International Conference on Advanced Manufacturing and Industry Applications (ICAMIA 2018)

IOP Conference Series: Materials Science and Engineering Volume 429

Sarawak, Malaysia
15 – 17 August 2018

ISBN: 978-1-5108-7508-1
ISSN: 1757-8981

Printed from e-media with permission by:

Curran Associates, Inc.
57 Morehouse Lane
Red Hook, NY 12571

Some format issues inherent in the e-media version may also appear in this print version.

Printed by Curran Associates, Inc. (2019)

For permission requests, please contact the Institute of Physics
at the address below.

Institute of Physics
Dirac House, Temple Back
Bristol BS1 6BE UK

Phone: 44 1 17 929 7481
Fax: 44 1 17 920 0979

techtracking@iop.org

Additional copies of this publication are available from:

Curran Associates, Inc.
57 Morehouse Lane
Red Hook, NY 12571 USA
Phone: 845-758-0400
Fax: 845-758-2633
Email: curran@proceedings.com
Web: www.proceedings.com

TABLE OF CONTENTS

AUTOMOTIVE TECHNOLOGIES

ENERGY AND THERMOFLUIDS

MANUFACTURING ENGINEERING AND TECHNOLOGIES

IOP Conf. Series: Materials Science and Engineering **429** (2018) 011001 doi:10.1088/1757-899X/429/1/011001

Preface

The International Conference On Advanced Manufacturing and Industry Applications (ICAMIA 2018) was held on 15-17 August 2018 in Kuching, Sarawak, one of the most popular tourist destinations in Malaysia. ICAMIA 2018 was organized by Malaysian Society of Instrument and Control Technology (MSICT), Mechanical Engineering Programme in School of Mechatronic Engineering, Universiti Malaysia Perlis and Technopreneur At UniMAP Sdn. Bhd (TUSB).

The ICAMIA 2018 aims to provide a platform for participants from diverse field of Mechanical and Manufacturing Engineering backgrounds to exchange ideas and findings to promote cooperation, networking as well as knowledge sharing in both engineering practice and research in higher education. The conference includes the fields of Applied Mechanics and Design, Automotive Technologies, Energy and Thermofluids as well as Manufacturing Engineering and Technologies. The conference received more than 120 papers from Malaysia and other countries over the world. After a rigorous peer-review process, 105 papers were accepted. The accepted papers that focused on theory, applications, design and experimental studies relevant to mechanical engineering field were invited to present in this conference.

I wish to record my gratitude to all reviewers, both local and international, for their time and effort in reviewing the papers. Their feedbacks to the authors have been excellent and useful to ensure the high quality of papers for the conferences. My sincere thanks also go to all sponsors for their generosity. Last but not least, I would like to thank also the organizing committee members for their hard work, cooperation and suggestions in ensuring the success of this conference.

Ir. Dr. Zuradzman Mohamad Razlan
Chairman
ICAMIA 2018

Peer review statement

All papers published in this volume of *IOP Conference Series: Materials Science and Engineering* have been peer reviewed through processes administered by the proceedings Editors. Reviews were conducted by expert referees to the professional and scientific standards expected of a proceedings journal published by IOP Publishing.

IOP Conf. Series: Materials Science and Engineering **429** (2018) 012001 doi:10.1088/1757-899X/429/1/012001

Development and implementation of FaRSLeSS approach at aircraft composite manufacturing industry : A case study

M S Ismail[1,a] , M I Hussain[1,b] , Z Mohd Zain[1,c] , N Muhammad[1,d]

[1]School of Manufacturing Engineering, University Malaysia Perlis, Pauh Putra Campus, 02600 Arau, Perlis, Malaysia

[a]mohdshihabudin@unimap.edu.my, [b]miqbal@unimap.edu.my, [c]zuraidah@unimap.edu.my, [d]noorhafiza@unimap.edu.my

Abstract. FaRSLeSS approach is integration between Lean and Six Sigma methods to reduce the failure rate in manufacturing production line. It represents 'Failure Rate Reduction and Sustainability through Lean and Six Sigma". Many manufacturing industries carry out corrective actions for their production line failure rate by using either lean or six sigma methods separately. This may prolong the failures or defects and sometimes even could not be solved effectively. FaRSLeSS approached was developed as guidance to manufacturing companies on the appropriate steps needs to be taken in order to solve the production line failure rate faster and effectively. Besides, this approach also suggests on how to sustain the performance once the failure rate was reduced to the ultimate target. During developing the FaRSLeSS approach, a case study was carried out at aircraft composite manufacturing industry (company A).

1. Introduction

The implementation of integration between lean and six sigma approaches should be properly organized in order to achieve the main objectives of any project. Six sigma methods alone are very hard to be implemented since involves data mining, analyzing and monitoring. In another words, it involves numbers. It is very difficult to get employees' commitment to involve in this type of approach. Therefore, the integration between six sigma and lean approach which involve more on how to reduce waste and how to ease the jobs and processes, expected to be accepted by the employee easier. The first main task need to be focused on by the researcher is how to create a lean and six sigma culture among the employees in the company. This culture is very important in order to ensure that all levels of employees from operator to manager level within the factory is speaking with a 'common language', which is 'lean six sigma language'. Every problems and actions taken need to be carried out with a guidance of lean and six sigma approaches, not with a traditional approach; doing daily 'fire fighting' to solve problems at the production line. The ultimate objective is to create a lean six sigma culture in the company.

2. Methodology of FaRSLeSS development, implementation and its sustainability

Figure 1 shows the process flow on the methodology to complete FaRSLeSS (Failure rate Reduction & Sustainability through Lean & Six Sigma) approach. The flow includes all the three main objectives of this study which are :

a) development and implementation of FaRSLeSS approach at company A,

b) enhancement of lean six sigma culture and
c) sustainability of the approach in the company.

For a development of FaRSLeSS, it starts with previous papers review and data collection/analysis from company A. Numbers of papers in a similar research study and scope were selected for the review purposes. The paper review process concluded that for aircraft composite panel industry, there is still lack of research done related to application of both lean and six sigma in solving the failure rate reduction at the production line, mainly in composite manufacturing production.

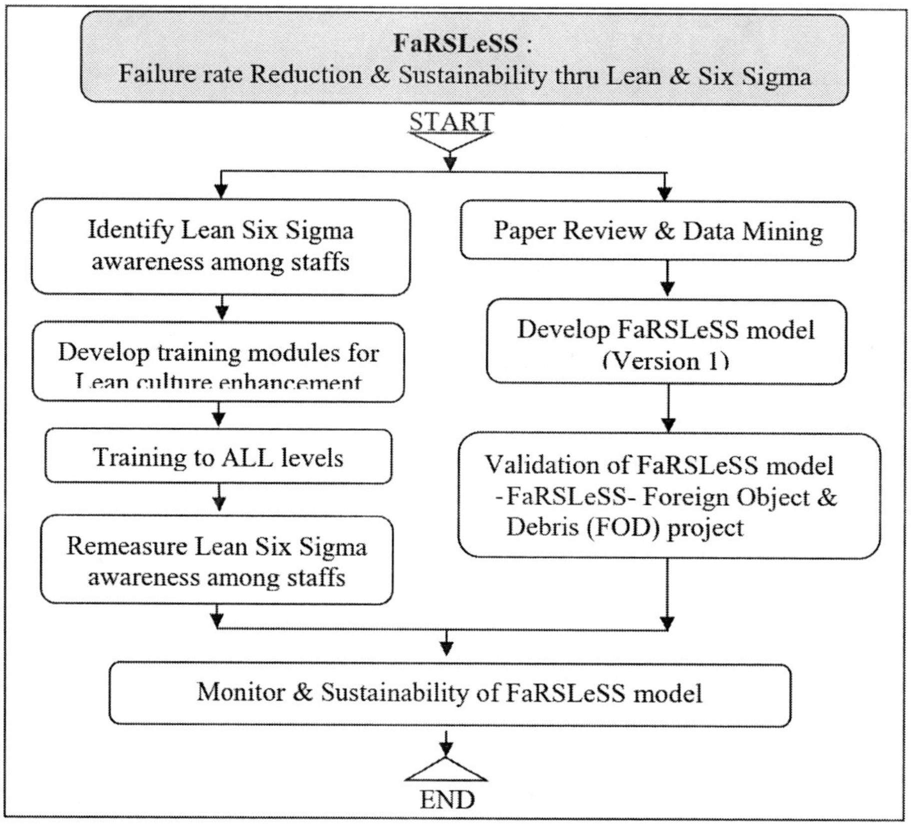

Figure 1. FaRSLeSS approach development

There is also no study so far on keeping the production yield at a desired value set by the management team for aircraft composite manufacturing. Besides that, the review also found that most of the research carried out at manufacturing production line was using either six sigma or lean tools separately. Therefore, FaRSLeSS is developed to come out with a new approach in order to consolidate both lean and six sigma tools for a better result, especially at aircraft composite panel manufacturing industry.

During the stage of FaRSLeSS approach development, data collection activities were carried out simultaneously in company A. From the overall failure rate pareto in one year as shown in figure 2, there were two most important processes in company A which contributed to 85% of the total failure items. The two processes were trimming (53.3%) and hand layup (38.2%). The most probable reason for these two processes with the highest failure rate is because of both of them mainly are manual processes by operators. Manual process will always exposed to high process variation and production waste. Therefore both trimming and lay up processes were selected for the study since it is expected to give a bigger impact to the overall result later, in term of failure rate reduction, production pass yield and productivity, which all of them will lead to reduce parts disposal and simultaneously increase profit to company A. The FaRSLeSS approach will look into two main items; firstly on how to reduce the failure rate for both processes and secondly how to sustain the failure rate after the reduction achieved.

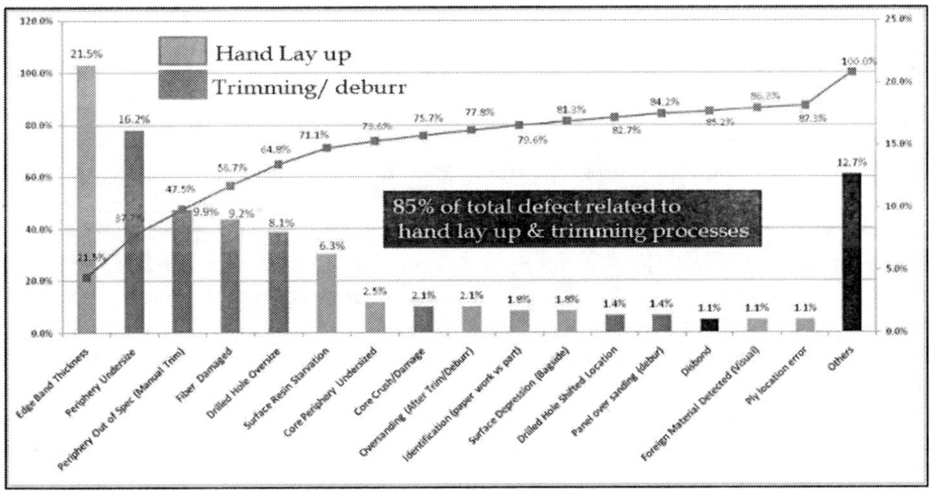

Figure 2. Overall failure rate at company A production line

3. FaRSLeSS Model

By referring to the situation above, a guided lean six sigma integrated with lean technique is required to be applied in aerospace composite panel manufacturing company. By this application, the manual process variation could be reduced statistically and unwanted production waste could be eliminated from the production line in order to give a higher productivity and quality of the product. FaRSLeSS model is then developed as shown in figure

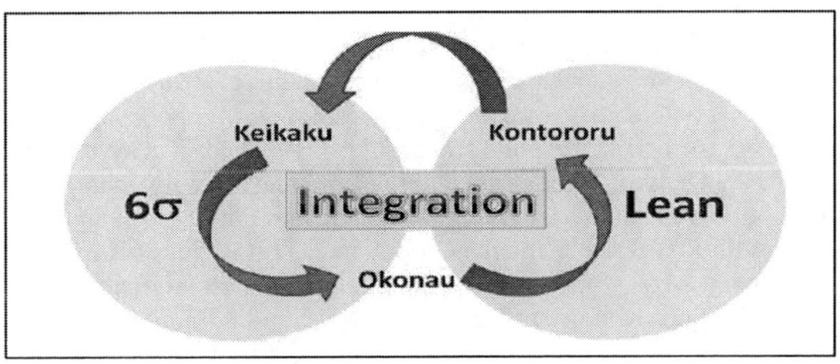

Figure 3. FaRSLeSS model

The integration between lean and six sigma approaches in FaRSLeSS model is guided by the KOK cycle which is from three Japanese words represents three main steps involved; *Keikaku* (Planning), *Okonau* (Executing) and *Kontororu* (Controlling). These three steps are in a cycle in order to complete the implementation and sustainability purposes.

3.1. Step 1 - Keikaku (Planning)

IOP Conf. Series: Materials Science and Engineering **429** (2018) 012001 doi:10.1088/1757-899X/429/1/012001

The FaRSLeSS model begins with Keikaku step. In this step, a proper plan need to be conducted by the team involves production engineers and top management. The problem needs to be identified, recognized and defined clearly and need to be transparent to the whole team members. Then, the problem statements, objectives and scopes of the problem need to be determined and written. It is a good practice if a project name to be allocated for the visibility of the project and easy monitoring. For instance in company A, one of the major issue rose in year 2014 is very high parts disposal cost. Therefore the parts disposal reduction by 50% was selected as a key performance index (KPI) to link with the lean and six sigma approaches research.

3.2. Step 2 - Okonau (Executing)

Okonau means executing the project by measuring, analyzing and improving activities. Measuring has two main purposes ; gather data to validate and quantify the problem (opportunity) and begin teasing out facts and numbers that offer clues about the causes of the problem. For example, the first year period part disposal data was gathered and monitored. From the company's parts failure by items data, overall scenario of the pareto for model X disposal parts need to be measured. After measurement completed, the data requires analyzing. One of the big challenges during analyzing data is to select right tools. When causes go deeper or when the relationship between the problem and other factors is complex and hidden, more advance statistical techniques may be required to identify and verify the cause.

Training to employees needs to be conducted in order to familiarize them with lean six sigma culture. The training could be conducted includes basic production innovation (PI), kaizen and waste (Muda) management and motion ergonomic. Then for lean, waste at production line and the overall factory need to be identified by using Value Streaming Mapping (VSM) and other related tools. As for six sigma, analysis need to be carried out for a second and third level pareto of finding a root cause for the part disposal issue. This could be achieved by using six sigma tools such as Cause and Effect diagram, Analysis of Variance (ANOVA), Failure Mode Engineering Analysis (FMEA) and Design of Experiment (DOE). The analysis step conclusion is then concluded before move to a next step. At this analysis step, the knowledge and theory about advance composite materials and its manufacturing processes need to be known and understood well by the researcher and team members. Discussion and brainstorming sessions among team members need to be carried out regularly. All the limitations such as specification by the customers need to be understood as a scope of the analysis.

A traditional method of improving and solving problem, the repairer or engineer tends to straight away try to look into a temporary solution or countermeasure without finding as actual root cause of the problem. This method also sometimes happens in company A. The disadvantage of this method is although the problem is solved, it may be repeated since the actual root cause is not solved. With systematic tools in a lean six sigma approach, the cause of a problem could be solved until the root of it. From a lean point of view, an improvement could be done on the continuous waste management practice. This includes Kaizen implementation, 5S and 3 M's (Muda, Muri, Mura) improvement. As for a six sigma view, production line processes need to be improved according to the statistical and non statistical result obtained from analysis step.

From the FaRSLeSS model developed in company A, the Okonau step could be specified more detail as in figure 4. This detail sub steps is also included in the FaRSLeSS model development for a proper sequence of lean and six sigma tools the management need to use for achieving a desired production failure rate target.

3.3. Step 3 : Kontororu (Controlling)

Control and sustain step is very important in order to make the improvement carried out in previous step could be remained. Otherwise the problem and the similar cycle of solving might be repeated again and again. Firstly, the improvement items need to be reviewed regularly. The reasonable frequency for review process is in weekly basis, depends on the production output quantity. If the result found diverting from its lower and upper control limit, action need to be taken immediately.

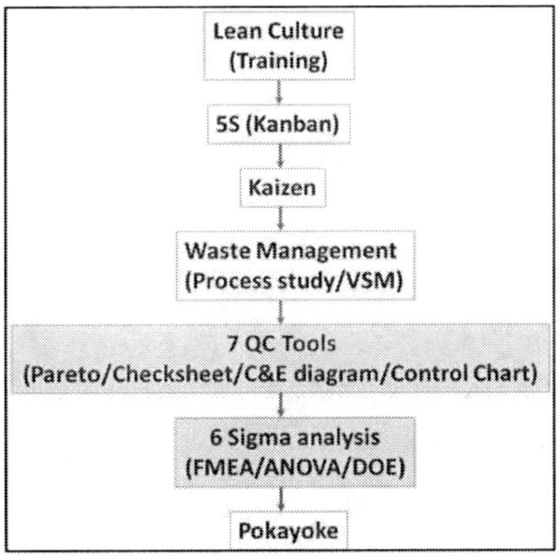

Figure 4. Okanou (executing) steps

A system standard document for process sustainability needs to be created by the team for easy monitoring. The most common lean six sigma tools to serve this purpose are Standard Operating Procedure (SOP) or Work Instruction (WI), Process Flow, Statistical Process Control Chart (SPC), continuous Kaizen activities and Continuous Quality Improvement (CQI). Specific control tasks that must be completed by the teams include developing a monitoring process to keep track of changes they have set out, creating a response plan for dealing with problems that may arise and helping focus management's attention on a few critical measures that give them current information on the outcomes of the project.

4. Conclusion

From figure 1, the FaRSLeSS model was validated at one of a project in company A which is called Foreign Object & Debris (FOD) failure reduction project. It is about foreign object contamination defects for aircraft composite panel. In year 2016, there are 28 cases of FOD failure in company A production line. The FaRSLeSS approach was implemented in early year 2017 for this FOD issue and monitored until end of the year. There is very significant reduction in the FOD failure rate after the approach implemented where only 3 cases recorded. This 90% reduction in FOD failure rate proven that the FaRSLeSS model could be accepted as a good guided approach for failure rate improvement at manufacturing production line. The continuous monitoring is carried out at company A. As a further confirmation, this approach should be validated at other manufacturing industries.

References

[1] Everton D. & Sergio E.G., 2013, *Lean, Six Sigma and Lean Six Sigma: an analysis based on operations strategy,* International Journal of Production Research

[2] Frida A., Astrid H., 2014, Design for Manufacturing of Composite Structures for Commercial Aircraft – the Development of DFM strategy at SAAB Aerostructures, CIRP Conference on Manufacturing System: **17**(2014) 362-367

[3] A. R, Chaple., S. S, Khedakar., S. R, Dharmadhikari., N. R, Chaple., 2015, *Newly Developed Automatic Lay-Up Process for Manufacturing of FRP Sheets,* International Journal Of Computational Engineering Research, **3**, pp.1

IOP Conf. Series: Materials Science and Engineering **429** (2018) 012002 doi:10.1088/1757-899X/429/1/012002

Finite Element Simulation on Railway Wheels under Various Loading

M A Che Zulkifli, K S Basaruddin*, M Afendi, W H Tan and E M Cheng

School of Mechatronic Engineering, Universiti Malaysia Perlis, Pauh Putra Campus, 02600 Arau, Perlis, Malaysia.

*Corresponding author: khsalleh@unimap.edu.my

Abstract: This study aims to predict the safety factor of railway wheel under various loading. A three dimensional finite element (FE) railway wheel model was developed to calculate the mechanical stress on railway wheel. In this study, the railway wheel contact was analyzed under static loading. The material used was steel A-900. The ANSYS Workbench was employed for the FE analysis. The results show that the highest safety factor was found under cornering load, compared to the analyses under vertical load, braking load and contact pressure.

1. Introduction

Railway wheel is commonly used to move the train that carrying passenger or cargo from one place to another destination. Nowadays, the new railway wheel was used so that it can sustain the stress due to rolling contact between wheel and rail. The old railway wheel that have high risk of accident and can happen at any time due to several problems such as excessive load from the train load and also the train handling that can cause unexpected types of loading on the wheel-rail contact. Continuous contact surface between the wheel and rail generates the kinetic friction. This rolling contact also can be defined as friction forces that exists between the rail and wheel of contact surface.

Analysis on the failure prediction of railway's wheel and rail under various condition has attracted many researchers. Most of them utilised finite element method (FEM). FEM has been established and widely used for many engineering problems even for complex geometry such as cellular solids [1], honeycomb [2], cancellous bone [3], [4] and many more. Kuna et al. [5] conducted the fracture mechanical assessment on railway wheels made of austempered ductile iron with graphite nodules under static and cyclic loads using FEM. They found that the critical crack sizes determined by their investigations can be diagnosed certainly and reliably by means of available non-destructive testing. Moreover, Arslan and Koyabasi [6] demonstrated a well-known FEA procedure on an example three-dimensional rail–wheel and axle model. The results obtained in 3D FE analysis show good agreement with real life problems experienced at both railway wheel and rail. A railway system containing wheel, rail, axle, and pads was modeled and analyzed by Aalami et al. [7]. They utilised ANSYS software for the FE analysis and used elastic-plastic materials, mapped meshing, and the rolling motion of the wheel contingent upon the up-to-date international railway systems that results in high accuracy in the solutions. The simulation results have suitable agreement with the real life experiences.

On the other hand, it is widely known that the residual stress is one of the crucial parameters in railway wheels. The designer will ensure the residual stresses in the wheels at minimum as possible. Nejad [8] estimated the residual stresses in the rail wheels due to heat treatment process of a railway wheels. The FE analysis was conducted by applying the elastic-plastic material properties for the rail wheels under thermal loads. The results obtained show good agreement with those achieved in field measurements. Ramanan et al. [9] also applying the thermal load due to braking during services in the FE analysis of the rail wheel. They found that stresses due to braking dominate over the payload stresses based the results obtained from the coupled temperature-displacement analysis. Hence, this paper aims to predict the safety factor of the railway wheels under various loads during services which is rolling contact, braking and cornering.

2. Method

The wheel profile was selected according to the information provided by the local railways company. The type of train used to perform the analysis was the wagon type. The information provided are the wheel diameter which is 965 mm (38 in), the vertical load (maximum design load of 78.48 kN), the material properties of wheel (yield strength of 765 MPa; Young's modulus of 205 GPa; friction coefficient of 0.3; material density of 7850 kg/m^3; rotational velocity of 24.6 rad/s; ultimate tensile strength of 1080 MPa) and the rail dimension with 1000 mm long.

Figure 1 shows the geometrical model and boundary conditions of the present analysis. The convergence test was performed in order to determine the optimum mesh size. Tetrahedral and hexahedral types of element were used in the meshing of the FE model. Four set of boundary condtions were applied on the rail-wheel models, which is vertical load, braking load, cornering load and contact pressure. The safety factor was then calculated based on the yield strength divided by the von-mises stress.

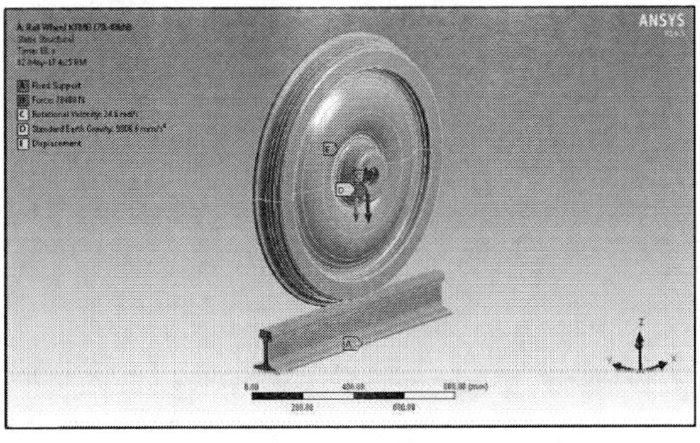

Figure 1. Geometrical model and boundary conditions setup.

3. Results and discussion

3.1. Response of railway wheel under vertical load

Figure 2 shows the contour distribution of von-mises stress and total deformation for railway wheels and rail under vertical load (one sample). Obviously, the high stress was found concentrated in the rolling contact between rail and wheel, whereas the high total deformation was subjected to the axle of the wheel. Safety factor of the wheels as found decreased with respect to the increase of vertical load as shown in Fig. 3. The wheel was still in safe condition under the vertical load up to 120 kN.

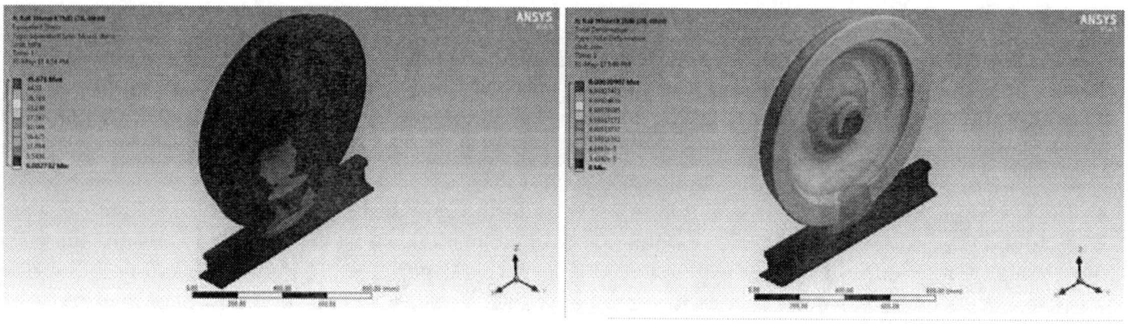

(a) (b)

Figure 2. Contour distribution under rolling contact. (a) von-mises stress. (b) total deformation.

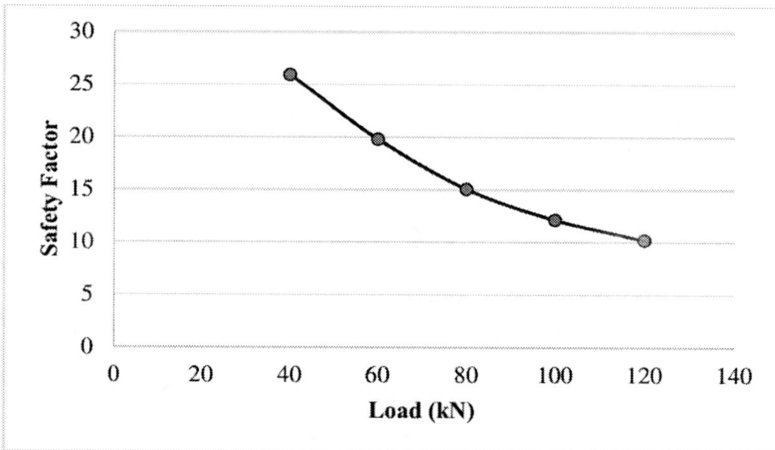

Figure 3. Safety factor of the railway wheels under vertical load

3.2. Response of railway wheel under braking load

The braking load was applied to wheel from the brake pad and the contact between rail and wheel generated the shear force. The stress distribution for (a) von mises and (b) shear stress on the rail-wheel is shown in Fig. 4. It shows that the critical area also concentrated on the rail-wheel contact area. Similarly, the safety factor was decreased when the braking force increased. But it is found that the safety factor of the wheel is slightly lower under braking load compared to vertical load.

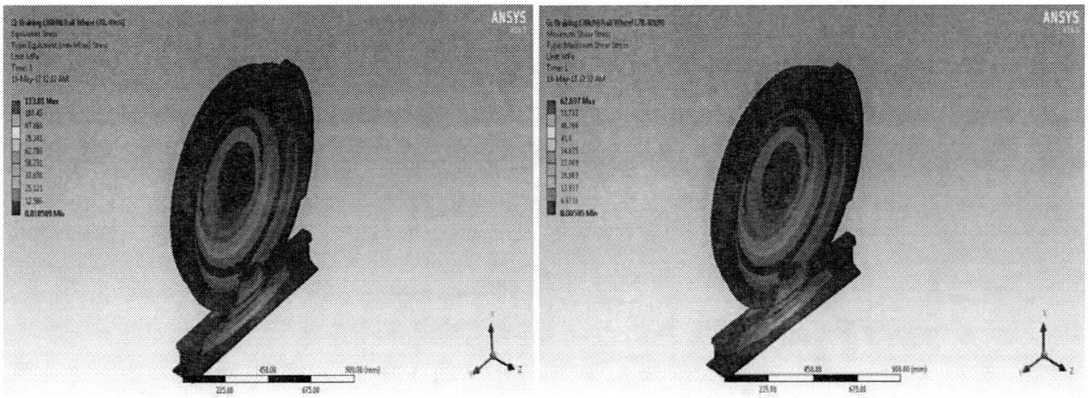

(a) (b)

Figure 4. Stress distribution under braking force. (a) von-mises stress. (b) maximum shear stress.

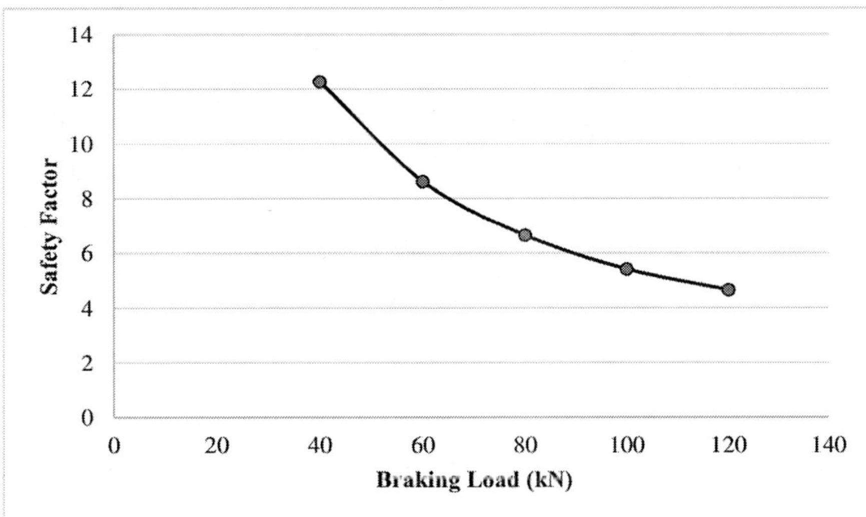

Figure 5. Safety factor of the railway wheels under braking load

3.3. Cornering Load

Figure 6 shows the safety factor of the wheel under cornering load. It shows the similar pattern as safety factor under vertical and braking load. However, the value is largely higher compared to the former conditions. The contact pressure contour between rail and wheel is shown in Fig. 7. The result shows that the pressure was concentrated on the central part of the contact area. This is the area where the facture would be initiated if the excessive load is applied.

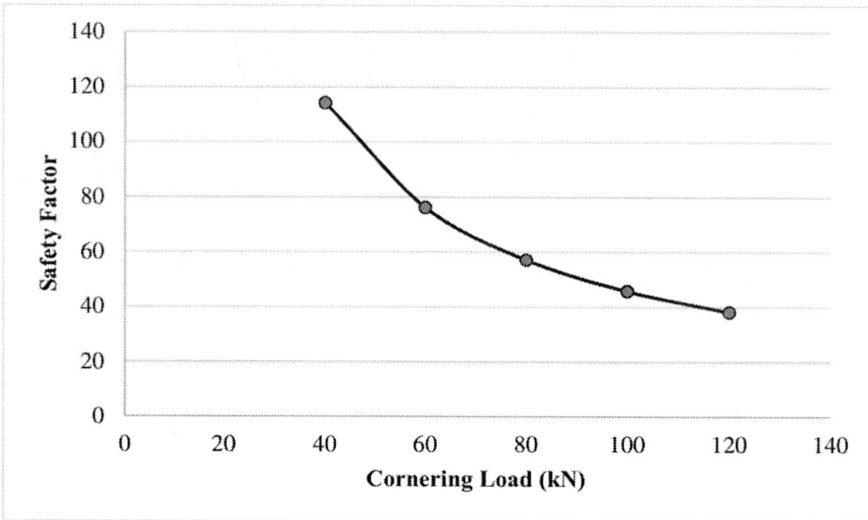

Figure 6. Safety factor of the railway wheels under cornering load

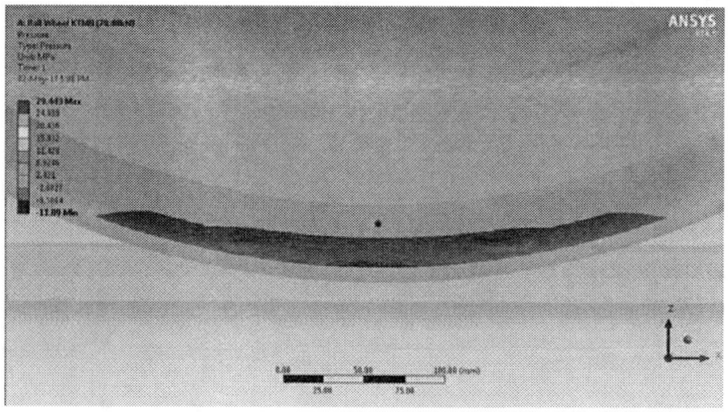

Figure 7. Contact pressure contour between rail and wheel.

4. Conclusion

The present study was carried out to analysed the safety factor of the railway wheel under service conditions which is vertical rolling contact, braking load and cornering load. The results show that the safety factor of the railway wheel was lowest under braking load compared to vertical load and cornering load. The railway wheel was safe under all simulated loads that up to 120 kN. The findings of present study could be used for designer to predict failure on the railway wheels.

Acknowledgement

This study was financially supported by Universiti Malaysia Perlis under Journal Incentive Grant (No. 9007-00214).

References

[1] Basaruddin, K. S., Kamarrudin, N. S. & Ibrahim, I. 2014. *Lat. Am. J. Solids Struct.* **11,** pp. 755–769.
[2] Halim, S. A., Basaruddin, K. S., Daud, R., Safar, M. J. A. & Rahman, A. S. A. 2015. *ARPN J. Eng. Appl. Sci.* **10,** pp. 9353–9358.
[3] Basaruddin, K., Takano, N., Yoshiwara, Y. & Nakano, T. 2012. *Med. Biol. Eng. Comput.* **50,** pp. 1091–1103.
[4] Yoshiwara, Y., Clanche, M., Basaruddin, K. S., Takano, N., Nakano, T. 2011. *J. Biomech. Sci. Eng.* **6,** pp. 270–285.
[5] Kuna, M., Springmann, M., Mädler, K., Hübner, P. & Pusch, G. 2005. *Eng. Fract. Mech.* **72,** pp. 241–253.
[6] Arslan, M. A. & Kayabaşi, O. 2012. *Adv. Eng. Softw.* **45,** pp. 325–331.
[7] Aalami, M. R., Anari, A., Shafighfard, T. & Talatahari, S. 2013. *Adv. Mech. Eng.* **2013,** pp. 1-9.
[8] Masoudi Nejad, R. 2014. *Eng. Fail. Anal.* **45,** pp. 449–455.
[9] Ramanan, L., Krishna Kumar, R. & Sriraman, R. 1999. *Int. J. Mech. Sci.* **41,** pp. 487–505.

IOP Conf. Series: Materials Science and Engineering **429** (2018) 012003 doi:10.1088/1757-899X/429/1/012003

Polyamide Nanofibers Reinforced Titanium Nanoparticles Composites for hydrophobic surfaces

K Salman[1,*], Z M Razlan[2]

[1]Department of Electro-mechanics Engineering, University of Technology, Baghdad-Iraq
[2]School of Mechatronic Engineering, Universiti Malaysia Perlis, Pauh Putra Campus, 02600 Arau, Perlis, Malaysia

*Corresponding author: drkhansaa_ds@yahoo.com

Abstract. In this study, samples of polyamide (6) (PA 6) nanofibers armed with TiO_2 nanoparticles were prepared using an electrospinning method for the preparation of nanocomposite fibres. The nanofibers were prepared under 0.15% weight, 25 kV applied voltage, 15 cm electrospinning distance, 0.48 mm tip diameter and 0.3 ml/hr injection rate to obtain nanofibers with a diameter of about 100 nm and then reinforced with (50 nm diameter) Titanium Nanoparticles. The properties and morphology of the surface were studied using the atomic force microscopy (AFM) scanner. The polyamide structure was also examined by the Fourier transform infrared radiation (FTIR) infrared device, and the contact angle was measured with water using a contact angle analyzer. The results showed that the diameter of the fibres ranged between 80-100 nm and contain some beads. The infrared test also showed that the intensity of the peaks decreases after the transformation of polyamide into nanofibric. The resulting surface has the hydrophobic properties, especially after the addition of Titanium nanoparticles.

1. Introduction

The ultra-hydrophobic material restores the water at a very high contact angle, allowing the surfaces to remain dry. These materials lead to many benefits such as ease of manufacturing, low cost, scalability, a small quantity of large area coats, eliminating existing water corrosion. [1]. this is the aim of this research. A number of methods have been used to disperse ultra-super hydrophobic surfaces, such as electrochemical deposition [2], electrophoresis [3], chemical etching [4], layer after self-assembly of layer [5], chemical vapor deposition [6]. Nanotechnology has entered into this field, in order to improve the hydrophobic surfaces, scientists deliberately to form micro and nano surface protrusions, that will increase the water droplets by hand (or even any other means), so as to protect the optimal surface, He walked in the term science material "lotus effect" refers to the hydrophobic surfaces [7]. There are many previous studies on the hydrophobic coatings such as Qian F. et. al (2013), who found the TiO_2/ high density poly ethylene (HDPE) nanocomposite have super hydrophobic properties without any chemical exposure, while it transfers for hydrophilic surface after exposure to UV-RAD. [8]. Also, Srinivasan A. et. al (2013) studied the fabricate Graphene-loaded TiO2 thin films on glass substrates by the spin-coating technique. They conclude the resulting fabric has inexpensive, transparent, conductive, super hydrophilic, and highly photo-catalytically active properties. [9].

2. E perimental Part

2.1. Materials

Polyamide-6 as white, transparent from Alibaba Company (China) to preparation nanofibers by electrospinning technique was used. Formic acid from Sigma-Aldrich as a solvent was used. C-Titanium (TiO_2) nanoparticles with (50 nm) diameter from Sigma-Aldrich offers Aldrich-791342, used as a reinforcement material.

IOP Conf. Series: Materials Science and Engineering **429** (2018) 012003 doi:10.1088/1757-899X/429/1/012003

2.2. Electrospinning set up

The electrospinning solution held in a syringe connected to the syringe pump to regulate fluid flow, while the positive high voltage power supply is connected to the metallic needle, and the negative energy supply is connected to the ground metal collector, as in Figure 1.

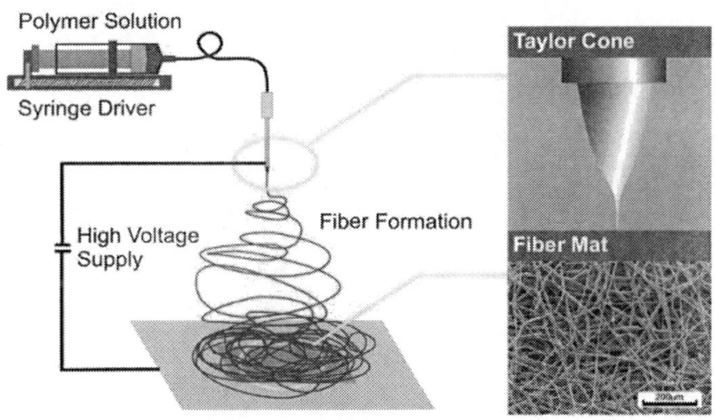

Figure 1. Set Up of Electrospinning

2.3. Fabrication of Super-Hydrophobic/ Super Hydrophilic Textile

20 ml of polyamide 6 (PA 6)/formic acid pumped by electrospinning system for (15 hr), and collected on aluminum foil (20 cm * 20 cm^2), then it cut according to a required test as in table (1).

Table 1. The dimensions of Samples According to Tests

Samples	Dimension, cm^2	Test
1	1	AFM
2	1	Contact angle
3	17.5	FTIR

3. Results and discussion:

3.1. FTIR analysis

Figure 2 (a) shows the infrared analysis of the polyamide, and Figure 2 (b) shows a FTIR analysis of polyamide nanofibres. Note that there are new links created after the transformation of polyamide from microstructure to nanostructures. Also observed from this nanoparticle structure of polyamide polycrystalline has more crystallization than the macroscopic structure, increasing the amorphous structure leads to increased crystalline peak amplitude and density. As nanostructures are highly crystallized, so the nanoparticle sample shows a weaker peak. This was confirmed by Zimba et al. [10], which observed these peaks more clearly in the amorphous PA structure of nanocrystalline structures.

IOP Conf. Series: Materials Science and Engineering **429** (2018) 012003 doi:10.1088/1757-899X/429/1/012003

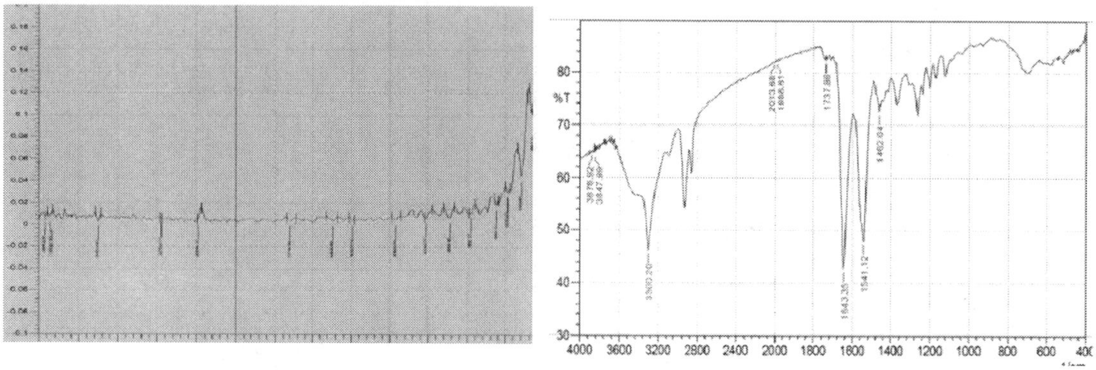

a b

Figure 2. FTIR Analysis of (a) Macromolecules PA6 (b) PA6 Nanofibers

3.2. Contact angle

Figure (4) show the relationship between the contact angle of (3a. foil aluminum, 3b. foil aluminum coated by PA nanofibers, and 3c. foil aluminum coated by PA nanofibers reinforced with TiO_2 nanoparticles). We show from this shape the contact angle of foil aluminum is smaller than 90º, this leads to hydrophilic behavior as in Figure (3a), also the contact angle changes with time dramatically. Figure (3b) shows the hydrophilic behavior of PA nanofibers, this is because the PA surface able to create a bond with water and the contact angle is smaller than 90°. Figure (3c) shows the hydrophobic behavior of (PA + TiO_2) nanofibers this is because the TiO_2 adding leads to hydrophobic behavior of surfaces because it leads to increasing of the roughness of surfaces. Also noticed that, the contact angle decrease with remaining time on the surfaces, with (PA + TiO_2) surface it decrease slowly because this surface is more roughness than other samples with (aluminum foil, & PANF), this is because the these surfaces have less roughness [11].

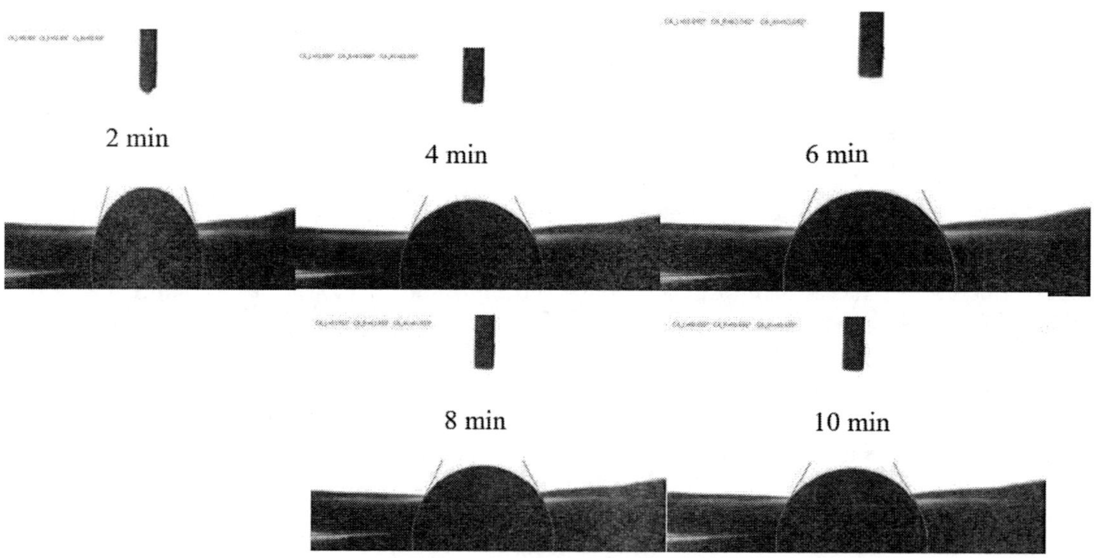

Figure 3a. Contact angle of aluminum foil with time

IOP Conf. Series: Materials Science and Engineering **429** (2018) 012003 doi:10.1088/1757-899X/429/1/012003

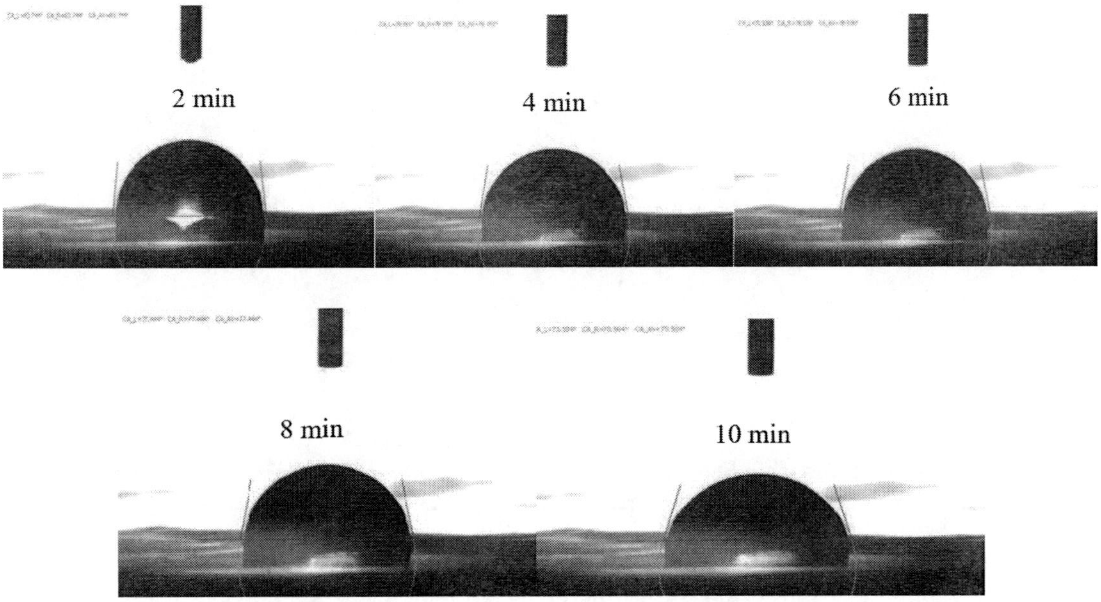

Figure 3b. Contact of (0.15w/v) PA6 surfaces nanofibers with time

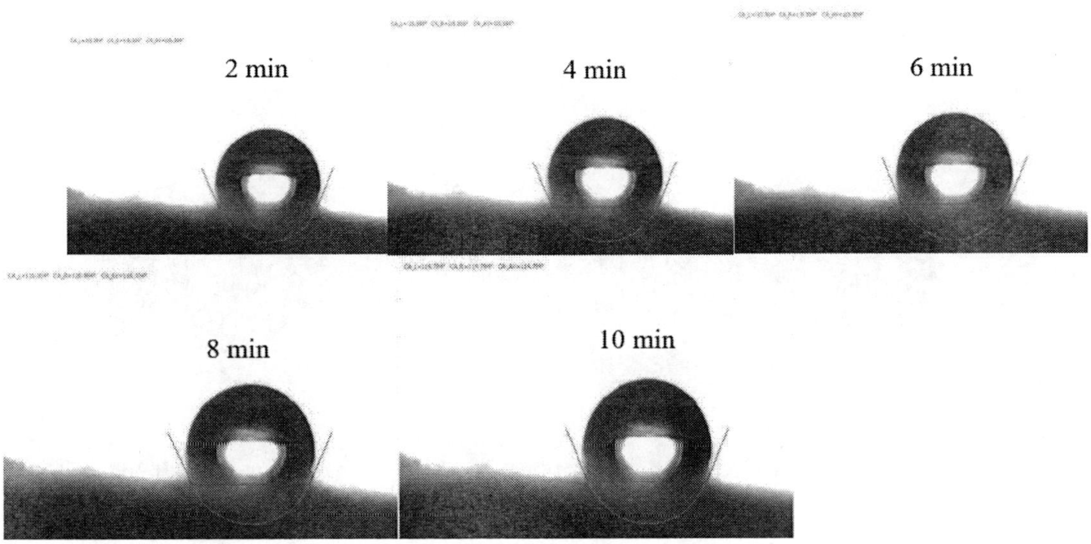

Figure 3c. Contact angle of (PA 6 + TiO_2) surfaces nanofibers with time

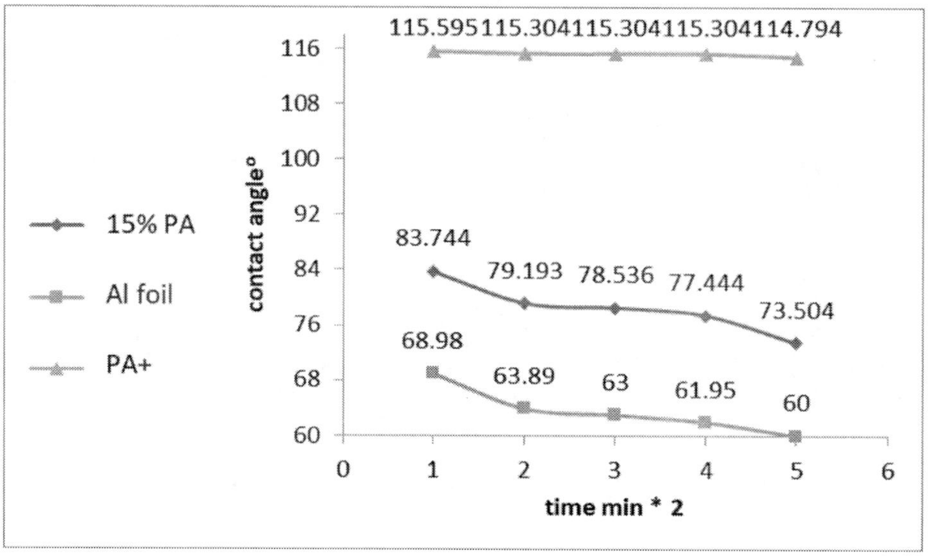

Figure . Show the relationship between the contact angle and remaining time of water droplet to 3a. Aluminum foil, 3b. PA nanofibers and 3c. (PA+TiO_2) nanofibers coating

3.3. AFM results

Figure (5 a,b) shows the AFM results of the morphology of 5a. PA 6 nanofibers surfaces and 5b. (PA 6 +TiO_2) surface.

Notice the PA 6 nanofibers surface has roughness less than (PA 6 +TiO_2) surface this is because the TiO_2 nanoparticles adding leads to increasing of the roughness of the surface and this leads to increasing of the hydrophobic behavior of surface.

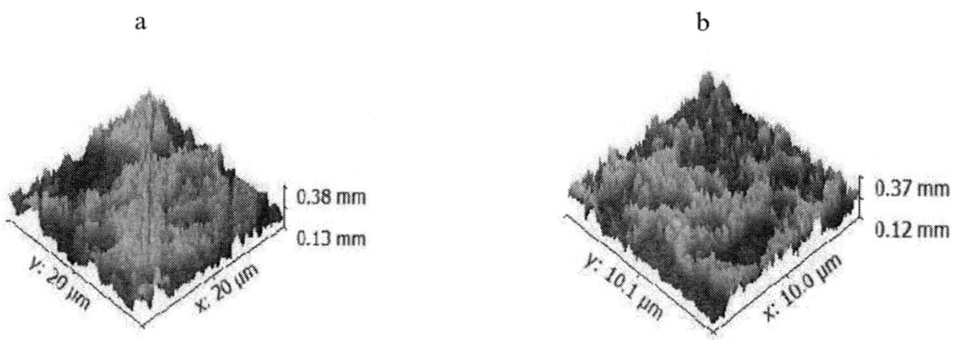

Figure . AFM of surface, a. PA 6 nanofibers and b. PA+TiO_2 nanocomposite nanofibers

. Conclusions
In this paper, we conclude that:
1. The polyamide 6 nanofibers have hydrophilic behavior with contact angle about (68.98°).
2. The adding of TiO_2 nanoparticles leads to increasing of the contact angle of result nanocomposites nanofibers (PA6 + TiO_2 nanoparticles) and hydrophobic behavior of the surface.
3. This leads to using of this nanocomposites membrane surface in prevention, packaging, protection, and limiting of pollution applications.

References
[1] Panagiotis N. Manoudis, Ioannis Karapanagiotis, Andreas Tsakalof§, Ioannis Zuburtikudis and Costas Panayiotou, " Superhydrophobic Composite Films Produced on Various Substrates", *Langmuir*, *24* (19), pp 11225–11232. 2008
[2] Xufeng Wu and Gaoquan Shi, "Production and Characterization of Stable Superhydrophobic Surfaces Based on Copper Hydroxide Nanoneedles Mimicking the Legs of Water Striders", *J. Phys. Chem. B*, 2006, 110 (23), pp 11247–11252
[3] Zuxin She, Qing Li, Zhongwei Wang, Longqin Li, Funan Chen, and Juncen Zhou," Novel Method for Controllable Fabrication of a Superhydrophobic CuO Surface on AZ91D Magnesium Alloy", ACS Appl. Mater. Interfaces, 2012, 4 (8), pp 4348–4356
[4] Zhijian Chen and Wei Li, " Introduction to the special issue of the 2009 International Symposium on Crystal Engineering & Drug Delivery Systems", Frontiers of Chemical Engineering in China, March 2010, Volume 4, Issue 1, pp 1–1
[5] Shunsuke Nishimoto*, Hitomi Sekine, Xintong Zhang, Zhaoyue Liu†, Kazuya Nakata, Taketoshi Murakami, Yoshihiro Koide and Akira Fujishima, " Assembly of Self-Assembled Monolayer-Coated Al_2O_3 on TiO_2 Thin Films for the Fabrication of Renewable Super hydrophobic–Superhydrophilic Structures", *Langmuir*, 2009, 25 (13), pp 7226–7228
[6] Ana Borras, Angel Barranco†, and Agustín R. González-Elipe, "Reversible Superhydrophobic to Superhydrophilic Conversion of Ag@TiO_2 Composite Nanofiber Surfaces", *Langmuir*, 2008, 24 (15), pp 8021–8026
[7] Genzer, J, Efimenko, K, ‘‘Recent Developments in Superhydrophobic Surfaces and Their Relevance to Marine Fouling: A Review.’’Biofouling,22(5) 339–360. 2006
[8] Qian Feng Xu, Yang Liu, Fang-Ju Lin, Bikash Mondal, and Alan M. Lyons," Superhydrophobic TiO2–Polymer Nanocomposite Surface with UV-Induced Reversible Wettability and Self-Cleaning Properties", ACS Appl. Mater. Interfaces, 2013, 5 (18), pp 8915–8924
[9] Srinivasan A., Tata N. Rao, Marappan S. Dinesh R., Itaru H., and Masahiro M., " Superhydrophilic Graphene-Loaded TiO2 Thin Film for Self-Cleaning Applications", ACS Appl. Mater. Interfaces, 2013, 5 (1), pp 207–212
[10] C. Zimba; J. Rabolt, " Raman spectroscopic characterization and molecular force field development of a synthetic polyamide: Nylon 66". Macromolecules 1989, 22(6), 2863-2867.
[11] Meng L-Y., Park S-J., "Effect of fluorination of carbon nanotubes on superhydrophobic properties of fluoro-based films". Journal of Colloid and Interface Science, 342, 559–563 .2010

Acknowledgments
Thanks to all the authors and those who contributed and helped in completing this paper.

Fracture risk prediction on children with Osteogenesis Imperfecta subjected to loads under activity of daily living

S B C Wanna[1], K S Basaruddin[1], M H Mat Som[1], A R Sulaiman[2], A Shukrimi[3], S F Khan[1], M S Abdul Majid[1], M J M Ridzuan[1]

[1]School of Mechatronic Engineering, Universiti Malaysia Perlis, Pauh Putra Campus, 02600 Arau, Perlis, Malaysia.
[2]Department of Orthopaedics, School of Medical Sciences, Universiti Sains Malaysia, 16150 Kota Bharu, Kelantan, Malaysia.
[3]Department of Orthopaedics, Kulliyyah of Medicine, International Islamic University Malaysia, 25710 Kuantan, Pahang, Malaysia

Abstract. Children with Osteogenesis Imperfecta (OI) often vulnerable to fracture even on their daily basic tasks. The unpredicted fracture leads to intramedullary surgery. This study aims to investigate the fracture risk in children with OI in daily basic tasks including standing, walking, and jumping. Ten models of OI finite elements were reconstructed based on the radiography CT image of enrolled OI patients. An actual body weight of OI patient was used as the main force acting vertically towards the femoral head and then the fracture risk was observed. In standing and walking configuration, all ten OI models were found able to sustain the acted load. Whereas in jumping configuration, all OI models suffered to become fracture. Early prediction of fracture risk could help medical personnel and OI patients from fall or injury.

1. Introduction

Osteogenesis Imperfecta (OI) is a hereditary disease which causes the bone fracture to the patient. The unpredictable occurrence of fracture always becomes major issues discussed among the orthopedic, researchers and patients. the lack of type I collagen which covers up almost 95% of the entire bone formation lead to abnormal structure collagen produced in the bone where it causes low bone mass and bone fragility which leads to a health condition such as OI [1,2,3]. To date, once the fracture occurs, OI patient will be treated either with a surgical procedure or physical therapy. Surgical procedure involved the intramedullary nail insertion into the bone to join the fractured bone. However, this also does not solve the issue as the occurrences of unpredictable fracture might repeat in future.

Assessment of fracture risk in OI bone had been conducted by several researchers in the past [4,5]. The bone failure has been widely investigated where could be due to the physical properties at micro level [6,7] or macro level [8]. The finite element model of OI bone was presented with variation of bowing angles to resemble real OI bone [9]. The findings suggested that the gait cycle and various of load impacts influenced the risk of fracture. In order to develop the OI bone in the finite element method to presents the reliable model, there are several aspects should be taken into consideration. Geometrical model, bone properties and boundary conditions are the main parameters to be emphasised in finite element analysis. Commonly, the difficulty encountered by researchers is to develop the geometry of the finite element models to serve as the OI bone. Bone material properties must also be clearly defined considering the randomness of microstructures [10, 11]. The fragility of the bone largely affected by the porosity, crosslinking profile, mineralization distribution, or the collagen content which influence the fracture risk. Fan et al. [8] suggested the Young's modulus for OI bone is 19 GPa with isotropic linear elastic. This value based on experimental conducted on OI bone sample through nanoindentation technique.

In this study, the fracture risk with respect to activities of daily living (ADL) which is standing, walking and jumping was predicted using finite element analysis. Ten OI finite element models were reconstructed based on the radiography image of the real OI patient. The actual body weight opposed by OI patients was used as load acting on the femoral head.

2. Method

2.1. Standard femur (SF)

The "Standardised femur" (SF) model was first obtained from Biomedtown website proposed by Viceconti et al. [12] which make it available for download. The length of SF is 409.4 mm. SF was imported to finite element software, ANSYS for simulation purpose. First, the process of separating SF into two parts that consists of femoral head and femoral shaft takes place since the SF was imported in one full part thus separating its solid design into two parts is required for the selection of force region in the femoral head. The procedures are as follows.

i. At the origin, a perpendicular line with 45° was drawn in Z-plane to divide the femoral head and femoral shaft.
ii. Once divided, the femoral head and femoral shaft were bonded with same mechanical properties.

Once completing the separating process, convergence test was conducted to determine the optimum element size to ensure the accuracy of numerical results. The SF meshed with 28,105 nodes and 12,745 elements of tetrahedron and hexahedron types chosen based on the convergence test.

2.2. Finite element modeling of OI bone

Reconstruction of OI bones takes place in ANSYS. Ten OI models were reconstructed to resemble real OI bone geometry. CT scan image of enrolled OI patient obtained from the Hospital Universiti Sains Malaysia with patient's consent was used as a guideline for this purpose. The bowing angles in sagittal plane were varied from 7.5° to 30.0° with increment of 2.5° for each model. The bone length was measured and divided by half, where this middle point was used as the reference point to deform the bone to represent the OI bone geometry. Figure 1 shows the radiography CT image from enrolled OI patient whereas Figure 2 shows the reconstruction of OI finite element model.

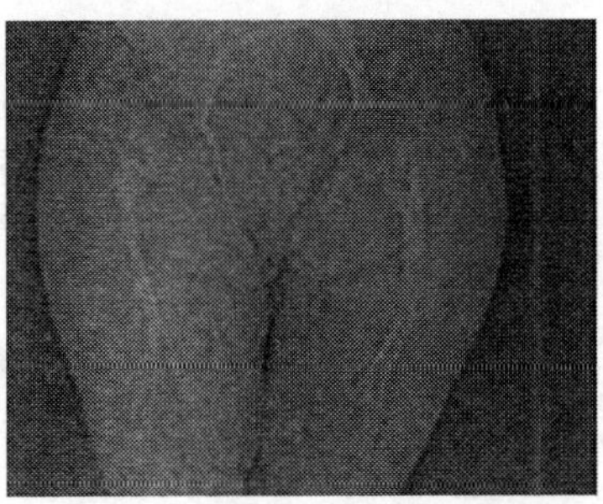

Figure 1. CT scan image of real OI patient.

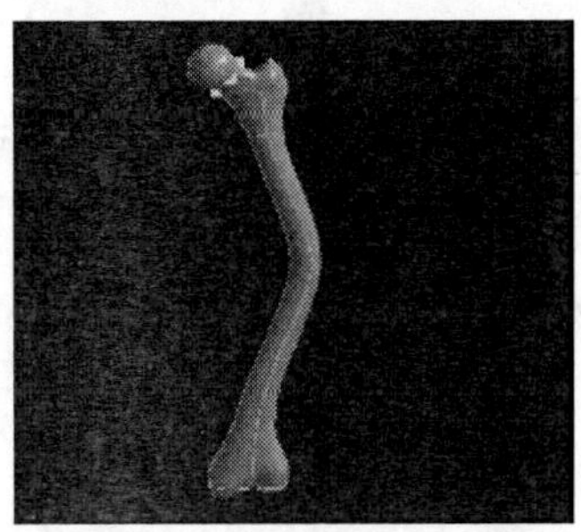

Figure 2. Reconstruction of finite element OI model.

2.3. Material properties

Material properties of the reconstructed OI bone was set as isotropic linear elastic with the Young's Modulus of 19 GPa and the Poisson ratio of 0.3 [13].

2.4. Boundary conditions

Boundary conditions were set to represent the standing, walking and jumping conditions. In standing, the compressive force was assumed as one-third of body weight. Walking takes approximately three to seven times of body weight, whereas jumping greatly affected on 15 to 20 times of body weight [14]. The body weight of real OI patient which is 37.5 kg was used as the main force acted upon the femoral head which indicated the children for age between 13 years old to 18 years old [15]. The constraint was set at the medial condyles, lateral condyles and intercondylar fossa of the femur. Figure 3 illustrates the force and constraint used throughout this study.

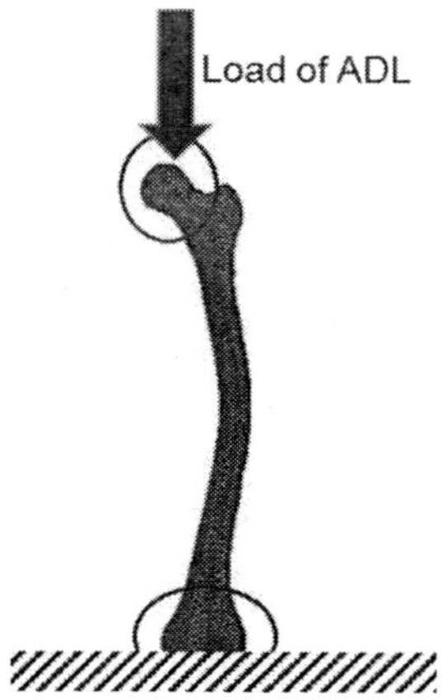

Figure 3. Setup of boundary conditions on femoral head (ADL load) and medial condyles (fixed).

2.5. Fracture risk analysis

Bone fracture risk is clinical risk used to determine the fragility of the bone. Fracture risk calculation was made based on formulation by Keyak et al. [16]. Fracture strength of 115MPa was considered into this study based on nanaindentation test performed by previous study [13]. Maximum principle stress is a direct value obtained through the simulation for all boundary conditions.

$$\text{Fracture risk} = \frac{\text{Fracture strength}}{\text{Maximum principle stress}} \quad (1)$$

3. Result & Discussion

Figure 4 presents the graph of fracture risk against ten OI models with variation of bowing angles. In Figure 4(a), for standing configuration, noticed here that the trend shows in decreasing pattern when the bowing angle increased. During the first three bowing angles which are 7.5°, 10.0° and 12.5°, fracture rik decrease as the OI bowing angles increase. When bowing angle reaches 15.0°, 17.5°, 20.0° and 22.5°, it is observed that there is a slight change of the trend which shows the constant pattern. However, at 25.0°, 27.5°and 30.0°, there is an increment in fracture risk. On the other hand, Figure 4(b) shows the fracture risk for walking and jumping configuration. The black line indicates the fracture risk level where the value less than one (FR<1), there is a fracture and vice versa. Discovered here that in both walking configuration, for minimum and maximum walking given the load possessed by OI patient, the result shows that all OI models have the ability to walk on their own without any aid as the fracture risk is above fracture risk value of 1. This result supported by previous study conducted by Fritz et al, in which during seven phase of gait cycle, all the OI patient able to perform gait cycle without fracture [5]. However, a precaution should be taken and the patient must be monitored even though the patients can perform this tasks on their own because any external impact from the surrounding might eventually cause a challenge to them which lead to fracture. For jumping configuration, the analysis pointed that for minimum jumping where involve 15 times on body weight can actually have accomplished by the OI models. On the other hand, the maximum jumping will definitely cause fracture as all OI models show the value of fracture risk below 1. Highlighted here, for walking and jumping, the trend was different compared to the standing configuration. The trend which fairly remains at constant was observed for walking and jumping configurations indicated that there are only a slight difference in fracture load between each OI models when performing walking and jumping. The compressive force acting on the femoral head during all configurations seem to affect mostly at the femoral shaft region represented by the red contour. The stress concentration accumulated at the middle of femoral shaft part at the bowing angles region. The stress distributed among the femur region and at the femoral head and medial condyles, lateral condyles and intercondylar fossa of the femur do not seem to be affected by the compressive force acted upon standing, walking either jumping. On the other, the middle femoral shaft region is great suffering from a fracture. This circumstance can be related to the bone geometry itself. Well aware that in OI bone, the progressive deformities of long bones are heredity disease which mainly affected the porosity of bone and hence influence the mechanical strength of OI bone [17]. Thus far, the bending region of OI bone is considered as the most fragile region and will eventually break once vulnerable to an excessive amount of force.

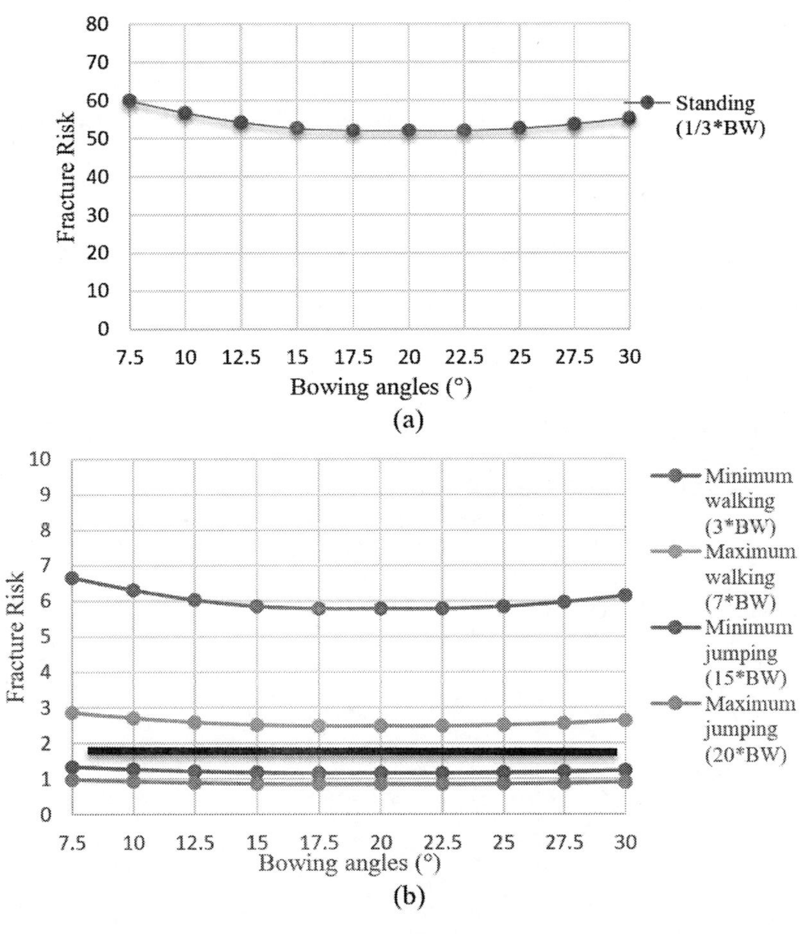

Figure 4. Fracture risk with variation bowing angles under loads of (a) standing, (b) walking and jumping.

4. Conclusion

In conclusion, the ability of OI models to perform the daily basic tasks were clearly simulated through finite element analysis in this study. The finite element OI models used to represent the real OI patients conducted in this study proposed that in standing and walking configuration, all level of OI models were able to perform the tasks without fracture given the body weight of 37.5 kg. On contrary, the jumping configuration will experience fracture given the same load. However, minimum jumping barely able to endure under this circumstance. Overall, this study presents the activities of daily basic greatly influence the mobility of OI patients and risk of fracture.

Acknowledgment

The authors would like to thanks for financial support from the Ministry of Education Malaysia through Fundamental Research Grant Scheme (FRGS/1/2016/TK03/UNIMAP/02/6).

References

[1] Fratzl-Zelman N, Barbara M, Klaus K, Klausing and Roschger P 2015 Wien. Med. Wochenschr.**165** pp 271–277
[2] Primorac D, D Anticević, I Barisić, D Hudetz and Alan I 2014 *Coll. Antropol.* **38** pp 767–72
[3] Marina J C 2011 *Kliegman RM, Behrman RE, Jenson HB (Ed.), Nelson.textb. Pediatr.* **19**
[4] Caouette C, Ikin N, Villemure I, Arnoux P-J, Rauch F and Aubin C-É 2017 *Med. Biol. Eng. Comput.* **55** pp 549-560
[5] Fritz J M, Guan Y, Wang M, Smith P A and Harris G F 2009 *Med. Eng. Phys.* **31** pp 1043–1048
[6] Yoshiwara Y, Clanche M, Basaruddin, K S, Takano N, Nakano T 2011 *J. Biomech. Sci. Eng.* **6** pp 270-285
[7] Basaruddin K S, Takano N, Yoshiwara Y, Nakano T 2012 *Med. Biol. Eng. Comput.* **50**, pp. 1091-1103
[8] Fan Z, Smith P, Reiners K, Hassani S and Harris G 2004 *Conf. Proc. IEEE Eng. Med. Biol. Soc.* **7** pp 4884–4887
[9] Wanna S B C, Basaruddin K S, Som M H M, Hashim M S M, Daud R, Majid M S A, Sulaiman A R 2017 *J. Physics: Conf. Series* **908**
[10] Halim S A, Basaruddin K S, Daud R, Safar M J A, Rahman A S A 2015 *ARPN J. Eng. Appl. Sci.* **10** pp 9353–9358.
[11] Basaruddin K S, Kamarrudin N S, Ibrahim I 2014 *Lat. Am. J. Solids Struct.* **11** 755–769
[12] Viceconti T A, Casali M, Massari B, Cristofolini L and Bassini S 1996 *J. Biomech.* **2** 1241
[13] Fan Z F, Smith P, Rauch F and Harris G F 2007 *Compos Part B Eng.* **3** p 411
[14] Dalla Pria Bankoff A 2012 *Hum. Musculoskelet. Biomech.* pp.61–86
[15] Graff K, Syczewska M, 2017 *Eur. J. Pediatr* **176** pp. 311–316
[16] Keyak J H, Rossi S A, Jones K A 1998 *J. Biomech* ***31*** p 125
[17] Zimmermann E A, Busse B, Ritchie R O 2015 *Bonekey Rep.* **4**, pp 1–13

The Effect of Multiple Surface Treatments on Oil Palm Empty Fruit Bunch (OPEFB) Fibre Structure

M K Faizi[1,2,a], Shahriman A B [1,2], M S A Majid[1], Shamsul B M T [3], A S M Nawawi[1], I Zunaidi[1], Ng Y G[3], Z M Razlan[1,2] and W K Wan[1]

[1]School of Mechatronic Engineering, Universiti Malaysia Perlis, Pauh Putra Campus 02600 Arau, Perlis, Malaysia.

[2]Motorsport Technology Unit, Universiti Malaysia Perlis, Pauh Putra Campus 02600 Arau, Perlis, Malaysia.

[3]Faculty of Medicine and Health Science,Universiti Putra Malaysia, 43400 UPM Serdang, Selangor Darul Ehsan.

[a]mkhairulfaizi@unimap.edu.my

Abstract. The key objective of this work was to study the effect of multiple surface treatments on the oil palm empty fruit bunch (OPEFB) fibres. To achieve this purpose, the OPEFB fibres were treated with the sodium hydroxide (NaOH) (3%), silane (2%), a combination of NaOH (3%) and silane (2%) (NaOH+silane), and NaOH (3%) prior silane (2%) (NaOH-silane). The soaking time was standardized for 7 hours. The multiple treated and untreated fibres were subjected to single fibre test and scanning electron microscopy (SEM). The single fibre tests were performed according to ASTM D3822-07 standard using INSTRON Micro tester while microscopy was observed by using Scanning Electron Microscope (SEM) TM3000. The result shows that the treated fibres with 3% concentration of NaOH have achieved the maximum ultimate tensile stress of 82.04 MPa compared with other treatment and untreated fibre.

Keywords: Natural fibre reinforced composites; natural fibres; alkali treatment, silane treatment, cellulosic.

1. Introduction

Due to the light weight, high quality to weight proportion and the decomposition necessary in building and designing areas, natural fibre composite is a promising future composite material [1]. The oil palm empty fruit bunch (OPEFB) is one of the potential natural fibres to be explored [2]. Well known that Malaysia has the successful oil palm industry that contributes to the major national financial development [3]. Found that, as much as 90% of the aggregate biomass delivered into the wastes whereby its consist of OPEFB, trunks, kernel shells and oil palm fronds and part of that only 10% of it delivered as oil [4]. In real field, certain waste like kernel shells and fronds can be used as boiler organic fuel and food for livestock animals respectively. However, most of it likes OPEFB and trunks tend to dispose and the disposal process is always exposed to the pollution issue; OPEFB and trunks are mostly incinerated. Based on the observations, the OPEFB fibre waste can be changed into useful products by utilising fibre based (natural fibre) composite as opposed to blazing it. The celluloses, hemicelluloses and lignin are the three primary constituent's structures of OPEFB. The modification of these constituents might result for a steady holding among matrix and fibre with increasing the mechanical properties of the composite. For this purpose, chemical surface treatment on fibre is the common approach for fibre constituent's modification in order to achieve a good consolidation between fibre and matrix [5]-[9]. The common chemicals such as sodium hydroxide (NaOH), acetyl acid, and silane were used for surface treatments [10], [11]. In this paper, elastic properties; tensile, modulus of elasticity and morphological structure of the fibres are discussed.

2. Materials and methods

The United Oil Palm Industries, Nibong Tebal, Pulau Pinang had supplied the oil palm empty fruit bunch (OPEFB) fibres for this research. Following the water retting process, the fibres were manually extracted hair by hair. The water retting process took about a week to remove the residual oil and cleaned using distilled water. Next, the fibres dried under the sun to ensure the removal of its moisture content. The OPEFB fibres were then treated with NaOH (3%), silane (2%), a combination of NaOH (3%) and silane (2%) (NaOH+silane), and NaOH (3%) prior silane (2%) (NaOH-silane). Entire treatments were standardized for 7 hours of soaking time at room temperature and maintaining the liquor ratio 40:1 to remove the hemicelluloses and surface impurities of the fibre. Finally, the fibres were cleaned using distilled water and dried at room temperature. The measurement of the length, and diameter were taken as shown in Table 1. The measurements were performed prior to the tensile test

Table 1. Physical properties of OPEFB fibres. (40 samples for each treatment).

Type of treatment	Average diameter (mm)	Fibre length (mm)
Untreated	0.475 ± 0.012	60
NaOH (3%)	0.375 ± 0.009	60
Silane (2%)	0.449 ± 0.015	60
NaOH (3%) + silane (2%)	0.323 ± 0.012	60
NaOH (3%) – silane (2%)	0.358 ± 0.007	60

The single fibre tests were performed to determine the tensile strength of the fibre according to ASTM D3822-07 standard using INSTRON Micro universal testing machine with a load cell of 2kN. The fibre was mounted onto a tab-shaped piece of paper with the gauge length 40 mm. The dimensions of the multiple treated and untreated fibres were measured using digital microscope and weighed using an analytical balance device. The fibre's surface morphological were observed using Scanning Electron Microscope (SEM) TM3000.

3. Result and discussion

Figure 1 shows the stress–strain responses of untreated and multiple treated of OPEFB fibers. The NaOH (3%) treated fibers exhibits a higher ultimate tensile stress following NaOH (3%)+silane (2%), silane (2%), NaOH (3%)-silane (2%), and untreated for decreasing pattern. The treated fibers exhibit the higher tensile stress due to changes in the cellulose crystallinity. The change process involves the removal of weak amorphous of the fibers. The removal enables the fibril to rearrange them in a more compact manner, thus enhancing the tensile strength of the fiber [8].

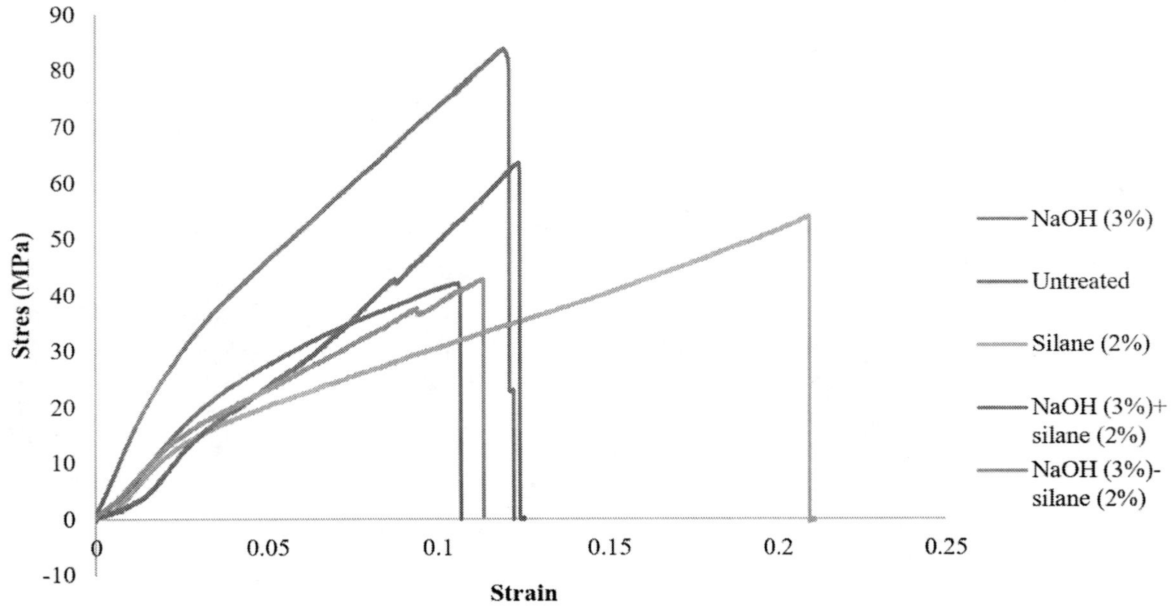

Figure 1. Stress–strain responses for untreated and multiple treated of OPEFB fibre.

A Figure 2 exhibits the clear view of the average tensile stress of fibres with 82.04 MPa is a higher value represents NaOH (3%) treatment followed by 61.47 MPa, 54.32 MPa, 42.1 MPa, and 39.99 MPa for NaOH (3%) +silane (2%), silane (2%), NaOH (3%) -silane (2%), and untreated respectively. This indicates that the NaOH(3%) had increased exceeded 100 % from the untreated fibre.

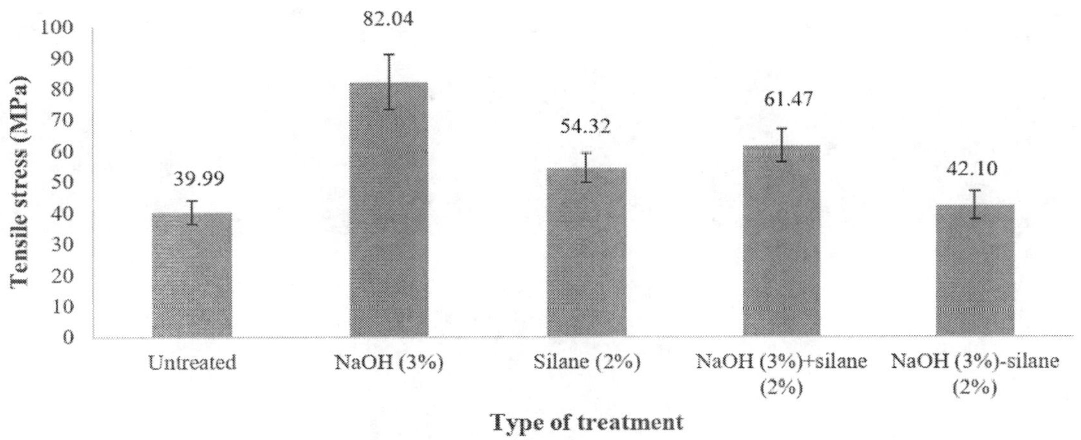

Figure 2. The average tensile stress of untreated and multiple treated of OPEFB fibres.

The highest average Young's Modulus of OPEFB fibres recorded is 1.05 GPa for silane (2%) while the lowest is 0.79 GPa for NaOH (3%)-silane (2%). Sighted these results in Figure 3, all the surface treatments improved the fibre elasticity compared with untreated except for the NaOH (3%)-

silane (2%). The NaOH (3%)-silane (2%) indicates a slight reduction in elasticity about 3.8%. Hence, the different modulus elasticity may result in the different strain to failure of specific fibre.

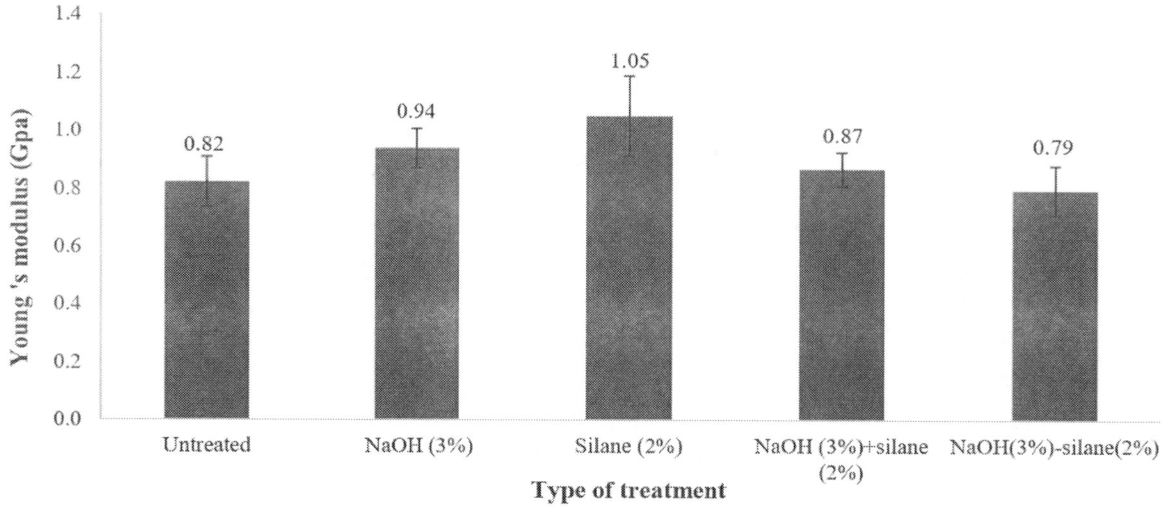

Figure 3. Modulus of Young for untreated and multiple treated OPEFB fibres.

The Scanning Electron Microscope (SEM) was used to learn about the surface morphology of the untreated and multiple treated OPEFB fibres and its exhibits in Figure 7. The untreated and NaOH (3%)-silane (2%) treated fibres depicts an almost similar physical condition whereby the impurities clearly existed on the fibres surface as shown in Figure 7(a), and (e). The NaOH (3%) and NaOH (3%)+silane (2%) were also depicted the similar physical condition which the rough surfaces existed as shown in Figure 7(b), and (d). Meanwhile, for silane (2%) in Figure 7(c), the surface was not very clean compared with others but it has fewer impurities and a rough surface existed. Due to fewer impurities and rough surface existed, the silane (2%) treated fibres registered as higher Young' Modulus.

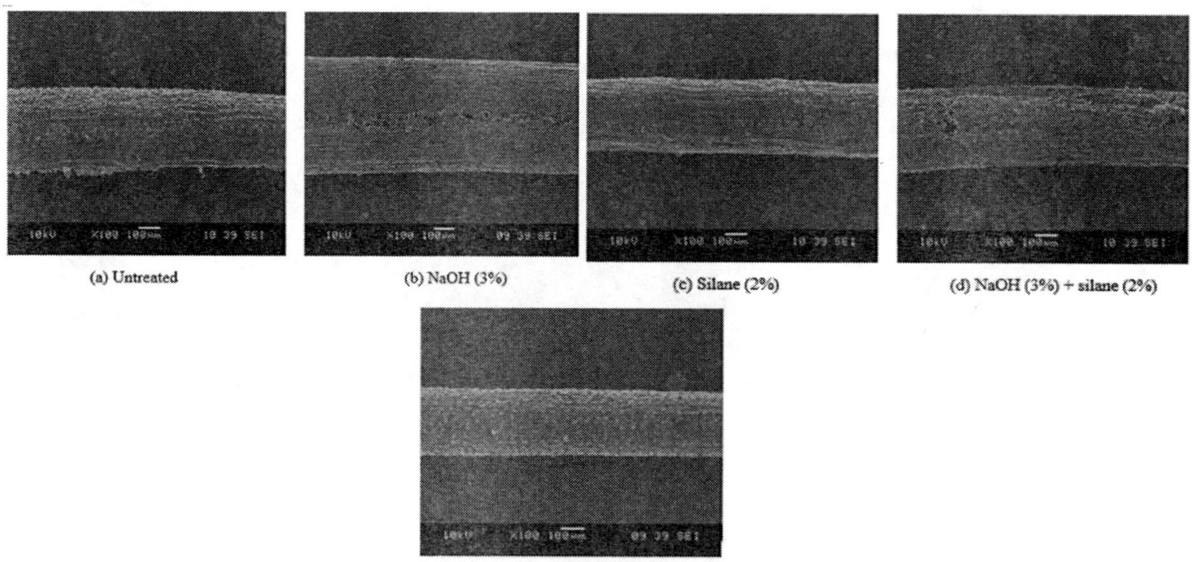

Figure 4. Surface morphologies for untreated and multiple treated of OPEFB fibres.

. Conclusions
The mechanical properties of the multiple treated and untreated OPEFB fibres were investigated in this paper. Found that, all treated OPEFB fibres shows the increasing of tensile strength compared with untreated fibres. The structure of the fibre also differences between the multiple treated and untreated fibre. The result shows that the NaOH (3%) treated fibre exhibited better tensile strength with 82.04 MPa compared with other treated and untreated. The surface observation through Scanning Electron Machine (SEM) shows the texture of the untreated and NaOH (3%)-silane (2%) fibres achieved the similar morphology with the impurities were significant while NaOH (3%) and NaOH (3%)+silane (2%) treated also have the similar condition with surface roughness existed. Finally, silane (2%) treated fibre shown different physical when no significant impurities and surface roughness existed. The study further supported the feasibility of utilising OPEFB fibres as reinforcing materials in polymer composites.

Acknowledgement
The authors would like to be obliged to Universiti Malaysia Perlis (UniMAP) for providing laboratory facilities, Motorsport Technology Unit (MOTECH) UniMAP for providing financial assistance, and Universiti Putra Malaysia (UPM) as collaboration partnership.

References
[1] R. K. and R. S. Sandeep Bhardwaj 2013 Natural Fibre Composites and Its Potential, vol. 1, no. 4, pp. 37–40.
[2] M. K. Faizi, Shahriman A.B., M.S. Abdul Majid, Shamsul B. M. T., S. N. Basah, E. M. Cheng, M. Afendi, Zuradzman M. R., Khairunizam WAN, and D. Hazry 2017 An overview of the oil palm empty fruit bunch (OPEFB) potential as reinforcing fibre in polymer composite for energy absorption applications, MATEC Web of Conferences 90.
[3] M. F. Awalludin, O. Sulaiman, R. Hashim, and W. N. A. W. Nadhari 2015 An overview of the oil palm industry in Malaysia and its waste utilization through thermochemical conversion, specifically via liquefaction, Renew. Sustain. Energy Rev., vol. 50, no. October, pp. 1469–1484.
[4] M. F. Awalludin, O. Sulaiman, R. Hashim, and W. N. A. W. Nadhari 2015 An overview of the oil palm industry in Malaysia and its waste utilization through thermochemical conversion, specifically via liquefaction Renew. Sustain. Energy Rev., vol. 50, pp. 1469–1484.
[5] M. M. Kabir, H. Wang, K. T. Lau, and F. Cardona 2013 Effects of chemical treatments on hemp fibre structure, Appl. Surf. Sci., vol. 276, pp. 13–23.
[6] W. Fatra, R. Sanjaya, H. Rionaldo, and Z. Helwani 2013 Alkaline Treatment of Oil Palm Frond Fibers by Using Extract of Oil Palm Empty Fruit Bunch Ash for Use in Natural Fiber Reinforced Composite, vol. 47, no. October, pp. 1–10.
[7] W. Fatra, H. Rouhillahi, Z. Helwani, Z. Zulfansyah, and J. Asmura 2016, Effect of Alkaline Treatment on the Properties of Oil Palm Empty Fruit Bunch Fiber-reinforced Polypropylene Composite, Int. J. Technol., vol. 7, no. 6, p. 1026.
[8] M. J. M. Ridzuan, M. S. Abdul Majid, M. Afendi, S. N. A. Kanafiah, and M. B. M. Nuriman 2015 Effects of alkaline concentrations on the tensile properties of Napier grass fibre Appl. Mech. Mater., vol. 786, no. June, pp. 23–27.
[9] M. K. Faizi, A.B. Shahriman, M. S. Abdul Majid, Z. A. Ahmad, B. M. T. Shamsul, and Y. G. Ng 2017 The effect of alkaline treatments soaking time on oil palm empty fruit bunch (OPEFB) fibre structure, IOP Conf. Series: Journal of Physics: Conf. Series 908.
[10] Li Xue, Tabil Lope G., Panigrahi Satyanarayan, "Chemical treatments of natural fibre for use in natural fibre reinforced composites: A review," Journal of Polymers and the Environment., vol. 15, February, pp. 25-33, 2017.
[11] Islam M.S., Pickering K.L., Foreman N.L., "Influence of alkali treatment on the interfacial and physic-mechanical properties of industrial hemp fibre reinforced polylactid acid composites," Composites Part A: Applied Science and Manufacturing., vol. 41, August, pp. 596-603, 2009.

IOP Conf. Series: Materials Science and Engineering **429** (2018) 012006 doi:10.1088/1757-899X/429/1/012006

The Effect of Alkaline Treatments with Various Concentrations on Oil Palm Empty Fruit Bunch (OPEFB) Fibre Structure

M K Faizi[1,2,a], Shahriman A B[1,2], M S A Majid[1], Shamsul B M T[3], I Zunaidi[1], Ng Y G [3], Z M Razlan[1,2] and W K Wan[1]

[1]School of Mechatronic Engineering, Universiti Malaysia Perlis, Pauh Putra Campus 02600 Arau, Perlis, Malaysia.

[2]Motorsport Technology Unit, Universiti Malaysia Perlis, Pauh Putra Campus 02600 Arau, Perlis, Malaysia.

[3]Faculty of Medicine and Health Science, Universiti Putra Malaysia, 43400 UPM Serdang, Selangor Darul Ehsan.

[a]mkhairulfaizi@unimap.edu.my

Abstract. The main goal of this work is to study the effect of different concentration level of alkaline solution for oil palm empty fruit bunch (OPEFB) fibre structure. As to achieve this objective, the OPEFB fibres were treated with 2 hours of soaking time with 3%, 5%, 7%, and 10% of sodium hydroxide (NaOH) concentration portion. The single test for treated and untreated fibers was then carried out according to the ASTM D3822-07 standard. Next, the surfaces of the fibres prior and after the treatment were observed with a scanning electron microscope (SEM) TM3000. The result shows that the 3% of the NaOH concentration exhibits better tensile strength compared to the other concentrations and untreated fiber.

Keywords: Natural fibre reinforced composites; natural fibres; alkali treatment, cellulosic.

1. Introduction

Recently, the utilising natural fibres as composite have expanded significantly in a couple of years. It is promising as future composite material due to the light weight, high quality to weight proportion and the decomposition necessary in building and designing areas [1]. One of the potential natural fibres to be explored is the oil palm empty fruit bunch (OPEFB) [2]. The oil palm trees can grow well in Malaysia atmosphere. It has extremely turned into the most critical rural product in Malaysia and has been most successful industry that contributes to the national financial development after petroleum [3]. However, only just 10% of the aggregate biomass delivered into oil and 90% of it delivered as wastes. The wastes consist of large amounts of lignocellulosic material, for example, OPEFB, trunks and oil palm fronds [4]. In order to dispose the waste, the process is always exposed to the pollution issue; OPEFB and trunks are mostly incinerated. Found that these agriculture wastes, particularly of OPEFB, can be changed into useful products by utilising fibre based (natural fibre) composite as opposed to blazing the OPEFB. The structure of OPEFB comprises of three primary constituents; celluloses, hemicellulose and lignin. The steady holding among matrix and fibre can increase the composites mechanical properties by modifying the constituents. Thus, to achieve a good consolidation between fibre and matrix, surface treatment on fibre is the common approach adopted via chemical treatments for fibre constituent's modification purposes [5]-[9]. The sodium hydroxide (NaOH), acetyl acid, and silane were the common chemicals that used for surface treatments [10]. However, this research is focused on the effect of NaOH concentration of the OPEFB fibre. The elastic properties; tensile, modulus of elasticity and morphological structure of the fibres are discussed.

2. Materials and methods

The fibres that obtained from United Oil Palm Industries, Nibong Tebal, Penang were manually extracted prior continue to water retting process for a week. The fibre was then cleaned using distilled water and dried under the sun for removal of the moisture content. Next, the processed of OPEFB fibre was treated with 3%, 5%, 7% and 10% aqueous solutions of NaOH at room temperature for 2 hours, maintaining the liquor ratio 40:1 to remove the hemicelluloses and surface impurities of the fibre. Finally, the distilled water was used to clean the fibres before it dried at room temperature. The physical properties of the treated and untreated fibres are presented in Table 1.

Table 1. Physical properties of OPEFB fibres. (20 samples for each NaOH concentration)

NaOH concentration (%)	Average diameter (mm)	Standard deviation	Average length (mm)
3	0.34	0.05	70
5	0.27	0.04	70
7	0.29	0.04	70
10	0.31	0.05	70
Untreated	0.47	0.06	70

ASTMD3822-07 standard was used in single fibre testing to determine the tensile strength of the OPEFB fibres. The treated and untreated fibres were weighed and physically measured by using an analytical balance device, and microscopic. Meanwhile the surface of the fibres was observed using Scanning Electron Microscope (SEM) TM3000. The fibres were dried by using oven to the temperature of 60°C to 100°C about 30 minutes to an hour before proceeding with the test. The fibre was mounted onto a tab-shaped piece of paper with the gauge length is 70 mm. The INSTRON 5848 Micro universal testing machine with a load cell of 2kN were used to determine the tensile strengths of the treated and untreated fibres.

3. Result and discussion

The stress–strain responses for untreated and alkali-treated OPEFB fibres are shown in Figure 1. The treated fibres exhibit a higher ultimate tensile stress due change in the cellulose crystallinity. The change process involves the removal of weak amorphous of the fibers. The removal enables the fibril to rearrange them in a more compact manner, thus enhancing the tensile strength of the fiber [8]. The result shows that the 3% of NaOH concentration have the highest ultimate tensile test. Following the 10%, 5%, 7% of the NaOH concentration, there is an increase in the tensile stress but a decrease in strain compared to untreated fibers.

As shown in Figure 2, the 3% treated fibres exhibit 111.8 MPa which is the highest tensile strength. The tensile strength initiates to decrease from 100.8 MPa, 68.4 MPa, 31.6 MPa, and 28.9 MPa with 7%, 5%, 10%, and 28.9% respectively. The alkali treatment causes fibrillation, which is a process that causes the fibre bundle to break into smaller bundles. Smaller bundles help to distribute the load applied throughout the fibres [9].

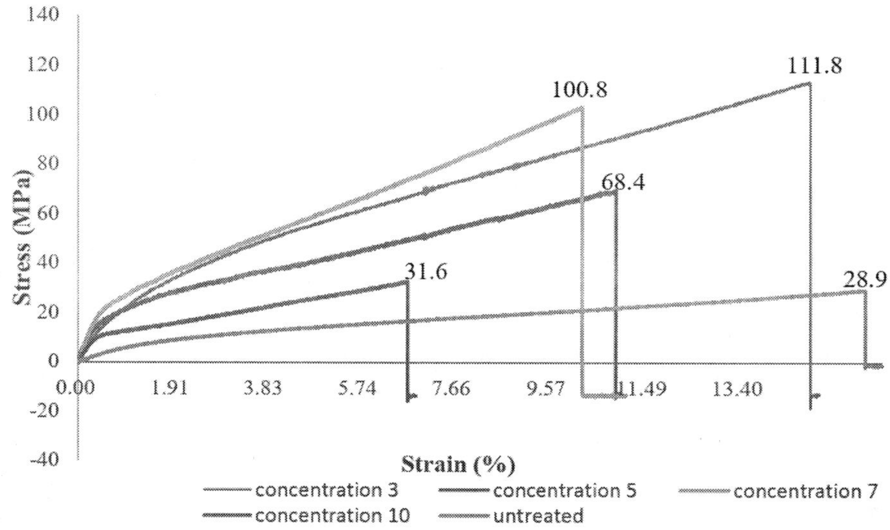

Figure 1. Stress–strain responses for untreated and treated of OPEFB fibres

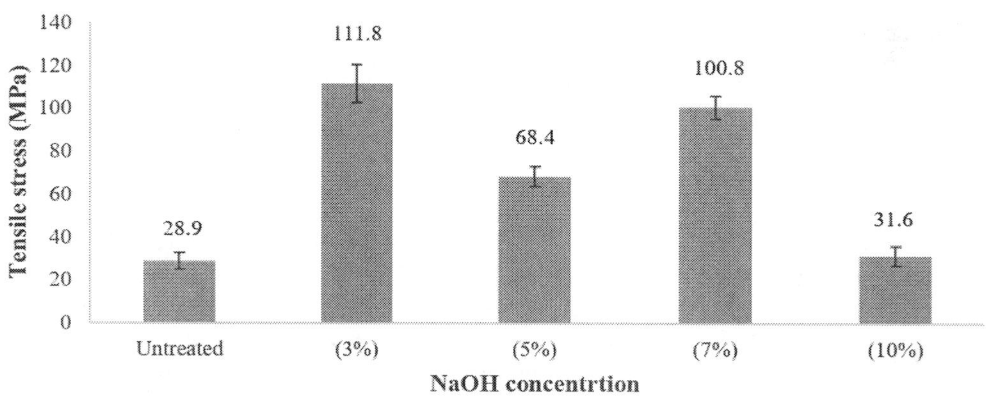

Figure 2. The average tensile stress of untreated and multiple treated of OPEFB fibres.

Based on Figure 3, the highest average Young's Modulus of OPEFB fibre recorded is 2.6 GPa 5% of NaOH concentration. While the lowest average modulus of elasticity is 3% with 0.6 GPa although previously it exhibited the highest tensile strenght. Seeing these results, the alkali treatment improved of the fibre elasticity compared with untreated fibres concerning the alkali concentration except for 3% of NaOH whereby reduced about 25%. However, after 5% treatment the modulus tremendously increased in the highest value before dropping for 1.8 GPa under 7% treatment. The modulus then indicates a slight increase for 2 GPa correlate with the increasing of alkali concentration in 10%. Hence, the different modulus elasticity may result in the different strain to failure of specific fibre.

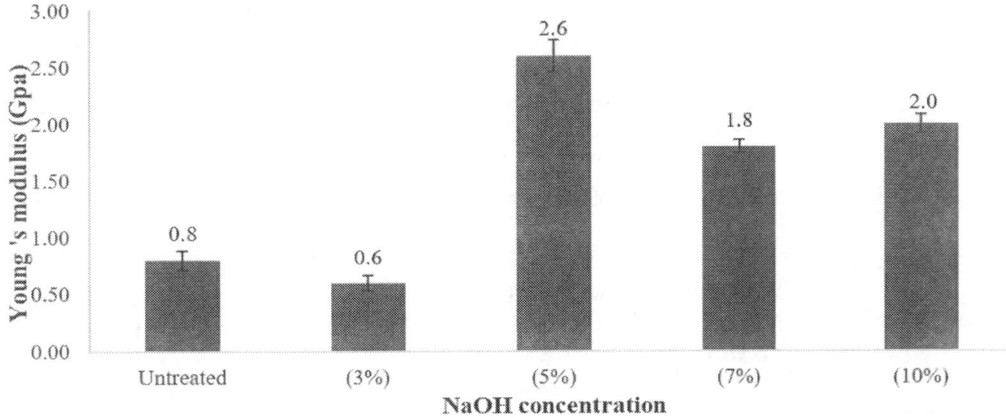

Figure 3. Young's Modulus of OPEFB fibres

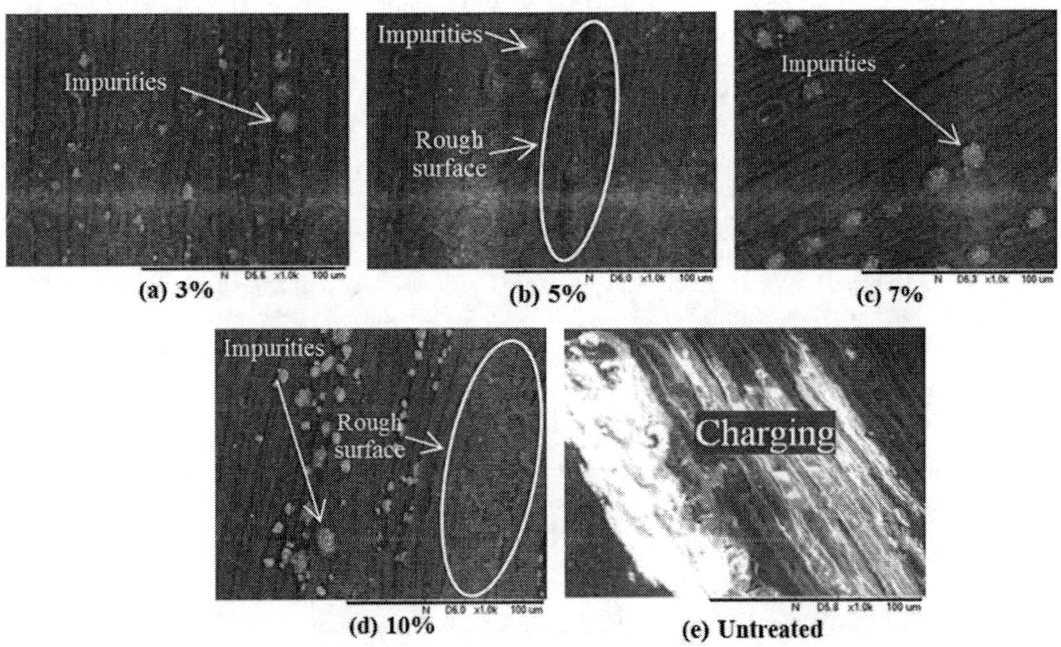

Figure 4. Surface morphologies for untreated and multiple treated of OPEFB fibres.

Figure 4 shows the SEM micrographs of the surface for untreated and treated OPEFB fibres. The treated fibres exhibit cleaner surfaces when the impurities reduced compared to untreated. The white spots in Figure 4 represent the presence of impurities on the surface of the fibres. However, the 5% and 10% NaOH concentration shows that the rough surfaces existed as depicted in Figure 4(b), (d) and its differ compare with 3% and 7% NaOH concentration which were smoother as shown in Figure 4(a), and (c). While Figure 4 (e) represents untreated fibers whereby clearly contain severe impurities, wax, fatty substances and globular pro- trusions called tyloses [11]. Based on this morphological observation, the alkali treatment caused the removal of impurities on the OPEFB fibres.

Conclusions

The mechanical properties of the treated and untreated OPEFB fibres with 2 hours soaking time accompanied by different NaOH concentrations were investigated in this paper. It is shown that the OPEFB fibres were increased in their tensile strength when treated in alkaline solution but slightly differ on Young's Modulus when only 3% NaOH concentration shows the reduction compare to untreated fibre. The 3% NaOH concentration shown the highest tensile strength compared to the other soaking time and untreated fibre. The surface observation through Scanning Electron Machine (SEM) shows the texture of the fibres achieved the best result after alkaline treating when the impurities removed from the fibres. These results are quite promising as for further study to support the feasibility of utilising OPEFB fibres as reinforcing materials in polymer composites.

Acknowledgement

The authors would like to be obliged to Universiti Malaysia Perlis (UniMAP) for providing laboratory facilities, Motorsport Technology Unit (MOTECH) UniMAP for providing financial assistance, and Universiti Putra Malaysia (UPM) as collaboration partnership.

References

[1] R. K. and R. S. Sandeep Bhardwaj 2013 Natural Fibre Composites and Its Potential, vol. 1, no. 4, pp. 37–40.

[2] M. K. Faizi, Shahriman A.B., M.S. Abdul Majid, Shamsul B. M. T., S. N. Basah, E. M. Cheng, M. Afendi, Zuradzman M. R., Khairunizam WAN, and D. Hazry 2017 An overview of the oil palm empty fruit bunch (OPEFB) potential as reinforcing fibre in polymer composite for energy absorption applications, MATEC Web of Conferences 90.

[3] M. F. Awalludin, O. Sulaiman, R. Hashim, and W. N. A. W. Nadhari 2015 An overview of the oil palm industry in Malaysia and its waste utilization through thermochemical conversion, specifically via liquefaction, Renew. Sustain. Energy Rev., vol. 50, no. October, pp. 1469–1484.

[4] M. F. Awalludin, O. Sulaiman, R. Hashim, and W. N. A. W. Nadhari 2015 An overview of the oil palm industry in Malaysia and its waste utilization through thermochemical conversion, specifically via liquefaction Renew. Sustain. Energy Rev., vol. 50, pp. 1469–1484.

[5] M. M. Kabir, H. Wang, K. T. Lau, and F. Cardona 2013 Effects of chemical treatments on hemp fibre structure, Appl. Surf. Sci., vol. 276, pp. 13–23.

[6] W. Fatra, R. Sanjaya, H. Rionaldo, and Z. Helwani 2013 Alkaline Treatment of Oil Palm Frond Fibers by Using Extract of Oil Palm Empty Fruit Bunch Ash for Use in Natural Fiber Reinforced Composite, vol. 47, no. October, pp. 1–10.

[7] W. Fatra, H. Rouhillahi, Z. Helwani, Z. Zulfansyah, and J. Asmura 2016, Effect of Alkaline Treatment on the Properties of Oil Palm Empty Fruit Bunch Fiber-reinforced Polypropylene Composite, Int. J. Technol., vol. 7, no. 6, p. 1026.

[8] M. J. M. Ridzuan, M. S. Abdul Majid, M. Afendi, S. N. A. Kanafiah, and M. B. M. Nuriman 2015 Effects of alkaline concentrations on the tensile properties of Napier grass fibre Appl. Mech. Mater., vol. 786, no. June, pp. 23–27.

[9] M. K. Faizi, A.B. Shahriman, M. S. Abdul Majid, Z. A. Ahmad, B. M. T. Shamsul, and Y. G. Ng 2017 The effect of alkaline treatments soaking time on oil palm empty fruit bunch (OPEFB) fibre structure, IOP Conf. Series: Journal of Physics: Conf. Series 908.

[10] Li Xue, Tabil Lope G., Panigrahi Satyanarayan, "Chemical treatments of natural fibre for use in natural fibre reinforced composites: A review," Journal of Polymers and the Environment., vol. 15, February, pp. 25-33, 2017.

[11] M. A. N. Izani, M. T. Paridah, U. M. K. Anwar, M. Y. M. Nor, and P. S. H, "Composites : Part B Effects of fiber treatment on morphology , tensile and thermogravimetric analysis of oil palm empty fruit bunches fibers," vol. 45, pp. 1251–1257, 2013.

Analysis of local thickening by beading and compression using FEM

Liyana Tajul[1], Quratul Aini [1], Hasnulhadi Jaafar[2]

[1] School of Mechatronic Engineering, Universiti Malaysia Perlis, Pauh Putra Campus, 02600 Arau, Perlis, Malaysia.
[2]School of Manufacturing Engineering, Universiti Malaysia Perlis, Pauh Putra Campus, 02600 Arau, Perlis, Malaysia.

Abstract. This paper analysed the local thickening of tailored blanks using Finite Element Method. In this process, local thickening was produced at the centre of the sheet by beading and compression. In the beading process, the sheet is beaded freely and no change in thickness occurs. The beaded area was then compressed with no change in width to form a thicker portion at the centre of the sheet. The amount of thickening formed increases with the increase of the beading height. However for beading height 8 mm folding occurred. These processes do not involve joining and hence stress concentration by the change of thickness can be prevented.

1. Introduction

The demand of lightweight vehicle is increasing to reduce fuel consumption and global emission. Tailored blanks having thickness distribution are attractive in automotive industry for the purpose of weight reduction. Generally, sheets having different thicknesses and strength are joined together by laser welding and it is called tailor-welded blanks. Tailored blanks control the thickness distribution of sheet metals, and thus material utilization is improved, strength is increased and weight is reduced. For example, area requiring high strength is made thicker, while low strength area is made thinner.

Although tailor blanks are advantages, the stress concentration from the change in thickness at the weld area affects the formability of the parts [1]. As the thickness ratio increases, the formability of the tailor welded blanks decreases. Rojek et al. [2] investigated the mechanical properties of the weld zone of tailor-welded blanks. Satya Suresh et al. [3] investigated the formability of tailor-welded blanks joined by TIG and laser welding. During forming, thinner area deforms more as compared to the thicker area and hence failure occurs at the thinner one. Tailored blanks without joining are also produced by rolling process [4]. The gap between rolls is controlled to produce blanks having thickness distribution. The stress concentration is reduced from the gradual change in thickness between thick and thin areas. Han et al. [5] investigated the production of tailor rolled blanks with thickness change in both longitudinal and transverse directions. For cold stamping, the thin area produced by rolling possesses higher strength as compared to the thicker part due to work hardening. Sun et al. [6] reported that tailor rolled blanks for automotive parts improves the energy absorption and crashworthiness. Wernicke et al. [7] produced local thickening of sheets by in-plane swaging. A roller is used to form an accumulation of material and hence thickness is increased.

During stamping of sheet metal, the corners of the stamped part undergo excessive thickness reduction causing the part to fracture. Tan et al. [8] reported that thickness reduction occurred at the inner corner of a wheel disk after stamping for a uniform thickness sheet. Increasing the whole thickness of the part to prevent thinning at a small area will add extra weight and cost. Therefore an optimum thickness distribution is needed to improve the formability of the stamped part without adding extra weight and hence the material usage can be improved. Sheets having local thickening are attractive for this purpose as the thickness and strength of the part can be controlled. This study analysed the local thickening of tailored blanks by beading and compression using Finite Element Method (FEM).

2. Approach of local thickening by beading and compression

Figure 1 shows the stages in producing sheet having local thickening by beading and compression. In the 1st stage, the sheet is beaded freely and the beading portion is produced at the centre of the sheet. The width of the sheet reduces while no change in thickness occurs during this stage. Then in the 2nd stage (compression), the beaded portion is compressed to produce local thickening. During compression, force is applied on the flange of the sheet using springs and blank holders to prevent wrinkling. Both ends of the sheet are constrained to prevent change in the width direction. Floating die aids in removing the sheet after the compression process.

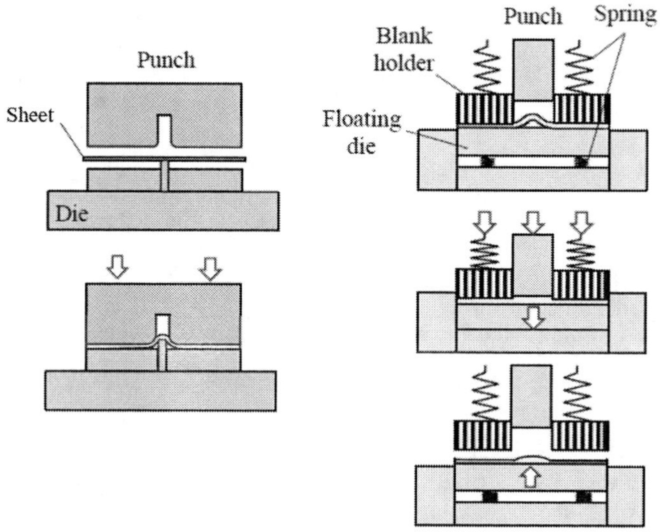

(a) Beading (b) Compression

Figure 1: Local thickening of sheet by beading and compression

The analysis of the two-stage forming process is conducted by Finite Element Method using Ansys Workbench software based on explicit-dynamic formulation. The width of the sheet is between 101 mm to 106.5 mm and the thickness is 1.0 mm. The width is varied according to the beading die height, h = 3, 5, and 8 mm. The dimension of the tools is given in Figure 2. Since the groove of the beading punch and the width of the compression punch is 20 mm, the width of the local thickening will be limited within 20 mm. The beading die is made flat with varying beading heights. The compression die is designed with flanges on both ends to constraint the width of the sheet. Figure 3 shows the finite element model of local thickening by beading and compression. The processes were conducted under one stroke. The material is Aluminium and the properties are as listed in Table 1. The tools are made rigid while the sheet is flexible. The meshing element of the sheet is quadrilateral /triangle.

IOP Conf. Series: Materials Science and Engineering **429** (2018) 012007 doi:10.1088/1757-899X/429/1/012007

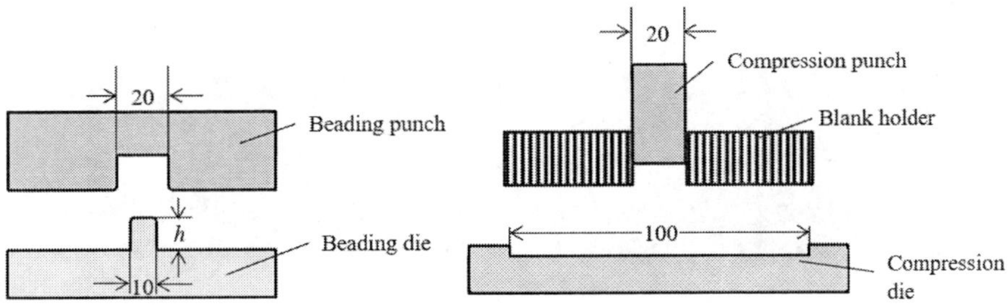

(a) Beading die (b) Compression die

Figure 2: Dimension of tools

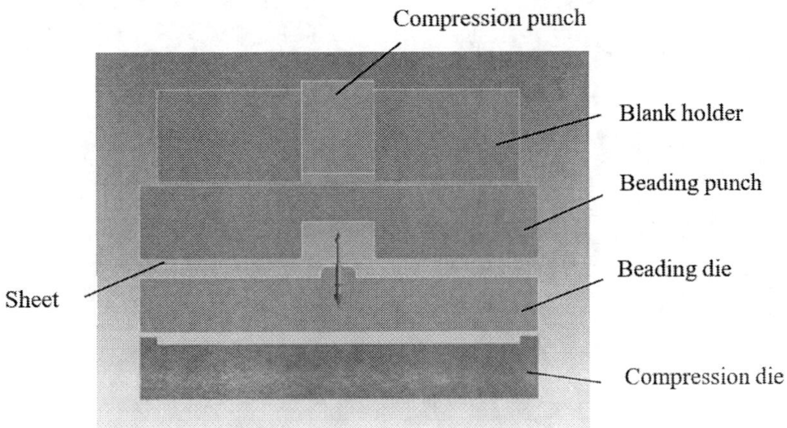

Figure 3: Finite element model of local thickening by beading and compression

Table 1: Material properties

Properties	Value
Density (kg/m^3)	2710
Young's modulus (GPa)	71
Yield Strength (MPa)	290
Poisson's ratio	0.3

3. Result & Discussion

Figure 4 shows the resulted sheet having thickness distribution after the beading and compression processes. The sheet follows the shape of the beading die of 180° and forms a beaded portion at the centre. No change in thickness occurs during beading. The compression die limits the change in the width direction to allow for thickening to form. Thickening are formed for beading height 3 mm and 5 mm while folding occurred for beading height 8 mm.

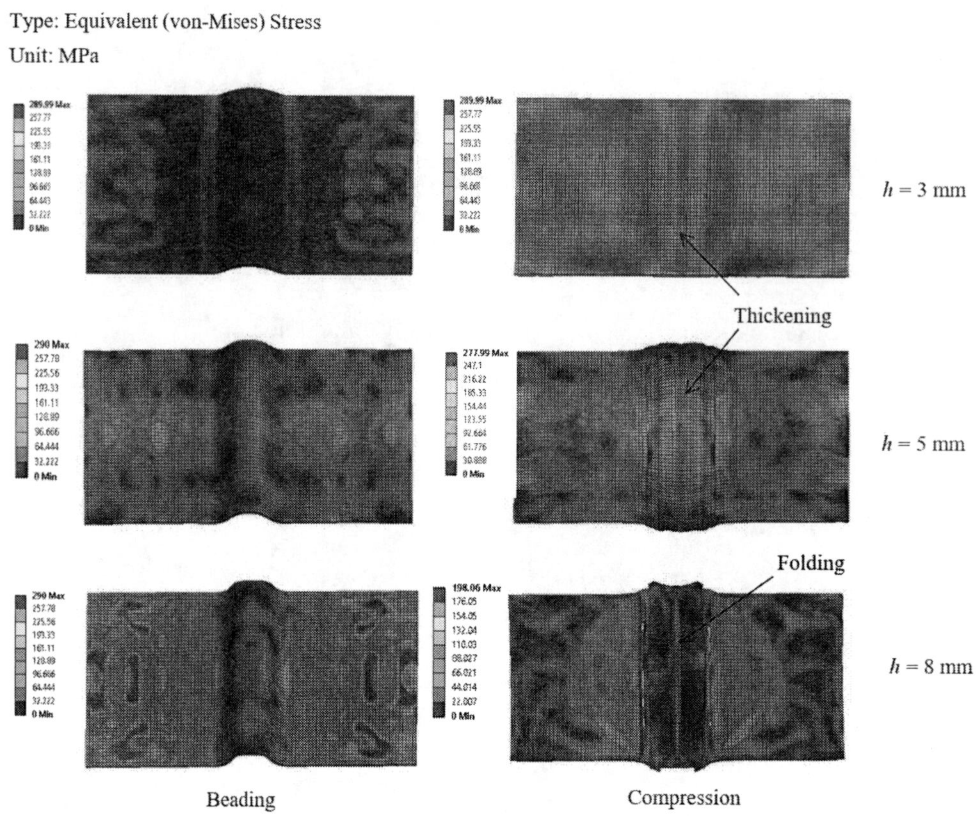

Figure 4: Sheets after beading and compression

The maximum change in thickness for beading height 3 mm and 5 mm are 5% and 70%, respectively (see Figure 5). As the beading height increases the maximum change in the thickness increases. A large beading height will produce large beaded area and hence more thickening are observed. However folding was observed for beading height, $h = 8$ mm. The shape of the beaded area affects the formation of thickening and folding. Figure 6 shows the cross-sectional view of sheets after beading and compression. The beaded area for beading height 3 and 5 mm is triangular while for beading height 8 mm the side walls are almost vertical. The vertical sidewall causes folding.

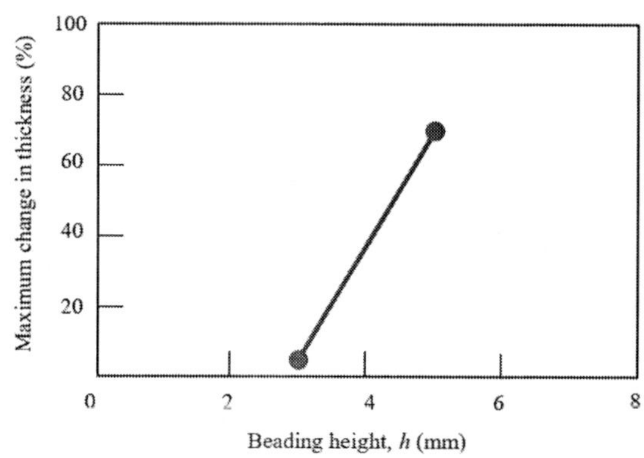

Figure 5: Maximum change in thickness after beading and compression

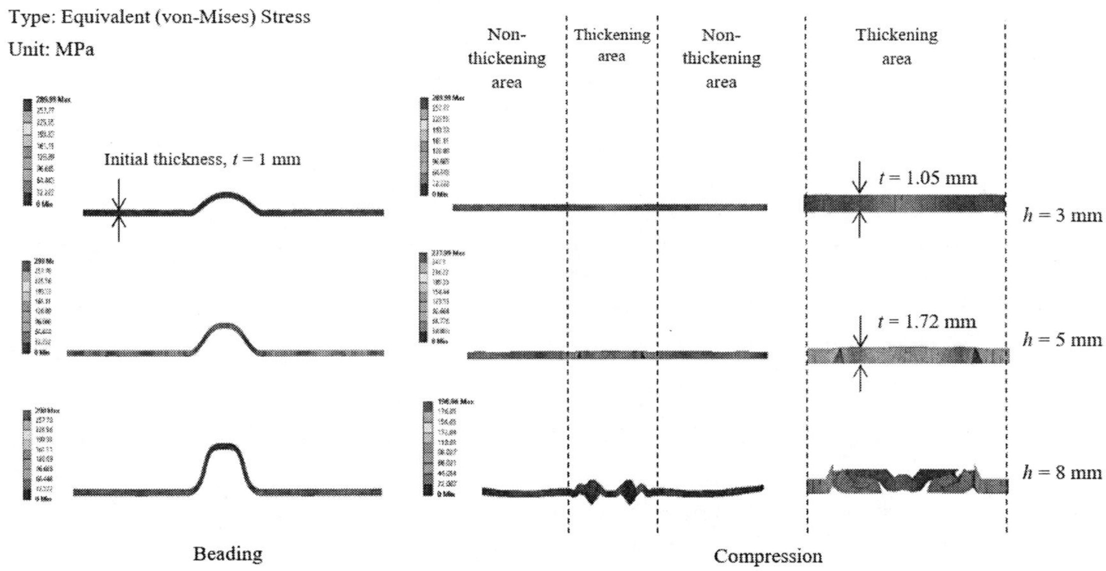

Figure 6: Cross sectional view of sheets after beading and compression

4. Conclusion

Local thickening by beading and compression has been analysed by using Finite Element Method. The sheets having thickness distribution were produced without joining. The amount of thickening formed can be controlled by varying the beading height. Beading height of 8 mm produced folding due to the vertical side walls formation after beading process. A formability test is to be conducted in the future to understand the effect of local thickening on the drawing depth and thinning compensation.

References

[1] Chan S M, Chan L C, Lee T C 2003 Tailor-welded blanks of different thickness ratios effects on forming limit diagrams *Journal of Materials Processing Technology* **132** 95–101.

[2] Rojek J, Hyrcza-Michalska M, Bokota A, Piekarska W 2012 Determination of mechanical properties of the weld zone in tailor-welded blanks *Archives of Civil and Mechanical Engineering* **12** 156-162.

[3] Satya Suresh V V N , Regalla S P, Gupta A K, Padmanabham G. 2015 Weld line shift in the case of tailor welded blanks subjected to differential strengths with respect to TIG and laser welding *Materials Today: Proceedings* 2, 3501-3510.

[4] Kopp R, Wiedner C, Meyer A 2005 Flexibly rolled sheet metal and its use in sheet metal forming *Advanced Materials Research* **6-8** 81–92.

[5] Han S, Hwang T, Oh I, Choi M, Moon Y H 2018 Manufacturing of tailor-rolled blanks with thickness variations in both the longitudinal and latitudinal directions *Journal of Materials Processing Technology* **256** 172-182.

[6] Sun G, Tian J, Liu T, Yan X, Huang X 2018 Crashworthiness optimization of automotive parts with tailor rolled blank *Engineering Structures* **169** 201-215.

[7] Wernicke S, Sieczkarek P, Martins P A F, Tekkaya A E 2016 Local sheet thickening by in-plane swaging *International Journal of Mechanical Sciences* **19** 59-67.

[8] Tan C J, Mori K, Abe Y 2008 Forming of tailor blanks having local thickening for control of wall thickness of stamped products *Journal of Materials Processing Technology* **202** 443-449.

IOP Conf. Series: Materials Science and Engineering **429** (2018) 012008 doi:10.1088/1757-899X/429/1/012008

Sound Transmission Loss Analysis for Laminated Glass and Perspex

W. H. Tan[1,*], E. A. Lim[2], E. M. Cheng[1], K. S. Basaruddin[1] and M. Hafifi Ramli[1]

[1]School of Mechatronic Engineering, Universiti Malaysia Perlis (UniMAP), Pauh Putra Campus, 02600 Arau, Perlis, Malaysia.
[2]Institute of Engineering Mathematics, Universiti Malaysia Perlis (UniMAP), Pauh Putra Campus, 02600 Arau, Perlis, Malaysia.

*whtan@unimap.edu.my

Abstract. The aim of this study is to analysis the sound transmission loss (STL) for the laminated glass and Perspex. In this study, the STL of laminated glass and Perspex are determined using the two-load transfer function method (TFM) impedance tube measurement setup. There are 5 mm and 8 mm thickness of glass and Perspex were used. Glass and Perspex will be laminated into double and triple layer in order to investigate its STL due to interlayer effect. All the specimens are prepared into round shape with the diameter of 33 mm that fit with the inner diameter of impedance tube. From the measurement, it is concluded that the STL almost similar for the Perspex and glass without the layer lamination at the frequency range below 3500Hz. Number of interlayer for the glass or Perspex lamination plays the major role as main parameters that affect the overall STL.

1. Introduction

Nowadays, there are many noise problems happened in our workplace and surrounding, such as noises come from busy traffic of transportation, industries machining, construction site and etc. that affecting our daily life. To solve these problems, it is crucial to understand the knowledge of acoustic, noise control and its prevention. Usually, to reduce noise level, it is compulsory to deal with frequency, amplitude and wavelength of the sound. The sensitivity of human being on the sound frequency is limited from 20 Hz to 20 KHz [1].

A laminated glass is the glass that consists of two or more layer of glasses with an interlayer that typically made using ethylene-vinyl acetate (EVA) or polyvinyl butyral (PVB) in the middle of the glasses [2]. Usually, laminated glass was used for automobile windshield, exterior store front, window and wall for the building. Laminated glass is considered one type of safety glass in protecting human from injuries as it can hold together when shattered. Thus, vehicle front windscreen normally installed with laminated glass. The stress applies on the laminated glass could cause it to deform elastically and demonstrate no plastic deformation before it fractures [3]. On the contrary, Perspex flexibility is higher than the glass and therefore, it is able to withstand high speed wind and pressures as it is used as building windows [4]. In general, Perspex was made using heating method, bending and pressure forming [5].

In this study, the acoustic property of laminated glass and Perspex were analysed. The sound transmission loss (STL) of the laminated glass and Perspex are measured using the two-load transfer function method (TFM) impedance tube [6] [7].

2. Methodology

In this study, the impedance tube was fabricated using acrylic tube. The inner diameter of the tube is 33 mm with length of 1.25 m [8]. The detectable frequency range for the impedance tube is in between 60 Hz to 6000 Hz. The schematic drawing of the impedance tube is shown in Figure 1, and Figure 2 shows the fabricate impedance tube for this study.

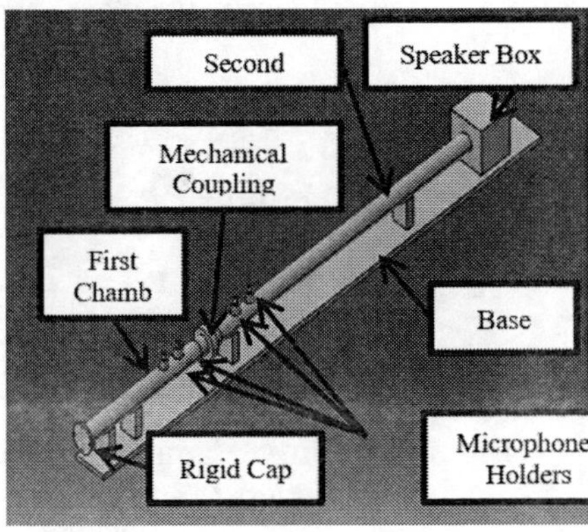

Figure 1. Design for impedance tube CATIA V5.

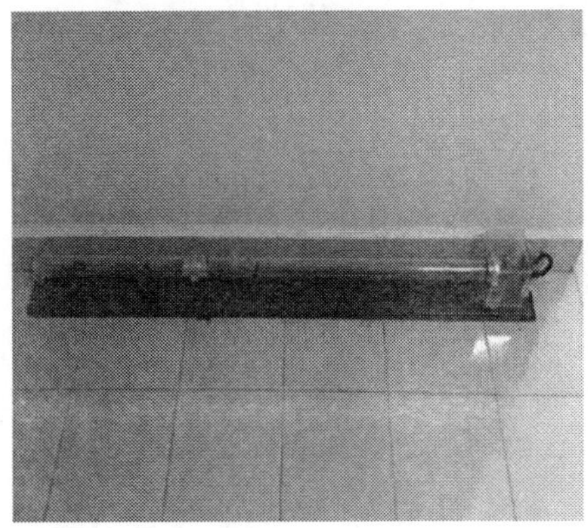

Figure 2. Fabricated impedance tube.

2.1. Specimen preparation

Table 1 listed the specimens that were used for the sound transmission loss (STL) measurement. There are 5 mm and 8 mm thickness of glasses and Perspex were used. Glass and Perspex are laminated into double and triple layer to investigate its STL via interlayer effect. All the specimens are prepared into round shape with the diameter of 33 mm that fit tightly with the inner diameter of impedance tube. For the laminated glass and Perspex specimen, the interlayer is seals using silicone SN 503 Silicon Sealant. The veneer clipper tool is the jig that is used to control the thickness of interlayer for the laminated glass and Perspex [10]. Figure 3–4 shows the prepared specimens that are used for the measurement. The cross section layouts of specimens are showed in Figure 5–6.

Table 1. Material selection with thickness and interlayer.

Material	Thickness	
	5mm (single interlayer)	8mm (double interlayer)
Glass	Single	Single
Perspex	Single	Single
Laminated Glass	Double	Triple
Laminated Perspex	Double	Triple

IOP Conf. Series: Materials Science and Engineering **429** (2018) 012008 doi:10.1088/1757-899X/429/1/012008

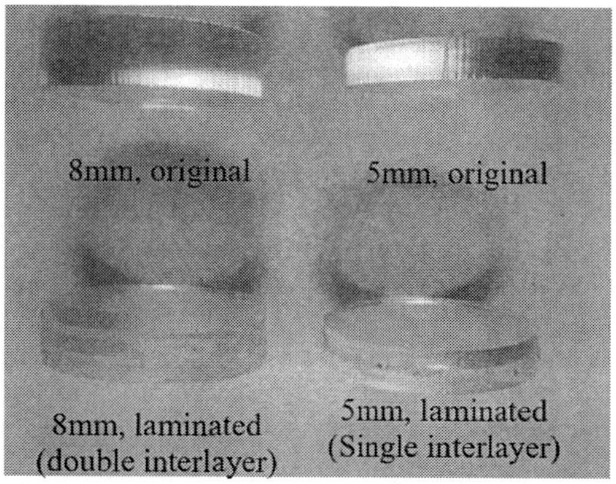

Figure 3. Perspex and laminated Perspex specimens.

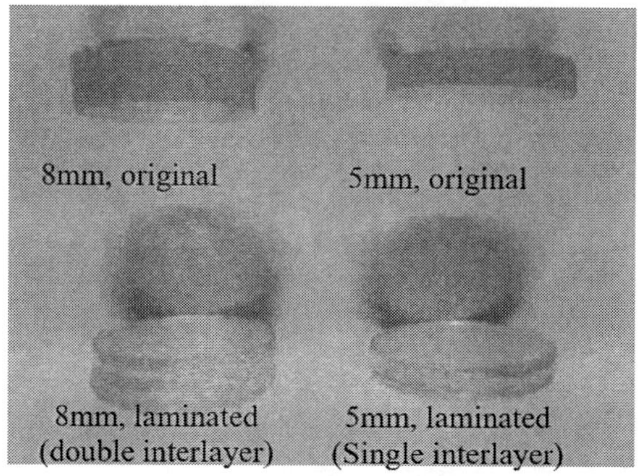

Figure 4. Glass and laminated glass specimens.

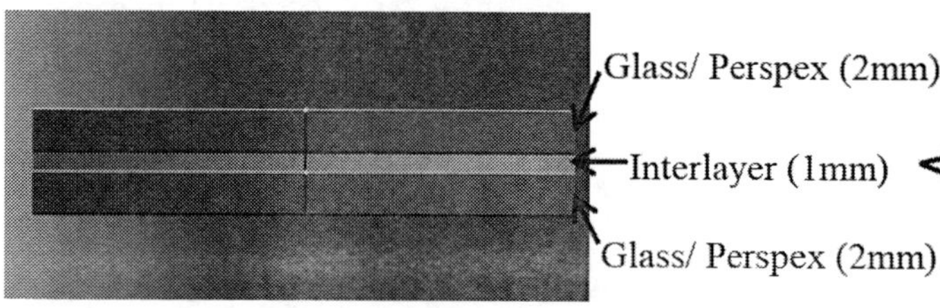

Figure 5. Double laminated layer with single interlayer (5mm thickness).

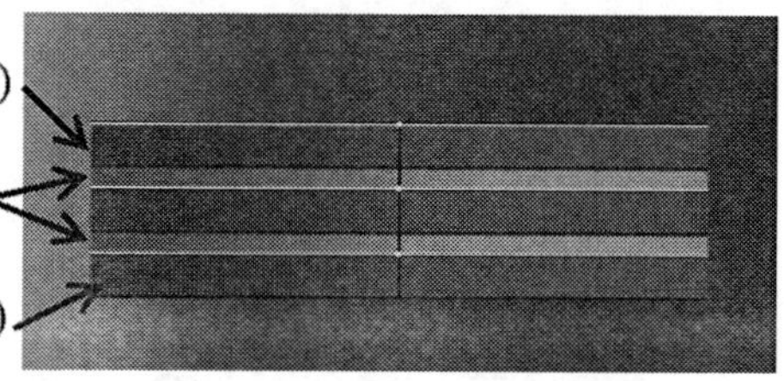

Figure 6. Triple laminated layer with double interlayers (8mm thickness)

2.2. Measurement setup

In this study, two-load transfer function method (TFM) impedance tube is used for the sound transmission loss (STL) measurement. Figure 7 showed STL measurement setup in the laboratory. For the measurement setup, the impedance tube required four microphones with each connected to the Data Acquisition (DAQ), where the sensor signals are synchronised in post-processing and stored in a computer where the measurement is done with two sensors simultaneously instead of one sensor in two-times [11].The speaker was connected to the output port of DAQ, that provides random sinusoidal noise signal generated by the DAQ using an analysis software known as LMS Test Lab 14A [12]. The rigid cap is used as the rigid termination, while the sponge works as the anechoic termination for the two-load TFM. All microphones were calibrated using microphone calibrator before the measurement. The measurement data is collected and analysed via DAQ coupled with the installation of analysis software in the computer. The STL measurement graph against frequency is then plotted [13].

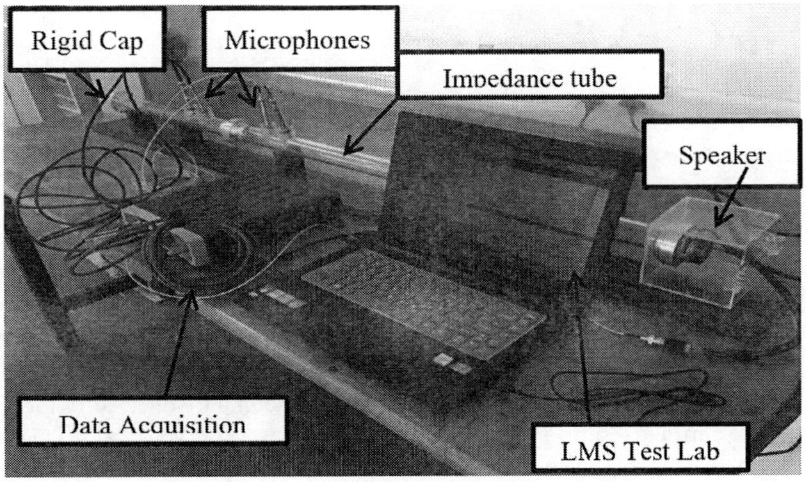

Figure 7. Setup of two-load transfer function method (TFM) impedance tube.

3. Result and discussion

In this section, the sound transmission loss (STL) measurement results are compared and discussed, in term of specimen thickness, materials, and interlayer thickness.

3.1. Effect of specimen thickness on STL

The sound transmission loss (STL) for the glass with the thickness of 5mm and 8mm is shown in Figure 8. The blue line indicates the STL of 5mm thick glass while the red line is 8mm thick glass. From the result, it is discovered that STL of 5mm thick glass is always higher than 8mm thick glass for the frequency range of 0Hz to 4500Hz. Overall, the STL of 5mm and 8mm thick glass are relatively low, typically below 10dB for the frequency range below 3500Hz. Meanwhile, for the frequency range 3500Hz and above, the STL shows much higher value for 5mm and 8mm glasses. Similar observation is found for the Perspex with the thickness of 5mm and 8mm as shown in Figure 9.

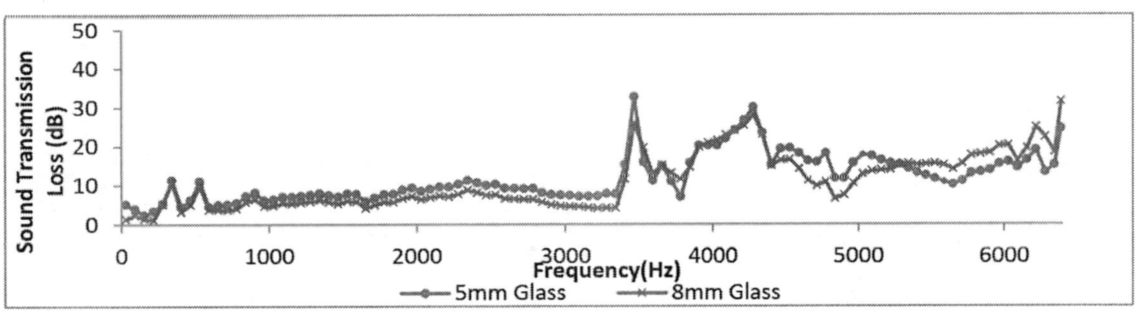

Figure 8. Sound transmission loss for 5mm and 8mm thick glasses.

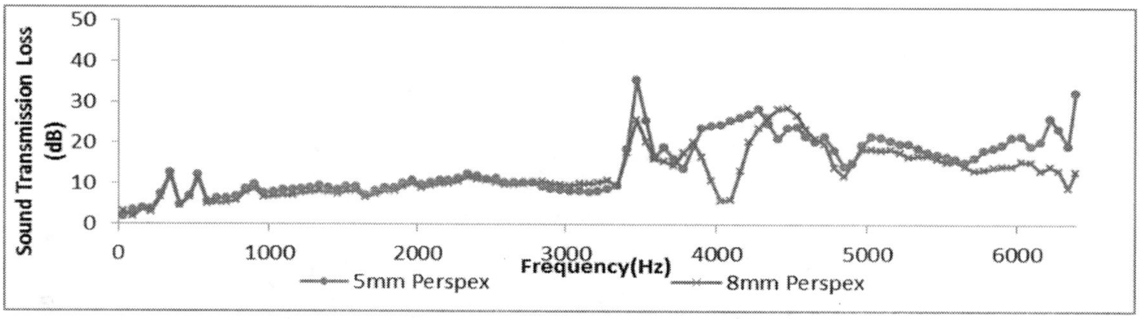

Figure 9. Sound transmission loss for 5mm and 8mm thick Perspex.

3.2. STL comparison between glass and Perspex

Figure 10 and 11 show the sound transmission loss (STL) comparison results for glass and Perspex with thickness of 5mm and 8mm. Based on Figure 10, it is observed that STL of Perspex always higher than glass with similar thickness of 5mm throughout the whole investigated frequency domain. Similar trend is observed for the 8mm thick Perspex with frequency below 3500Hz as illustrated in Figure 11. Hence, it is concluded that Perspex gives higher insulation than glass for same thickness based on the obtained STL results in this study.

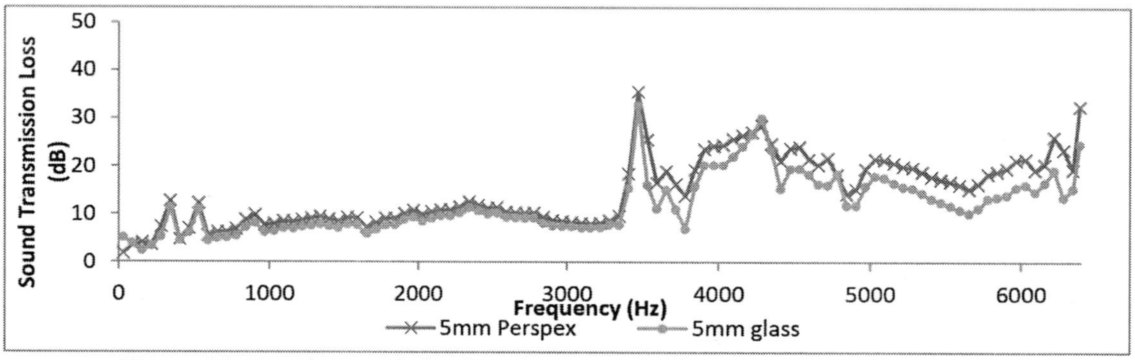

Figure 10. Sound transmission loss for Perspex and glass with thickness of 5mm.

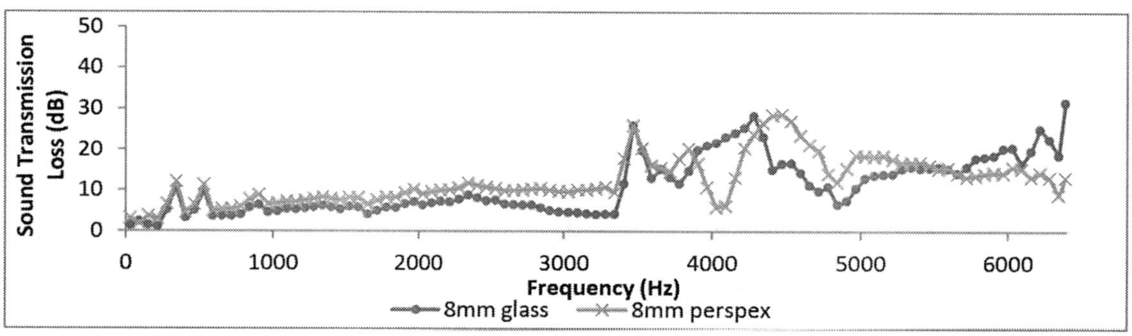

Figure 11. Sound transmission loss for Perspex and glass with thickness of 8mm.

3.3. Comparison of laminated layer with single and double interlayer on STL

Figure 12 and 13 shows the comparison sound transmission loss (STL) graph for laminated glass and Perspex with single and double layers respectively. According to Figure 12, it is observed that STL for laminated Perspex and glass is almost similar with small differences of STL for the frequency range from 0Hz to 3500Hz. In contrary, when the interlayer increases to double layers, the STL is also improved significantly compared with single interlayer. This phenomenon is proven in Figure 13 with double interlayers. Simultaneously, it is also observed that laminated Perspex with double interlayers obtained better STL throughout the investigated frequency domain in compares to laminated glass with double interlayers. Therefore, it can be concluded that laminated Perspex with double interlayers delivers higher STL as the interlayer is considered as an insulation material where it can absorb more noise when the interlayer increases [14]. This phenomenon is proven in Figure 14 and 15, where the comparison is done using similar thickness for laminated glass or Perspex, and the results shown the laminated glass or Perspex is able to provide higher STL compared to non-laminated glass or Perspex.

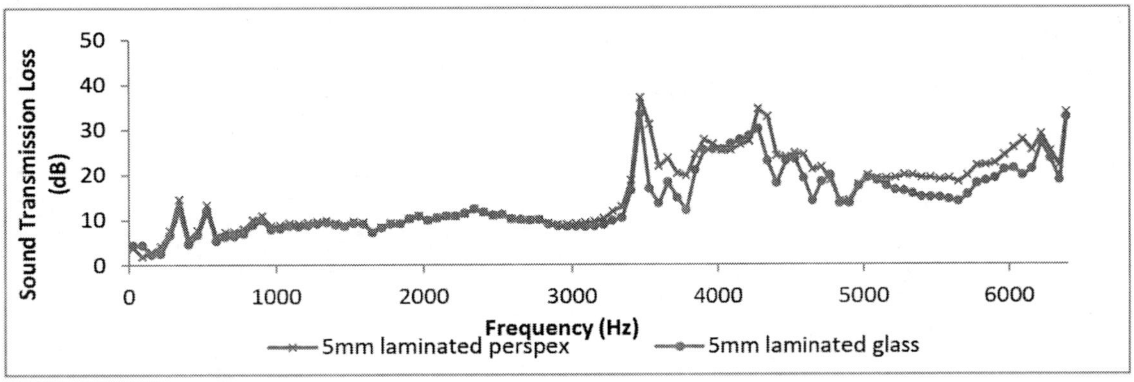

Figure12. Sound transmission loss for laminated Perspex and glass with single interlayer (5mm thickness).

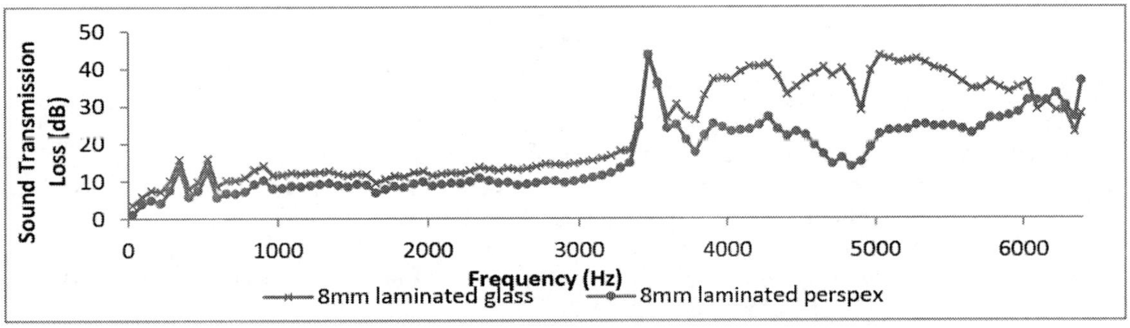

Figure 13. Sound transmission loss for laminated Perspex and glass with double interlayers (8mm thickness).

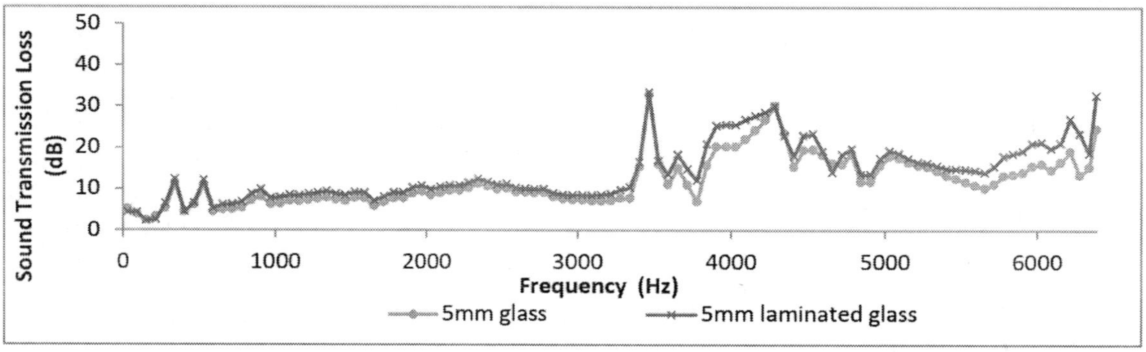

Figure 14. Sound transmission loss for 5mm thick glass and 5mm thick laminated glass with single interlayer.

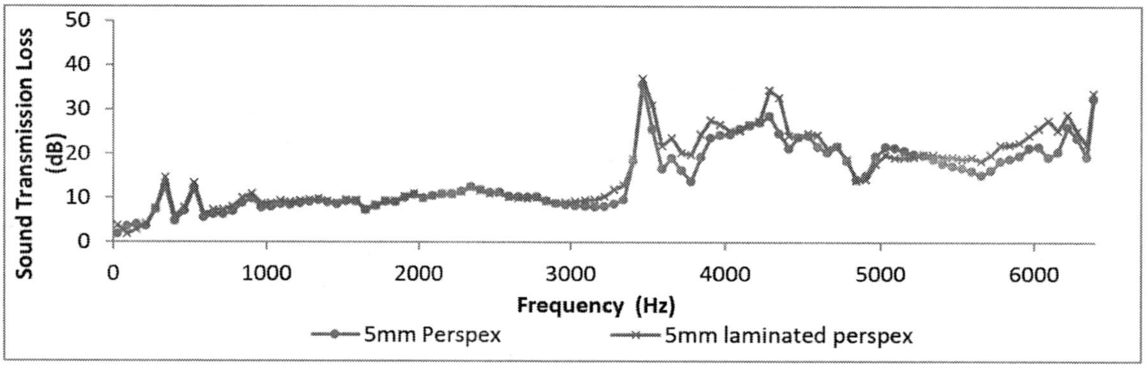

Figure 15. Sound transmission loss for 5mm thick Perspex and 5mm thick laminated Perspex with single interlayer.

4. Conclusion

This study involves exploring, observing, and understanding the sound transmission loss (STL) for the laminated glass and Perspex. From the STL results, it is concluded that the STL almost similar for the Perspex and glass without the layer lamination. Simultaneously, one of the significant findings observed in this study is when the interlayer increases for laminated Perspex or glass; the STL is also increasing respectively. Thus, the number of interlayer for the glass or Perspex lamination plays the major role in parameters that affect the overall STL.

Acknowledgement

The authors are grateful for the lab assistance, colleagues who are willing to share their opinion and School of Mechatronic Engineering, Universiti Malaysia Perlis (UniMAP) for providing the equipment and environment during this research.

References

[1] N. Robert B., Noise Control for Building. Valley Forge: CertainTeed Corporation, 2011.

[2] "Glass Manufacturing Process", Aisglass.com, 2017. [Online]. Available: https://www.aisglass.com/glass-manufacturing-process. [Accessed: 16- Jun- 2017].

[3] J. Schimmelpenningh and J. Schimmelpenningh, " Acoustic Interlayers for Laminated Glass – What makes them different and how to estimate performance ," pp. 2–8, 2012.

[4] F. Bill, Working with Acrylic. Albuquerque: Craftics, Inc., 1992.

[5] W. Practice, "Forming PLEXIGLAS ® Guidelines for Workshop Practice," Form. PLEXIGLAS ® Guidel. Work. Pract., pp. 1–32.

[6] G. D. Considine, "Transmission Loss," Van Nostrand's Sci. Encycl., vol. 9, no. 79, 2005..

[7] J. Kunio and R. W. H. Laboratories, "A Comparison of Two and Four Microphone Standing Wave Tube Procedures for Estimating the Normal Incidence Absorption Coefficient," Internoise 2009, 2009.

[8] B. and Kjaer, "Transmission Loss Measurement in a Standing Wave Tube," Tech. Rev., no. 1, pp. 1–50, 2007.

[9] B. Yousefzadeh, M. Mahjoob, N. Mohammadi, and A. Shahsavari, "An experimental study of Sound Transmission Loss (STL) measurement techniques using an impedance tube," Acoust. 08 Paris, pp. 965–968, 2009.

[10] W. Practice, "Forming PLEXIGLAS ® Guidelines for Workshop Practice," Form. PLEXIGLAS ® Guidel. Work. Pract., pp. 1–32.

[11] H. V. Hendrix, "Acoustic Impedence Measurements," in Standing Waves (Microflown Ebook), vol. 4c2, U.S.A: Borgo Press, 2006, pp. 1–65.

[12] C. V. Hrncir, Bill, "LMS Test . Lab Introduction," LMS international, Leuven, Belgium, pp. 1–8.

[13] R. V. Lopes, "Design And Construction Of A Low Cost Impedance Tube For Sound Absorption Coefficients Measurements," no. Cobem, pp. 105–115, 2013.

[14] E. A. Khidir, "A Comparative Study of Sound Transmission Loss Provided by Glass, Acrylic and Polycarbonate," J. Teknol., vol. 60, pp. 1–4, 2013.

IOP Conf. Series: Materials Science and Engineering **429** (2018) 012009 doi:10.1088/1757-899X/429/1/012009

High-cycle fatigue of rotating cantilever round copper tube under load-controlled condition

Z L Gan and C J Tan

Department of Mechanical Engineering, Faculty of Engineering, University of Malaya, 50603 Kuala Lumpur, Malaysia.

Abstract. A cantilever rotating bending testing machine accepting only rod specimen with a fixed clamping diameter of 9 mm is modified for tube specimens using two developed adapters. One end of a ϕ15 mm copper tube is expanded into ϕ18 mm and forms the tube specimen. Two adapters fitting for ϕ15 mm and ϕ18 mm tube ends respectively are designed and fabricated. The ϕ15 mm adapter is connected to a shaft subject to an applied load and the ϕ18 mm adapter is connected to a motor, respectively through two collets. Both tube ends are adhered to the inner wall of the adapters with superglue to prevent slippery under the rotation. Finite element analysis is used to estimate the maximum applied stress in the tube specimen i.e. adjacent to the transition for a known counter weight. Only one fatigue test result i.e. at 2.2×10^6 cycles for applied stress of 205 MPa is successfully obtained from the experiment. The collets failed to clamp the adapters for applied stress higher or lower than 205 MPa due to the present of oxide powders resulting from the wear of the collect as it is a 15-year-old machine.

1. Introduction

Copper tubes have been widely used in the plumbing and the Heating Ventilation Air-Conditioning (HVAC) industries due to its excellent heat & electrical conductivity, high corrosion resistant and good formability. These tubes are often bent into small corner radii in the condenser units to have more compact piping design. However, the bent tubes are subject to cyclic bending stress due to the vibration originated from the rotating fan, compressor etc. leading to fatigue failure, particularly at the bent corners [1]. Key components for air conditioners such as evaporator / condenser coils and heat exchangers etc are made of copper. The evaporator tubes directly transfer the heat through a compressor to a condenser unit, and the heat is being removed in this process. In this application, the Copper tubes must hold at certain pressure and be able to perform well at low temperatures. Any failures in the tubes may cause leaking in the cooling system, leading to low cooling efficiency [2]. S-N curve has been used as a criterion for in-house piping stress assessment in the industries. However, this curve is only obtained from rod specimen, but not from the cold drawn copper tube. In the application, the tubes are subject to multi-axial loading. It may not be tension-compression in nature. In the study of copper-phosphorous multi bent tubes in outdoor air-conditioning units, Lim, Mohammad, & Yapreported that the mode shapes of the vibration response of such tubes are not limited to compression-tension. In fact, bending also present hugely in all mode shapes [3]. Martin developed a formula to estimate the endurance limit of the ferrous metal specimen based on the factors such as size, surface finish, loading, temperature, reliability etc [4]. However, it is not applicable for non-ferrous metals such as Copper as it does not have an endurance limit i.e. it will eventually fail for increase in rotating cycles. Therefore, it is crucial to obtain the fatigue property of the copper tubes particularly at high-fatigue cycle for more accurate in-house piping stress assessment. The study aims to investigate the fatigue property of cold drawn copper phosphorous alloy tube using a modified cantilever rotating bending fatigue test machine for HVAC applications.

2. Material & method

The mechanical and the chemical properties of the tube specimens are shown in Table 1 and Table 2. The minimum tensile strength of 250 MPa is given for annealed tubes. It may increases to 380 MPa due to the cold tube drawing process, resulting into low percentage of total elongation. The copper tube has an outer diameter of 15 mm and wall thickness of 0.7 mm (KEMBLA BS EN 1057 Table X). It is a half-hard temper copper tube with a designation of Cu-DHP or CW024A following European Committee for Standardization (CEN), 2006. It is a commercially-pure copper which has been deoxidized with Phosphorus to leave a relatively high residual content. It is not susceptible to hydrogen embrittlement. The conductivity is relatively low on account of high Phosphorus content.

Table 1. Mechanical Property

Temper Designation (according to EN 1173)	R250
Tensile Strength (min) (MPa)	250
Total Elongation (min) (%)	30
Hardness (HRB)	75 to 100
Maximum Working Pressure (KPa)	5800

Table 2. Chemical Composition (wt %)

Cu	P
> 99.90	0.015 to 0.04

The schematic and the photo of the cantilever rotating bending fatigue test machine (Hi-Tech Scientific HSM-19 Mark 3) are shown in Figure 1. In the cantilever rotating bending test, non-uniform bending moment is generated along the specimen length with constant load amplitude. The machine has an electronic cycle counter to record the number of rotating cycles during the test. Counter weights are added for obtaining different applied bending stresses in the tube specimens. Originally, it is designed for solid machined specimen with a fixed outer diameter of 9.0 mm at its both ends and a fixed length of 63.7 mm. The solid specimen is mounted onto the machine though a pair of collets. Since the size of the tube specimen in this study is different from the standard solid machined specimen, an adapter is needed to mount the tube specimen to the existing collets. Since it is an old and obsolete machine, collets with other diameters are no longer available in the market, and it is costly and time consuming to fabricate new ones for this study.

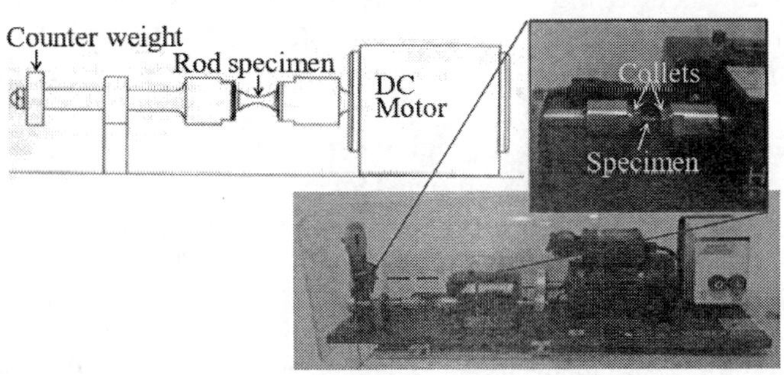

Figure 1. Cantilever rotating bending fatigue test machine (Hi-Tech Scientific HSM-19 Mark 3)

The exploded and the assembly drawings for the tube specimen, adapters and collets are shown in Figure 2. One end of the copper tube is expanded from diameter of 15 mm to 18 mm for stress concentration in the smaller hollow section adjacent to the transition. Two adapters are used to join expanded tube specimen to the existing collets at its both ends. Therefore, it has a 9 mm diameter protrusion in one end to be inserted to the existing collet, and another end to be attached to the specimen in either 15 mm or 18 mm diameter. Both tube ends are adhered to the side wall of the adapters with super glue (Cyanoacrylates) to transfer the rotational movement during the test. Since the join are not exerted with tension and compression force but rather a small amount of torque, which is around 1N according to the user manual [5], it can be firmly attached together with the super glue. However, curing time of at least 24 hours is needed to attain the required adhesive strength before performing the fatigue test. After inserting the protrusion of the adapter to the collet, the collet is clamped to prevent slippery under the rotation. To minimize the strain of the glue due to any loading, the adapter is machined to fit into the copper tube end with a tight tolerance. This might remove any movement of the specimen inside the adapter during loading, resulting into the low accuracy of the test.

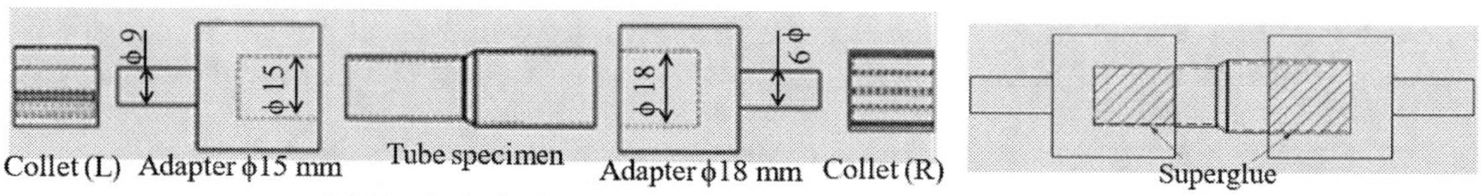

(a) Exploded view (b) Assembly view

Figure 2. Exploded and assembly views of tube specimen, adapters and collets

The detailed design of the adaptor for tube end ϕ15 mm is shown in Figure 3. It has a ϕ9 mm protrusion with 20 mm length at one end to be inserted to the existing collet. The tube specimen is inserted and glued to the other end of the adapter with an inner diameter of 15 mm and depth of 20 mm.

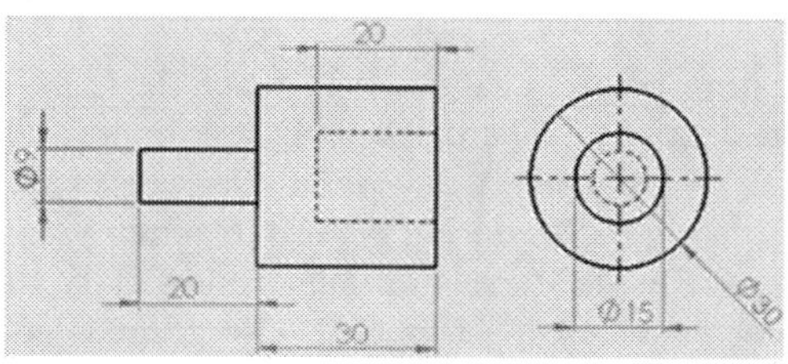

Figure 3. Detailed design of adaptor for tube end ϕ15 mm

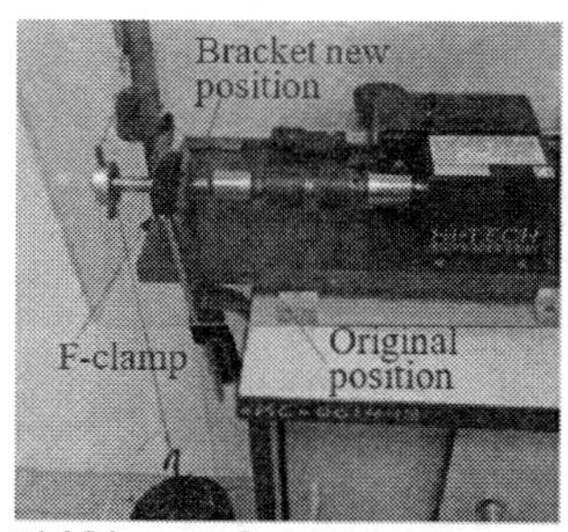

Figure 4. Shifting of bracket and F-clamp of cantilever rotating bending machine to accommodate for insertion of two adapters

Shifting of the bracket and the F-clamp of the cantilever rotating bending machine to accommodate for the extended length between the two collets after the insertion of the two adapters is illustrated in Figure 4. Two new holes are drilled in the base plate at approximately 60 mm from original holes away the motor as a new mounting location for the bracket. These modifications are reversible and can be restored to its original structure.

The detailed designs of tube expander and the photo of the tool are shown in Figure 5. The inner diameter of the tube specimen is expanded from 13.6 mm to 16.7 mm with an expansion ratio of 22.6% using a 10-ton manual hydraulic press. Barreling is observed along the sidewall of the expanded section after piercing the expander into the tube end. An outer die or ring with a clearance of 0.85 mm between the die and the expander is used to flatten the sidewall and to maintain the cylindricity of the expanded section. The cylindricity of the expanded section relative to the original section is measured with a dial gauge. In the measurement, the unexpanded section of the specimen is clamped to a lathe machine and the cylindricity of the expanded section is measured with the dial gauge under rotation. A cylindricity value of less than 0.2 mm is obtained in the measurement. Buckling is observed along the unexpanded section for die clearance of less than 0.85 mm.

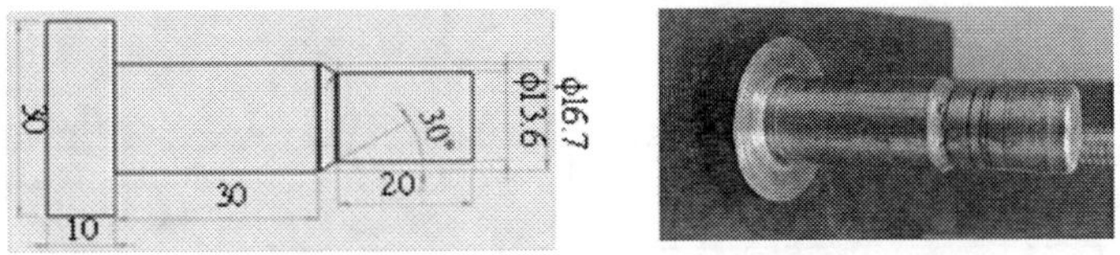

Figure 5. Detailed designs and photo of tube expander

Finite element analysis is performed using ANSYS to estimate the applied bending stress in the specimen under the controlled load during the fatigue test. In the simulation, the tube specimen is divided with tetrahedral mesh into 37413 elements. The elastic modulus of the tube is 120 GPa. The expanded section is fully constrained. A bending moment of 5880 N.mm is applied at the tube end of the unexpanded section for a counter weight of 40 N. The distance between the applied load and the collet attached to the motor is 147 mm.

3. Results & discussion

The comparison between the location of the fatigue crack in the test and the stress concentration zone in the simulation is shown in Figure 6. The location of the fatigue crack in the test is well predicted in the simulation. The crack is located in the unexpanded section adjacent to the transition. A maximum stress of 204.6 MPa is obtained from the simulation for a counter weight of 40 N.

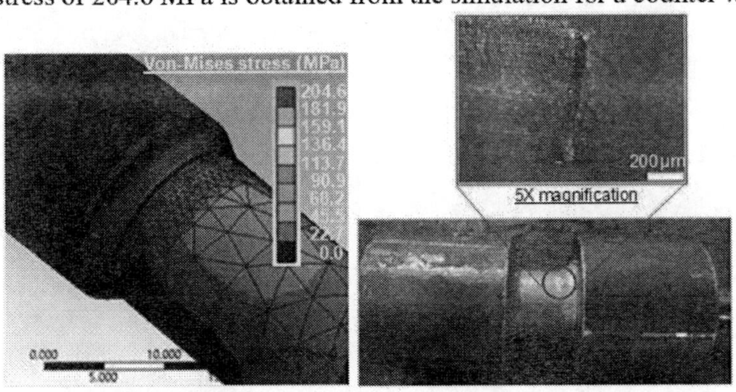

(a) FEA results (b) Location of crack

Figure 6. Comparison between FEA and experimental results

Figure 7. Wear of Collet for counter weight of 40 N after 2.2 x 10^6 cycles

In the experiment, two types of counter weight blocks are available for the test i.e. 20 N and 2 N. Although it is possible to generate a maximum bending stress of up to 300 MPa in tube specimen with the combination of these weight blocks, the clamp between the collet holder and the collet tend to loose due to the wear of the collet. Oxide powder generated from the collet resulting from the friction and wear for counter weight of 40 N after 2.2 x 10^6 cycles is shown in Figure 7. The clamp failed after thousands of rotating cycles for counter weight of more than 40 N. It also failed after millions of cycles for counter weight lower than 40 N. Therefore, only one set of data could be obtained with the existing setup. Since the collet has been used for more than 15 years in the test, the clamping force exerted onto the protrusion is getting weaker resulting from the wear. It is also difficult to fabricate new collets due to the lack of accurate and detailed dimensions of the worn part. There is also no spare part available in the market as the machine is obsolete. However, Stephens et al. reported that S-N curves from torsional or bending load-control test tend to have no data at lower fatigue life, i.e.: 10^3 or 10^4 cycles due to the significant plastic deformation [6].

The comparison of fatigue strength among different test methods for pure copper is shown in Table 3. Zainuddin et al.reported that the fatigue strength of 320 MPa in a 3-point bending fatigue test of copper tube at 2.2 x 10^6 cycles [7]. Farshbaf & Habibi reported the fatigue life of more than 10 million cycles for bending stress of 200 MPa in rotating bending fatigue test of solid machined specimen. At around 2.2 million cycles, the bending stress is around 225MPa [8].

Table 3. Comparison of fatigue strength among different test methods for pure copper

Test method	Specimen	Fatigue strength @ 2.2 x 10^6 cycles (MPa)
3-point bending [7]	Round tube	320
R-R Moore rotating bending [8]	Solid specimen	225
This study	Round tube	205

4. Conclusion

In this study, the conventional cantilever rotating bending fatigue test machine accepting only rod specimen with a fixed clamping diameter of 9 mm is modified and the fatigue life of an expanded copper tube specimen with a diameter of 15 mm is successfully obtained. The diameter of one tube end is expanded from 15 mm to 18 mm and straightened for stress concentration adjacent to the transition in the smaller section. Two adapters are designed and fabricated in this modification for mounting the expanded tube specimen to the machine through the collets. The fatigue strength of the tube specimen at 2.2 x 10^6 cycles is 205 MPa using a counter weight of 40 N. The strength is lower than the other reported results for the same material at 2.2 x 10^6 cycles. Due to loose of the clamp resulting from the wear of the collet, only the fatigue test for bending stress of 205 MPa is successfully completed. The clamp tends to loose for counter weight of more than 40 N or for fatigue life of more than millions of cycles using counter weight of less than 40 N.

5. Reference

[1] Elbel S, Duggan M, LaGrotta T, Raj S and Hrnjak P 2016 *16th Int. Refrigeration and Air Conditioning Conf. Purdue* Paper 2438

[2] Rex M and Mark R M 2004 *Air conditioning Home & Commercial Fifth Edition* (Danvers: John Wiley & Sons Inc.)

[3] Lim D Z, Mohammad H F and Yap K H 2014 *J. Eng. Sci. Technol.*40

[4] Martin J 1962 *Mechanical Behavior of Engineering Materials*. ed Englewood Cliffs N J (Prentice-Hall)

[5] Roberts N P 2001 *Instruction Manual HSM19 Mk3 Rotating Fatigue Machine* In H.-T. Scientific (Ed.) UK pp 17

[6] Ralph I, Stephens A F, Robert R S, Henry O F 2001 *Metal Fatigue in Engineering (Second ed.)* (New York: John Wiley & Sons Inc.)

[7] Zainuddin S, Amirhossein B, Muhammad F M and Junaidi S 2014 *Adv. Mat. Res.* **896** 4

[8] Farshbaf P and Habibi F 2015 *4th Int. Engineering Materials and Metallurgy Conf. (iMat 2015)*

Three Dimensional Finite Element Analysis on Railway Rail

M A Zulkifli[1], K S Basaruddin[1], Y Abdul Rahim[2], M Afendi[1], P Gurubaran[1], I Ibrahim[1]

[1]*School of Mechatronic Engineering, Universiti Malaysia Perlis, Pauh Putra Campus, 02600 Arau, Perlis, Malaysia.*
[2]*School of Manufacturing Engineering, Universiti Malaysia Perlis, Pauh Putra Campus, 02600 Arau, Perlis, Malaysia.*

**Corresponding author: khsalleh@unimap.edu.my*

Abstract: A prediction of fracture risk on railway rail using finite element analysis (FEA) was carried out in this paper. A three dimensional model was developed using CATIA to investigate the mechanical stress distribution under various loadings. The analysis was performed using ANSYS software to determine the stress response under loading of rolling contact, braking and cornering. In this study, the interaction between rail and wheel at some point was analyzed under static loading. The fracture risk via safety factor was then calculated based on the fracture strength of the wheel's material. The results showed that the braking load condition contributes to the lowest possibility of fracture on the surface of rail compared to the rolling contact and cornering loads.

1. Introduction

Railroads has a critical influence on the transportation framework of a nation and was assumed as an imperative part in managing a solid economy. Thus, railroad is adapted to redesign and overhaul its framework to take care of future demand of developing activity. The rolling contact between wheel and rail can be influenced by certain factors such as train speed, loading, wheel/rail profiles, and the response of the track parts. Railway track can be divided into several structures that play an important role in ensuring the smooth running of the train. Rail track commonly was classified into two main parts; substructure and superstructure [1]. Railway track structure comprises of rails, rail pads, sleepers, fastenings, ballast, sub-ballast, and subgrade. Therefore, two subsystems of a ballasted track structure can be recognized: the superstructure, made out of rails, sleepers, stabilizer and sub-ballast, and the substructure made out of an arrangement layer and the ground. An investigation on mechanical stress distribution is important in order to predict fracture or safety factor of the railways rail.

Numerous studies [2–4] have investigated the mechanical stress distribution on the railways rail. Ertz and Knothe [3] reported that the stress distribution is an important factor for the rail-wheel contact interface analysis. It clearly shows that two materials contacting at rolling interface, are highly dependent on the geometry of the contacting surfaces, loading conditions and material properties of the railway system [5]. Analysis of rail–wheel contact was studied by many researchers in different context such as different rail/wheel geometries to calculate the Hertzian contact behavior and determine the stress distributions at the area of interest. As a consequence, damage mechanisms such as surface cracks, plastic deformation, wear and operating rolling noise were also investigated [4]. For braking load condition [3], most of the previous studies investigated the thermal analysis between wheel and the brake pad. For instance, a modelling thermal effects in the braking systems of railway vehicles was studied by Arslan [6]. It is necessary to derive an analytical model of a thermal analysis that describes the heating transfer of the heat generated by friction at surfaces which are in contact, between a railway wheel and braking blocks, through the wheel and blocks, as well as heat outflow of the whole braking system due to cooling of the surrounding air [7]. Vo et al. [3] studied a nonlinear cornering model of a bogie. The nonlinearities, such as the ability to model stick-slip (or strictly roll-slip) between a wheel and a rail, are found to have a very significant influence on the development of corrugation in cornering. In particular, the periodic occurrence of complete sliding of the wheels of the leading axle makes possible corrugation growth associated with a lateral mode of vibration of the track that is not significantly excited until this sliding occurs [8].

On the other hand, most of the studies on the mechanical stress of railways rail have utilised the finite element method (FEM). Applications of FEM in structural analysis have been used extensively for macro and micro levels [9,10]. Even the use of FEM was not limited to regular microstructure but also for random microstructures [11,12]. Hence, this study aims to develop a finite element model of railway rail in order to predict the safety factor under various loading during service. Investigation on mechanical stress distribution was performed under three types of loading which is the rolling contact loading, braking loading and cornering loading.

2. Methodology

2.1 Development of railway rail model

A rail with 1000 mm length was created using CATIA. The assembly model of the railways rail-wheel and its half symmetrical model are shown in Figures 1 and 2, respectively. The three dimensional model was exported to ANSYS workbench for finite element analysis. The rail was made of 900A steel. The present analysis was tested by using a standard of 60 kg/m Rail Profiles-Types UIC 60 which provided by local train company.

Figure 1: Rail-wheel assembly in CATIA.

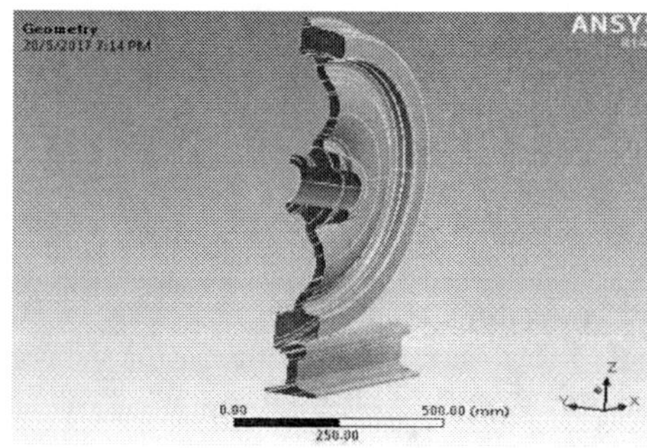

Figure 2 : Cross section of rail-wheel contact in ANSYS.

2.2. Assign material properties

The information regarding to the railways wheel and rail was provided by the local train company, including the wheel diameter of 965 mm (38 in), the vertical load (maximum design load = 78.48 kN), and the mechanical properties of wheel (yield strength = 438 MPa; Young's modulus = 205 GPa; friction coefficient = 0.3; material density = 7850 kg/m^3; rotational velocity = 24.6 rad/s; ultimate tensile strength = 1080 MPa).

2.3. Meshing

The meshed models of both rail and wheel were generated in ANSYS with the default mesh as shown in Figure 3. A selection of meshing size of rail was checked by verifying the quality of mesh which is available in ANSYS software. The contact in mesh between wheel and rail is depicted in Figure 4. A total of 197,530 nodes and 173,200 elements (including structural and contact elements) have been generated in the present models.

2.4. Boundary conditions

The boundary conditions were applied based on real conditions of the railway system during service. Figure 5 shows the boundary conditions that was applied to the rail-wheel in contact surface. In order to investigate the relationship between the variable vertical loading towards the stress response and fracture risk of railway rail, a set of different vertical loads were applied in the finite element analysis.

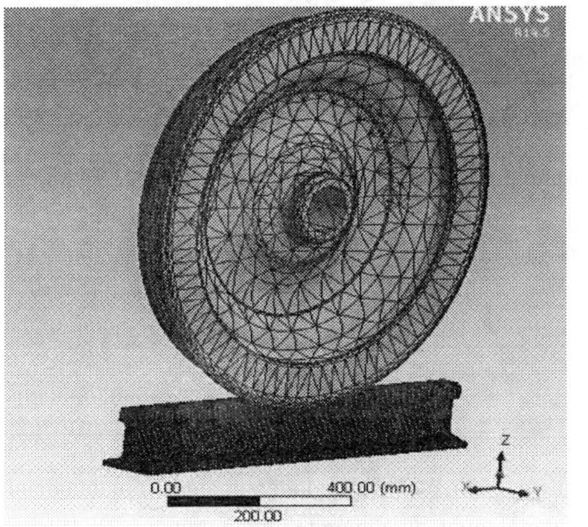

Figure 3: Meshing model of rail-wheel

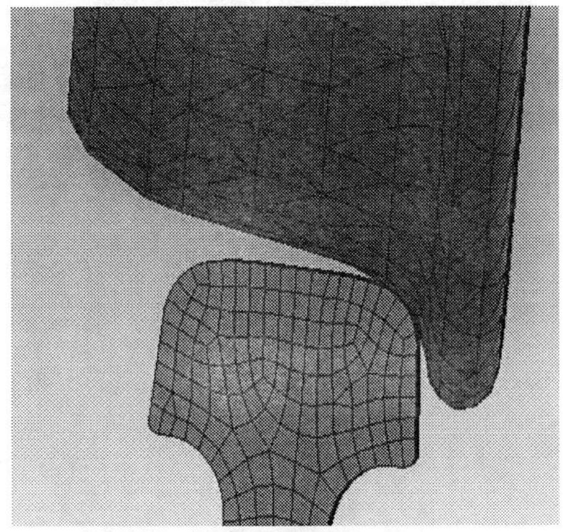

Figure 4: Meshing types for rail-wheel

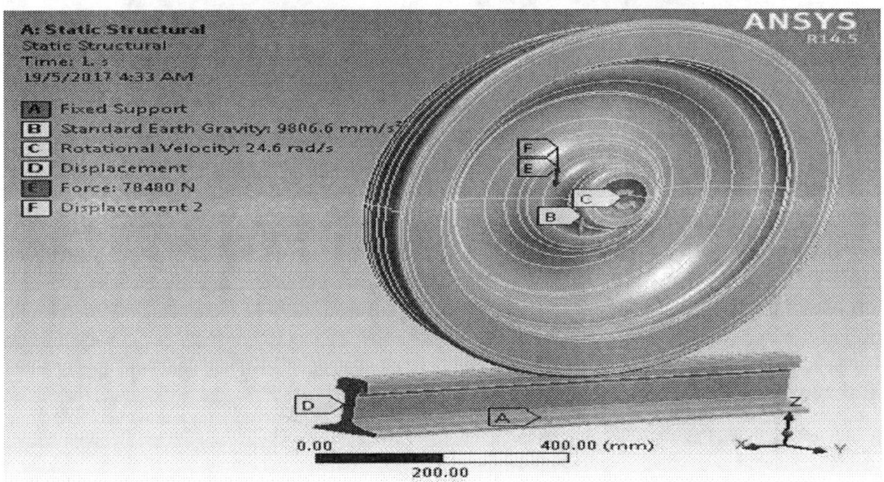

Figure 5: Boundary conditions

2.5. Prediction of fracture risk

The maximum equivalent von-misses stress obtained in the analysis was considered so that the risk of fracture on the rail can be predicted. The fracture risk also kwon as safety factor, was calculated based on eq. (1).

Fracture risk = Yield Strength/Maximum Equivalent Von-Misses stress Eq. (1)

3. Results and discussion

Finite element models were simulated successfully to visualize stress, deformation and strain levels. Static load analysis was performed in order to predict the fracture risk of railways rail. In Figure 6, distribution of Von Mises stress on the rail was shown. The maximum stress was found at the outer surface of the rail-wheel contact. Figures 7 and 8 show that the maximum of Von Misses stress distribution under rolling contact and cornering load conditions, respectively.

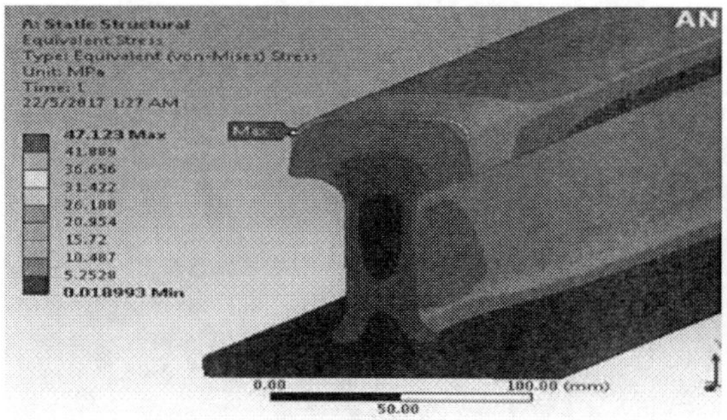

Figure 6 : Maximum Von Mises stress distribution under braking load

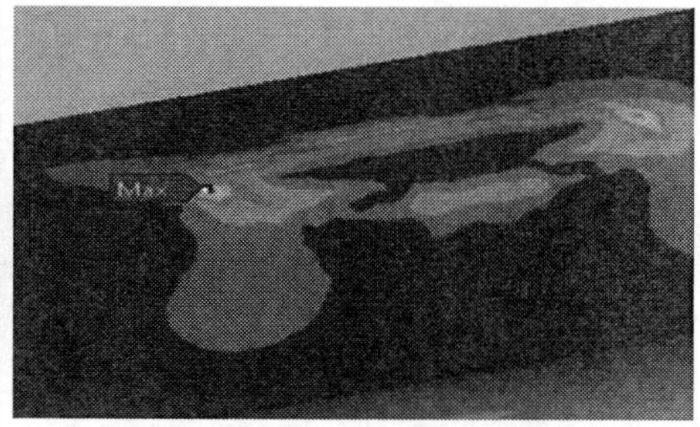

Figure 7: Maximum Von Mises stress distribution under rolling contact

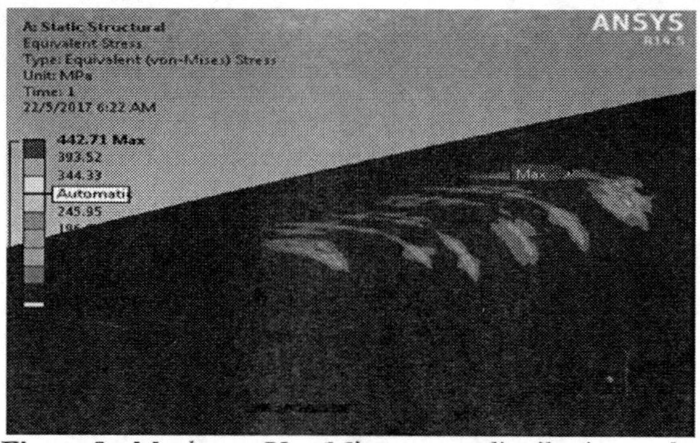

Figure 8 : Maximum Von Mises stress distribution under cornering load

In Figure 9, the graph indicates decrement in fracture risk of rail under rolling contact load when the increase of load was applied. As shown in Figure 10, when the load reached 100 kN, there is possibility of fracture going to happen because the fracture risk below than one. However, the rail was found safe under braking load if the load applied between 20 to 100 kN as shown in Figure 10. In contrary, under cornering load as depicted in Figure 11, the possibility of fracture going to occur on the rail if the load reached at 80 kN.

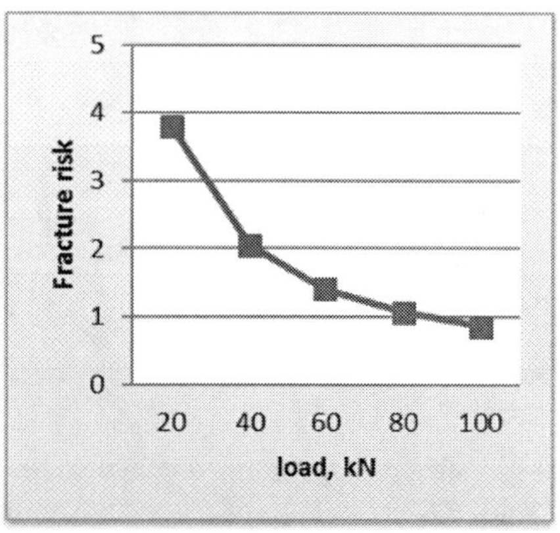

Figure 9 : Fracture risk under rolling contact load

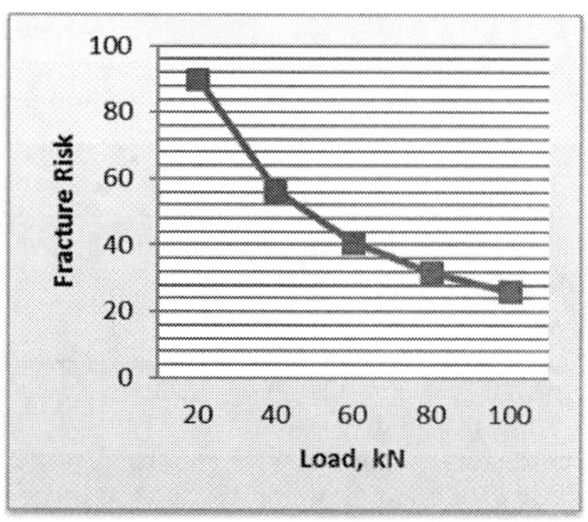

Figure 10 : Fracture risk under braking load.

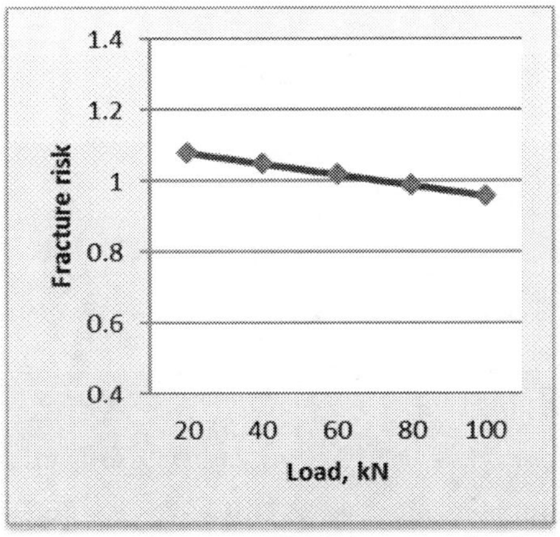

Figure 11 : Fracture risk under cornering load

4. Conclusion

This study was carried out to investigate the mechanical stress distribution and the fracture risk of the railway rail subjected to various loadings. Variable loads were applied to represents the actual load during service. Three type of load (rolling, brake and corner) were applied. The results showed that the highest fracture risk was obtained under braking load whereas more the critical conditions were found during loads of cornering and rolling.

5. References

[1] Kaewunruen, S. & Remennikov, A. 2008. in *New Research on Acoustics* 197–220.
[2] Aalami, M. R., Anari, A., Shafighfard, T. & Talatahari, S. *Adv. Mech. Eng.* **2013,** pp. 1–9.
[3] Ertz, M. & Knothe, K. A. 2002. *Wear* **253,** pp. 498–508.
[4] Vo, K. D., Zhu, H. T., Tieu, A. K. & Kosasih, P. B. 2015. *Wear* **323,** pp. 61–75.
[5] Sladkowski, A. & Sitarz, M. 2005. *Wear* **258,** pp. 1217–1223.
[6] Arslan, M. A. & Kayabaşi, O. 2012. *Adv. Eng. Softw.* **45,** pp. 325–331.
[7] Cai, W., Wen, Z., Jin, X. & Zhai, W. 2007. *Eng. Fail. Anal.* **14,** pp. 1488–1499.
[8] Donzella, G. & Petrogalli, C. 2010. *Int. J. Fatigue* **32,** pp. 256–268.
[9] Halim, S. A., Basaruddin, K. S., Daud, R., Safar, M. J. A. & Rahman, A. S. A. *ARPN J. Eng. Appl. Sci.* **10,** pp. 9353–9358.
[10] Basaruddin, K. S., Kamarrudin, N. S. & Ibrahim, I. 2014. *Lat. Am. J. Solids Struct.* **11,** pp. 755–769.
[11] Yoshiwara, Y., Clanche, M., Basaruddin, K. S., Takano, N. & Nakano, T. 2011. *J. Biomech. Sci. Eng.* **6,** pp. 270 285.
[12] Basaruddin, K., Takano, N., Yoshiwara, Y. & Nakano, T. 2012. *Med. Biol. Eng. Comput.* **50,** 1091–1103.

Experimental investigation of workpiece deformation due to WEDM

Adel Saleh Warregh [1] and Mohd Zahiruddin bin Md. Zain [2]

[1]School of manufacturing, University Malaysia Perlis, Kampus Tetop U/u Pauh, Jalan Arau- Changlun 02600 Arau Perlis, Malaysia

E-mail: [1]adelsaleh201313@yahoo.com, [2]zahiruddin@unimap.edu.my

Abstract. Deformation of stainless steel AISI 316 which machined by wire electrical discharge machining (WEDM) was studied in this work. First, the temperature distribution in a workpiece due to WEDM single discharge was calculated using the finite element method, and the result was used to estimate residual stress. It was found that the residual stress in the radial direction, σ_r (z) along the center axis of the single discharge crater was distributed non- linearly in the thickness direction from the top surface of the workpiece. Then, in structural analysis, the result of σ_r (z) was imposed within web part model. This study aims to investigate the deformation of AISI316 material during WEDM discharging by taking a discharge current, pulse on time, pulse off time, and heat input to a workpiece as input parameters. As temperature increases due to increase in discharge current during the process caused residual stresses induced by the machining lead to plastic deformation. To explain the deformation of the workpiece, additional models of the uniform rectangular beam were built. As a result, the non- linear distribution of residual stress was found to be the main reason that causes the non- uniform curvature of the workpiece.

Keywords:
WEDM, thermal stress, residual stress, surface quality, deformation.

1. Introduction

Electrical discharge machining (EDM) has the ability to produce micro components with complex features on any electrically conductive materials regardless of hardness. However, deformation of EDM workpiece at certain sizes and geometries influences component's dimensional stability and accuracy which limits further miniaturization of the process. There are two kinds of residual stress which might cause the deformation. One is residual stress that exists within the workpiece prior to machining [1], and the other one is residual stress caused by EDM itself [2]. The stress that exists prior to machining is the result of plastic deformation processes such as forging. This kind of residual stress can be released by annealing before the machining operation [1]. On the other hand, during EDM only part of the zone on the workpiece where the temperature is above melting point is removed and the remaining molten zone is left in the crater to form a recast layer. While the area near crater is in a molten state, no residual stress will be formed due to free movement. However, at the end of discharge, the recast layer solidifies and the cooling process begins. At the cooling zone, the tensile residual stress is formed, causing the zone near the crater to contract [2] [3] [4]. In wire EDM, fabricated fins started to bend when the fin thickness was less than 100 μm in rough cutting condition, and it was considered that the occurrence was due to the EDM [2].

For thin part components, WEDM might be a solution. However, few researches have focused on the thermal deformation phenomenon which usually appears when machining the thin-walled components in WEDM. The thermal deformation cannot be neglected due to its considerable influence on the mechanical performance and microstructure of the components during service life [5–7], especially machining the thin-walled components. One reason for thermal deformation is always neglected is due to non-macro cutting force during the WEDM process. However, as a matter of fact, thermal effects play the dominant role in WEDM except for ultrashort discharges. The thermal deformation may be caused by two kinds of residual stress [8]. One is the residual stress exists in the workpiece before machining which can be released by annealing in the pre-processing, while the other one is the residual stress generated by the WEDM process which is not yet explained clearly.

Finally, the influence of residual stress on workpiece deformation has never been calculated by other researchers. Consequently, in this work thermal and structural analysis are studied in order to understand the deformation. Residual stress due to a single discharge is studied by using the model of heat conduction within the workpiece. Then, the studied residual stress is imposed within the web part workpiece model in structural analysis to understand the non- uniform deformation of the web part.

2. Literature Review

Since the early seventies, researchers worldwide tried to investigate the workpiece deformation by wire electrical discharge machining and search out the residual stress that causes the workpiece deformation in the material AISI 316.

G. spur et al. [2] studied the influence of the maximum discharge current and the pulse on time on the dimensional stability of the teeth and found the tooth edge deviation occur at a tooth width of 50 μm. This shape deviation results from the melting of the basic material under tensile residual stress, so that the elastic limit of the stressed tooth cross- section is exceeded and a plastic deformation.

M. Zahiruddin et al. [9] studied 2D surface machining by micro EDM to investigate the analysis of micro-fin deformation after calculating the temperature distribution in a workpiece due to micro EDM single discharge and using the result to calculate the residual stress. It was found that the residual stress in the radial direction, $\sigma_r(z)$ along the center axis of the single discharge crater was distributed non-linearly in the thickness direction from the top surface of the workpiece. Then, in structural analysis, the result of $\sigma_r(z)$ was imposed within micro fin model. In addition, it was found that the summation of calculated deflections after three sequential discharges was almost equal to the measurement result. As a result, the non- linear distribution of $\sigma_r(z)$ was found to be the main reason that causes the non- uniform curvature of micro fin. From obtained result the fin straight near the tip and curvature is larger near the root. On the other hand, workpiece deformation occurs at the tip of the beam workpiece in the case of WEDM is comparatively different than micro EDM.

J. P. Kruth, Ph. Bleys et al. [10] observed up to a thickness of 0.4 mm, the beams show no deflection because of a sufficient ratio of thickness to stress penetration, which prevents deformation. On the other hand, the beam of 0.2 mm and 0.1mm show warping to the left, i.e. towards the last machining path. It must be noted that the beam of 0.1mm seems to bend to the right, but this phenomenon is caused by dielectric flushing combined with an undercut in the corners, severely weakening the structure. In addition, he found the measurement of residual stress profiles for rough and finish WEDM of tool steel show that the maximum tensile stress is reduced with increasing number of finishing steps.

3. Experiment

3.1. Workpiece preparation

Stainless steel AISI316 is selected as the workpiece material and the standard size from suppliers was $200mm \times 100mm \times 3mm$ as shown in Figure 1. The designed web beam for cutting operation as shown in Figure 2. The brass wire electrode diameter (0.25 mm) was used as the tool electrode.

Figure 1. Workpiece material used in Standard size

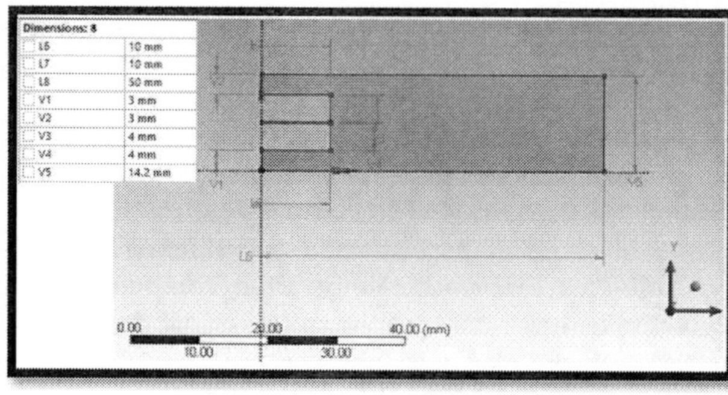

Figure 2. The designed web thickness

3.2. Machine

The workpiece was machined using AQ327 WEDM machine of Sodik Europe ltd as shown in Figure 3 with an isopulse generator and deionized water was used as dielectric liquid with Sodik's new lp control, it can import parasolid (solid models) files directly into the controller. This unique feature maximizes cutting efficiency, reduces the work flow and reduces programming. Figure 4 illustrates the setup of EDM wire cut machine.

Figure 3. EDM wire cut machine

Figure 4. Setup of EDM wire cut machine

3.3. Material properties

The chemical composition for the stainless steel grade AISI 316 is indicated in table 1. Also the thermal and mechanical properties are set out in table 2.

Table 1. Chemical composition of AISI 316 at room temperature

Chemical Composition								
Element	Mn	Si	P	S	Cr	Mo	Ni	N
Content %	2	0.75	0.045	0.03	18	3	14	0.10

Table 2. Thermal and mechanical properties

Thermal properties	
Thermal conductivity	17W/mK
Co-efficient of Thermal Expansion	$18*10^{-6}$
Specific Heat	530 J/kgK
Density	800Kg/m^3
Melting Temperature	1673k
Mechanical Properties	
Modulus of Elasticity	193GPa
Poisson's Ratio	0.275
Tensile Yield Strength	515MPa

3.4. Machining conditions used

Table 3 demonstrates the values of the input parameters which used in the experiment, and cover the reasonable range of the input parameters available according to the characteristics of WEDM. Other parameters such as frequency constant, discharge voltage and spark radius were kept constant.

Table 3. Process parameters used for modeling

Sr. no	Parameters	Values
1	Discharge Voltage. V	45 V
2	Spark radius, R	100μm
3	Workpiece polarity	Anode
4	Frequency constant, Kf	2.4
5	Material	AISI 316
6	Dielectric medium	water
7	Reference (ambient) temp. To	298K
8	Discharge Current, I	6, 25, 40A
9	Heat input to workpiece, P	0.15, 0.2, 0,25 kw
10	Pulse on time, T_{on}	100, 210, 260 μs
11	Pulse on time, T_{off}	160, 200, 265μs
12	Electrode	Brass

3.5. Method to cut the workpiece

Figure 5 shows a method to cut the thick plate workpiece. The plate workpiece was held horizontally. In order to explain the method, Figure 5a. is referred. First, the tool was moved downward to the starting position in the direction of cutting depth, cd; approximately 10000μm (10 mm) for the workpiece which was held horizontally. Then the tool was relatively moved towards the plate side surface perpendicularly and cutting begins. In this work, the tool electrode was moved during the thick plate cutting so that excessive too electrode wear does occur at only one side of the tool electrode surface which may severely influence the cutting progress. After the first slot cutting, the tool electrode was returned to the starting position, and then it was moved in the y- direction relative to the plate. Then, the tool was moved again towards the plate side surface to begin the 2nd slot cutting. Here, movement in the y- direction is important to control the part thickness, *pt*. Workpiece with plate thickness, T*h* equal to 3000μm (3mm). In this work, the web part with size of 10mm × 3mm × 0.2mm was obtained by using the cutting method in this section.

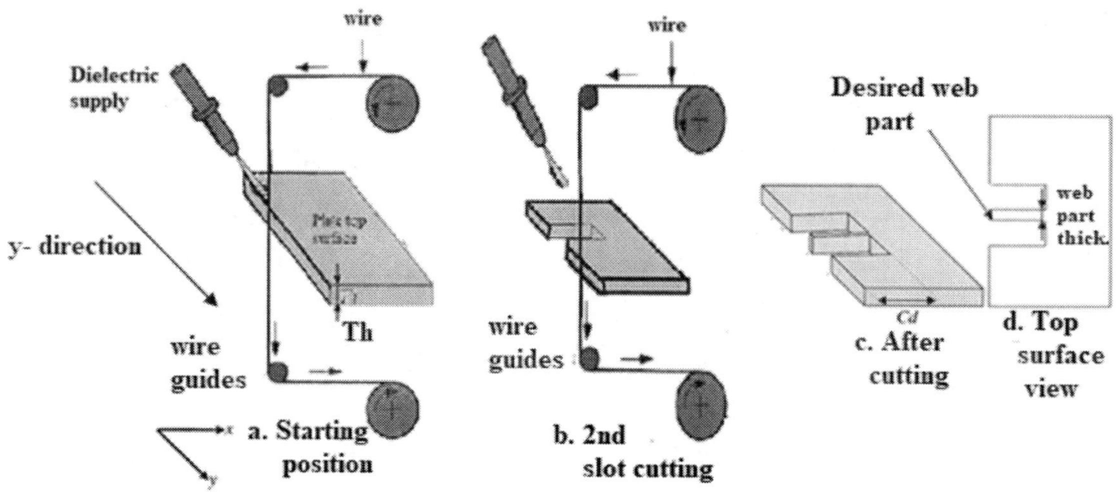

Figure 5. Method to cut the workpiece

4. Methodology

4.1. Theory of the formation of residual stress and workpiece deformation

Figure 6a. tensile stress is formed and remains at both sides of the 1st slot after the first cutting. During the 2nd slot cutting, the stress of part surface at the adjacent area of the 1st slot is released. At the same time, in the region on the part surface at the adjacent area of the 2nd slot, tensile stress is formed. When the part width, p_w is sufficiently small, the tensile stress at the adjacent area of the 2nd slot will cause the part to be bent toward this slot as shown in figure 6c. To obtain quantitatively the part bending due to WEDM, it is necessary to conduct the structural analysis to calculate the workpiece deformation caused by the thermal stress. [9].

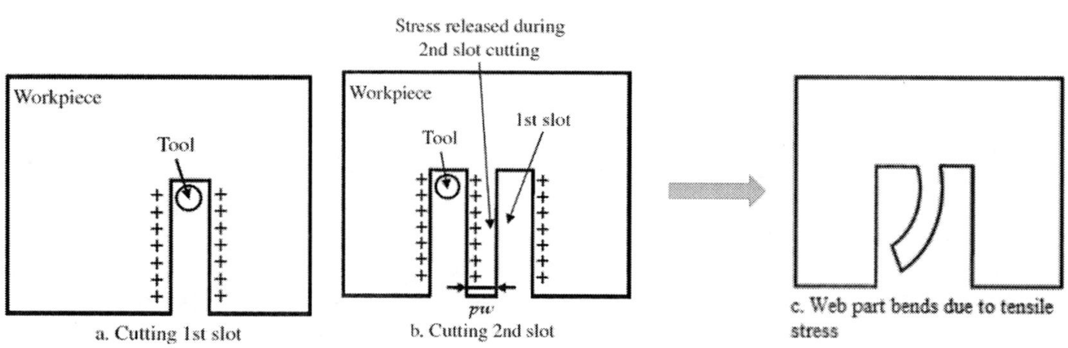

Figure 6. Theory of the formation of residual stress and workpiece deformation

4.2. Residual stress due to single discharge

The power input of EDM single discharge is required to build the heat conduction model. In WEDM, discharge energy per unit time distributed at workpiece under each unit area of circular surface heat source so-called power density, P_e was used as a measure of the power input. It is calculated as follows:

$$P_e = X \cdot i_e \cdot u_e / (\pi \cdot d^2/4) \qquad (1)$$

Where:

- i and u_e are discharge current and discharge voltage, respectively which are measurable during discharging.
- X is the ratio of energy distributed into workpiece with respect to the total discharge energy, and it can be expected between the case of micro and macro EDM [9].
- d is plasma diameter. In macro EDM, plasma diameter expands about 5 times of crater diameter (dm = 68.4μm) within the first 2μs of t_e [10]. In this work t_e was 100 μs which is significantly longer than 2μs. Thus, d in this work was assumed to be 342 μm [11].

All the parameters that are required to determine P_e in WEDM are shown in table 4, and in this work, they will be used for the thermal stress analysis. The physical model of heat conduction to estimate the residual stress was built using ANSYS; finite element analysis software. In this work, stainless steel AISI316 was used as the workpiece because its high thermal conductivity makes it suitable to function as a tool workpiece in EDM wire cut machine. However, though the toughness of stainless steel is relatively high, stainless steel workpiece are vulnerable to thermal deformation. The energy input as shown in Table 4 was used to model the power density on the workpiece. Figure 7 shows an overview of the axisymmetric model, with 10 mm in radius and 3 mm in thickness. All surfaces were assumed adiabatic except the discharge area.

Table 4. WEDM power input

Parameters	Values
Discharge current, i_e [A]	12
Discharge voltage, u_e [V]	20
Discharge duration, t_e [μs]	100
Energy distribution ratio into workpiece, x [%]	34
Plasma diameter at end of t_e, d [μm]	342 [11]
Power density on workpiece, $p_e(t_e)$ [Gw/m^2]	0.7
Crater diameter, dm [μm]	68.4

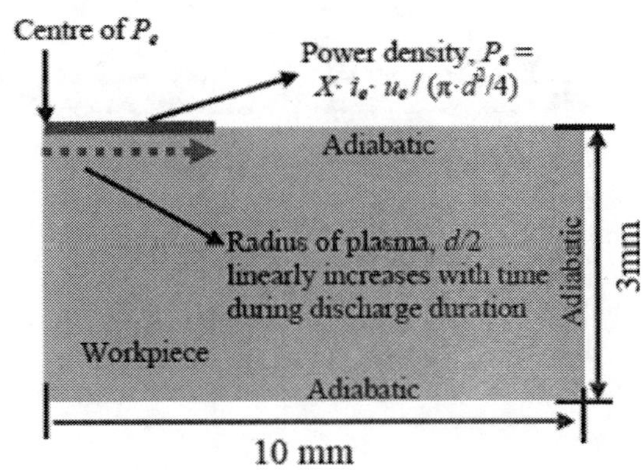

Figure 7. Overview of heat conduction model of satinless steel workpiece

Figure 8 shows the magnification of the discharge zone. The zone in the workpiece where the temperature exceeds boiling point was assumed to vaporize immediately, and thus each element in the zone was removed from the model. Accordingly, the power density location was changed as shown in Figure 8.

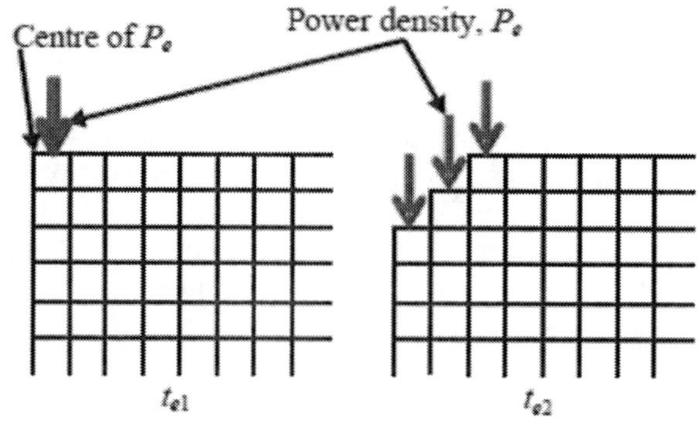

Figure 8. Magnification of discharge zone

After the end of discharge, the simulation was continued to allow cooling before obtaining the residual stress. Figure 9 shows the temperature distribution at the end of cooling (t = 265 μs after the ignition of discharge). As shown in the figure, the maximum temperature of the workpiece model drops to 100 ˚C which is only 75 ˚C above the initial temperature. This can be considered sufficiently low for the residual stress calculation.

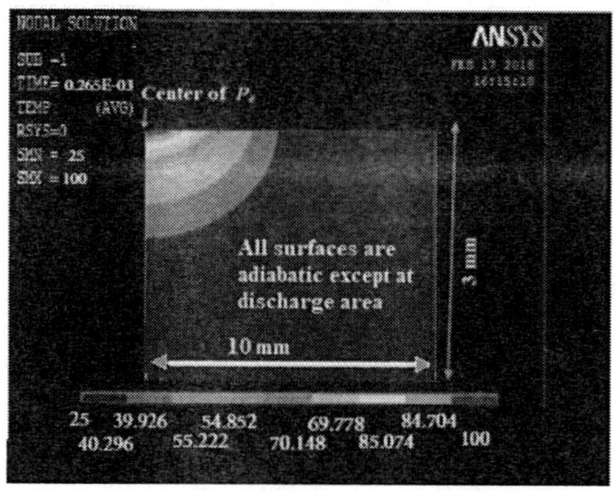

Figure 9. Temperature distribution at the end of cooling time

From this model, the residual stress distribution along the center axis of a crater due to WEDM single discharge was measured. Figure 10 shows measured residual stress and penetration depth for three different machining situations. As shown in the figure, the distribution of residual stress is non- linear, coinciding with the results obtained by other researchers [3] [4]. In later section the influence of non-linear distribution of residual stress on part beam deformation will be observed.

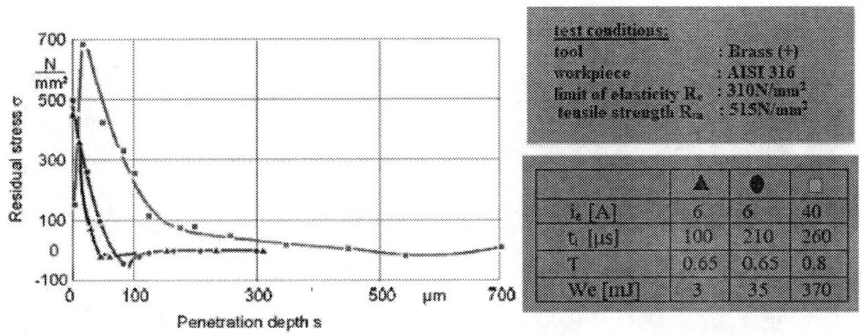

Figure 10. Measured residual stress vs. depth for three different machining situations.

4.3. Criterion of workpiece deformation

An exploratory experiment was carried out to observe the workpiece deformation in WEDM. Stainless steel plate with standard size from supplies was 200mm × 100mm × 3mm as shown in Figure 1. The workpiece was cut off with fixed dimensions as shown in Figure 2. The part beam was machined using WEDM with taking a discharge current, pulse on time, pulse off time and heat input to a workpiece as input parameters, the applied machined path is shown in the figure 11a. In order to study the complex phenomenon accurately, the part beam with size of 10mm × 3mm × 0.2mm were selected as experiment specimen. Moreover, the criterion of workpiece deformation is performed. Generally, the beam deflection could be used to express the deformation behaviour. Through the magnitude of deformation during the WEDM process is micron scale, the beam deflection was measured directly by the KEYENCE VH-Z500R digital microscope. Therefore, the beam deflection is selected as the criterion of workpiece deformation as shown in Figure 11b.

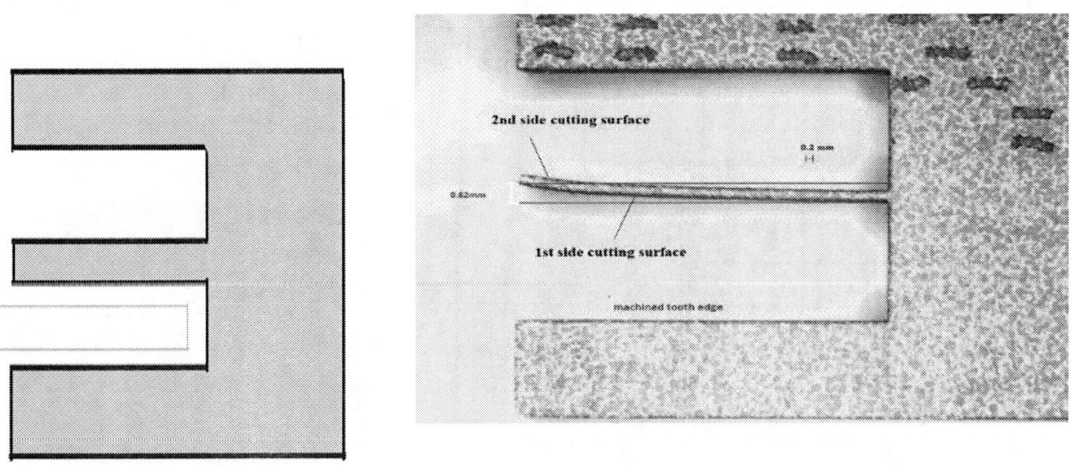

a. Workpiece with cutting direction **b**. picture of the machined stainless steel workpiece

Figure 11. Method to produce stainless steel beam workpiece

5. Results and discussions

5.1. Preliminary experiment

IOP Conf. Series: Materials Science and Engineering **429** (2018) 012011 doi:10.1088/1757-899X/429/1/012011

In order to get the beams deformation after machining WEDM, beams thickness was selected from 0.3 mm to 0.1 mm. Figure 12 shows the problem of deformations when machining very thin parts by WEDM.

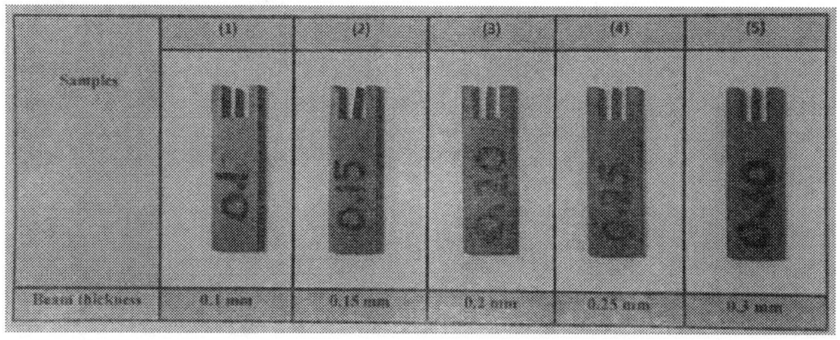

Figure 12. WEDM of thin parts results in considerable deformation

Up to a thickness of 0.25 mm the beams show no deflection, because of a sufficient ratio of thickness to stress penetration, which prevents deformations. The damage observed at the tip of 0.1 mm beam. The beams of 0.2 mm and 0.1 mm on the other hand show warping to the right, i.e. towards the last machining path. It is clear this makes further finishing of the part impossible, thus limits the thickness of structures that can be machined first by WEDM before finishing. It must be noted that the beam of 0.1 mm seems to bend to the right, but this phenomenon is caused by dielectric flushing combined with an undercut in the corners, severely weakening the structure (figure 12, sample 1). Figure 12 should not lead to the interpretation that the structures shown on the figure cannot be correctly machined by WEDM.

5.2. Deformation after machining WEDM

Experiments were run with a stainless steel AISI 316 as used in conventional EDM and a wire diameter of 250 µm. The experiments served to examine the influence of the maximum discharge current ($\hat{i}_e$) and pulse on time (t_e) on the dimensional stability of the part beam. The generator parameters were varied for these investigations in such a way that the total machining spectrum could be examined. At the first, the settings were selected in a way that the share of open- circuit impulses, predominated during machining. Later, the settings were gradually varied until further machining became impossible due to the high portion of short circuits.

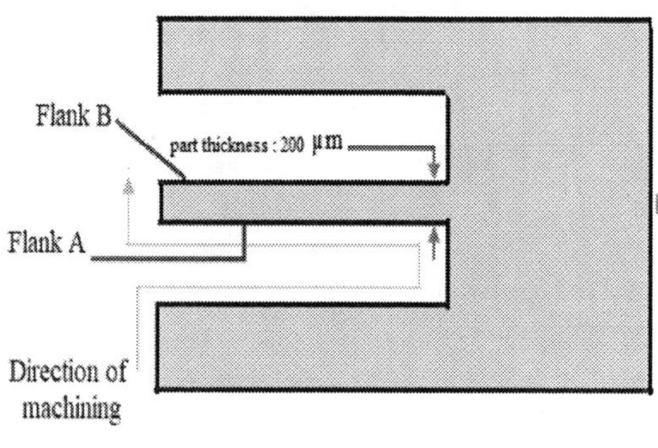

Figure 13. Part beam workpiece

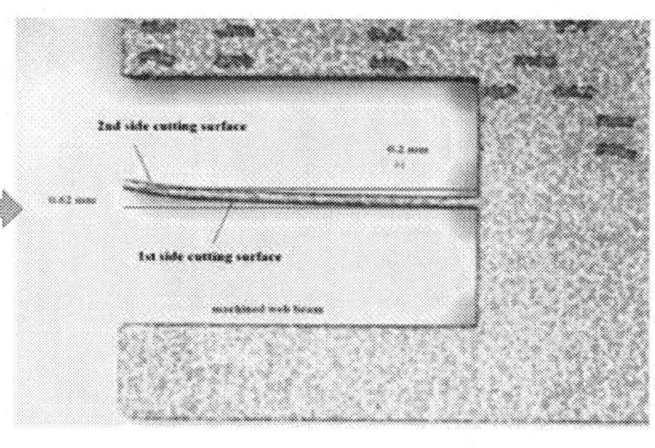

Figure 14. Part beam deformation

Figure 13 shows that a decreasing part thickness leads to a deflection in shape from the programmed workpiece contour. The tensile residual stresses induced during machining of the increasing part beam flank and their depth of indentation into the sub-surface layer exert a major influence on the intensity of the deflection.

The deflection was observed clearly in figure 14, where λ = 0.62 mm, while the basic body of the sample prevents a deformation during machining of the increasing flank (figure 13, flank A), the part beam over during machining of decreasing flank (figure 13, flank B). The shape deflection results from the melting of the basic material under tensile residual stress, so that the elastic limit of the stressed part beam cross- selection is exceeded and a plastic deformation. However, the deflection was observed clearly in figure 14, where λ = 0.62 mm. Figure 15 shows the measured deflection of the part workpiece with respect to its thickness, where the deflection of the workpiece increased when the workpiece thickness was reduced.

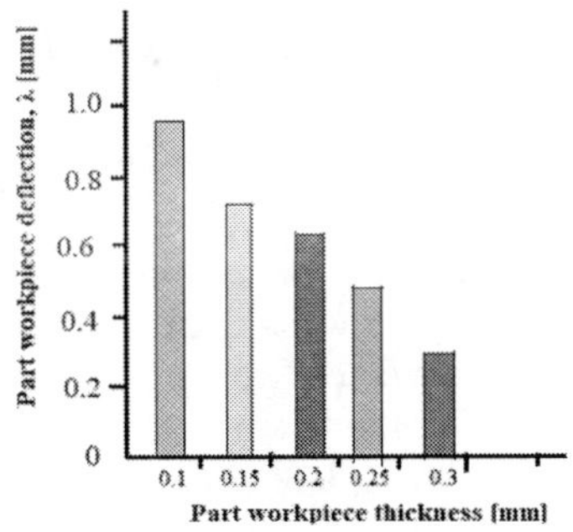

Figure 15. Measured deflection of part workpiece with respect to its thickness

5.3. Effect of residual stress on the workpiece deformation

First, it is fully understood that the larger residual stress accompanies with the larger deformation. In fact, since the workpiece material is heated, molten and removed at the high temperature during the pulse discharge process, the thermal stress appears in the workpiece surface and leads to thermal deformation. In the heating process, the elastic deformation appears at first. Once if the thermal stress within the material is beyond the yield strength of materials, the plastic deformation takes place. Then cooling process comes after the pulse discharge process, and the recast layer begins to form in the cooling process so that the tensile residual emerges in the recast layer surface. It is mentioning worthy that the residual stress distributed at the top of the thin part beam should balance with that force on the bottom. If not, the thin part beam will keep on deforming until the residual stress is balanced by the internal force of materials.

Finally, the results of experiment in this work as shown previously in Figure 14. During the experiment, one side of stainless steel AISI 316 workpiece was rigidly supported, and WEDM results in non- linear residual stress distribution within the workpiece as shown in Figure 10. The non- uniform deformation as shown in Figure 17 can be observed.

5.4. Non- linear deformation in thin part beam

There is a non- linear deformation phenomenon in the WEDM processing the thin part beam component as shown in Figure 14. This deformation is placed non- linear from fixed end to free end along the beam direction. The non- linear deformation phenomenon needs to be further analysed to determine the deformation characteristics. Therefore, the deformation data are measured at each interval of 50 microns from the fixed end to free end of the cutting beam. As shown in Figure 16, it can be clearly seen that there exists the maximum deformation as well as the maximum deflection angle at the tip of the beam. With the increase of the beam thickness along the cutting direction, the deformation decreases rapidly and presents the strong non-linear characteristic, which is exactly consistent with the research results of Zahiruddin and Kunieda [9]. This might be attributed to that when the wire tool is beginning to cut into the workpiece, the pulse discharge status hardly keeps stable as well as more the ratio of abnormal discharge (namely arc discharge, transition discharge, short circuit, open circuit) increases dramatically.

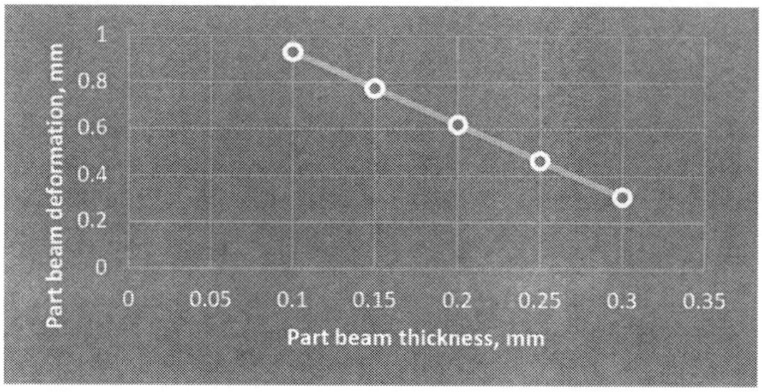

Figure 16. The relationship between part beam thickness and deformation

Additionally, the debris has been flushed away at a high speed in the first cutting so that the debris has little influence on the pulse discharge process, which leads to much more strong input discharge energy directly acting on the workpiece surface. The reasons which mentioned above probably produce the large thermal deformation at the tip of the beam. Moreover, another probable reason is that the tip of the beam is a free end without any constraints, therefore the residual stress in the workpiece material needs to be balanced by enough strong inner force from the larger deformation generated. On the contrary, after the wire tool cutting into the workpiece for some distance away from the free end, the pulse discharge status comes to be relatively stable. Correspondingly, the residual stress of the material decreases and uniformly distributes along the cutting direction. On the other hand, the tail of the beam is a fixed end which can constrain the deformation and balance the residual stress.

6. Conclusion

Workpiece deformation due to WEDM has been studied and the work can be concluded as follows:

1. Up to a thickness of 0.25 mm, the beams show no deflection because of a sufficient ratio of thickness to stress penetration, which prevents deformation.
2. On the other hand, the beams of 0.2 mm and 0.1 mm show warping to the right, i.e. towards the last machining path.
3. The curvature was observed at the tip of the beam workpiece.
4. Non-linear distribution of residual stress was found to be the main reason that causes the non-uniform curvature of the workpiece.
5. The deflection of workpiece increased when the workpiece thickness was reduced.
6. Imposed residual stress was non- linearly distributed in the thickness direction below WEDM crater.
7. As a consequence, a non- uniform bending of web part workpiece along its length due to WEDM can be observed.
8. To understand the workpiece deformation, it must be comparison the curvature of machined web beam workpiece with the curvature of simulated web beam workpiece and also, a comparison between the distribution of residual stress generated by multi- discharges in the depth direction and distribution of residual stress generated by single discharge.
9. As temperature increases due to increase in discharge current during the process caused residual stresses induced by the machining lead to a shape deflection of contour elements.
10. The deflections was measured, where $\lambda = 0.62$ mm. This result indicates that the deflection of the workpiece increased when the workpiece thickness was reduced.

ACKNOWLEDGEMENT

The authors acknowledge the School of Manufacturing Engineering, University Malaysia Perlis for the lab facilities special thanks to those who contributed to this project directly or indirectly.

References

[1] Kawakami T and Kunieda M 2005 Study on Factors Determining Limits Of Minimum Machinable Size in Micro EDM Annals of the CIRP, Vol. **54**, 1, pp. 167-170.

[2] Spur G, Uhlmann E, Doll U and Daus N 1999 WEDM of Microstructured Component Parts – Heat Conduction Model IJEM, No.**4**, pp. 41-46.

[3] Yang Y and Mukoyama Y 1996 Three-dimensional analysis of residual stress in EDM process IJEM, No. **1**, pp. 27–33.

[4] Das S, Lots M and Klocke F 2003 EDM simulation: finite element-basedcalculation of deformation, microstructure and residual stresses J. of Materials Processing Technology, **142**, pp. 434-451.

[5] Cheng X, Yang X, Zheng G, Huang Y and Li L. 2014 Fabrication accuracy analysis of tools with complicated geometries by wire EDM J. Mech. Sci. Technol; **2** (6):2329–35.

[6] Sanchez JA, Plaza S, Ortega N, Marcos M and Albizuri J 2008 Experimental and numerical study of angular error in wire-EDM taper-cutting Int. J. Adv. Manuf. Technol; (12):1420–8.

[7] Sanchez JA, Rodil JL, Herrero A, De Lacalle LL and Lamikiz A 2007 On the influence of cutting speed limitation on the accuracy of wire-EDM corner-cutting. J. Mater. Process. Technol; **1 2**(1):574–9.

[8] Kawakami T and Kunieda M 2005 Study on factors determining limits of minimum machinable size in micro EDM CIRP Ann. – Manuf. Technol; (1):167–70.

[9] Zahiruddin M and Kunieda M 2016 Analysis of micro fin deformation due to micro EDM proc. CIRP, **42**, pp. 569-574.

[10] Kruth J P and Bleys Ph 2000 Measuring residual stress caused by wire EDM of tool steel International Journal of Electrical Machining, No **5**, pp. 23- 28.

[11] Kojima A, Nastu M and Kunieda M 2008 Spectroscopic measurement of arc plasma diameter in EDM Annals of the CIRP, Vol. **57**, pp. 203-207.

Behaviour of GRE composite pipes after fire exposure under dry and wet internal conditioning

Abrar, F.S.B., M.S. Abdul Majid*, M.J.M. Ridzuan, M.K. Faizi, K.S. Basaruddin, N.A.M. Amin, E.M. Cheng.

School of Mechatronic Engineering, Universiti Malaysia Perlis, Pauh Putra Campus 02600 Arau, Perlis, Malaysia.

*shukry@unimap.edu.my

Abstract. This paper presents the behaviour of GRE composite pipe after fire exposure under dry and wet internal conditioning. Numerous studies has been conducted in the past, however there is lack of research in the residual strength, and the damage mechanisms of the pipe due to fire exposure. Therefore, this will be the focus of the research study. The specimens were exposed to fire for a period of 10s to 90s under dry and wet internal conditioning, followed by undergoing a monotonic burst test and scanning electron microscopy (SEM) analysis. The results show that the residual strength in terms of axial stress decreases with an increase in fire exposure time. Water is found to absorb thermal energy from the fire source, hence reducing the rate of matrix breakdown and char formation of the pipe under the wet internal conditioning.

Keywords: Composite pipes, GRE pipes, fire exposure, residual strength, microstructure

1. Introduction

Glass-reinforced epoxy (GRE) is one of the composites that are widely used in low and high-pressure piping to transport fluids. This can be seen in offshore industries where GRE pipes to transport oil and water over long distances because it is easy to handle and repair [1-4]. GRE composite pipe consists of high strength fibres which are bonded to a matrix with well-defined interfaces between them. The matrix and fibres are able to retain their existing chemical and physical properties while producing a better combination of properties that cannot be achieved with either component acting alone.

An intrinsic issue with composite materials is their poor fire performance. The reason for this is because of the flammable property of the composite material organic matrix and the associated risk from fire as mentioned by Mouritz and Mathys [5]. Polyester, epoxies and vinyl ester are the most common resin systems used in composite pipes, and these polymers lose much of their stiffness and strength at temperatures close to glass transition temperature which often lies between 80°C - 150°C. Higher temperatures, more often in excess of 400°C will result in decomposition and products that may burn. These products will subsequently release heat, limiting visibility hindering escape and thus constituting a further hazard and making fire-fighting a dangerous task [6-7]. Furthermore, the behaviour also raised concerns about the integrity of the composite pipe due to degradation in the mechanical properties.

Despite its combustibility, polymer composites have shown positive results when exposed to fire such as excellent thermal insulation and slow burn-through effect based on past research [8]. The reason for this is due to low diffusivity and conductivity of the composites, endothermic effect of the matrix decomposition, convective cooling flow of the volatiles to hot face and low thermal conductivity of residual glass depleted resin content. The most important factor supporting the composites integrity in fire is the heat absorption property due to the endothermic pyrolysis of the polymeric matrix.

Thus the current study investigates the residual strength of GRE composite pipes after being exposed to fire and pressure loading under dry and wet internal conditioning. The investigation focuses on the residual strength of GRE composite pipes after fire exposure and monotonic burst test. This takes into account the process of cutting the composite pipe into smaller specimens which were exposed to fire. The heat flux of the fire is kept constant at 112kW/m2 while varying the fire exposure time ranging from 10s to 90s to simulate in service environments. Then the burnt specimen was subjected to internal pressure loading from the monotonic burst pressure test to determine the deformation behaviour of the composite pipe.

2. Materials and method

2.1 Heat flux calibration

The heat flux of the fire source is set as a controlled variable in this study while manipulating the exposure time on the specimen. Hence, a suitable method is needed to calibrate the fire testing process. The solution is to use the copper block heat flux meter. Figure 1 shows the layout of the apparatus to calibrate the heat flux. The apparatus consists of a copper block with 3 thermocouples placed at different points within the block [9]. This will ensure a more accurate temperature sensing as the average of the temperatures is needed to obtain the heat flux. Besides that, it is important to take note that the exterior of the block is insulated with kaowool, bound together by a calcium silicate board frame. A small circular surface is left exposed at the front for heat to penetrate through during the calibration.

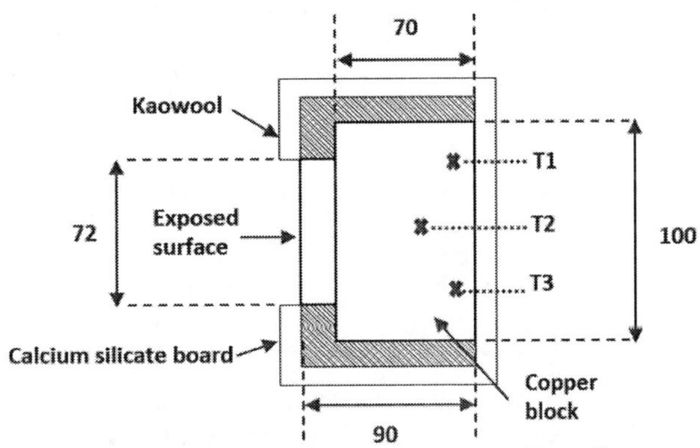

Figure 1. Schematic diagram of the copper block heat flux meter (in mm)

The calculation to obtain the heat flux is based on the following formulation:

$$Q=V.\rho.Cp.\partial T\partial t \qquad (1)$$

where, Q = Energy Input (J/s)
V = Volume of the copper block ($m3$)
ρ = Density of copper
Cp= Specific heat of copper (J/kg.°C)
$\partial T/\partial t$ = Temperature change with respect to time (°C/s)

2.2. Fire exposure
This test utilises a propane fuel source to burn a pipe test specimen to simulate actual fire damage situation. The fire test rig consists of a 14kg LPG cylinder, two support stands, burner array and GRE composite pipe specimen [10]. The setup configuration of the test is shown in Figure 2. The pipe specimen is supported parallel to the burner rows and above the centre axis of the burner assembly. The pipe specimen is fixed at a 12.5 cm height from the burner array whereas the propane gas pressure is fixed at 0.25 bar.

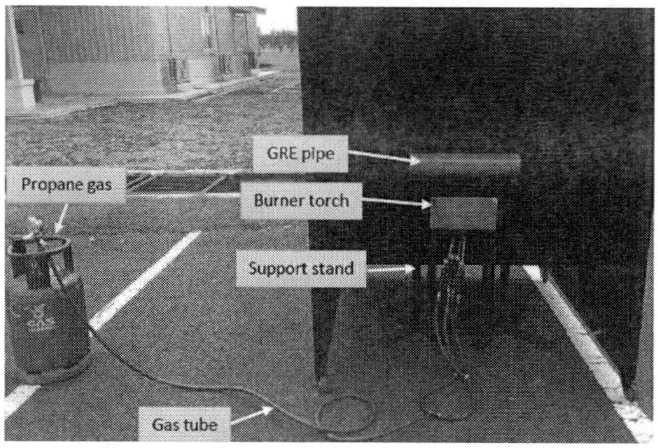

Figure 2. Propane burner conditioning setup

2.3. Monotonic burst test
The burst test is conducted on the composite pipe in 2 conditions that are before and after fire exposure. This test is done investigate the burst strength and functional failure of the specimens. The test utilises a pressure test rig developed based on ASTM D1599 (2014) standard [11]. The test rig was presented in a previous publication by the authors [12]. The concept of this test is to continuously pressurise the specimen until failure occurs. To lock the pressure in the pipe, custom designed end fittings that are based on the pipe dimensions were used. O rings are used together with the end fittings to ensure pressure leaks do not occur during the test. Besides that, tapered sleeves are placed together with the end fittings to limit external movement, acting as a mechanical lock. As a safety mechanism, clamps will be used at both ends during the testing process.

3. Results and discussion

3.1. Heat flux calibration
The heat flux calibration is performed to obtain a value of 113.6 ± 11.4 kW/m^2 based on ISO 14692. The average heat flux was calculated from the results is 111.92 kW/m^2 which is within the desired heat flux range. The calibration found that a gas pressure of $P = 0.25$ bar is required for the burner while the distance between the burner and pipe surface is maintained at $D = 125$ mm.

3.2 Monotonic burst tests
Next, the monotonic burst tests were conducted on the GRE composite pipes after exposure to fire at different time configurations under dry and wet internal conditions. This test aims to determine the pipes' burst strength after subjecting them to fire and different internal conditioning. The burst strength refers to the axial stress caused by the internal burst pressure. Table 1 presents a summary of the fire damaged specimens.

It can be observed that there is a decrease in burst strength when fire exposure duration increases. This is because of the drop in the mechanical strength of the GRE composite pipes following the breakdown of the epoxy matrix, which deteriorates the interfacial fibre-matrix interface. For the dry condition, the pipe started off with burst strength of 123.88 MPa with 10s of fire exposure. There was a drastic decline in the burst strength to 29.79 MPa and 32.51 MPa with the 20s and 30s fire exposure respectively, indicating a strength reduction of more than 70%. The dry 30s specimen reported a lower burst pressure than dry 20s specimen although the burst strength of the dry 30s is still slightly higher, showing that the burst strength did not undergo a rapid decline from 20s to 30s of fire exposure as seen from 10s to 20s. This is due to the formation of char which has low thermal conductivity compared to the virgin composite material, which slows down thermal energy transfer and reduces matrix decomposition rate [13].

Table 1. Summary of monotonic burst tests on various condition and exposure time

Condition	Fire exposure time (s)	Max Burst Pressure (MPa)	Axial Stress (Mpa)	Hoop Stress (Mpa)	Failure Type	Failure Region
Dry	10	16.47	123.88	247.76	Weepage	Near end caps
	20	4.85	29.79	59.57	Weepage	Fire exposed area
	30	4.09	32.51	65.02	Weepage	Fire exposed area
Wet	10	16.69	129.81	259.63	Weepage	Near end caps
	20	16.81	134.24	268.48	Weepage	Near end caps
	30	15.49	125.36	250.71	Weepage	Near end caps
	90	6.17	48.69	97.39	Weepage	Fire exposed area

Besides that, limited oxygen is being transported to the reaction zone as char covers the area and limit oxygen supply causing a slowing down in matrix decomposition. For the wet internal condition, the pipe showed burst strength of 129.81 MPa at 10s of fire exposure which is slightly higher than the same fire exposure duration in the dry condition. After 90 s of fire exposure, the burst strength reduced to 48.69 MPa. The results show that the burst strength of the wet 90 s specimen is still higher than the dry 20s specimen, indicating that the presence of water in the pipe is capable of prolonging functionality of the pipe. This is due to the water acting as a heat conduction medium by absorbing the thermal energy from the fire source, slowing down the matrix decomposition rate [14].

3.3 Residual burst strengths

The evaluation of the residual strength of fire exposed pipes was conducted by slowly pressurising the GRE pipes until failure by weepage occurred. The tests were stopped once the rate of weepage equals the delivery from the pump. The criterion to determine the increased damage upon the residual pressure capacity of the pipes is the residual strength capability that is expressed in percentage with the following formulation:

$$\text{Residual Strength Capability} = \frac{\text{Axial Stress (Fire damaged specimen)}}{\text{Axial Stress (Nondamaged specimen)}} \times 100\% \quad (2)$$

Figure 3 presents the relationship between the residual strength capability and fire exposure time of the GRE composite pipe. The residual strength of the pipes decreases with increasing fire exposure time under both dry and wet internal conditioning. The trends for these two conditions are similar with the main difference being the rate at which the residual strength is decreasing. For the dry internal conditioning, the plot is characterised into two distinct regions whereby there is a rapid reduction of residual strength by 76-80% within the first 20s of fire exposure. This is followed by a slower depletion of residual strength by another 20% until no structural integrity is present. The trend indicates that the pipe will lose total functionality at 39.4s. The behaviour is caused by the rapid degradation of the polymer matrix and fibre bonding at the outer pipe surface which then spreads towards the inner pipe surface which explains the rapid reduction for the first 20s. The matrix decomposition produces char which continues to build up from the outside. The char acts as a thermal insulation layer because the thermal conductivity is much lower compared to the virgin composite material. This effect slows the matrix decomposition rate, hence reducing the damage to the internal matrix layer of the pipe from 20 s to 30 s as seen in the trend.

In the wet conditioning samples, the first region is from 0s to 30s where the pipe maintains a residual strength capability of above 90% with very small decrement. The trend decreases gradually from 30 s onwards until no structural integrity is present. The trend indicates the pipe will lose total functionality by 112.7s when there is water present in the pipe during exposure to fire. The water absorbs thermal energy by heat conduction through the pipe thickness, thus very effectively protecting the matrix-fibre bonding on the inside of the pipe during the first 30s of fire exposure. As water continues to absorb thermal energy, the water temperature increases rapidly and slowing down the energy absorption process [15]. The temperature within the inner matrix-fibre layers starts to increase to above 70 °C, exceeding the T_g of the epoxy which begins the decomposition process of the matrix.

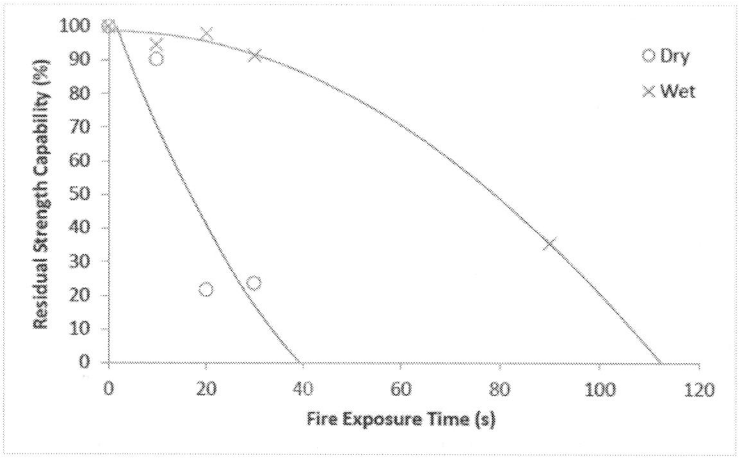

Figure 3. Residual strength capability of fire damaged specimen at all conditions

4. Conclusion

The paper presents the experimental investigation of monotonic burst test performed on fire exposed composite pipes under dry and wet internal conditions. The conclusions are as follows;

1. The average heat flux from the calibration is 111.92 kW/m^2, which is within the desired heat flux range based on ISO14692. The calibration found that a gas pressure of P = 0.25 bar with the distance between the burner and pipe surface of D = 125 mm are required to obtain the required heat flux.

2. The burst strength decreases when fire exposure time increases. For the dry internal conditioning, the plot is characterised into two distinct regions whereby there is a rapid reduction of residual strength in the first 20s due to the quick degradation of the epoxy matrix and fibre bonding at the outer pipe surface. This is followed by a slower reduction of residual strength from 20s to 30s due to the accumulation of char which acts as a thermal insulation layer. This slows down both the matrix decomposition rate and residual strength reduction.

3. In the wet conditioning samples, the first region is from 0s to 30s where the pipe maintains high residual strength capability with very small decrement. The trend decreases gradually from 30s onwards until no structural integrity is present. The pipes under wet internal conditioning exhibit stronger residual strength compared to the dry condition of the same fire exposure time due to the water absorbing thermal energy from the heat source, reducing the temperature distribution along the thickness of the composite laminate, thus slowing down the matrix decomposition rate of the epoxy.

References

[1] Pranesh Krishnan, M.S. Abdul Majid, M. Afendi, S. Yaacob, A.G. Gibson, "Effects of hydrothermal ageing on the behaviour of composite tubes under multiaxial stress ratios", *Compos. Struct.* vol. 148, pp. 1–11, July 2016.

[2] S.N. Fitriah, M.S. Abdul Majid, M.J.M. Ridzuan, R. Daud, A.G. Gibson, T.A. Assaleh, Influence of hydrothermal ageing on the compressive behaviour of glass fibre/epoxy composite pipes, *Compos. Struct.,* vol. 159 January 2017, pp. 350–360.

[3] M.S. Abdul Majid, A. G. Gibson, M. Hekman, M. Afendi, and N. A. M. Amin, "Strain response and damage modelling of glass/epoxy pipes under various stress ratios," Plast. Rubber Compos., vol. 43, no. January 2016, pp. 290–299, 2014.

[4] P. Krishnan et al, "Effects of hydrothermal ageing on the behaviour of composite tubes under multiaxial stress ratios", Composite Structures, vol.148, pp1-11, 2016.

[5] A. P. Mouritz and Z. Mathys, "Post-fire mechanical properties of glass reinforced polyester composites," *Composites Science and Technology*, Vol. 61, no. 4, pp. 475-490, March 2001.

[6] A. G. Gibson and A. P. Mouritz, Fire Properties of Polymer Composite Materials, Dordrecht, Netherlands: Springer, 2006

[7] A. G. Gibson et al, "Modelling composite high temperature behaviour and fire response under load," Journal of Composite Materials, Vol. 46, no. 16, pp 2005-21, 2012.

[8] A. Anjang, "Post-fire mechanical properties of sandwich composite structures," Composite Structures, vol. 132, pp.1019-28, 2015.

[9] T. N. A. Browne, "A Model for the Structural Integrity of Composite Laminates in Fire", Ph.D. dissertation, University of Newcastle upon Tyne, United Kingdom, 2006.

[10] United States. Department of Transportation, Coast Guard Office of Engineering, Logistics and Development, "Fire Endurance Testing of Fiberglass Piping", 1992. [Online]. Available: http://www.dtic.mil/dtic/tr/fulltext/u2/a268638.pdf.

[11] ASTM D1599, Standard Test Method for Resistance to Short-Time Hydraulic Pressure of Plastic Pipe, Tubing and Fittings, 2014.

[12] Pranesh Krishnan, M.S. Abdul Majid*, Ang Jia Yi, M. Afendi, S. Yaacob, A.G. Gibson, An Automated Portable Multiaxial Pressure Test Rig for Qualifications of Glass/Epoxy Composite Pipes, Science and Engineering of Composite Materials, DOI: https://doi.org/10.1515/secm-2016-0036

[13] R.J. Bansal et al, "Thermal stability and degradation studies of polyester resins", Journal of Applied Polymer Science, vol. 37, pp. 1901-1908, 1989.

[14] S. Boryniec, W. Przygocki, "Polymer Combustion Processes.3. Flame Retardants for Polymeric Materials", Progress in Rubber and Plastics Technology, vol. 17, no. 2, 2001

[15] R.N. Walters, "A microscale combustion calorimeter study of gas phase combustion of polymers", Combustion and Flame, vol. 162, no.3, pp.855-863, 2015.

IOP Conf. Series: Materials Science and Engineering **429** (2018) 012013 doi:10.1088/1757-899X/429/1/012013

Thermoplastic Elastomer composite using Benzyl Chloride Treatment on Kenaf Core Powder Mixing with Polypropylene and Virgin Acrylonitrile Butadiene Rubber

Z Mustaffa [1], S Ragunathan [1], N S Othman[1], A A Ghani[2], W A Mustafa[2], P C Aswarya [1], I Zunaidi[3], W K Wan[3], Z M Razlan[3] and Shahriman A B[3]

[1]School of Environmental Engineering, Kompleks Pusat Pengajian Jejawi, Universiti Malaysia Perlis, 02600 Arau, Perlis, Malaysia

[2]School of Engineering Technology, Kampus Sungai Chuchuh, Universiti Malaysia Perlis, 02100 Padang Besar, Perlis, Malaysia

[3]School of Mechatronic, Universiti Malaysia Perlis.

Email: mustaffazainal@gmail.com,

Abstract. The thermoplastic elastomer composite based tensile properties and morphology of kenaf core power (KCP) treated with benzoyl chloride filled polypropylene (PP)/ virgin acrylonitrile butadiene rubber (vNBR) composites were aim and investigated. The composites with different KCP loading from 0 wt % to 30 wt % with constant PP and vNBR (70 wt% and 30 wt%) were prepared using heated two-roll mill by melt mixing at temperature of 180 °C . Tensile properties of composites which is tensile strength, Young's Modulus and elongation at break were investigated. The morphology of filler and composites were investigated using scanning electron microscope (SEM). The result shown, the tensile strength, elongation at break and Young's modulus of treated composites have higher by 4 - 14 %, 4 - 13 % and 0.5 - 3 % respectively compare untreated composite. These findings were supported by micrograph diagram from the morphological study. KCP filler treated with benzoyl chloride has improved the adhesion and gave strong interfacial bonding between KCP filler and PP/vNBRr matrices which results in good tensile strength of composites.

1. Introduction

Kenaf is relatively commercially available and economically reasonably-priced among different natural fibre reinforcing material. Kenaf is usually denoted as industrial kenaf because of its first rate in interest for the manufacturing of industrial raw materials. Kenaf is a hardy, strong and tough plant with a fibrous stalk, proof to insect harm and calls for tremendously fewer amount of or no pesticides. There are three kinds of fibre which are bast, core and pith constitutes the kenaf plant [1]. On the other hand, kenaf is characterized by two distinct fibres bast and core which are 35% and 65% respectively [2]. The bark and core fibre are taken into consideration as two distinct sorts of raw material that can be distinguished by the way of their chemical composition anatomical appearances. The pith is comprising totally of parenchymatous cells, which are in form of polygonal not typically prismatic [3]. The kenaf bast fibres possess putting mechanical properties that lead them as a replacements to glass fibres in polymer composites as reinforcing elements by according to researchers [4].

There are various chemical treatments and modifications being studied to improve the compatibility of matrix and fibre [5-8]. Pre-treatment the filler using alkali (sodium hydroxide) is a traditional way. This Alkali pre-treatment is still relevant to clean filler surface to prepare the filler with a better site for interaction until now. Then, this alkali pre-treatment will be followed by another type of chemical treatment, usually a coupling agent, using organic and inorganic chemical to enhance the tensile, water absorption and thermal properties of the composites [9,10]. Apart from the alkali pre-treatment, there are few more treatments have been applied to the fibre in order to improve the interfacial adhesion between fibre and matrix such as silane, isocyanate, and peroxide chemical treatment. Each of the chemical treatments has their own advantages depending on the type of filler and the matrix being used [11,12].

In this research, the kenaf core have been treated with benzoyl chloride as coupling agent for PP/vNBR composite. The surface modification of kenaf core have been analysed using Scanning Electron Microscopy (SEM). The untreated and treated kenaf core are mixed with PP and vNBR in order to develop the composite. The effects of benzoyl chloride treatment on tensile and morphological properties of PP/vNR/KCP composites have been studies.

2. Material and Methods

Polypropylene (PP) Grade 6331 was obtained from Titan Pro Polymers (M) Sdn Bhd Johor with a melt flow index of 14 g/10 min at 230 °C and a density of 0.9 g/cm^3. The vNBRr was obtained from Juara One Resources Sdn Bhd, Penang. Coupling agent benzyl chloride was supplied by A.R Alatan Sdn Bhd. Kenaf core is collected from Malaysia National Tobacco Board. The KC were grinded into powder. The KC powder was then sieved at a particle size of 150 µm. The KC powder was dried at 80 °C for 24 hour in the oven to remove the water content. Meanwhile the vNBR was cut into small piece with size 0.5 – 1 mm.

2.1. Chemical Modification of Kenaf Core powder.

35 g of Kenaf core powder merges in 6 % NaOH solution for 30 min at room temperature. Then KCP raised with distilled water. The treated KCP soaked in 10 % NaOH solution with added with 50 ml benzoyl chloride and agitated vigorously. After 15 min, the mixture filtering and washing with water. Then the treated powder was drying at room temperature. The KCP were then merges in ethanol for 1 hour to remove the unreacted benzoyl chloride. Finally, the KCP washed with water and dried in an oven at 80 °C for about 24 hour.

2.2. Sample Preparation PP/vNBR/KCP

Formulation of PP/NBRr/KC composites is given in table 1. The mixing process was carried out by using heated two roll mill machine at 180 °C. PP was allowed to melt for 4 min, followed by vNBR and adding of KC filler after 6 min with total mixing time of 9 min. The PP/NBRr/KC composites are prepared in 1 mm sheets by using compression-moulding. The hot-press machine was preheated for 6 min at 180 °C, 3 min of compressing and 2 min of cooling down the sample composite. The total preparation time is 11 min. After the compression process, the composite moulded into square shape with the thickness of 1 mm thin. The moulded composite was cut into dumbbell shape according to ASTM 412 using a Wallace die cutter S6/1/6A.

Table 1. The formulation of PP//vNBR/KCP. *

Material	Composition (wt %)
PP	70
vNBR	30
KC	5,10,15,20,25, 30

(wt%)-weight percent.

*Similar composition of material for treated KCP

2.3. Tensile Properties

The tensile tests were conducted on 5 samples for each formulation according to ASTM D638 standard by using a Universal testing (intron 3366) machine. The testing was carried out at a crosshead speed of 5 mm/min of dumbbell specimens and was performed at room temperature. Tensile strength, young's modulus, and elongation at break will be measured.

2.4. Scanning Electron Microscopy (SEM)

SEM is used to study the surface morphology sample. SEM is a scientific instrument which utilizes a focused beam of electrons to inspect objects with much higher magnification and resolution. Scanning electron microscopy (SEM) model JOEL JSM-6460LA was used in this research. All samples were coated with platinum by using an auto fine coater model JFC-1600 from Japan to avoid electrostatic charging and poor image resolution during the testing. The morphological studied, analysed tensile fracture surface of composites for untreated and benzoyl chloride treatment. The observed samples were taken from the 1 mm dumbbell fracture surface samples from the tensile test. In this analysis, the magnification used was 300x magnifications.

3. Result and discussion

3.1. Tensile Properties

Figure 1-3 shown the tensile strength, elongation at break, E_b and Young's modulus of PP/vNBR/KCP composite. The tensile strength of composites decreasing with the increasing of KCP loading. The tensile strength of the composites decreased due to the inability of the filler to support stress transferred from the PP matrix [13,14]. Poor interfacial adhesion between KCP and the matrices raise the stress concentration area and decrease the tensile strength of composite [15,16]. Meanwhile, the tensile strength of treated PP/vNBR/KCP composites shows higher tensile strength when compared with the untreated composites. Results indicated that the PP/vNBR/KCP composite with 5 wt% kenaf loading exhibits tensile strength increase 1.7 MPa more compare untreated composite. The improvement in tensile strength after treatment indicate that the treatment enhancement the KCP-PP/vNBR interaction.

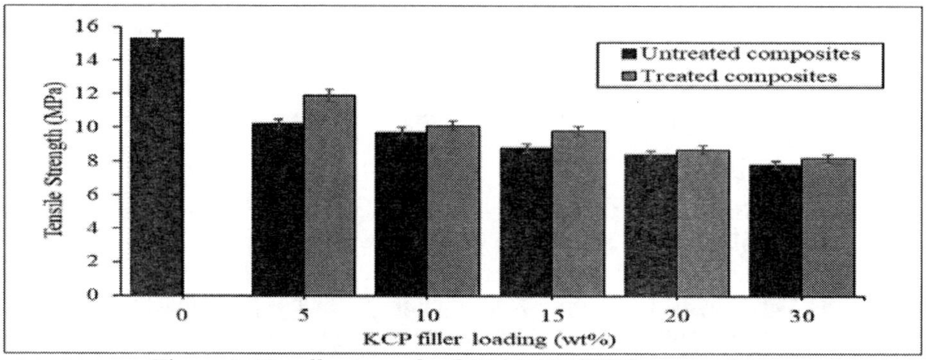

Figure 1: Tensile strength of PP/vNBR/KCP composites.

Figure 2 shows the elongation at break, E_b of PP/vNBR/KCP composites. The E_b drop with the increasing of KCP loading. This due to the raise in stiffness of the compound with the adding of KCP filler. The PP/vNBR/KCP composites with benzoyl chloride treatment, shows an improvement of E_b. The improved are due to a good filler distribution and better adhesion between KCP and PP/vNBR [17].

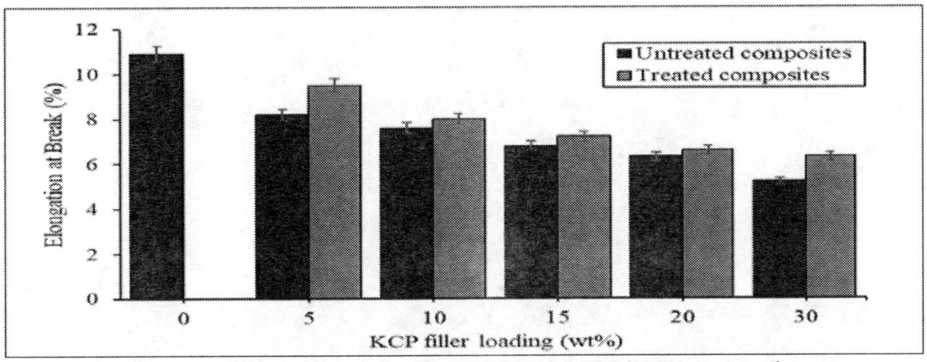

Figure 2: Elongation at break, E_b of PP/vNBR/KCP composites

Figure 3 shows Young's modulus of PP/vNBR/KCP composites. The Young's modulus shows a slight increased with the increase of KCP loading. This make the soft matrix become rigid with additional of KCP filler. KCP was able to minimise the movement of segmental PP/vNBR matrices. PP/vNR/KCP composites with the benzoyl chloride treatment show higher Young's modulus values than the untreated composites [18]. The improvement on Young's modulus was due to a good interfacial adhesion between treated KCP with PP/vNBR matrices [15].

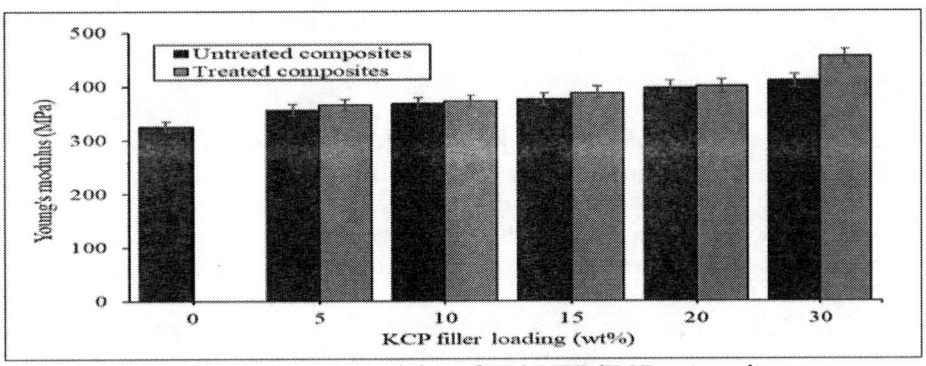

Figure 3: Young's modulus of PP/vNBR/KCP composites

3.2. Morphological Properties

Figure 4 (a-b) shows the tensile fractured surface morphology of PP/vNBR/KCP composites with KCP loading of 30% untreated and treated composites. Based on figure 4 (a), it can be clearly seen that the filler is pulled out from the matrix which is primarily caused by weak interfacial adhesion. This happens because the matrix-cellulose interface does not present homogeneity due to opposite nature of the components (hydrophobic PP/vNBR versus hydrophilic cellulose) [8,19]. Thus, the stress transferred across the interface is very poor. However, Figure 4 (b), adding coupling agent to composite enhances the dispersion in the matrix and enables direct contact between cellulose fibre and PP/vNBR matrices [20]. The PP/vNBR/KCP composite with benzoyl chloride treatment showed an enhancement in interfacial adhesion. The less filler pulled out was observed. The particles are seem to be embedded strongly in the matrix. This is as result of an agreement with the increase of tensile strength of PP/vNBR/KCP with benzoyl chloride treatment [15].

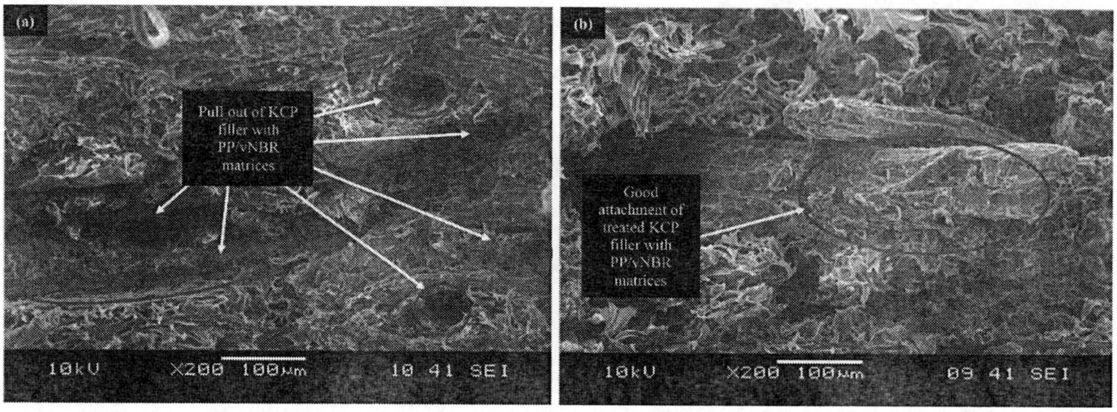

Figure 4: SEM micrographs of the fractured surface morphology of PP/vNBR/KCP composites with KCP loading of 30% (a) Untreated composite (b) benzoyl chloride treated composite.

4. Conclusion

In this research, the effect surface modification using benzoyl chloride as coupling agent on KCP filler mixing with PP/vNBR based on tensile and morphology properties are investigated. Based tensile properties, the tensile strength and elongation at break of untreated PP/vNBR/KCP composites decreases by 4-23% and 7-36% respectively with increasing of KCP filler loading. Meanwhile, the tensile modulus increases by 2-13% with increasing filler loading. However, the treated PP/NBRv/KCP composite have higher tensile strength, elongation at break and Young's modulus. The increasing by 4-14% of tensile strength, 4-13% of elongation at break and 0.5-3% of Young's modulus of treated composites compare with untreated composite. The chemical treatments improved the interfacial bonding between the KCP filler and the thermoplastic polypropylene matrix in the fabricated composites and thereby resulted in greater tensile strength, elongation at break and Young's modulus values. The benzoyl chloride which act as coupling agent causes changes in the linkages between fibre and matrix part of the compound. The pre-treatment of sodium hydroxide ionizes the hydroxyl group to the alkoxide, thus increases the surface roughness in addition to mechanical interlocking. This have been supported by SEM of surface morphology of KCP filler. The SEM of fractured surface of treated composites, show enhancement in interfacial adhesion between KCP filler and PP/VNBR matrices and have less pull out compare to untreated composite.

Reference

[1] Karimi S, Tahir P, Karimi A, Dufresne A and Abdulkhani A 2014 Kenaf bast cellulosic fibers hierarchy : A comprehensive approach from micro to nano *Carbohydr. Polym.* **101** 878–85

[2] Abdul Khalil H P S, Yusra A F I, Bhat A H and Jawaid M 2010 Cell wall ultrastructure, anatomy, lignin distribution, and chemical composition of Malaysian cultivated kenaf fiber *Ind. Crops Prod.* **31** 113–21

[3] Md. Tahir P, Ahmed A B, SaifulAzry S O A and Ahmed Z 2011 Retting process of some bast plant fibres and its effect on fibre quality: A review *BioResources* **6** 5260–81

[4] Faruk O, Bledzki A K, Fink H P and Sain M 2012 Biocomposites reinforced with natural fibers: 2000-2010 *Prog. Polym. Sci.* **37** 1552–96

[5] Mustafa W A, Saidi S A, Zainal M and Santiagoo R 2018 A Proposed Compatibilizer Material on Banana Skin Powder (SP) Composites Using Different Temperatue *J. Adv. Res. Fluid Mech. Therm. Sci.* **43** 121-7

[6] Huda M S, Drzal L T, Mohanty A K and Misra M 2008 Effect of fiber surface-treatments on the properties of laminated biocomposites from poly(lactic acid) (PLA) and kenaf fibers *Compos. Sci. Technol.* **68** 424–32

[6] Sreekala M S, Kumaran M G, Joseph S, Jacob M and Thomas S 2000 Oil palm fibre reinforced phenol formaldehyde composites: influence of fibre surface modifications on the mechanical performance *Appl. Compos. Mater.* **7** 295–329

[7] Mohanty A K, Misra M and Drzal L T 2002 Sustainable Bio-Composites from renewable resources: Opportunities and challenges in the green materials world *J. Polym. Environ.* **10** 19–26

[8] George J, Sreekala M S and Thomas S 2001 A review on interface modification and characterization of natural fiber reinforced plastic composites *Polym. Eng. Sci.* **41** 1471–85

[9] Akil H M, Omar M F, Mazuki A A M, Safiee S, Ishak Z A M and Abu Bakar A 2011 Kenaf fiber reinforced composites: A review *Mater. Des.* **32** 4107–21

[10] Lai S and Li P 2017 Effect of Thermoplastic Polyurethane-Modified Silica on Melt- Blended Poly (Lactic Acid) (PLA) Nanocomposites *Polym. Polym. Compos.* **25** 583–92

[11] Ragunathan S, Mustaffa Z, Kamarudin H, Sam S T and Ismail H 2017 The effect of polypropylene maleic anhydride on polypropylene/(recycled acrylonitrile butadiene rubber)/(sugarcane bagasse) composite *J. Vinyl Addit. Technol.* **23** 228–233

[12] Zainal M, Santiagoo R, Ayob A and Mustaffa W A 2017 Mechanical properties and chemical reaction of 3-aminopropyltriethoxysilane of polypropylene, recycle acrylonitrile butadiene rubber and sugarcane bagasse composites *Int. J. Microstruct. Mater. Prop.* **12** 55

[13] Majid R A, Ismail H and Taib R M 2016 Benzoyl Chloride Treatment of Kenaf Core Powder: The Effects on Mechanical and Morphological Properties of PVC/ENR/kenaf Core Powder Composites *Procedia Chem.* **19** 803–9

[14] Ragunathan S, Ismail H and Hussin K 2014 Comparison of processing and mechanical properties of polypropylene/recycled acrylonitrile butadiene rubber/rice husk powder composites modified with silane and acetic anhydride compound *J. Thermoplast. Compos. Mater.* **27** 1651–66

[15] Nor N A M and Othman N 2016 Effect of Filler Loading on Curing Characteristic and Tensile Properties of Palygorskite Natural Rubber Nanocomposites *Procedia Chem.* **19** 351–8

[16] Blackley D C 1983 *Synthetic Rubbers: Their Chemistry and Technology* (London: Applied Science Publishers)

[17] Ismail H, Ragunathan S and Hussin K 2011 Tensile properties, swelling, and water absorption behavior of rice-husk-powder-filled polypropylene/(recycled acrylonitrile-butadiene rubber) composites *J. Vinyl Addit. Technol.* **17** 190–7

[18] Zainal M, Aihsan M Z, Mustafa W A and Santiagoo R 2018 Experimental Study on Thermal and Tensile Properties on Polypropylene Maleic Anhydride as a Compatibilizer in Polypropylene / Sugarcane Bagasse Composite *J. Adv. Res. Fluid Mech. Therm. Sci.* **43** 141–8

[19] Santiagoo R, Ismail H and Hussin K 2012 Effects of Acetic Anhydride on the Properties of Polypropylene(PP)/Recycled Acrylonitrile Butadiene(NBRr)/Rice Husk Powder(RHP) Composites *Polym. - Plast. Technol. Eng.* **51** 1505–12

The effect of heat treatment on the tensile strength and ductility of pure titanium grade 2

LDK Catherine [1, a], **Darulihshan Bin Abdul Hamid** [1, b]

[1] *Universiti Kuala Lumpur (MIDI)*

[a] *louis.denis06 @s.unikl.edu.my,* [b] *darulihsan@ unikl.edu.my,*

Abstract. The commercially pure titanium (CP) and titanium alloy are widely used in medical, aerospace and automotive industries due to their attractive mechanical properties especially because of its strength-to-weight ratio that is the highest among all known metal. A thorough understanding of the mechanical properties of titanium is very important so that an optimize machining parameters can be selected. Pure titanium has a higher level of corrosion resistance compared to titanium alloy due to the presence of an oxide layer on its surface. In this paper, heat treatment especially the annealing process was used prior to machining in order to study on the variation of its tensile strength and ductility at two different annealing conditions. The three conditions that were used in this experiment are untreated (T1), heat-treated at 700°C for 1 hour (T2) and 900°C for 1 hour (T3). In this work, a higher level of oxidation was observed for T3 compared to T2 due to the exposure to a temperature that is above the beta transus temperature that triggered the formation of acicular martensite. It was also observed that there was a decrease in the tensile strength when treated at T2 and a recovery of the strength was noticed at T3. The ductility properties were more significant at T2.

1. Introduction

The commercially pure (CP) titanium and its alloy are widely used for industrial applications due to its strength-to-weight ratio, high corrosion resistance and high temperature resistance [1]. Pure titanium differ itself from its alloy counterparts due to its high biocompatibility with human body environment and free from toxic elements such as vanadium and aluminum that are found in titanium grade 5. In the dental manufacturing industry, most of them make use of pure titanium and titanium alloys with a treated surface in order to optimize the contact between alveolar bone and the device surface that is also known as osseointegration [1]. According to ASTM F67, the CP titanium are classified into four distinct grades, G1-G4. On the other hand, CP Ti are seldom use in application that are subject to a high stress such as orthopaedic prostheses. In addition, some of the main disadvantage of CP Ti over the alloys are higher young modulus, low mechanical strength, poor wear resistance and a limitation to improve the mechanical properties. The mechanical properties of CP Ti are generally determine by the weight percentage of nitrogen, oxygen and carbon. These three key elements are kept at a lower percentage if a high toughness is required. Heat treating titanium are carried out for three main reasons, namely to reduce residual stress that are developed during fabrication, to customize the material to produce a better combination of ductility and machinability, to increase the strength and finally to optimize special properties such as fracture, toughness, fatigue strength and high temperature creep strength [2]. Titanium and its alloys exist in three structural forms: alpha (α), beta (β) and alpha-beta. In titanium alloy, aluminium is the alpha-phase stabilizer whereas vanadium is the beta-phase stabilizer. CP Ti undergoes an allotropic transformation at 882,5°C and changes from its initial structure of alpha to beta phase that is from an hexagonal close pack (HCP) crystal structure to body centered cubic (BCC) [3]. In a study, the authors investigated the effect of heat treatment on three different metal namely mild steel, titanium grade 2 and stainless steel. The annealing temperature applied for mild steel and stainless steel were set at 200 °C and hold for 120 mins, then furnace cool. In addition, Titanium grade 2 was annealed at 450°C and also furnace cool. They concluded that the ductility decreases significantly for stainless steel but increases for titanium and mild steel [4, 5].

2. Materials and methods

The titanium grade 2 round bar which originate from Japan was manufactured according to ASTM F67 and was purchased from a local distributor that is E-Steel Sdn. Bhd. The length of the round bar was one meter and 15 mm in diameter. A total of nine tensile specimens were prepared according to ASTM E8M standard as shown in Figure 1 for small size specimens. The choice for selecting smaller size specimens in this study was mostly motivated base on the limited budget allocated for this research and also saving the amount of material. The chemical composition and mechanical properties of the titanium grade 2 provided by the manufacturer are listed in Table 1 and 2. The tensile test was perform on an Instron Universal Testing Machine that has a maximum load of 100 kN.

Table 1. Chemical composition of pure titanium grade 2

Element	Fe	O	C	N	H	Ti
Content (Wt.%)	1.143	0.285	0.286	0.593	0.286	0.593

Table 2. Mechanical properties of pure titanium grade 2

Properties	Values
Tensile strength (Mpa)	344
Yield strength (Mpa)	375
Elongation (%)	20
Poisson ratio	0.37
Modulus of elasticity (Mpa)	105
Hardness (HRB)	80

2.1 Fabrication of tensile specimens

The nine tensile specimens were fabricated according to the ASTM E8M dimension shown in Figure 1. The CNC Mazak nexus 200 turning machine as shown in Figure 2 was used for the fabrication of the specimens. A preliminary turning was performed using mild steel as raw material to verify the uploaded machining program and few adjustment were done for the turning parameters. A few samples made of mild steel were damaged due a high cutting speed. Given that the gage diameter is 6 mm in diameter, a lower cutting speed was more suitable to achieve the required dimension for the tensile specimen. The final tensile specimen is as shown in Figure 3.

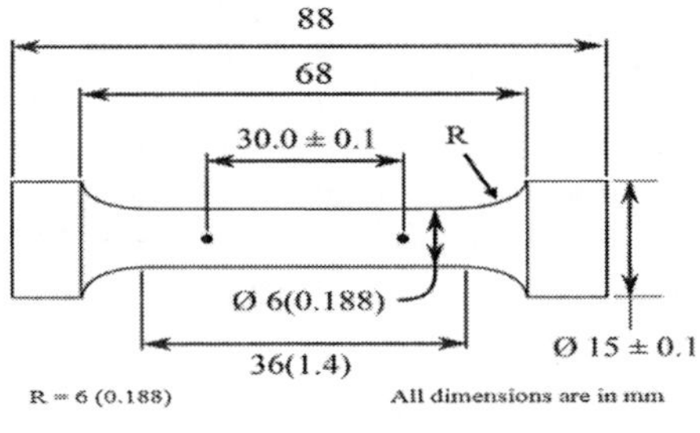

Figure 1. Tensile test specimen according to ASTM E8M

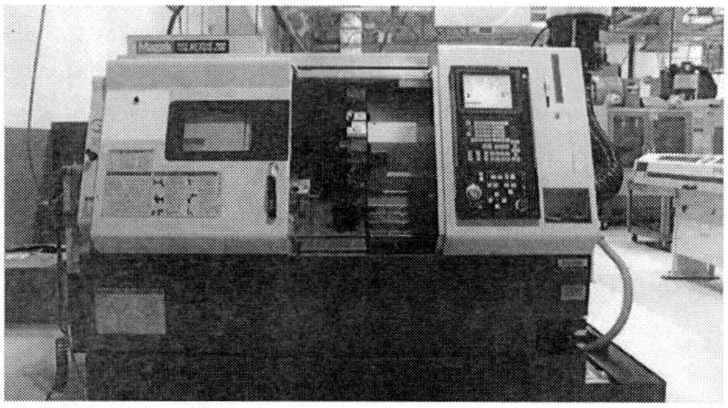

Figure 2 . CNC turning Mazak nexus 200

Figure 3. Machined specimen

2.2 Heat treatment of tensile specimens

The type of heat treatment that was used in this study was annealing. Three batches of tensile specimens namely T1 for untreated specimens, T2 for specimens annealed at 700°C for 1 hour and T3 for specimens that was annealed at 900°C for 1 hour. All the specimens were cool down in the furnace. The heating rate was set at 25°C/min until it reach the required annealed temperature and hold for one hour. In addition, the cooling rate was set at 12°C / min to allow a more suitable rearrangement of the grain. The annealing process was performed in a furnace as shown in Figure 4. The selection of the annealing temperature were done base on the literature reviews where it was stated that above the annealing temperature of 883°C, phase transformation start to occur. In the case of titanium grade 2, the phase transformation is from alpha to beta-phase. However, in a study it was stated that at 700°C it was observe that the Vickers hardness was among its lowest value [6]. In addition, the second annealing temperature of 900°C was selected among others because at this specific temperature the microstructure is fully transform into a beta-phase and the maximum hardness was reached [6]. Furthermore, mechanical properties such as yield stress, strength coefficient and strain demonstrated a significant variation [6].

Figure 4. Furnace

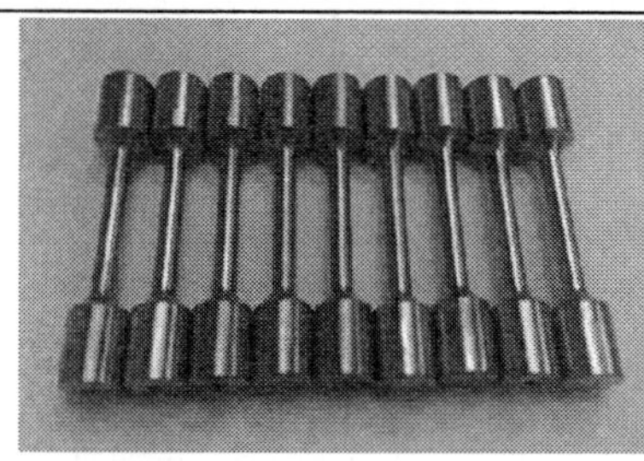

Figure 5. Nine tensile specimens

3. Results and discussion

The change in physical appearance of the specimens after the different heat treatment was related to the phenomenon of oxidation that occur to the tensile test specimens as shown in Figure 6. At T2, a dark scale was form on the sample suggesting an oxidation that occur due to the interaction with atmosphere in the furnace. In addition, T3 exhibit a higher level of oxidation and a light brownish colour scale was formed on the specimens. The oxidation product of titanium during exposure to air is known as TiO_2(titania) which is a tetragonal rutile crystal structure that allow titanium to be one of the most corrosion resistant material [7]. During this process, the high affinity of titanium to oxygen and the high solid solubility of oxygen in titanium that is around 14.5% results in the formation of scale that varies in colours according to the exposed high temperature [8]. Moreover, this oxygen rich layer that is observe on the specimens are widely known as α-case. It adheres strongly to the titanium substrate surface and its strength is fully dependent on three components namely the oxidation temperature, thickness of oxide layer and the influence of nitrogen on the oxidation in air [8]. It was also reveal in a study that the strength of the oxide layers to the substrate decreases with the decrease in the layer thickness as well as the oxidation temperature increases [9, 10]. Figure 7 shows the specimen after that

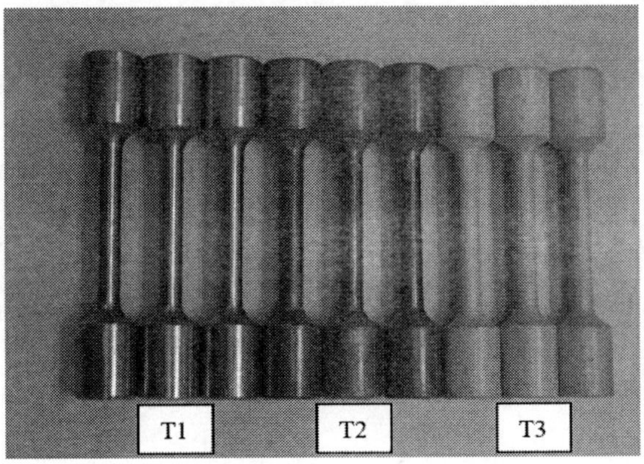

Figure 6. Titanium grade 2 specimen after heat treatment.

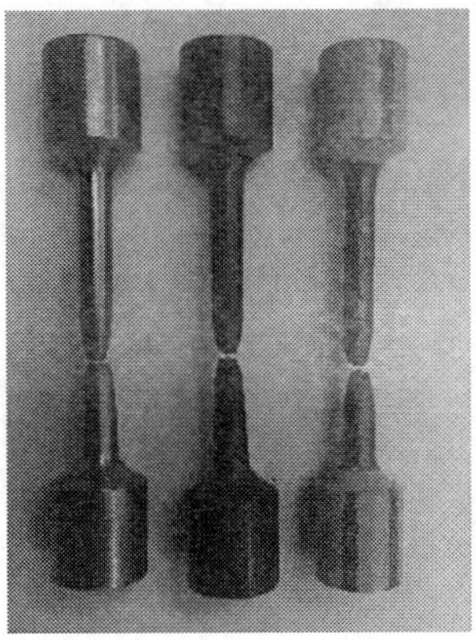

Figure 7. tensile specimen

the tensile test was conducted. Although that the different heat treatment specimens exhibit percentage of elongation, the location and trend of the necking behaviour is same for all of them. The generated results from the tensile test shown in Figure 8 and summarize in table 3 indicate that there is a decrease in the ultimate tensile strength at T2 compare to the other conditions. On the other hand, T2 has the highest ductility with a strain of 50 %. The high ductile behaviour of T2 and T3 can be attributed to the ease of plastic deformation when subject to an annealing condition. In the first case, T2 when heated at 700°C that is a temperature below the transformation to beta-phase, is slightly more ductile than T3 with less tensile strength. Moreover, at T3 that is above the transformation temperature of 883°C to beta phase for Ti grade 2, the ductility increases and the tensile strength bounce back to the original value of 376 Mpa.

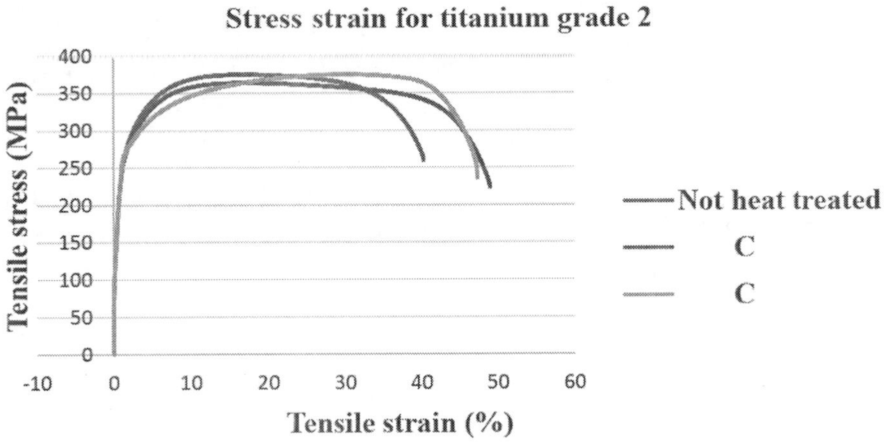

Figure 8. Engineering stress-strain curves of specimens

Table 3. Tensile data of Ti grade 2 for three different conditions.

Sample	TS (Mpa)	Elongation (%)	Area reduction (%)
T1	375	41	75
T2	360	50	82
T3	376	47	65

4. Conclusion

Heat treatment has been used mainly to harden and also to increase the strength of material. In this study, it was applied to observe the behaviour of the mechanical properties on CP Ti grade 2. It was found that oxidation layer on the specimens occur at different rate when subject to higher temperature. The two different heat treatment on CP Ti grade 2 had demonstrated that there is a possibility of softening the material at 700°C that is below the transformation temperature. In addition, the values of the elongation and the percentage in reduction of the area have also confirmed that the ductility for T2 as well as for T3 have increased. It was also revealed that the tensile strength decreases at T2 but recover to its initial strength at T3.

References

[1] Elias CN, Lima JH, Valiev R, Meyers MA. Biomedical applications of titanium and its alloys. Jom. 2008 Mar 1;60(3):46-9.

[2] Smith WF. Structure and properties of engineering alloys. Structure and properties of engineering alloys.2nd edition McGraw-Hill Series in Materials Science; 1993.

[3] Madhukar S, Sai PV, Kumar S, Prakash DJ. Experimental Investigation on Effect of Heat Treatment on Mechanical Properties of Titanium Alloy (Ti-6Al-4V). International Journal of Current Engineering and Technology. 2017 Feb; 7(1).

[4] Madhukar S, Reddy AV, Reddy NV, Reddy PS. Experimental Investigation on Effect of Heat Treatment on Mechanical Properties of Steels and Titanium.

[5] ASTM International. ASTM F67-06-Standard Specification for Unalloyed Titanium, for Surgical Implant Applications (UNS R50250, UNS R50400, UNS R50550, UNS R50700).

[6] Jiang CP, Huang ZH. Grain Size Effect on Mechanical Properties of Titanium Alloy. InKey Engineering Materials 2015 (Vol. 626, pp. 548-552). Trans Tech Publications.

[7] Serrao P, Nayak J. Effect of Surface Mechanical Attrition Treatment on High Temperature Oxidation of CP Titanium. American Journal of Materials Science. 2016; 6(4A):20-4.

[8] Wen M, Wen C, Hodgson P, Li Y. Thermal oxidation behaviour of bulk titanium with nanocrystalline surface layer. Corrosion science. 2012 Jun 1;59:352-9.

[9] Lu G, Bernasek SL, Schwartz J. Oxidation of a polycrystalline titanium surface by oxygen and water. Surface Science. 2000 Jun 20; 458(1-3):80-90.

[10] Wen M, Wen C, Hodgson P, Li Y. Thermal oxidation behaviour of bulk titanium with nanocrystalline surface layer. Corrosion science. 2012 Jun 1;59:352-9.

IOP Conf. Series: Materials Science and Engineering **429** (2018) 012015 doi:10.1088/1757-899X/429/1/012015

Tensile and morphological studies of polypropylene/ empty fruit bunch composite: Effect of maleic anhydride-grafted polypropylene

N S Othman[1], R Santiagoo[1], M Zainal[1], W A Mustafa[2], V Sanapan[1], I Zunaidi[3], Z M Razlan[3], W K Wan[3] and Shahriman A B[3]

[1]School of Environmental Engineering, Universiti Malaysia Perlis, Pusat Jejawi 3, 02600 Arau, Perlis, Malaysia

[2]Faculty of Engineering Technology, Kampus Sg. Chuchuh, Universiti Malaysia Perlis, 02100 Padang Besar, Perlis, Malaysia

[3]School of Mechatronic, Universiti Malaysia Perlis, 02600, Arau Perlis.

Email: wanienurul91@yahoo.com

Abstract. The effect of maleic anhydride-grafted polypropylene (PPMAH) as compatibilizer on the tensileand morphological properties of polypropylene (PP)/ empty fruit bunch (EFB) composite were investigated. The composite was melt mixed using a heated two roll mill at 180 °C with speed of 15 rpm for seven different EFB filler loadings (70/30, 75/25, 80/20, 85/15, 90/20, 95/5 and 100/0 wt%). From the analysis, tensile strength and elongation at break decreased with increasing EFB filler loading. However, it is interesting to observe that the addition of PPMAH has resulted in5 to 10 % better tensile strengthcompareto without PPMAHcompatibilizer. The morphological studies from scanning electron microscopy (SEM) of tensile fracture surfaces showed thatPPMAH compatibilizercomposites also indicates improved dispersion and adhesion of EFB fibre on the PP matrix.

Keywords. Maleic anhydride-grafted polypropylene, empty fruit bunch, polypropylene, tensile properties, compatibilizer

1. Introduction

Recently, with the increasing environmental consciousness and new government rules, the motive for manufacturing product has to be more suitable to the environment[1].For over a decade, the application of biocomposite that consists of natural fibre and polymer has grabbed their attention due to its characteristic. Natural fibre like kenaf, jute, hemp and empty fruit bunch has been replacing from the conventional fibre such as glass and carbon[2–4]. Natural fibre has been a great solution for efficient and a very cheap absorbing material. They are also an ideal optiondue to their renewable nature and eco-friendly properties. Production of the composite using this fibre gives a low cost, low density, biodegradable, good thermal and insulating properties[5–7].The usage of natural fibres as reinforcing fibres in thermoplastic and thermoset matrix composite produce a good environmental benefit along with ultimate disposability and uses raw material. These fibre have high toughness, acceptable specific strength and good thermal properties compared to glass fibres and mica[1].

IOP Conf. Series: Materials Science and Engineering **429** (2018) 012015 doi:10.1088/1757-899X/429/1/012015

Usually, natural fibre such as empty fruit bunch (EFB) is either burned or dumped in the plantation area for natural decomposition[8].The uses of these fibres in a composite can reduce the amount of polymer used which eventually protect the environment and less polymer waste to be disposed[9]. Besides, usage of natural fibre is suitable as reinforcement in polymer matrix as it is available in abundance, requires easy processingand cheap[10]. However,natural fibres arenaturally hydrophilic and thermoplastic have hydrophobic properties.The incompatibility of this two material will create poor interfacial adhesion and affect the performance of the composites [7, 8].Compatibilizer is eventually used to enhance the interfacial adhesion between the fibre and matrix.Maleic anhydride-grafted polypropylene (PPMAH) is one of the compatibilizers, produced by letting the PP chain to link with maleic anhydride and form covalent bonds at the interface between fibres with PPMAH [13].

Thus, this study is aimed to investigate the tensileproperties, which consist of tensile strength, Young's modulus, elongation at break, and morphological properties of the PP/EFB composite with and without the PPMAHcompatibilizer. The effect of EFB filler loading on the composite isalso taken into consideration.

2. Material and method

2.1. Material Preparation

PP grade 6331 as a matrix was supplied from Titan Pro Polymers (M) Sdn. Bhd. EFB fibre from United Oil Palm SdnBhd was cleaned to remove all the impurities such as dried leaves and dirt.EFB was then cut and resized into small pieces. Next, the cleaned EFBwas dried in the oven with a temperature of 90°C for 24 hours to remove the moisture content. It wasgrinded into powder formand sieved into the particular size less than 150 μm. The EFB powder dried again in the oven to remove excessive moisture content at the same temperature as earlier for 24 hours. PPMAH was supplied from Sigma Aldrich chemical.

2.2. Mixingcompounding and sample preparation

The materials were prepared and weighed according to the formulation as shown in table 1. PP and EFB were mixed using a heated two roll-mill machine at 180° C with a rotor speed of 15 rpm. The PP was first charged into the machine and followed by EFB powder at first four minutes. The total mixing time is 7 minutes. For the compounding with PPMAH compatibilizer, it was added simultaneously with PP. Then, all the compounds were compressed using a hot press machine at 180° C. The compounds were preheated for three minutes prior to being compressed for four minutes. The 1 mm thin sheet was allowed for cooling for another three minutes. It was cut into dumbbell shape using Wallace die cutter model S6/1/6.A. Each sample was prepared with five duplications.

Table 1. Formulation for PP/EFB composite with and without PPMAH compatibilizer.

Materials	PP/EFB (wt%)							PP/EFB/PPMAH (wt%)						
PP	100	95	90	85	80	75	70	100	95	90	85	80	75	70
EFB	0	5	10	15	20	25	30	0	5	10	15	20	25	30
PPMAH[a]	-	-	-	-	-	-	-	3	3	3	3	3	3	3

[a]the weight percent of PPMAH is dependent on PP

2.3. Testing and characterization properties

2.3.1. Tensileproperties

Tensile tests were performed on the samples using an Instron Universal Testing Machine. Tensile tests were conducted according to ASTM 638 Standard Test Method to determine tensile strength, Young's Modulus and elongation at break. The thickness and width of each dumbbell specimens were recorded using a vernier caliper and the initial jaw separation distance on the tensile tester was set to 50mm. All dumbbell specimens were tested with a crosshead speed of 5mm/min. The sample was clamped at both ends with one end fixed to the machine and the other end at the removable clamp. Loads were applied until the samples were broken. All five tests were conducted for each type of sample and the average values are reported.

2.3.2. Morphological properties

Scanning electron microscope (SEM) was used to identify the surface morphologies of the fracture tensile specimens. Prior to the SEM testing, an auto fine coater was used to coat all the specimens with platinum to avoid charging effect. The tensile fracture and cross-sectional micrographs were taken to study the fracture mechanisms and interface adhesion of the composites. The SEM creates various images by focusing a high beam of electrons onto the surface of a sample and detecting signals from the interaction on the incident electrons with the surface of the sample.

3. Result and discussion

3.1. Tensile properties

The tensile strength of PP/EFB composites and PP/EFB/PPMAH composites is shown in figure 1. It can be observed that tensile strength decreases with increasing of EFB filler loading. This could be due to the incompatibility of the non-polar PP matrix with polar EFB which results in poor adhesion at the interfaces [14]. As the amount of fiber loading increase, low PP incorporation could result in insufficiency of the matrix to hold the EFB fiber. Thus, this occurrences affected the ability of the matrix to absorb and transfer the stress to the fibers[15]. However, PP/EFB/PPMAH composites have an improved tensile strength for almost 5 to 10 % compared composite without PPMAH compatibilizer. It is proved that the compatibilizer can improve the matrix-filler distribution in the composite. Addition of PPMAH has enhanced the rate of stress transmitted from the matrix to the EFB, which eventually result ina better connection at the interfacial region[16]. Addition of PPMAH has increased the adhesion by forming the C=C group's reaction with the PP matrix[14].

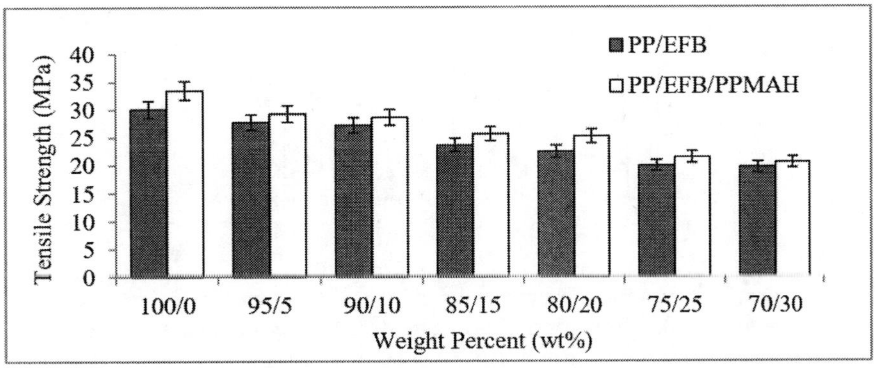

Figure 1. The tensile strength of PP/EFB composites with and without PPMAH compatibilizer.

The Young's modulus of PP/EFB composite with and without the addition of PPMAH is shown in figure 2. Young's modulus was found to be increased with increasing of EFB filler loading[17]. This could be due to the addition of EFB filler into the PP matrix has improved the stiffness of the composite. Lignocellulosic filler in common has its natural stiffness, which is expected higher than the matrix [18]. Composite with the presence of PPMAHcompatibilizerhasa higher modulus than those with the absence of PPMAH. The modulus increases as the amount of filler increased and this indicates that the increased adhesion at the interfacial area might add the stiffness of the composite [19]. Addition of compatibilizer has strengthened the interfacial adhesion between the EFB filler and PP matrix, hence makes the bonding stiffer [20].

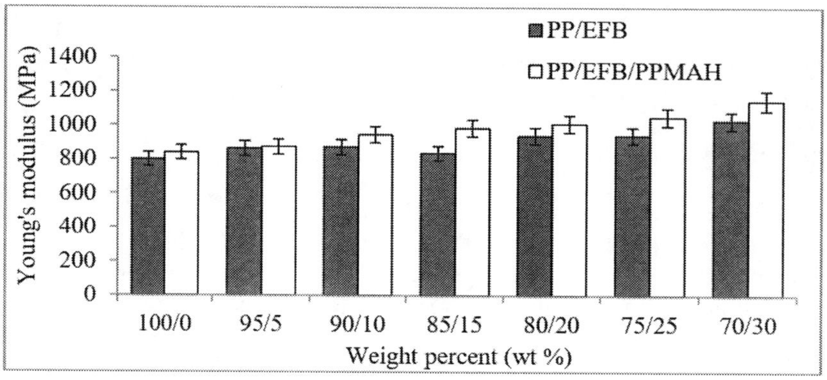

Figure 2. Young's modulus of PP/EFB composites with and without PPMAH compatibilizer.

Figure 3 shows the elongation at break of PP/EFB composites with and without the addition of PPMAH. The elongation at break indicates a decreasing pattern with increasing fibre loading. The break of the sample at a lower value of deformation might due to the effect of rigid fibre disturbing the movement of the polymer network [21]. A decreased of elongation at break with a higher EFB filler loading may be due to the increased stiffness of the composite[18]. However, at similar EFB filler loading, PP/EFB/PPMAH composites have a lower elongation at break. Addition of PPMAH increased the interfacial adhesion resulting more rigid composite. Thus, it will reduce the elasticity or flexibility of the polymer chain[13].

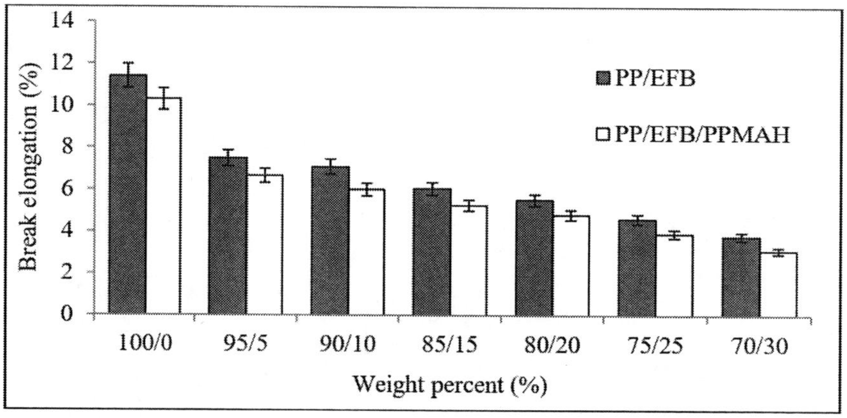

Figure 3. Elongation at break of PP/EFB composites with and without PPMAH compatibilizer.

3.2. Morphological Properties

Figure 4and figure 5 show micrographs of PP/EFB composite without and with PPMAH compatibilizer at EFB filler loading of 15 wt%, respectively. Figure 4 has revealed some of the EFB fibrepull-outsand created holes in the PP/EFBcomposite. The fiberpull-outswere due to the poor fibre-matrix interaction. This interaction will impair the properties of the composite such as poor tensile strength[22].However, for PP/EFB/PPMAH composite as shown in figure 5, a better dispersion and adhesion is detected in the form of a more cohesive interface between the matrix and EFBfibre.The EFB fibre was nicely embedded in the PP matrix. Thus, these results serve as clear evidence that the compatibility between the EFB and PP is significantly enhanced upon reaction with PPMAH[23].

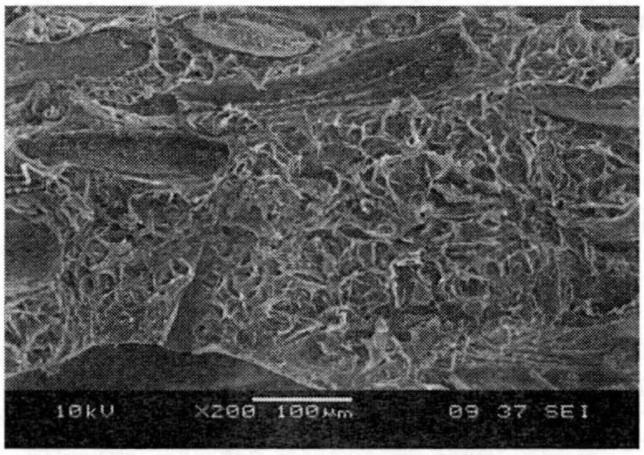

Figure 4. SEM micrograph of PP/EFB composite at 15 wt% fibre loading with 200X magnification.

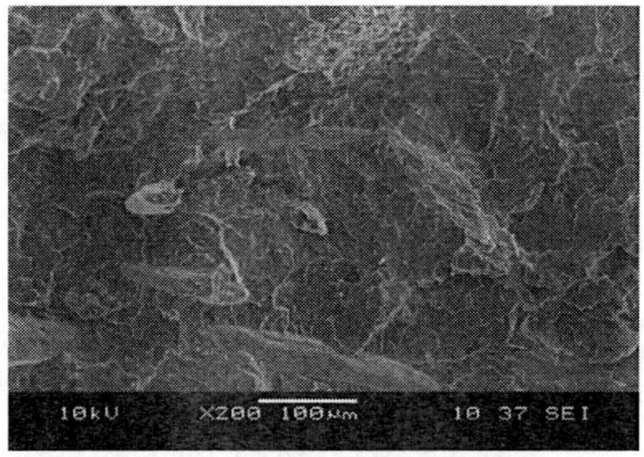

Figure 5. SEM micrograph of PP/EFB/PPMAH composite at 15 wt% fibre loading with 200X magnification.

4. Conclusion

In short, the tensile strength and elongation at break decreased but Young's Modulus increased with increasing of EFB filler loading. PP/EFB/PPMAH composite showed better tensile properties compare to PP/EFB composite. These findings were well supported by micrographs from morphological studies. The good adhesion and interfacial bonding between EFB fibre and PP matrix clearly observed with the addition of PPMAH compatibilizer.

5. References

[1] Mohanty, A.K., Misra, M. &Hinrichsen, G. 2000 Biofibres, biodegradation polymers and biocomposites: An overview *Macromol. Mater. Eng.***276-277** 1–24

[2] Mohammed L, Ansari M N M, Pua G, Jawaid M and Islam M S 2015 A Review on Natural Fiber Reinforced Polymer Composite and Its Applications *Int. J. Polym. Sci.***2015**

[3] Mustafa W A, Saidi S A, Zainal M and Santiagoo R 2018 Experimental Study of Composites Material Based on Thermal Analysis *J. Adv. Res. Fluid Mech. Therm. Sci.***43** 37–44

[4] Mustafa W A, Saidi S A, Zainal M and Santiagoo R 2018 A Proposed Compatibilizer Materials on Banana Skin Powder (BSP) Composites Using Different Temperature *J. Adv. Res. Fluid Mech. Therm. Sci.***43** 121–7

[5] Faizi M K, Shahriman A B, Majid M S A, Shamsul B M T and Ng Y G 2017 An overview of the Oil Palm Empty Fruit Bunch (OPEFB) potential as reinforcing fibre in polymer composite for energy absorption applications *MATEC Web Conf.***01064**

[6] Zainal M, Aihsan M Z, Mustafa W A and Santiagoo R 2018 Experimental Study on Thermal and Tensile Properties on Polypropylene Maleic Anhydride as a Compatibilizer in Polypropylene / Sugarcane Bagasse Composite *J. Adv. Res. Fluid Mech. Therm. Sci.***43** 141–8

[7] Zainal M, Santiagoo R, Ayob A and Mustaffa W A 2017 Mechanical properties and chemical reaction of 3-aminopropyltriethoxysilane of polypropylene, recycle acrylonitrile butadiene rubber and sugarcane bagasse composites *Int. J. Microstruct. Mater. Prop.***12** 55–65

[8] Ooi, C.H., Cheah, W.K., Sim, Y.L. &Pung S Y 2017 Conversion and characterization of activated carbon fiber derived from palm empty fruit bunch waste and its kinetic study on urea adsorption *J. Environ. Manage.***197** 199–205

[9] Song, J. H., Murphy, R. J., Narayan, R., & Davies G B H 2009 Biodegradable and compostable alternatives to conventional plastics *Philos. Trans. R. Soc. B Biol. Sci.***364** 2127–39

[10] Jawaid, M., Khalil, H.P.S.A. &Bakar, A.A. 2010 Mechanical performance of oil palm empty fruit bunches / jute fibres reinforced epoxy hybrid composites *Mater. Sci. Eng. A***527** 7944–9

[11] Khatoon, H. & Ahmad S 2017 A review on conducting polymer reinforced polyurethane

composites *J. Ind. Eng. Chem.***53** 1–22

[12] Zainal M, Santiagoo R, Ayob A and Mustaffa W A 2017 Mechanical properties and chemical reaction of 3-aminopropyltriethoxysilane of polypropylene, recycle acrylonitrile butadiene rubber and sugarcane bagasse composites *Int. J. Microstruct. Mater. Prop.***12** 55

[13] Shinoj S, Visvanathan R, Panigrahi S and Kochubabu M 2011 Oil palm fiber (OPF) and its composites: A review *Ind. Crops Prod.***33** 7–22

[14] Rozman H D, Tay G S, Kumar R N, Abusamah a, Ismail H and Ishak Z a M 2001 Polypropylene ± oil palm empty fruit bunch ± glass ® bre hybrid composites : a preliminary study on the ¯ exural and tensile properties *Eur. Polym. J.***37** 1283–91

[15] Anuar, N.I.S., Zakaria, S., Harun, J. & Wang C 2017 Kenaf/PP and EFB/PP: Effect of fibre loading on the mechanical properties of polypropylene composite. *Mater. Sci. Eng.***217**

[16] Othman N S, Santiagoo R, Abdillahi K M and Ismail H 2017 Effect of polypropylene maleic anhydride (PPMAH) on mechanical and morphological properties of polypropylene (PP)/recycled acrylonitrile butadiene rubber (NBRr)/empty fruit bunch (EFB) composites *AIP Conf. Proc.***1865** 1–5

[17] Farsi M 2012 Thermoplastic Matrix Reinforced with Natural Fibers: A Study on Interfacial Behavior *Some Critical Issues for Injection Molding* ed J Wang

[18] Tay, G. S., Zaim, J. M. &Rozman H D 2010 Mechanical Properties of Polypropylene Composite Reinforced with Oil Palm Empty Fruit Bunch Pulp *J. Appl. Polym. Sci.***116** 1867–72

[19] Khalid, M.,Ratnam, C.T., Chuah, T.G., Ali, S. &Choong T S Y 2008 Comparative study of polypropylene composites reinforced with oil palm empty fruit bunch fiber and oil palm derived cellulose. *Mater. Des.***29** 173–8

[20] Nasution, H., Pandia, S., Maulida&Sinaga M S 2015 Impact Strength and Thermal Degradation of waste Polypropylene (wPP) / Oil Palm Empty Fruit Bunch (OPEFB) Composites: Effect of Maleic-g-polypropylene (MAPP) Addition *Procedia Chem.***16** 432–7

[21] Marquès, E.F., Méndez, J.A.,Pèlach, M.A., Vilaseca, F., Bayer, J. &Mutje P 2011 Influence of coupling agents in the preparation of polypropylene composites reinforced with recycled fibers *Chem. Eng. J.***166** 1170–8

[22] Mohsin, M. E. A., Ibrahim, A. N., Arsad, A. Rahman, M.F.A. &Alothman O Y 2015 Effect of Polypropylene, Ethylene Vinyl acetate and Polyamide-6 on Properties of Recycled Polypropylene / Empty Fruit Bunch Composites. *Fibers Polym.***16** 2359–67

[23] Rozman H D, Saad M J and Ishak Z A M 2002 Modification of Oil Palm Empty Fruit Bunches with Maleic Anhydride: The Effect on the Tensile and Dimensional Stability Properties of Empty Fruit Bunch/Polypropylene Composites *J. Appl. Polym. Sci.***87** 827–35

Fabrication and Properties of Polypropylene and Kenaf Fiber Composite

Z Mustaffa[1], S Ragunathan[1], N S Othman[1], A A Ghani[2], W A Mustafa[2], A M Farhan[1], I Zunaidi[3], Z M Razlan[3], W K Wan[3] and Shahriman A B[3]

[1] School of Environmental Engineering, Kompleks Pusat Pengajian Jejawi, Universiti Malaysia Perlis, 02600 Arau, Perlis, Malaysia
[2] Faculty of Engineering Technology, Kampus Sungai Chuchuh, Universiti Malaysia Perlis, 02100 Padang Besar, Perlis, Malaysia
[3]School of Mechatronic, Universiti Malaysia Perlis, 02600, Arau Perlis.

Email: mustaffazainal@gmail.com

Abstract. The effect polypropylene maleic anhydride (PPMAH) as a compatibilizer in mixing of polypropylene (PP) with kenaf fiber (KP) has been studied. Seven (7) different PP/KF composites ratio (100/0, 95/5, 90/10, 85/15, 80/20, 75/25 and 70/30) where compatibilizer use for PP/KF composite with fixed 3 wt% of maleic anhydride grafted polypropylene were carried out. In this project, all compound preparation was done by using a heated two-roll mill at a temperature of 180 °C for 7 minutes total of the mixing process. Then, the mixture was moulded by using a hot press machine at a temperature of 180 °C for a total 12 minutes to form 1 mm sheet and then cut into dumbbell shape. This sample was tested for mechanical properties include tensile strength, young modulus and elongation at break and morphology of the fracture surface of PP/KF composites. The result show PP/KF composite with compatibilizer has a higher tensile strength and young modulus compare to uncompatibilizer PP/KF composite. The adhesion between fiber and matrix has improved with addition of PPMAH compatibilizer. It make the PP/KF composite increase the stiffness and tensile strength. The scanning electron microscopy (SEM) study of tensile fracture surfaces of the PP/KF composites also indicates improved adhesion of kenaf fiber with polypropylene matrix with the presence of PPMAH compatibilizer.

1. Introduction

In past few years, many studies have been carried out on uses of natural fiber for polymer composites which are develop rapidly to meet varied end use such as transportation, economical building and other construction material [1,2]. It is because there was an abundance of natural fiber that can substitute with synthetic fiber such as glass and carbon because of synthetic material generally expensive, not environmentally friendly and not renewable [3,4].

Nowadays, kenaf research was more focused on the production of biofuel [5]. Many studies have been carried out which focusing on fabricating lighter and tougher PP/KF composites. The PP/KF composite also having potential in order to improve fuel consumption for automotive part [6]. The beginning of kenaf history in Malaysia beginning in 1999 where it was initiated by the National Economic Action Council (NEAC) which formed a committee to identify the potential of expanding kenaf in Malaysia [7]. The most important thing to be a concern in the research of kenaf was to evaluate kenaf adaptation with climate in Malaysia [8]. As the commercial utilization of kenaf keeps on differentiating from its chronicled part to its various application like paper product, materials for building, and food for livestock [9]. Besides that, reinforced of natural fiber with polymer of PP/KF composite have concern as concern as innovative materials in several applications, however their potentiality is limited for certain industries [10].

Even though most of the natural fiber suitable to replace synthetic fiber, most of them are hydrophilic in nature which absorbs water and causes problems in between fiber and matrix [11]. So it will need some modification to gain better properties of PP/KF composite. In order to overcome the incompatibility between natural fiber and a hydrophobic polymer, the chemical modification was carried out to surmount the problem. The incompatibility may cause of polypropylene is non-polar and hydrophobic while lignocellulosic fiber material is polar due to the hydroxyl groups in the cellulose which will results in poor adhesion [12]. Maleic anhydride-grafted polypropylene (PPMAH) compatibilizer use to improve the interfacial adhesion between fiber and non-polar polymer [13–15]. In general, by using the plantation waste, it can decrease petroleum-based fuels products. So, kenaf has gain attention nowadays to use in the production of natural fiber PP/KF composite which is a combination of kenaf as fiber and polymer that is polypropylene as a matrix.

The main objectives of this research were to fabricate a PP/KF composite to evaluate the tensile and morphological properties of polypropylene and kenaf fiber PP/KF composite as well to study the maleic anhydride-grafted polypropylene effect of the polypropylene and kenaf PP/KF composite.

2. Material and Methods

Polypropylene (PP) Grade 6331 was provided by Titan Pro Polymers (M) Sdn Bhd Johor with a melt flow index of 14 g/10 min at 230 °C and a density of 0.9 g/cm^3. Kenaf fiber (KF) is collected from the Malaysia National Tobacco Board. The KF were grinded into powder and sieved at a particle size of 150 μm. Then it was dried at 85 °C for 24 hours in the oven to remove the water content [16]. Polypropylene grafted maleic anhydride (PPMAH) was obtained by Sigma Aldrich Chemical with a melt flow index of 11.5 g / 10 min. 3 wt% of PPMAH was used as compatibilizer in this formulation.

Material	Function	Supplier
Polypropylene (PP)	Matrix	Titan PP(M) Sdn Bhd
Kenaf Fiber (KF)	Fiber	Lembaga Kenaf dan tembakau
Maleic Anhydride-grafted polypropylene (PPMAH)	Compatibilizer	Sigma Aldrich Chemical

2.1. Sample Preparation PP/KF Composite.

The melt mixing process at temperature 180 °C with rotor speed 15 rpm using a The heated two roll mill machine was used to prepared PP/KF composites. PP was melting followed by PPMAH and melted for 4 min. Then, KF was added continued for another 3 min and the compound was taken out. The total time for mixing was 7 min. The compound was prepared into 1 mm sheets using compression moulding. The hot-press machine was preheating at temperature 180 °C for 7 minute, compressing for 3 minute and cooling down the sample composite for 2 minute. The Wallace die cutter model S6/1/6.A. was used to cut composites into dumbbell shapes.

Table 1. The formulation for PP/KP composite with PPMAH as a compatibilizer in weight percentage.

Material	PP/KF composites composition (wt%)						
PP	100	95	90	85	80	75	70
KF	0	5	10	15	20	25	30
PPMAH	3	3	3	3	3	3	3

2.2. Tensile Properties

Tensile properties test followed ASTM 638 Standard Test Method to determine the tensile strength, Young's Modulus and elongation at break. Tensile tests were conducted using using an Instron Universal Testing Machine 5569 on the PP/KF composite samples. Before that, the thickness and width of each dumbbell specimens were recorded by using Vernier Calliper. Then all dumbbell specimens were tested with a 50 mm initial jaw and crosshead speed of 5 mm/min.

2.3. Scanning Electron Microscopy (SEM)

SEM is a scientific instrument, which utilizes a focused beam of electrons to inspect objects with much higher magnification and resolution. Platinum powder used to coated sample by using an auto fine coater model JFC-1600 from Japan to avoid electrostatic poor image resolution and charging during the testing. The morphological studied analyzed the tensile fracture surface of PP/KF composite. The observed samples were taken from the 1 mm dumbbell fracture surface samples from the tensile test. In this analysis, the magnification used was x200 magnifications.

3. Result and discussion

3.1. Tensile Properties

Figure 1showns the tensile strength and Young's module of PP/KF composite uncompatibilizer and with PPMAH compatibilizer. The tensile strength was decreased with increasing of KF loading. This is maybe caused by the inability of fiber to support stress and distribute it to polymer [12]. On the other hand, the tensile strength of PPMAH compatibilizer PP/KF composite showed higher tensile strength compared to the uncompatibilizer PP/KF composite. The improvement by 3.98 – 18.50 % compared uncompatibilize composites. The presence of PPMAH in the PP/KF composite have better bonding between KF fiber and PP matrix [17,18].

Meanwhile, the Young's Modulus was increased with increasing of KF loading. This is may be due to the addition of fiber into the matrix has increased the stiffness of the PP/KF composite [19]. Furthermore can be seen Young's modulus of PP/KF composite is higher compared to uncompatibilizer PP/KF composite. It due to the replacement of a polymer matrix with smaller fiber particle, which improves the modulus of PP/KF composites [20,21]. The increase in modulus was also factor that deformation and matrix mobilization were limited by present of fiber particle that form mechanical barriers [22].

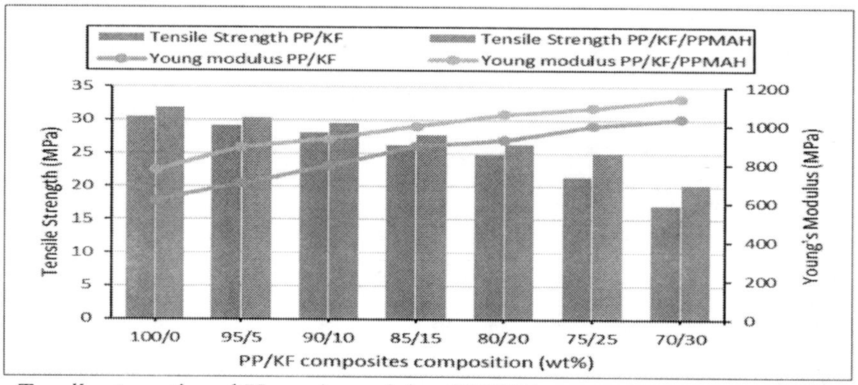

Figure 1: Tensile strength and Young's module of PP/KF composite uncompatibilizer and with PPMAH compatibilizer.

Figure 2 shows elongation at break, E_b of PP/KF composite uncompatibilizer and with PPMAH compatibilizer. It can be seen that uncompatibilize rPP/KF composite show a higher value of elongation compared to the compatibilizer PP/KF composite and this is maybe due to compatibilizer itself where the presence of hydroxyl group of PPMAH that improved the interfacial adhesion between polypropylene and kenaf which resulting the decreasing in elongation at break [23]. In addition, it can be seen that with increasing of fiber show the decreasing of elongation at break and as explain before, this could be due to the crack transmit through lack of interfacial region and caused the PP/KF composite to become brittle as well resulting low value of elongation [24]. The elongation at break decreased with a higher fiber loading cause the increased stiffness of the PP/KF composite.

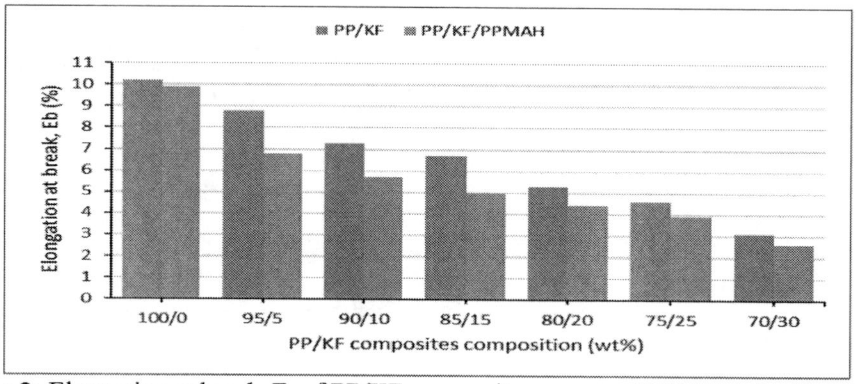

Figure 2: Elongation at break, E_b of PP/KF composite uncompatibilizer and with PPMAH compatibilizer

3.2 Morphological Properties

Figure 3 (a-b) shows SEM micrographs of the fractured surface morphology of PP/KF composite with KF loading of 70/30 wt% (a) uncompatibilizer (b) compatibilizer PP/KF/PPMAH composite. Based on Figure 3 (a) the detachment site of KF can clearly see. This due to PP matrix and KF have poor adhesion between them. Furthermore, the hole is observed in uncompatibilize composite. This because of poor interfacial adhesion between non-polar polymer and fiber [13].

Figure 3 (b) shows a better attachment of kenaf fiber on the PP/KF/PPMAH composites as the fiber has been encapsulated by the matrix. It shows improvement of attachment between kenaf fiber and matrix by using a PPMAH compatibilizer. This was proven that the compatibilizer was effective in improving the mechanical and morphological properties for the PP/KF composite. The less pull out fiber from the matrix and the fiber to the presence of compatibilizer PPMAH that improve the bonding of fiber in composite [25].

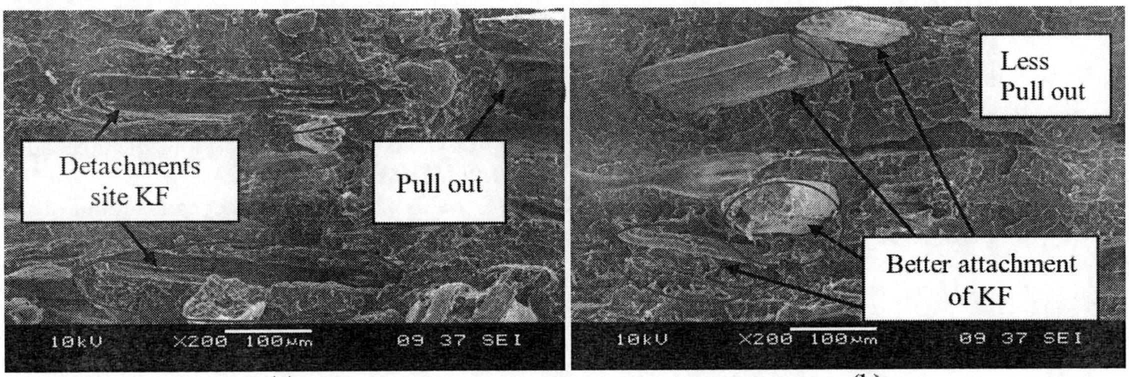

Figure 3: SEM micrographs of the fractured surface morphology of PP/KF composite with KF loading of 70/30 wt% (a) Uncompatibilizer (b) compatibilizer PP/KF/PPMAH composite.

4. Conclusion

In conclusion, PPMAH composites show an improvement in tensile strength by 3.98 – 18.50 % and Young's Modulus by 9.32 – 26.79 % compared to uncompatibilizer composites. Meanwhile, the elongation at break, E_b decrease by 2.94 – 25.37 %. Besides that, morphology properties for PPMAH composite show good interfacial adhesion and less pull out of kenaf fiber.

Reference

[1] Sanjay M R, Arpitha G R, Naik L L, Gopalakrishna K and Yogesha B 2016 Applications of Natural Fibers and Its Composites: An Overview *Nat. Resour.* **07** 108–14

[2] Mustafa W A, Saidi S A, Zainal M and Santiagoo R 2018 A Proposed Compatibilizer Materials on Banana Skin Powder (BSP) Composites Using Different Temperature *J. Adv. Res. Fluid Mech. Therm. Sci.* **43** 121–7

[3] Thiruchitrambalam M, Athijayamani A, Sathiyamurthy S and Syed Abu Thaheer A 2010 A review on the natural fiber-reinforced polymer composites for the development of roselle fiber-reinforced polyester composite *J. Nat. Fibers* **7** 307–23

[4] Mustafa W A, Saidi S A, Zainal M and Santiagoo R 2018 Experimental Study of Composites Material Based on Thermal Analysis *J. Adv. Res. Fluid Mech. Therm. Sci.* **43** 37–44

[5] Saba N, Jawaid M, Hakeem K R, Paridah M T, Khalina A and Alothman O Y 2015 Potential of bioenergy production from industrial kenaf (Hibiscus cannabinus L.) based on Malaysian perspective *Renew. Sustain. Energy Rev.* **42** 446–59

[6] Anuar H, Hassan N A and Mohd Fauzey F 2011 Compatibilized PP/EPDM-Kenaf Fibre Composite Using Melt Blending Method *Adv. Mater. Res.* **264–265** 743–7

[7] Mohd H A B, Arifin A, Nasima J, Hazandy A H and Khalil A 2014 Journey of kenaf in Malaysia: A Review *Sci. Res. Essays* **9** 458–70

[8] H`ng P S, Khor B N, Tadashi N, Aini A S N and Paridah M T 2009 Anatomical Structure and Fiber Morphology of New Kenaf Varieties *Asian J. Sci. Res.* **2** 161–6

[9] Webber III C L, Bhardwaj H L and Bledsoe V K 2002 Kenaf production: fiber, feed, and seed *Trends in new crops and new uses* ed J Janick and A Whipkey (Alexandria) pp 327–339

[10] Babaei I, Madanipour M, Farsi M and Farajpoor A 2014 Physical and mechanical properties of foamed HDPE/wheat straw flour/nanoclay hybrid composite *Compos. Part B Eng.* **56** 163–70
[11] Zainal M, Santiagoo R, Ayob A and Mustaffa W A 2017 Mechanical properties and chemical reaction of 3-aminopropyltriethoxysilane of polypropylene, recycle acrylonitrile butadiene rubber and sugarcane bagasse composites *Int. J. Microstruct. Mater. Prop.* **12** 55
[12] Santiagoo R, Ismail H and Hussin K 2012 Effects of Acetic Anhydride on the Properties of Polypropylene(PP)/Recycled Acrylonitrile Butadiene(NBRr)/Rice Husk Powder(RHP) Composites *Polym. Plast. Technol. Eng.* **51** 1505–12
[13] Yang H S, Wolcott M P, Kim H S, Kim S and Kim H J 2007 Effect of different compatibilizing agents on the mechanical properties of lignocellulosic material filled polyethylene bio-composites *Compos. Struct.* **79** 369–75
[14] Ragunathan S, Mustaffa Z, Kamarudin H, Sam S T and Ismail H 2017 The effect of polypropylene maleic anhydride on polypropylene/(recycled acrylonitrile butadiene rubber)/(sugarcane bagasse) composite *J. Vinyl Addit. Technol.* **23** 228–233
[15] Affandi R D, Ragunathan S, Ismail H and Sam S T 2018 A study on mechanical and thermal properties of polypropylene/recycled acrylonitrile butadiene rubber/palm kernel shell composites with maleic anhydride grafted polypropylene as compatibilizer *J. Vinyl Addit. Technol.* **24** E125–32
[16] Seng O H and Teik T S 2006 Drying of Kenaf in a Flat Bed Box Dryer for Animal Feed *J. - Inst. Eng.* **67** 17–21
[17] Kim H S, Lee B H, Choi S W, Kim S and Kim H J 2007 The effect of types of maleic anhydride-grafted polypropylene (MAPP) on the interfacial adhesion properties of bio-flour-filled polypropylene composites *Compos. Part A Appl. Sci. Manuf.* **38** 1473–82
[18] Santiagoo R, Affandi R D, Noraishah S, Ismail H and Hussin K 2016 The compatibilizing effect of polypropylene maleic anhydride (PPMAH) on polypropylene (PP)/acrylonitrile butadiene rubber (NBR)/palm kernel shell (PKS) composites *ARPN J. Eng. Appl. Sci.* **11** 1666–72
[19] Amri F, Husseinsyah S and Hussin K 2013 Effect of sodium dodecyl sulfate on mechanical and thermal properties of polypropylene/chitosan composites *J. Thermoplast. Compos. Mater.* **26** 878–92
[20] Fu S Y, Feng X Q, Lauke B and Mai Y W 2008 Effects of particle size, particle/matrix interface adhesion and particle loading on mechanical properties of particulate-polymer composites *Compos. Part B Eng.* **39** 933–61
[21] Othman N S, Santiagoo R, Abdillahi K M and Ismail H 2017 Effect of polypropylene maleic anhydride (PPMAH) on mechanical and morphological properties of polypropylene (PP)/recycled acrylonitrile butadiene rubber (NBRr)/empty fruit bunch (EFB) composites *Proceeding of the 3rd International Conference of Global Network for Innovative Technology 2016 (3rd IGNITE-2016)* vol 1865 pp 1–5
[22] Leong Y W, Abu Bakar M B, Mohd Ishak Z A and Ariffin A 2004 Characterization of talc/calcium carbonate filled polypropylene hybrid composites weathered in a natural environment *Polym. Degrad. Stab.* **83** 411–22
[23] Wang N, Zhang J, Fang Q and Hui D 2013 Influence of mesoporous fillers with PP-g-MA on flammability and tensile behavior of polypropylene composites *Compos. Part B Eng.* **44** 467–71
[24] Ragunathan S, Nurul S O, Abdillahi K M and Ismail H 2016 Effect of Maleic Anhydride-Grafted Polypropylene on Polypropylene/Recycled Acrylonitrile Butadiene Rubber/Empty Fruit Bunch Composite *J. Vinyl Addit. Technol.* **On Press**
[25] Zainal M, Aihsan M Z, Mustafa W A and Santiagoo R 2018 Experimental Study on Thermal and Tensile Properties on Polypropylene Maleic Anhydride as a Compatibilizer in Polypropylene / Sugarcane Bagasse Composite *J. Adv. Res. Fluid Mech. Therm. Sci.* **43** 141–8

IOP Conf. Series: Materials Science and Engineering **429** (2018) 012017 doi:10.1088/1757-899X/429/1/012017

Relationship Between Number and Size of Required Steel Reinforcement to Beam Dimension of Standard Public School in Malaysia

S M Razak[1,a)], M A Azizan[1,b)], A Adnan[2,c)], R K Wong[1,d)], K W C Tan[1,e)], A N Rizalman[3,f)], and M Ezdiani[4,g)]

[1]Faculty of Engineering Technology, Universiti Malaysia Perlis, Kampus Uniciti Alam, Sungai Chuchuh, 02100 Padang Besar, Perlis, Malaysia.

[2]Engineering Seismology & Earthquake Engineering Research (e-SEER), Institute of Noise & Vibration, Universiti Teknologi Malaysia, 81310 UTM Skudai, Johor, Malaysia.

[3]Faculty of Engineering, Universiti Malaysia Sabah, 46667 Petaling Jaya, Selangor, Malaysia.

[4]Faculty of Computing, Universiti Teknologi Malaysia, 81310 UTM Skudai, Johor, Malaysia

*Email: skmuiz@unimap.edu.my

Abstract. Overdesign of structural components may result in unnecessary wastage of materials and contribute towards higher construction cost. One structural component most frequently overdesigned is the Reinforced Concrete beam where the amount of steel reinforcements used are way much higher than required to cater for intended design loading. This paper scrutinizes the relationship between the size of beam towards number and size of the required steel bar of the general public school, the number and size of the tension reinforcement bar needed due to the variation of beam size in each ratio based on the general public school and the number and size of the shear reinforcement bar needed due to the variation of beam size in each ratio based on the general public school to become the guideline for the beam design.

1. Introduction

Beam is a component of a building structure which attached to both another components of structure which is slab and column. Usually beams are responsible to withstand tensile strength from loading which transferred to slab and the beam accepted the loading from slabs [1].

Study shows most of the construction project faced the problem of cost overrun due to the less economical design of the building from the aspect of specification, and material usage. This subsequently leads to the loss of profits to the investors who invest their money to construct the building [2]. Sagging moment usually happen at spans of the beam, where hogging moment usually happen at spans of the beam [3].

To investigate the number and size of the tension and compression reinforcement bar needed and the number and size of the shear reinforcement bar needed due to the varies of beam size in each ratio based on the public school. In this study, the beam is limited as continuous beam with three spans. The specification of materials is in accordance to the Standard of Department of Work.

As the tension stress build-up in the concrete, the concrete will need to relieve the tension stress inside the reinforced concrete beam, thereby causing the formation of cracks in the process to lower down the tension stress of the whole structure [4]. In principle, tensile strain is directly proportional to tensile load applied on the reinforced concrete beam. The highest stress accumulated at the bottom region of the beam's mid-span [5].

Shear force is summarised as Vc. For Vd, which is called dowel shear force, it is an action of the rebars to ends up the cracking propagated towards the support component such as the interconnection between beam and column [6,7]. Deflection in reinforced concrete beam is inversely proportional to flexural durability of the beam. Deflection of the beam may cause creep effect and shrinkage on the beams due to the loading on the slabs and the span of the beam [8].

2. Methodology

The design of beam has been separated into three parts which is flexural and shears reinforcement. The beam size for the general school standard is 200x600, 250x500, 240x600,225x500, so in this study, the beam size has been distributed into four ratio of beam size, which is 0.3, 0.4, 0.45 and 0.5. The calculation was done using the spreadsheet in Microsoft Excel. The result was then tabulated in table to shows the number and the size of tension reinforcement bar and shear link. The distance between every shear link was also tabulated in the table of result.

3. Results and Discussion

Table 1 shows that the tension steel reinforcement bar needed due to the beam size with ratio of 0.45 in slab spanning 7 m. The table indicated that the minimum tension steel reinforcement bar needed in the beam size are 200 mm x 450 mm, 250 mm x 550 mm, 275 mm x 600 mm, 300 mm x 650 mm and 325 mm x 700 mm which are capable to avoid from cracking and deflection.

3.1. Parameters Results

Table 1 The number and size of the tension steel reinforcement bar needed due to the beam size with ratio of 0.45 in 7m span of slab

SPAN 7m				
Ratio: 0.45			H16	
b	h	Support	Outside span	Middle span
200	450	7H16	6H16	5H16
250	550	6H16	5H16	4H16
275	600	6H16	4H16	3H16
300	650	6H16	4H16	3H16
325	700	5H16	4H16	3H16
Ratio:	0.45		H20	
b	h	Support	Outside span	Middle span
200	450	5H20	4H20	3H20
250	550	4H20	3H20	3H20
275	600	4H20	3H20	2H20
300	650	4H20	3H20	2H20
325	700	4H20	3H20	2H20
Ratio: 0.45			H25	
b	h	Support	Outside span	Middle span
200	450	3H25	3H25	2H25
250	550	3H25	2H25	2H25
275	600	3H25	2H25	2H25
300	650	3H25	2H25	2H25
325	700	2H25	2H25	2H25
Ratio: 0.45			H26	

IOP Conf. Series: Materials Science and Engineering **429** (2018) 012017 doi:10.1088/1757-899X/429/1/012017

b	h	Support	Outside span	Middle span
200	450	3H26	3H26	2H26
250	550	3H26	2H26	2H26
275	600	3H26	2H26	2H26
300	650	2H26	2H26	2H26
325	700	2H26	2H26	2H26
Ratio: 0.45			H28	
b	h	Support	Outside span	Middle span
200	450	3H28	2H28	2H28
250	550	2H28	2H28	2H28
275	600	2H28	2H28	1H28
300	650	2H28	2H28	1H28
325	700	2H28	2H28	1H28

Table 2 shows that the shear steel reinforcement bar needed due to the beam size with ratio of 0.45 in slab spanning 7 m. The table indicated that the minimum distance between shear steel reinforcement bar needed in the beam size are 200 mm x 450 mm, 250 mm x 550 mm, 275 mm x 600 mm, 300 mm x 650 mm and 325 mm x 700 mm which afford to avoid shear failure at the support of beam.

Table 2 The size of shear link and the spacing between shear link due to the beam size with ratio 0.45 in 7 meter span of slab.

SPAN 7m

Ratio: 0.45			H6	
Beam Size		Outside Support	Outside span	Middle span
200	450	H6-100	H6-80	H6-160
250	550	H6-120	H6-90	H6-130
275	600	H6-130	H6-100	H6-120
300	650	H6-140	H6-110	H6-110
325	700	H6-150	H6-110	H6-100
Ratio:	0.45		H8	
Beam Size		Outside Support	Outside span	Middle span
200	450	H8-120	H8-140	H8-290
250	550	H8-220	H8-170	H8-230
275	600	H8-240	H8-180	H8-210
300	650	H8-250	H8-190	H8-190
325	700	H8-270	H8-200	H8-180
Ratio: 0.45			H10	
Beam Size		Outside Support	Outside span	Middle span
200	450	H8-120	H8-140	H8-290
250	550	H8-220	H8-170	H8-230
275	600	H8-240	H8-180	H8-210
300	650	H8-250	H8-190	H8-190
325	700	H8-270	H8-200	H8-180
Ratio: 0.45			H12	
Beam Size		Outside Support	Outside span	Middle span
200	450	H12-300	H12-300	H12-300
250	550	H12-380	H12-370	H12-380
275	600	H12-410	H12-400	H12-410
300	650	H12-450	H12-420	H12-430
325	700	H12-490	H12-450	H12-400

In the design of shear link and the spacing between the shear links, the outside support and middle support has the different percentage of shear force exerted in the location. The outside support shows 0.45 F and middle support shows 0.55 F and 0.6 F. The amount of shear force in the middle part of each span of beam is minimum compared to those at outside and inside supports. Thus, the minimum shear link should be installed at the span part.

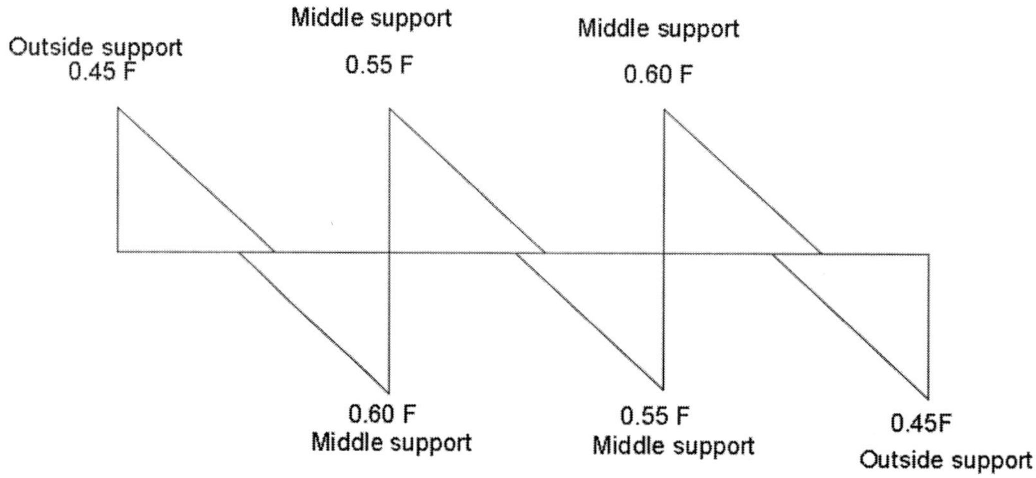

Figure 1 Shear force diagram for a three-span continuous beam

From Table 2, the middle support shows the higher requirement of the shear link than the outside support. This is due to the middle support has the higher percentage of shear force exerted on it compared to the outside support. The more amounts of shear link and the short distance between the shear link may increase the capacity to withstand the shear force.

The larger the size of beam shows the less requirement to the shear link which is little amount of shear link and the far distance between the shear link. This is due to the larger amount of concrete increase the compressive strength of the beam which may afford larger amount of shear force due to loading. The minimum shear link is suitable to place at the span part of the beam due to the low shear force on that part.

4. Conclusions

Furthermore, after the results of the number and size of the tension reinforcement bar needed due to the varies of beam size in each ratio based on the public school are tabulated, it can be concluded as the number of tension reinforcement bar needed which is over 6 reinforcement bars is considered as unreasonable. The number of tension reinforcement bar needed should be maintained in the number around 2 to 6 for tension reinforcement. Last but not least, the conclusion for the results of the shear reinforcement bar needed due to the varies of beam size in each ratio based on the general public school are the larger the size of beam shows the less requirement to the shear link which is little amount of shear link and the far distance between the shear link.

References

[1] Han H, Cao D, Liu L 2017 Green's function for forced vibration analysis of bending-torsion coupled Timoshenko beam *Applied Mathematical Modelling* 45 621-635
[2] Betlejewska R S, Potkany M 2015 Construction costs analysis and its importance to the economy. Procedia Engineering **34** 35-42
[3] Mosley B 2007 Reinforced Concrete Design to Eurocode 2. New York. Palgrave Macmillan Group.
[4] Bhatt P 2006 Reinforced Concrete Design theory and examples. London and New York. Taylor and Francis Group
[5] Albitar M, Mohamed Ali, M S Visinthin, P Drechler M. 2017 Durability Evaluation of Geoploymer and Conventional Concretes *Construction and Building Materials* **136** 374-385.

[6] Franke S, Franke B, Harte A.M 2015 Failure Modes and Reinforcement Techniques for Timber Beams – State of Art *Construction and Building Materials* **97** 2-13
[7] Zagon R, Matthys S, Kiss Z 2016 Shear behaviour of SFR- UHPC I shaped beams *Construction and building Materials* **124** 258-268
[8] Pagani A 2017 Large Deflection and Post Buckling Analyses of Laminatd Composite Beams by Carrera Unified Formulation *Composite Structures* **170** 40-52

On-Site Structural Design Capacity Checker for Reinforced Concrete Beams

S M Razak[1,a), M A Azizan[1,b), A Adnan[2,c), C H Ing[1,d), K W C Tan[1,e), N Z Zainol[1,f) and N Yahya[1,g)

[1]Faculty of Engineering Technology, Universiti Malaysia Perlis, Kampus Uniciti Alam, Sungai Chuchuh, 02100 Padang Besar, Perlis, Malaysia.
[2]Engineering Seismology & Earthquake Engineering Research (e-SEER), Institute of Noise & Vibration, Universiti Teknologi Malaysia, 81310 UTM Skudai, Johor, Malaysia.

*Email: skmuiz@unimap.edu.my

Abstract. This paper presents a structural design capacity checker that has been developed by using Microsoft Excel in order to carry out the calculations of simply supported beam as based on Eurocode 2. The purpose of this research is to develop an Excel User Interface and to perform structural design adequacy check of simply-supported beam under construction at the site. Besides, the required formulations for main reinforcement, shear, deflection and cracking checks are also programmed. The results of maximum shear force, maximum bending moment, tension and compression bars provided in Excel User Interface and STAAD Pro are compared. There are four main parts in Excel User Interface which are input, output, detailed calculation and data sheet. Input and output are the visible parts whereas detailed calculation and data sheet are hidden. In the output sheet, graph of shear force and bending moment, tension and compression bars provided, adequacy of main and shear reinforcement, deflection and cracking checking are shown. Another design of simply supported beam is designed by using STAAD Pro. The results that obtained from Excel User Interface and STAAD Pro is compared. This research helps users to carry out a structural capacity checking of simply supported beam in simple, precise and quicker.

1. Introduction

In the building construction, substructure and superstructure are the two-basis part of the building. Superstructure is the part of structure that located above the ground level. The structure which is known as superstructures include column, beam, slab, brick wall and roof [1]. Thus, every single important step in the building design and construction need to be measured so that the structure of the building is in the safety limits.

The criteria in designing the current capacity of structure are the strength and service ability. The strength of the structure is the adequacy of the structure to withstand the applied load without beyond the limit of any part of the structural member system. For the serviceability of the structure, it is normally defined as deflection which meant structure's fitness which used for specific design loadings. There are allowable limits for all the structural elements. It is a risk when it is required to find the present capacity of the structure [2].

IOP Conf. Series: Materials Science and Engineering **429** (2018) 012018 doi:10.1088/1757-899X/429/1/012018

The safety of the structural buildings and other facilities are the most concern issues for the clients, contractors and engineers. It is a human tragedy which includes high human suffering and monetary loss such as injuries, death, suffering and loss of productivity. In the construction site, safety issues due to the failure of structure need to be taken in order to eliminate the destruction of building during construction time [3]. Eurocode 2 (EC2) is the standard of structural design of concrete structure which is proposed by European. There are 9 separate documents which is related to concrete design in Eurocode 2 and it is published by the British Standard Institution [4]. EC2: Part I is about the building's design. The principle of Eurocode 2 stated any alternatives are not allowed and the designer must comply with the standard [5].

2. Methodology

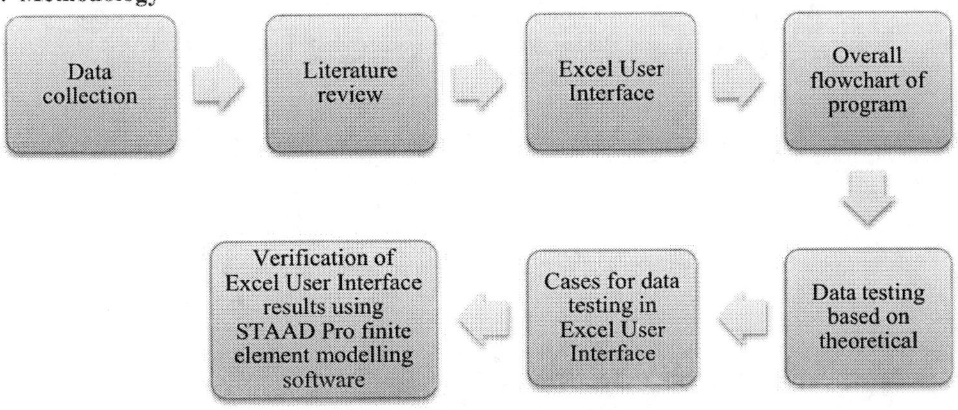

Fig.1: Overall process of flow chart

Data Collection. The software that is used to develop the user interface is by using Microsoft Excel 2013. Data collection such as all the function library of Microsoft Excel 2013 is determined. There are hundreds of functions in Excel but only some of the function is used which depend on the type of data that are needed. Excel Easy, Excel Exposure, YouTube such as Excel Tutorial and ExcellsFun are the medium that provide tutorial of using Microsoft Excel on the network. Besides, Function Library is on the Formula tab and it is a place which can explore the functions of Excel.

Literature Review. Once the data collection has been completed, online database is familiarized and identified in the field of study. Google Scholar, e-resources and Science Direct are the sources which are used to search the relevant databases about the topic of the study. Firstly, the articles are identified and located before starting to write the literature review. The objective and content of the articles are analyzed. The articles are arranged in categories. Then, the literature in table or idea of flowchart is summarized. An outline of the final overview is developed and written the review of literature. Finally, a coherent essay is developed. The reference of databases which is used in the thesis is also written.

Excel User Interface. Excel User Interface was a program for reinforced concrete simply supported beam design as per Eurocode 2. The main purpose was to help user who do or do not have any structural background but need to do structural calculation of checking the capacity of simply supported beam. Thus, the program is established with simple and easy to use interface. User only need to key or select the data of simply supported beam to input and the results of calculations is showed on the output sheet. The user interface was the program designed by using Microsoft Excel 2013. Excel is the most easy and simplest program to use by the user. Therefore, good user interface is produced.

IOP Conf. Series: Materials Science and Engineering **429** (2018) 012018 doi:10.1088/1757-899X/429/1/012018

Overall Flowchart of Program. A program which is used to check the capacity of the simply supported beam is developed by using Microsoft Excel 2013. In the excel user interface, formulation flow for the simply-supported beam is designed as per Eurocode 2 (Design of Reinforced Concrete Structure). The excel user interface is divided into 3 main parts which are input, detailed calculation and output while data sheet was the subpart of the program. Some of the function of input is depended to the detail value of data sheet while the function of detailed calculation is connected by the input which is chosen by the users and value of data sheet which related to the input of data. Lastly, the output is only connected to the detailed calculation part. Anybody can use the user interface included user who do not have any structural knowledge. When the data of the simply supported structure is put into the input sheet of excel user interface and the results of calculations is shown on the output sheet automatically.

In Excel User Interface, input sheet and output sheet were visible by users whereas detailed calculation and data sheet were not visible by users. The worksheets of both detailed calculation and data sheet were make it very hidden by using Microsoft Visual Basic. It was to prevent hidden worksheet appearing in hidden sheet dialog. The worksheet and the workbook of Excel User Interface is protected the structure of Excel User Interface with a password. There are two processes in worksheet protection. The first step was to unlock cells that needed to be editable and then the worksheet is protected with a password. The workbook protection was to prevent users from viewing hidden worksheet, adding, moving or deleting worksheets of Excel User Interface. Figure 3.2 showed the overall flowchart for the excel user interface.

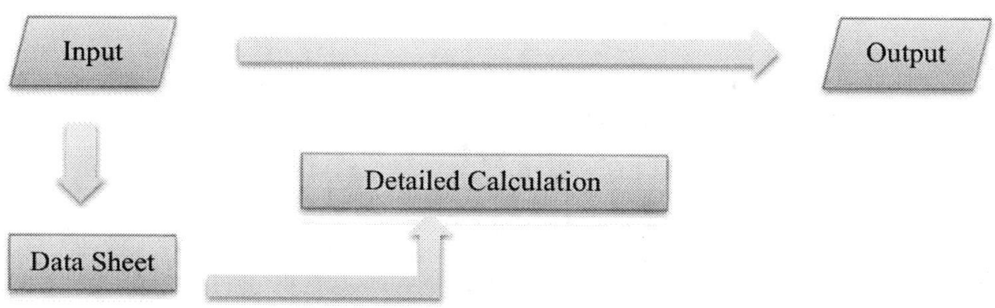

Fig. 2: Overall flowchart of Excel User Interface

Data Testing Based on Theoretical. Data testing based of Excel User Interface is tested based on the theoretical with my supervisor. All of the formulation in the Excel User Interface is checked in order to ensure that the accurate results and readings are given to the user.

Cases for Data Testing in Excel User Interface. There are 8 possible cases that can be used for the simply supported beam in the Excel User Interface, considering maximum of two beam on each size of the simply supported beam being considered.

Table 1: Cases for data testing in Excel User Interface

Case 1: Simply Supported Beam Subjected to 4 Point Loads and Distributed Load	Case 2: Simply Supported Beam Subjected to 3 Point Loads and Distributed Load Simply Supported Beam Subjected to 3 Point Loads and Distributed Load

<table>
<tr>
<td>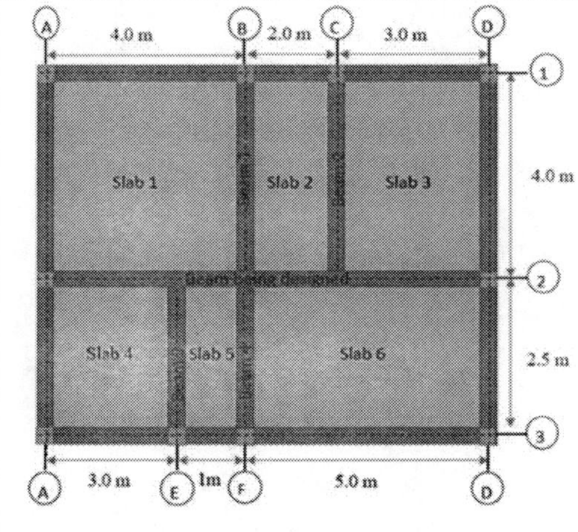
</td>
<td>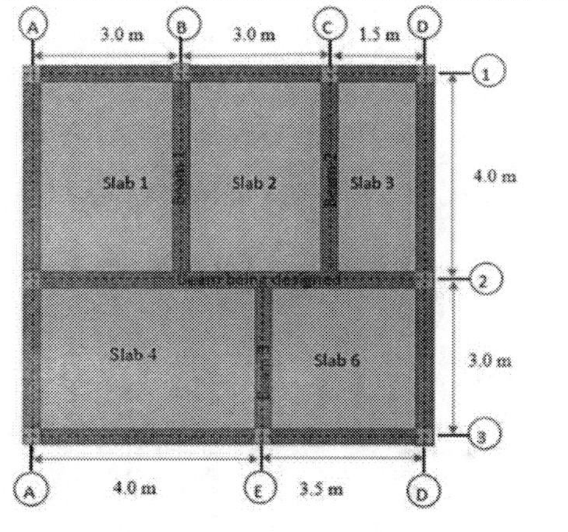
</td>
</tr>
<tr>
<td>Case 3: Simply Supported Beam Subjected to 2 Point Loads at the Same Side and Distributed Load</td>
<td>Case 4: Simply Supported Beam Subjected to 2 Point Loads from Two Different Sides and Distributed Load</td>
</tr>
<tr>
<td>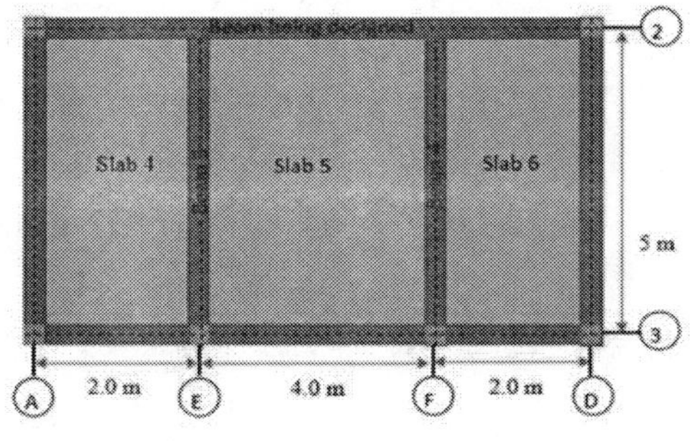
</td>
<td>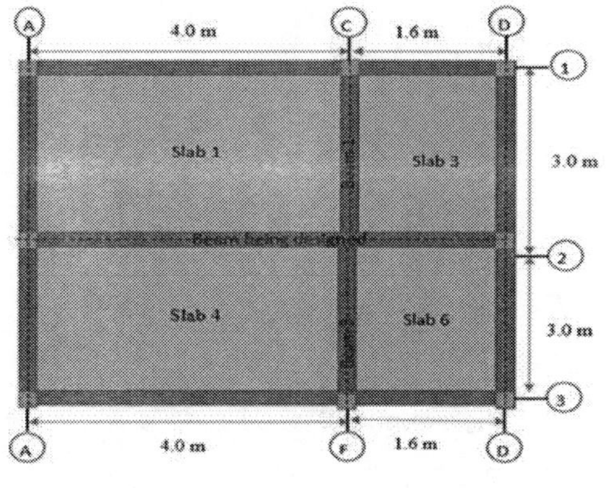
</td>
</tr>
</table>

IOP Conf. Series: Materials Science and Engineering **429** (2018) 012018 doi:10.1088/1757-899X/429/1/012018

Case 5: Simply Supported Beam Subjected to 1 Point Loads and Distributed Load from 2 Different Sides

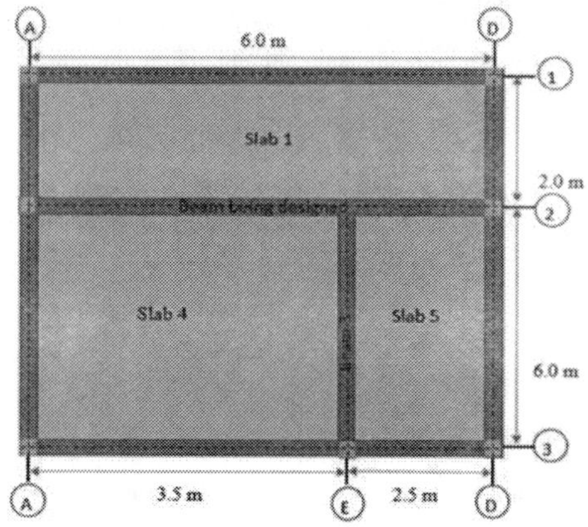

Case 6: Simply Supported Beam Subjected to 1 Point Loads and Distributed Load from Same Side

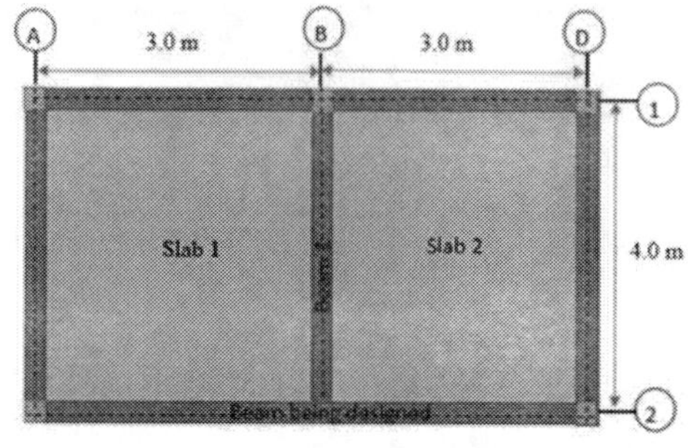

Case 7: Simply Supported Beam Subjected to Distributed Load with 2 Slabs

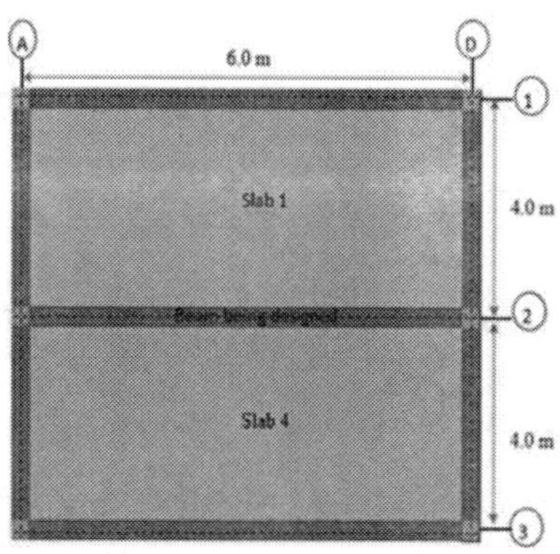

Case 8: Simply Supported Beam Subjected to Distributed Load from 1 Slab

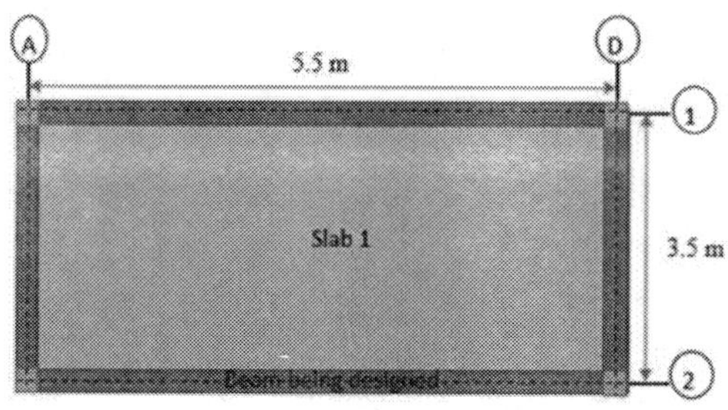

Verification of Excel User Interface Results Using STAAD Pro Finite Element Modelling Software. Lastly, data verification is performed by comparing the output that Excel User Interface and results of STAAD Pro. By performing the checking, functionability of the Excel User Interface can be determined and the analysis is done in order to check the accuracy of the Excel User Interface.

3. Results and Discussions

3.1. Maximum Shear Force

Table 2 showed the maximum shear force and percentage of difference which was obtained from Excel User Interface and STAAD Pro.

Table 2: Maximum shear force

Maximum shear force (KN)

Case	Excel User Interface	STAAD Pro	Percentage of difference (%)
1	374.272	375.650	0.3668
2	363.499	363.498	0.0003
3	153.464	153.465	0.0007
4	197.157	197.156	0.0005
5	157.893	157.894	0.0006
6	134.818	134.817	0.0007
7	93.085	93.086	0.0011
8	78.686	78.684	0.0025

From Table 2, it was clearly showed that there was only small percentage of difference in the value of maximum shear force which was obtained from the Excel User Interface and finite element software STAAD Pro. For the maximum shear force, all the percentage of difference for both Excel User Interface and finite element software STAAD Pro for 8 cases was below 1%. The highest difference percentage of maximum shear force was case 1 which was about 0.3668% while the lowest difference percentage of maximum shear force was case 2 which was about 0.0003%. The percentage of difference for maximum shear force in 8 cases were quite low and thus the results of maximum shear force which obtained from Excel User Interface were quite accurate when compared with the STAAD Pro.

3.2. Maximum Bending Moment

Table 3 showed that the maximum bending moment and distance of maximum bending moment that obtained from the Excel User Interface and STAAD Pro. The percentage of difference for both values that obtained were calculated.

Table 3: Maximum bending moment

		Excel User Interface		STAAD Pro		
Case	Length of simple supported beam (m)	Maximum bending moment (KNm)	Position of maximum bending moment (m)	Maximum bending moment (KNm)	Position of maximum bending moment (m)	Percentage of difference (%)
1	9.0	-959.764	4.056	-956.707	4.500	0.3195
2	7.5	-734.634	3.833	-734.398	3.750	0.0321
3	8.0	-293.608	3.461	-306.930	4.000	4.3404
4	5.6	-264.988	3.337	-264.872	3.270	0.0438
5	6.0	-288.356	3.500	-288.359	3.500	0.0010
6	6.0	-255.525	3.000	-255.524	3.000	0.0004
7	4.0	-93.085	2.000	-93.086	2.000	0.0011

8	6.0	-118.027	3.000	-118.026	3.000	0.0008

In the maximum bending moment, there was only one of the percentage difference was above 1% and the others was below 1%. Case 3 had the highest difference percentage of maximum bending moment (4.304%) while case 6 had the lowest difference percentage of maximum bending moment (0.0010%). The distance of maximum bending moment for first 4 cases which obtained from Excel User Interface and finite element software STAAD Pro were different. This was because bending moment is obtained from shear force in Excel User Interface. However, in STAAD Pro, the shear force is obtained through finite element matrices. The values of shear force along the distance different to manual calculation. Therefore, there had some difference on the maximum bending moment and the position of the maximum bending moment between Excel User Interface and the finite element software STAAD Pro. However, the percentage of difference for both maximum bending moment and distance of maximum bending moment occur are not much different. It concluded that the results of maximum bending moment and distance of maximum bending moment that shown in Excel User Interface are accurate.

3.3. Tension Bars

Table 4 showed that the tension bars provided and difference number of bars provided between the Excel User Interface and finite element software STAAD Pro.

Table 4: Tension bars provided

Case	Excel User Interface	STAAD Pro	Difference of tension bars provided
1	8H32	7H32	1H32
2	7H32	5H32	2H32
3	4H25	3H25	1H25
4	5H25	3H25	2H25
5	5H25	4H25	1H25
6	7H20	5H20	2H20
7	5H16	3H16	2H16
8	4H20	3H20	1H20

In Table 4, the tension bars that provided by Excel User Interface and finite element software for 8 different cases were different. It showed that the tension bars provided by Excel User Interface had higher number of tension bars provided than the results obtained by the software. This is because additional longitudinal reinforcement due to shear were included in the calculation of Excel User Interface while STAAD Pro limitation in considering the additional longitudinal reinforcement on simply supported beam. Thus, Excel User Interface was more emphasis on the safety factor of design. Excel User Interface took the higher number of tension bars provided in order to consider the worst-case scenario.

3.4. Compression Bars

Table 5 showed that the compression bars provided and difference number of bars provided between the Excel User Interface and finite element software STAAD Pro.

Table 5: Compression bars provided

Case	Excel User Interface	STAAD Pro	Difference of compression bars provided
1	2H25	2H25	0
2	2H25	2H25	0
3	-	-	-
4	-	-	-
5	3H16	3H16	0
6	-	-	-
7	-	-	-
8	-	-	-

From Table 5, it showed that there were same number of compression bars provided between Excel User Interface and finite element software STAAD Pro in case 1, case 2 and case 3. There was not any difference of compression bars provided for case 1, case 2 and case 3. Thus, it can be concluded that the results of compression bars provided which obtained by Excel User Interface was same with the results that obtained from STAAD Pro. The results of compression bars provided in Excel User Interface were accurate.

4. Conclusions

Excel User Interface has been developed for everyone with or without the structural background. The main purpose of this program is to help users with or without the structural background to perform structural design adequacy check of simply supported beam under construction at the site based on Eurocode 2.

In Excel User Interface, the formulation of main reinforcement, shear reinforcement, deflection and cracking checks are programmed. It is a simple program and it provides some basic information to users in the input and output sheets. The maximum concentrated forces that subjected to simply supported beam is four. It is a worst-case scenario when the simply supported beam is subjected with 4 points loads and uniform distributed loads. There are 8 possible cases for checking the adequacy of simply supported beam. An "ERROR" is showed when the incorrect input used in the calculation. Excel User Interface is produced an output about the graph of maximum shear force and maximum bending moment, adequacy of main reinforcement and shear reinforcement provided and lastly checking the deflection and cracking of the simply supported beam.

Lastly, the results of maximum shear force, maximum bending moment, tension and compression bars provided in Excel User Interface and STAAD-Pro are compared. The results that obtained between Excel User Interface and finite element software STAAD Pro had highest percentage of accuracy. Therefore, Excel User Interface had higher safety factor of design.

In conclusion, it showed that the results that obtained from Excel User Interface were accurate. Excel User Interface is a user-friendly interface and it was safe to be used by everyone. Everyone can design simply supported beam based on Eurocode 2 in a simple and quick manner.

References

[1] J Kevin, Carpentry and Joinery 2008 London: Pearson Education Limited 75-81
[2] S. Nolan 2002 New load for old structure *Journal of Entertainment Service and Technology Association*

[3] M. Usmen and M. Vilnitis 2015 Evaluation of Safety, Quality and Productivity in Construction *Materials Science and Engineering*
[4] Eurocode 2 2004 Design of Concrete Structures: Part 1-1: General Rules and Rules for Buildings
[5] F Shaffi, W Omar, S Mohammwas and A M Makhtar 2001 Standardization of Structural DeSIGN: A Shift from Bristish Standards to Eurocodes *Journal Technology* 21-30

IOP Conf. Series: Materials Science and Engineering **429** (2018) 012019 doi:10.1088/1757-899X/429/1/012019

Effect of Solvent Vapor Treatment on Electrospun Nylon 6,6 Nanofiber Membrane

N S A Halim[1], M D H Wirzal[1*], M R Bilad[1], A R M Yusoff[2], N A H M Nordin[1], Z A Putra[1], J Jaafar[3]

[1]Chemical Engineering Department, Universiti Teknologi PETRONAS, Bandar Seri Iskandar, 32610 Perak, Malaysia.
[2]Faculty of Science, Universiti Teknologi Malaysia, 81310 Skudai, Johor, Malaysia.
[3]Higher Institution Centre of Excellence (HICoE), Block N29a, Advanced Membrane Technology Research Centre (AMTEC), Faculty of Chemical & Energy Engineering (FCEE), Universiti Teknologi Malaysia, 81310 Skudai, Johor, Malaysia.

*Email: mdzulhakim.wirzal@utp.edu.my

Abstract. This paper highlights the effect of solvent vapor treatment on characterization of electrospun nanofiber membrane (ENM). Fouling issue in membrane separation has becoming a debate since it reduces the lifetime of the membrane. In order to solve this issue, ENM has been recommended due to its great performance in terms of filtration, permeability and stability. However, ENM has several drawbacks such as low mechanical strength and loosely packed layers of nanofibers. Thus, solvent vapor treatment was used to solve this issue. In this study, nylon 6, 6 was fabricated as ENM and was exposed to formic acid as solvent vapor treatment. Solvent vapor treatment helps to increase the bonding of fiber by promoting crosslinking at the junction of fiber thus making the membrane mechanically durable. By controlling the solvent vapor exposure time, excessive fusion between the fibers can be prevented as over-crosslinking would reduce membrane permeability. Solvent vapor treatment also reduces the number of layers of ENM, by making it more closely packed and reduces the risk of membrane delamination. All treated ENM were found to be more hydrophilic and has lower surface roughness, which would make them more resistant towards fouling than the untreated one.

1. Introduction

Water scarcity is a serious threat that is faced by most population in the world [1]. It gives huge impact and may cause fatality if it is not treated properly since water is one of the necessities for living things. The demand for clean freshwater keep increasing everyday due to climate change, advance in industrial development and increase in world population. To address this issue, numerous water treatment technologies have been implemented and one of them would be membrane technology [2].

Membrane technology has emerged as one of the reliable and efficient remedy for water treatment. It requires small footprint and gives stable quality of effluent. Nevertheless, membrane separation always faces the threat of fouling since it reduces flux and efficiency of membrane [3]. These are due to formation of unwanted macro-void around the cross-section of the membrane, consistent distribution of pore size and pores geometrical structure [2].

For better quality of water treatment, nanofiber membrane (NFM) has great potential to be used as a filter especially for pre-treatment before nanofiltration or reverse osmosis processes [4]. NFM can be fabricated via melt blowing [5], template synthesis [6], phase separation [7], self-assembly [8] and the most common one; electrospinning [9].

Electrospinning is one of the favorable and advanced method to produce nanofibers with diameter range of nanometers to micrometers. Electrospinning produces membranes that have huge surface area to volume ratio, high porosity, great water permeability and flexibility [10-12]. ENM is broadly used in air filters [13], membrane separation [14], conductive fibers [15], tissue scaffolds [16], catalysts [17], drug delivery [18] and protective clothing [19]. Moreover, electrospinning also is able to control the formation of pore structure during fabrication [2].

However, electrospun nanofiber membrane has poor mechanical strength and loosely packed layers of nanofibers. It is because of the random orientation within fibers and weak bonding at junctions of the fibers [20, 21]. In order to enhance the mechanical properties of the electrospun nanofiber membrane, post treatment technique can be applied apart from blending of polymer [22] or incorporate nanofillers into the fiber [23].

Solvent vapor exposure is preferable to be prepared for post treatment of NFM due to more controlled nanoscale phase separation and better thermodynamic stable morphology of active layer [24]. Moreover, it would not make great change to the morphology or dimension of the membrane and less aggressive for fibers welding. This is because the residue solvent will ease the fusion between the fibers [21].

In this study, nylon 6,6 polymer is chosen to be fabricated as NFM since it has great hydrophilicity surface, stable material and has high resistance to chemicals. Materials that have good hydrophilicity are more resistant to fouling problems [25]. For solvent vapor treatment, formic acid vapor will be exposed to nylon 6,6 NFM under different exposure time. Later, the membranes will be tested with water permeability together with the characterization of membrane where surface morphology, porosity, contact angle and surface roughness will also be studied.

2. Materials and Method

2.1. Preparation of nylon 6,6 solution

The materials used were formic acid, FA (98-100%, MERCK), glacial acetic acid, AA (99.85%, VWR Chemicals) and nylon 6, 6 pellets (Sigma Aldrich). In order to prepare nylon 6,6 solution, mixture of FA and AA with a ratio of 1:1 was used.

2.2. Electrospinning

Nylon 6,6 solution was filled inside a 5 mL syringe and it was attached with 0.6 mm inner diameter of needle tip. The feeding rate was set at 0.4 mL hr^{-1}. The voltage used was 26.0 kV and the distance from needle tip to a metal screen collector was set to 15 cm.

2.3. Membrane Characterization

The membranes were characterized based on their surface and cross-sectional morphology, porosity, water contact angle and surface roughness. Scanning Electron Microscopy (SEM) was used to study the morphology of the membrane. Porosity was calculated by measuring the volume of water absorbed and volume of dry membrane. Water contact angle was measured by using goniometer and for surface roughness, Atomic Force Microscopy (AFM) was used.

2.4. Solvent vapor treatment

Electrospun nylon 6, 6 NFM was exposed to FA vapor inside a vacuum chamber at room temperature for 5, 12, 24 and 48 hours as shown in Figure 1.

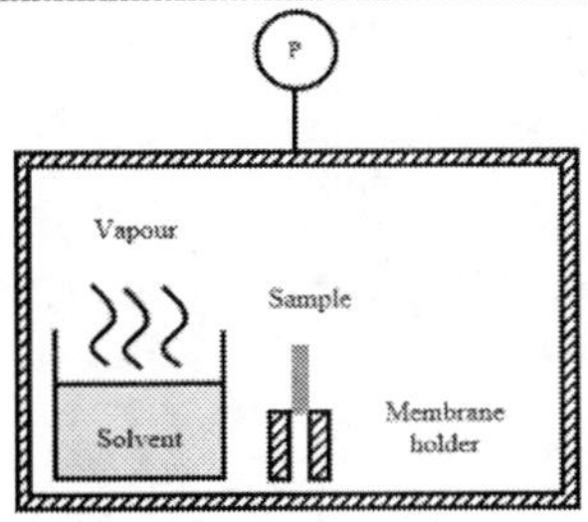

Figure 1. Solvent vapor treatment set-up where the samples were exposed under different period of times (5, 12, 24 and 48 hours) in vacuum chamber at room temperature.

2.5. *Permeability Analysis*
Permeability of NFM for pure water was measured using cross-flow testing unit with pressure at 0.1 bar. The permeate was measured every 10 minutes until it reached a stable value.

3. Results and Discussion

3.1. *Membrane Characterization*
Figure 2 shows the SEM images with magnification of 10,000 x for electrospun nylon 6,6 NFM which has been treated with formic acid vapor under different exposure time. For a certain period of time, there was a change in morphology of nanofiber between the untreated and treated membrane. As the exposure time increased from 5 h to 12 h, no obvious changes could be seen at the membrane surface morphology. However, the membrane started to swell as the exposure time was set to 24 hours and swelled completely when it was exposed for 48 hours. As predicted, when the membrane was exposed to formic acid vapor for a long period of time, too much cross linking would occur [4]. Swelling occur as the membrane fiber fused with each other where the polymer of the membrane dissolved and melted with the help of solvent vapor [21, 26]. High volatility of formic acid also contributed to the cross-linking of the fibers. Figure 3 and 4 show the cross-sectional of untreated and 5 h treated membranes. As the untreated membrane was prepared for the SEM testing, the layers of membrane were easily deformed and delaminated due to unbind layers of nanofibers where crosslinking was not being performed [4]. The 5 h treated membrane layers were seen to be closely packed and dense after underwent exposure to formic acid vapor. From here, it can be conclude that solvent vapor exposure does help to strengthen up the membrane integrity by fusing the fibers at the contacting points of the fiber and reducing the number of mat layers of NFM.

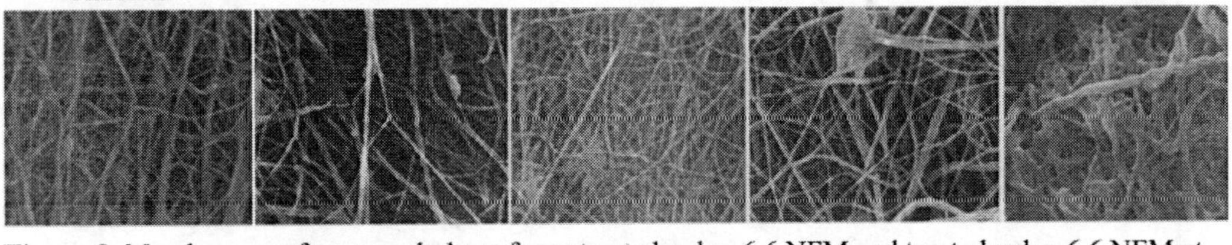

Figure 2. Membrane surface morphology for untreated nylon 6,6 NFM and treated nylon 6,6 NFM at treatment times of 5, 12, 24 and 48 h. All images share the same scale-bar.

Figure 3. Cross-sectional of untreated nylon 6,6 with magnification of 300x.

Figure 4. Cross-sectional of 5h treated nylon 6,6 with magnification of 300x.

Table 1. Summary of membrane properties

Sample name	Treatment Duration (hours)	Mean Pore Size (µm)	Porosity (%)	Contact Angle (°)
S-0	0	2.115	73.67 ± 0.93	50.0 ± 10.5
S-5	5	0.028	69.97 ± 0.73	50.0 ± 9.0
S-12	12	0.012	68.75 ± 0.45	50.0 ± 2.2
S-24	24	0.010	64.40 ± 0.50	49.0 ± 5.5
S-48	48	0.009	62.10 ± 0.70	44.0 ± 14.8

Table 1 shows the summary of membrane properties. As can be seen from the table, untreated nylon 6,6 NFM has the highest average porosity and contact angle of 73.67% ± 0.93% and 50.0° ± 10.5°. The mean pore size, porosity and contact angle show a decreasing trend as the treatment duration increases. The mean pore size and porosity decreases as the exposure time increases were due to the melting of fibers in which the fiber fused with each other. Hence, the pore size of the membrane was reduced, which correlated with Figure 2 [21]. Moreover, ENM has water contact angle in the range between 44-50°. Since the water contact angle was less than 90°, it can be conclude that nylon 6,6 NFM has great hydrophilic properties [27].

Figure 5 shows the surface roughness results of S-0, S-5, S-24 and S-48 membranes acquired from AFM testing. It was found that the surface roughness for S-5, S-24 and S-48 were lower than S-0 membrane. S-24 has the lowest surface roughness with Ra of 80.63 nm while S-0 has the highest surface roughness with Ra of 155.6 nm. Surface roughness is one of the factors that affect membrane fouling. Lower surface roughness is preferable since it has low fouling tendency [28]. It is because smooth surface will slower down the growth of deposit particles as the feed flows through the membrane [29].

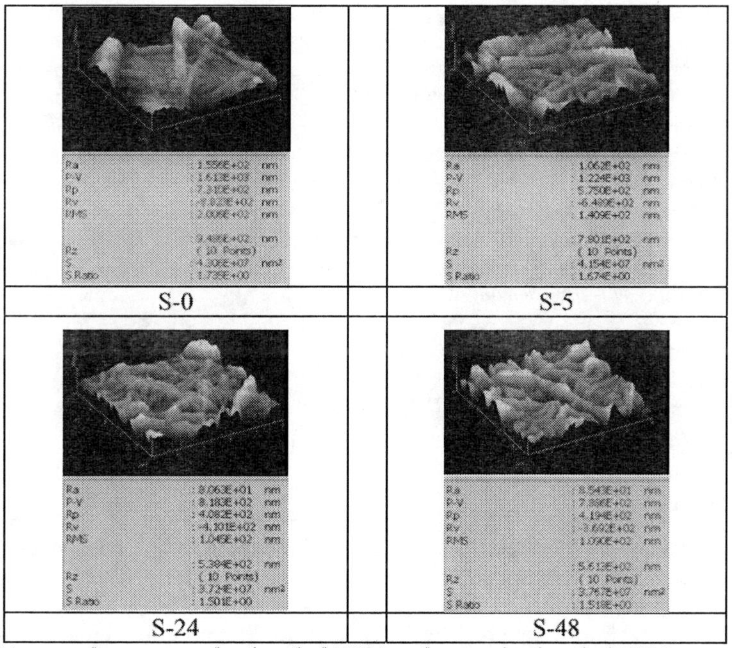

Figure 5. *AFM images for untreated nylon 6, 6 NFM and treated nylon 6, 6 NFM at treatment times of 5, 24 and 48 h.*

3.2. Permeability analysis

Figure 6 shows the steady-state permeability of pure water for untreated and treated nylon 6,6 NFM. As treatment duration of NFM increases, permeability of pure water decreases. S-0 membrane has the highest steady-state water permeability while S-48 membrane has the lowest permeability. This relates with the SEM images (Figure 2) where it shows that as the treatment duration increases, the pore size of the membrane become smaller. This is due to excessive fusion of fiber, which eventually will reduce the porosity and water permeability of the membrane. As for S-5 membrane, the permeability is relatively similar with the S-0 membrane. Ideally, a shorter period of time is preferable to promote fusion of fibers at the contacting points. Only some parts of the nanofiber surfaces are dissolving which are just adequate to form physical crosslinks between the fibers. From this, membrane porosity and permeability would not be affected by at the same time, the membranes fibers can be strengthen via crosslinking [4].

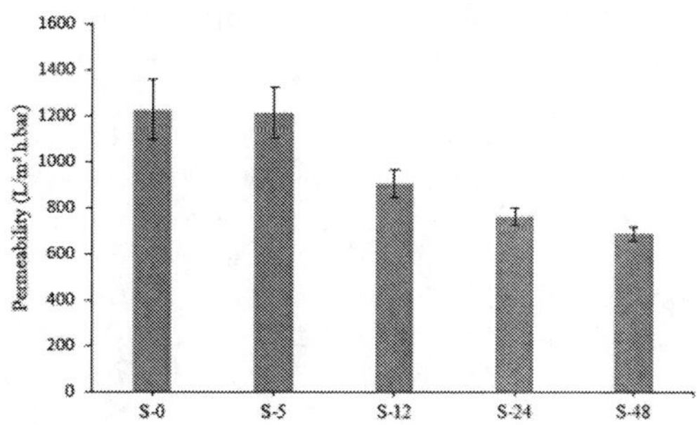

Figure 6. Steady-state pure water permeability for untreated nylon 6,6 NFM and treated nylon 6,6 NFM with pressure at 0.1 bar.

4. Conclusion

In this research, the effect of solvent vapor treatment on electrospun nylon 6,6 NFM for water treatment was studied. It is found that solvent vapor treatment promotes crosslinking at the junction of the fibers which prevents the stacked layers of nanofibers to be loosed and make them become more closely packed. This proves that the mechanical strength of NFM can be enhanced by treatment of solvent vapor under controlled exposure time. Solvent vapor treatment also increases hydrophilicity and decreases surface roughness of the membranes. S-24 membrane has the lowest surface roughness with Ra of 80.63 nm. However, excessive exposure to solvent vapor reduces porosity and permeability of NFM. From the study, it is recommended to conduct solvent vapor treatment within 5 hours in order to maintain the performance of the membrane while still can reduce the number of layers of NFM. Apart from that, it can be conclude that treated electrospun nylon 6,6 NFM has great potential to be applied as a filter especially for water treatment.

Acknowledgement

We would like to acknowledge Universiti Teknologi PETRONAS and Unversiti Teknologi Malaysia for providing grants (UTP grant number (YUTP): 0153AA-H29) and (UTM grant number (HICOE): A. J090301.5300.07092) for conducting the research activities.

References

[1] Ibrahim N A, Wirzal M D H, Nordin N A H and Halim N S A 2018 Development of Polyvinylidene Fluoride (PVDF)-ZIF-8 Membrane for Wastewater Treatment *IOP Conference Series: Earth and Environmental Science* **140**(1) 012021.

[2] Bae J, Baek I and Choi H 2016 Mechanically Enhanced PES Electrospun Nanofiber Membranes (ENMs) for Microfiltration: The Effects of ENM Properties on Membrane Performance *Elsevier* **105** 406–412.

[3] Eliseus A, Bilad M R, Nordin N A H M, Putra Z A and Wirzal M D H 2017 Tilted Membrane Panel: A New Module Concept to Maximize The Impact of Air Bubbles for Membrane Fouling Control in Microalgae Harvesting *Bioresour. Technol.* **241**(Supplement C) 661–668.

[4] Liu C, Li X, Liu T, Liu Z, Li N, Zhang Y and Feng X 2016 Microporous CA/PVDF Membranes Based on Electrospun Nanofibers with Controlled Crosslinking Induced by Solvent Vapor. *Elsevier* **512** 1–12.

[5] Raghavan B, Soto H and Lozano K 2013 Fabrication of Melt Spun Polypropylene Nanofibers by Forcespinning *Journal of Engineered Fibers and Fabrics* **8**(1).

[6] Wan Y, Zheng M, Lu H, Feng S, Ji G and Cao J 2010 Template Synthesis of Carbon Nanofibers Containing Linear Mesocage Arrays *Nanoscale Research Letters* **5**(6) 913–916.

[7] Ramakrishna S 2005 An Introduction to Electrospinning and Nanofibers *World Scientific.*

[8] Rolandi M and Rolandi R 2014 Self-assembled Chitin Nanofibers and Applications *Advances in Colloid and Interface Science* **207**(Supplement C) 216–222.

[9] Khajavi R and Abbasipour M 2012 Electrospinning as a Versatile Method for Fabricating Coreshell, Hollow and Porous Nanofibers *Scientia Iranica* **19**(6) 2029–2034.

[10] Azizo A S, Wirzal M D H, Bilad M R and Yusoff A R M 2017 Assessment of Nylon 6, 6 Nanofibre Membrane for Microalgae Harvesting *AIP Conference Proceedings* **1891**(1) 020032.

[11] Bilad M R, Westbroek P and Vankelecom I F J 2011 Assessment and Optimization of Electrospun Nanofiber-Membranes in a Membrane Bioreactor (MBR) *Journal of Membrane Science* **380**(1) 181–191.

[12] Wang X, Yu J, Sun G and Ding B 2016 Electrospun Nanofibrous Materials: A Versatile Medium for Effective Oil/Water Separation. *Materials Today* **19**(7) 403–414.

[13] Wang N, Mao X, Zhang S, Yu J and Ding B 2014 Electrospun Nanofibers for Air Filtration. *In Electrospun Nanofibers for Energy and Environmental Applications* pp. 299–323 Springer Berlin Heidelberg.

[14] Obaid M, Tolba G M K, Motlak M, Fadali O A, Khalil K A, Almajid A A and Barakat N A M 2015 Effective Polysulfone-Amorphous SiO2 NPs Electrospun Nanofiber Membrane for High Flux Oil/Water Separation *Chemical Engineering Journal* **279** 631–638.

[15] Cong Y, Liu S and Chen H 2013 Fabrication of Conductive Polypyrrole Nanofibers by Electrospinning.

[16] Manoukian O S, Matta R, Letendre J, Collins P, Mazzocca A D and Kumbar S G 2017 Electrospun Nanofiber Scaffolds and Their Hydrogel Composites for The Engineering And Regeneration Of Soft Tissues *Methods in Molecular Biology (Clifton, N.J.)* **1570** 261–278.

[17] Coşkuner Filiz B and Kantürk Figen A 2016 Fabrication of Electrospun Nanofiber Catalysts and Ammonia Borane Hydrogen Release Efficiency *International Journal of Hydrogen Energy* **41**(34) 15433–15442.

[18] Hu X, Liu S, Zhou G, Huang Y, Xie Z and Jing X 2014 Electrospinning of Polymeric Nanofibers for Drug Delivery Applications *Journal of Controlled Release* **185**(Supplement C) 12–21.

[19] Gorji M, Bagherzadeh R and Fashandi H 2017 Electrospun Nanofibers in Protective Clothing. *In M. Afshari (Ed.), Electrospun Nanofibers* (pp. 571–598) Woodhead Publishing.

[20] Ahmadi A, Qanati O, Seyed Dorraji M S, Rasoulifard M H and Vatanpour V 2017 Investigation of Antifouling Performance A Novel Nanofibrous S-PVDF/PVDF and S-PVDF/PVDF/GO Membranes Against Negatively Charged Oily Foulants *Journal of Membrane Science* **536**(Supplement C) 86–97.

[21] Huang L, Manickam S S and McCutcheon J R 2013 Increasing Strength of Electrospun Nanofiber Membranes for Water Filtration Using Solvent Vapor *Journal of Membrane Science* **436**(Supplement C) 213–220.

[22] Zhao Y, Qiu Y, Wang H, Chen Y, Jin S and Chen S 2016 Preparation of Nanofibers with Renewable Polymers and Their Application in Wound Dressing.

[23] Dai X, Cao Y, Shi X and Wang X 2016 The PLA/ZIF-8 Nanocomposite Membranes: The Diameter and Surface Roughness Adjustment by ZIF-8 Nanoparticles, High Wettability, Improved Mechanical Property, and Efficient Oil/Water Separation *Wiley Online Library* 1–6.

[24] Zheng Y, Li S, Zheng D and Yu J 2014 Effects of Different Polar Solvents for Solvent Vapor Annealing Treatment on The Performance Of Polymer Solar Cells *Organic Electronics* **15** 2647–2653.

[25] Bilad M R, Azizo A S, Wirzal M D H, Jia Jia L, Putra Z A, Nordin N A H M and Yusoff A R M 2018 Tackling Membrane Fouling in Microalgae Filtration Using Nylon 6,6 Nanofiber Membrane. *Journal of Environmental Management* **223** 23–28.

[26] Drioli E, Giorno L and Fontananova E 2017 Comprehensive Membrane Science and Engineering. *Elsevier.*

[27] Muppalla H 2011 Highly Hydrophilic Electrospun Fibers for The Filtration of Micro and Nanosize Particles Treated with Coagulants JNT University.

[28] Hobbs C, Taylor J and Hong S 2006 Effect of Surface Roughness on Fouling of RO and NF Membranes During Filtration of a High Organic Surficial Groundwater *Journal of Water Supply: Research and Technology – Aqua* **55**(7–8) 559–570.

[29] Kukulka P, Kukulka D and Devgun M 2007 Evaluation of Surface Roughness on the Fouling of Surfaces.

IOP Conf. Series: Materials Science and Engineering **429** (2018) 012020 doi:10.1088/1757-899X/429/1/012020

The Effectiveness of Non-Destructive Bat Control Measures in Residential Building

N Z Zainol[1,a)], **F A Ibrahim**[1,b)], **S N A Mohd Noor**[1,c)], **M A Azizan**[1,d)], **M M Ahmad**[1,e)]

[1]Faculty of Engineering Technology, Universiti Malaysia Perlis, Malaysia

Email: norrazman@unimap.edu.my

Abstract. Bat management is uncommon among the Malaysian even though the said wildlife pest possesses a quite alarming threat to human especially in the suburban. The bats have losses their natural habitat to development, consequently, it adapts to co-exist among the human by residing in man-made structures. Many studies have been done on bat control measures especially for residential building and plenty of methods have been developed and executed in controlling bat from entering building either destructive or non-destructive method. In most countries, non-destructive is preferable such as bat proofing structures, bat proofing materials, artificial roost, and temporary outside roost but how far the method successfully and efficiently containing the problem remains questionable. Therefore, the purpose of this study is to determine the effectiveness of bat control measures available in the market nowadays. Applying Technology Acceptance Model (TAM) as the underlying theory to determine the most reliable and efficient bat control method, the research was accomplished through a survey among the pest management companies. A total of 30 respondents with vast experiences in pest control had completed the questionnaires. Results indicated that TAM successfully fit to be applied for this research. Generally, the best and effective bat control method was determined by perceived usefulness and perceived ease of use of each selected control methods. Both independent variables have a strong positive relationship with the effectiveness of bat control measures. The best control that perceived to be the most effective non-destructive bat control method is bat house; follows by repellent, bat proofing structure, bat- friendly house and lastly the fumigant. Each of these methods could be used in preventing the bat from entering the building but, it is suggested to the pest control companies to consider a non-destructive method which will not harm the bat and provides an alternative dwelling similar to its natural habitat.

1. Introduction

Wildlife pest are animals that have little of fear for living close to human and most commonly, in the residential buildings. Most of the pest stays away from human sights, unfortunately; these animals possess a threat to the health and well-being of people by spreading diseases, pollute the surrounding with their essences, droppings and to some extends, damaging the property. The animals that are normally considered as wildlife pest are birds, bats, rat, squirrel, mole, crow, and raccoon [1]. Pest management required effective control measures to contain the infestation as well as to eliminate the threat. Common approaches are combination of habitat management, sanitation, repellents and also removal. For an example, trapping, baiting and relocation are common practices for controlling wildlife pest [2].

Bats are small in size thus; it is able to enter the buildings through small opening, cracks or gaps. Unfortunately, despite their uniqueness, bats can spread disease such as rabies and skin rashes to human and animals [1]. Bats' urine and guano (i.e. bad droppings) are acidic that potentially corrodes metal or sandstone for some building, eventually, damaging the structures. It also leaves stains at the building wall [3]. Hence, it is important to take control measures in preventing bats from roosting in residential building.

Plenty of studies on bat control measures for residential building [1,2,4,5] to put away bat from entering building such as bat proofing structures, bat proofing materials, artificial roost, and temporary outside roost. Although, there are many methods in preventing bats from roosting and entering a building, it is still a need in finding the effective method or control measures of bats from entering and roosting in the building [1,5]. Besides that, the approaches in controlling bats need to be non- destructive because bats are legally protected in few countries such as in the United Kingdom [5]. Thus, this research is to determine the effectiveness of non-destructive bat control measure in residential building by measuring the perceived of usefulness and ease of use that will influence the effectiveness of bat control measures in Malaysia.

2. Literature review

Globally, natural environment is diminishing because of development and land usage. Consequently, wildlife which losing their natural home habitats are seeking for alternative to settle as dwelling. Habitat-like and safer, bat preferred structures as roosting sites such as buildings, bridges and tunnels which provides secure, dry place for it to rest and reproduce. The potential roost sites for bats in the structure of buildings are in the walls, attic and roof [4]. However, the presence of bats in the building can affect the health and property of the occupants, for example the urine and bat guano can corrode metal or sandstone for some building. Bat crawls, thus, leaves stains at the building wall and some spreads disease such as rabies and skin rashes to human [3]. Colonization may take years but once happened, chasing the bats out will be difficult. Eradication normally last for months before the colonization reoccurs. Importantly, the best prevention method should be non-destructive to avoid damage to the bat population.

There are five methods in controlling bats from entering and roosting in the building. Bat proofing structure is one of the methods by sealing the entrance of bats with bat proofing materials [1]. Repellents such as fans and electric lights might be useful in bat control measure [4]. Next, providing alternative dwelling space likes bat houses that resemble similar conditions of tree cavities [6] will attract bat to house the space. The bat friendly house is the buildings that had been abandoned by humans and were renovated so that it would be preferred by bats to dwell the building [3]. Lastly, fumigants is the method of pest control that completely fills an area with gaseous in order to evict the bats from the building [7].

Table 1 Common locations of bats roost [1]

Natural or rural environments	Urban environments
Caves	Attics
Cavities and crevices created in trees	Barns and sheds
Mines	Bridges
Palm tree fronds	Culverts
Rock crevices	Roofing tiles

2.1. *Technology Acceptance Model*

The Technology Acceptance Model (TAM) was an adaptation to the Theory of Reasoned Action. Technology Acceptance Model was designed to predict information technology acceptance and usage on the job. This theory has been widely used to a diverse set of technologies and users in order to better explain intention to use a technology. Later, the theory has been simplified and named TAM 2. TAM 2 is consists of "perceived usefulness" (PU) as the primary driver of usage intentions and "perceived ease of use" (PEU) even though, has been proved less influential but remained primarily overlooked by further researchers [8].

2.2. Theory of Perceived Usefulness and Perceived Ease of Use (PUPEU)
Perceived usefulness is "the degree to which a person believes that using a particular system would enhance his or her job performance." This follows from the definition of the word useful: "capable of being used advantageously." Hence, a tool high in perceived usefulness is one for which a user believes in the existence of a positive use-performance relationship [8].

Perceived ease of use, refers to "the degree to which a person believes that using a particular system would be free of effort." This follows from the definition of "ease": "freedom from difficulty or great effort". A tool that is easy to use is more likely to be accepted by users [8].

Perceived ease of use and perceived usefulness are probably two of the most widely cited correlates of technology acceptance and usage. Thus, by using the theory perceived usefulness and perceived ease of use in this study as an instruments or tools it can provide the important results for assessing the easiness and usefulness of the bat control methods [9].

3. Methodology
This research use quantitative survey method to determine the effectiveness of non-destructive bat control measures in residential building in Northern State of Malaysia. The survey was performed to collects data from experienced workers of pest management companies. The representatives were as ked on their opinion and perception on usefulness and ease of use for the available bat control measures. The selected respondents were based on the pest control company that had registered under Pest Control Association of Malaysia and a total of 30 respondents had participated in this research.

Originally basing on TAM [8], the original framework theory of PUPEU was constructed to provide important information to researchers that seeks the statistically test of relationships among these perceived of usefulness and perceived ease of use [10]. Adapting this framework, a conceptual framework as in Figure 1 were constructed in order to demonstrates the relationship between perceived ease of use and perceived usefulness and based on the theory of perceived of usefulness and perceived ease of use. Figure 1 shows the proposed research model with different kinds of variables and the relationship of each variable in order to achieve the effectiveness in bat control measures. There were three research objectives; first, to determine whether perceived of usefulness has positive relationship with effectiveness of bat control measure in residential building, and secondly, to determine whether perceived ease of use has positive relationship with the effectiveness of bat control measure in residential building. Ultimately, the last research outcome is to find the best control measure to prevent bats from entering the building especially the houses.

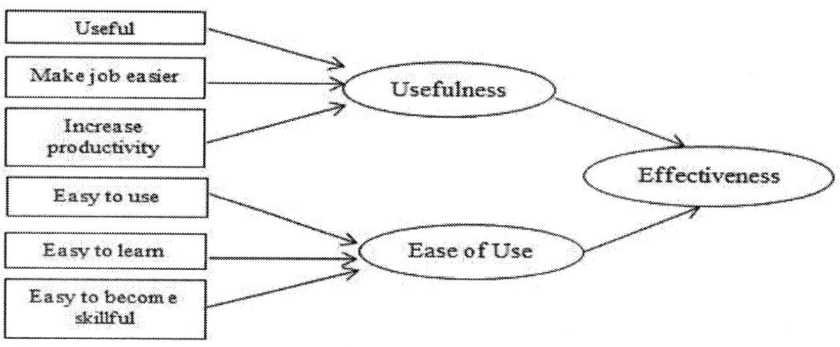

Figure 1: Framework of Perceived Usefulness (P.U) and Perceived Ease of Use (P.E.U) [10]

4. Result and Discussion

4.1. Descriptive

The research sample consists of 30 respondents working in pest control companies in Northern State of Malaysia. The selection of potential respondent was based on gender, age, working position and working experience. The job was dominated by men with 80% of the respondents were male (24 respondent) and the other 20% were female (6 respondent). Pest control companies claimed that the work of pest management required them to do odd jobs which not suitable for most of the female such as climbing up the attic or roof of a building to inspect existence of pests or spraying smelly pesticide. Table 4.1 shows more than 80% of the respondents were age more than 30 years while the rest, below than 30 years. Half of the respondents are supervisor (15 respondents), one third was working as technician (10 respondents) and the remaining as a fumigator (4 respondents) and a helper. Majority have a vast experience in the field, i.e. more than 5 years. The frequencies verified that respondents have vast experience, skilful, and knowledgeable in pest control management and the collected data could reveal the insight on bat control measures and the most effective method that has highest potential to be effective in preventing the bats' colonization.

Table 4.1 Frequency analysis for 30 respondents

Profile	Frequency	Percentage (%)
Gender:		
Male	24	80
Female	6	20
Current Position:		
pest control supervisor	15	50
pest control assistant	1	3.3
pest control technician	10	33.3
fumigator	4	13.3
Working experience:		
Below 1 year	1	3.3
1 year and above	6	20.0
5 years and above	8	26.7
10 years and above	15	50

4.2. Reliability

The reliability test, Cronbach's Alpha (CA) (α) was established to test the stability and consistency of the variables. Cronbach's Alpha (α) are used to test how well the items in an instrument set are positively correlated to one another and will be accepted if the value were 0.6 and above [11]. The result from CA shows that the values of α ranging from 0.649 to 0.911, thus, all the tested variables in the framework were accepted. Hence, the result shows that the questionnaires are reliable measurement instrument.

4.3. Correlation

This test was to determine the direction and significance of the bivariate relationships of the variables used in the study (the relationship between any two variables among the variables tapped in the research). A Pearson Correlation (r) matrix will provide this information, that is, it will indicate the direction, strength and significance of the bivariate relationships of all the interval or ration variables in the study. From table 4.3.1 the results show that the relationship between the two independent variables and the dependent has a mixed result. The perceived usefulness and perceived ease of use had a moderate relationship to very high relationship to the effectiveness of bat control measure in residential building.

Overall results show that the relationship between independent variables and dependent for each of the bat control methods were significantly positive. For the perceived usefulness the highest score is bat proofing structure with 0.744 with significant 0.000 and the weakest relationship is fumigant with 0.431 with significant of 0.018. The theory postulated that the user believes that using particular system would enhance his/her job performance [8]. Therefore, bat proofing structure has high potential to be the correct method to prevent bat colonization in a house.

Next, for the perceived ease of use, the highest score is bat house with 0.759 significant is 0.000 and the lowest score is the bat proofing structure is 0.598 with significant of 0.000. The theory of perceived ease of use states that, a tool that is easy to use is more likely accepted by users [8] thus, the method of bat house is the method that is easy to use and more likely accepted by the users.

The first hypothesis stated that, there is a positive relationship between perceived of usefulness towards effectiveness of bat control measure in residential building. From table 4.3.2, the Pearson correlation for this relationship was valued at 0.666, meaning, it has a moderate relationship between the perceived usefulness and effectiveness. For the second hypothesis which proposed a positive relationship between perceived ease of use towards effectiveness of bat control measure in residential building, the Pearson correlation value were 0.766. Therefore, there is a strong positive relationship between the perceived ease of use and effectiveness. These variables have significant relationship as the significant value is less than 0.01 values. Hence, the hypothesis is accepted because there is significant relationship.

Table 4.3.1 Correlation Analysis

		Effectiveness				
		Bat Proofing Structure	**Repellent**	**Bat House**	**Bat Friendly House**	**Fumigant**
Perceived Usefulness	Pearson Correlation	.744**	.638**	.681**	.566**	.431*
	Sig.	.000	.000	.000	.001	.018
Perceived Ease of Use	Pearson Correlation	.598**	.708**	.759**	.654**	.699**
	Sig.	.000	.000	.000	.000	.000

** Correlation is significant at the 0.01 level (2-tailed)
* Correlation is significant at the 0.05 level (2-tailed)

Table 4.3.2 Correlation Analysis Total

		Effectiveness Total
Perceived Usefulness Total	Pearson Correlation	.666**
	Sig.	.000
Perceived Ease of Use Total	Pearson Correlation	.766**
	Sig.	.000

** Correlation is significant at the 0.01 level (2-tailed)

5. Conclusion

As a conclusion, the perceived of usefulness and ease of use tested in this study have been substantiated to have significant relationship with effectiveness of bat control measure in residential building. Yet again, modified TAM is proven to test the efficient and effectiveness of technology that being employed in the industry. Out of five tested bat control methods, the most effective method is bat house by providing alternative dwelling space that resemble similar conditions of the bat's natural habitat. The other non-destructive method also preferable but, fumigant method which uses gaseous to evict the bat is the least desirable since the method might harm the bat.

6. Acknowledgement

The author would like to acknowledge the support from the Fundamental Research Grant Scheme (FRGS) under a grant number of FRGS/1/2017/SS03/UNIMAP/03/1 from the Ministry of Higher Education Malaysia.

References

[1] T Salmon, D Whisson and R Marsh 2006 *Wildlife Pest Control around Gardens and Homes*, 2nd ed. California: University of California Agriculture and Natural Resources Publication

[2] J N Langan 2012 Elsevier Health Sciences *Integrated pest management in Fowler's Zoo and Wild Animal Medicine* **7** 51–59

[3] C C Voigt, K L Phelps, L F Aguirre, M Corrie Schoeman, J Vanitharani and A Zubaid 2016 Bats and Buildings *The Conservation of Synanthropic Bats*

[4] A M Glassey and J F Karlik 2009 *Bats in Integrated Pest Management for Home Gardeners and Landscape Professionals* California: Pest Notes Publication

[5] R Coles 2015 Bat Building Inspection Urban Green United Kingdom

[6] T H Kunz and S D Reynolds 2003 Bat colonies in buildings *Monitoring Trends in Bat Population of the United States* 91–102

[7] A K Taylor, F J Mazzotti and C N Huegel 2002 Bats *Information for the Florida Homeowner*

[8] F Davis Perceived usefulness, perceived ease of use, and user acceptance of Information technology *MIS Quarterly* **13** 3 1989.

[9] A M Aladwani 2002 The development of two tools for measuring the easiness and usefulness of transactional Web sites European Journal of Information System 11 223-234

[10] H Segars, Albert and V Grover 1993 Re-Examining Perceived Ease of Use and Usefulness *A Confirmatory Factor Analysis* **17** 517–25

[11] U Sekaran 2003 *Research method for business: A skill building approach* 4th ed.

The Effect of Concentration Staging on Performance of Produced Water treatment using Forward Osmosis

S M Hizam[1], N A H M Nordin[1*], M R Bilad[1], M D H Wirzal[1], Z A Putra[1], N S Sambudi[1], M M E Nasef[1]

[1]Chemical Engineering Department, Universiti Teknologi PETRONAS, Bandar Seri Iskandar, 32610 Perak, Malaysia.

*Email: nahadi.sapiaa@utp.edu.my

Abstract. During oil and gas production, a large amount of produced water (PW) is generated, which would be harmful to the environment if not treated properly. Treatment of such wastewater requires multi-processing units and typically includes fouling prone pressure driven membranes. Forward osmosis (FO) seems promising due to its high salt rejection, high total dissolved solid removal, more reversible fouling type, does not require external pressure to operate and potentially can be used as a standalone unit with minimum pretreatment. However, FO still suffers from low permeate flux, high reverse solute flux (RSF) and membrane fouling, although not as severe as in pressure driven membrane. Since FO depends largely on osmotic pressure difference, the effect of concentration staging was investigated. It was found that multi-stage concentration was able to achieve up to 80% water recovery due to sufficient osmotic pressure.

1. Introduction

Produced Water (PW) is wastewater from oil and gas production partly from water flooding process in during enhance oil recovery (EOR) [1]–[3]. PW is the major waste stream produced from oil and gas industry [4]. PW generally consists of various hydrocarbon, organic and inorganic matters, dispersed oil and various solutes [4]. It must be treated properly due to the high level of contaminants present in it.

Conventional treatments offer its own advantages and disadvantages. For example, while adsorption is able to remove heavy metals, benzene, toluene, ethylene and xylene and dispersed oil, suspended particles may plug the adsorbent media hence reducing its efficiency overtime [5]–[7]. Regeneration of absorbent media requires usage of chemicals in which requires an extensive treatment for the chemicals used. Another conventional treatment includes evaporation pond which has been used for both onshore and offshore treatment. However, it requires a relatively large space to function efficiently [8]. In short, conventional PW treatment faces a few challenges such as high treatment cost, usage of toxic chemicals, requires large installation space and generation of secondary waste [7]. To solve these problems, membrane separation technologies have been continuously researched.

Common membrane based technology includes the usage of pressure driven membrane such as microfiltration (MF), ultrafiltration (UF), nanofiltration (NF) and reverse osmosis (RO). These membranes use differential pressure as driving force to facilitate mass transport. The degree of particle separation depends on pore size of the membrane. Example, MF has the largest pore (0.1-3μm) and is used to remove suspended solids while RO has the smallest pore (about 0.1nm) and is used to remove salts [7], [8]. However, pressure driven membrane are prone to membrane fouling [9], [10]. The usage of high pressure tends to have an irreversible fouling effect on the membrane [11].

Another alternative for membrane separation technology is forward osmosis (FO). FO is an emerging technology and have been investigated in PW treatment [12]–[14]. In contrary to pressure driven membrane, FO relies on osmotic pressure difference between two solutions; the feed solution (FS) and the draw solution (DS) where the DS has the highest osmotic pressure to facilitate mass transport. FO results in dewatering and concentrating the feed while diluting the DS.

In the treatment of PW, FO offers a variety of advantages. Firstly, FO does not require any external pressure to operate unlike in RO since it utilizes osmotic pressure difference between two solutions. The absence of high external pressure promotes lower energy consumption and reduce fouling tendency [15]. It also renders the fouling type to be more reversible [11] and therefore a simpler cleaning method such as osmotic backwashing can offer complete flux recovery [16]. As such, FO can potentially operate as a stand-alone unit without the use of pretreatment, unlike NF and RO [15]. Moreover, FO is reported to have high salt rejection up to 96% while operating at ambient condition [15]. FO can also treat hyper saline wastewater (>65 bars) in which RO is unable. However, FO also have a number of setback such as lower water flux (depending on osmotic pressure difference) [17], reverse salt flux due to concentration polarization [18] and membrane fouling (despite of its reversibility), although it is not as severe as in NF or RO [10 - 11].

Since FO relies much on osmotic pressure difference, this study aims to investigate the effect of concentration staging performance on the treatment of PW via FO. Concentration staging will be compared based on single-stage and multi-stage concentration. Rejection analysis was also done for both single-stage and multi-stage concentration.

2. Materials and Method

2.1. List of Chemicals and Materials
PW feed solution (FS) was obtained from a petroleum production well in peninsular Malaysia. The characteristic of the PW sample is summarized in **Table 1**. Note that the conductivity of the feed sample is 2 mS/cm. Control feed was also used as to mimic the PW in terms of osmotic pressure. To produce such solution, sodium chloride (NaCl) was added to deionized (DI) water to obtain around 2000 µS/cm. The control feed was used to measure the performance of FO without any fouling agent while still having similar osmotic pressure of the feed sample. The control feed was used mainly in the study on effect of staging. Seawater was used as the DS and was obtained locally from Teluk Batik, Perak, Malaysia. It has a conductivity of 58 mS/cm and a pH of 6.5. The large difference in conductivity between the feed and the DS ensures sufficient driving force for FO.

2.2. Experimental Setup
Figure 1 shows the schematics for the experimental setup. The feed is in direct contact with the DS through the FO membrane where both are kept under ambient condition. The membrane was kept on AL-

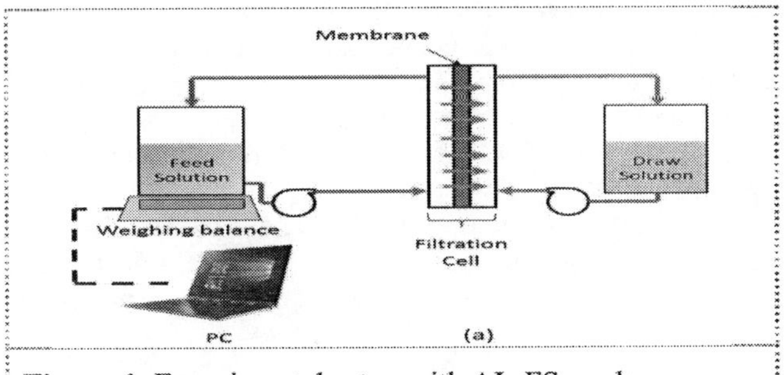

Figure 1. Experimental setup with AL-FS mode

IOP Conf. Series: Materials Science and Engineering **429** (2018) 012021 doi:10.1088/1757-899X/429/1/012021

Table 1. Summary of Produced Water Characteristics

Parameter	Amount
Conductivity (μS/cm)	2000
pH	6
COD (mg/L)	2010
TOG (mg/L)	11.27
Total Nitrogen (mg/L)	30.6
Total Phosphorus (mg/L)	10.28
Turbidity (NTU)	79.4

FS (active layer facing feed solution) mode to reduce the effect of fouling [19]. The membrane area was 3.4×10^{-3} cm^2. Two Seko PR7 Peristaltic Dosing Pumps were used to recirculate both FS and DS to the membrane module at 150 ml/min. The FS was kept on a Mettler Toledo MS 3002S /01 weighing balance to record the changes of the weight which was later used to determine the flux according to equation (1). The weigh balance was connected with a computer for data logging through Mettler Toledo Serial-Port-to-Keyboard software. All of the data were logged and computed in Microsoft Excel.

$$J = \frac{\Delta V}{A\Delta t} \quad (1)$$

where J is permeate flux in L/m^2h, V the volume of the feed (L), A the area of the membrane (m^2) and t the time (h).

2.3. *Effect of Staging: Single-stage concentration vs Multi-stage Concentration*

For both experiments, a TFC-PA membrane from Toray was used. The objective was to concentrate PW from 300 ml to 60 ml. The initial ratio of feed to DS was kept at 1:1. Particularly, for single-stage concentration, the same DS (taken at once) was used throughout the experiment. However, for the multi-stage concentration, the replenishment of DS was introduced. This was done stage by stage i.e.; for each 60 ml decrement of PW FS, the DS is replenished with fresh 300 ml DS. The replenishment of DS was necessary to provide enough driving force for FO process. The process continues until PW reached 60 ml, leading to 4 stages of DS replenishment.

2.4. *Rejection Analysis*

The rejection analysis was conducted for both experiment. Several parameters were gauged which were conductivity, pH value, total oil and grease (TOG), turbidity, chemical oxygen demand (COD), total nitrogen (TN) content, total phosphorus (TP) content. These parameters were measured by using conductivity meter (Eutech Instrument CON 700), pH meter (Eutech Instrument pH 700), total oil in water analyzer (Turner Design TD-500D), turbidity meter (Hach 2100Q Turbidity Meter), and Hach-Lange kits for COD, TN and TP respectively.

3. Results and Discussion

3.1. *Single-stage concentration vs Multi-stage Concentration*

Based on **Figure 2**, for the single-stage concentration, the flux for control feed was in the range of started from 16.27 L/m^2h and declined to 6.07 L/m^2h. The final conductivity value for both control feed and DS were 8.14 mS/cm and 48.7 mS/cm respectively. However, when using real PW as the feed, the flux was declining much further from 9.49 L/m^2h to 0.39 L/m^2h. Furthermore, the amount of water that was able to recover was only up to 56.67%, instead of 80% when using control feed. At the end of the experiment, it was found that the conductivity for PW and draw solution were 31.4 mS/cm and 31.8 mS/cm. The finding suggested that the draw solution was not able to draw water from the feed. Prolonged operation resulted in balanced water transport. This effect can also be explained by the effect

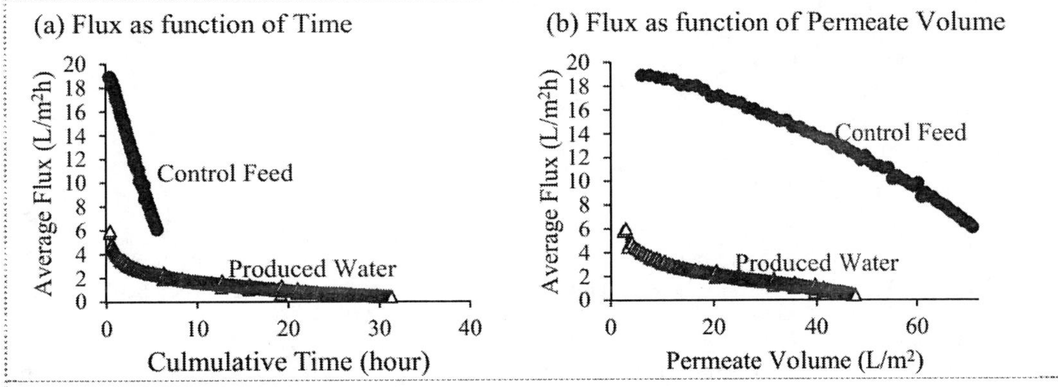

Figure 2. Single-stage concentration performance plotted as (a) a function of time and (b) function of permeate volume.

fouling, more specifically cake enhanced concentration polarization (CECP) [20]. The gap of flux between the control feed and PW suggested that fouling occurred on the membrane, thus reducing the effective area of the membrane. Moreover, RSF occurred due to concentration polarization. Therefore, the osmotic pressure shifted towards the membrane side that was facing the feed. Consequently, the osmotic pressure differences reduced and permeate flux decreased.

For multi-stage concentration, replenishment of draw solution was done for every 60 ml reduction of the feed. Based on **Figure 3**, the flux for control feed started from 20.15 L/m^{2}h and declined to 8.01 L/m^{2}h, which was more than that of single-stage concentration. The final conductivity for control feed and draw solution were 8.46 mS/cm and 48.4 mS/cm respectively. When real PW was used as the feed solution, the permeate flux obtained was decreased from 15.48 L/m^{2}h. to 1.72 L/m^{2}h. Indeed, fouling also occurred in multi-stage concentration, which explained the flux gap between control feed and PW besides osmotic dilution of the draw solution. However, multi-stage concentration was able to achieve 80% water recovery due to sufficient osmotic pressure difference provided by replenished draw solution.

3.2. Summary of Rejection Analysis

Table 2 shows the summary of the rejection analysis done for both single-stage concentration and multi-stage concentration. Note that for multi-stage concentration, the rejection analysis was done at the

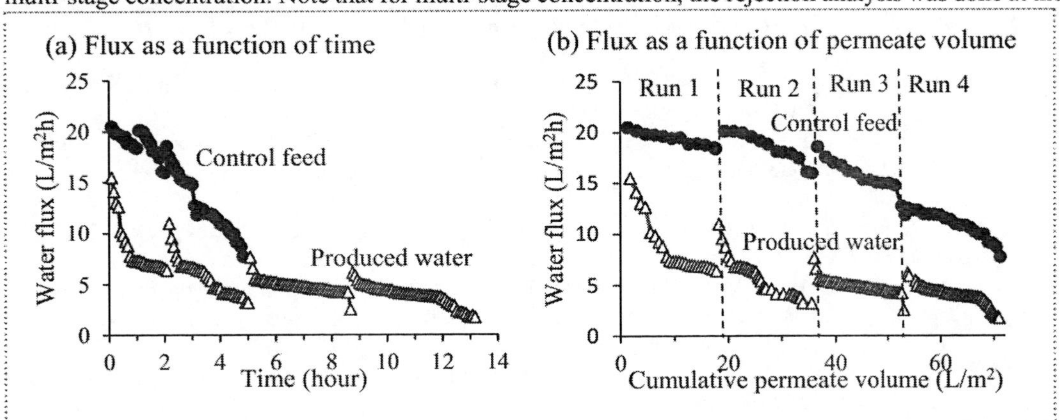

Figure 3. Multi-stage concentration performance plotted as (a) a function of time and (b) function of permeate volume.

Table 2. Summary of Rejection Analysis on Draw Solution

Parameter	Single-stage	Multi-stage (n=4)
Conductivity (μS/cm)	31876	47121
pH	6.5	6.5
COD (mg/L)	230	220
TOG (mg/L)	7	7
Total Nitrogen (mg/L)	0	0
Total Phosphorus (mg/L)	1.3	1.2
Turbidity (NTU)	1.47	1.13

final stage (n=4). The parameters of the DS for both single-stage concentration and multi-stage concentration were almost similar and both have meet the demands of standard B discharge effluent.

4. Conclusion

This study investigates the effect of concentration staging where multi-stage concentration outperformed single-stage concentration. However, fouling still persist despite the claims of FO membrane having lower fouling tendency. Further study on membrane fouling control strategy is thus required.

Acknowledgement

We acknowledge Universiti Teknologi PETRONAS for providing grants (grant number: 0153AA-F58 and 0153AA-F59) and facilities for conducting the research activities.

References

[1] S Deng *et al* 2002 Produced water from polymer flooding process in crude oil extraction: characterization and treatment by a novel crossflow oil–water separator *Sep. Purif. Technol.* **29** 207–216

[2] A D Horn 2009 Breakthrough Mobile Water Treatment Converts 75% of Fracturing Flowback Fluid to Fresh Water and Lowers CO2 Emissions *SPE Americas E&P Environmental and Safety Conference.*

[3] D L Shaffer, L H Arias Chavez, M Ben-Sasson, S Romero-Vargas Castrillón, N Y Yip and M Elimelech 2013 Desalination and Reuse of High-Salinity Shale Gas Produced Water: Drivers, Technologies, and Future Directions *Environ. Sci. Technol.* **47** 9569–9583

[4] J A Veil, M G Puder, D Elcock and R J Redweik 2004 A White Paper Describing Produced Water from Production of Crude Oil, Natural Gas, and Coal Bed Methane Argonne National Lab., IL (US), ANL/EA/RP-112631

[5] F R Spellman 2013 *Handbook of Water and Wastewater Treatment Plant Operations* 3rd ed. CRC Press

[6] Colorado School of Mines 2009 Technical Assessment of produced water treatment technologies *Integr. Framew. Treat. Manag. Prod. Water RPSEA Proj. 07122-12 Colo.* 8–128

[7] A Fakhru'l-Razi, A Pendashteh, L C Abdullah, D R A Biak, S S Madaeni and Z Z Abidin 2009 Review of technologies for oil and gas produced water treatment *J. Hazard. Mater.* **170** 2 530–551

[8] E T Igunnu and G Z Chen 2014 Produced water treatment technologies *Int. J. Low-Carbon Technol.* **9** 3 157–177

[9] E-S Kim, Y Liu and M Gamal El-Din 2011 The effects of pretreatment on nanofiltration and reverse osmosis membrane filtration for desalination of oil sands process-affected water *Sep. Purif. Technol.* **81** 3 418–428

[10] I Sutzkover-Gutman and D Hasson 2010 Feed water pretreatment for desalination plants *Desalination* **264** 289–296

[11] S Lee, C Boo, M Elimelech and S Hong 2010 Comparison of fouling behavior in forward osmosis (FO) and reverse osmosis (RO) *J. Membr. Sci.* **365** 1 34–39

IOP Conf. Series: Materials Science and Engineering **429** (2018) 012021 doi:10.1088/1757-899X/429/1/012021

[12] K L Hickenbottom *et al.* 2013 Forward osmosis treatment of drilling mud and fracturing wastewater from oil and gas operations *Desalination* **312** 60–66
[13] B D Coday *et al.* 2016 Evaluation of the transport parameters and physiochemical properties of forward osmosis membranes after treatment of produced water *J. Membr. Sci.* **499** 491–502
[14] M Al-Furaiji 2016 Hyper-saline produced water treatment for beneficial use
[15] B D Coday *et al.* 2014 The sweet spot of forward osmosis: Treatment of produced water, drilling wastewater, and other complex and difficult liquid streams *Desalination* **333** 23–35
[16] Y Yu, S Lee and S K Maeng 2016 Forward osmosis membrane fouling and cleaning for wastewater reuse *J. Water Reuse Desalination*
[17] A Seppälä and M J Lampinen 2004 On the non-linearity of osmotic flow *Exp. Therm. Fluid Sci.* **28** 4 283–296
[18] N T Hancock and T Y Cath 2009 Solute Coupled Diffusion in Osmotically Driven Membrane Processes *Environ. Sci. Technol.* **43** 17 6769–6775
[19] C Y Tang, Q She, W C L Lay, R Wang and A G Fane 2010 Coupled effects of internal concentration polarization and fouling on flux behavior of forward osmosis membranes during humic acid filtration *J. Membr. Sci.* **354** 1 123–133
[20] E A Bell, T E Poynor, K B Newhart, J Regnery, B D Coday and T Y Cath 2017 Produced water treatment using forward osmosis membranes: Evaluation of extended-time performance and fouling *J. Membr. Sci.* **525** 77–88

IOP Conf. Series: Materials Science and Engineering **429** (2018) 012022 doi:10.1088/1757-899X/429/1/012022

Advanced Aeration Using Finned Panel Spacer For Fouling Control In Produced Water Membrane Filtration

R.Rahmawati, M.R Bilad*, N.A.H.M Nordin

Chemical Engineering Department, Universiti Teknologi PETRONAS, Bandar Seri Iskandar, 32610 Perak, Malaysia

*Corresponding author: mroil.bilad@utp.edu.my

Abstract. Treatment of produced water (PW) is a challenge largely due to its form as stable oil/water emulsion. Membrane technology has shown promise in treating PW by efficiency removing micron-size oil droplets, but still suffers from major obstacle of membrane fouling. Many approaches have been introduced to mitigate membrane fouling including employing air bubbles (aeration) which improve hydrodynamics for scouring off the foulants. Despite offering some advantages, it still constricted by poor contact and distribution toward membrane surface. Thus, those limitations became highly concerned, and will be handled in this study by applying new finned spacer. The spacer encourages air bubbles flow path homogenously covering whole membrane surface. The filtration tests were performed using lab-made Polysulfone (Psf) flat-sheet membrane assembled into a flat-sheet panel. The results show excellent effect of aeration using finned panel spacer by performing higher flux and rejection. The bubbles are well distributed and homogeneously projected on the membrane surface by the finned spacer. Thus, the objective to mitigate fouling can be obviously achieved.

1. Introduction

Oil and gas industries are paramount significance in modern civilization. During extraction and processing, they generate large volume of oily waste water which contains various organics and inorganics [1]. Produced water (PW) is defined as water extracted from subsurface containing oil and gas. In old wells, its volume can reach of about 10 times of the produced oil [2] and its fraction is larger as the aging of the wells. It has different characteristics by well to well [3]. It consists of around 70% of total waste water during oil production [3]. In PW, oil is dispersed in water, stabilized by surfactant [4], which makes its treatment challenging. PW also contains suspended solids and dissolved organics [4].

Various methods have been used to treat PW, including both of conventional and advanced methods. One of the advanced methods is membrane technology that has long been implemented for treatment of PW [5,6]. Although it brings promises, it still suffers from membrane fouling [7]. Currently, all membrane based process require rigorous pre-treatment before used as polishing step in PW treatment [5]. Polymer material that commonly used to prepare MF/UF are polysulfone (Psf), polyethersulfone (PES), PVDF, PAN and cellulose acetate (CA) [8]. Among them, Psf has high mechanical stability over wide range of temperatures from -100 ^{0}C to160 ^{0}C. It has excellent chemical and physical properties e.g thermal and chemical stability, mechanical strength and oxidative resistance. Thus, Psf becomes preferred membrane material especially for water treatment in ultrafiltration system [9]. Moreover, Psf also offers many advantages: (i) highly resistant to mineral acids, alkali and electrolytes in pH ranging

2 to 13, (ii) resistant to oxidizing agent, so they can be cleaned by bleaches (iii) resistant to surfactants and hydrocarbon oil (iv) soluble in solvent (DMAc and NMP), make it applicable for conventional phase inversion method (v) good mechanical strength and permeability (vi) easily to modify properties through polymer blending.

Air bubbling has become a standard method for pressure driven membrane fouling control [10–12]. The air bubbles help to restrict foulant deposition and suppress cake layer build-up to maintain permeability [11,13]. Despite many advantages, air bubbles scouring still has few limitations: producing weak shear rates and the impact reaches plateau over at certain air supply rate [11,14]. This work focuses on maximizing the impact of air bubble for membrane cleaning, preferably under moderate velocity [15]. Bubble size and its number/amount must be optimized to give desirable hydrodynamic conditions and at the same time to minimize energy consumption [16]. Another factor considered is the effect of turbulences in membrane process. Turbulence flow regime can suppress concentration polarization. It can be built up by employing two-phase flow involving air bubbles [16]. Thus, it needs to be evaluated and optimized thoroughly.

Recently, an attempt has been proposed via introduction of spacer in the module assembly to enhance the effect of air bubbles for membrane fouling control. The spacer promotes bubble oscillation and random movement which effectively decline membrane fouling by improving turbulences, diminishing concentration polarization and increase mass-transfer rate [17]. Such system was found effective, especially when coupled with elevated liquid cross flow velocity. The bubble flow paths also highly depend on cross flow of the feed liquid [17]. This method is quite practical but exhibits higher energy compared to the standard aerated panel (without spacer), because the need of high liquid cross-flow velocity to encourage air bubble distribution. An attempt to maximize contact of air bubbles and membrane has also been introduced via a tilted panel [10]. The system is found to be effective for one sided membrane, low in foot-print but requiring an impractical switching mode to enable intermittent aeration to both sides of panel. The most preferred is indeed to use a conventional of vertical panel membrane due to its easy handling and small footprint.

Introducing of spacer in membrane module assembly is found effective to distribute air bubble but still highly depends on feed cross-flow velocity that can inflate energy consumption. In this study, a new system is proposed to maximize the role of air bubble (aeration) in membrane fouling control by employing a novel finned spacer. The fins are placed in between two adjacent vertical panels that help to encourage air bubble flow path in order to get excellent bubble flow trajectory distribution on the membrane surface and better bubble-surface contacts. This project proposes to optimize the role of air bubbling for membrane fouling control on vertical membrane panel via introduction of finned spacer.

2. Material and method

Membrane preparation, characterization and module assembly

Polysulfone (Psf) (BASF, Mn 22,000) polymer membrane, was prepared via wet phase inversion method. Polyethylene glycol (PEG) (Sigma-Aldrich Mw 1,0000) as additive, DMAc (Sigma-Aldrich) as solvent and Deionized water as non-solvent. The dope solution composition was 15%, 1% and 84% respective to Psf, PEG and DMAc. The mixture was mixed up by magnetic stirrer at 60 ^{0}C for 36 hours to obtain homogeneous solution. The homogenous solution then was degassed for 24 hours to release entrapped air bubbles. The solution was casted with casting knife at 0.22 mm wet-thickness followed by directly soaked in non-solvent. The produced Psf membrane was kept in wet until assembled into a filtration panel. The prepared membrane will be characterized by contact angle to measure its hydrophilicity.

Membrane porosity (δ) was measured using gravimetric method and the value was estimated using (Eqs 1&2).

$$\delta = \frac{Vpore}{Vmembrane} \times 100 \qquad (1)$$

$$Vpore = \frac{Mwet-Mdry}{\rho wetting\ agent} \qquad (2)$$

Where, V_{pore} is volume of membrane pores, M_{wet} and M_{dry} indicate wet membrane weight and dry membrane weight, the applied wetting liquid was n-butanol with $\rho = 0.92\ g/cm^3$

To be practically functional, membrane was assembled into a membrane filtration panel with effective area 10 cm x 14 cm (140 cm^2) (Figure 1). Two panels were required to build up one module system. One membrane sheet was glued on frame using oil-resistant glue to prevent any leaking due to the use of PW as feed liquid. After passing leaking test, two panels were arranged in parallel, sandwiching tightly a finned spacer between them. The packing was set tight to prevent air bubble lost from the module system. Thus, from that arrangement air bubbles will totally distribute inside the module.

Produced water

PW sample was collected from one of oil production unit in Malaysia. Since the properties characteristic of PW changes over time and to preserve the emulsion, PW sample was collected at once and was kept at around 4^0C in dark room. PW characteristic is provided in Table 1.

Finned spacer

Finned spacer was made using PMMA/acrylic having 73^0 contact angle which means having hydrophilic properties. Fins were arraged horizontally on spacer stand. Finned Spacer was placed in between two adjacent membrane panel.

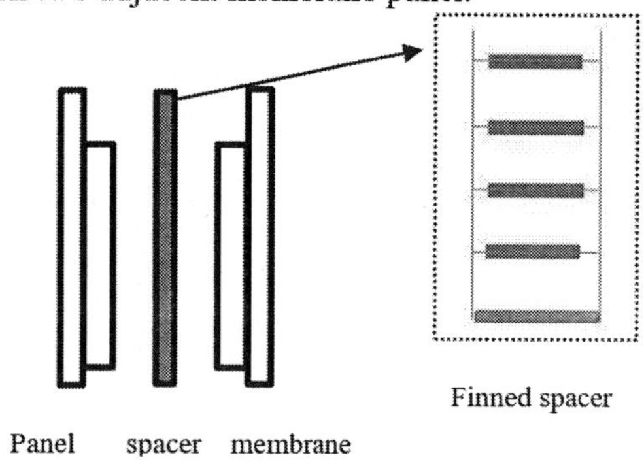

Figure 1. Membrane module

Table 1. PW characteristic

Items	Amount
Conductivity	2000 μ s cm^{-1}
pH	6
Turbidity	79.4 NTU
COD	2060 mg L^{-1}
Total oil and grease (TOG)	11.27 mg L^{-1}
Total nitrogen (TN)	36 mg L^{-1}
Total phosphor (TP)	12.8 mg L^{-1}

Experimental set-up

The membrane filtrations were run in submerged and constant-pressure system (Figure 2). A vacuum pressure pump also be used to exert vacuum condition and drive the membrane filtration. The ΔP was kept constant at -0.1 bar by adjusting control valve. The permeate has been collected semi-batch wise for every operation cycle (9.5 min filtration and 0.5 min relaxation). After permeate volume was measured, the permeate was returned back into the tank to maintain feed liquid level.

The permeance flux (J) was calculated from permeate flow rate measured every 10 min for 12 cycles. Permeability (L) can be attained from flux (Eqs 3&4). The permeability profile indicates the effectivity of membrane operation and detects fouling occurrence.

$$J = \frac{V}{A\,\Delta t}\ (\mathrm{L/m^2h}) \quad (3)$$

$$L = \frac{J}{\Delta P}\ (\mathrm{L/m^2hbar}) \quad (4)$$

Where, V is permeate volume (l), A effective membrane area (m^2), t filtration time (h), and ΔP trans-membrane pressure (bar).

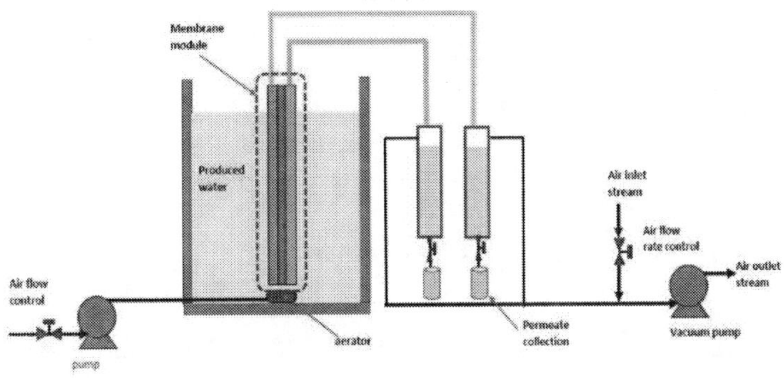

Figure 2. Experimental Set-up

Rejection was calculated using Eqs 5 :

$$R\,(\%) = \left(1 - \frac{Cp}{Cf}\right) \times 100 \quad (5)$$

Where, C_p and C_f is permeate and feed parameters.

3. Result and discussion

Membrane characteristics

Membrane characteristics in Table 2 indicate that it is suitable for PW treatment. Contact angle test employs water drops on membrane surface then at certain magnification it will get the degree of water drop toward membrane surface. Angles of less than 90^0 means hydrophilic, while more than 90^0 means hydrophobic. This membrane is hydrophilic with 67.9^0 of contact angle. Hence, the membrane can facilitate water transport and rejecting other hydrophobic matters.

Based on surface pore (Figure 3), the membrane is in ultra-filtration (UF) which is the most used in water treatment especially in PW treatment [18]. UF membrane has been specialized to remove color, odor, viruses and colloidal organic matter [19,20] also known effectively in oil removal compared to traditional methods in PW treatment [18,21].

Psf membrane is asymmetric, comprising selective layer (up layer) and support layer (bottom layer). The core advantage of this morphology is having strong structure supported by second layer, which can maintain both of high selectivity and high flux.

Table 2. Membrane Characteristic

Items	Amount
Contact angle	67.9^0
Nominal pore size	<0.01 µm (UF)
Thickness	0.26 cm
Surface porosity	51.2%
Pore morphology	Asymmetric
Specification	Ultrafiltration

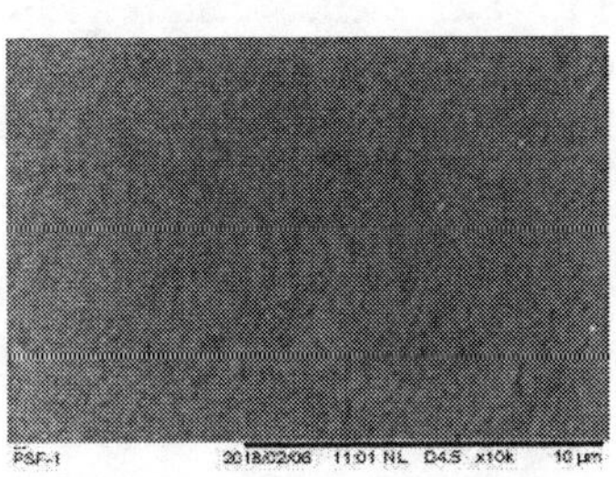

Figure 3.Surface SEM image of 15%Psf-1%PEG

Hydraulic Membrane Performance

A. Relative permeability

Figure 4 shows the permeability trend over the time for membrane 1 and 2. Those data were obtained under different parameters: using spacer and aeration 1.5 L/min (spacer+bubble), using aeration 1.5

L/min but without spacer (no spacer+bubble) and using neither spacer nor aeration (no spacer+no bubble).

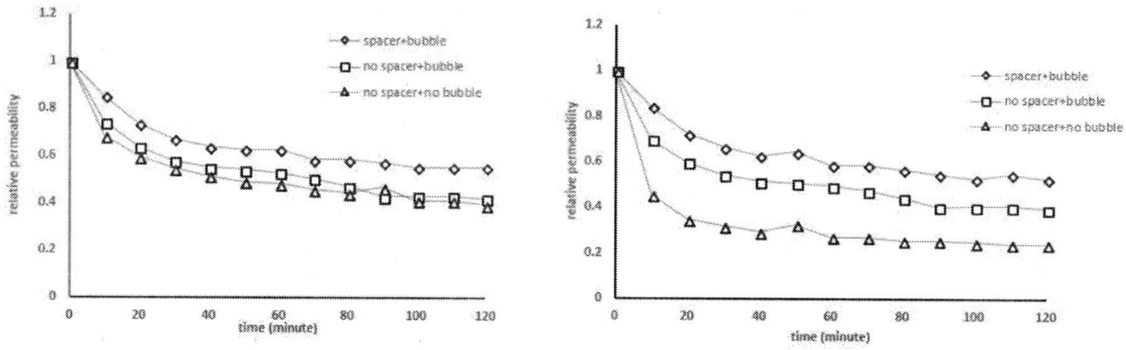

[a] [b]

Figure 4. Relative permeability: [a] membrane 1 ; [b] membrane 2

As shown in Figure 4, spacer+bubble system has the highest relative permeability. The no-spacer + bubble system shows better performance than no-spacer + no-bubble system. This finding shows effectiveness of spacer and bubble enhance membrane permeability. Both graphs show about similar of decreasing permeability over time. Permeability decreases rapidly within first 40 min operation and gradually reach plateau for the last 30 min. Sharp decline on permeability on the first 40 min is due to high initial flux (low initial membrane mass transfer resistance) [22,23]. Overtime, membrane mass transfer resistance increases due to pore clogging, cake layer formation, concentration polarization, etc [23,24]. All of them contribute on membrane fouling. Lastly, permeability reaches steady value due to balance between convective flow of foulant approaching membrane surface and its back diffusive flow away from the membrane surface.

B. Effect of air bubble (aeration)

The effect of air bubble in membrane filtration, can be seen by comparing the permeability profile of variable different set-ups: no-spacer+bubble and no-spacer+no-bubble in Figure 4. Both filtration systems operated under air bubbling show higher permeability than no-bubble operation. This suggests that air bubbles can maintain membrane permeability due to their roles in removing foulant from membrane surface. Moreover, air bubbles can help to lift the oil droplets like ones normally seen in dissolved air flotation process. The bubbles have buoyancy force to float those particulates and bring them onto feed liquid surface.

The role of air bubble flotation in waste water treatment has long been known and this phenomenon is utilized in dissolved air flotation (DAF) process. DAF removes dissolved and dispersed oil from produced water, hence can reduce oil and turbidity [25,26]. However, stand-alone DAF is still constricted by the higher energy consumption of generating micro-size bubbles. Employing air bubble in membrane filtration, on another hand, not only gives significant impact to reduce fouling propensity.

C. Effect of Spacer

The effectiveness spacer in membrane filtration can be seen by the relative value of permeability between spacer+bubble and no-spacer+bubble. Significantly higher permeability was found for spacer+bubble than no-spacer+bubble. This result suggests the role of spacer to maximize the air bubbles contact toward membrane surfaces. Better bubble and membrane contacts leads to get better fouling control.

It is still not clear enough to see the bubble pathway encouraged by spacer. Roughly, Figure 4 shows that the bubble distribution tends to membrane 2, implied by higher permeability and significant improvement. We can conclude that the use of spacer in membrane module is to get excellent bubble distribution.

D. Oil Rejection

To get more precise TOG analyzing, each variable has done by three times measurement (Table 3). The amount of oil rejection can be sequenced from the worse to the best one: no-spacer+no-bubble, no-spacer+bubble and spacer+bubble.

Combination of spacer+bubble gives the best performance in oil rejection due to the role of spacer to maximize the bubble reducing the fouling. Membrane 1 can obtain more than 60% of oil rejection while membrane 2 rejects 55% oil. Meanwhile, combination of no-spacer+bubble doesn't perform oil rejection as much as spacer+bubble due to the uncertain bubble distribution in term of the spacer absence. On the other hand, it still performs better rejection compared to no-spacer+no-bubble.

Table 3. Oil Rejection

variable	TOG (mg/L)		%rejection	
	M1	M2	M1	M2
No spacer + no bubble	7.2	8.9	35.82	21.03
	7.4	9		
	7.1	8.8		
average	7.23	8.9		
No spacer + bubble	5.7	7.9	50.01	27.83
	5.7	8.2		
	5.5	8.3		
average	5.63	8.13		
Spacer + bubble	4.3	5.3	60.96	55.04
	4.4	5		
	4.5	4.9		
average	4.4	5.1		

Meanwhile, combination of no-spacer+bubble doesn't perform oil rejection as much as spacer+bubble due to the uncertain bubble distribution in term of the spacer absence. On the other hand, it still performs better rejection compared to no-spacer+no-bubble.

The combination of spacer+air bubble to suppress fouling proves the significant effect in oil rejection. Better permeance is indicated by higher oil rejection. Moreover, it can be concluded that air bubble can decrease fouling propensity by doing both scouring off the foulant from membrane surface and floating those particulates onto PW liquid surface (DAF). Spacer also gives the significant impact to distribute bubbles.

4. Conclusion

Membrane fabrication using polymer blending of Psf and PEG produces membrane ultrafiltration (UF) which is suitable for PW treatment. Membrane performance over time gets declined permeability due to fouling effect. The using of advanced aeration occupied finned spacer performs better fouling control. It functions by removing those foulants and gives significant impact in permeability profile and oil rejection.

5. Acknowledgements

We acknowledge Universiti Teknologi Petronas – Malaysia for providing grants and facilities for conducting the research activities.

References

[1] F Ahmadun, A Pendashteh, and L Chuah 2009 Journal of Hazardous Materials **1** 530–51
[2] K L Benko and J E Drewes 2008 Produced Water in the Western United States: Geographical Distribution, Occurrence, and Composition *Environ. Eng. Sci.* **2** 239–46
[3] M Çakmakce, N Kayaalp, and I Koyuncu 2008 Desalination of produced water from oil

production fields by membrane processes *Desalination* **222** 176–86
[4] J M Dickhout, J Moreno, P M Biesheuvel, L Boels, R G H Lammertink, and W M de Vos 2017 Produced water treatment by membranes: A review from a colloidal perspective *J. Colloid Interface Sci.* 523–34
[5] S Alzahrani and A W Mohammad 2014 Challenges and trends in membrane technology implementation for produced water treatment: A review *J. Water Process Eng.* 107–33
[6] D P Leta 2009 (19) United States (12) **1**
[7] A Bahadori, G Zahedi, S Zendehboudi, and M Bahadori 2013 Estimation of air concentration in dissolved air flotation (DAF) systems using a simple predictive tool *Chem. Eng. Res. Des.* **1** 184–90
[8] M Padaki, R Surya Murali, M S Abdullah, N Misdan, A Moslehyani, M A Kassim, N Hilal, and A F Ismail 2015 Membrane technology enhancement in oil-water separation. A review *Desalination* **3** 197–207
[9] H L Huang and S Yang 2006 Filtration characteristics of polysulfone membrane filters *J. Aerosol Sci.* **3** 1198–208
[10] A Eliseus, M R Bilad, N A H M Nordin, Z A Putra, and M D H Wirzal 2017 Bioresource Technology Tilted membrane panel : A new module concept to maximize the impact of air bubbles for membrane fouling control in microalgae harvesting *Bioresour. Technol.* **2** **1** 661–8
[11] T Hwang, Y Oh, B Kim, and J Han 2015 Bioresource Technology Dramatic improvement of membrane performance for microalgae harvesting with a simple bubble-generator plate *Bioresour. Technol.* **1** 343–7
[12] P Le-Clech, V Chen, and T A G Fane 2006 Fouling in membrane bioreactors used in wastewater treatment *J. Memb. Sci.* **2** 17–53
[13] S Chang and A G Fane 2001 <Chang, Fane - 2001 - The effect of fibre diameter on filtration and flux distribution — relevance to submerged hollow fibre modules.pdf> **1** 221–31
[14] M R Bilad, G Mezohegyi, P Declerck, and I F J Vankelecom 2011 Novel magnetically induced membrane vibration (MMV) for fouling control in membrane bioreactors *Water Res.* 63–72
[15] F Meng, F Yang, B Shi, and H Zhang 2008 A comprehensive study on membrane fouling in submerged membrane bioreactors operated under different aeration intensities *Sep. Purif. Technol.* 91–100
[16] G Accary, D Morvan, and S Me 2008 The Human Line-1 Retrotranspsons Creates DNA Double Strand Breaks *Fire Saf. J.* **3** 173–8
[17] C Fritzmann, M Hausmann, M Wiese, M Wessling, and T Melin 2013 Microstructured spacers for submerged membrane filtration systems *J. Memb. Sci.* 189–200
[18] E T Igunnu and G Z Chen 2014 Produced water treatment technologies *Int. J. Low-Carbon Technol.* 157–77
[19] J Drewes, T Cath, P Xu, and J Graydon 2009 An Integrated Framework for Treatment and Management of Produced Water *Tech. Assess. Prod. water Treat. Technol.* 8–128
[20] R Han, S Zhang, D Xing, and X Jian 2010 Desalination of dye utilizing copoly(phthalazinone biphenyl ether sulfone) ultrafiltration membrane with low molecular weight cut-off *J. Memb. Sci.* **3** 1–6
[21] Y He and Z W Jiang 2008 Technology review: Treating oilfield wastewater *Filtr. Sep.* 14–6
[22] V Chen, A G Fane, S Madaeni, and I G Wenten 1997 Particle deposition during membrane filtration of colloids: Transition between concentration polarization and cake formation *J. Memb. Sci.* **12** 109–22
[23] D J Miller, S Kasemset, D R Paul, and B D Freeman 2014 Comparison of membrane fouling at constant flux and constant transmembrane pressure conditions *J. Memb. Sci.* 505–15
[24] M Taniguchi, J E Kilduff, and G Belfort 2003 Modes of natural organic matter fouling during ultrafiltration *Environ. Sci. Technol.* **3** 1676–83
[25] J M Younker and M E Walsh 2014 Bench-scale investigation of an integrated adsorption-

coagulation-dissolved air flotation process for produced water treatment *J. Environ. Chem. Eng.* **2** 692–7

[26] S Jiménez, M M Micó, M Arnaldos, E Ferrero, J J Malfeito, F Medina, and S Contreras 2017 Integrated processes for produced water polishing: Enhanced flotation/sedimentation combined with advanced oxidation processes *Chemosphere* **1** 309–17

The Possible Challenges of Green Procurement Implementation

S L A Rais[1*], Z A Bidin[1], A A M Bohari[1], M M Saferi[1]

[1]Faculty of Architecture, Planning and Surveying, Universiti Teknologi MARA, Cawangan Sarawak, Kota Samarahan, Sarawak, Malaysia

*Email: syamimi6893@sarawak.uitm.edu.my

Abstract. The construction industry is one of the industries which significantly contribute to the negative impact on the environment, natural resources, and public health. To tackle this issue, the Malaysian government has suggested the green procurement initiative under *MyHijau* Programme to be implemented in construction projects. Green procurement is an effective tool to minimise the environmental impacts and has been widely applied in many countries. However, in Malaysia, the implementation of green procurement is still at introduction phase. Thus, this study aimed to identify the challenges faced by stakeholder in the implementation of green procurement for construction project. This research has conducted extensive literature review on green procurement and its challenges and later, validated through focus group discussion among experience practitioners in construction industry and green project. This study revealed that lack of knowledge and lack of standard guideline are the main challenges for implementing green procurement. The research findings could possibly be a basis of reference to assist the stakeholder to implement green procurement for the project and take necessary measure to tackle the green procurement challenges.

1. Introduction

Construction activities are one of the major contributors to environmental issues such as air pollution, waste pollution, noise pollution, and water pollution [1]. Besides, it also negatively impact natural resources since construction activities consume about half of renewable and half of non-renewable resources and materials [2]. In 2010, Malaysia was identified to have the highest GHG emission in Asia with 7.7 metric tons per capita compared with Indonesia, Singapore, China, and India [3]. Thus, the Malaysia government takes initiative to encourage the adoption of sustainable and green construction to tackle this issue. The implementation of sustainable construction concept in the construction industry can positively contribute to the environment [4]. In 2012, MyHIJAU Programme was implemented by the Ministry of Energy, Green Technology and Water (KeTTHA) and the Malaysia Green Tech Corporation (MGTC) to support the development of green technology. This programme is executed to encourage manufacturers, suppliers, and producers to produce more green products and services [5]. One of the initiatives under the MyHIJAU Programme is green procurement which is a tool to manage environmental problems. The public and private organisations should be implementing green procurement to reduce environmental problem [6]. However, in Malaysia, the implementation of green procurement in the private sector is higher compared to the public sector [7]. The implementation of green procurement is still in its introduction phase and stakeholder faces with various of challenges such as lack of knowledge and awareness on green procurement concept [8,9]. This paper aims to provide an overview of green procurement and identify its possible challenges faced by stakeholder in the implementation of green procurement. The paper is based on a review of literature which examines green procurement and its challenges and later, the findings extracted from the literature review will be validated through a focus group discussion among experienced practitioners.

2. Green Procurement for Construction Project

Since the 1990s, green procurement has been known as an effective way in product production and consumption to reduce the negative impact towards the environment. Currently, green procurement has been recognised as an effective tool to minimise negative environmental impacts and has been applied widely in many countries throughout the world [8]. The term, 'green', refers to recognition, integration, and implementation of environmental practices to reduce negative environmental impact. Meanwhile, 'procurement' is defined as purchasing or chain management which is considered as a support service for companies to achieve their aims [10]. As stated by the Ministry of Energy, Green Technology, and Water (KeTTHA), green procurement is "procurement activities of products, services and works considering environmental criteria and standards that conserve the natural environment and resources which minimises the negative impact of human activities". Green Council [11] defined green procurement as the purchase of services and products which reduce and provide a positive impact on the environment. In the construction industry, green procurement is very important to manage the environmental degradation problems which are a common occurrence in the construction industry. Currently, there is no specific standard guideline for green procurement in the construction industry [12]. Therefore, green procurement as the act to obtain sustainable goods and services for a construction process from the commencement to the end of the project [13]. It considers all lifecycle construction process from the extraction of raw material, transportation, manufacturing, packaging, storage, handling, to the disposal of materials. Other than that, the important part of green procurement includes appointing the contractor and setting the environment requirement in the contract. In tender assessments, environmental consideration must be included and the criteria of procurement process must be evaluated such as the preliminary design or architectural competition stage, tendering for contracts, and building services.

Green procurement aims to reduce negative environmental impact compared with other products or services that serve the same purpose or products that meet certain predefined environmental criteria [10]. The adoption of green procurement provides benefits to the environmental aspect and corporate aspect [14]. In terms of environmental aspect, the benefits can be achieved by reducing negative impacts on the surrounding humans and the environment. For the cooperate aspect, the benefits are obtained by carrying business benefit to the company by improving reusable content, improve environmental performance, and reduce environmental handling cost. The implementation of green procurement in the construction industry improves the performance of the building and construction process [14]. The performance of the project is measured based on quality, time, and cost and currently the project performance includes green performance. Green performance consists of three aspects that include environmental, social, and economic [12]. The aim of environmental performance is to reduce the environmental impact such as reducing greenhouse gas emissions and pollution. Social performance relates to the clients' level of satisfaction in terms of cost and design performance while economic performance relates to the value for money. Adham [8] highlighted that the challenges of green procurement vary according to countries due to the differences in socioeconomic, demographic and cultural. There are seventeen possible challenges of green procurement implementation that can be identified from the literature review. The exhaustive explanation on possible challenges will be discussed in the finding and analysis section and the list of author and possible challenges of green procurement are listed in Table 1 and the categories are listed in Table 2.

3. Methodology

An intensive literature review was conducted to explore information related to green procurement. The aim of this paper is to highlight the challenges of green procurement implementation. The study begins with a review of the nature of green procurement and its benefits in the construction industry. A list of possible challenges implementation of green procurement was presented and discussed. The literature review was carried out to analyses the possible challenges of green procurement implementation that can assist stakeholder to take necessary measures to tackle the challenges. The information was obtained from fourteen journal articles which contain the 'green procurement' and 'challenges' as keywords. Other related research materials were also referred.

4. Analysis and Findings: The Possible Challenges of Green Procurement Implementation

The implementation of green procurement is still in its introduction phase and faces various challenges. As highlighted by Adham & Siwar [8], Bohari & Xia [13], Bouwer et al., [15], and Fischer [16], lack of knowledge among stakeholders is a key challenge of green procurement implementation. This is supported by Buniamin et al., [9], stating that the low level of understanding and knowledge among people working in public client organisation and stakeholder's organisation are obstructing the implementation of green procurement. Therefore, it might influence other staff not to appreciate the benefits of green procurement due to the minimal environmental principle knowledge and understanding. Besides, the lack of training for procurement officers has resulted in the inadequate knowledge and information on green procurement [15]. Therefore, the number of qualified staff to manage green procurement is still insufficient [9]. This may be attributed to the lack of awareness of the green procurement concept [7, 14, 17,18]. Low level of awareness of the green procurement is not only faced by workers in public client organisations but also in stakeholder organisations. In addition, the enforcement by the government in implementing green procurement is still poor and caused hesitancy among clients and developers adopt green practices. As studied by Wong et al., [14] in China, green procurement is not mandatory and their government only provide a macro level of control for construction projects to reduce the environmental effects. Besides, there are insufficiencies in terms of policies and regulation to promote green procurement [15,17]. In order to mitigate this issue, more mandatory requirements are needed to provide better sustainability.

As highlighted by Wong et al., [14] poor commitment by its top management hampers to implement green procurement. This is supported by Adham & Siwar [8], Sourani & Sohail [17], Buniamin et al., [9], McMurray et al., [7], Min & Galle [19], and Mensah & Ameyaw [20] who emphasised that the lack of management commitment in terms of money and time are slowing down the implementation of green practices. A study by Buniamin et al., [9] identified that the most common challenge of green procurement implementation is the lack of proper guideline which is agreed by Bohari & Xia [13], Sourani & Sohail [17], and Gunther [21]. Currently, the guideline is fragmented as it is developed by different organisations for different purposes. Thus, there is no standardised guideline that can be followed. Without a standard guideline, there is no clarity on how and when to apply and by whom. These unanswered questions create confusion and difficulties to stakeholder [17]. As mentioned by Bouwer et al., [15] and Sourani & Sohail [17], other challenges are lack of practical tools and information such as handbook or internet tools. The existing tools are not comprehensive to evaluate the criteria of sustainability. Thus, they need to create more comprehensive tools to deal with a situation where sustainability needs to be evaluated [17]. Other contributing challenges include the poor market demand for recyclable construction materials [14] and a limited supply of green products that hinder the implementation of green procurement [9,22]. However, to increase the number of green products, the government authority should create a demand for it. Thus, supplier and clients should create a good communication and coordination channel. According to Wong et al., [14] and Sourani & Sohail [17], there are insufficient of integration and link-up in the industry among clients, construction team, and supplier. Thus, a few opportunities for obtaining innovative and sustainable solution might be missed [17]. Another challenge of green procurement is the process and procedure to implement green procurement is that it is time-consuming [9,17]. The public client might have not sufficient time to address the suitability requirements, thus, they prefer to adopt the practices that they are currently used.

Table 1. List of possible challenges of green procurement implementation

Authors [7]McMurray et al., (2014), [8]Adham & Siwar (2012), [9]Buniamin et al., (2016), [13]Bohari & Xia (2015), [14]Wong et al., (2016), [15]Bouwer et al., (2005), [16]Fischer (2010), [17]Sourani & Sohail (2010), (2016), [18]Ruparathna & Hewage (2015), [19]Min & Galle (2001), [20]Mensah & Ameyaw (2015), [21]Gunther (2003), [22]Varnas et al., (2009), [23]Carlsson & Waara (2006)														
Possible Challenges	[7]	[8]	[9]	[13]	[14]	[15]	[16]	[17]	[18]	[19]	[20]	[21]	[22]	[23]
Lack of knowledge		✓	✓	✓		✓	✓							
Lack of training for procurement officers						✓								
Insufficient qualified staff to handle green procurement			✓											
Lack of awareness on the green procurement concept	✓				✓			✓	✓					
Lack of enforcement by government to implement green procurement			✓		✓									
Insufficient policies and regulation promoting green procurement						✓		✓						
Lack of top management commitment (including money and time)	✓	✓	✓		✓			✓		✓	✓			
Lack of proper guideline in implementing green procurement			✓	✓				✓				✓		
Lack of practical tools and information						✓		✓						
Poor market demand for recyclable material					✓									
Insufficient integration and link up in the industry					✓			✓						
Procedure to implement green procurement are time consuming			✓					✓						
Lack of incentive for companies to implement green procurement	✓		✓		✓			✓	✓					
Limited supply for green product			✓										✓	
High cost of green product and services			✓								✓			✓
Perception that green product would be more expensive						✓		✓		✓			✓	
Insufficient research and development					✓			✓						

As highlighted by Wong et al., [14], McMurray et al., [7], and Ruparathna & Hewage [18], the challenges that hamper the implementation of green procurement are lack of incentives for

IOP Conf. Series: Materials Science and Engineering **429** (2018) 012023 doi:10.1088/1757-899X/429/1/012023

companies. This is supported by Sourani & Sohail [17] who pointed out that there are inadequate incentives for government departments and restrictions on expenditure. These might hinder organisation to invest in sustainability. The failure to provide environmentally friendly products to government agencies is due to inadequate incentives given to enterprise managers [9]. Another challenge in the implementation of green procurement is that it requires high capital cost [9, 20, 23]. Thus, approaches to cutting the cost might be adopted to reduce the total cost of procurement. There is no necessary increased capital cost in sustainable construction procurement [17]. However, it is contributing to the perception of organisation towards green product are would be more expensive and sustainability always leads to incurring high capital cost [15, 17, 19,22]. As mentioned by Sourani & Sohail [17] and Wong et al., [14], other challenges include insufficient research and development. The inadequate research and development are related to funding issues where there is limited research regarding facilities' operational cost and the relationship between the capital cost and operational cost [17]. The list of possible challenges in green procurement implementation from the literature review is listed in Table 1.

TABLE 2. Possible challenges of green procurement implementation categorization based on table 1

No.	Categories	Possible challenges	Coding
1.	Knowledge	Lack of knowledge Lack of practical tools and information	(K1) (K2)
2.	Training	Lack of training for procurement officers Insufficient qualified staff to handle green procurement	(T1) (T2)
3.	Awareness	Lack of awareness on the green procurement concept	(A1)
4.	Policy	Lack of enforcement by government to implement green procurement Insufficient policies and regulation promoting green procurement Lack of proper guideline in implementing green procurement Insufficient research and development	(P1) (P2) (P3)
5.	Commitment	Lack of top management commitment including money and time	(P4) (C1)
6.	Demand	Poor market demand for recyclable material	(D1)
7.	Integration	Insufficient integration and link up in the industry	(I1)
8.	Time	Procedure to implement green procurement are time consuming	(T1)
9.	Cost	Lack of incentive for companies to implement green procurement High cost of green product and services Perception that green product would be more expensive	(CO1) (CO2) (CO3)
10.	Availability	Limited supply for green product	(A1)

5. Conclusions

Green procurement is still a relatively new concept in Malaysia. Thus, green procurement needs to be introduced to the public and private stakeholder since the environmental issues are increasing. This paper describes the result of the literature review in identifying a list of possible challenges in implementing green procurement. The outcome of this paper could help stakeholder to take necessary measures to tackle the green procurement challenges. This paper has identified seventeen [17] possible challenges of green procurement implementation [refer Table [1] and the challenges were further classified under ten [10] categories [refer Table 2] including knowledge, training, awareness, policy, commitment, demand, integration, time, cost, and availability. Thus, these ten [10] categories of possible challenges will be explored in the next phase of this research which is the group discussion. The focus group discussion will be conducted among experienced practitioners in the construction industry and green construction to validate the ten categories of possible challenges. Also, the findings from this paper can serve as a basis for further research in the future.

References

[1] Shen L Y and Tam V 2002 Implementation of environmental management in the Hong Kong Construction Industry *International Journal of Project Management* **20** 7 535-543

[2] Zolfagharian S, Nourbakhsh M, Irizarry J, Ressang R and Gheisari M 2012 Environmental Impacts Assessment on Construction Sites *Construction Research Congress 2012 © ASCE 2012*

[3] Zaid S M, Myeda N E, Mahyuddin N and Sulaiman R 2015 Malaysia's Rising GHG Emissions and Carbon 'Lock-In' Risk: A Review of Malaysian Building Sector Legislation and Policy *Journal of Surveying, Construction and Property (JSCP)* **6** 1-13

[4] Abidin N Z 2009 Sustainable Construction in Malaysia – Developers' *Awareness Proceedings of World Academy of Science, Engineering and Technology*

[5] MyHijau Programme *Sustainable Consumption and Production Malaysia,* 2013 Web 25 January 2018.

[6] Bohari A A M and Xia B 2015 Developing Green Procurement Framework for Construction Projects in Malaysia *The 6th International Conference on Engineering, Project, and Production Management*

[7] McMurray A J, Islam M M, Siwar C and Fien J 2014 Sustainable Procurement in Malaysia Organization: Practices, Barriers and Opportunities *Journal of Purchasing and Supply Management* **20** 3 195-207

[8] Adham K N and Siwar C 2012 Empirical Investigation of Government Green Procurement (GGP) Practices in Malaysia *OIDA International Journal of Sustainable Development*

[9] Buniamin S, Ahmad N, Rauf F H A, Johari N H and Rashid A A 2016 Green Government Procurement Practices (GGP) in Malaysian Public Enterprises *7th International Economics and Business Management Conference*

[10] Mosgaard M A 2014 Improving the Practices of Green Procurement of Minor Items *Journal of Cleaner Production*

[11] Green Council 2010 Report of The Research Study on The Current Status and Direction for Green Purchasing in Hong Kong *Green Council*

[12] Bohari A A M, Skitmore M, Xia B and Teo M 2017 Green Oriented Procurement for Building Projects: Preliminary Findings from Malaysia *Journal of Cleaner Production* **148** 690-700

[13] Bohari A A M and Xia B 2015 Green Procurement Framework for The Malaysian Construction Industry *7th International Conference of SuDBE2015*

[14] Wong J K W, Chan J K S and Wadu M J 2016 Facilitating Effective Green Procurement In Construction Projects: An Empirical Study of The Enablers *Journal of Cleaner Production*

[15] Bouwer M, de Jong K, Jonk M, Berman T, Bersani R, Lusser H, Nissinen A, Parikka K and Szuppinger P 2005 Green Public Procurement in Europe2005 – status overview

[16] Fischer E A 2010 Green Procurement: Overview and Issues for Congress *Congressional Research Service (CRS) Report for Congress*

[17] Sourani A and Sohail M 2010 Barriers to Addressing Sustainable Construction in Public Procurement Strategies *Engineering Sustainability*
[18] Ruparathna R and Hewage K 2015 Sustainable Procurement In The Canadian Construction Industry: Challenges and Benefits *Canadian Journals of Civil Engineering*
[19] Min H and Galle W P 2001 Green Purchasing Practices of US Firm *International Journal of Operations & Production Management* **21** 9 1222-1238
[20] Mensah S and Ameyaw C 2012 Sustainable Procurement: The Challenges of Practice in The Ghanaian Construction Industry *Conference: West Africa Built Environment Research (WABER)* 871-880
[21] Gunther P A 2013 Hurdles in green purchasing - method, findings and discussion of the hurdle analysis. In C. Erdmenger (ed.), Buying into the Environment: Experiences, Opportunities and Potential for Eco Procurement *Greenleaf Publishing* United Kingdom
[22] Varnas A, Balfors B and Faith-Ell C 2009 Environmental Consideration in Procurement of Construction Contracts: Current Practices, Problems and Opportunities in Green Procurement in The Swedish Construction Industry *Journal of Cleaner Production*
[23] Carlsson L and Waara F 2016 Environmental concerns in Swedish local government procurement *Academic Press*

IOP Conf. Series: Materials Science and Engineering **429** (2018) 012024 doi:10.1088/1757-899X/429/1/012024

Green Procurement for Construction Project: The Roles of Stakeholder Values

Munira Mohd Saferi[1*], Asmah Alia Mohamad Bohari[1], Zafikha Aida Bidin[1], and Syamimi Liyana Amat Rais[1]

[1]Faculty of Architecture, Planning and Surveying, Universiti Teknologi MARA, Cawangan Sarawak, Kampus Samarahan, Kota Samarahan, Sarawak, Malaysia.

*Email: munira749@sarawak.uitm.edu.my

Abstract. Construction development at a large scale is a matter of concern to the society. Green construction is introduced to minimize the problem of environment degradation. Green procurement is perceived as an effective method to improve the environmental performance of a green construction. However, green procurement is at its initial stages and it is still not been widely adopted by the stakeholders in the construction development. Acknowledging the importance of stakeholder, thus, extensive literature review was conducted to study the concept of stakeholder values in adoption of green procurement. A thorough literature review was conducted by summarizing the relevant journal from the Journal of Cleaner Production. The aim of this paper is to study the roles of stakeholder values for the success of green procurement implementation. It was found that there are five categories of stakeholder values identified for the implementation of the green procurement practice. These values serve as a guideline for future research and references for the success of green procurement adoption.

1. Introduction

Issues regarding environment degradation are increasing due to the vast development in the country [1]. The construction industry is under constant pressure to improve environmental performance. A new policy is vital to fulfil the requirements of the continuous development by considering the importance of the environment. Thus, the need for 'sustainable development' concept to be implemented. Energy efficiency is employed on construction project with either the design, management or quality of the construction itself as part of the sustainable development agenda [2]. Thus, sustainable construction or green construction is established to achieve sustainable development.

Green construction comprises procuring environmentally friendly materials. Environmentally friendly materials include natural and recycled material and taking into consideration the design that provide optimum energy efficiency [3]. Green construction is the coordinated structure of outline, which integrates the whole process of construction from design to the demolition of the project that incorporate the environmental value of the construction project [4]. The construction is conducting by ensuring preservation of nature and in the importance of the development [5]. In recent years, emphasis has been on the global movement on promoting green buildings. Whereby, the Tenth (10) Malaysia Plan 2011-2015 focuses on the adaptation strategies in ensuring a sustainable growth in the development of the construction industry. This has been pursued under the Eleventh (11) Malaysia Plan 2016-2020, in line with Malaysia's general plan of promoting sustainable and environmentally friendly development.

The construction process requires a higher level of integration among partners within a construction project as manpower, materials, and equipment within a project must be coordinated [6]. Thus, Green Procurement is introduced to enhance the effectiveness of green construction development. As Zsidisin and Siferd [7] mentioned, green procurement is measured as integrated policies, process and relationships that are associated with natural environment. In construction projects, stakeholders play an important role in determining the path and critical success factors of green procurement establishment [8]. Each stakeholder has an influential power to enhance the effectiveness of green procurement. Stakeholder competencies and commitment have critical and direct effect on green procurement undertakings in green construction [9].

Therefore, this paper provides a preliminary study of stakeholder values in the construction development and analyse their influence towards the success of the adoption of green procurement in Malaysia, by employing paper review technique. The development of green construction in Malaysia is discussed followed by a brief review of the stakeholder values and participation in green procurement adoption. A qualitative method measuring stakeholder values that focus on the green procurement adoption is conducted.

2. Green Procurement

2.1. Construction Procurement

A construction project is known as a complex process whereby the project stakeholders integrate in achieving their goals to construct the facility [10]. According to Cox and Townsend [11], there are differences between the supply of the construction industry that involves engineers, architects, contractors, builders, surveyors, suppliers, and laborers with the demand such as project owners and representatives. Therefore, procurement acts as a link of both the supply and demand of the construction industry.

Procurement is the most important part in a construction project as it incorporates all activities in the construction project such as knowledge, manpower, equipment and materials sourced for management services required in the accomplishment of project objectives [12]. A study initiated by Rowlinson and McDermott [13] indicate that a procurement process should comprise both functional (i.e. contract strategy, client, law, conditions of contract, and performance) and vivid part (i.e culture, sustainability, economics, political environment, learning, leadership, satisfaction, self-esteem and motivation). It is highlighted that procurement has great potential in improving the project management, as each construction project will undergo a procurement phase.

The rule of operation, roles and contractual relations between the parties are defined by the procurement method chosen. According to Ruparathna and Hewage [14], construction procurement represents two segments of the purchasing transaction in obtaining a constructed facility and the process of fulfilling the need for constructing the facility of a potential client, which should be a focus for each study pertaining procurement activities.

2.2. Green Procurement for Construction Industry

According to Ran and Tang [15], green development can enhance nature of life and save energy, with lower maintenance and operational cost. Green construction as part of green development initiative, promotes a concept that is healthy, safe, comfortable and Green concept is a primary theme selection in the current modern construction [16]. Organizations may also be considering the execution of environmental friendly construction activity [17] that include transport, storage and disposal of the purchased materials [18].

Green procurement includes the purchase of products, policies and directives with minimal or positive environmental impacts [9]. The attributes of green procurement may differ with each organization purpose as organizations may consider purchase of materials from green supplier that include environment measures, such as using recycled material [19]. The essential value of green procurement enhances the condition and financial execution of buildings and other developments [20].

IOP Conf. Series: Materials Science and Engineering **429** (2018) 012024 doi:10.1088/1757-899X/429/1/012024

There are two benefits of implementation of green procurement which are i) limiting the negative consequences of environment and surroundings of human being and ii) it helps in obtaining business benefits to the company by reducing environmental handling costs [18].

According to green procurement practices, the environmental criteria should be considered from the early stage of planning [8] and tender assessment where the evaluation will be conducted during the progress of procurement [21]. This procedure can assist client to control the use of energy, wastage, carbon footprint and minimize risk to the environment [22]. Green procurement described by Bouwer et. al. [23], where it integrates every stages of procurement process with environmental criteria to encourage the environmentally friendly products and technologies of construction development, in minimizing impact towards environment. The Ministry of Energy, Green Technology and Water Malaysia (KeTTHA) [24] described the term environmental criteria as the construction process that reduces risk to the environment, enhance the environment condition, have low greenhouse gas (GHS) discharge, preserve the consumption of energy and resources, and encourage the reuse of resources throughout the process.

2.3. *Role of stakeholder in Construction Industry*

Turner and Zolin [25] indicate that the apparent outcomes of the project are affected by the perception of stakeholders towards the project and influence the accomplishment of the goals set by the association [26]. Furthermore, perceptions of stakeholders over project outcomes also change occasionally [25]. Previous study was conducted to identify the influencers of the project that are the stakeholders that portray the relationship between these parties and their designate concern [27]. It is essential to identify the stakeholder interests, assessing their interactions and measures their influence towards the construction project [28]. In order to act, they need to be aware on the significance of the sustainable construction and its need [5].

The study by Carmelli and Tishler [29] shows that values are considered as an influence to stakeholders work-related behaviour. As stated by Olander and Landin [30], construction project implementation can be obstructed by the negative attitude of the stakeholders. Loch and Kavadias [31] support this notion whereby based on their observations, the project failed due to the unstated issues among the project stakeholders. Stakeholder values implicit depends on their expectations on the project that worship to their commitments [32].

In the construction context, green performance depends on stakeholder values whereby their commitment indicate a huge impact to the green performance [9]. Study by Eskerod and Huemann [33] shows that it is necessary to identify and manage the stakeholder values in sustainable construction as a future demand in green construction development. Distinctive stakeholders in construction projects are Sustainability Consultant, Contractor, Employee, Client, Engineers, Trade Subcontractor, Archaeologist, Development Manager, Local Government, Design Coordinator, Regulatory Agency, Managing Director, Technical Director, Conservationist, Environmentalist, Project Manager, Area Manager, Material Supplier, Subcontractor, Architect and Quantity Surveyor and other specialist consultants [34].

3. Methodology

In accordance with the objective of this paper, the literature review was done in depth to establish some knowledge pertaining to the green procurement and relevant background in Malaysia. The first part of the literature review focused on the nature of procurement that include the project participant involvement and scope of works. The second part explored in depth the current green procurement practice, issues and role of stakeholders in green procurement adoption. A thorough literature review was conducted to analyse the stakeholder values that can assist in the development of green procurement in green construction process. For the completion of literature review, information was obtained from several sources such as published books, textbooks, articles in journals, technical papers, guideline and procedure available at the respective practice in the industry, information from website and some others published and unpublished research work. The keywords used in searching the article are "green procurement", "sustainable procurement" and "green purchasing". These key word used because the term green and sustainable is interchangeable. A review of the literature identifies and highlights the important variables and documents that provide significant findings from that will serve as a foundation on which theoretical framework for the current investigation can be built.

4. Analysis and Findings

According to Bullen and Love [35], green building implementation is unsafe due to the traditionalist behaviour of the stakeholders in the construction industry. In order to ensure the success of green construction development, there is a need for an organized action that involves all relevant stakeholders in the building sector starting from designers, architects, workers, investors, legal representatives of governmental bodies and decision makers, to the final users. Thus, green procurement will not only lay out the framework but also incorporates the structure of responsibilities and authorities of each stakeholder for green construction development [8]. Based on extensive LR conducted, summary of stakeholder values categorization is listed in Table 1. Five main categories of stakeholder values are identified, which are knowledge, awareness, capability, affordability and commitment.

TABLE 1. Stakeholder values categorization by varies authors

Stakeholder values	Authors	Statement
Knowledge	[36]	• Unfamiliarity with green technologies. A project manager has to deliver the project with the required performance specified by the client (Ling, 2003), and unfamiliarity with the performance of green technologies may affect the performance outcome.
	[37]	• Lack of knowledge on green technology and materials as barriers to green construction
	[2]	• The lack of knowledge on green technology and the durability of green materials is a significant barrier preventing the construction industry from implementing the strategies and specifying green construction.
Awareness	[37]	• Regional ambiguities in the green concept • Conflicts in benefits with competitors • Dependence on promotion by government
	[8]	• The stakeholder is not guided by any specific definition but more from the need for compliance
	[38]	• The need for improved awareness of green building incentives among owners
Affordability	[39]	• Affordability was the biggest barrier to sustainable construction and would indicate that sustainable construction is more expensive to execute compared to standard practices.
	[2]	• The utilization of green techniques such as high-performance insulation protection, water and energy saving equipment often increase the capital cost.
	[40]	• The extra cost occurred is the fundamental barrier for implementing green construction
	[41]	• Green materials are often expensive and the conflict of interests among stakeholders can result in the uncertainties and inadequate trust relationship

	[42] [43] [44]	• the greatest obstacle to green construction is the higher investment cost that may incur and the risk of unforeseen cost
Commitment	[39]	• Lack of client demand as a second barrier to sustainable construction
	[8]	• The current challenge is to shift the status quo of current practices.
	[45]	• The inclination of designers to adopt financial and time-effectiveness criteria for environmental elements during the design process has hindered green procurement development
Capability	[2]	• Immature green supply chain and a lack of trust on suppliers. It indicates that an integrated industrial supply chain is crucial for the success of green construction. • Any degradation of aesthetic appearance derived from the adoption of green construction technologies was a concern to most stakeholders.

Stakeholder knowledge is the main factor contributing to green procurement adoption. Lack of knowledge in terms of green construction will lead to failure in promoting this green procurement initiative [46]. Suggested by Bohari et al. [8], they need to understand the basic knowledge pertaining green procurement criteria and procedure. For instance, the materials involve in the project and the performance outcome of each technology used in green construction adoption. This is supported by Min and Galle [47] in which the stakeholders need to be familiar with green purchasing, thus education and training should be provided. The training can the procedure for tendering, supplier requirement and material specification.

Awareness among stakeholders pertaining green and sustainability is one of the main factors that help the successful implementation of green initiative. According to Olubunmi [38], awareness can trigger intention to concern with factors that promote the adopting of green initiative by building stakeholders. Awareness can be a push factor that drive the industry to wide-spread the interest to explore the potential of green practices. Thus, this study recognizes awareness among stakeholder as important values for adopting green practices.

Often, green approach requires complicated techniques and construction processes [48]. Thus, the transition towards green construction demands the organization to have capability in terms of technical knowledge on green practices and approach. Stakeholder education is important to demonstrate the benefits of such green practices [49]. This paper agreed that the clients and other stakeholder as the key driver towards sustainable construction [50]. Eisenberg et al. [51] posed two reasons, as obstacles are insufficient knowledge or technical expertise and unfamiliarity with the products, materials, system, or design. Ling [52] supported by stating that unfamiliarity with the performance of green technologies may affect the performance outcome.

Green construction is recognized to be righteous for the environment, but the stakeholders, especially the developer is not interested as it is known to be high in cost from the material purchasing to the technology used. They believed that it is not worth to invest in green material as it is costlier [53], thus, the reason to the least favourite compared to the conventional method of construction [54]. Apart from that, once decided to adopt green project, the green technology will be involved therefore will lead to higher capital cost as stated by Shi et al. [2]. People also perceive green products to be costly and not worth buying [53]. Financial incentives would help to drive demand from stakeholders.

According to Zollo and Winter [55], commitment is developed when organization had their goals and the individual who work on it share the same belief with that goals. Qi et al. [56] highlighted that top-level management commitment affecting the green practice level of adoption. Besides, Adetunji et al. [57], the client as one of the most important stakeholders normally trigger the needs to shift towards greener approach and demand for projects green requirement. As highlighted by Qi et al. [56], the commitment also must include by the external stakeholder such as the government and community. The research conducted by Zhang et al. [58] shows those pressures from the community play significantly positive roles in engaging firms in improving environmental management performance. Hwang and Ng [36] highlighted two important factors to call for the stakeholder commitment, which shall refer to communication and nurture the interest from all the team members including the supplier and the sub-contractors.

Figure 1. Stakeholder values categorization

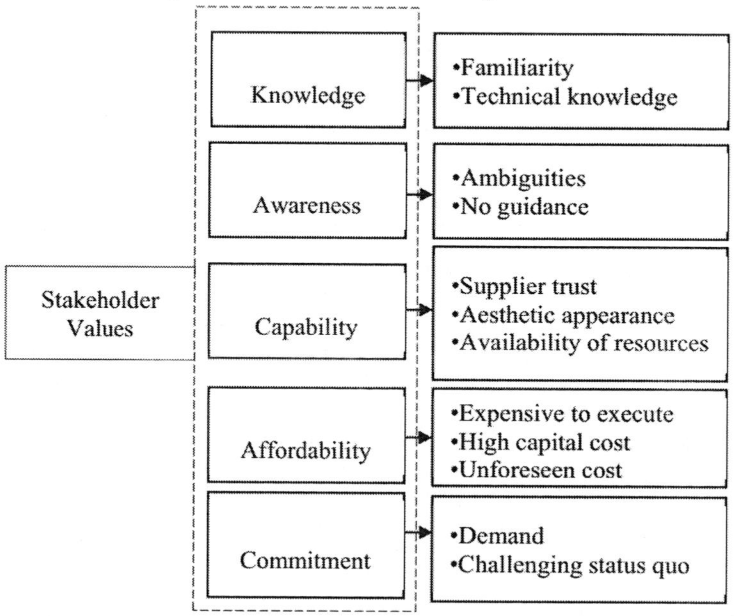

5. Conclusions

Green procurement is regard as an important tool to promote sustainability for construction project. Project stakeholder in other hands, posed as project stakeholder is the key that will determine project success. The influence of stakeholder has the impact on the adoption of green procurement practices. However, currently, project stakeholder values are least being discussed in the literature and motivate this research to be carried out. This paper reviewed possible stakeholder values that deemed to be important and employed thematic analysis, there are five stakeholder values identified. The five stakeholder values that deemed to be important include knowledge, awareness, capability, affordability and commitment Thus, this paper provides a basis to further investigate the impact of stakeholder values on the green procurement practices implementation. The list of stakeholder values gathered in this paper will later be used to develop a survey instrument in the next phase of this research.

References

[1] R Saffuan, J Ariffin, and Z Amin 2012 Planning Park for Liveable Cities: Green Technology Design Practice for Tasik Biru Kundang, Kuang, Selangor, Malaysia *Procedia - Soc. Behav. Sci.* **35** 705–12

[2] Q Shi, J Zuo, R Huang, J Huang, and S Pullen 2013 Identifying the critical factors for green construction - An empirical study in China *Habitat Int.* **40** 1–8

[3] S Kubba 2010 Defining "Green" and "Sustainability" *LEED Pract. Certif. Accredit. Handb.* 1–18

[4] F Li, T Yan, J Liu, Y Lai, S Uthes, Y Lu, and Y Long 2014 Research on social and humanistic needs in planning and construction of green buildings *Sustain. Cities Soc.* **12** 102–9

[5] L Stojanovska-Georgievska, I Sandeva, and H Spasevska 2017 An empirical survey on the awareness of construction developers about green buildings in Macedonia *2017 2nd Int. Multidiscip. Conf. Comput. Energy Sci. Split. 2017*

[6] N Čuš-Babič, D Rebolj, M Nekrep-Perc, and P Podbreznik 2014 Supply-chain transparency within industrialized construction projects *Comput. Ind.* **65** 345–53

[7] G A Zsidisin and S P Siferd 2001 Environmental purchasing: A framework for theory development *Eur. J. Purch. Supply Manag.* **7** 61–73

[8] A A M Bohari, M Skitmore, B Xia, and M Teo 2017 Green oriented procurement for building projects: Preliminary findings from Malaysia *J. Clean. Prod.* **148** 690–700

[9] J K W Wong, J K S Chan, and M J Wadu 2016 Facilitating effective green procurement in construction projects: An empirical study of the enablers *J. Clean. Prod.* **135** 859–71

[10] A S Oyegoke, M Dickinson, M M A Khalfan, P McDermott, and S Rowlinson 2009 Construction project procurement routes: An in-depth critique *Int. J. Manag. Proj. Bus.* **2** 338–54

[11] A Cox, P Ireland, and M Townsend 2018 *Managing in Construction Supply Chains and Markets Reactive and Proactive Options for Improving Performance and Relationship Management*

[12] P Eskerod and T Larsen 2018 Advancing project stakeholder analysis by the concept 'shadows of the context' *Int. J. Proj. Manag.* **36** 161–9

[13] S Rowlinson and P Mcdermott 1999 *Procurement Systems A Guide to Best Practice in Construction*

[14] R Ruparathna and K Hewage 2015 Review of Contemporary Construction Procurement Practices *J. Manag. Eng.* **31** 04014038

[15] J Ran and M Tang 2018 Passive cooling of the green roofs combined with night-time ventilation and walls insulation in hot and humid regions *Sustain. Cities Soc.* **38** 466–75

[16] A Darko and A P C Chan 2016 Critical analysis of green building research trend in construction journals *Habitat Int.* **57** 53–63

[17] K Otsuki 2011 Sustainable partnerships for a green economy: A case study of public procurement for home-grown school feeding *Nat. Resour. Forum* **35** 213–22

[18] M A Salam 2008 An empirical investigation of the determinants of adoption of green procurement for successful green supply Chain management *Proc. 4th IEEE Int. Conf. Manag. Innov. Technol. ICMIT* 1038–43

[19] Q Zhu, Y Geng, and J Sarkis 2013 Motivating green public procurement in China: An individual level perspective *J. Environ. Manage.* **126** 85–95

[20] Anon 2006 Status and Development of Green Purchasing in Hong Kong Background of Green Council *Chief Exec.*

[21] L Preuss 2007 *Buying into our future: Sustainability initiatives in local government procurement* vol 16

[22] A H I Lee, H Y Kang, C F Hsu, and H C Hung 2009 A green supplier selection model for high-tech industry *Expert Syst. Appl.* **36** 7917–27

[23] M Bouwer, K de Jong, M Jonk, T Berman, R Bersani, H Lusser, A Nissinen, K Parikka, and P

Szuppinger 2005 Green Public Procurement in Europe 2005 - Status overview *Virage Milieu Manag.* 107
[24] M Ministry of Energy, Green Technology and Water 2009 National Green Policy **1** 22
[25] R Turner and R Zolin 2012 *Forecasting Success on Large Projects: Developing Reliable Scales to Predict Multiple Perspectives by Multiple Stakeholders Over Multiple Time Frames* vol 43
[26] R Freeman and J Mcvea 2001 *A Stakeholder Approach to Strategic Management*
[27] A Gupta 1995 A stakeholder analysis approach for interorganizational systems *Ind. Manag. {&} Data Syst.* 2–8
[28] J Yang, P Q Shen, L Bourne, C M F Ho, and X Xue 2011 A typology of operational approaches for stakeholder analysis and engagement *Constr. Manag. Econ.* **29** 145–62
[29] A Carmeli and A Tishler 2004 *The Relationships between Intangible Organizational Elements and Organizational Performance* vol 25
[30] S Olander and A Landin 2005 Evaluation of stakeholder influence in the implementation of construction projects *Int. J. Proj. Manag.* **23** 321–8
[31] C Loch and S Kavadias 2011 *Implementing strategy through projects*
[32] B Cornell and A C Shapiro 1987 Corporate Stakeholders and Corporate Finance *Financ. Manag.* **16** 5–14
[33] P Eskerod and M Huemann 2013 Sustainable development and project stakeholder management: what standards say *Int. J. Manag. Proj. Bus.* **6** 36–50
[34] M Bal, D Bryde, D Fearon, and E Ochieng 2013 Stakeholder Engagement: Achieving Sustainability in the Construction Sector *Sustain.* **5** 695–710
[35] P A Bullen and P E D Love 2010 The rhetoric of adaptive reuse or reality of demolition: Views from the field *Cities* **27** 215–24
[36] B G Hwang and W J Ng 2013 Project management knowledge and skills for green construction: Overcoming challenges *Int. J. Proj. Manag.* **31** 272–84
[37] P T I Lam, E H W Chan, C K Chau, C S Poon, and K P Chun 2009 Integrating Green Specifications in Construction and Overcoming Barriers in Their Use *J. Prof. Issues Eng. Educ. Pract.* **135** 142–52
[38] O A Olubunmi, P B Xia, and M Skitmore 2016 Green building incentives: A review *Renew. Sustain. Energy Rev.* **59** 1611–21
[39] M Pitt, M Tucker, M Riley, and J Longden 2009 Towards sustainable construction: Promotion and best practices *Constr. Innov.* **9** 201–24
[40] G Ofori and H L Kien 2004 Translating Singapore architects' environmental awareness into decision making *Build. Res. Inf.* **32** 27–37
[41] H Li, E W L Cheng, and P E D Love 2000 Engineering , Construction and Architectural Management Article information : To cite this document : Users who downloaded this article also downloaded : About Emerald www.emeraldinsight.com Partnering research in construction *Constr. Archit. Manag.* **7** 76–92
[42] L B Robichaud and V S Anantatmula 2010 Greening project management practices for sustainable construction **27** 48–57
[43] T Häkkinen and K Belloni 2011 Barriers and drivers for sustainable building *Build. Res. Inf.* **39** 239–55
[44] N Zainul Abidin Nazirah 2010 Investigating the awareness and application of sustainable construction concept by Malaysian developers *Habitat Int.* **34** 421–6
[45] F Testa, E Annunziata, F Iraldo, and M Frey 2016 Drawbacks and opportunities of green public procurement: An effective tool for sustainable production *J. Clean. Prod.* **112** 1893–900
[46] A Sourani and M Sohail 2011 Barriers to addressing sustainable construction in public procurement strategies *Proc. Inst. Civ. Eng. - Eng. Sustain.* **164** 229–37
[47] H Min and W P Galle 2001 Green purchasing strategies of U.S firms *Int. J. ofOperations Prod. Manag.* **21** 1222–38
[48] L Shen, Z Zhang, and Z Long 2017 Significant barriers to green procurement in real estate

development *Resour. Conserv. Recycl.* **116** 160–8

[49] A Abuzeinab, M Arif, M A Qadri, and D Kulonda 2018 Green business models in the construction sector: an analysis of outcomes and benefits *Constr. Innov.* **18** 20–42

[50] S Renukappa, A Akintoye, C Egbu, and S Suresh 2016 *Sustainable procurement strategies for competitive advantage: An empirical study* vol 169

[51] D Eisenberg, R Done, and L Ishida 2002 Breaking down the barriers: challenges and solutions to code approval of green building *Dev. Cent. Appropr. Technol.* 32

[52] J Ung Ling 2003 *The project manager's personal characteristic, skills and roles in local construction industry*

[53] S Brammer and H Walker 2011 *Sustainable procurement in the public sector: An international comparative study* vol 31

[54] M Mosgaard, H Riisgaard, and R Dorothea Huulgaard 2013 *Greening non-product-related procurement - When policy meets reality* vol 39

[55] M Zollo and S G Winter 2002 Deliberate Learning and the Evolution of Dynamics Capabilities *Forthcom. Organ. Sci.* 47

[56] G Y Qi, L Y Shen, S X Zeng, and O J Jorge 2010 The drivers for contractors' green innovation: An industry perspective *J. Clean. Prod.* **18** 1358–65

[57] I Adetunji, A Price, P Fleming, and P Kemp 2003 Sustainability and the UK construction industry—a review *Proc. Inst. Civ. Eng. - Eng. Sustain.* **156** 185–99

[58] D liang Zhang, G Huang, Y Xu, and Q Gong 2015 *Waste-to-Energy in China: Key Challenges and Opportunities* vol 8

IOP Conf. Series: Materials Science and Engineering **429** (2018) 012025 doi:10.1088/1757-899X/429/1/012025

Assessment of Carbohydrate Contents in Perlis Harumanis Mango Leaves during Vegetative and Productive Growth

N. H. Hazis[1], A. A. Aznan[1] , M. N. aafar[2], F. A. Azizan[1], R. Ruslan[1], I. H. Rukunudin[1]
[1]School of Bioprocess Engineering, Universiti Malaysia Perlis, Perlis, Malaysia.
[2]Institute of Sustainable Agrotechnology, University Malaysia Perlis, 02100 Perlis, Malaysia.

*Email: aimiathirah@unimap.edu.my

Abstract - This study was aim to assess one of the flower inducing parameter of mango plant for further research and development to produce premium Perlis Harumanis mango in off-season. The carbohydrate content in mango leaves were known as the flower inducing parameter but there was no scientific research on the concentration variation of carbohydrates especially for Perlis Harumanis mango grown in the greenhouse. In this study, the elements of carbohydrate content such as glucose and starch concentration level in the leaves were analyzed using UV spectrophotometer observed during the vegetative and productive growth. The results of the study showed a higher concentration value of glucose and starch during productive growth. Meanwhile for the total carbohydrate content, the highest concentration value was obtained during the vegetative growth.

1. Introduction

Mango (*Mangifera Indica*) is one of the famous fruits grown in tropical and sub-tropical area especially in South East Asia countries. Malaysia is known as one of the mango producer and one of the iconic mango is known as the Perlis Harumanis mango. The Perlis Harumanis mango is very popular because of its aroma, rarity and sweet rich taste. The local demand for Perlis Harumanis mango is very high as the fruits usually fully booked even before it is ripen especially on the beginning of its season. The increasing demand for Perlis Harumanis mango also has been proven by the agreement on exporting the premium quality Perlis Harumanis mangoes into Japanese market by both government of Malaysia and Japan [1]. As the Perlis Harumanis mango is a seasonal fruit, thus the off-season fruit production may contribute to the high potential of extra income for the farmers as well as to the local community who involved in the agro-tourism sector.

The vegetative growth period for Perlis Harumanis mango trees is approximately from July to December. The productive growth period usually occurs on January to February. Meanwhile, during March to April, the Perlis Harumanis mango trees started to bear fruits and the harvesting season will started on April until June. The pre-flowering growth period is recognized as one of major stages in the physiologic phenomenon of a mango plant. The growth of Perlis Harumanis mango tree can be affected by the environmental and biological factors. It is very crucial that the final stems of the trees are let to be break for a period of time in the dormancy period after the last vegetative development so that the new reproductive shoots can be generated. This dormancy period is crucial to provide enough time for the stem to develop the perfect whorl, length and diameter [2].

During the vegetative and productive growth, flowering period is the transition phase from the vegetative to the productive growth of a tree [3]. Carbohydrates is the energy source that is the product of photosynthesis process are stored in different parts of the trees such as in the leaves. The higher level of carbohydrates concentration is desired during the floral development [4]. Thus, the understanding of carbohydrate content variation in the leaves is important to develop a guideline of floral induction procedure for the local farmers but there is lack of scientific report on the variation of carbohydrate content in the leaves especially for Perlis Harumanis mango trees. Thus, the objective of this study was to assess the variation in carbohydrate content in the Perlis Harumanis mango leaves during vegetative and productive growth.

2. Materials and Method

2.1 Study Area
This study was carried out in a greenhouse at Institute of Sustainable and Agrotechnology (INSAT), Universiti Malaysia Perlis (UniMAP), Padang Besar, Perlis. The greenhouse is planted with 212 Harumanis mango trees located at the GPS location of 6°39'45.439N and 100°19'17.994''E.

2.2 Leaves sample collection
The leaves samples were collected from 21 healthy branches from Harumanis mango trees where 9 branches were in a vegetative growth and 12 branches from productive growth in 4 weeks. The leaves sample for vegetative growth was collected from four different stages of vegetative growth and was labelled as V1, V2, V3 and V4. For productive growth stage, the leaves samples were collected at a flowering branch started from budding until the flowers were fully bloomed and was labelled as F1, F2, F3, and F4. The sample images of each stage for vegetative and productive growth is as shown in Figure 1 (a-b). All of the samples were sealed in plastic bags to be analysed in the laboratory. Table 1 shows the number of collected leaves samples for vegetative and productive growth in the greenhouse. In the laboratory, the leaves samples were composited with the respective growth stage were dried in the oven at 60°C for 24 hours. The dried samples were ground into powder form and kept in a cold room at 4°C.

Figure 1 (a b) Stages in vegetative and productive growth

Table 1 Leaves samples for vegetative and productive growth collected inside the greenhouse for 4 weeks

Growth	Stage	Number of Sample
Vegetative	V1	3
	V2	12
	V3	15
	V4	6
Reproductive	F1	3
	F2	12
	F3	18
	F4	15

2.3 *Determination of Total Carbohydrate Content*

The total carbohydrate content was determine using anthrone method. Leaves sample of 0.05g was added into a boiling tube. It was hydrolysed in a boiling water bath for 3 hours using 5mL 2.5 N hydrochloric acid. After 3 hours, it was let to be cooled to room temperature. Then, the solution was neutralised with solid sodium carbonate until the bubble formed. The solution was added with distilled water until the volume reached 50mL before it was centrifuged for 10 min. Then, the supernatant were collected and 1mL of the aliquot were taken for analysis. Meanwhile, the standard glucose was prepared by adding 0.1g of glucose in 100mL water as stock solution. The working standard was made from 10mL of stock solution that was diluted to 100mL water. The standard was prepared by taking 0, 0.2, 0.4, 0.6, 0.8 and 1mL of the working standard and let '0' serves as blank. The volume was made up to 1mL in all the tubes including the sample tubes by adding distilled water. Then, 4mL of Anthrone reagent was added into the tubes and was heated for 8 minutes in a boiling water bath. After that, the solution were cooled rapidly before it was placed in the UV spectrophotometer (Thermo Fisher Scientific Orion AquaMate 7000) and the light absorption was recorded at 630nm [4].

A standard graph was developed by plotting the concentration of the standard glucose on the X-axis versus absorbance on the Y-axis in order to determine the concentration of glucose of each test sample. From the graph, the amount of carbohydrate present in the sample tube was calculated using Equation 1:

$$\begin{aligned} &Amount\ of\ carbohydrate\ content\ present\ in\ 100\ mg\ of\ the\ sample \\ &= \frac{mass\ of\ glucose\ (mg)}{Volume\ of\ test\ sample} \times 100 \end{aligned} \qquad \text{(Equation 1)}$$

2.4 *Determination of Glucose and Starch Content*

To determine the total glucose and starch content in Perlis Harumanis mango leaves, 1g of finely ground sample was put into 50mL centrifuge tube and 10 mL of hot 80% ethanol were added. The tubes were let to vortex for 2 min and it was centrifuged at 3000 rpm for 10 min in a bench top centrifuge. After centrifuging, the supernatant were decanted. These steps were repeated until the solution did not give any changes in colour when tested with anthrone reagent. As the solution gave no reaction in term of colour changes the residue was cooled in ice water. A 6.5 mL of 52% perchloric acid were added to the residue while stirring the content and then it was centrifuge at 4°C in refrigerated centrifuge. The supernatant were used and made up the volume to 100mL with distilled water. The 0.2mL supernatants were pipetted out and the volumes were made up to 1 mL with distilled water for sample solution. The standards glucose solutions was prepared by taking 0.2, 0.4, 0.6, 0.8 and 1mL of the working standard and the volume were increased to 1 mL with water in each tube. A 4 mL anthrone reagent were added into each tube and were heated up for 8 min in a boiling water bath. After that, the tubes were cooled under running tap water and the sample was read at 630 nm light absorption using the UV spectrophotometer. A standard curve was prepared to determine the glucose content in each samples. The obtained value was calculated using Equation 2 and 3 to convert the glucose into starch concentration value [5].

$$\begin{aligned} &\text{starch value in sample solution} \\ &= 0.9 \times \text{glucose absorbance of sample} \end{aligned} \qquad \text{(Equation 2)}$$

$$\begin{aligned} &\text{starch value in 1 g of powdered sample} \\ &= \text{1g of sample} - \text{starch value in sample solution} \end{aligned} \qquad \text{(Equation 3)}$$

3. Results and iscussion

Total carbohydrate content, glucose and starch level for Perlis Harumanis mango leaves consisted of 84 samples based on different stages of vegetative and productive growth. Figure 2 and Figure 3 illustrates the average of total carbohydrate content, glucose and starch concentration levels for vegetative and productive growth respectively.

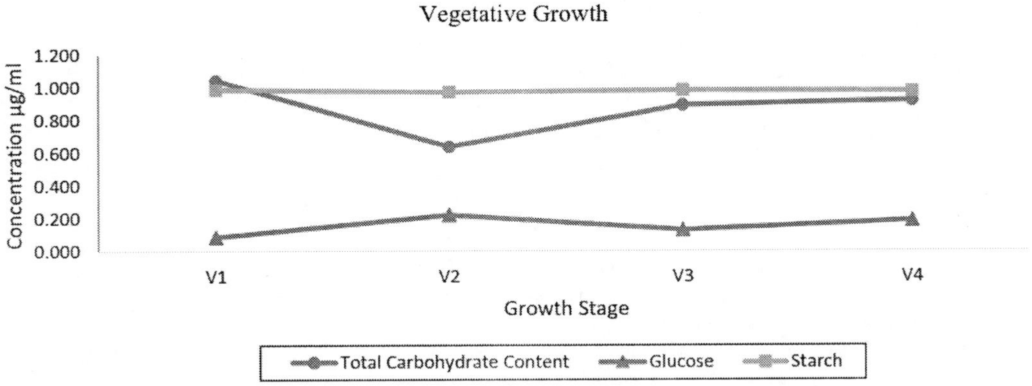

Figure 2 Total carbohydrate content, glucose and starch concentration levels at vegetative growth

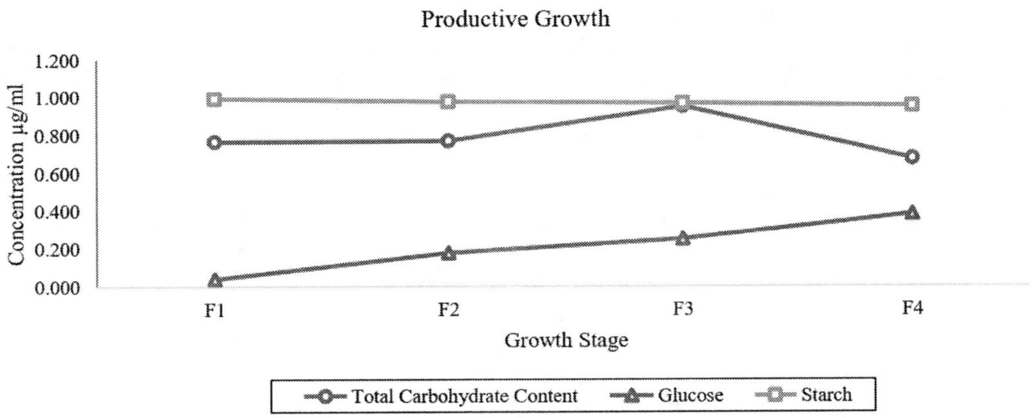

Figure 3 Total carbohydrate content, glucose and starch concentration levels at productive growth

The total carbohydrate content during the vegetative growth was decreased at V1 to V2 and steadily increased at V2 to V4. Meanwhile for productive growth, the total carbohydrate content were steadily increased at F1 to F3 and decreased back from F3 to F4. The highest concentration level of total carbohydrate content for vegetative and productive growth was 1.042 µg/ml and 0.950 µg/ml respectively. The total carbohydrate content was assumed to be high during the vegetative growth. As supported by Helaly [6], the high level of differences in carbohydrate content may be caused by the water stress in plant thus it show the changes of carbohydrate content throughout the stages in vegetative growth.

The glucose concentration during vegetative growth increased from V1 to V2 and decreased back from V2 to V3. However, the glucose values started to increase back from V3 to V4. Meanwhile, for productive growth, the glucose values showed a steady increment from F1 until F4. From the graph, the highest glucose values were obtained during productive growth which was 0.383 µg/ml as compared to 0.220 µg/ml in vegetative growth. These differences may be due to the movement of sugars from leaf to the tissue as the result from requirement of higher photosynthesis activity process as the trees require more energy during productive growth. The findings was conform with the study conducted by Sandip [7], where the high energy consumption are required in the form of high soluble sugars and it was important in floral activity during productive growth.

The starch concentration trend in Figure 2 does not showed significant different for both vegetative and productive growth. The highest starch concentration obtained in productive and vegetative growth were at the concentration of 0.991 µg/ml and 0.986 µg/ml respectively. The starch value in the productive growth is higher as compared to vegetative growth although there a small difference in the concentration value. It is may due to the reduction in photosynthesis rate [8].

. Conclusion

This study showed that there were differences of concentration level for carbohydrate, glucose and starch content in Perlis Harumanis mango leaves. The finding of this study provides a reference for further research to identify the possible flower inducing parameter for off-season Harumanis mango production in the greenhouse.

References

[1] F A S Syahir, A y m Shakaff, A Zakaria, M Z Abdullah and A H Adom 2012 Bio-inspired vision fusion for quality assessment of harumanis mangoes *Proc. - 3rd Int. Conf. Intell. Syst. Model. Simulation, ISMS 2012* 317–324

[2] R S M Farook *et al* 2013 Harumanis Mango Flowering Stem Prediction using Machine Learning Techniques *Res. Notes Inf. Sci.* **13** 46–51

[3] P Shivakrishna, K Ashok Reddy and D Manohar Rao 2016 Effect of PEG-6000 imposed drought stress on RNA content, relative water content (RWC), and chlorophyll content in peanut leaves and roots *Saudi J. Biol. Sci.* 8–12

[4] J E Hedge and B T Hofreiter 1962 Carbohydrates chemistry *Biochem. Methods* 19

[5] K Mesa, S Serra, A Masia, F Gagliardi, D Bucci and S Musacchi 2016 Seasonal trends of starch and soluble carbohydrates in fruits and leaves of 'Abbé Fétel' pear trees and their relationship to fruit quality parameters *Sci. Hortic. (Amsterdam).* **211** 60–69

[6] M N Helaly, H El-Hoseiny, N I El-Sheery, A Rastogi, and H M Kalaji 2017 Regulation and physiological role of silicon in alleviating drought stress of mango *Plant Physiol. Biochem.* **11** pp. 31-44

[7] M Sandip, A N Makwana, A V Barad and B D Nawade 2015 Physiology of Flowering- The Case of Mango **1** 11 1008–1012

[8] P Wu, C Wu and B Zhou 2017 Drought Stress Induces Flowering and Enhances Carbohydrate Accumulation in Averrhoa Carambola *Hortic. Plant J.* **3** 2 60–66

IOP Conf. Series: Materials Science and Engineering 429 (2018) 012026 doi:10.1088/1757-899X/429/1/012026

Groundwater Assessment Using Geophysical Survey at Insat, Perlis, Malaysia

Azizan F. A.[1*], **Aznan A. A.**[1], **Ruslan R.**[1], **M. Nazari M.** [1], and **Jaafar M. N.**[2]

[1]Faculty of Chemical & Energy Engineering, Universiti Malaysia Perlis, Malaysia
[2]Institute of Sustainable Agrotechnology (InSAT), Perlis, Malaysia.

*Email: fathin@unimap.edu.my

Abstract. In this study, a 2D geo-electrical resistivity survey was conducted at Institute of Sustainable Agrotechnology (InSAT) Padang Besar, Perlis to locate potential groundwater zone. A set of ABEM Terrameter SAS 400 was used to measure resistivity and chargeability value of 400-meter survey line. Then, the measured data was analysed using RES2DINV software to produce resistivity and chargeability imaging profiles. It was found that, there is one groundwater reservoir detected at a distance of 180 m from left end at distance 0 m. The groundwater reservoir zone was located at 50 m depth below ground surface. The resistivity and chargeability values of this zone varied between 500 to 5000 Ωm and 0 to 2 ms, respectively.

1. Introduction

Sufficient water supply is a vital process throughout the growing stages in all plants. Even though Malaysia has high annual rainfall of approximately 2600 millimeters, which is above the global average, the distributions of the rainfall are uneven [1]. This has caused certain areas to experience limited water supply problem. In addition to that, there are many inter-related changes to the weather due to climate change. Thus, it affects the pattern of rainfall and vegetation system. The changes in amount and distribution of water especially may modify the environment [2].

In Malaysia, surface water is the main source for fresh water supply including for irrigation. This source however is limited due to above problems. The increasing in water demand and pollution of this source also contributes to the problems. Therefore, a clean, low cost and yet sustainable alternative should be investigated. Thus, the most suitable and readily available alternative source to the surface water is groundwater.

Groundwater forms as an alternative source to fulfill the water demand specifically in the states of Kelantan and Perlis. In 2011, the statistic showed that one-third of the water supply in Perlis comes from groundwater sources [3]. Groundwater in Malaysia is an important resource that is yet to be exploited on a bigger scale to meet the increasing demand for various uses. The utilization of groundwater can also help to solve water shortage in the areas where surface water is limited.

Nowadays, groundwater has become an important source of water to meet the increasing requirement for domestic, industrial and agricultural needs. With the new technology on geophysical technique approach, Malaysia will be able to tap this resource effectively and in sustainable manner, and at the same time be able to protect it from any adverse effect [4].

In order to explore and determine the potential location of this alternative resource, geophysical technique approach was applied in this study. The geophysical technique such as electrical resistivity survey has increased the possibility to find the location of groundwater resources [5-7]. The resistivity imaging profile of the survey line represents the subsurface profile. This subsurface profile shows the geological formation of that particular area which may then indicate the suitable layers that hold groundwater. Therefore, this study was conducted to detect the potential groundwater zone based on resistivity imaging profile using geo-electrical resistivity method.

2. Material and Materials

2.1. Study Area

This study was carried out at Institute of Sustainable Agrotechnology (INSAT), Universiti Malaysia Perlis. The area covers 300,000 m2 which consists of a main office, laboratories and more than 50 greenhouses. The study location lies on latitude and longitude of 6°39′15.25 ″N and 100°15′50.37″E, respectively and was marked 'X' as in Figure 1. The area is relatively flat at 53m above mean sea level. The average annual temperature and precipitation of study area is 32°C and 1869 mm, respectively.

Figure 2 shows geological map of Perlis. Generally, the state composed of six geological formations; Kubangpasu Singa, Setul, Chuping Limestone, Alluvium, Granite and Bukit Arang. Based on this map, the study area is composed of a Kubang Pasu - Singa formation. The Kubang Pasu – Singa formation consists of shale, siltstone and sandstone. There is also slightly thin layer of chert. Arenitic rocks mostly have poor sorted such as expansive rocks and can also be composed of archos and sub-archos [8]. Argillite rocks consist of shale of various colors, from dark gray, gray, red to brown. The laminate in the shale is clear. In addition to shale, there are also siltstone. In the northeastern part of Perlis, Yap [8] divided the sequence of the Kubang Formation to three facies; Passage beds facies, Dominant sandstone facies, and Argilitary rock facies.

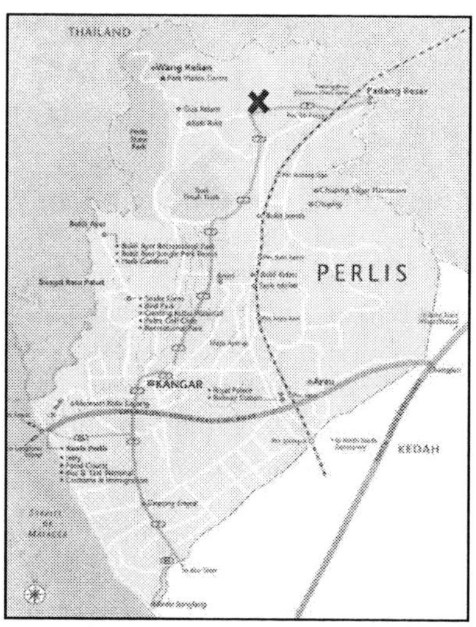

Fig. 1. Study location

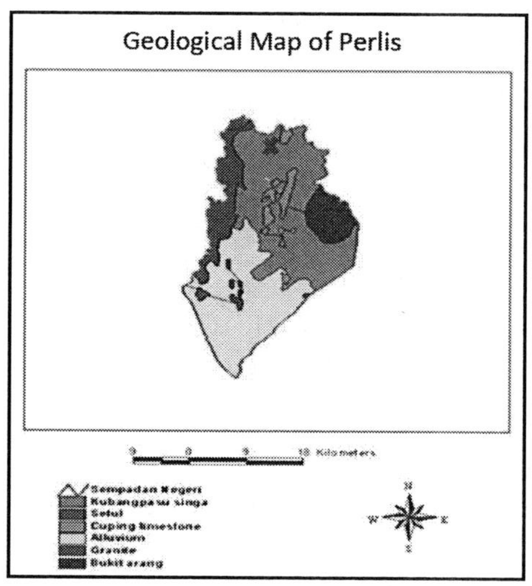

Fig. 2. Geological Formation of Study Area [9]

2.2. Geophysical Method

A 2D geo-electrical method that consists of electrical resistivity and induced polarization survey were conducted using ABEM Terrameter SAS4000 equipment set. A resistivity survey line was spread over the length of 400 m as suggested in the instruction manual [10]. Figure 3 shows field arrangement of resistivity survey while Figure 4 shows exact location of the survey line in study area. The line started from point A to point C (where B is a centre) extending from south-east to west-east direction.

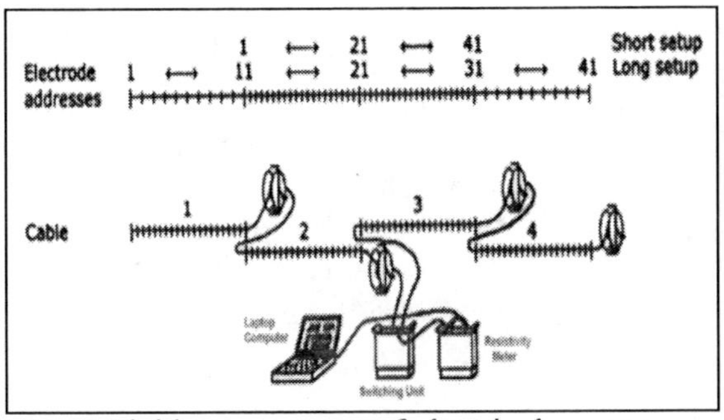

Fig. 3. Field arrangement of electrical survey conducted [9]

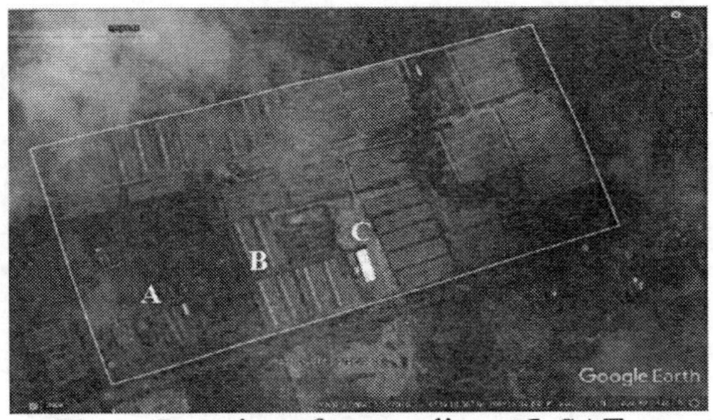

Fig. . Location of survey line at InSAT

A total of 41 electrodes were used in the survey with electrode spacing of 5 m. Pole-dipole array was selected in this study as it has relatively good horizontal coverage [11]. In addition, the array has a significantly higher signal strength as compared to dipole-dipole array and it is not sensitive to telluric noise as the pole-pole array. The induced polarization and resistivity data were generated and recorded automatically by the resistivity meter. These data later were extracted and processed using RES2INV software [12] to convert the apparent resistivity data to true resistivity by inversion method. The resistivity meter measures apparent resistivity from which pseudosections were developed and subsequently inverted to true resistivity 2D sections [13].

In this survey, the focus was directed to the contrast of the resistivity values which is expected to reflect the geological structures. The chargeability values also have been analysed for the determination of an expected groundwater potential zone at the study area.

3. Results and Discussions

Figure 5 and 6 present the respective resistivity and chargeability profiles of study area. The resistivity values obtained were ranging from 0 Ωm to 10,000 Ωm while the chargeability values were ranging from 0 ms to 500 ms. The maximum depth of imaging profile is 135.5 m beneath ground surface. It can be seen that, the resistivity patterns in Figure 5 can be divided into two sections; A-B and B-C. This is due to different pattern appeared. The trend clearly shown resistivity values increased with the increased of depth for A-B cross section resistivity profile. In addition, the A-B resistivity profile was identified with three underlying zones. The zones are Zone 1 or top layer, Zone 2 or intermediate layer and Zone 3 or bottom layer. However, the three underlying zones was overlapped by zone 4.

It is believed that, zone 1 composed of clay soil as the resistivity values was ranging from 10-100 Ωm. Palacky [14] has concluded that the clay soil has resistivity within the same range. This finding has been supported by Keller and Frishcknecht [15] and Daniels and Albert [16]. Their study have found that clay soil falls within the resistivity range of 1-100 Ωm. Other study also concluded to the same interpretation [17]. Meanwhile, Zone 4 is expected to be the zone of true residue soil due to low resistivity value. It might be clay soil that extending from zone 1 or other rocks. It is difficult to confirm this geological body as resistivity values decreases when the rocks are saturated with water. Zone 2 and 3 were observed to be composed of sedimentary rocks due to high resistivity values. Zone 2 was expected to compose of moderate weathered rocks (100-500 Ωm) while zone 3 was possibly compose of completely into highly weathered rocks (500-5000 Ωm). Based on the resistivity range, the rocks might be shale, siltstone or/and sandstone. This assumption was aligned with geological map of Perlis as in Figure 2 as the formation of this study area is Kubang Pasu – Singa Formation. Keller and Frishcknecht [15] indicated that resistivity values for shale and sandstones are within 20 - 2000 Ωm and 8 - 4000 Ωm, respectively. In addition, Telford et al [18] proposed the range of siltstone to be in between 1.5x104 to 5.6x108 Ωm depending on water content.

From chargeability imaging profile in Figure 6, half of the profile area on the right side is consists of materials that have a very high chargeability measurement. The measurement was ranging between 10 ms and 500 ms and represented by the colour scheme between brown to purple. The high chargeability reading is probably due to highly contaminated area [19-20]. However, through the chargeability profile of Figure 6, there is a possibility of freshwater reservoir that might presents. The area is suspected to contain groundwater due to its low chargeability values [18]. The area is on left side of Figure 6 filled with greenish-blue colour. The chargeability values of this area were ranged between 0 – 2 ms. The expected freshwater reservoir can be found at 180 m from point A with a depth of 50 m below ground surface. Based on groundwater zone that have been detected, a water well can be constructed in the suggested location later.

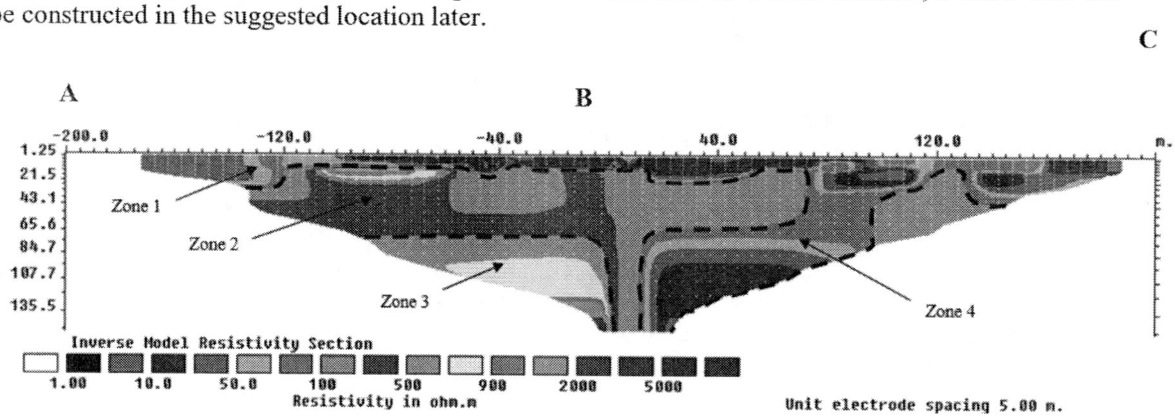

Fig. . Resistivity imaging analysis result

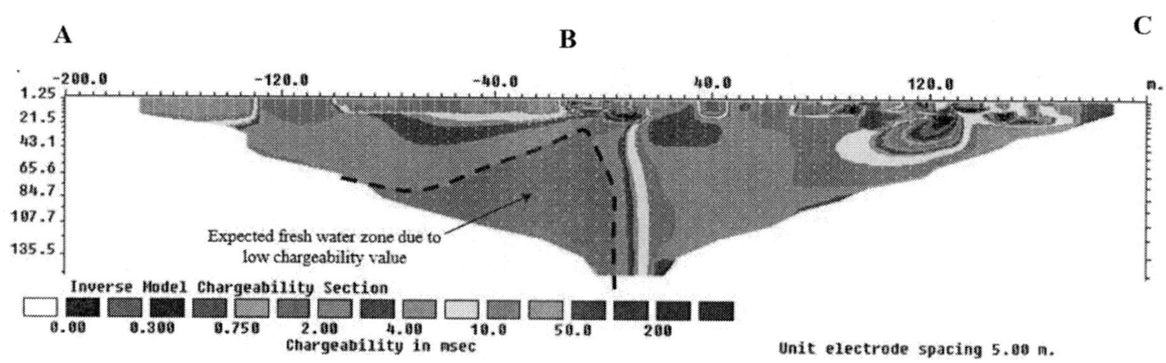

Fig. . Chargeability imaging analysis result

4. Conclusions

In this study, the results showed that geo-electrical resistivity was proven as a successful method as it is widely applied in finding the potential groundwater zone. It was found that, the area consists of three layers with overlapping formation of carbonate rocks. The three layers were identified as clay soil, moderate weathered rocks, and highly weathered rocks respectively based on resistivity value. The chargeability image revealed that, there is one potential zone was detected to contain groundwater based on low chargeability value. The zone was located 180 m from point A at 50 m below ground surface. The results obtained have laid an important platform from which groundwater well can be constructed at suggested zone of groundwater later.

Acknowledgements
The author would like to acknowledge support from the Short Term Grant under a grant number of 9001-00556, Universiti Malaysia Perlis (UniMAP).

References

[1] M Telipot 1998 Pertanika *Journal of Science and Technology* **6** 1 59 – 70
[2] Z A Mokhtar, A S Yahaya, F Ahmad, S Suri and M H Halim 2014 Trends for daily rainfall in Northern and Southern Region of Peninsular Malaysia *Journal of Civil Engineering Research* **4** 3A 222-227
[3] M I A Kamal 2011 Identification of Groundwater Prospective Zones in Perlis, Malaysia using Remote Sensing and Geographic Information System
[4] A K M Hatta, M A Azmer, A B Rashid 2000 An Overview of the Status of Groundwater Utilisations in Peninsular Malaysia Ministry of Natural Resources and Environment
[5] K M Adel 2016 Egypt Arabian Journal of Geosciences **9** 11 1-9
[6] D R Rahmawati, Supriyadi, N P Aryani, M A Naufal 2018 Groundwater potential prediction by using geoelectricity method a case study in Simpang Lima and around it *Journal of Physics: Conference Series* **983**
[7] Z A M Hazreek, A G A Raqib, M Aziman, A T S Azhar, A T M Khaidir, Y M Fairus, S Rosli, I M Fakhrurrazi and R A Izzaty 2017 Preliminary Groundwater Assessment using Electrical Method at Quaternary Deposits Area *IOP Conference Series: Materials Science and Engineering* **226**
[8] K F Yap 1991 Geologi Am Kawasan timur laut Perlis Universiti Kebangsaan Malaysia
[9] W I Ruslan 2008 Timah Tasoh Reservoir, Perlis, Malaysia
[10] ABEM Terrameter SAS 4000 /SAS 1000 Instruction Manual 2010
[11] Samouëlian, I Cousin, A Tabbagh, A Bruand and G Richard 2005 Electrical resistivity survey in soil science: A review *Soil and Tillage Research* **83** 2 173-193
[12] M H Loke, I Acworth, T Dahlin 2003 A comparison of smooth and blocky inversion methods in 2D electrical imaging surveys *Exploration Geophysics* **34** 182–187
[13] M H Loke, R D Barker 1996 Rapid Least-Squares Inversion of Apparent Resistivity Pseudosections by a Quasi-Newton Method *Geophysical Prospecting* **44** 131–152
[14] G J Palacky 1988 Investigations in Geophysic **3**
[15] G V Keller, F C Frischknecht 1966 Electrical Methods In Geophysical Prospecting Pergamon Press Inc. Oxford
[16] F Daniels, R A Alberty 1996 Physical Chemistry John Wiley and Sons Inc.
[17] M H Z Abidin, R Saad, F Ahmad, D C Wijeyesekera and M F T Baharuddin 2014 Correlation Analysis Between Field Electrical Resistivity Value (ERV) and Basic Geotechnical Properties (BGP) *Soil Mechanics and Foundation Engineering* **51** 3 117-125
[18] W Telford, L P Geldert, R E Sheriff and D A Keys 1990 *Applied Geophysics* Cambridge UK
[19] P Vaudelet, M Schmutz, M Pessel, M Franceschi, R Guérin, O Atteia, A Blondel, C Ngomseu, S Galaup, F Rejiba, P Bégassat 2011 Mapping of contaminant plumes with geoelectrical methods A case study in urban context *Journal of Applied Geophysics* **75** 4 738-751
[20] J Deceuster, O Kaufmann 2012 Improving the delineation of hydrocarbon-impacted soils and water through induced polarization (IP) tomographies: a field study at an industrial waste land *Journal of Contaminant Hydrology* **136** 25-42

IOP Conf. Series: Materials Science and Engineering **429** (2018) 012027 doi:10.1088/1757-899X/429/1/012027

Validation of Groundwater Potential Zone Based on Imaging Profiles using Different Array and Lines of Survey Position

Azizan F. A.[1*], Sah S. S.[1], Mohamed Nawawi M. A.[2]

[1]School of Bioprocess Engineering, Universiti Malaysia Perlis, Malaysia
[2]Department of Agricultural and Biosystems Engineering, Faculty of Engineering, Universiti Putra Malaysia, Malaysia

*Email: fathin@unimap.edu.my

Abstract. Two resistivity and induced polarization surveys were carried out in this study to validate the groundwater potential location. These 2D geo-electrical survey lines were perpendicular to each other and intersected at point E. The survey lines used a different array configuration; pole-dipole and Wenner-Schlumberger for AB and CD lines, respectively. The surveys were conducted using set of ABEM Terrameter SAS 400 which resulted a resistivity and induced polarization imaging profiles after an analysis in RES2DINV software. The results revealed that, both induced polarization profiles have a low chargeability values of between 0 – 10 ms at intersect point, E even though the resistivity values varied. Thus, the results have verified the location to be a highly potential groundwater reservoir. The reservoir may be found around 50 m depth below ground surface.

1. Introduction

Groundwater in Malaysia is an important resource that is yet to be exploited on a bigger scale to meet the increasing demand for various uses [1]. The utilisation of groundwater can help solve water shortage in areas where surface water is limited [2]. The capacity and limitation of groundwater in Malaysian aquifer should be determined in knowing the groundwater storage available in the aquifer.

An increase of cropped area and population in Malaysia has contributes to a rapid increase of groundwater exploration for irrigation purposes [3,4]. The water shortage by caused of the drought season especially in several states including Perlis, Malaysia need this alternative resources. This resources of water have also been proven to be a reliable resource that farmers exploit to offset droughts and shortfalls in surface water irrigation supplies [5].

In Malaysia, farmers are normally use surface water as source of irrigation. Nevertheless, the study area has no nearby water body, thus fully depended on water pipes with huge cost. To reduce the dependency on pipe water, shallow groundwater resources can be assessed. Groundwater location should be studied to discover the potential aquifer area. Many methods have been proven to treasure groundwater and the most popular and non-destructive method is geophysical exploration.

As existing techniques and methods of geophysical exploration can be modified to adapt the environmental and engineering investigation, the development of the new one is no use [6]. Most cases of environmental and engineering interest are at shallow depth of subsurface, therefore near-surface study can be applied. The near-surface study techniques of geophysical investigation can be classified into several categories which includes 2D geo-electrical method. A 2D geo-electrical method compose of two surveys; resistivity and induced polarization. Many researchers [7, 8, 9] have used resistivity and induced polarisation survey in their study and evidenced that this technique is reliable to detect groundwater based on the imaging result profile.

However, this type of survey has several electrode configurations that can be used for measuring an electric current. The electrode configuration or arrangement is called array. Each array has their own advantage and disadvantage based on four characteristics which are (i) the sensitivity of the array to vertical and horizontal changes in the subsurface resistivity, (ii) the depth of investigation, (iii) the horizontal data coverage and (iv) the signal strength [10]. Besides, the resolutions, depth of investigation and noise sensitivity of these arrays are distinctive and recognized [11, 12].

As different array has different conditions, the imaging profiles results will be slightly differed from each other. This differentiation might lead to inaccurate interpretation. Therefore, this study was conducted to validate the potential zone of groundwater by using two different arrays and position of survey lines.

2. Materials and Methods

2.1. *Study area*

The study area is located at Institute of Sustainable Agrotechnology (INSAT), Universiti Malaysia Perlis, Perlis, Malaysia. The area lies between 6° 39′ 07″ to 6° 39′ 26″ North and 100° 15′ 34″ to 100° 16′ 03″ East. Besides 50 greenhouses of Harumanis mango tree, the study area was also built up with offices and laboratories. The average temperature is about 32°C at daytime and 24°C at night, with annual rainfall of 1869 mm. The study area formation is Kubang Pasu - Singa formation. This formation makes up of shale, siltstone and sandstone. In the north-eastern part of Perlis, Yap [13] divided the sequence of the Kubang Formation to three facies; Passage beds facies, Dominant sandstone facies, and Argilitary rock facies.

2.2. *Survey Technique*

In this study, two electrical resistivity and induced polarization surveys were carried out using ABEM Terrameter SAS4000 equipment set. The survey lines; namely AB and CD lines were overlapped at 190 m from each left-end as shown in Figure 1. The lines were intersected at point E as the location has a potential groundwater zone under the ground surface. Each survey line in study area has 400 m length and was arranged as standard stated in instruction manual [14]. Figure 2 shows field arrangement of 400 m survey line.

Fig. 1. Location of survey line at InSAT, UniMAP

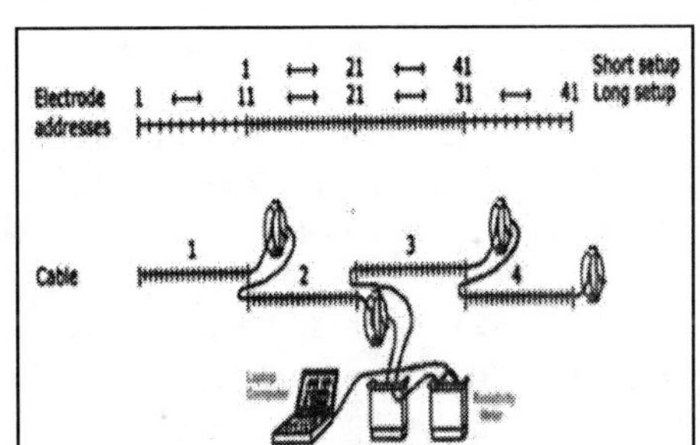

Fig. 2. Field arrangement of survey conducted [14]

The resistivity measurements are normally made by injecting current into the ground through two current electrodes, C1 and C2, and measuring the resulting voltage differences at two potential electrodes, P1 and P2, as in Figure 3. From the current and voltage values, an apparent resistivity, ρ_a value is calculated by multiplying k (geometric factor) and R (resistivity). This geometric value depends on the arrangement of four electrodes which are named array or protocol. Arrays types of Pole-Dipole and Wenner-Schlumberger were chosen for AB and CD survey lines, respectively. The arrangement of these array was shown in Figure 4. The field data were recorded automatically by the resistivity meter and processed using RES2INV software [15] to produce imaging profile.

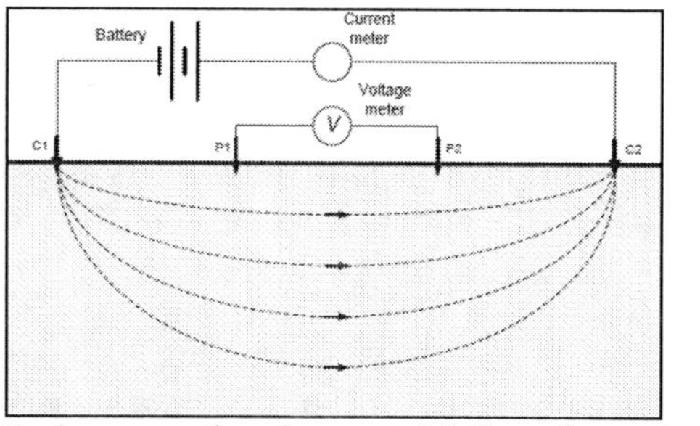

Fig. 3. A conventional array with four electrodes

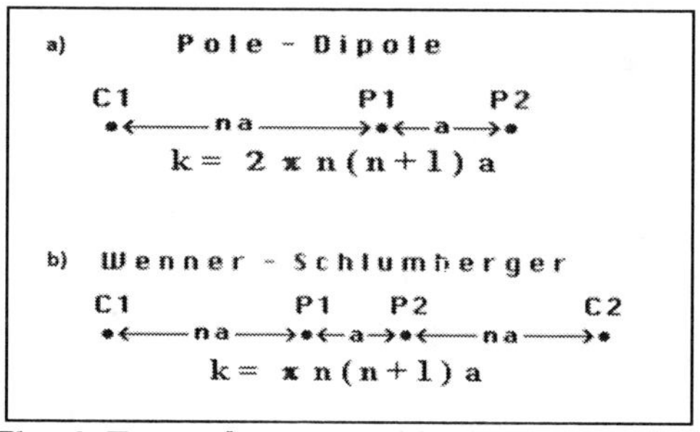

Fig. 4. Type of array used in this study [20]

In this study, AB survey line were performed and result of imaging profiles were analysed before the position of survey line CD was carried out. This to ensure the intersection of these lines are lies on potential groundwater location. The lines were position in perpendicular in order to re-evaluate the resistivity and chargeability value of the intersection point. In this kind of survey, geological structure of study area is reflected by resistivity imaging profiles while the groundwater is reflected by low chargeability value in induced polarization profiles. The results and discussion of this paper are focused on validating the potential groundwater location of point E in term of both imaging profiles produced from geo-electrical survey.

3. Results and Discussions

The resistivity imaging profiles for AB and CD lines are shown in Figure 5 and 6, respectively. The profiles have an overall depth of 135.5 m for Pole-Dipole array and 84.7 m for Wenner-Schlumberger array. The results of resistivity values were ranging between 1 Ωm to 5,000 Ωm for AB lines whereas the values were extending to 10,000 Ωm for CD lines.

It can be seen that; Figure 5 has more depth compared to Figure 6. The depth of each particular survey will be depended on maximum electrode spacing, depth factor and geometric factor [16]. It is also appeared that, Figure 5 and 6 has three underlying zones; Zone 1, Zone 2, and Zone 3 as labelled in the figures. These zone are exist as horizontal structures. However, in Figure 5 there is Zone 4 overlapping these three zones in vertical way. Pole-dipole has better resolution for vertical structures whereas Wenner-Schlumberger array has better resolution for horizontal structures [17].

Zone 1 was detected to compose of clay soil as the resistivity values was ranging from 10 to 100 Ωm. Several researchers have interpreted and proved that, this resistivity value ranging was belonging to clay type of soil [18-20]. Zone 4 was believed to have similar composition as Zone 1. The second underlying zone, Zone 2 has a resistivity value between 100 to 500 Ωm and is expected to compose of moderate weathered rocks. Zone 3 has the highest range of resistivity value of 500 to 10,000 Ωm. It is possible compose of completely into highly weathered rocks. Both zones were perceived to be composed of sedimentary rocks of shale, siltstone and/or sandstone due to high resistivity values [19, 21]. These rocks are a type of rocks in Kubang Pasu – Singa Formation [22].

Figure 7 and 8 show induced polarization imaging profiles for AB and CD survey lines, respectively. The chargeability values in induced polarization profiles in Figure 7 can be divided into two sections; half of the profile on left side has low chargeability measurement while right side of the profiles having a high chargeability value. Through profile in Figure 7, there is a possible freshwater reservoir indicated by greenish-blue colour fill. The values were ranged between 0 – 2 ms. In Figure 8, there is two regions of low chargeability values of less than 10 ms exists. This region is separated by high chargeability structures. All regions (region 1, 2 and 3) located in Figure 7 and 8 are suspected to hold groundwater due to its low chargeability values [21].

IOP Conf. Series: Materials Science and Engineering **429** (2018) 012027 doi:10.1088/1757-899X/429/1/012027

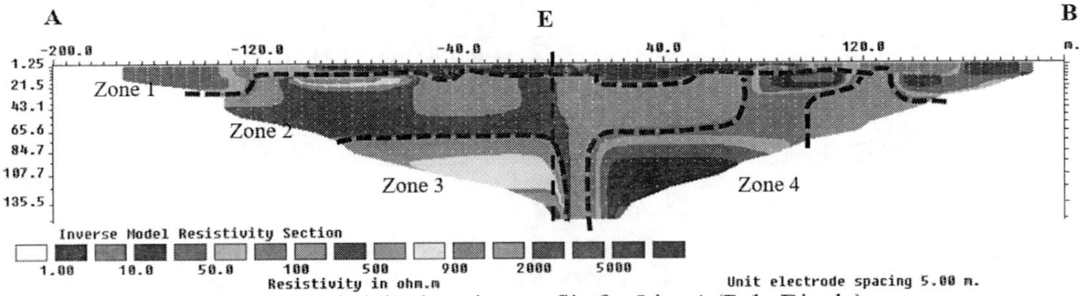

Fig. 5. Resistivity imaging profile for Line 1 (Pole-Dipole)

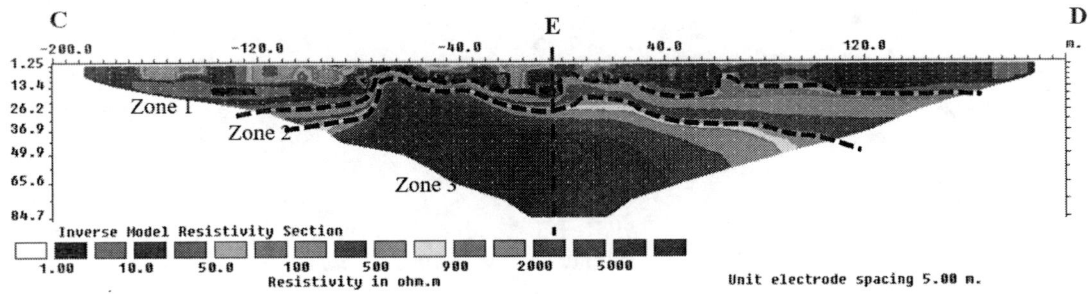

Fig. 6. Resistivity imaging profile for Line 2 (Wenner-Schlumberger)

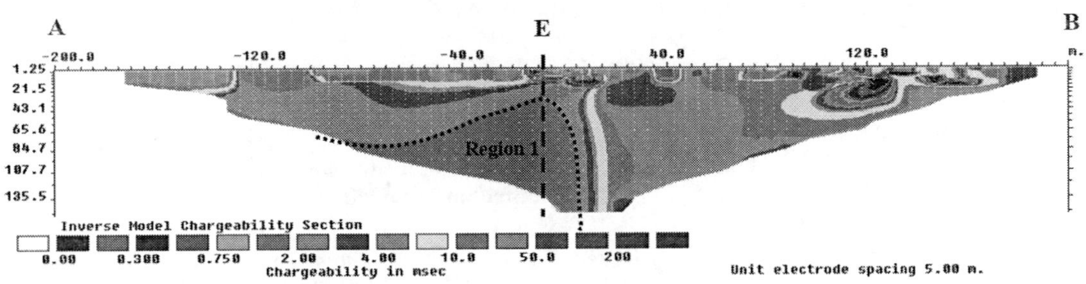

Fig. 7. Induced Polarization imaging profile for Line 1 (Pole-Dipole)

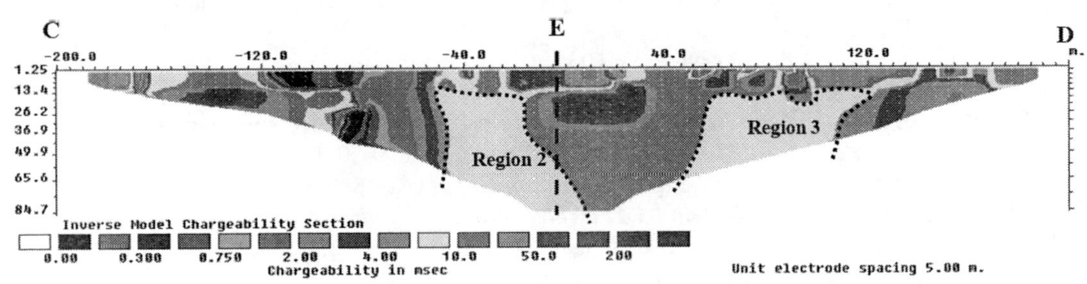

Fig. 8. Induced Polarization imaging profile for Line 2 (Wenner-Schlumberger)

Table 1 shows the results of intersection (point E) based on parameters investigated. Despite the resistivity values are much varied for both survey profiles, the chargeability values are within 0 – 10 ms which is fall in groundwater range. Therefore, this results may validate the existing of groundwater reservoir at that location. The depth of reservoir from AB survey lines is around 50 m below ground surface and for CD lines the depth also fall on same depth.

TAB E 1. Results of several parameters at intersect point E.

Parameters	Survey Line	
	AB (Pole-dipole)	CD (Wenner-Schlumberger)
Depth (m)	≈ 120	≈ 80
Resistivity value (Ωm)	250 – 900	800 – 10,000
Chargeability value (ms)	3 – 4	6 – 8

4. Conclusions

This study is helpful in validating the potential groundwater location at study area. The potential location of groundwater is located at point E at depth of 45 m below ground surface. The results also show that despite of different resistivity value of geology formation at point E, both lines show low conductivity values of the range 0 to 10 ms. This conductivity range is verified to contain groundwater.

Acknowledgments

The author would like to acknowledge support from the Short Term Grant under a grant number of 9001-00556, Universiti Malaysia Perlis (UniMAP).

References

[1] F Mohamed, W Z Wan Yaacob, M R Taha and A R Samsudin 2009 Groundwater and soil vulnerability in the Langat basin Malaysia *European Journal of Scientific Research* **27** 4 628-635
[2] T A Mohammed and A H Ghazali *Pertanika Journal of Sciences and Technology* **17** (1), 33 (2009).
[3] N Dor, S Syafalni, I Abustan, M T A Rahman, M A A Nazri, R Mostafa and L Mejus 2011 *Journal of Water Resources Management* **25** 11 2837– 2853
[4] X M Song, F Z Kong and C S Zhan 2011 *Journal of Water Resources Management* **25** 3 857-873
[5] L C Chang, C C Ho, M S Yeh and C C Yang 2011 *Journal of Water Resources Management* **25** 1 29-78
[6] P V Sharma 1997 Environmental and Engineering Geophysics Cambridge University Press
[7] M A Koukadaki, G P Karatzas and M P Papadopoulou 2007 *Water Resources Management Journal* **21** 1881-1898
[8] P Sikandar and E W Christen 2012 *Water Resources Management Journal* **26** 1201-1215
[9] M S E Juanah, S Ibrahim, W N A Sulaiman and P A Latif 2013 *Arab Journal Geoscience* **4** 4129-4144
[10] M H Loke 2000 Electrical imaging surveys for environmental and engineering studies: A practical guide to 2-D and 3-D survey
[11] R D Barker 1989 *Geophysics* **54** 1031-1037
[12] T Dahlin and B Zhou 2004 *Geophysical Prospecting* **52** 379-398
[13] K F Yap 1991 *Geologi Am Kawasan timur laut Perlis* Universiti Kebangsaan Malaysia
[14] ABEM Terrameter SAS 4000/SAS 1000 Instruction Manual 2010
[15] M H Loke, I Acworth and T Dahlin *Exploration Geophysics* **34** 182–187
[16] M H Loke 2001 *Tutorial: 2-D and 3-D electrical imaging surveys*
[17] M H Loke, P B Wilkinson and J E Chambers 2010 *Computers & Geosciences* **36** 11
[18] G J Palacky 1988 *Investigations in Geophysic* **3**
[19] G V Keller, F C Frischknecht 1966 *Electrical Methods in Geophysical Prospecting* Pergamon Press Inc., Oxford
[20] F Daniels and R A Alberty 1996 *Physical Chemistry* John Wiley and Sons Inc.
[21] W Telford, L P Geldert, R E Sheriff and D A Keys 1990 *Applied Geophysics* Cambridge, UK
[22] W I Ruslan 2008 *Timah Tasoh Reservoir, Perlis, Malaysia*

IOP Conf. Series: Materials Science and Engineering **429** (2018) 012028 doi:10.1088/1757-899X/429/1/012028

Determining the Suitability of POFA as Partial Replacement to Cement in Concrete for Acoustic Properties: Sounds Transmission Loss and Sounds Absorption Assessment

Abdul Wafi Razali[1,a)], Mohd Asraf Ayob[2,b)], Mohammad Nabil Fikri Saaid[2,c)], Awang Ihsan Awang Yunus[2,d)] and Khairul Anuar Maarof[2,e)]

[1]Faculty of Engineering, Universiti Malaysia Sarawak, Jalan Datuk Mohammad Musa, 94300 Kota Samarahan, Sarawak
[2]Faculty of Architecture, Planning and Surveying, Universiti Teknologi Mara Sarawak Branch, Jalan Meramek 94300 Kota Samarahan, Sarawak

*Email: rawafi@unimas.my

Abstract. Abundance of palm oil fuel ash (POFA) produced by the process of burning fibers and fruit bunches as fuel has contribute to several problems related to the space and environmental control. By having an alternative to manage and control the POFA as agricultural waste, it can give great benefits in construction industry for replacing the dependence of Ordinary Portland Cement (OPC) as binder in concrete. The properties of POFA itself are close to the OPC as binder in concrete yet there are some other unique characteristic of POFA in providing a better acoustical performance in a concrete. In this study shown that by replacing some part of cement with POFA in concrete mixture, it contribute to greater acoustical performance due to the high sound transmission loss and sound absorption found in this study. This study focus on single concrete mix with water cement ratio of 0.45 for concrete grade 25 and can be concluded that higher percentage of POFA replacement can enhance the acoustic properties of a concrete block but may reduce the strength of the concrete.

1. Introduction

Palm Oil Fuel Ash (POFA) is a kind of agricultural waste produced by the combustion process in steam boiler at the Palm Oil Mill and the waste were not really useful to industry as they contain not enough nutrients and left dumped near the palm oil mill [1]. Palm oil industry is the biggest agricultural industry in Malaysia and it produces approximately 41% of world palm oil production and 47% of world exports as stated by Malaysian Palm Oil Council (MPOC) and for every 100 tons of fresh fruit bunches processed, approximately 52 tons of nut shells, fibers and empty bunches discharged from the mills [2]. Several studies and experiments were conducted to determine the effect of using agricultural waste as cement replacement in concrete [1][2]. POFA will act as replacement to cement in concrete mix these ashes have been precisely studied on its properties and proved that it has cementitious properties which called as pozzolanic materials but the capabilities of the ashes must be examined before being commercialized [1].

As the biggest agricultural industry in Malaysia, palm oil mills produced huge number of waste in solid form [3]. The alternatives to take the benefit of POFA as substitute or replacement for cement in concrete mixture leads to optimize the utilization of oil palm waste and also to produce low cost mineral admixture sufficiently pozzolanic material that could be used in pozzolanic cement. 10% of ultra-fine sand (UFS) replacement generates higher sound absorption coefficient due to excessive amount of micro silica which resulting in higher amount of pore formation hence increased the void in the concrete [4]. The finer particle indicates the best results where the finer particle can easily vibrate compared to the coarse particle. This happened because the low inertia made the particles vibrate smoothly and drive the energy away. Vibrated rubber can dissipate energy easily. The frequency and the content of the rubber can dissipate the elastic energy. Thus, since the rubber has a good ability to dissipate sound energy, it produced a better airborne sound insulation performance against the cork granulates in gypsum where the rubber has a great elasticity [5]. Higher fineness of POFA exhibit higher resistance towards sulphate attack compared to less fineness of POFA where pozzolanic reaction in concrete mix form secondary C-S-H gel in shorter time [6].

Nowadays, designing building with good acoustical properties has become popular as the regulations are more demanding [7]. This study is about to assess the acoustical performance for concrete block with POFA in the mixture. There are two acoustical aspects been studied which are the sound transmission loss and sound absorption coefficient. An improvement of sound transmission loss in building element can help to reduce the interior noise from the transmission path [8].

Today, porous materials such as fibers are famously used as sound absorptive materials for a better acoustical performance [9]. There was a study on the crumb rubber replacement in concrete resulting the sound energy can be absorbed easily when the percentage of crumb rubber replacement increased causing incremented in air void in the crumb rubber concrete [10]. The amount of recycle aggregate in the porous concrete had less effect towards the absorption characteristic. The main factors contribute to the higher absorption coefficient is the surface area of the voids. As the surface area of the voids increased, the absorption coefficient also increased [11]. It is good if a concrete block with some part of POFA replacement in cement can act as a partition in an enclosed space having a lightweight characteristic and great in acoustical performance. This can help to benefit the surplus of POFA as agricultural waste to something useful in construction industry.

A field measurement of apparent airborne sound reduction index for interior wall was conducted by using software dB Bati 32 [12] and another field measurement for floor sound insulation impact have been carried out according to ISO 140-7 standard and by using software of analysis and elaboration dB BATI 32 [13]. By referring to ISO 140-8, the impact sound reduction for each of floor covering were measured and tested using software of analysis and elaboration data dB BATI 32 [12].

2. Methodology

In assessing the acoustic properties of Concrete with POFA-cement replacement for this study, there were 2 tests conducted which were the sound transmission loss test and sound absorption test. Materials for concrete such as Ordinary Portland Cement (OPC), fine aggregates (<600 μm), coarse aggregates and water are the main materials prepared and being used for all the concrete block samples. POFA used for this study was supplied from Palm oil mill at Samarahan belongs to the Sarawak Land Consolidation and Rehabilitation Authority (SALCRA). The design mix for the study is using concrete grade 25 with water cement ratio 0.45. There were 5 different samples of concrete mix design consists of control sample (the normal mixture without replacing part of cement with POFA in the mixture), 2.5% POFA-cement replacement, 5%, 7.5% and 10%. There were 100 concrete blocks size 240mm x 190mm x 100mm and tested on 7, 14 and 28 days for the sound transmission loss test and sound absorption test. Same materials compositions were used for casting the concrete blocks to conduct the tests.

2.1. Sound Transmission Lost Test

Concrete block size 240mm x 190mm x 100mm with different mixtures with POFA added according to the different percentage for every sample tested were prepared. This part of BS EN ISO 140-3:1995 lays down specifications towards laboratory test facilities for sound insulation measurements of building elements. The laboratory test facility consists of two adjacent reverberant rooms with a test opening between them in which the test specimen is inserted. When measuring walls or floors that the specimen should cover a total partition wall or ceiling of the test room, i.e. the test opening should extend from wall to wall. Reverberant chamber used for this study has opening size 1m x 1m in area. Twenty units of concrete blocks needed to enclose and cover the whole opening. To prevent any sound infiltration, blocks were sealed with lime mortar. Three tests on acoustical aspects conducted to measure the sound transmission losses which were the reverberation time (RT), Background Noise and Emi/Rec.

First test conducted was the reverberation time for every samples. Then, the measurement of background noise levels shall be made to ensure that the observations in the receiving room are not affected by noise from outside the test room, electrical noise in the receiving system, or electrical cross-talk between the source and the receiving systems. Then the results for transmission loss can be determined by using Buildings Acoustic Software (dbBati 32-bit Version 4.532).

2.2. Sound Absorption

Sound absorption test was tested at the receiver's room only. The entire test specimen (concrete blocks) were placed on the floor and the opening was sealed with steel plate to prevent sound transmission loss. The test must be conducted four times with different position of the microphone. Concrete blocks were placed on the floor under almost the same conditions of temperature and humidity. The reason is to ensure the air absorption not differ significantly and the microphone was mounted on a tripod and placed on the concrete blocks which are at least 1.5 m apart, 2 m from any sound source and 1 m from any room surface and the tested samples. After the measurement done, the microphone is rotated in 4 different directions (clockwise). Concrete block was covered with aluminum foil at all side. The sound in the reverberation room was generated from sound source which is noise source with an omni-directional radiation pattern. Using Buildings Acoustic Software (dbBati 32-bit Version 4.532).

3. Results and Discussions

3.1. Sound Transmission Loss

There are 5 different samples of concrete blocks mix with POFA. The cement content of the concrete mixture was replaced with 2.5%, 5%, 7.5% and 10% POFA. For each sample, there were 20 blocks produced and tested on 7 days, 14 days and 28 days. Total concrete blocks for sound transmission loss test are 100 units. **Table 1** shows the results of sound reduction for 5 different samples with different ages of concrete blocks. The range of frequency used to determine the transmission losses towards the concrete blocks are from 100Hz to 5,000Hz. Control samples show 37dB, 37dB and 39dB transmission loss recorded for 7, 14 and 28 days of concrete ages tested respectively. Transmission loss increased for 2.5% POFA replacement samples as results 45dB, 45dB and 47dB were recorded for 3 different ages of concrete respectively and positively increase for 5% POFA replacement samples as the results for 3 different days of concrete age's recorded 48dB, 48dB and 49dB sound transmission loss. 7.50% POFA replacement show almost similar results as 48dB sound transmission loss for 7 days and 49dB sound transmission loss for 14 and 28 days were recorded. Highest sound transmission loss was recorded by 10% POFA replacement as the amount of 49dB for 7 and 14 days and 50dB were recorded for the last 28 days of concrete age.

Higher number of sound transmission loss indicates a good acoustic performance due to the sound wave and sound energy that will less transmit through the concrete block. This will reduce the amount of noise to be transmitted from outside into the enclosed space in the building. Less amount of air void will prevent he sound wave passed through the concrete easily. And by adding up of higher amount of silica fume the reduction in sound transmission loss (STC) can be partially recover [11] It can best be concluded that the best POFA replacement for achieving high sound transmission loss is by 10%.

TABLE 1. SOUND TRANSMISSION LOSS RESULT FOR ALL SAMPLES

	7 Days (dB)	14 Days (dB)	28 Days (dB)
Control Sample	37.0	37.0	39.0
2.50%	45.0	45.0	47.0
5.0%	48.0	48.0	49.0
7.50%	48.0	49.0	49.0
10.0%	49.0	49.0	50.0

3.2. Sound Absorption Test

There are 5 different samples of concrete blocks mix with POFA. The cement content of the concrete mixture was replaced with 2.5%, 5%, 7.5% and 10% POFA. For each sample, there were 20 blocks produced and tested on 7 days, 14 days and 28 days. The total concrete blocks for sound transmission test is 100 units. **Table 2** shows sound absorption coefficient for three different concrete sample's ages against the percentage of POFA replacement in samples. The sound absorption coefficient for control samples are 0.05, 0.02 and 0.02 for 7, 14 and 28 days. According to the Table 2, when 5.0% POFA replacement samples were tested for 7, 14 and 28 days of ages, the results show that the sound absorption coefficients are 0.05, 0.04 and 0.06. It continued increased when 5.0% POFA replacement tested as 0.05, 0.06 and 0.07 of sound absorption coefficient were recorded. The results for 7.5% POFA replacement show inconsistency results due to the inconsistency preparation of concrete samples mixture affecting the concrete become weak in sound absorption. Highest sound absorption coefficient recorded for 10% POFA replacement as 0.05 sound absorption coefficient recorded for 7 days, 0.1 for 14 days and 0.08 for 28 days age of concrete tested. Sound absorption coefficient for concrete blocks with POFA replacement give better results due to the higher fiber content in the concrete blocks.

Result shows that the sound absorption coefficient increased over the percentage of POFA replacement. Higher fiber or POFA in the concrete can improve the sound absorption characteristic in concrete. The void exists inside concrete blocks due to the existence of POFA can absorb the sound energy that transmitted through the surface of concrete blocks. The main factors contribute to the higher absorption coefficient is the surface area of the voids. As the surface area of the voids increased, the absorption coefficient also increased [14]. It can be concluded that 10% POFA replacement is best for sound absorption coefficient in concrete.

TABLE 2. SOUND ABSORPTION COEFFICIENT RESULTS FOR ALL SAMPLES

	7 Days	14 Days	28 Days
Control Sample	0.05	0.02	0.02
2.50%	0.05	0.04	0.06
5.0%	0.05	0.06	0.07
7.50%	0.1	0.02	0.07
10.0%	0.05	0.1	0.08

4. Conclusion

Higher percentage of POFA increased the sound transmission loss and sound absorption coefficient. By looking at the results for transmission loss, it clearly stated that 10% of the POFA replacement produces better sound insulation for the concrete blocks which is 50 dB. In the mean time, higher POFA replacement also exhibited better sound absorption where the absorption coefficient for 10% replacement is 0.08. This indicates that higher amount of POFA in concrete can produce a better acoustic performance rather than control sample with 0% POFA replacement. Also to be concluded, higher amount or higher percentage of POFA replacement in concrete can improve and increase the acoustic performance due to the air voids in the concrete where the sound energy can be absorbed and reflected when passing through the concrete blocks mixed with POFA.

It requires further research and analysis to dig into the relationship between concrete properties and acoustical properties of concrete blocks with POFA replacement as alternative for cement as binder. In a positive perspective, higher POFA replacement in concrete can contribute to the greater acoustical performance in a concrete. Both sound insulation and sound absorption were in a good condition when replacing cement with POFA as alternative. In conclusion, higher percentage of POFA can be an alternative for providing a partition with greater acoustical performance.

References

[1] J H Tay and K Y Show 1995 Use of ash derived from oil-palm waste incineration as a cement replacement material *Resour. Conserv. Recycl.* **13** 27–36
[2] S Bamaga, M A Ismail, M W Hussin, and U Teknologi 2010 Chloride Resistance of Concrete Containing Palm Oil Fuel Ash *Concr. Res. Lett.* **1** 158–66
[3] S F S H Chea Chandara, Etsuo Sakai, Khairun Azizi Mohd Azizli, Zainal Arifin Ahmad 2010 The effect of unburned carbon in palm oil fuel ash on fluidity of cement pastes containing superplasticizer *Constr. Build. Mater.* **24** 1590--1593
[4] S S Shebl, H S Seddeq, and H A. Aglan 2011 Effect of micro-silica loading on the mechanical and acoustic properties of cement pastes *Constr. Build. Mater. - CONSTR BUILD MATER* **25** 3903–8
[5] S Herrero, P Mayor, and F Olivares 2013 Influence of proportion and particle size gradation of rubber from end-of-life tires on mechanical, thermal and acoustic properties of plaster–rubber mortars *Mater. Des.* **47** 633–642
[6] A S M A Awal and S Ibrahim Abubakar 2011 Properties of concrete containing high volume palm oil fuel ash: Ashort-term investigation *Malaysian J. Civ. Eng.* **23**
[7] S M Hasheminejad, A Hakimi, and H Keshavarzpour 2018 Broadband sound transmission loss enhancement of an arbitrary-thick hybrid smart composite plate using multi-objective particle swarm optimization–based active control *J. Intell. Mater. Syst. Struct.* **29** 1724–47
[8] Y Zhang, X Xiao, D Thompson, G Squicciarini, Z Wen, Z Li, and Y Wu 2016 Sound transmission loss of windows on high speed trains *J. Phys. Conf. Ser.* **744**
[9] F Shahani, P Soltani, and M Zarrebini 2014 The Analysis of Acoustic Characteristics and Sound Absorption Coefficient of Needle Punched Nonwoven Fabrics *J. Eng. Fabr.* **9** 84–92
[10] B Mohammed, K Hossain, J Ting Eng Swee, G Wong, and M Abdullahi 2011 Properties of crumb rubber hollow concrete block *J. Clean. Prod. - J CLEAN PROD* **23**
[11] S B Park, D Seuk Seo, and J Lee 2005 Studies on the sound absorption characteristics of porous concrete based on the content of recycled aggregate and target void ratio *Cem. Concr. Res. - CEM CONCR RES* **35** 1846–54
[12] M Goretti and F Cotana 2014 Sound insulation of identical building elements in different buildings *21st International Congress on Sound and Vibration 2014, ICSV 2014* vol 6 pp 1–8
[13] M G Franco Cotanaa 2005 Tools for building acoustics design and experimental performances
[14] H K Kim, J H. Jeon, and H K. Lee 2012 Workability, and mechanical, acoustic and thermal properties of lightweight concrete with a high volume of entrained air *Constr. Build. Mater. - CONSTR BUILD MATER* **29**

Achieving Sustainable Affordable Housing Scheme from the Perspective of Multi Eco-System

Siti Hamidah Abdul Hamid[1], Shariffah Zatil Hidayah Syed Jamaludin[2], Siti Akhtar Mahayuddin[3]

[1,2]Faculty of Architecture Planning & Surveying, Universiti Teknologi MARA, Sarawak Branch, 94300 Kota Samarahan, Sarawak, Malaysia. Malaysia
[3] Faculty of Architecture Planning & Surveying, Universiti Teknologi MARA, Seri Iskandar Branch, 32600 Seri Iskandar, Perak, Malaysia

Email: hamidah059@sarawak.uitm.edu.my

Abstract. Sustainable development and affordable housing are some of the challenges Malaysia and many other developing countries are experiencing. Majority of the population which comes from the middle-income group found difficulties in purchasing an affordable house and living in sustainable communities. Hence, in response to the Malaysian government's effort in the Eleventh National Plan (2016-2020) for every Malaysian to have equal access to affordable housing with sustainable approaches, the drifts of integrating affordability criteria in sustainable housing development seems vital. It is essential to simultaneously tackle sustainability and affordability housing issues. Thus, the aim of this study is to determine the Critical Success Factors (CSFs) in achieving Sustainable Affordable Housing Scheme from the economic, environmental, and social perspectives for the middle-income group. This study examines issues based on a comprehensive review of journal articles and official reports. Relevant literature resources on the CSFs of sustainable affordable housing to meet the needs and ensure satisfaction of end users are reviewed. It is expected that the findings will elicit several strategies for the construction industry players to enhance the quality of life for sustainable communities via the Sustainable Affordable Housing Scheme.

1. Introduction

Housing is an essential element for the economic growth and development of a country. An effective and efficient housing provision would not only contribute to the rapid economic growth but also provide a link between the corporeal growths of a community-built environment and its social outcomes [1]. This is due to the fact that the socio-economic stability of a country is always dependent on the housing affordability of the residents in the country. Malaysia is of no exception when it comes to deal with housing development. Through the National Housing Policy, extensive laws covering sustainable development of physical, economic, environmental and social aspects in housing were created [2]. The housing policy in Malaysia has evolved over the years through a number of national development plans. The objectives of the policy are to provide affordable and adequate housing, mainly to the low-income group. Nevertheless, most of the citizens are excluded from the housing assistance programme as they are classified under the middle-income group [3].

The middle-income group, which accounts for 40% of the Malaysian population, has found difficulties in getting their own house due to the significant increment in the housing prices since most of the developers in Malaysia tend to build high-cost houses more than medium-cost houses [4]. In this scenario, the developers played a vital role to realise the sustainable house concept by implementing sustainable aspects to the medium-cost housing. Although sustainable housing incorporates energy and water efficiency, better indoor environmental quality, and optimises material and resources used, the developers are bound to bear the extra cost of the new technologies involved [5]. Thus, redirecting the extra cost to consumers will increase the housing selling price [6].

1.1. Sustainable Affordable Housing and the Multi Eco-System

According to [7], sustainable affordable housing is a type of housing that is facilitated by the government via various housing programmes. The house needs to be built with environmental friendly criteria with the used of sustainable materials and able to sustain economically, environmentally and socially without escalating the life-cycle cost. The multi eco-system believes that many contemporary vital issues are directly or indirectly influenced by a comprehensive system consisting of economic development, natural environment, and social system [8]. The multi eco-system is composed of three subsystems: economic, environmental, and social which are accomplished through an integrated development among the subsystems on the basis of the sustainability of each subsystem. This is because the provision of sustainable affordable housing is always constrained by the need to consider economic, environmental and social implications [9]. [10] define sustainability as the observation of balancing between three concepts, namely, economic development, environmental protection and social equity. It needs an orderly operation of economic, environment, and social subsystems to guarantee the sustainable development of housing projects [11]. Considering the characteristics of sustainable affordable housing and the concept of multi eco-system, the sustainable affordable housing can be defined as a developmental level of sustainable housing project to meet the housing needs of present medium-low income group.

1.2. Research Methodology

An extensive literature review was conducted to explore the current scenario of Malaysian construction industry on sustainable affordable housing implementation. This paper specifically aims to emphasise the critical success factors (CSFs) in achieving the sustainable affordable housing schemes in many developed and developing countries. Various published sources such as journal articles, proceeding papers, and other related research materials were thoroughly examined to highlight the CSFs. The study starts with a review of the sustainable housing criteria and questions why sustainable house is not affordable, especially for the middle-income group of people. In that case, the CSFs of achieving sustainable and affordable housing scheme were gathered and selected based on the contribution of each factor to reduce the cost of a project, consequently assisting to increase the middle-income group's ability to own a sustainable affordable house. The CSFs were divided into multi eco-systems related to the aspects of sustainable development, which are economic, environmental and social. In order to achieve sustainable affordable housing scheme, all issues relating to the three subsystems need to be addressed holistically. In this paper, these issues are addressed using Critical Success Factors (CSFs) of sustainable affordable housing which are necessary to reach a project's goal [12].

1.3. Economic Critical Success Factors

Table 1.0 shows twenty-one (21) CSFs of sustainable affordable housing scheme as highlighted by various researchers. The CSFs were categorised under three subsystems; economic, environmental, and social. The economic CSFs were identified based on a review of literature consisting of eight (8) success factors, which illustrate the top three CSFs in high frequency of occurrence in reducing the cost of sustainable housing. These are; (i) **subsiding the cost of provision, initial purchase, operational, rent and mortgage loan rates,** (ii) **efficient management to minimise life-cycle cost, project risk management and problem-solving abilities,** and (iii) **effective housing policy and framework**.

TABLE 1. The Critical Success Factors (CSFs) for Sustainable Affordable Housing Scheme from the perspectives of economic, environmental and social

Critical Success Factors (CSFs)	Sources									
	Oyebanji et al. (2017)	Dave et al. (2017)	Said et al. (2016)	Copiello (2015)	Gan et al. (2015)	Xia et al. (2014)	Yates (2014)	Mc Murray et al. (2014)	Ihuah et al. (2014)	Wallbaum et al. (2012)
Economic										
Subsidising the cost of provision, initial purchase, operational, rent and mortgage loan rates etc.	√	√		√	√		√	√	√	√
Efficient management to minimise life-cycle cost, project risk management and problem-solving abilities	√	√			√	√	√	√	√	√
Effective housing policy and legal framework	√			√	√		√	√	√	√
Excellent governance in promoting economic growth	√			√	√			√	√	
Adequate funding to facilitate developers in providing sustainable affordable housing	√				√		√		√	
Economic design that leads to reducing upcoming maintenance and operation cost.	√	√				√	√			√
Appropriate technology to allocate for refurbishment	√	√		√	√					√
Great economic planning considering infrastructure services	√		√	√	√				√	√
Environmental										
Appropriate materials which are sustainable and environmental friendly	√	√		√	√	√		√	√	√
Appropriate land use plan for avoiding misuse of land and financial resources.	√		√	√					√	
Energy efficient fixtures, appliances, and equipment		√		√		√				√
Environmental protection by means of penalties to polluters.	√						√			
Whole building energy assessment		√				√				
Social										
Stakeholder's participation in the development process	√		√		√	√			√	
Enhance awareness of sustainable lifestyle	√				√			√	√	
Good quality social housing that creates the sense of place to live	√				√		√			√
Promote equity through equal distribution, social justice, gender equity, women empowerment and etc.	√									
Provides skills acquisition and job opportunities	√									√
Support community development and social services	√						√			√
Provide health and recreational facilities	√						√			√
Integration of supply chain								√		√

To ensure a successful implementation of sustainable affordable housing construction, the government should subsidise the cost of provision, initial purchase of the house, operational, rent and mortgage loan rates to ensure buyers from the middle-income group to meet their housing needs with ease [7]. For instance, the high value of house price, in particular, sustainable housing, in Malaysia causes buyers to either delay owning a house, forcing them to select a house other than their preference or suffer from high housing loan [13]. The initial investments are generally for the purpose of technology to protect the environment, noise barrier system, staff training, and human resources [14]. With the aid of government, the sustainable housing price is within the buyers' financial ability while meeting other essential basic living costs [15].

Apart from that, they also need to form an efficient management team, as they will have the best available information on the materials, method of construction, tools, and equipment to achieve sustainable affordable housing projects [16]. An efficient management should be concerned with the research and design process and pays higher attention to the resources, quality of product entering the market, absence of related standards, and certification to minimise life–cycle cost and risks [17]. It is notable that wastage of resources due to inefficient management processes causes an approximate increase of 12% in investment costs [18]. Support from industry players to the government will enable the creation of an effective housing policy relating to sustainable construction practices. Government agencies should embark on applicable policies that could provide critical support to ensure the feasibility of sustainable housing [19; 3]. Effective housing policy and legal framework are vital in terms of sustainability cost cutting, ensuring standardisation of materials, specifications and construction techniques which benefits the residents and population. These include reducing greenhouse gas emission, durability, and resilience to climate change and health benefits [20].

1.4. Environmental Critical Success Factors

Another common aspect of sustainable affordable housing scheme is concerned for the environment. The environmental CSFs highlighted in the literature review include 5 variables. The top three CSFs are **(i) appropriate materials which sustainable and environmental friendly (ii) appropriate land use plan for avoiding misuse of land and financial resources,** and **(iii) energy efficient fixtures, appliances and equipment.** A study by [18] clearly indicates that highly industrialised construction techniques such as steel frames, polystyrene cladding, and metal panel do not represent optimal solutions for the sustainable affordable housing problem as compared to timber, bamboo, and other energy-efficient materials. The decreasing cost of energy-efficient materials and techniques will simultaneously decrease the overall cost of housing construction. In addition, the use of local and recyclable, supply energy from renewable sources like solar or wind will be able to facilitate sustainable affordable housing construction. Meanwhile, the system of modularisation and flexibility in floor plan changes need to be adapted according to the different needs of each location and culture [18]. Thus, a future expansion cost of a housing unit will be better satisfying the owner.

An appropriate land use plan cannot be overlooked if sustainable affordable housing provision is to be met. A proper plan and process of acquiring certificate of Occupancy of land will reduce the financial cost in an attempt to achieve the delivery of sustainable affordable housing [21]. Land conservation should be conducted in accordance with the construction plan, abandoned sites are in rational use, and cleaning up the construction waste must be completed in time [11]. The use of energy efficient fixtures, appliances and equipment also contributed to the success of sustainable affordable housing implementation. For instance, tighter building envelop leads to higher thermal performance and energy efficiency, hence, greater construction quality and cost control can be attained[20]. A survey by [11] on Public Rental Housing project in China suggests that the air-conditioning system is controlled by the owner and type of balcony wall and also solar water heaters provide centralised heat for individual heat storage. A timed lighting system was also used in the area with better daylighting such as halls and stairs for full use of renewable resources and reduce energy consumption.

1.5. Social Critical Success Factors

A total of 8 CSFs were highlighted in the social subsystem based on the literature review. The three most common CSFs are **(i) stakeholder's participation in the development process, (ii) enhancing public knowledge and awareness of sustainable lifestyle,** and **(iii) good quality social housing that creates the sense of place to live.** The impact of high awareness and knowledge on the success of sustainable construction among project stakeholders has been reported in the literature [22; 16]. The success of sustainable construction is highly dependent on the desire of stakeholders in the construction industry to be committed to change and working towards congruent goals and objectives which are focused on reducing unnecessary cost, avoid overdesign, and zero defect in a project.

Despite stakeholders' participation in sustainable affordable housing implementation, enhancing public knowledge and awareness is one of the CSFs discovered. The residents should be well-educated to accept and live in a sustainable lifestyle. The long-term social benefits of sustainable affordable house are improving welfare and quality of life of the residents as well as social integration, lower health cost, increased individuals performance and productivity [23]. Good quality of social housing that creates the sense of place to live is also one of the CSFs under the social subsystem. A livable social home has to cater the various needs of existing and future residents such as housing that is inclusive, well-planned, and offers equal opportunity and good services for all [15]. For instance, a housing area which have freedom for social interaction, save and secure environment, access for spiritual development, public health facilities, and natural resources assist the financial expenditure of the family [24]. Hence, it creates satisfaction for the community members to live and work at present and in the future.

2. Conclusions

As an original contribution, the study focuses on CSFs of achieving sustainable affordable housing scheme from the perspectives of three subsystems, namely, economic, environmental, and social. There is a need to have an efficient management to minimise life-cycle cost, project risk management and problem-solving abilities, subsidising the cost of provision, initial purchase, operational, rent and mortgage loan rates and also effective housing policy and legal framework on the economic aspect. These factors made it evident that the government and housing authorities in Malaysia need to play a significant role in providing the scheme. From the environmental aspect, it can be clearly seen that the stakeholders including developer, designer, and contractor are responsible for using appropriate materials that are sustainable and environmentally friendly for reducing maintenance and life-cycle costs. They also need to ensure appropriate land use plan for avoiding misuse of land, human and financial resources and exploit energy efficient fixtures, appliances and equipment in delivering sustainable affordable housing. By successful adopting all the elements, it will reduce the cost of a project, hence, increasing the affordability of sustainable housing. As for the social context, the review confirmed that the stakeholder's participation in the development process, enhancing public knowledge and awareness of sustainable lifestyle, and good quality social housing that creates the sense of place to live are among the CSFs. The stakeholders need to be proactive in their decisions and work as a team in the provision of sustainable affordable housing. Meanwhile, professional and educational bodies could raise the level of understanding and awareness among Malaysians on the benefits of living in sustainable affordable housing. This paper has identified the pathway of future research which needs to execute an integrated sustainable affordable strategies for the used of industry players in achieving the goal of housing scheme that is both sustainable and affordable.

3. Acknowledgement

The authors would like to extend the utmost appreciation to Universiti Teknologi MARA (Sarawak Branch) toward the success of this research. This research is fully funded by Research Management Institute (RMI) under Excellent Fund UiTM.

References

[1] Ihuah P W, Kakulu II, and Eaton D 2014 A review of Critical Project Management Success Factors (CPMSF) for sustainable social housing in Nigeria *International Journal of Sustainable Built Environment*
[2] Said R, Majid R A, Alias A, Adnan Y M and Najib Razali M 2016 Sustainable housing affordability in Sabah *Journal of the Malaysian Institute of Planner*
[3] Baqutaya S, Ariffin A S, and Raji F 2016 Affordable housing policy: issues and challenges among middle-income groups *International Journal of Social Science and Humanity*
[4] Chiali K C and Choon A 2014 http://www.starproperty.my/index.php/articles/property-news/peoples-perception-of-affordable-housing
[5] Deng Y, Li Z, and Quigley J M 2012 Economic returns to energy-efficient investments in the housing market: evidence from Singapore *Regional Science and Urban Economics*
[6] Myeda N E , Kamaruzzaman S N and Zaid S M 2016 Sustainable Housing: Demographic Analysis Of Customers'demands In Klang Valley *Journal of Building Performance*
[7] Oyebanji A O, Liyanage C, and Akintoye A 2017 Critical Success Factors (CSFs) for achieving sustainable social housing (SSH) *International Journal of Sustainable Built Environment*
[8] Ma S J, and Wang RS 1984 Social-economic-natural complex ecosystem *Chinese Academy of Sciences*

[9] Salama A M and Alshuwaikhat H 2006 A trans-disciplinary approach for a comprehensive understanding of sustainable affordable housing *Global Built Environment*

[10] Drexhage J and Murphy D 2010 Sustainable development: from Brundtland to Rio 2012. Background paper prepared for consideration by the High Level Panel on Global Sustainability at its first *International Institute for Sustainable Development*

[11] Dezhi L, Yanchao C, Hongxia C, Kai G, Chi Man Hui E and Yang J 2016 Assessing the integrated sustainability of a public rental housing project from the perspective of complex eco-system *Habitat International*

[12] Wai S H, Yusof A M , and Ismail S 2012 Exploring success criteria from the developers' perspective in Malaysia *International Journal of Engineering Business Management*

[13] Goh A T, and Yahaya A 2011 Public Low-Cost Housing in Malaysia: Case Studies on PPR Low-Cost Flats in Kuala Lumpur *Journal of Design and Built Environment*

[14] Shen L, Tam V W Y, Tam L, and Ji Y 2010 Project feasibility study: the key to successful implementation of sustainable and socially responsible construction management practice *Journal of Cleaner Production*

[15] Wiesel I, Davison G, Milligan V, Phibbs P, Judd B and Zanardo M 2012 Developing sustainable affordable housing: A project level analysis *Australian Housing and Urban Research Institute (AHURI)*

[16] Ofori G 2015 Nature of the construction industry, its needs and its development: A review of four decades of research *Journal of Construction in Developing Countries*

[17] Zhang X, Wu Y, Shen L, and Skitmore M 2014 A prototype system dynamic model for assessing the sustainability of construction projects *International Journal of Project Management*

[18] Wallbaum H, Ostermeyer Y, Salzer C and Escamilla E Z 2012 Indicator based sustainability assessment tool for affordable housing construction technologies, *Ecological Indicators*

[19] Ametepey O , Aigbavboa C , and Ansah K 2015 Barriers to successful implementation of sustainable construction in the Ghanaian construction industry *Procedia Manufacturing*

[20] Dave M, Watson B and Prasad D 2017 Performance and perception in prefab housing: An exploratory industry survey on sustainability and affordability *Procedia Engineering*

[21] Aluko O E 2011 Sustainable housing, population growth and poverty: The implications on Lagos mega city *Journal of Sustainable Development*

[22] Hakkinen T and Belloni K 2011 Barriers and drivers for sustainable building *Building Research & Information*

[23] United Nations Environmental Programme (UNEP) 2013 *http://wedocs.unep.org/handle/20.500.11822/8607*

[24] Pattinaja A M and Putuhena F J 2010 Study on the requirements for sustainable settlement development for low income community in Indonesia *Journal of Environmental Science Engineering*

[25] Copiello S 2015 Achieving affordable housing through energy efficiency strategy *Energy Policy*

[26] Gan X, Zuo J, Ye K, Skitmore M and Xiong B 2015 Why sustainable construction? Why not? An owner's perspective *Habitat International*

[27] Yates J K 2014 Design and construction for sustainable industrial construction *Journal of construction engineering and management*

[28] McMurray A J, Islam M M, Siwar C and Fien J 2014 Sustainable procurement in Malaysian organizations: Practices, barriers and opportunities *Journal of Purchasing for Supply Management*

The Mechanical Properties of HFRC Under Elevated Temperature Exposure

C. Petrus[1,a)], H. A. Azhar[2,b)], L. D. Goh[3,c)], A. Alisibramulisi[4,d)], I. A. Abu Bakar[5,e)]

[1]Faculty of Civil Engineering, Universiti Teknologi MARA, Cawangan Sarawak. Malaysia
[2,3]Faculty of Civil Engineering, Universiti Teknologi MARA, Pulau Pinang, Jalan Permatang Pauh, 13500 Permatang Pauh, Pulau Pinang, Malaysia
[3,5]Faculty of Civil Engineering, Universiti Teknologi MARA, Shah Alam, Malaysia

*Email: clotilda166@uitm.edu.my

Abstract. This paper discusses the effect of exposure to elevated temperature on the mechanical properties of hybrid fibre reinforced concrete (HFRC). Literature review indicates that high strength concrete is prone to sudden explosion particularly, when exposed to elevated temperature. Polypropylene fibres (PPF), play an active role in improving spalling resistance of concrete induced to elevated temperature. However, it has less significant effect on the strength of concrete. The proposed HFRC by incorporating Steel Fibre (SF) and Polypropylene Fibre (PPF) with certain ratios is expected to improve the mechanical properties of concrete. The experimental program examined the mechanical properties of HFRC at a temperature of 800°C with four different heating time. The compression and flexural test were conducted on thirty (30) cubes, twenty (20) cylinders and ten (10) beams. From the experimental results, it was found that the HFRC with 0.5% PPF and 0.5% SF has effectively improved the interfacial bonding, lowering the tendency of spalling and increased both the compressive and tensile strength at high temperature.

1. Introduction

Exposure of concrete to fire or to extreme heat can cause severe effects on concrete's strength and caused massive cracking and spalling failure due to loss of cement paste-aggregate bond [1]. High strength concrete tends to experience pressure build up because of its low permeability compared to normal strength concrete [2]. The strength of concrete depends most on the hydration products (calcium silicate hydrate gel, calcium hydroxide) formed during hydration reaction between water and cement components. During concrete exposure to fire, evaporation occurs at lower elevated temperatures that cause free water to escape from the concrete. With increasing temperature, dehydration process and loss of chemically bonded water might occur. Previous researchers indicate that aggregates dissociated when heated to temperature over 800°C. At this point, pore size and porosity of hydrated matrix will increase which reduce the volume and consequently the strength of concrete will be weakened [3].

Recently, many researchers studied the mechanical properties of the HFRC incorporating with different kind of fibres; such as carbon fibres and PPF [4], glass fibre and PPF [5], carbon fibre and glass fibre [6], or SF and PPF [7-8]to concrete. In general, they reported that, by incorporating concrete with fibres, proven to overcome concrete weakness and also increase in tensile strength, strain capacity, toughness and durability. However, information on the behaviour of HFRC when exposed to extreme temperature is still limited. Therefore, in this study, the mechanical and microstructure of high strength concrete with the addition of SF and PPF at extreme temperature is investigated. These two complementary fibres can enhance mechanical properties and improve the residual strength of the concrete [3], [10–15]. Polypropylene can resist the initial crack and shrinkage while steel fibre can increase the strength [9]. The mechanical properties investigated in this study are compression and flexural strengths of HFRC exposed to high temperature with different heating time.

2. Experimental Detail

High strength concrete with 50 MPa and 60 MPa characteristic strengths made of Ordinary Portland Cement (OPC) with proportion of 0.3:1:0.11:0.35 and 0.3:1:1:1.9 respectively were used in this study. The aggregates used were sand and crushed granite of particle sizes range from 5 mm to 20 mm. The ratio of both hooked SF and PPF used were 0.5% of concrete volume respectively. The diameter of PPF is 22 μm, the length is 15 mm, the elastic modulus is 8 GPa and the tensile strength is 800 MPa. The length of SF is 50 mm with a diameter of 0.7 mm, the elastic modulus is 200 GPa and the tensile strength is 1500 MPa.

2.1. *Sample preparations*

Specified amount of bulk SF was kept in water for an hour to make sure the bulk SF losses into single SF. The mixing process started with mixing the dry cement, coarse and fine aggregates for 1 min; then water is added and mixed for another 3 min. After the mixing process, the specified amount of single SF and PPF was added to the wet concrete. The mixture was mixed for about 3 min to ensure that the fibres can evenly disperse throughout the concrete mixture. For this study, 30 cubes (100 mm *x* 100 mm *x* 100 mm), 20 cylinders (150 mm *x* 300 mm) and 10 beams (500 mm *x* 100 mm *x* 100 mm) were prepared. All specimens then removed from their mould after 24 hours and then cured for 28 days.

2.2. *Test methods*

Prior to the experimental testing, the specimens left at room temperature for 4 hours. The specimens were then placed in the muffle furnace and heated to 800°C with increment of 5°C/min. The heating time divided into two phases; namely phase 1 for 0-30 minutes and phase 2 for 30-60 minutes for data analysing. After the temperature reaches 800°C, the temperature was allowed to cool at the rate of 2°C/min. The specimens were left in the muffle furnace until the muffle furnace had completely cooled, before conducting the compressive and flexural tests. The compressive strength was tested on cube and cylinder specimens with a continuous load increment of 6.8 kN/sec and 5.3 kN/sec respectively until the specimens failed. The flexural test was conducted using ASTM C78/C78M – 10e1 [13] for the flexural toughness and the first crack strength of HFRC [17-18]. A constant rate of 0.2 kN/sec loaded on the beam specimens.

3. Results and Discussions

3.1. *Compressive strengths*

Exposure to high temperature resulted in significant strength loss for HFRC. FIGURE. 1 and FIGURE. 1 show the compressive strength of the HFRC for cube and cylinder specimens after exposed to the temperature of 800°C at four different heating times. The results show significant reduction of compressive strength especially at phase 1. However, when the heating time increases to phase 2, the residual strength rose to 8.63% for concrete grade C60 at 45 minutes before gradually drops to 6.28%. The compressive strength decreased with increasing heating time. This occurs due to the high elastic modulus of fibers as SF bridging effects that mitigate the initiation and expansion of micro-

defects on HFRC. In additional, phase 1 may also be regarded as a significant heating time for the strength loss of HFRC because, denser specimens with first exposure may experience hydration in this phase. Due to first exposure to extreme temperature on concrete with high moisture content, transportation of water vapour inside the concrete matrix is difficult to release [3].
Table 2 shows the compressive strength reduction for all specimens. The highest strength reduction occurred at heating time in phase 2 which is recorded about 10.94%. However, the total strength reduction percentage of all specimens showed a small increment at 45 minutes heating time. This small increment occurred due to the fiber's tensile resistance behaviour to an extreme temperature that maintains the residual strength of HFRC. This lowest reduction happened when HFRC exposed to an extreme temperature, HSC-C-S creates high vapour pressure due to the inner moisture of concrete. Since there are no micro-channels to release the pressure that will caused more damage and cracking in the concretes. This phenomenon also explains why there is a sudden-collapse of a structure during a fire. The sudden-collapse of a structure is due to the exposure to extreme heat that causing the inner moisture trapped within concrete matrix. Therefore, adding PPF and SF to HFRC mitigate the deterioration process of HFRC after exposed to extreme temperature.

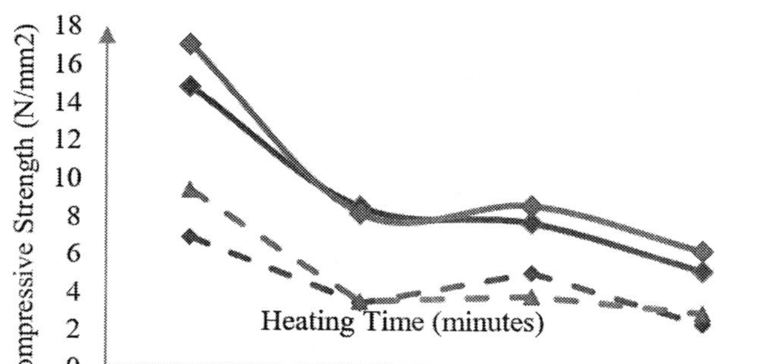

	4	15	45	60
C50 cube sample	14.85	8.53	7.72	5.21
C60 cube sample	17.07	8.19	8.63	6.28
C50 cylinder sample	6.93	3.45	5.06	2.41
C60 cylinder sample	9.41	3.54	3.84	3.00

FIGURE. 1. Compressive strength for cube and cylinder specimens.

TABLE 1 Reduction of strengths in HFRC

Time (min)	Reduction of Strength (%)						Total Strength Reduction (%)
	Cube		Beam		Cylinder		
	C50	C60	C50	C60	C50	C60	
Phase 1 0-30	6.32	8.88	1.86	2.24	3.67	4.49	8.16
Phase 2 30-60	9.64	10.79	1.87	2.27	4.52	6.42	10.94

3.2. Flexural strength

Figure 2 shows the decrement of flexural strength of HFRC beam specimens after exposed to the temperature of 800°C up to a duration of 60 minutes. As expected, the strengths of HFRC beam specimens of C60 at phase 1 improved than the specimens at C50 as compared to the residual strength at phase 2. However, the percentage difference of the flexural strength between both concrete grades are 1.86% and 2.24%. It was observed to be insignificant for beam specimens with C50 as stated in TABLE 1 with a minimum average of 0.38% and the behaviour of both grades were almost identical. The reason is likely due to failure of well-bonded PPF that supposed to provide post-crack ductility that refine the pores inside the concrete [19].

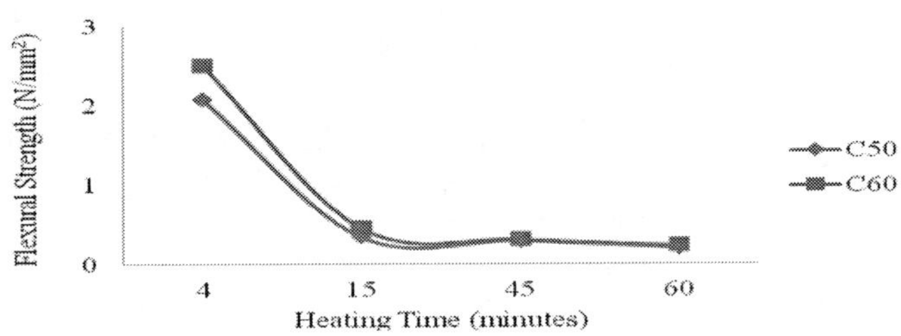

FIGURE. 2 Flexural strength for beam specimens.

4. Conclusions

Based on the scope of this study, it is concluded that incorporating 0.5% SF and PPF respectively of the volume to HFRC enhanced the mechanical and microstructure properties. Although the concrete specimens of HFRC were exposed to an extreme temperature of 800°C, the concrete specimens were able to provide advantageous results on the compressive and flexural strengths.

- The residual compressive strength for C60 remains higher than the residual compressive strength for C50 where it decreases until reached 60 minutes steadily. This behaviour shows that higher concrete grade improves the residual compressive strength for cylinder samples.
- The percentage difference of the flexural strength between both concrete grades are 1.86% and 2.24%. It was observed to be insignificant for beam specimens of C50 with a minimum average of 0.38%. This is due to the failure of well bonded PPF that supposed to provide post-crack ductility which refine the pores inside the concrete and the SF that will improves in plastic restraint by bridging effect.
- Generally, it is found that the HFRC with 0.5% PPF and 0.5% SF has effectively improved in the residual strength at 45-60 minutes (in phase 2) wheby the total strength reduction has minimised by 8.53%.

Acknowledgement
This research was funded by Grant of Universiti Teknologi Mara (UiTM) [600-RMI/FRGS 5/3 (42/2012)]. The authors would like to express their appreciation to the Research Management Institute (RMI) of University Teknologi Mara (UiTM) and the Ministry of Education (MOHE). Special thanks also dedicated to Mr. Ahmad Hawari and Mr. Salleh, for their assistance to complete this research.

References

[1] A Çavdar 2012 A study on the effects of high temperature on mechanical properties of fiber reinforced cementitious composites *Journal of Composites Part B: Engineering* **43** 2452–2463

[2] P Kalifa, G Chéné, and C Gallé 2001 High-temperature behaviour of HPC with polypropylene fibres - From spalling to microstructure *Cem. Concr. Res.* **31** 1487–1499

[3] Erjola Reufi and IlirjanaBoci 2016 Optical Microscopic Characterization Of Polypropylene And Steel Fiber Reinforced Concrete At Elevated Temperatures *Journal of Multidisciplinary Engineering Science and Technology (JMEST)* **3** 4747-4752

[4] J Li, L Yan, and Y Xing 2013 High-temperature mechanical properties and microscopic analysis of hybrid-fibre-reinforced high-performance concrete *Mag. Concr. Res.*, **65** 139–147

[5] M Kakemi 1995 Mathematical model for tensile behaviour of hybrid continuous fibre cement composites *Journal of Composites* **26** 637–643

[6] R Ismail, F A A Zakwan, C Petrus, N A Marzuki, N H Hashim, and M F Mustafa 2014 Compressive Behavior of Steel Fiber Reinforced Concrete After Exposed to High Temperatures *Proceedings of the International Civil and Infrastructure Engineering Conference* 731–740

[7] C Petrus, H A Azhar, G L Dee, R Ismail, A Alisibramulisi 2016 Compressive Strength of Concrete With Fibres At Elevated Temperature *Jurnal Teknologi* **78** 71-74

[8] G Asok and S George 2016 Investigation on Hybrid Concrete Using Steel and Polypropylene Fiber *International Journal of New Technology and Research (IJNTR)* **2** 61–64

[9] E T Dawood and M Ramli 2012 Durability of high strength flowing concrete with hybrid fibres *Constr. Build. Mater.* **35** 521–530

[10] H A Azhar C Petrus L D Goh R Ismail N H Hashim F A A Zakwan 2015 The Effect of High-Temperature on The Strength and Performance of Hybrid Fibre Concrete *International Conference on Advances in Civil and Environmental Engineering*

[11] H A Azhar, R Ismail, C Petrus, N H Hashim, F A Ahmad Zakwan, N A Marzuki, 2014 reinforced cementitious composites *Journal of Composites Part B: Engineering* **43** 2452–2463

[12] S P Singh 2011 Fatigue strength of hybrid steel-polypropylene fibrous concrete beams in Study flexure *Procedia Eng.* **14** 2446–2452

[13] C X Qian and P Stroeven 2000 Development of hybrid polypropylene-steel fibre-reinforced concrete *Cem. Concr. Res.* **30** 63–69

[14] E T Dawood and M Ramli 2012 Durability of high strength flowing concrete with hybrid fibres *Constr. Build. Mater.* **35** 521–530

[15] J T Cochrane 2003 Flexural Creep Behaviour of Fiber Reinforced Concrete Under High Temperature **Thesis (M.A.Sc.)** Dalhousie University

[16] C X Qian and P Stroeven 2000 Development of hybrid polypropylene-steel fibre-reinforced Performance Fiber Reinforced Concrete (UHPFRC) with steel fiber *Compos. Struct.* **92** 61–71

[17] B de Rivaz 2009 Steel fiber reinforced concrete (SFRC): The use of SFRC in precast segment for tunnel lining *Water Energy Int.*, **66** 75–84

[18] N Banthia, F. Majdzadeh, J. Wu, and V. Bindiganavile 2014 Fiber synergy in Hybrid Fiber Reinforced Concrete (HyFRC) in flexure and direct shear *Cem. Concr. Compos.* **48** 91–97

[19] A N Noumowe, R Siddique and G Debicki 2009 Permeability of high-performance concrete subjected to elevated temperature (600°C) *Constr. Build. Mater.* **23** 1855–1861

[20] M Ozawa and H Morimoto 2014 Effects of various fibres on high-temperature spalling in high-performance concrete *Constr. Build. Mater.* **71** 83–92

[21] M. Hsie, C. Tu, and P. S. Song 2008 Mechanical properties of polypropylene hybrid fiber-reinforced concrete *Mater. Sci. Eng. A.* **494** 153–157

IOP Conf. Series: Materials Science and Engineering **429** (2018) 012031 doi:10.1088/1757-899X/429/1/012031

Performance of Sand Reinforced with White Coir Fibre

J D Nyuin[1,a)], **N S Khalil**[2,b)] and **C Petrus**[1,c)]

[1]Faculty of Civil Engineering, Universiti Teknologi MARA Cawangan Sarawak, Kampus Kota Samarahan 2, 94300 Kota Samarahan, Sarawak, Malaysia
[2]Perunding Azizi, No. 3 Tingkat 2, Taman Arasu, Jalan Raja Omar, 32000 Sitiawan, Perak

*Email: joenyuin@uitm.edu.my

Abstract. Coconut fibre or more commonly known as coir fibre is obtained from the outer shell of coconut. The production of coconut in Malaysia is the fourth important industrial crop after oil palm, rubber and paddy field. However, wastes generated by the coconut are almost fully utilised as by-products and reinforcing material for soil. In fact, the availability of the coconut waste in Malaysia makes it suitable for soil reinforcing material purposes. Problem that sometimes occurs is to dispose this big, heavy and hard coconut at landfill area. It may require more space to go through landfill method. Disposal of this waste material is vital as it can affect the environment. Therefore, by using part of the coconut component can help to overcome this problem. In this research, white coir fibre was utilized as the reinforcing material to sandy soil to suit the backfill material for earth retaining structure that is related to road infrastructure. The lengths of the white coir fibre were fixed at some length according to its orientation such as random, vertical and horizontal. Shear strength parameter were tested using Direct Shear Test. In this research, inclusion of white coir fibre in sandy soil shows enhancement of the shear strength parameter such internal friction angle of soil regardless of its orientation. However, there is no clear relationship between the effect of white coir fibres orientations to the shear strength of sandy soil.

1. Introduction

The concept of reinforcing soil with natural fibre materials originated in ancient times [1]. Nowadays, soil reinforcement or soil stabilisation is no longer odd in the research field. Soil reinforcement is defined as a technique adopted to improve the engineering characteristics of soil in order to improve the soil properties such as shear strength, compressibility, density; and hydraulic conductivity [2,3]. Many cases develop due to unsuitable backfill material of retaining earth structure. Nevertheless, sandy soil is usually used to suit the backfill material. Hence, to increase the shear strength of this soil, suitable materials should be utilised. Other soil reinforcing technique such as the commonly used soil nailing is expensive. In order to substitute this method, utilisation of soil reinforcement using waste materials are utilized. The most important concept of reinforcement is the development of friction between the soil and the reinforcement leading to the transfer of the built-up loads of the soil to the reinforcement [4,5].

There are two types of materials that can be utilized for the purpose of soil reinforcing which are manmade and natural materials. Manmade materials examples are Polypropylene (PP) fibres, glass

IOP Conf. Series: Materials Science and Engineering **429** (2018) 012031 doi:10.1088/1757-899X/429/1/012031

fibres, steel fibres and others. Meanwhile, natural fibres that are widely used by the researchers are coconut fibre, palm fibre, jute, bamboo, cane and sisal.

Inclusion of coir fibre into the soil as soil reinforcement is highly efficient. There are two types of coir fibre which is brown coir fibre and white coir fibre. Mature coconuts produced brown coir fibres while immature coconuts produced white coir fibres. This research focused on the use of white coir fibre as reinforcing material as it is easy handling. With the high degree of cultivation of coconut in Malaysia, the coir fibre can be put to use for reinforcing material. This shall prove to be a highly economic measure of strengthening of soil and with high degree of efficiency [6]. The previous study that has been carried out by previous researchers is using brown coir fibre as soil reinforcing material. Brown coir fibre is used because these are thick, coarse, strong, durable and have high abrasion resistance. Due to its high lignin content, coir degradation takes place much more slowly than in other natural fibres. So, the fibre is also very long lasting with infield service life of 4-10 years [7]. Fibres were placed at different specific orientations with respect to the shear plane. The fibre content, orientation of fibres, and modulus of fibres were found to influence the contribution of fibres to the shear strength development [8].

The main objective of this research was to determine the shear strength of sand reinforced with white coir fibre. Furthermore, the effect of white coir fibre orientation in reinforced sand was investigated. From this research, the performance of white coir fibre will be evaluated and compared with the brown fibre.

2. Materials and Methods

In this research, two types of sands were used i.e. fine sand (0.2mm to 0.063mm in diameter) and coarse sand (2.0mm to 0.6mm in diameter). Coconut wastes were collected from a commercial centre at Jalan Permatang Pauh, Penang (Figure1). The coconut wastes were kept in a tight container to preserve the moisture content and natural properties before the coir were extracted.

FIGURE 1. Coconut waste and the extracted coir

A series of direct shear test according to BS 1377: Part 7: 1990 were conducted to determine the shear strength of sand reinforced with coir fibre. Shear box of 60 mm x 60 mm dimension was used, and the rate of displacement was kept at 0.5 mm/min. In order to ensure uniform specimen density, dry pluviation method was implemented whereby sand was dropped at the height of 30 cm from the shear box plate using suitable funnel. The coir fibres were placed in three different orientations; vertical, horizontal and random, as shown in Figure 2.

IOP Conf. Series: Materials Science and Engineering **429** (2018) 012031 doi:10.1088/1757-899X/429/1/012031

FIGURE 2. Orientation of coir fibres for direct shear test

Summary of test parameters for the direct shear test is shown in Table 1 below.

TABLE 1. Summary of test parameters.

Factors	Orientation		
	Horizontal	Vertical	Random
Length of coir fibres(mm)	50	25	50
Gap between coir fibres(mm)	5	10	-
Number of coir fibres	11	25	11

3. Results and Discussions

This research investigates the effect of coir fibre orientations to the shear strength behaviour of sand. The results of direct shear tests for both fine and coarse sand without coir fibres are compared with sand with coir fibres. The results of the direct shear test are shown in Table 2.

TABLE 2. Summary of direct shear test results

Specimen	Coir Orientation	Maximum stress (kPa)					
		Normal stress (kPa)					
		100	(Diff)%	200	(Diff)%	300	(Diff)%
Unreinforced Fine Sand	-	117.833	-	214.625	-	345.083	-
Fine Sand with Coir Fibres	Vertical	155.708	(+32)	290.375	(+35)	366.125	(+6)
	Horizontal	143.083	(+21)	223.042	(+4)	357.708	(+3)
	Random	138.875	(+17)	265.125	(+23)	391.375	(+14)
Unreinforced Coarse Sand	-	122.042	-	189.375	-	328.250	-
Coarse Sand with Coir Fibres	Vertical	198.583	(+62)	273.542	(+44)	328.250	0
	Horizontal	185.167	(+51)	366.125	(+93)	446.083	(+35)
	Random	227.250	(+86)	336.667	(+77)	483.950	(+47)

Specimens reinforced with coir fibres for all orientations are found to show higher maximum stress values compared to specimens without coir fibres. This indicates improvement to the reinforced specimens. The highest improvement is found to be in coarse sand with random orientation of coir fibres which shows 86% improvement. However, there is no clear relationship between the effect of coir fibres orientation to the amount of enhancement.

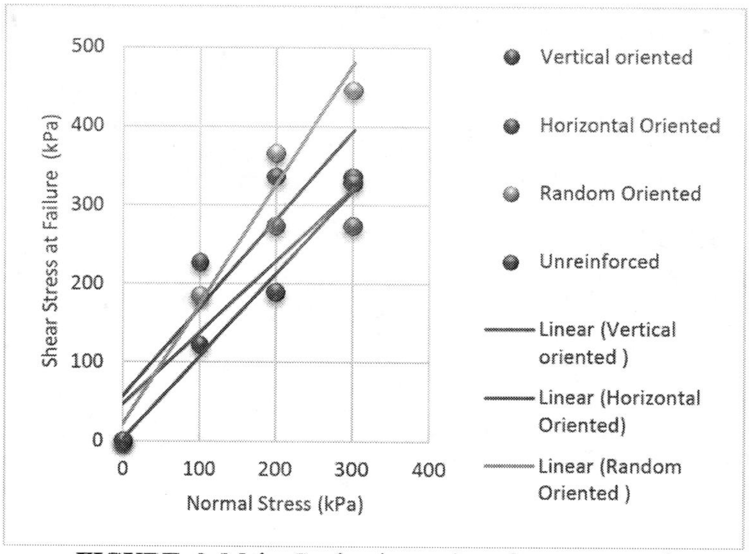

FIGURE 3. Mohr-Coulomb envelope for coarse sand

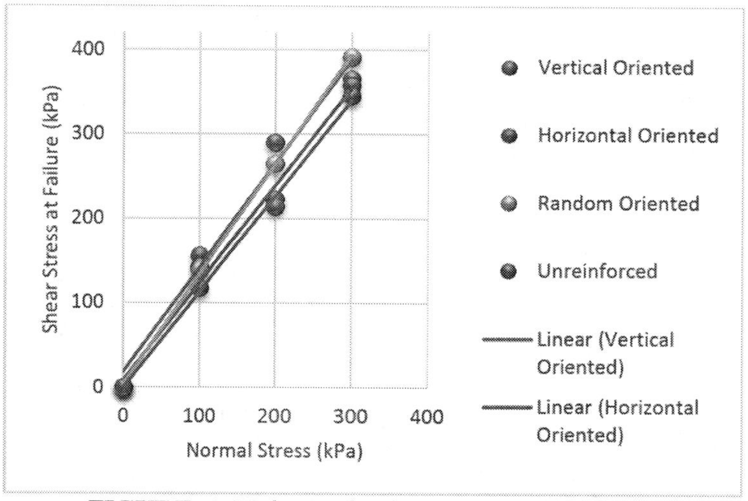

FIGURE 4. Mohr-Coulomb envelope for fine sand

TABLE 3. Summary of direct shear test results

	Coarse Sand				Fine Sand			
	Without Coir	Vertical	Horizontal	Random	Without Coir	Vertical	Horizontal	Random
Angle of friction (°)	46.45	42.27	56.64	57.36	48.54	50.96	49.07	52.44
(Diff.) (%)	-	(-9)	(+22)	(+23)	-	(+5)	(+1)	(+8)

4. Conclusions

This study aims to determine the performance of sandy soil reinforced with white coir fibres. The effect of coir fibres orientation to the soil shear strength is also investigated. Based on the study, the additional of coir fibres is found to improve the shear strength of sandy soil. The followings are the conclusions of the study:

- Sand reinforced with white coir fibres show improvement compared to sand without coir fibres.
- There is no clear relationship between the effect of white coir fibres orientations to the shear strength of sandy soil.
- Based on the study, there is a good potential for white coir fibres to be used as soil reinforcement material especially as backfilling material for retaining walls.

Acknowledgements

The authors acknowledge the Faculty of Civil Engineering, Universiti Teknologi MARA Cawangan Pulau Pinang for the lab facilities. Special thanks to Encik Mohamad Azrul Aswad Bin Muhamad Nor and those who contributed to this study directly or indirectly.

References

[1] T Yetimoglu and O Salbas 2003 A study on shear strength of sands reinforced with randomly distributed discrete fibers *Geotext. Geomembranes*

[2] Harish Parimi, Lavanya Punnam, Haragopal Peddinti 2018 A Modern and Experimental Study on stabilization of Marine Clay by using coir fibre for Foundation *Int. J. Trend Sci. Res. Dev.*

[3] S Kazemian, B B K Huat, a Prasad, and M Barghchi 2011 A state of art review of peat: Geotechnical engineering perspective *Int. J. Phys. Sci.*

[4] K Rao and P P R T Anusha 2011 STABILIZATION OF EXPANSIVE SOIL WITH RICE HUSK ASH LIME AND GYPSUM – AN EXPERIMENTAL STUDY *Int. J. Eng. Sci. Technol.*

[5] M K Rahman, S Rehman, and O S B Al-Amoudi 2011 Literature Review on Cement Kiln Dust Usage in Soil and Waste Stabilization and Experimental Investigation *IJRRAS*

[6] D Das, D Kaundinya, R Sarkar, and B Deb 2016 Shear Strength Improvement of Sandy Soil Using Coconut Fibre *Int. J. Civ. Eng. Technol.*

[7] S M Hejazi, M Sheikhzadeh, S M Abtahi, and A Zadhoush 2012 A simple review of soil reinforcement by using natural and synthetic fibers *Constr. Build. Mater.*

[8] G M Ayininuola and P O Oladotun 2016 Geotechnical Properties of Coconut Coir Fiber Soil Mixture *J. Civ. Eng. Res.*

IOP Conf. Series: Materials Science and Engineering **429** (2018) 012032 doi:10.1088/1757-899X/429/1/012032

Phytoremedial potential of aquatic macrophytes for the removal of copper in contaminated waters

C Marajan[1*], M Neging[1], N A Akbar[2], F Kennedy[2], N H Johnny[2]

[1]Faculty of Civil Engineering. Universiti Teknologi MARA Sarawak, 94300 Kota Samarahan, Sarawak, Malaysia.
[2]Faculty of Civil Engineering, Universiti Teknologi MARA Pulau Pinang, 13500 Permatang Pauh, Pulau Pinang, Malaysia.

*Email: carolinemarajan508@uitm.edu.my

Abstract. The aim of this study is to identify the potential aquatic macrophytes to be used for phytoremediation and to compare the reduction of copper by two species of aquatic plants. Two aquatic plants; *Centella asiatica* and *Eichhornia crassipes*, the effectiveness were evaluated for their capabilities of copper removal from copper solution. The aquatic macrophytes were placed for 7 days in 6 liters of solution that contained 1 mg/L, 3 mg/L and 5 mg/L of copper. Atomic absorption spectrometer (AAS) was used to conduct analyses of heavy metals contents. It was shown in the results that *Centella asiatica* achieved complete removal of 1 mg/L copper concentration after the third day of treatment. *Centella asiatica* also displayed maximum removal of 92% of 3 mg/L copper concentration and 62% for 5 mg/L copper concentration. The peak efficiency at the end of the experiments were at almost similar for *Eichhornia crassipes*, 83% for the 1 mg/L concentration, 80% for the 3 mg/L concentration and 83% for the 5 mg/L concentration. Overall, *Centella asiatica* exhibit faster removal rate of copper compared to *Eichhornia crassipes*. In terms of effectiveness, the percentage removal in solution was higher using *Centella asiatica* at lower concentration of copper. However, at higher concentration of copper (5 mg/L), *Eichhornia crassipes* was found to be more effective in removing copper compared to *Centella asiatica*.

1. Introduction

The increasing rate of urbanization all over the world has led to the increase of economic and industrial activities, which in turn would most certainly increase the amount of harmful contaminants and pollutants discharged into the surrounding environment of those expanding urban areas. Depending on how the contaminants and pollutants are being handled, the level of pollution on the surrounding environment may vary and discharged into various mediums namely the air, water and soil. Heavy metals contaminant, in particular, is the by-product of mostly industrial and agriculture activities, which could seep into the soil and may make its way into nearby water bodies such as rivers, lakes and underground water wells. Contamination of agricultural soils with several heavy metals including lead (Pb), cadmium (Cd), copper (Cu) and chromium (Cr) has been widely reported [1,2].

Contamination of heavy metals in water source used for human consumption would have a detrimental effect to humans. For the sake of human safety and health in the area of contamination of heavy metals, the presence of heavy metals need to be reduced or better, removed to ensure the water on the site of contamination would not be a health hazard to the people who are living in and around the vicinity of that area. There are a number of ways to remediate a contaminated water body, which can be categorized into three methods namely, biological treatment, chemical treatment and physical treatment. Each method has its own advantages and disadvantages in terms of practicality, costs of treatment process and its effect on the surrounding landscape. Therefore, the selection of method for remediating a water body are determined by many factors such as the scale of the pollution, the size of water body, the type of water body, the funding for the remediation process and also the cause of pollution [2,3].

Phytoremediation is an approach using aquatic macrophytes to treat water contaminated by organic and inorganic substances including heavy metals. Currently, phytoremediation using the aquatic macrophytes are used to overcome the environmental issues especially water and soil pollution. Phytoremediation takes advantage of the natural process in which the aquatic macrophytes undergo a process of breaking down inorganic compounds to less potent compounds [3,4]. Inorganic compounds such as heavy metal and synthetic substance cannot be biodegraded but they can be transformed through the body of the macrophytes where it is transformed into compounds, which have less mobility or toxicity than their original form. Aquatic macrophytes have great potential to accumulate heavy metals inside the plant bodies. In this study, two aquatic macrophytes, pennywort (*Centella asictica*) and water hyacinth (*Eichhornia crassipes*) were used to determine their potential in removing copper (Cu) of various concentrations. The effectiveness of the plants was determined based on copper content analyses in the solution.

2. Materials and Methods

2.1. Selection of aquatic macrophytes

Centella asiatica and *Eichhornia crassipes* were collected from Sungai Kelang Ubi, Bukit Mertajam, Pulau Pinang. These aquatic macrophytes were placed in a tank containing tap water for two-weeks, before using the macrophytes for the removal of copper.

2.2. Preparation of copper contaminated solution

Synthetic copper solution at 1 mg/L, 3 mg/L and 5 mg/L concentration were prepared using the compound copper (II) sulphate, Cu (II) SO_4 and distilled water. It was weighted accordingly using the analytical balance to determine the correct amount to obtain the desired concentration of copper (Cu). The calculation for the copper (Cu) concentrations is calculated using the dilution method formula, which is shown in equation 1.1.

$$m_1 v_1 = m_2 v_2 \tag{1.1}$$

Where:

m_1 = Concentration of the stock copper (Cu) solution (mg/mL)
v_1 = Volume of the stock copper (Cu) solution needed to be added in to the container to achieve the required concentration (mL)
m_2 = Required concentration of the copper (Cu) solution (mg/mL)
v_2 = Required volume of copper (Cu) solution (mL)

2.3. Experimental set-up

The aquatic macrophytes, which were placed in tap water for two weeks, were washed with no tissue removed. To minimize the variations of the macrophytes tested, it is important that the plants have the same background, looked green and fresh, have approximately same size and have approximately the same number of leaves to avoid big surface differences. The criteria of the plant that were taken into account in choosing it for the experiment was its size, weight, root length and number of stems/leaves on the plant. On average, the weight of all the macrophytes were approximately 27 g, the roots length were about 8 cm and the number of stem/leaves were 5 or 6 for each plant. The criteria of the macrophytes would affect its functionality to absorb the copper (Cu) in the solution; therefore it is crucial to use plants of similar physical criteria to obtain a consistent and accurate data as possible. The experiment is set-up in 3 sets of varying copper (Cu) solution concentrations with 3 containers containing the same concentration of copper (Cu) solution. The aquatic macrophytes were placed in 6 liters of solution containing of 1 mg/L, 3 mg/L and 5 mg/L of copper. The macrophytes were placed in 2 out of the 3 containers in each set, where the container without any macrophytes inside of it acts as the control container. The study was conducted for 7 days where copper (Cu) solution is assessed by comparing the daily readings for copper (Cu) concentration in the solution where the macrophyte is placed in a container containing the copper (Cu) solution. Samples of the copper (Cu) solution are taken daily to check for the concentration of copper (Cu) using atomic adsorption spectrometer (AAS) throughout the period of the experiment. At the end of the experiment, the reduction of copper (Cu) in the solution is converted into efficiency in form of percentage.

3. Statistical Analysis

The data were presented as mean values of duplicates. Statistical analysis using the one-way ANOVA by Excel 2010 (Microsoft Office) was measured to assess significant differences between the two aquatic macrophytes. The comparisons of mean using the least significant different test were calculated for P-values. A value of $P < 0.05$ was considered significant.

4. Results and Discussions

Figure 1(a) and Figure 1(b) shows the mean final concentrations of copper solution. For containers containing *Centella asiatica*, the aquatic macrophytes was able to reduce the concentration of copper from 5 mg/L to 1.90 mg/L, 3 mg/L to 0.23 mg/L and complete removal for 1 mg/L copper concentration as shown in Figure 1(a). Comparable trend of removal was also observed for containers containing *Eichhornia crasippes*, where the concentration of copper decreases from 5 mg/L to 0.87 mg/L, 3 mg/L to 0.59 mg/L and 1 mg/L to 0.17 mg as shown in Figure 1(b). As reported by [3], *Centella asiatica* and *Eichhornia crassipes* used in the removal of copper was effective treating copper contaminated water with initial concentration of 5.5 mg/L where copper concentration was reduced to 0.92 mg/ and 2.1 mg/L by pennywort (*Centella asiatica*) and water hyacinth (*Eichhornia crasippes*), respectively. Similar study by [4], also reported effective removal of copper with initial concentration of 0.35 mg/L, which was reduced to 0.02 mg/L within 5 days.

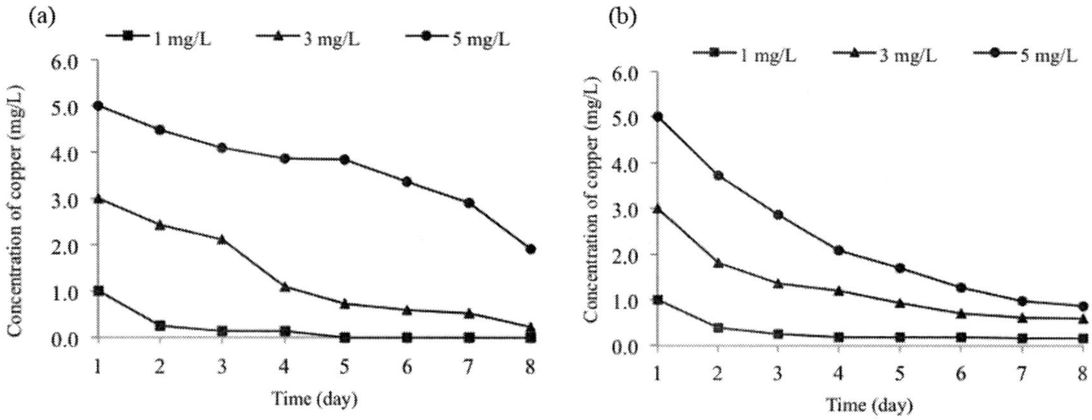

FIGURE 1. Final concentration of copper by (a) *Centella asiatica* and (b) *Eichhornia crassipes*

Figure 2(a) and Figure 2(b) shows the percentage removal of copper within seven days for *Centella asiatica* and *Eichhornia crassipes*. Results showed that *Centella asiatica* achieved complete removal of 1 mg/L copper concentration after the third day of treatment as shown in Figure 2(a). *Centella asiatica* also exhibited maximum removal of 92% of 3 mg/L copper concentration and 62% for 5 mg/L copper concentration. Figure 2(b) showed that *Eichhornia crassipes* removed 83% of copper at 1 mg/L, 80% of copper at 3 mg/L and 83% of copper at 5 mg/L within 7 days. Apparently as shown in Figure 2(a) and 2(b), the highest removal is obtained after the first day and the percentage removals for all concentrations reached the maximum after 5 days. Figure 2(b) also shows that copper removals by *Eichhornia crassipes* are the highest for the concentration of 1 mg/L and the least copper removals were at 83% for *Eichhornia crassipes. Centella asiatica* was able to removed 62% of copper at the concentration of 5 mg/L. As reported by [3], the reason why *Eichhornia crassipes* and *Centella asiatica* could perform extremely well at low concentration of copper and less effective at higher concentration of copper is due to that the loading effect where sorption sites were saturated by copper ions is at the highest concentration.

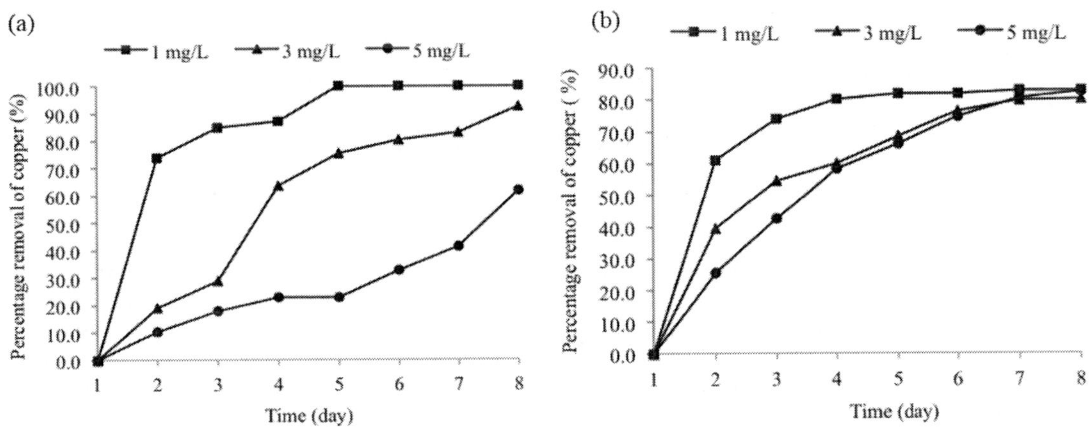

FIGURE 2. Percentage removal of copper by (a) *Centella asiatica* and (b) *Eichhornia crassipes*

Table 1 shows the comparison of removal efficiencies of *Centella asiatica* and *Eichhornia crassipes* with previous studies done by other researcher using the same aquatic macrophytes. It is found that *Centella asiatica* in this study was slightly lower than the study conducted by [3]. Meanwhile for *Eichhornia crassipes* displayed better removal capability than the study conducted by [3]. This study also exhibited similar capability in removing copper when compared to [4] and [5] using *Eichhornia crassipes* where removal efficiency was between 83% to 89%.

TABLE 1. Comparison of removal efficiency of copper between the selected aquatic macrophytes with other studies

Reference	*Centella asiatica*	*Eichhornia crassipes*
This study	62%	83%
[3]	83%	62%
[4]	-	83%
[5]	-	89%

The analysis of variance (ANOVA) test done for the reduction percentage differences of the copper removal between *Centella asiatica* and *Eichhornia crassipes*. Table 2 shows the statistical analysis for the performance of *Centella asiatica* and *Eichhornia crassipes* treatment system. The ANOVA analysis showed that no significant difference ($P > 0.05$) at lower concentration of copper solution (1 mg/L & 3 mg/L) by both *Centella asiatica* and *Eichhornia crassipes* where both aquatic macrophytes have similar capability in removing copper. The removal efficiency at 5 mg/L copper concentration for both aquatic macrophytes was proved to be staticsally significant ($P<0.05$) where *Eichhornia crassipes* was found to be more effective in removing copper at the concentration of 5 mg/L compared to *Centella asiatica.*

TABLE 2. ANOVA table showing the performance of aquatic macrophytes in term of average percentage reduction of copper

Source of Variance	df	SS	MS	F	P
1 mg/L					
Between macrophytes	1	646.43063	646.43063	0.6558717	> 0.05
Error	14	13798.474	985.60527	-	-
Total SS	15	14444.904			
3 mg/L					
Between macrophytes	1	16.382256	16.382256	0.0170599	> 0.05
Error	14	13443.881	960.27719	-	-
Total SS	15	13460.263			
5 mg/L					
Between macrophytes	1	3026.1551	3026.151	4.951	< 0.05
Error	14	8556.86	611.20428	-	-
Total SS	15	11583.015			

5. Conclusions

The results of the present study showed that *Centella asiatica* and *Eichhornia crassipes* were effective in removing copper in contaminated water. In addition, *Centella asiatica* and *Eichhornia crassipes* successfully removed 1 mg/L copper concentration within the permissible amount allowed (0.2 mg/L) for industrial effluent discharge standard for Malaysia. The highest percentages of copper removal were 83% for *Eichhornia crassipes* and complete removal for *Centella asiatica* were obtained at concentrations of copper solution of 1 mg/L. Overall, *Centella asiatica* exhibit faster removal rate of copper compared to *Eichhornia crassipes*. In terms of effectiveness, the percentage removal in solution was higher using *Centella asiatica* at lower concentration of copper. However, at higher concentration of copper (5 mg/L), *Eichhornia crassipes* was found to be more effective in removing copper compared to *Centella asiatica*. Hence, it can be proposed that phytoremediation can be applied using both of the aquatic macrophytes for the removal of copper in contaminated water.

Acknowledgement

The authors would like to thank the Faculty of Civil Engineering, Universiti Teknologi MARA Pulau Pinang for the providing the laboratory facilities, and the laboratory technicians for providing technical assistance.

References

[1] Vigliotta G, Matrella S, Cicatelli F, Guarino F and Castiglione S 2016 *Journal of Environmental Management* **179** 93–102
[2] Rezania S, Ponraj M, Talaiekhozani A, Mohamad S E, Md Din M F, Taib S M and Sairan F M 2015 *Journal of Environmental Management* **163** 125–133
[3] Mokhtar H, Morad N and Ahamd-Fizri F F 2011 *International Journal of Environmental Science and Development* **2** 205-210
[4] Swain G, Adhikari S and Mohanty P 2014 *International Journal of Agricultural Science and Technology* **2** 1 – 7
[5] Keith C, Borazjani H, Diehl S V, Su Y and Baldwin B S 2006 *36th Annual Mississippi Water Resources Conference*, Mississippi
[6] Sharma H K, Dogra P, Sharma N and Sharma S 2013 *Research Journal of Recent Sciences* **2** 261 – 267

IOP Conf. Series: Materials Science and Engineering **429** (2018) 012033 doi:10.1088/1757-899X/429/1/012033

Composite Cathode of NCM Particles and Li_3PS_4-LiI Electrolytes Prepared using the SEED Method for All-Solid-State Lithium Batteries

R Matsuda[1*], E Hirabara[1], N H H Phuc[1], H Muto[1], A Matsuda[1]

[1]Toyohashi University of Technology, 1-1 Hibarigaoka, Tempaku, Toyohashi, Aichi, 441-8580, Japan

*Email: matsuda@ee.tut.ac.jp

Abstract. $Li_7P_2S_8I$ (LPSI) electrolytes were directly produced on a cathode active material $LiNi_{1/3}Mn_{1/3}Co_{1/3}O_2$ (NMC) using the new liquid process which is named 'SEED method'. The formation of electrolytes on NMC was confirmed using scanning electron microscopy and energy dispersive X-ray mapping. The ionic conductivities of the LPSI electrolyte prepared by the SEED method were evaluated and the LPSI electrolyte showed 2.5×10^{-4} Scm^{-1} at room temperature which is almost equal to that of the LPSI electrolyte prepared by the previous liquid-phase shaking process. All-solid-state cells composed of [90MNC·10LPSI composite | Li_3PS_4 | In-Li] were fabricated and the electrochemical properties of the cells were evaluated. The cell fabricated using this composite as cathode material exhibited enhanced performance in comparison with the cell using the cathode material prepared by hand mixing LPSI with NMC.

1. Introduction

Design of high ion conducting paths and good electrochemical contacts between active materials and the solid electrolytes is crucial for fabrication of all-solid-state lithium batteries. The sulfide-based solid electrolytes are currently well known as a promising candidate for all-solid-state batteries and extensively studied due to their high lithium ionic conductivities and suitable plasticity for good contacts at the interface.[1,2] Various methods for composite fabrication of cathode active materials and solid electrolytes have been extensively studied.[3,4,5] However, the volume change in the composites during charge-discharge cycles causes in the gap increment between active materials and electrolytes, which results in a decrease in the capacity of the all-solid-state battery. Thus, it is essential to design the composite electrodes which can adjust and/or follow the gap accordingly to the volume change during the charge-discharge cycling.

Recently our group had developed a new process, liquid-phase shaking (LS) method [6] for preparation of electrolytes used in the Li2S-P2S5 system. The electrolytes prepared by LS method can be mixed with active materials homogeneously due to the fine particle sizes of the resultant electrolytes. [6] We propose a new preparation method named as 'SEED method' to fabricate composites of active materials and solid electrolytes based on the liquid-phase process. In this method, the solid electrolytes were prepared from Li2S using as a seed materials, which are nuclei precipitated on the surface of the active materials that act as scaffold for the composites preparation. The composite of electrolyte Li3PS4 (LPS) and LiNi1/3Co1/3Mn1/3O2 positive electrode active material (NMC) which is a high-potential oxide active material was obtained using the SEED method [7]. In this study, Li7P2S8I (LPSI) solid electrolyte in the Li3PS4-LiI system, which shows higher ionic conductivities than LPS[8], [9], was prepared by the SEED method on the surface of NMC. The NMC-LPSI composites were obtained and their battery characteristics were evaluated.

2. Experimental

Figure 1 shows the preparation procedure of NMC-LPSI composite by the SEED method. Li_2S coating on active material NMC reacts with $Li_3P_2S_6I$ ethyl propionate (EP) solution and then the composites of NMC and electrolyte LPSI with molar ratio of Li_2S:P_2S_5:LiI =3:1:1 were obtained. The illustration shown in Figure 1 is the concept of the SEED method. Two-thirds of Li_2S that was required for preparation of LPSI was coated on active materials NMC while the remaining required one-third of Li_2S was dissolved in the solution to containing '$Li_3P_2S_6I$ with a molar ratio of Li_2S:P_2S_5:LiI =1:1:1. The resultant solution was transparent and homogeneous.[10]

The detailed preparation procedure of NMC-LPSI composite is as follow. Firstly, the NMC was coated with Li_2S. Li_2S (99.9%, Mitsuwa) was added to ethanol (EtOH, Super Dehydrated (99.5%, Wako) and dissolved by stirring treatment for 1 h. NMC (1 wt% $LiNbO_3$ coated) was immersed in the as-obtained Li_2S EtOH solution and then EtOH was evacuated at 170°C for 2 h under reduced pressure to form the NMC-Li_2S composite. Secondly, Li_2S, P_2S_5 (99%, Merck) and LiI (99%, Aldrich) were mixed and dissolved in EP by stirring treatment for 30 min. The obtained EP solution ($Li_3P_2S_6I$ in EP) was transparent and pale yellow. The NMC-Li_2S composite was then added to the '$Li_3P_2S_6I$ in EP' and agitation treatment with stirring was carried out for 6 h at 50°C. The precursor of NMC-LPSI was heat treated at 170°C for 2 h under reduced pressure after evaporated the solvent EP at room temperature. The gray powders of NMC-LPSI composites were obtained.

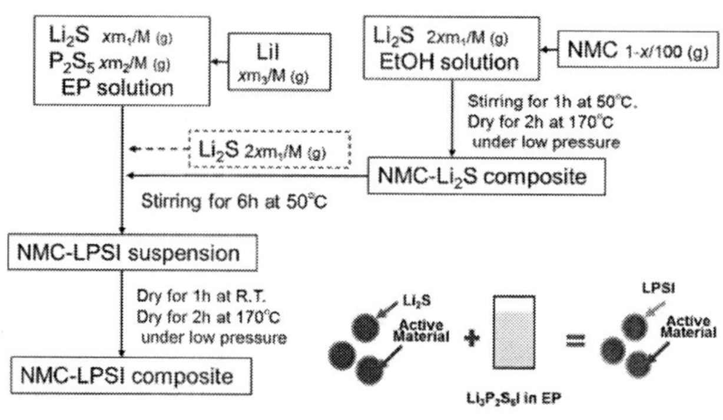

Figure 1: The preparation procedure of NMC-LPSI composites. The weight ratio of NMC: LPSI is (100-x): x. In the Figure, m_1, m_2, m_3 and M (=$3m_1$+m_2+m_3) are the molar weight of Li_2S, P_2S_5, LiI and LPSI, respectively.

The prepared composite was characterized using scanning electron microscopy (SEM; S4800, Hitachi). Elemental mapping was taken using an energy dispersive X-ray spectroscopy (EDXS, Horiba) detector attached to the SEM system. The X-ray diffraction patterns of LPSI were obtained using Ultima IV (Rigaku).

All-solid-state cells were fabricated using the prepared composite as a cathode material, LPS prepared by the liquid-phase shaking method [6] as an electrolyte layer and indium foil (diameter about 8 mm) as a counter electrode. To form electronic conducting paths to the composites, vapor grown carbon fiber (VGCF) was hand mixed with the composites. The cathode composites consist of 90NMC-10LPSI-3VGCF (wt. %). The composite electrodes for electrochemical measurements were prepared by uniaxial pressing (about 80 mg of sample dried at 170°C) into pellets of approximately 0.6 mm in thickness and 10.0 mm in diameter at a pressure of 330 MPa at room temperature. The prepared pellet was then placed in a holder made from polyetheretherketon (PEEK) and switched between two stainless steel rods. The charge-discharge tests of the cells were conducted at a constant current at 0.1 C. In order to compare the effect of composite by the SEED method, the similar tests were also carried out on all-solid-state cells fabricated using positive electrodes composite prepared by hand mixing of NMC and LPSI. All processes for preparation the composites and fabrication the all-solid-cells were performed in an Ar-filled dry box. The cells were placed in a glass tube filled with Ar during the electrochemical characterization measurements.

3. Results and Discussion

Figure 2 shows an SEM image and EDXS elemental mappings of P, S and I of the composites composed of NMC and LPSI produced by the SEED method. There are some sheet-like solid on the surface of NMC particle. From the EDXS mappings, it can be observed that P, S and I are dispersed on the whole particle surface. This indicates that the thin film of LPSI electrolyte was generated on the surface of the particle and followed by growth of sheet-like solid LPSI grew on the NMC.

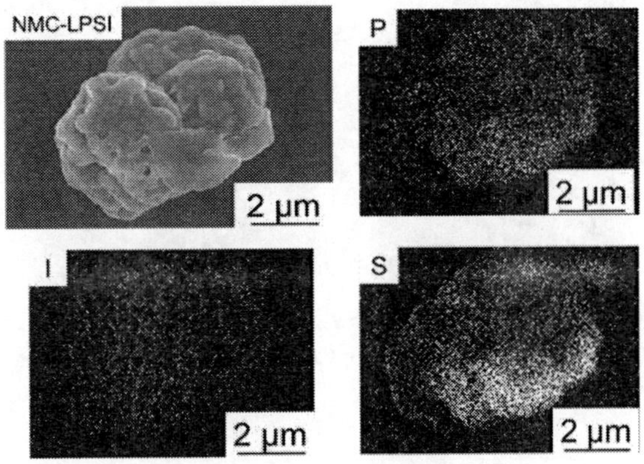

Figure 2: An SEM image of the NMC-LPSI composite particle and corresponding EDXS elemental mappings of P, S and I.

According to the procedure shown in Figure 1, the LPSI electrolyte was prepared using the SEED method without NMC particles to confirm the properties of the composite electrolytes obtained by this method. The Li_2S was added to the $Li_3P_2S_6I$ EP solution instead of the NMC-Li_2S composite that is shown with dotted line in Figure 1. Figure3 shows XRD patterns and SEM images of electrolytes prepared by SEED method and the LS method. Both electrolytes are exhibited almost similar XRD patterns assigned to the crystalline $Li_7P_2S_8I$. It is noteworthy that $Li_7P_2S_8I$ crystals were precipitated from the $Li_3P_2S_6I$ in EP by adding Li_2S where a molar ratio of Li_2S:P_2S_5:LiI was 3:1:1. The SEM image of the electrolyte obtained by SEED method shows sheet-like structure and irregular-shaped particles compared with the electrolyte formed by the LS method. This electrolyte formed by SEED method exhibited almost same profile of the temperature dependence of ionic conductivities as shown in Figure4. The ionic conductivity of the electrolyte formed by SEED method at room temperature was 2.5×10^{-4} Scm^{-1} which is higher than the ionic conductivity of LPS but it still slightly lower than that of LPSI formed by conventional LS method. Thus described that the LPSI on the active material composite produced by SEED method is considered to possess almost similar properties as the LPSI prepared by the previous method.

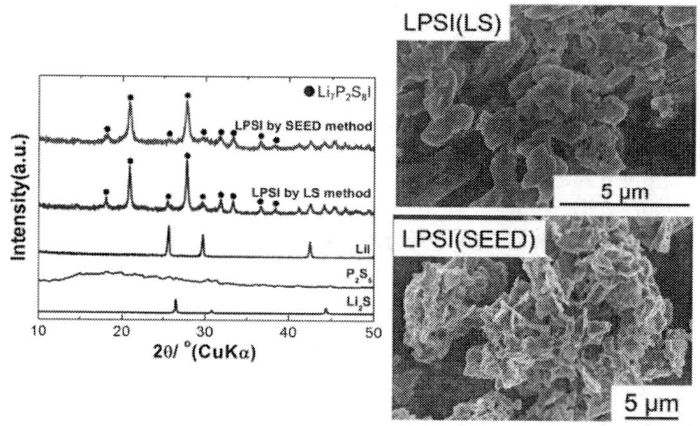

Figure 3: XRD patterns and SEM images of $Li_7P_2S_8I$ prepared by the SEED method and the LS method.

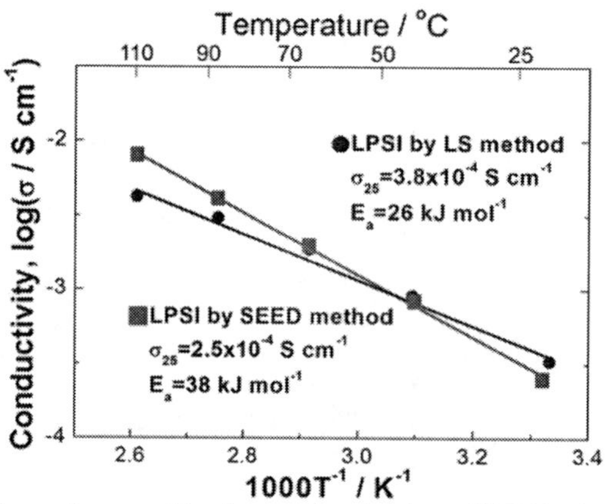

Figure 4: The temperature dependence of ionic conductivities of LPSI electrolytes prepared by SEED method(■) and LS method(●).

The all-solid-state cell using as-obtained active material NMC-electrolyte LPSI composite by SEED method was fabricated. The weight ratio of the composite used was NMC:LPSI =90:10. The cell with hand-mixed active material consisting of NMC and LPSI was also fabricated for comparison. The weight ratio was the same as that of the composite formed by SEED method. LPSI used in the hand-mixed active material was prepared by the LS method. Figure.5 shows the initial charge-discharge curves of the cells at 0.1C. The discharge capacity of the cell using the hand mixed active material as cathode was only 27 $mAhg^{-1}$(dash line in Figure,5) because of the amount of electrolyte content was too low to be mixed homogenously with NMC. On the other hand, the discharge capacity of the cell using NMC-LPSI composite formed by SEED method in the present study exhibited more than 90 $mAhg^{-1}$ (solid line in Figure.5) and good performance compared to the cell formed by the hand-mixing although the amount of electrolyte content was same. This result indicates that the NMC particles were coated homogenously with the LPSI electrolyte by using SEED method.

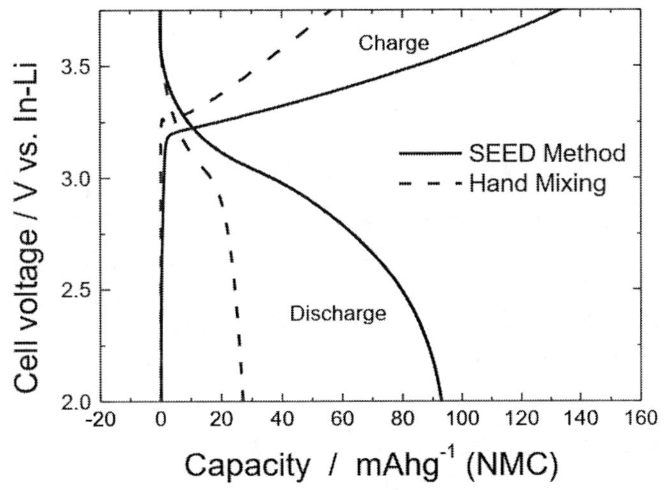

Figure 5: Charge-discharge voltage profiles in the first cycle of the all-solid-state cells fabricated using 90NMC-10LPSI-3VGCF |LPS (LS method)|In at 30°C, 0.1C. Cathode electrode composites were prepared by hand mixing (dash line) and SEED method (solid line).

4. Conclusions

The cathode composite was successfully prepared using the SEED method. Crystalline $Li_7P_2S_8I$ electrolyte was directly produced on the active material NMC by the SEED method. The charge-discharge profiles of the all-solid-state cells show that the composite interface between LPSI and the active material formed by SEED method has good contacts. The SEED method developed in this study can be applied not only to cathode composites but also anode composites using Li metal or carbon.

Acknowledgements

This study was supported by the Advanced Low Carbon Technology Specially Research for Innovative Next Generation Batteries (JST-ALCA-SPRING) program of the Japan Science and Technology Agency. The authors thank Prof. Tatsumisago, Prof. Hayashi and Dr. Sakuda (Osaka Prefecture University) for their assistance and useful discussions.

References

[1] Hayashi A, Hama S, Morimoto H, Tatsumisago M and Minami T 2001 *J. Am. Ceram. Soc.* **84[2]** 477-479
[2] Kamaya N et al 2011 *Nat. Mater* **10,** 682-686
[3] Sakuda A, Hayashi A and Tatsumisago M 2013 Sci. Rep. **3** 2261
[4] Yubuchi S et al 2015 *J. Power Sources* **293** 941-945
[5] Kim D H et al 2017 *Nano Lett. 17* **(5)** 3013–3020
[6] Phuc N H H, Morikawa K, Totani M, Muto H and Matsuda A 2017 *Ionics* **23** 2061-2067
[7] Hirahara E et al 2018 *56th The Ceramic Society of Japan Basic Meeting* 1E10
[8] Ujiie S, Hayashi A and Tatsumisago M 2012 *Solid State Ionics* **211** 42-45
[9] Phuc N H H et al 2017 *J. Power Sources* **365** 7-11
[10] Phuc N H H and Matsuda A 2018 *Advances in Chemistry Research* **42** Chapter 8.

IOP Conf. Series: Materials Science and Engineering **429** (2018) 012034 doi:10.1088/1757-899X/429/1/012034

Noise-Induced Hearing Loss: Engineering Control at Industry and Clinical Audiology Approach at Hospital Level

Ahmad Anas Nagoor Gunny[1], Rafidah Hassan Mydin[2] and Sriyana Abdullah[1]
[1] Department of Chemical Engineering Technology, Faculty of Engineering Technology, University Malaysia Perlis, Kampus UniCITI Alam, Sungai Chuchuh, 02100 Padang Besar, Perlis, Malaysia
[2] Department of Otorhinolaryngology,Tuanku Fauziah Hospital, Perlis, Malaysia

E-mail: ahmadanas@unimap.edu.my

Abstract. In today's increasingly complex industrial design and size, noise is one of the serious hazards to hearing. Loss of hearing as a result of long term exposure to high level noise at industrial sector is become one most potential harmful for workers. In-line with this concern, engineering control design has been adapted at industrial level as prevention approach and as a part of occupational safety and health culture at workplace. While hearing impairment of workers due to high level of noise exposure require proper hearing rehabilitation strategy at clinical level. This paper discusses briefly some of the engineering control adapted at industrial scale as a prevention strategy of hearing impairment and audiological assessment and management for the workers in risk for hearing impairment at clinical level.

1. Introduction

Noise is any unwanted or unpleasant sound that has potential to cause injury and affect the health and quality of life of a person. Excessive exposure to noise, especially at work place, will cause damage to hearing structure, that's lead to noise induced hearing loss (NIHL). Exposure to noise depends on many factors, such as duration of exposure, the magnitude of sound pressure, spectral content and the character of the noise either continuous, varying, intermittent, or impulsive.

Hearing loss is associated to accoupation and nonaccoputaion causes. Occupational hearing loss is one of the most common work-related illness in the world. NIHL associated with occupational can be caused by a one-time exposure to an intense "impulse" sound, such as an explosion in industries, or by continuous exposure to high intensity sounds over an extended period of time, such as noise generated in a manufacturing process.

Most of the industrial sectors associated with this occupational disease are forestry, fishing, mining, transportation, quarry, manufacturing and construction. Exposure to excessive noise in these industires becoming the the most common preventable cause of hearing loss in workers of every country in the world [1] where estimated 16% of the disabling hearing loss in adults in the world is associated to occupational noise[2].

IOP Conf. Series: Materials Science and Engineering **429** (2018) 012034 doi:10.1088/1757-899X/429/1/012034

The Department of Occupational Safety & Health, Malaysia in 2014 reported 2648 cases of work diseases, of which 78.1% were attributed to NIHL [3]. While in Great Britain, it is estimated that 20,000 workers suffer from noise-induced hearing loss caused by their working environment based on data from the Labour Force Survey (three year average period 2013/14-2015/16), equating to a rate of 62 cases per 100,000 workers [4]. In Germany, 4–5 million people which 12–15% of the workforce are due to exposore to high magnitued of noise which defined as hazardous by WHO [5].

A study conducted by Crandell et al. (2004) estimate slightly higher, 11 million Americans exhibiting some degree of permanent noise-induced hearing loss and 40 million in at-risk associated to their working environment [6]. However, in comparing to developed countries, the effects of the exposure to occupational noise are reported higher in the developing regions [7].

Therefore, it is important to reduce the risk of hearing loss at the place or *in situ* by adopting the engineering noise control measure. Audiological assessment and management practice at hospital level aims to reduce the hearing impairment due to workplace conditions . Hence, this paper discuss briefly the engineering control strategies practice in the industry to reduce noise level to prevent mass damage to hearing , and also some of the audiological assessment and management for the workers at risk for NIHL at hospital level.

2. Engineering controls

Increasing size and types of industries increased the demand for strategies in order to reduce the negatives impact to environment and people. Similar to other types of hazard, noise induced hearing loss is preventable through a hierarchy of controls. The use of engineering controls is prioritized over administrative controls and personal protective equipment (PPE). Ear plug and other hearing protection devices are form of PPE generally used in industries to deal with noise pollution issue. However, the problem with this approach is PPE only protect the wearer but not the entire workplace that exposed the problem. On the other hand, Engineering control involve modifying, replacing workplace in order to reduce or eliminate any potential hazard in the wider scope.

Assessing the applicability of engineering controls is a sophisticated process. There are several stepsneed to follow in evaluation the applicabality of engineeringcontro. First, identiy and define the noise problem which include measuring the magnitued of noise levels and level of exposure on employee and the need for noise reduction. Next, identification of individual noise sources and an assessment of their affects to the overall noise levels. Following these identification and assessment, selection and development of appropiate engineering control. However the selection criteria will be influenced, to some extent, by the cost of purchasing, operating, servicing, and maintaining the control [8]

This type of control measurement involved the protection in first phase or known as design phase and at control the hazard at its potential source which cover more large space, personnel and more effective. Prevention through design is found to be effective in reducing the number of injury and illness cases reported in the workplace and minimizing cost associated with these cases [5].

With regards to noise hazard in industries, Engineering controls require physical changes to the workplace such as redesigning equipment at the noise source or along the transmission path to reduce the exposure of worker to high magnitude of noise. For example, re-design the continuous mining machine which is used in coal mining industries to reduce excessive noise that might cause the hearing impairment of mine operators. The flight bars and the conveyor belt tail rotor of the machine were coated with special synthetic crystalline compound that able to reduce noise and improve the life span of the equipment [9]. Liu (2003) reported an innovative silencer known as duct resonator array was installed at the diffuser in large centrifugal compressor that effectively reduced the sound level of the machine. There are several companies that prefer to buy quiet equipment rather than modifying their equipment to reduce noise pollution [10]. For example, in one of the successful noise control collaborations between Mining companies and NIOSH, the mining companies are committed to "buy quiet' equipment [11].

Another form of engineering control in design phase is by modifying the workplace or facility-related noise at the source to reduce risk related to hearing impairment. This may involve placing an acoustical shield between the noise source and employee and enclose or isolate the noise source. A complete enclosure that equipped with a closed top improved noise reduction as compare to partial enclosure. However, the design still needs some opening area for personal and equipment flow [11]. In a Practical Guide manual for preventing occupational hearing loss published by NIOSH, often a noise-related machine may be completely enclosed which allow the operator who operating the machine remain outside of the enclosure [8]. The noise reduction design of workplace manifests advantages not only to the operator but also to all the adjacent areas.

In addition to modification and design of equipments and facilities, modification of process design also can be adopted under engineering control program to reduce the negative effect of noise in industries. Most of industries which have the risk and exposure to both noise and chemicals include painting, printing,, boat building, construction, metal products, chemicals, petroleum, leather products, furniture making, agriculture, and mining. [12-13].

Studies showed clinical testing indicates that combine exposure to noise and solvents such as toluene, ethylbenzene, trichloroethylene induces synergistic adverse effects on hearing. [12,9]. The adverse effects of these chemicals and other ototoxicity chemicals can be seen as damage to cochlear hair cells, central nervous system, or both [14]. Thus, substitution of chemical that would not produce negative synergistic effect with noise in a process production is preferable in a workplace.

Apart from the design phase, proper maintenance procedure could reduce and minimise the noise from the source (ie.equipment). Reduction of noise resulting from friction between metal parts could be reduced by use lubrication or use of soft elastic interspacing on the parts. Other example is reduction of mechanical shock between parts by using conveyor belts instead of chutes to avoid noisy falls, covering metal tables, metal wheels, etc. with a material, such as rubber [15].

As explain above, engineering control has been adopted using different approaches to limit the risk of hearing loss at industries. On the other hand, at hospital level, initiating appropriate health surveillance program and continuous periodic audiological evaluation are regarded as the key to reduce the damage of hearing loss [16] whereby Metidieri et al., (2013) [16] found these audiological management can lead to rehabilitation through individual actions and group therapies.

3. Assessment of hearing: Pure tone Audiometry

Audiometric evaluation is crucial to the success of the hearing loss prevention program, since it is the only way to determine whether occupational hearing loss is being prevented. Management must allocate sufficient time and resources to the audiometric program to allow accurate testing; otherwise, the resulting audiograms will be useless.[8]

According to Factories and Machinery (Noise Exposure) Regulations, (1989), "Hearing status of each individual working in noisy environment should be tested in yearly basis" [17]. The test must

IOP Conf. Series: Materials Science and Engineering **429** (2018) 012034 doi:10.1088/1757-899X/429/1/012034

be preceded by a period of quiet of at least fourteen hours to eliminate the potential of temporary threshold shift (TTS) in the test results [16]. Pure tone audiometry is a standard test used to test hearing, where the softest intensity level of sound perceived by individuals at certain frequencies will be plotted and displayed in an audiogram.

The pattern of an audiogram begins with an elevation in hearing threshold around the 3- 6 kHz region of the audible spectrum leading to a 'notch' in the audiogram, may suggest presence of NIHL. In classic presentation, a notched sensorineural hearing loss is noted at 4000 Hz that should be relatively symmetric, as displayed in Figure 1. The degree of noise induced hearing loss that occurs is dependent on the level of noise, how long someone is exposed to it, and to some extent on individual susceptibility [14].

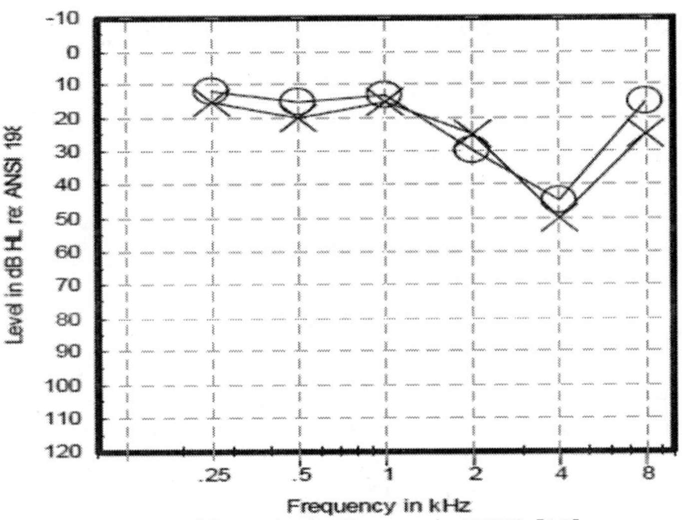

Figure 1: Audiogram in NIHL [19]

Patient presence with NIHL will commonly complaint of tinnitus which is the phantom perception of sound in the absence of other physical sound [19]. Studies carried out by a group reseachers on relationship between tinnitus and hearing loss found that 85 to 90% of the individuals with tinnitus present some level of hearing loss[20]. The perception of sound can be in the form of buzzing, ringing, or hissing, and can be continuous or intermittent. Tinnitus could be severe enough causing stress, hearing and sleep disturbance, and impact individual daily life.

4. Clinical approach; Diagnosis, Counseling & Recommendation

As part of audiological assessment approach, detailed history on patient current and past workplace, duration of work, also their past medical history is documented. Patient present with the pattern of NIHL, will be notified and a proper counseling conveyed. Counseling given on the aspects of effects of hearing loss, that result in major communication difficulties due to loss of sensitivity and clarity of high pitched sounds and inability to discriminate speech sounds particularly in the presence of background noise [21]. These difficulties will result in substantial physical and psychological distress, and lead to social isolation and depression. Explanation on the consequences of NIHL used to create awareness, and motivation for patient to protect their hearing by proper use of earmuffs, earplugs and other protective devices. Patient strictly advised, for consistent and correct use of hearing protection device and limit daily activities that expose them to high noise. As mentioned by Berger [22], effective training has motivated individuals to adopt hearing loss prevention behaviors.

Those patient that severely affected and bothered by their tinnitus problem, a thorough counseling using cognitive behavioral therapy (CBT) method given to them. This approach is to gradually distract their mind from thinking or focusing on the presence of tinnitus and engaged themselves to their daily activity. Apart from that, sound therapy, or hearing aid is preferential choice given to help them overcome tinnitus issues.

In some cases, if indicated, patient will be given a support letter about their current hearing status, potential damage and consequences they may face if continuously exposed to noise at their workplace. Administrative control of noise exposure may be suggested to employer, to lay out quiet workstations, and limiting employee exposure time by implementing shift work schedules, as restricted in Factories and Machinery (Noise Exposure) Regulations 1989 [17].

Individual detected with the permanent sensorineural hearing loss with significant difficulty in hearing ability will be offered for hearing aids. Hearing aid is an electronic device that will amplify the sound to and improved listening and speech comprehension of those with hearing loss [23]. Apart from that, a yearly hearing monitoring will be scheduled for patient as this will be considered as continuous yearly assessment which will provide insight on the workers motivation and awareness level for hearing protection, whereby the information not only useful for counseling, but also for any compensation claim later.

5. Conclusion

Hearing loss at industry is regarded as one of the serious problem that should be taken into consideration by the industries which carry the potential risk. Implementing engineering control at initial design stage should be prioritized by the employer as part of NIHL prevention strategy. Besides, healthcare professional, like audiologist should plays vital roles in reducing prevalence of NIHL within the industrial community and protect hearing health clinically with proper audiological management and assessment, in order to enhance the quality of life of millions of people.

Acknowledgement

This research project has been supported by University Malaysia Perlis under Short Term Grant (STG 9001-00507) and Faculty of Engineering Technology, University Malaysia Perlis, Malaysia.

References

[1] Alberti, P.W., Symons, F. and Hyde, M.L., 1979. Occupational Hearing Loss The Significance of Asymmetrical Hearing Thresholds. *Acta oto-laryngologica*, **87**(3-6), pp.255-263.
[2] Nandi, S.S. and Dhatrak, S.V., 2008. Occupational noise-induced hearing loss in India. *Indian journal of occupational and environmental medicine*, **12**(2), p.53
[3] Sam, W.Y., Anita, A.R., Hayati, K.S., Haslinda, A. and Lim, C.S., 2017. Prevalence of Hearing Loss and Hearing Impairment among Small and Medium Enterprise Workers in Selangor, Malaysia. *Sains Malaysiana*, **46**(2), pp.267-274
[4] Noise induced hearing loss in Great Britain, 2016, accessed 19 January 2017, *http://www.hse.gov.uk/statistics/causdis/deafness/index.htm*

[5] Concha-Barrientos, M., Steenland, K., Prüss-Üstün, A., Campbell-Lendrum, D.H., Corvalán, C.F., Woodward, A. and World Health Organization, 2004. Occupational noise: assessing the burden of disease from work-related hearing impairment at national and local levels.

[6] Crandell, C., Mills, T.L. and Gauthier, R., 2004. Knowledge, behaviors, and attitudes about hearing loss and hearing protection among racial/ethnically diverse young adults. *Journal of the National Medical Association*, **96**(2), p.176.

[7] Nandi, S.S. and Dhatrak, S.V., 2008. Occupational noise-induced hearing loss in India. *Indian journal of occupational and environmental medicine*, **12**(2), p.53.

[8] Franks, J.R. and Merry, C.J. eds., 1996. *Preventing occupational hearing loss: a practical guide* (No. 96-110). US Dept. of Health and Human Services, Public Health Service, Centers for Disease Control and Prevention, National Institute for Occupational Safety and Health, Division of Biomedical and Behavioral Science, Physical Agents Effects Branch.

[9] Smith, A.K., Kovalchik, P.G., Alcorn, L.A. and Matetic, R.J., 2009. A dual sprocket chain as a noise control for a continuous mining machine. *Noise Control Engineering Journal*, **57**(5), pp.413-419.

[10] Liu, Z., Dresser-Rand Co, 2003. *Acoustic liner and a fluid pressurizing device and method utilizing same*. U.S. Patent 6,550,574.

[11] Davies, P., Maling Jr, G.C. and Tocci, G.C., 2009. National Academy of Engineering "Technology for a quieter America" project—impact on "green acoustics"?. *The Journal of the Acoustical Society of America*, **125**(4), pp.2502-2502.

[12] Campo, P., Morata, T.C. and Hong, O., 2013. Chemical exposure and hearing loss. *Disease-a-month*, **59**(4), pp.119-138.

[13] Parent-Thirion, A., 2007. Fourth European working conditions survey. Office for official Publ. of the European Communities.

[14] Kirchner, D.B., Evenson, E., Dobie, R.A., Rabinowitz, P., Crawford, J., Kopke, R. and Hudson, T.W., 2012. Occupational noise-induced hearing loss: ACOEM task force on occupational hearing loss. Journal of occupational and environmental medicine, **54**(1), pp.106-108.

[15] Hansen, C.H. and Bies, D.A., 1995. Engineering noise control. London: Spon.

[16] Metidieri, M.M., Rodrigues, H.F.S., Oliveira Filho, F.J.M.B., Ferraz, D.P., Almeida Neto, A.F.D. and Torres, S., 2013. Noise-Induced Hearing Loss (NIHL): literature review with a focus on occupational medicine. International archives of otorhinolaryngology, **17**(2), pp.208-212.

[17] Mohtar, M., 1999. Factories and machinery (Noise exposure) regulations 1989 (No. L-0537). MOPGC.

[18] Reid, N., Thorne, P., Ameratunga, S., Williams, W., Purdy, S. and Dodd, G., 2006. Best practice in noise-induced hearing loss management and prevention: A review of literature, practices and policies for the New Zealand context. Accident Compensation Corporation of New Zealand.

[19] Han, B.I., Lee, H.W., Kim, T.Y., Lim, J.S. and Shin, K.S., 2009. Tinnitus: characteristics, causes, mechanisms, and treatments. Journal of Clinical Neurology, **5**(1), pp.11-19.

[20] Sanchez, T.G., Guerra, G.C.Y., Lorenzi, M.C., Brandão, A.L. and Bento, R.F., 2002. The influence of voluntary muscle contractions upon the onset and modulation of tinnitus. Audiology and Neurotology, **7**(6), pp.370-375.

[21] Tyler, R.S. and Baker, L.J., 1983. Difficulties experienced by tinnitus sufferers. Journal of Speech and Hearing disorders, **48**(2), pp.150-154.

[22] Berger, E.H., 2000. Noise control and hearing conservation: why do it. The noise manual. Fairfax, VA: American Industrial Hygiene Association.

[23] Dillon, H., 2008. Hearing aids. Hodder Arnold.

IOP Conf. Series: Materials Science and Engineering **429** (2018) 012035 doi:10.1088/1757-899X/429/1/012035

The Effect of Paddy Straw Content on Properties of Regenerated Cellulose Biocomposite Films

H. Farah Norain[1], A. W. M. Kahar[1,2] and H. Ismail[3]

[1]School of Materials Engineering, Universiti Malaysia Perlis (UniMAP), Jejawi, 02600 Perlis, Malaysia
[2]Center of Excellence Geopolymer and Green Technology, School of Materials Engineering, Universiti Malaysia Perlis (UniMAP), Kompleks Taman Muhibah, Jejawi 2, 02600, Perlis, Malaysia.
[3]School of Materials and Mineral Resources Engineering, Universiti Sains Malaysia (USM), Engineering Campus, 14300 Nibong Tebal, Penang, Malaysia.

*Email:kaharwahab@unimap.edu.my

Abstract. Regenerated cellulose (RC)/paddy straw (PS) biocomposite films was prepared by addition of PS powder in partially dissolving microcrystalline cellulose into an 8% lithium chloride/N,N-dimethylacetamide (LiCl/DMAc) solution. The effect of paddy straw on tensile properties and crystallinity index (CrI) of RC/PS biocomposite films were studied. The results obtained revealed that the tensile strength and modulus of elasticity of RC/PS biocomposite films compared with pure regenerated cellulose was significantly improved at 5 wt% PS loading but decrease with the increment of PS. The increase in crystallinity index at 5 wt% proved the enhancement affected by the incorporation of PS.

1. Introduction

The development of biopolymer from renewable resources has become the objective to substitute the non-biodegradable petrochemical synthetic polymer [1,2]. Cellulose is a promising resource for developing the sustainable, biodegradable and environmentally friendly product as it is a dominant component in the vast majority of plant form with outstanding properties such as high flexibility, good thermal and chemical stability, biocompatibility, non-toxicity and low cost [3,4]. The utilization of cellulose not only protects the environment but also helps to curb the dependence on fossil fuels while maintaining the consumption of it [5biocam].

However, cellulose is difficult to process because of its crystalline form and the close packing and numerous number of hydrogen bonds makes it neither melt nor soluble in water or any common solvent [6]. Currently, several solvent have been used for regeneration of cellulose process. Most studies related to regenerated cellulose composite describe use of LiCl/DMAc as the solvent system for the dissolution of cellulose. There are some other non-derivatising cellulose solvent such as mineral acids, various molten hydrates and ionic liquid that also have been used in the processing of regenerated cellulose composite [6-8]. Incorporation of filler into polymer matrix has been proved to increase the polymer properties. Rice is the main agriculture crop that generate huge amount of agriculture waste in the form of straw accounting around more than 100 MT annually in India. Globally, paddy waste is the major crop left over annual generation of more than 770 MT [9]. Paddy straw being a lignocellulose predominantly contains cellulose (32-47%), hemicellulose (19-27%) and lignin (5-24%) which are associated with each other through complex bonding [10].

In this study, the effect of PS powder loading on tensile properties and crystallinity index (CrI) of RC/PS biocomposite were investigated.

2. Eperimental

2.1. *Materials*

Paddy straw (PS) was obtained from Kilang Beras Bernas Kuala Perlis, Perlis. The collected agro residue was thoroughly washed and dried at 50 °C. The dried paddy straw was ground into powder and sieved. The average size of PS powder was 53 µm as measured by a Malvern particle size analyzer. The DMAc and LiCl were supplied by Merck and Acros Organic, while microcrystalline cellulose (MCC) was supplied by Sigma-Aldrich.

2.2. *Preparation of RC/PS Biocomposite Films*

Microcrystalline cellulose requires activation before dissolution in LiCl/DMAc. Activation procedure requires MCC to be immersed in distilled water followed by acetone and DMAc for 1 hour at room temperature. Activated MCC was dissolve in 8% (w/v) LiCl/DMAc. The cellulose solution was stirred for 1 hour until transparent solution was obtained. The viscous cellulose solution was casted on the glass plate for regeneration process to take place. The RC/PS biocomposite film was washed with distilled water and dried at room temperature. The formulation of RC/PS biocomposite films with different PS loading is shown in Table 1.

Table 1. Formulation of RC/PS biocomposite films

Materials	RC biocomposite films	RC/PS Biocomposite films
MCC (wt.%)	3	3
PS (wt.%)	-	5, 10, 15, 20

2.3. *Tensile Testing*

Tensile properties such as tensile strength, modulus of elasticity and elongation at break can be obtained from tensile test. Tensile testing was conducted using Instron 5569 Universal testing system by referring to ASTM D 882. A cross-head speed of 10 mm/min was used and test was carried out at room temperature. Five samples were tested with size of 50 x 15 mm was cut from each of RC biocomposite films.

2.4. *X-Ray Diffraction*

Crystallinity index (CrI) of RC biocomposite film samples was determined by X-ray diffraction method by using Bruker Advance model system. The pattern with Cu $K\alpha 1$=1.5406Å at 30kV and 10mA were recorded from 2° to 35°. CrI as defined by Segal was used as a measurement of crystallinity.

$$CrI\ (\%) = (I_{002}-I_{AM})/I_{002} \times 100 \quad (1)$$

Where I_{002} is the maximum intensity of the (022) lattice diffraction peak and I_{AM} is the intensity scattered by the amorphous part of the sample. The diffraction peak for plane (002) is located at the diffraction angle of around 2θ=21°-22° and the intensity scattered by amorphous part was measured at lowest intensity at a diffraction angle around 16°

3. Result and Discussion

3.1. Tensile Properties

Figure 1 and 2 shows the tensile properties of RC/PS biocomposite films containing different paddy straw weight content. The tensile strength and modulus of elasticity are significantly increase with the incorporation of PS powder up to 5 wt% and decreases with the continuous addition of PS in RC biocomposite films. This could be attributed to high aspect amount of PS ratio in the system that leads to agglomeration causing poor dispersion in the RC matrix. The improvement in mechanical properties of RC/PS biocomposite film over pure RC film could be due to better filler interaction thus enhanced the properties for stress transfer. The elongation at break of RC/PS biocomposite films is shown in Figure 2. As the PS content further increase, the elongation at break reduced. This is due to the restriction of RC chain movement by the PS powder during deformation.

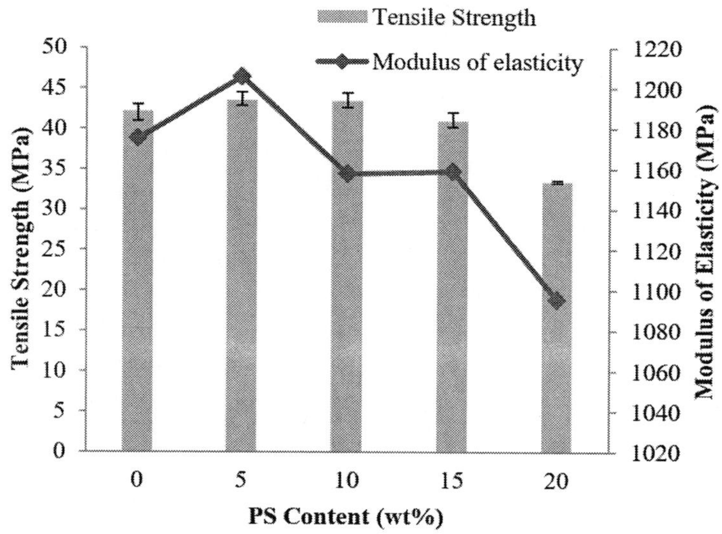

Figure 1: The effect of PS on tensile strength and modulus of elasticity of RC/PS biocomposite films

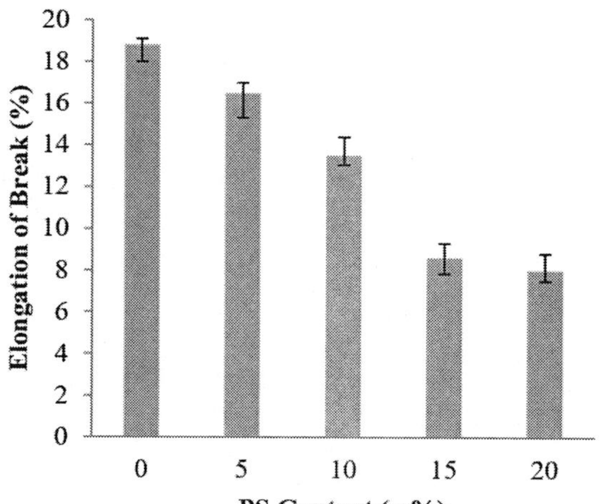

Figure 2: The effect of PS on the elongation at break of RC/PS biocomposite films

3.2. X-Ray Diffraction

XRD patterns of RC and RC/PS biocomposite films with different PS content are shown in Figure 3. The regenerated cellulose films exhibit characteristic peaks at $2\theta = 11°$-$12°$ which correspond to (110) plane, while $2\theta = 20°$ and $2\theta = 22°$ correspond to (110) and (020) planes. The diffraction peaks of RC/PS5 and RC/PS15 biocomposite films are very much similar to the pure RC film. This is show that PS was distinctly formed by the regenerated cellulose chain. The RC/PS biocomposite films with 15 wt% of PS show highest CrI as presented in Table 2. This could be attributed to the further increment of PS in the RC matrix increases the crystalline materials in the biocomposite films. Good interaction between PS and RC matrix also promotes the increment in crystallinity index of RC/PS biocomposite films.

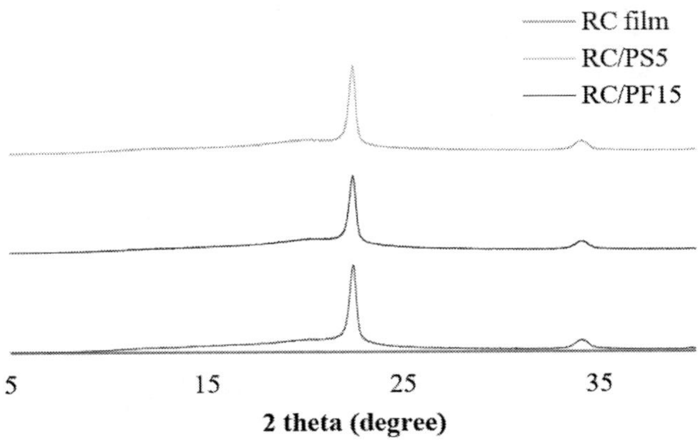

Figure 3: The effect of PS on the crystallinity index (CrI) of RC/PS biocomposite films

Table 2:

RC/PS biocomposite films	**Crystallinity index (CrI)**
RC/PS film 0 wt%	45.9
RC/PS film 5 wt%	46.0
RC/PS film 15 wt%	47.3

4. Conclusion

The incorporation of paddy straw in the regenerated cellulose biocomposite films improves the tensile strength and modulus of elasticity at 5 wt% of PS and decreased at higher PS loading. The elongation at break decrease with the increasing amount of PS loading in the biocomposite films. Crystallinity index (CrI) increase with the increment of PS content.

Acknowledgement

This research was supported by the Ministry of Higher Education (MOHE). The authors would like to thank the Fundamental Research Grant Scheme (FRGS), 9003 00601 for financial support.

References

[1] Soheilmoghaddam M Wahit M U 2013 Development of regenerated cellulose/halloysite nanotube bionanocomposite with ionic liquid *International Journal of Biological Macromolecules* **58** 133-139.

[2] Silva R D Vongsanga K Wang X Byrne N 2015 Development of a novel regenerated cellulose composite material *Carbohydrate Polymers* **121** 382-387
[3] Soheilmoghaddam M Wahit M U Mahmoudian S Hanid N A 2013 Regenerated cellulose/halloysite nanotube nanocomposite films prepared with an ionic liquid *Materials Chemistry and Physics* **141** 936-943
[4] Soheilmoghaddam M Wahit M U Whye W T Akos N I Pour R H Yussuf A A 2014 Bionanocomposites of regenerated cellulose/zeolite prepared using environmentally benign ionic liquid solvent *Carbohydrate Polymers* **106** 326-334
[5] Pullawan T Wilkinson A N Eichhorn S J Discrimination of matrix-fibre interaction in all-cellulose nanocomposite *Composites Science and Technology 70* **16** 2325
[6] Duchemin B J C Newman R H Staiger M P 2009 Structure-property relationship of all cellulose composites *Composite Science and Technology* **69** 1225-1230
[7] Huber T Mussig J Curnow O Pang S Bickerton S Staiger M P 2012 A critical review of all-cellulose composites *Journal of Materials Science* , **47**(3), 1171-1186.
[8] Kauldhar B S Yadav S K 2018 Turning waste to wealth: A direct process for recovery of nano-silica and lignin from paddy straw agro-waste *Journal of Cleaner Production* **194** 158-166.
[9] Kaur K Phutela U G 2016 Enhancement of paddy straw digestibility and biogas production by sodium hydroxide-microwave pretreatment *Renewable Energy* **92** 178-184.

Effect of Graphite Loading and Addition of Ferric Chloride ($FeCl_3$) on the Properties of Poly (Vinyl Chloride)/Poly (Ethylene Oxide)/Graphite Polymer Films

M D Siti Hajar[1*], P L Teh[1], A W M Kahar[1], M I M Yazid[1], H N Hamzah[1]

[1]School of Materials Engineering, Kompleks Pusat Pengajian Jejawi 2, Taman Muhibbah, Universiti Malaysia Perlis (UniMAP), Jejawi, 02600, Arau, Perlis, Malaysia.

*Email: sitihajarmohddin@gmail.com

Abstract. In this study, films consisting of a blend of poly (ethylene oxide)/ poly (vinyl chloride) (PEO/PVC) and a conductive filler, graphite with the addition of $FeCl_3$ as a doping were prepared and characterized for their mechanical and electrical properties. Solid polymer blend films based on PVC/PEO (50/50 wt%/wt%) with different graphite loading were prepared by using solution casting technique. Electrical conductivity results discovered the conductivity increased with increasing of filler loading and addition of $FeCl_3$. However, the addition of $FeCl_3$ in graphite loading led to a decreased in tensile strength of PEO/PVC/Graphite polymer films. The dispersion of graphite and mechanism of conductive path in the polymer films were also investigated by scanning electron microscopy (SEM). The morphology of the PEO/PVC/Graphite polymer films shows that agglomeration occurred to complete the connection of conductive path, thus improving the conductivity behavior of the polymer films.

1. Introduction

Basicly, polymeric materials either organic or inorganic are electrical insulators in their nature. Therefore limits its potentials as smart material under electrical conducting application. Efforts to promote from insulating polymer to conductive polymer composites bring to the rising of nano-carbons filler like carbon nanofibers [1], carbon nanotubes [2], and graphene [3]. Due to their excellent mechanical, thermal and electrical properties in composites materials leads to new research area in polymer matrix composites (PMC). There are a lot of advantages by using conductive polymers composites compared to metallic conductors. For instance, they can be easily converted into low cost technologies, light weight, provide corrosion resistance and also offer a wide variety of electrical conductivities.

Graphite is crystal form of a pure carbon much resembling of mica-sheets of strongly linked atoms, with very weak bonds between the sheets. Graphite generally used as electronically conducting filler because it is economical and abundant naturally [4]. It has excellent electrical properties and well dispersion in polymer matrix [5]. In this study, the electrical properties, mechanical properties and morphology of PVC/PEO blends containing graphite-$FeCl_3$ filler were investigated.

2. Experimental

2.1. Materials
PVC powder with molecular weight of 220 000 g/mol and PEO powder with molecular weight of 100 000 g/mol as the matrix for conductive polymer film was supplied by AR Alatan Sdn. Bhd., Alor Star, Malaysia. Graphite with molecular weight of 12.02 g/mol and particle size of <50 μm was supplied by R&M Chemicals. Plasticizer used in this research was Dioctyl terepthalate with molecular weight of 390 g/mol and tetrahydrofuran with molecular weight of 72.11 g/mol was provided from AR Alatan Sdn. Bhd.

Figure 1: Morphology of graphite particle.

2.2. Composites Preparation
The PEO/PVC/Graphite polymer films were processed by solution casting method. Initially, PEO and PVC are dissolved in tetrahydrofuran solution (THF) and were blended together to get homogeneous solution. The DOTP as was incorporated at the same amount for all composition. After that, conductive filler were added in the solution. The PEO/PVC/Graphite polymer films solutions were poured and casted on a glass mould after 4 hours and let it to evaporate in a fume cupboard at room temperature. The PEO/PVC/Graphite polymer films were formed when THF are fully dried. The same process was repeated to produce films which consist of different graphite content.

2.3. Preparation of Secondary Doping of Graphite with $FeCl_3$
Graphite were dissolved with 6 wt% of ferric chloride in toluene and then stirred for 24 hours. Next, the doped mixture was washed with distilled water and dried in an oven for 5 hours at 60°C. Finally, the doped graphite powders were grinded to fine powders and ready to preserve for sample preparations.

2.4. Scanning Electron Microscopy (SEM)
The tensile properties of the samples were assessed by using Universal Testing Machine Instron 5569 according to ASTM D638. Rectangular shaped specimens were tested at relative humidity (30 ± 2) °C and ambient temperature (25 ± 3) °C with the crosshead speed of 30 mm/min. The tensile strength and Young's modulus data were acquired from the average values of five samples for each samples.

2.5. Electrical Conductivity Test
The conductivity measurement of PEO/PVC/Graphite polymer films was conducted using I-V measurement systems. The conductivity was tested by Keithley Model 4200 Semiconductor Characterization System. The samples were conducted with voltage varied from 0 V to 10 V. The

electrical conductivity was determined using its relationship with resistivity and conductivity as shown in Equation 1and Equation 2 below:

$$\rho = R.\left(\frac{w \times t}{l}\right) \quad (1)$$

Where R = resistance of the films, w = width, t = thickness, l = length between the probe contact.

$$\sigma = \frac{1}{\rho} \quad (2)$$

3. Results and Discussion

Resistance, R of PVC/PEO/Graphite and PVC/PEO/Graphite-$FeCl_3$ conductive polymer films with filler content 5-25 wt% prepared by solution casting method was measured at room temperature by two point probe. Their conductivities, σ are plotted against filler loading as shown in Figure 2. Electrical conductivity of both PVC/PEO/Graphite and PVC/PEO/Graphite-$FeCl_3$ polymer films are increased with further increase in the amount of filler. Poor electron mobility in the polymer film was observed due to lower graphite loading. At this point, the graphite filler is covered by PVC/PEO matrix where the polymer film does not form conductive pathways for electron mobility to reach percolation limit.

The addition of $FeCl_3$ content in PVC/PEO/Graphite-$FeCl_3$ polymer films results in improvement in the electrical conductivity. The conductivity of PVC/PEO/Graphite-$FeCl_3$ is higher than PVC/PEO/Graphite conductive polymer films and was increased by numerous orders of magnitude. The electrical conductivity of PVC/PEO/Graphite polymer films increased to 10^{-2} S/cm compared to 10^{-4} S/cm at 25 wt% of graphite loading with the presence of $FeCl_3$ in the polymer films. From the previous study by J Jin et al. [6], addition of LiCl in Graphite/polyethersulphone composites has increased the electrical conductivity due to the formation of lithium ion-graphite intercalation compound (Li-GIC). The densities of holes in the composites are greater than pure graphite. So, the high density and the high mobility of holes contributed to the high electrical conductivity of the doped composites.

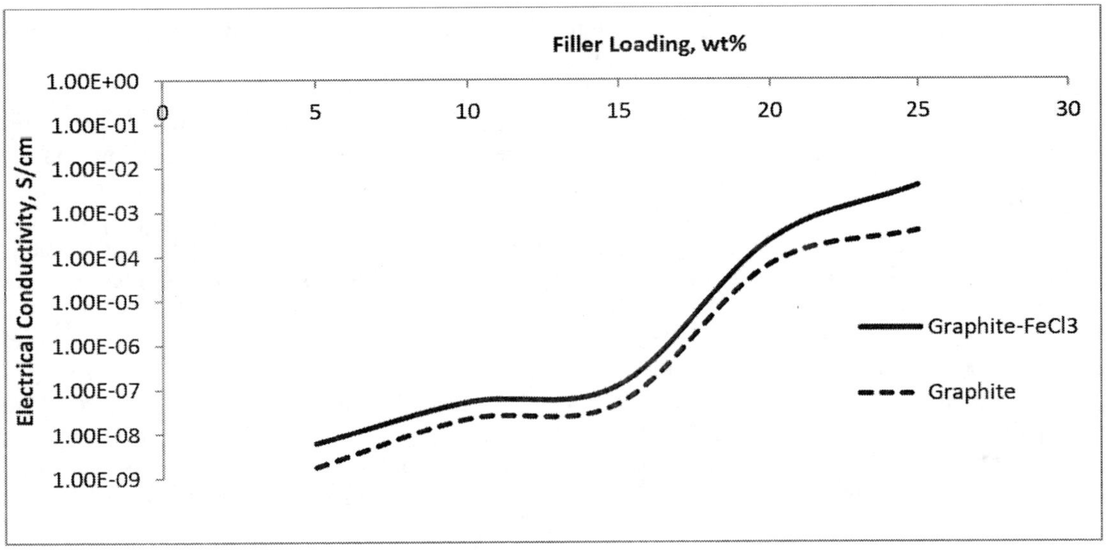

Figure 2: Electrical conductivity of PVC/PEO/Graphite and PVC/PEO/Graphite-$FeCl_3$ polymer films at different graphite loading.

The tensile strength of PVC/PEO/Graphite and PVC/PEO/Graphite-$FeCl_3$ polymer films is shown in Figure 3. From the Figure 3, it can be observe that the tensile strength decreased gradually with the increasing of graphite content. The addition of graphite fillers attributes a restrictive effect over the PEO/PVC polymer films by deformation of their molecular chains and restricting the motion. This can be explained by the weak filler-matrix interaction between the graphite and PVC/PEO matrix in the polymer films. Too high concentration of conductive filler could lead to material redundancy and dentrimental mechanical properties. At similar filler loading, PVC/PEO/Graphite polymer films with the presence of $FeCl_3$ showed lower tensile strength compared to PVC/PEO/Graphite polymer films. The additions of $FeCl_3$ have given negative effect in mechanical properties on PVC/PEO/Graphite polymer films.

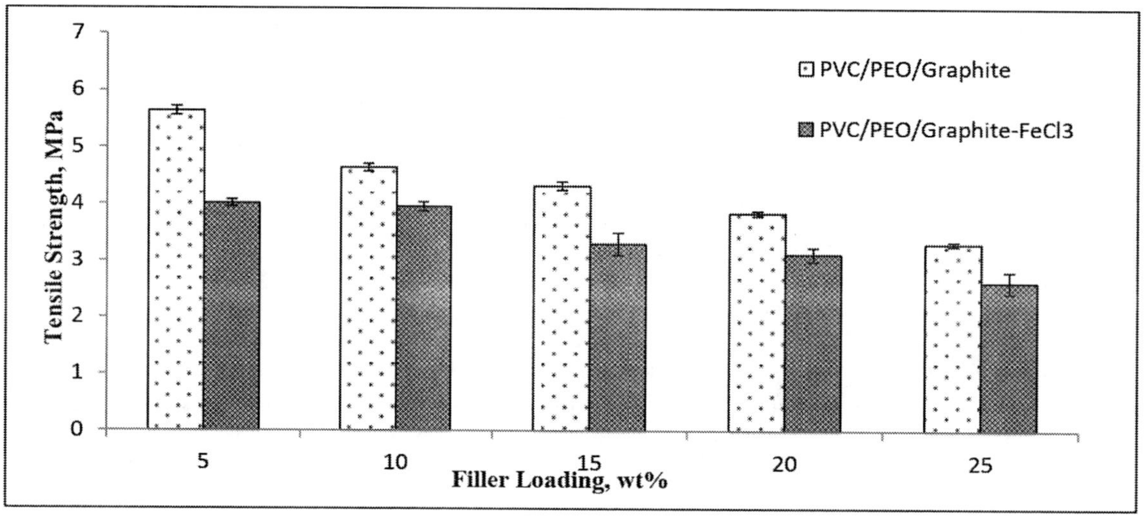

Figure 3: Tensile strength vs. graphite loading of PVC/PEO/Graphite polymer films with and without presence of $FeCl_3$.

The SEM morphologies of PVC/PEO/Graphite and PVC/PEO/Graphite-$FeCl_3$ polymer films are represented in Figure 4. The analysis was conducted to visualize the effect of graphite loading and presence of $FeCl_3$ on the properties of the films. Figure 4 (a)-(c) shows the surface morphology of PVC/PEO polymer films with addition of 5 wt%, 15 wt% and 25 wt% of graphite contents respectively. The surface morphology appears in the SEM images to be almost entirely filled at high graphite content as it surface are become rough and polymer matrix become darker colour. The effect of addition of $FeCl_3$ on the PVC/PEO/Graphite polymer films are shown in Figure 4 (d)-(f). The surface morphology of PVC/PEO/Graphite polymer films containing $FeCl_3$ exhibited poor surface morphology compared to polymer films with the absence of $FeCl_3$. Furthermore, the presence of $FeCl_3$ displayed agglomeration of the filler as a results of $FeCl_3$ ion reducing the PVC/PEO/Graphite miscibility and weaken the interaction between them. This proves that the tensile strength results are poor with the addition of $FeCl_3$ in the polymer films. However, the existences of agglomeration contributed to the high electrical conductivity as the electric charge are easy to transfer between graphite filler.

IOP Conf. Series: Materials Science and Engineering **429** (2018) 012036 doi:10.1088/1757-899X/429/1/012036

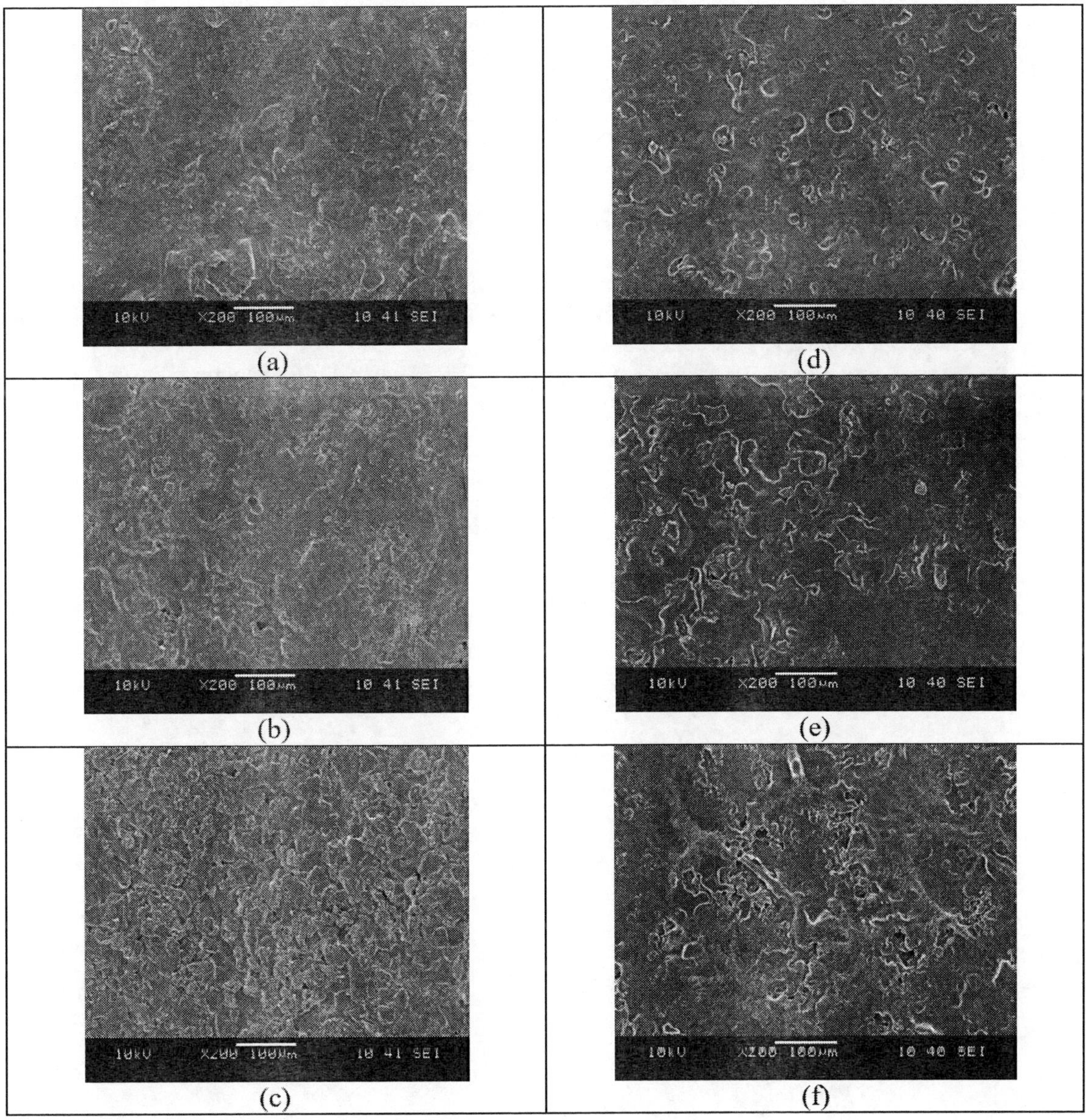

Figure 4. SEM micrograph of PVC/PEO/Graphite polymer films with and without presence of $FeCl_3$.

4. Conclusion

PVC/PEO/Graphite and PVC/PEO/Graphite-$FeCl_3$ polymer films were prepared by solution casting technique. The electrical conductivity of the films were found to be increased with higher loading of graphite and further improved with the introduction of $FeCl_3$. It was observed that the tensile strength of the films was reduced with the addition of graphite. At a similar loading, the presence of $FeCl_3$ reduced the tensile strength of the films. On the other hand, SEM morphology of PVC/PEO/Graphite and PVC/PEO/Graphite-$FeCl_3$ polymer films supported the tensile results by showing more voids and agglomeration as more graphite content incorporated as well as when the $FeCl_3$ were introduced into the films.

References

[1] Sridhar V, Jeon J and Oh I 2010 Microwave extraction of graphene from carbon fibers *CarbonN. Y.* **49** 222–6

[2] Li M, Wan Y, Gao Z, Xiong G, Wang X, Wan C and Luo H 2013 Preparation and properties of polyamide 6 thermal conductive composites reinforced with fibers *Mater. Des. 51* 257–61

[3] Kuila T, Bose S, Mishra A K, Khanra P, Kim N H and Lee J H 2012 Effect of functionalized graphene on the physical properties of linear low density polyethylene nanocomposites *Polym. Test. 31* 31–8

[4] Ahlatciotlu Ozerol, E., Senkal, B. F., and Okutan, M. (2015). Preparation and characterization of graphite composites of polyaniline. *Microelectronic Engineering*, **146**, 76–80

[5] Narimissa E, Gupta R K, Choi H J, Kao N and Jollands M 2012 Morphological, mechanical, and thermal characterization of biopolymer composites based on polylactide and nanographite platelets *Polym. Compos.* **33** 1505–15

[6] Jin J., Leesirisan S. and Song M. 2010 Electrical Conductivity of Ion-doped Graphite/Polyethersulphone Composites, *Composites Science and Technology*.

IOP Conf. Series: Materials Science and Engineering **429** (2018) 012037 doi:10.1088/1757-899X/429/1/012037

Entylene Vinyl Acetate/Carbonized Wood Fiber Composites: The Effect of Zinc Chloride Content on Tensile Properties, Electrical Properties and Thermal Degradation

M P M Hanif[1,2*], A R Rozyanty[1,2], M D S Hajar[1], S J Tan[1,2], A G Supri[3]

[1]School of Materials Engineering, Universiti Malaysia Perlis (UniMAP), Kompleks Taman Muhibah, Jejawi 2, 02600 Arau, Perlis, Malaysia.
[2]Center of Excellence Geopolymer and Green Technology, School of Materials Engineering, Universiti Malaysia Perlis (UniMAP), Kompleks Taman Muhibah, Jejawi 2, 02600 Arau, Perlis, Malaysia.
[3]Faculty of Engineering Technology, Universiti Malaysia Perlis (UniMAP), Aras 1, Blok S2, Kampus UniCITI Alam, Sungai Chuchuh, 02100 Padang Besar, Perlis, Malaysia.

*Email: anepisal@gmail.com

Abstract. Challenges faced to develop conductive composites comprised of poor electro-active stability, control of the mobility, mechanical properties, presentation and concentration of bioactive molecules. Carbonized wood fiber (CWF) was successfully produced from waste wood fiber by using conventional furnace machine at 700 °C in air atmosphere for 4 hours. While, CWF coated $ZnCl_2$ was carried out surface modification process via ethanol solution. In this study, the effect of different $ZnCl_2$ content coating CWF with the fixed CWF filler and ethylene vinyl acetate (EVA) composites was studied. The samples were prepared by using Brabender Plasticoder at160 °C and 50 rpm rotor speed. The result indicated that increasing $ZnCl_2$ content showed tensile strength, tensile modulus, electrical conductivity, and thermal stability compared to unmodified composites.

1. Introduction

Ethylene vinyl acetate (EVA) copolymer is an insulating material and converting it to conductive material is quite challenging. Nowadays, researcher proved that EVA material properties change when combine to another conductive filler into composites. According to Asakura et al.,[1] had prepared the modelling of dielectric properties for EVA/polyaniline (PAni) composites. They found that increased of filler content will increase the electrical conductivity. The use of filler in composites creates new materials simply known as polymer composites. Though, the practice of using synthetic fibers as reinforcement filler initiated health problems issues to the manufacturers and the consumers such as skin irritations and lung cancer due to the content of hazardous ingredients [2, 3]. Hence, many researchers start to explore new alternative reinforcing filler using natural fibers due to the abundance of natural fibers and ease of processability. The use of natural fibers as reinforcing filler in thermoplastic composites offers significant disposability and positive environmental benefits. The

major advantages of natural fibers over synthetic fibers include biodegradability, recyclability, lower specific weight, high specific strength and stiffness, good thermal properties, non-corrosive nature, and lower in cost [4-6].
Converting natural fiber to carbon, carbonization is a particular form of that process in chemical technology called pyrolysis that is the breakdown of complex substances into simpler ones by heating. Khalil et al.,[7] derived from bamboo stem (BS), coconut shells (CNS) and oil palm empty fiber bunch (EFB) were obtained by pyrolysis of fiber at 700 °C to produce carbon black (CB). Generally, wood fibers are hydrophilic in nature whereas synthetic polymers are hydrophobic. The hydrophilic characteristics also known as polar molecule of natural fibers leads to a high moisture absorption, causing dimensional change in the fiber and resulting in the swelling of manufactured composites [8]. This is because water is also a polar covalent molecule, which means it has oppositely charged regions. The oppositely charged regions of the water molecule and the hydrophilic polar molecule are able to attract one another, which is why hydrophilic molecules are generally soluble in water.

Many study involving carbonized wood fiber (CWF) composites have been largely performed by researchers. While, different waste wood, method, temperature, time produces different yields and characteristics of products. Furthermore, this will affected their application at end product. Generally, some researcher aims to focus on chemical properties of elemental analysis percentages after carbonation process. The average weight percentages of the chemical elements forming wood material are about 50 wt% of carbon (C), 43 wt% of oxygen (O), 6 wt% of hydrogen (H), and 1 wt% of nitrogen (N)[9]. Furthermore, Tillman[10] has been reported that hardwood species or grade A, average C content around 47-50 wt%, and for softwood species or grade C, around 50-53 wt%, due to the varying lignin and extractives content. All wood species contain about 6 wt% of H content. The O content ranges from 40-44 wt%, sulphur (S) is less than 0.1 wt% and N ranges from 0.1-0.2 wt%.

Works concerning about the use of different modification content to improve the compatibility between fibers and matrices have been revealed. Zinc chloride ($ZnCl_2$) was used, because zinc has the tendency to donate an electron higher than iron in electrochemical series. Advantages of metal salt catalyst such as ease of volatilization, availability, non-toxicity, low price, and easy in handling, even hold in combination with the toluene precursor, which is less toxic than benzene[11]. In this research, the compatibility of EVA filled CWF using different content of $ZnCl_2$ as a surface modification agent was studied.

2. Experimental

2.1. Materials

The EVA grade 2518 CO with contains 18.1% of vinyl acetate (VA) was supplied by The Polyolefin Company (Singapore) Pte Ltd. The waste wood fiber types of Pulverised Wood Filler with mixed wood grade was obtained from PolyComposite Sdn. Bhd. Kajang, Selangor. Besides that, zinc chloride ($ZnCl_2$) and ethanol were supplied from AR Alatan Sdn. Bhd. Alor Setar, Kedah.

2.2. Methodology

CWF was pyrolysized using a small-scale reactor at 700 °C with a residence time of 3 hours in a muffle furnace by using conventional heating machine with air comprise. Then, CWF filler was modified with different content of $ZnCl_2$ in ethonal solution under room temperature. The CWF was mixed with $ZnCl_2$ by using mechanical stirred model 5430 with rotor speed of 500 rpm for 24 hours. Then, CWF_{ZnCl2} filler was filtered and left to dry in the oven at 50°C for 24 hours to evaporate ethanol residue. The modified CWF_{ZnCl2} were used as filler in EVA composites. The EVA/CWF_{ZnCl2} composites with different content of $ZnCl_2$ were blended by using Brabender Plasticorder at temperature of 160 °C and rotor speed of 50 rpm for 10 minutes, respectively. Then, the hydraulic hot press machine (GT 7014A) was used form plate shape with a thickness of 1 mm. The formulations of EVA/CWF_{ZnCl2} composites are shown in Table 1. Table 2 show mass percentages of CWF and CWF_{ZnCl2} fillers by using energy dispersive X-ray (EDX) spectroscopy analysis.

Table 1 Formulations of EVA/CWF_{ZnCl2} composites at different $ZnCl_2$ content.

Materials	Ratio
EVA (phr)	100
CWF (phr)	10
$ZnCl_2$ (wt%)	0~8

Table 2 Mass percentages from EDX spectra.

Element	Mass (%)	
	CWF	CWF_{ZnCl2}
Carbon (C)	90.66	89.32
Oxygen (O)	9.34	8.20
Chloride (Cl)	-	1.54
Zinc (Zn)	-	0.92

2.3. Characteristic

Tensile properties in which tensile strength and tensile modulus were tested based on ASTM D638 standard by using Universal Testing Machine Instron 5569 with a capacity of the 5 kN, crosshead speed 50 mm/min and gauge length of 50 mm. Electrical properties of conductive polymer were measured by four-probe measurement system model Keithley Model 4200 Semiconductor Characterization System with voltage varied from 0~10 V. The resistivity (**ρ**) and electrical conductivity (σ) calculated by using equation 1 and 2;

$$\rho = \left(\frac{\pi}{ln2}\right) t \left(\frac{V}{I}\right) \quad (1)$$

$$\sigma = \frac{1}{\rho} \quad (2)$$

where; t is sample thickness, V is voltage difference in charge between two points and I is the rate at which charge is flowing. Thermal degradation analysis was carried out by using a Perkin Elmer Pyris TGA 7 thermogravimetric analyzer (TGA) according to ASTM D3850-2000. Samples of about 10 mg were decomposed from 30~900 °C with a heating rate of 10 °C/min and 50 ml/min of nitrogen gas flow to prevent thermal oxidation process of sample.

3. Results and Discussion

3.1. Tensile Properties

Fig. 1 presents the effect of $ZnCl_2$ content on a tensile strength of EVA/CWF_{ZnCl2} composites. The tensile strength increased as the $ZnCl_2$ content increased on EVA/CWF_{ZnCl2} composites. During surface modification of CWF, particle size of CWF particle found to reduce compared to unmodified CWF and this was due to corrosion of $ZnCl_2$ on the surface of CWF filler. When the particle size of CWF was decreased, and the specific surface areas increased during chemical modification. However, this chemical modification affected the chemical bonding between EVA and CWF_{ZnCl2} and their dispersion in the EVA matrix during compounding. These chemical bonding caused the an improvement of the interfacial interaction of filler and matrix[12]. Good interface interaction between CWF_{ZnCl2} particles and EVA copolymer help to tranfer stress from the matrix to filler, which contributed to the higher tensile strength of the composites. When content of $ZnCl_2$ up to 8 wt%, CWF_{ZnCl2} particles tended to agglomerate and lead to poor distribution of CWF_{ZnCl2} in EVA matrix and weak interface interaction between CWF_{ZnCl2} filler and EVA matrix. Thereby, the tensile strength of CWF_{ZnCl2} filled EVA composites decreased as the CWF_{ZnCl2} increased up to 8 wt%[13]. This phenomenon is further explained by the nature characteristics of CWF, which are easily agglomerated in its primary and secondary structure, which has led to poor dispersion and distribution into the polymer matrix [14]

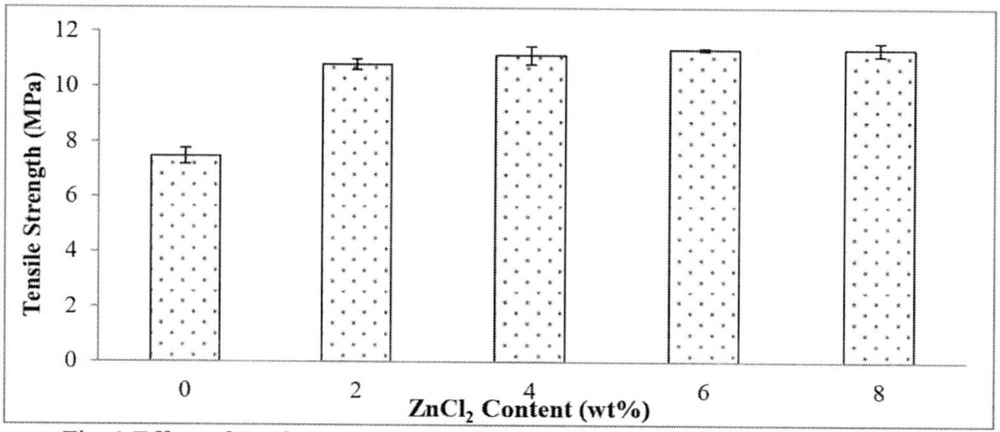

Fig. 1 Effect of $ZnCl_2$ content on tensile strength of EVA/CWF_{ZnCl2} composites.

Fig. 2 shows the effect $ZnCl_2$ content on a tensile modulus EVA/CWF_{ZnCl2} composites. The increment in tensile modulus with respect to $ZnCl_2$ content was due to the incorporation of rigid $ZnCl_2$ particle into the composites and hence imparted greater stiffness to the EVA/CWF_{ZnCl2} composites[15]. This was due to the stiffening effect caused by the good interaction between the $ZnCl_2$ coated CWF and the EVA matrix thus improved the stiffness of the composites. Besides, the polymer chain mobility was restricted by the addition of $ZnCl_2$ resulted from the improvement in rigidity of the composites[16]. It is suggested that the percolated network of CWF can help to improve the tensile modulus in the ternary composites and to stabilize the improvement with the introduction of $ZnCl_2$[17].

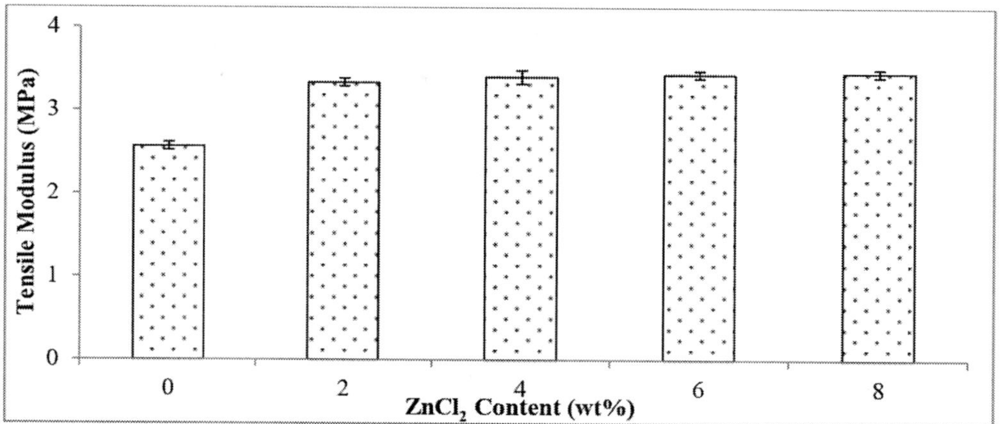

Fig. 2 Effect of $ZnCl_2$ content on tensile modulus of EVA/CWF_{ZnCl2} composites.

3.2. Electrical conductivity

The electrical conductivity of EVA/CWF composites at with different content of $ZnCl_2$ is shown in Fig. 3. The electrical conductivity of composites are increased with increasing of $ZnCl_2$ content. At low $ZnCl_2$ content, the electrical conductivity of the EVA/CWF_{ZnCl2} composites is slightly higher than that EVA/CWF composites due to the CWF coated $ZnCl_2$ particles are isolated from each other by the insulating matrix. As the content of CWF_{ZnCl2} into EVA matrix increased, the CWF_{ZnCl2}particles begin to contact with each other, and a continuous conductive pathway is formed throughout the entire volume of the composites which allowed the electron to travel. Furthermore, introduced $ZnCl_2$ into CWF filler through surface modification caused particle size of filler reduced. Thus, reduction particle size will increase the surface area of filler in matrix and electron can easily jump from one orbital to another orbital[18].

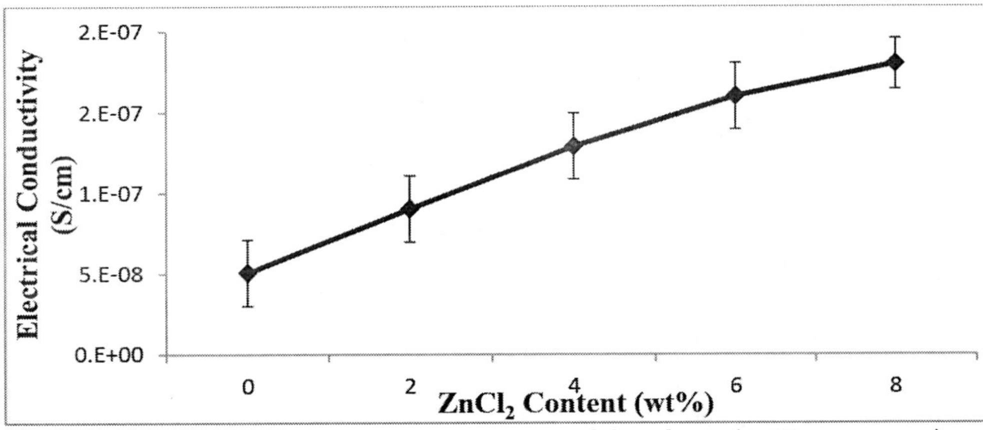

Fig. 3 Effect of $ZnCl_2$ content on electrical conductivity of EVA/CWF_{ZnCl2} composites.

3.3. *Thermal degradation*

The thermogram analysis (TGA) and differential thermogravimetric (DTG) curve of selected EVA/CWF_{ZnCl2} composites at different content of $ZnCl_2$ are presented in Fig. 4, meanwhile Table 3 recorded the initial decomposition temperature (IDT), final decomposition temperature ($T_{-Max\%wt}$) and residual mass of the composites. Addition of various $ZnCl_2$ content demonstrated weight loss through two-stage decomposition process at temperature around 315 to 390 °C and main stage decomposition at temperature 410 to 500 °C due to EVA phase and similar to demonstrated weight loss EVA/CWF composites. The IDT reduced with incorporation $ZnCl_2$ content and similar decrement with thermal stability at $T_{-Max\%wt}$ and residual mass compared to without surface modification. This was due to different particle size of CWF and CWF_{ZnCl2} filler[19]. With a smaller particle size, the CWF_{ZnCl2} fillers were well distributed and dispersed within the EVA matrix, this will form a good interfacial bonding between the filler and matrix. The CWF_{ZnCl2} fillers with the high aspect ratio increased the surface contact area and formed good interfacial bonding with the matrix [20].

Work by Yue[21] reported that weight loss of pure $ZnCl_2$ started at about 400 °C and had a maximum weight loss rate at 587 °C with a heating rate of 30 °C/min in N_2. Based on Table 3, increasing content of $ZnCl_2$ into EVA/CWF_{ZnCl2} composites, the IDT, $T_{-Max\%wt}$ and residual mass increased compared to unmodified composites. These results suggested that additions of $ZnCl_2$ contents will improve the thermal stability and residual mass of the EVA/CWF_{ZnCl2} composites because $ZnCl_2$ acts as a dehydrating agent, eliminating water and inducing the aromatization of carbon[21]. This higher thermal stability of polyacrylonitrile (PAN) owing to the metal powder of $ZnCl_2$ will affect the physical bonding which enhanced the interaction betweem filler to the matrix, thus created a barrier that slowed down the volatilization of the EVA/CWF_{ZnCl2} composites during thermal decomposition process[13]. Supported by Rusu[22], vicat softening temperature increases with the increase zinc powder content in high density polyethylene (HDPE) composites.

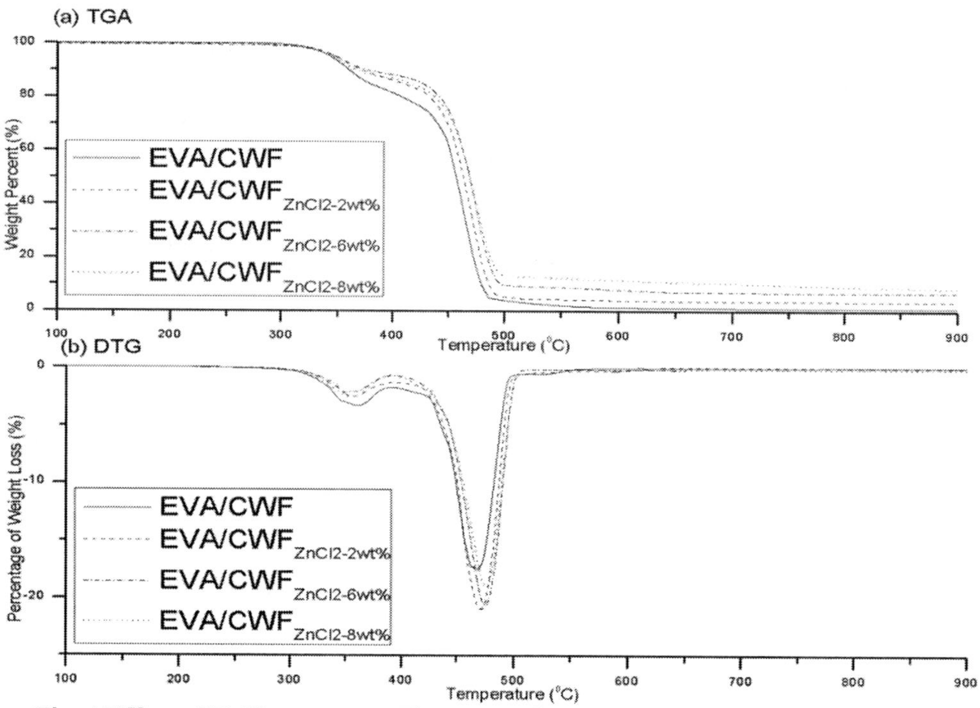

Fig. 4 Effect of $ZnCl_2$ content on Thermogravimetric of EVA/CWF_{ZnCl2} composites.

Table 3 Data of IDT, $T_{-Max\%wt}$ and residual mass of EVA/CWF_{ZnCl2} composites at different $ZnCl_2$ content.

Composite Codes	IDT (°C)	$T_{-Max\%wt}$ (°C)	Residual mass (%)
EVA/CWF	348.12	475.59	4.55
EVA/$CWF_{ZnCl2-2wt\%}$	346.18	474.75	3.53
EVA/$CWF_{ZnCl2-6wt\%}$	346.81	475.13	4.15
EVA/$CWF_{ZnCl2-8wt\%}$	347.14	476.61	4.34

4. Conclusion

The effect of $ZnCl_2$ content as surface modifer in terms of mechanical, electrical, and thermal properties at 10 phr of filler loading coated $ZnCl_2$ at different contents was investigated. The incorporation of CWF coated with $ZnCl_2$ in EVA matrix increased tensile strength, tensile modulus, electrical conductivity, thermal stability, and residual mass of the composites. The percolation threshold was achieved at 6 wt% of CWF_{ZnCl2} in EVA composites showed improvement in conductivity range from the insulator to semi-conductor properties.

Acknowledgements

The authors gratefully acknowledge the financial support for this research from Skim Latihan Tenaga Pengajar Akademik (SLTPA) under UniMAP and the Ministry of High Education Malaysia (MOHE).

References

[1] R. Asakura, M. Morita, K. Maruyama, H. Hatori, and Y. Yamada, "Preparation of fibrous activated carbons from wood fiber," vol. 39, ed: Springer, 2004, pp. 201-206.

[2] S. Kalia, B. Kaith, and I. Kaur, *Cellulose fibers: bio-and nano-polymer composites: green chemistry and technology*: Springer Science & Business Media, 2011.

[3] M. Nosonovsky and B. Bhushan, *Green tribology: biomimetics, energy conservation and sustainability*: Springer, 2012.
[4] P. Ghosh, *Fibre science and technology*: Tata McGraw-Hill Education, 2004.
[5] V. Matsagar, *Advances in structural engineering*: Springer, 2015.
[6] P. N. Khanam and M. AlMaadeed, "Improvement of ternary recycled polymer blend reinforced with date palm fibre," *Materials & Design,* vol. 60, pp. 532-539, 2014.
[7] H. P. S. A. Khalil, P. Firoozian, I. O. Bakare, H. M. Akil, and A. M. Noor, "Exploring biomass based carbon black as filler in epoxy composites: Flexural and thermal properties," vol. 31, ed: Elsevier, 2010, pp. 3419-3425.
[8] J. Mirbagheri, M. Tajvidi, J. C. Hermanson, and I. Ghasemi, "Tensile properties of wood flour/kenaf fiber polypropylene hybrid composites," *Journal of Applied Polymer Science,* vol. 105, pp. 3054-3059, 2007.
[9] D. Fengel, G. Wegener, H. S. Jojo, and S. Prawirohatmodjo, *Wood: chemistry, ultrastructure, reactions*: Gajah Madah Universiti Press, 1989.
[10] D. A. Tillman, *Wood Combustion: Principle, Processes, and Economics*: Elsevier, 2012.
[11] H. Hou, A. K. Schaper, Z. Jun, F. Weller, and A. Greiner, "Large-scale synthesis of aligned carbon nanotubes using FeCl3 as floating catalyst precursor," vol. 15, ed: ACS Publications, 2003, pp. 580-585.
[12] H. Pi, S. Guo, and Y. Ning, "Mechanochemical improvement of the flame- retardant and mechanical properties of zinc borate and zinc borate–aluminum trihydrate- filled poly (vinyl chloride)," *Journal of Applied Polymer Science,* vol. 89, pp. 753-762, 2003.
[13] C. Kim, B. T. N. Ngoc, K. S. Yang, M. Kojima, Y. A. Kim, Y. J. Kim, *et al.*, "Self- Sustained Thin Webs Consisting of Porous Carbon Nanofibers for Supercapacitors via the Electrospinning of Polyacrylonitrile Solutions Containing Zinc Chloride," *Advanced materials,* vol. 19, pp. 2341-2346, 2007.
[14] N. Roy, R. Sengupta, and A. K. Bhowmick, "Modifications of carbon for polymer composites and nanocomposites," *Progress in Polymer Science,* vol. 37, pp. 781-819, 2012.
[15] A. Gungor, "Mechanical properties of iron powder filled high density polyethylene composites," *Materials & design,* vol. 28, pp. 1027-1030, 2007.
[16] S. Ramli, C. Ratnam, S. Ahmad, and N. Athirah, "Study on thermal properties and crystallization behavior of electron beam irradiated ethylene vinyl acetate (EVA)/waste tyre dust (WTD) blends in the presence of polyethylene graft maleic anhydride (PEgMAH)," in *AIP Conference Proceedings*, 2013, pp. 146-152.
[17] J. Che, K. Wu, Y. Lin, K. Wang, and Q. Fu, "Largely improved thermal conductivity of HDPE/expanded graphite/carbon nanotubes ternary composites via filler network-network synergy," *Composites Part A: Applied Science and Manufacturing,* vol. 99, pp. 32-40, 2017.
[18] C. Zhang, X. Yuan, L. Wu, Y. Han, and J. Sheng, "Study on morphology of electrospun poly (vinyl alcohol) mats," *European polymer journal,* vol. 41, pp. 423-432, 2005.
[19] T. Mani, P. Murugan, J. Abedi, and N. Mahinpey, "Pyrolysis of wheat straw in a thermogravimetric analyzer: effect of particle size and heating rate on devolatilization and estimation of global kinetics," *Chemical engineering research and design,* vol. 88, pp. 952-958, 2010.
[20] Z. Anwar, A. Kausar, I. Rafique, and B. Muhammad, "Advances in epoxy/graphene nanoplatelet composite with enhanced physical properties: A review," *Polymer-Plastics Technology and Engineering,* vol. 55, pp. 643-662, 2016.
[21] Z. Yue, C. L. Mangun, and J. Economy, "Preparation of fibrous porous materials by chemical activation: 1. ZnCl2 activation of polymer-coated fibers," *Carbon,* vol. 40, pp. 1181-1191, 2002.
[22] M. Rusu, N. Sofian, and D. Rusu, "Mechanical and thermal properties of zinc powder filled high density polyethylene composites," *Polymer Testing,* vol. 20, pp. 409-417, 2001.

Effect of Fly Ash Content and Applied Load On Wear Behaviour of AA6063 Aluminium Alloy

Alaa Mohammed Razzaq [1,2,*], Dayang Laila Majid[1], Nor Hafizah Manan[1], Mohamad R. Ishak[1] and Uday M. Basheer [3,4]

[1]Department of Aerospace Engineering, Faculty of Engineering, Universiti Putra Malaysia,
43400 UPM Serdang, Selangor Darul Ehsan, Malaysia; alaaupm@gmail.com (A.M.R.);
mohdridzwan@upm.edu.my (M.R.I.)
[2]Oil Products Distribution Company, Iraq Ministry of Oil, 10022 Al Doura, Baghdad, Iraq
[3]Faculty of Mechanical Engineering, Universiti Teknologi Malaysia, 81310 Johor Bahru, Johor, Malaysia; uday@utm.my
[4]UTM Centre for Low Carbon Transport in Cooperation with Imperial College London, Institute for Vehicle Systems and Engineering, Universiti Teknologi Malaysia, 81310 Skudai, Malaysia; uday@utm.my

*Email: alaaupm@gmail.com

Abstract. The AA6063 alloy reinforced with 0, 2, 4 and 6 wt. % of fly ash (FA) particulate composites were produced by the compocasting technique. In this study, the effect of FA content and sliding speed on the wear behaviour of AA6063-FA composite was investigated. The wear was carried out by using pin on disc apparatus against rough steel counterface. These wear tests were conducted at applied load of 24.5, 49 and 73.5 N respectively with the constant sliding speed (150 rpm) and sliding time (10 minutes). In the results, it has been observed that upon addition of FA particles, the wear resistance of AA6063-FA composites was enhanced. The wear rate reduced with the increase in the content of FA particles. While the wear rate increased with increase in the applied load.

1. Introduction

The aluminium alloys are widely used in automobile industries as components of internal combustion engines such as cylinder heads, pistons and cylinder blocks due to light weight, high strength to weight ratio, and high thermal conductivity. The applications of aluminium alloys are limited because of poor wear resistance at room and elevated temperatures[1]. Aluminium metal matrix composites (AMMCs) are advanced materials, which aluminium metal combine as matrix material with the ceramic reinforcement such as SiC Al_2O_3, B_4C, TiB_2 etc. to produce composite materials. AMMCs have a various application in different fields like airframes, driveshafts, pistons and submarine. For the past few decades, there has been a sharp demand for AMMCs due to their light weight and higher strength to weight ratio, wear resistance, higher thermal conductivity. These composites can be fabricated in different processes such as solid state (powder metallurgy) and liquid state processes (stir casting, compocasting, squeeze casting, in situ casting routes)[2].

The homogeneous mixing during the fabrication process between the metal matrix and different reinforcement shapes plays an important role determination of mechanical properties of these composites. Therefore, it's one of the most significant mechanical properties that may be affected are the wear resistance. Generally, several factors are effected in the wear properties such as applied load, sliding distance, sliding velocity, friction temperature, and counterface hardness [3]. As well as the reinforcement contents, type, size and shape of materials are most factors that strongly influence in the wear properties [4, 5]. Various theories have been presented to give the influence of different parameters on wear mechanism of pure materials. But the wear mechanism in composite materials might be fundamentally different from the wear mechanism in pure metals. It is therefore difficult to determine the wear parameters required for these composite materials (AMMCs) unless these kind of parameters are closely studied by research.

Zhiqiang et al. [6] discovered that the wear weight loss of both the pure aluminum matrix and their silicon particle composites increased with increasing the applied load and sliding speed ~~application~~. The author's highlighted to the lower frictional coefficient of the aluminum silicon composites and weight loss related with the unreinforced alloy samples more than 65N applied load value and vice versa. The transition of wear conditions of aluminum matrix occurred at the applied load value of 50 N. While their composite specimens demonstrated to enhance the wear resistance. However, under high loads (73.5 N), the protective layer of the reinforcement particles cannot stable longer remain under the compact, micro-cut action and ploughing and the abrasion grooves formed are more distinguished. This may be linked to the presence of large plastic strain in the deformed layers, which resulted in void nucleation, subsurface crack initiation and propagation. The surface material was removed from the wear surface and the cracks moved near to the surface. This led to the surface layers removal by delaminating wear [7, 8]. Rajmohan et al. [9] have highlighted that the wear rate increased with increasing value of applied load for the hybrid aluminum matrix composite specimens. The results of the ANOVA analysis indicated that load is the important parameter, which effects the wear loss of aluminum matrix composites followed by mass fraction of ceramic particles. It was also observed that the wear mechanism of the composites transformed from abrasion to delamination.

In this work, AMCs consisting of reinforced AA6063 alloy with FA particles produced by compocasting and the wear behavior of AA6063 alloy and AA6063-FA composites with 0-6 wt. % FA.

2. Material and methods

2.1. Matrix and Reinforcement Materials

In the present investigation, AA6063 alloy was selected as a matrix material. 6063 alloy is an aluminum based alloy often used in the automotive and aerospace industry. Chemical composition of AA6063 alloy consists of 0.008 wt. % Ti, 0.4 wt. %Mn, 0.01 wt. %Cr, 0.01 wt. % Zn, 0.01 wt. % Cu, 0.19 wt. % Fe, 0.51 wt. % Mg, 0.8 wt. % Si, 98.4 wt. % Al, was supplied by Kamco Aluminium Sdn Bhd, Malaysia. The FA particles was selected as a reinforcement material. The chemical composition of FA powder consists of 0.01 wt. %CuO, 0.34 wt. %Na_2O, 0.38 wt. %MgO, 0. 63 wt. % CaO, .09 wt. % K_2O, 1.14 wt. % TiO_2, 2.78 wt. % Fe_2O_3, 19.09 wt. % Al_2O_3, and 95.98 wt. % SiO_2. The particle size is 44μm, it was obtained from Tanjung Bin Power plant in pontian, Johor, Malaysia.

2.2. Manufacturing the Composite Material

The proper selected parameters of manufacturing process is important to product the composite by compocasting technique such as molten temperature, stirrer speed, preheating temperature of reinforcement etc. preparation of the composite was carried out according to the following procedure:

AA6063 alloy was heated up to molten state in graphite crucible by induction furnace. The temperature of the melt was about 800°C, and start stirring process by stainless steel stirrer (Figure.1). Then reduce the temperature to 600°C to get semisolid state of the molten AA6063 alloy aluminium and add the preheated FA particles (900°C for 2 hours). Then increase the temperature with stirred the mixture 300 rpm. After completing the addition of the FA powder into the melt, the AA6063-FA composite was poured into the preheated mould about 250°C and allowed to cool in atmospheric temperature. The cast was then detached from the mould and the samples of wear test machined with heat treatment with different weight fractions of FA reinforcement (0-6%) in step 2%.

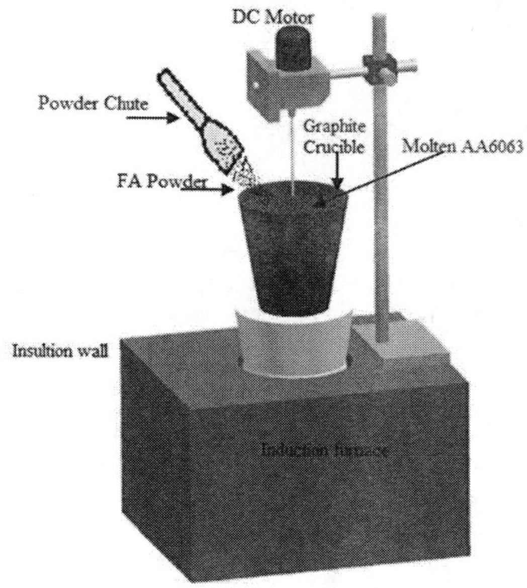

Figure.1. Schematic diagram of casting technique (Induction Melting Furnace)

2.3. Wear Test

The dry sliding wear behaviour test of AA6063-FA composites was carried out under ambient condition using a Pin on Disc apparatus (DUCOM's wear friction Monitor, Model-TR-20LE, Bangalore, India). Ducom Rotary Tribometer complies with ASTM G99-05 standard [10]. Figure.2 demonstrates the wear test conducted in the Tribology Laboratory at Universiti Teknologi Malaysia (UTM), under varying applied load (24.5, 49, and 73.5N) and at a fixed sliding speed of 150 rpm up to sliding time 10 min against hardened steel disc EN32 with hardness 60HRC. Wear test specimens were cut into rectangular bares to the dimensions of 8mm×8mm×24mm as shown in Figure. 2 [11, 12]. The test surface was adequately polished using different grades of silicon carbide papers to ensure the proper contact with steel disc. During sliding test, the load is applied on the specimen through cantilever mechanism and the specimens brought in intimate contact with the rotating disc at a track radius of 80 mm (Figure.2), the specimens were cleaned with acetone. The wear rate was calculated from the volumetric loss per unit sliding distance. An investigation graph shows the height loss or wear in μm against slippery time in sec (which is proportional to sliding distance) taken from wear testing machine is shown in Figure.3.

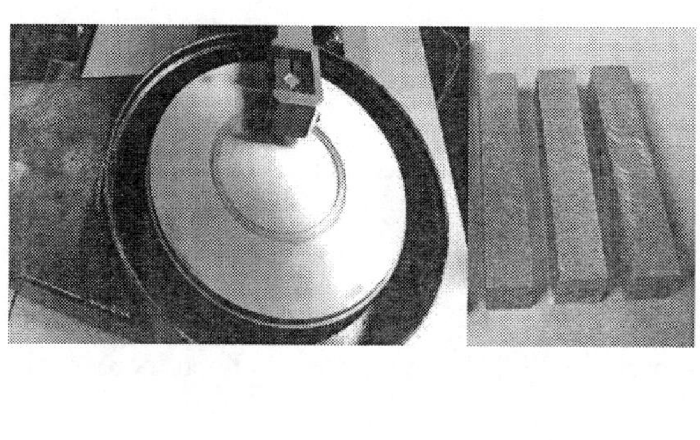

Figure.2. Top view of the pin on disk type wear testing machine with the photograph of wear test specimens.

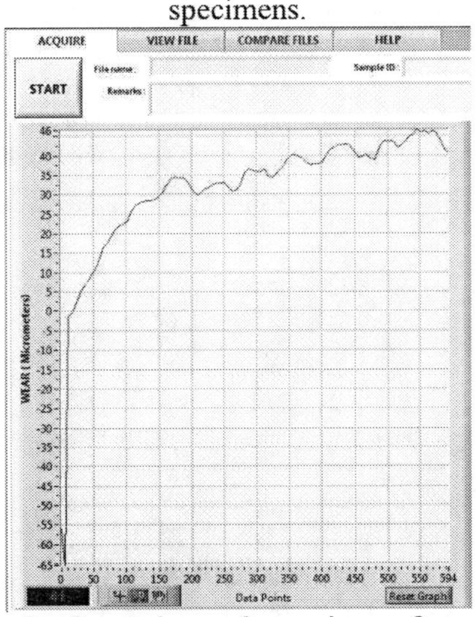

Figure.3. typical graphical result of experimental wear in μm from pin-on-disc type wear testing machine different weight fraction reinforced of AA6063 alloy at sliding speed 150 rpm.

3. Results and discussion

The wear test results of the AA6063 alloy and AA6063-FA composites under dry sliding conditions are shown in figure 4. The overall slippery time of 10 minutes at sliding speed 150 rpm under varying weight fraction of FA (0- 6%) in step 2% were studied. The wear rate of these test was obtained from the weight loss is plotted against applied load as shown in figure 4 for different FA content. The incorporation of FA particles was beneficial to the enhancement of wear resistance. The increased wear resistance was attributed to the higher hardness of the AA6063-FA composite according to increased FA content [13]. The results of added FA to composite were improved hardness offer resistance to the cutting action of the counterface asperities. The thermal mismatch between the AA6063 alloy and the FA particles created dislocations during solidification, those strain fields strengthen the subsurface. The homogenous distribution of FA particles led to Orowan strengthening [14]

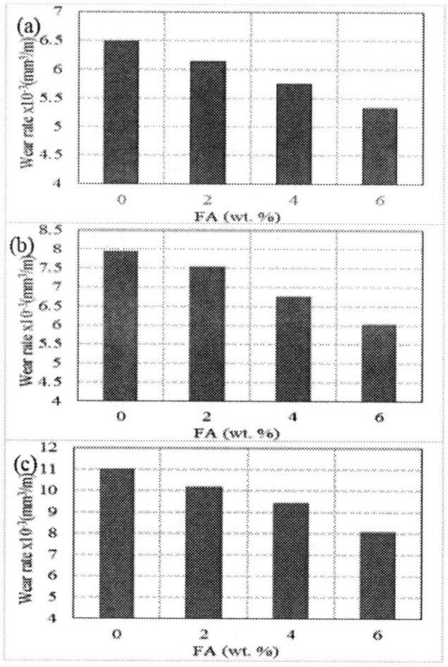

Figure.4. Wear rate of 6-FA composites at 150 rpm as a function of wt. % FA with variant load: (a) 24.5N, (b) 49N, and (c) 73.5N.

The wear weight loss of the AA6063-FA composite increases with increasing in the applied load. From the figure 5, it is clear that wear loss of unreinforced AA6063 alloy increases higher than AA6063-FA composite. An increased applied load on the sliding specimens increased the pressure on the sliding specimen. These results are in agreement with the well-known Archard's wear expression equation 1 [15, 16] Archard's law steady state wear was proportional to contact stress or pressure and sliding distance for these metallic composites.

$$W=KP_n/H \qquad (1)$$

Where H is the indentation hardness of the softer surface, K is the wear coefficient of Archard which is dimensionless (always less than unity) and P_n is the normal pressure applied between the surface.

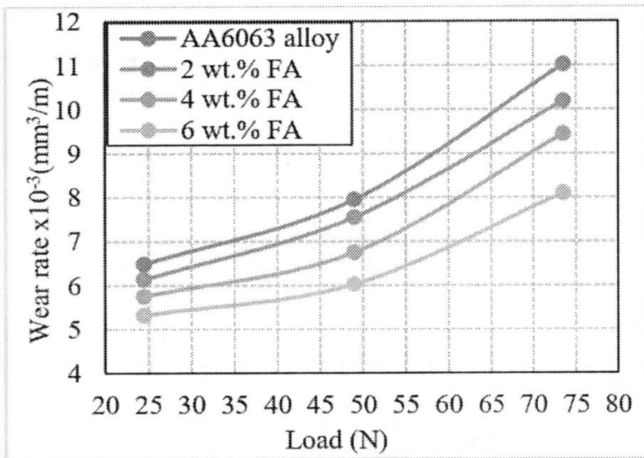

Figure. 5. Wear rate of AA6063 alloy-based AMCs as a function of applied load

4. Conclusion

In the present work, AA6063-FA composites were prepared by the compocasting technique. The effect of reinforced FA content and applied load on the wear behaviour of AA6063 alloy were investigated. The reinforcement of AA6063 alloy with FA particles has a significant effect on the wear behavior. The wear rate decreases with increasing the weight content of FA reinforcements. The results of this investigation are indicated about increasing the wear resistance when increases the FA content at constant sliding speed. As well as the wear rate decreases with increase in applied load.

Acknowledgements

The authors would like to thanks the Ministry of Education Malaysia (MOE), Aerospace Department in Faculty of Engineering at University Putra Malaysia (UPM) and Mechanical Engineering Faculty, UTM-Centre for Low Carbon Transport, Institute for Vehicle Systems and engineering at Technology University Malaysia (UTM) for providing the research facilities. This research work has been supported by the Ministry of Education Malaysia (MOE) for the FRGS Grant (R.J130000.7824.4F723), UPM Research Grant (9487900) and UTM research grant Q.J130000.2524.17H83.

References

[1] H M Rajan S Ramabalan I Dinaharan and S Vijay 2014 Effect of TiB2 content and temperature on sliding wear behavior of AA7075/TiB2 in situ aluminum cast composites *archives of civil and mechanical engineering* **14** pp 72-79

[2] C Jawalkar A S Verma and N Suri 2017 Fabrication of aluminium metal matrix composites with particulate reinforcement: a review *Materials Today: Proceedings* **4** pp 2927-2936

[3] J Singh and A Chauhan 2016 Overview of wear performance of aluminium matrix composites reinforced with ceramic materials under the influence of controllable variables *Ceramics International* **42** pp 56-81

[4] Q X Mingwu Bai Xiue Wang Yong Wan Weiming Liu 1995 Wear mechanism of SiC whisker-reinforced 2024 aluminum alloy matrix composites in oscillating sliding wear tests *Wear* **185** pp 197-202

[5] F B A Alahelisten M Olsson S Hogmark 1993 On the wear of aluminium and magnesium metal matrix composites *Wear* **165** pp 221-226

[6] S Zhiqiang Z Di and L Guobin 2005 Evaluation of dry sliding wear behavior of silicon particles reinforced aluminum matrix composites *Materials & Design* **26** pp 454-458

[7] V S K Ramachandran P V Ananthapadmanabhan KP Sreekumar 1998 Microstructure adhesion microhardness abrasive wear resistance and electrical resistivity of the plasma sprayed alumina and alumina–titania coatings *Thin Solid Film* **315** pp 144-152

[8] N Suh 1977 Delamination wear of dispersion-hardened alloys *Journal of Engineering for Industry* p 289
[9] T Rajmohan K Palanikumar and S Ranganathan 2013 Evaluation of mechanical and wear properties of hybrid aluminium matrix composites *Transactions of nonferrous metals society of China* **23** pp 2509-2517
[10] A Standard G99-05 2005 Standard test method for wear testing with a pin-on-disk apparatus *Annual Book of ASTM Standards* G99-05 3
[11] R Uyyuru M Surappa and S Brusethaug 2007 Tribological behavior of Al–Si–SiC p composites/automobile brake pad system under dry sliding conditions *Tribology International* **40** pp 365-373
[12] R Uyyuru M Surappa and S Brusethaug 2006 Effect of reinforcement ume fraction and size distribution on the tribological behavior of Al-composite/brake pad tribo-couple *Wear* **260** pp 1248-1255
[13] A Mohammed Razzaq D L Majid M R Ishak and U M Basheer 2017 Effect of Fly Ash Addition on the Physical and Mechanical Properties of AA6063 Alloy Reinforcement Metals **7** p 477
[14] G Gautam N Kumar A Mohan R Gautam and S Mohan 2016 Strengthening mechanisms of (Al3Zrmp+ ZrB2np)/AA5052 hybrid composites *Journal of Composite Materials* **50** pp 4123-4133
[15] J Archard 1953 Contact and rubbing of flat surfaces *Journal of Applied Physics* **24** pp 981-988
[16] I D J Allwyn Kingsly Gladston N Mohamed Sheriff J David Raja Selvam 2017 Dry sliding wear behavior of AA6061 aluminum alloy composites reinforced rice husk ash particulates produced using compocasting *Journal of Asian Ceramic Societies*

IOP Conf. Series: Materials Science and Engineering **429** (2018) 012039 doi:10.1088/1757-899X/429/1/012039

Development of Biopolymer film with different ratios of Gelatine to Chitosan reinforced with Zinc Oxide Nanoparticles for food covering/preservation

DY Yi, BM Siddique*, JC Lai

School of Chemical Engineering and Science, Faculty of Engineering, Computing and Science, Swinburne University of Technology Sarawak Campus, Kuching, Malaysia.

E-mail: msiddique@swinburne.edu.my

Abstract
In this research project, gelatine and chitosan have been chosen to form a composite film and the zinc oxide nanoparticles have been used to increase the properties of the composite film. In the experiment, the ratio of gelatine to chitosan as well as the amount of zinc oxide nanoparticles have been manipulated. These composite film with the gelatine (G) to chitosan (Ch) ratio of 40G:60Ch, 50G:50Ch, 60G:40Ch have incorporated with 0 mg, 1 mg and 10 mg of ZnO NP in each film. A layer of smooth film with 30 μm thickness were produced. From the experiment, the higher the ratio of chitosan in the composite have resulted a stronger adhesive properties of the film. Due to the high adhesive properties, ratio of 50G:50Ch and 60G:40Ch composite films were not able to be peeled off from the dish. Hence, only ratio of 40G:60Ch composite film with different amount of zinc oxide nanoparticles were evaluated. The transparency of the film decreased while increasing the amount of zinc oxide nanoparticles. Despite that, 10 mg zinc oxide loaded film has 10 % decrease in its water solubility as compared to the film that without loaded with nanoparticles. Moreover, the film loaded with 10 mg zinc oxide have a better ability to withstand in an acidic condition. When contact with 1 % acetic acid solution, the film could maintain its original shape but for the film loaded with 0 mg and 1 mg nanoparticles have rolled up when contacted with the acid solution. All the test samples were dissolved completely in acid solution after 24 hours.

1. Introduction

The properties of the films are a key requirement for an organic product to extend their shelf life after harvested. In contrast with petrochemical-based film, a biopolymer-based film has provided advantages not only for the environment but also to consumers. Biopolymer film possesses the biodegradable properties thus self-decompose after certain period of time. The biodegrading properties of the film have greatly reduced the environmental waste. Hence, the organic that wrapped with biopolymer film has contributed to a higher quality of the product.

In the past decades, edible films and coating such as wax were introduced and used for various fruits to prevent loss of moisture while create a glossy surface for decorative purposes. These practices are still carried out in the present day even though their researchers and associated chemistries were understood and accepted long before. According to Pavlath & Orts [1 p.1], the edible film had been only related to food applications in the past 50 years. The terms, film and coating had been used interchangeably to show the surface of a food is covered by a thin layer of material of certain composition. The only difference between film and coating are the notion that film is a stand-alone wrapping material, whereas a coating is applied and form directly contact on the surface of the food. In the early years on 1967, the idea of edible films are not popular and have been limited mostly to the wax layer that used on the fruits [1 p.1].

In the years of 1986, the products that wrapped or coated with edible films have offered only by nearly 10 companies. During intervening years, it shows significant changes in business who applied the concept of edible films, whereas it resulted in more than 600 companies offering such products in 1996 [1 p.1]. Nowadays, edible film uses have expanded widely in order to retain the quality of various horticultural product. Edible film wrapping technique has been developed rapidly because it extended the period of fresh horticultural products after harvest and also extend the useful marketing distances. Edible films have greatly reduced the economic waste in advanced and developed countries.

Nanotechnology is a popular research and has gained momentum in enhances 21st century. Due to their distinctive properties, nanoparticles have been widely used in various field. Among these metal nanoparticles, zinc oxide nanoparticles are most important as they have been used as biomedical, cosmetic, agriculture, and etc. The unique properties of zinc oxide nanoparticles against bacteria have attracted the attention of researchers toward food preservation [2 p.4-5]. The relatively low cost of zinc oxide as compared to silver or gold nanoparticles also have promoted the usage of zinc oxide nanoparticles. Besides that, zinc oxide nanoparticles have less harmful to human health as they have been used as an additive in food products such as breakfast cereals.

The constraints of natural polymers food packages can be recovered by applying composite polymer, which combined two or more biopolymers to form a single food package or implanting nanoparticles that used to strengthen the packages without causing any harm toward humans. The main aim of this research project was to study the properties and the effectiveness of the films which was loaded with zinc oxide nanoparticles and with various ratios of gelatine to chitosan content.

2. Methodology

2.1 Materials

A medium molecular weight of Chitosan with 75-85 % deacetylated, 30 g of gelatine powder (Type-A) from cold water fish, 20 mg of zinc oxide nanoparticles with average particle diameter of 30 nm, 15 g of glycerol and 1 % acetic acid were purchased.

2.2 Films Preparation

2.2.1 Preparation of Gelatine Solution

The gelatine solution was prepared by dissolving 0.5 g, 0.75 g and 1 g of gelatine powder in different beaker. Each of the powder in different beaker was dissolved by 50 ml of distilled water with a percent concentration of 1 % (w/v), 1.5 % (w/v) and 2 % (w/v) respectively. The mixture was stirred continuously in room temperature for 30 min.

2.2.2 Preparation of Chitosan Solution

The chitosan solution was prepared by dissolving 0.75 g of chitosan powder with 5 ml of glacial acetic acid. The solution filled up with distilled water to obtain a total volume of 50 ml mixture. The mixture was stirred overnight at room temperature and with a percent concentration of 1.5 % (w/v). The impurities will be filtered by using gravity filtration method.

2.2.3 Preparation of Gelatine-Chitosan Film Loaded with Zinc Oxide Nanoparticles

The film formation are based on Hosseini et al. [3 p.1491] but with slightly modification. The gelatine and chitosan film forming solution was prepared by adding 1 % (w/v), 1.5 % (w/v) and 2 % (w/v) gelatine solution with 1.5 % (w/v) chitosan solution in a proportions of 1:1 (v/v) to obtain different

IOP Conf. Series: Materials Science and Engineering **429** (2018) 012039 doi:10.1088/1757-899X/429/1/012039

proportions respectively (solution 40G:60Ch, 50G:50Ch, 60G:40Ch). All the mixture was stirred for 15 minutes to obtain a good blend. Then, glycerol (0.3 g/g gelation plus chitosan) was added to each mixture as a plasticiser and the solution was again stirred for 15 min. After that, 10 ml of each mixture was distributed into three beakers for the addition of 0 mg, 1 mg, 10 mg zinc oxide nanoparticles. All the solution were stirred for 5 min and homogenised for 10 min with 500 rpm.

A 3ml of each aliquots film forming solution was pipetted and pour onto the petri dishes. The solution was separated evenly until the solution has covered the surface of the dish. The dishes that filled with the solution were dried in an incubator at 35 °C for 72 hr. The dried films were manually peeled off and stored in the incubator at 35 °C for further analysis.

2.3 Film Morphology

The optical microscope (Nikon Eclipse 1500) equipped with capture system was used to observe the surface of the film. A magnification of 20×, 50× and 100× were used on microscope. The surface of the film has been captured by the installed image capture system.

2.3.1 Thickness

The optical microscope (Nikon Eclipse 1500) equipped with capture system was used to measure the thickness. A magnification of 5x will be used on microscope and the image will be capture by using the capture system.

2.3.2 Light Transmission and Transparency

The light transmission and transparency was determined according to the method reported by Hosseini et al. [3 p.1491-1492]. By using UV-Vis Spectrometer (Lamda-35), the wavelength between 200 nm and 800 nm (200, 300, 400, 500, 600, 700, 800) have been selected to measure the barrier properties of films against ultraviolet (UV) and visible light. Each sample was cut into a portion of 7 mm × 40 mm and placed into the cuvette. An empty cuvette has been placed into the spectrophotometer which used as the reference. The transparency of the film can be calculated by using the following equation. According to the equation, high transparency indicates opaque.

$$Transparency\left[\frac{A_{600}}{mm}\right] = -\frac{-\log T}{x}$$

$T = Fractional\ Transmittance$; $x = Thickness\ of\ the\ film\ (mm)$

2.3.3 Moisture Absorption

The sample films with 20 × 20 mm^2 were first stored in 0 % relative humidity (RH) (calcium sulphate) for 24 hr. The sample films were conditioned in a desiccator containing calcium nitrite saturated solution at 20-25 °C to ensure a relative humidity of 55 % after weighed the films. At desired intervals of time, the sample films were weighed until the equilibrium state was reached. The moisture absorption of the films was calculated as follows:

$$Moisture\ Absorption\ (\%) = \left[\frac{(W_t - W_0)}{W_0}\right] \times 100$$

$W_t = Weights\ of\ the\ sample\ after\ t\ time\ at\ 55\%\ RH;\ W_0 = Initial\ weight\ of\ the\ sample$

2.4 Water Solubility

The solubility of the film in distilled water was determined according to the method reported by Benbettaïeb *et al.* [4 p.2411] with slightly modification. Each of the film was dried in the incubator at 60 °C for 2 hr. After drying, each of the film has been cut into 3 mg in pieces. The dried film was immersed in 5 ml of distilled water under room temperature. After 24 hr immersion, film pieces were taken out from the distilled water and re-dried the film in the incubator at 60 °C for 2 hr to measure the weight of dry matter, which was not solubilized in water. The final dried weight of the film was weighed and recorded. The water solubility of the film (% FS_w) was calculated using the following equation:

$$FS_w(\%) = \left(\frac{W_i - W_f}{W_i}\right) \times 100$$

$FS_w = Film\ solubility\ in\ water\ (\%)$
$W_i = Initial\ dry\ weight\ (mg); W_f = Final\ dry\ weight\ (mg)$

2.5 Acid Solubility
The solubility of the film in acetic acid was determined according to the method reported by Cardoso *et al.* [5 p.88] with modification. Each of the film was dried in the incubator at 60 °C for 2 hr. After drying, each of the film has been cut into 3 mg in piece. The dried film was immersed in 5 ml of 1 % of acetic acid under room temperature. After 24 hr immersion, film pieces were taken out and re-dried the film in the incubator at 60 °C for 2 hr. The final dried weight of the film was weighed and recorded. The acid solubility of the film (% FS_A) was calculated using the following equation:

$$FS_A(\%) = \left(\frac{W_i - W_f}{W_i}\right) \times 100$$

$FS_A = Film\ solubility\ in\ acid\ (\%)$
$W_i = Initial\ dry\ weight\ (mg)\ ; W_f = Final\ dry\ weight\ (mg)$

3. Results & Discussion
From the experiment, 40G:60Ch, 50G:500Ch and 60G:40Ch composite films with each addition of 0 mg, 1 mg and 10 mg of zinc oxide nanoparticles have been produced. But only 40G:60Ch composite film loaded with 0 mg, 1 mg and 10 mg have been characterized. The 50G:50Ch and 60G:40Ch composite films unable to be peeled off from the petri dish due to the high adhesive properties which laminated on the petri dish. In can be observed that, increases the ratio of gelatine in the composite film have increase the adhesive properties of the film.

3.1 Physical Appearance & Thickness
The texture of the film is an important requirement for them to be used in product packages. From the experiment, the samples film which prepared was transparent and flexible. Figure 1 shows the appearance of the film and its clear from the image that it was transparent. The surface of the films was smooth especially the side that exposed to the air while drying on the petri dish. The other side of the sample contacted with the glass petri dish remained shining but it possesses an adhesive property which allow the films to stick on the surface of the food product. Apart from that, the acidic smell of the film was very slight or non-discernible. The appearance and texture of the films were in accordance with the description reported by Sukkunta [6 p.43] for the film which made by gelatine and chitosan. The levels of ZnO NPs added has no significantly effect on the appearances of the film.

Figure. 1 Physical appearance of the film

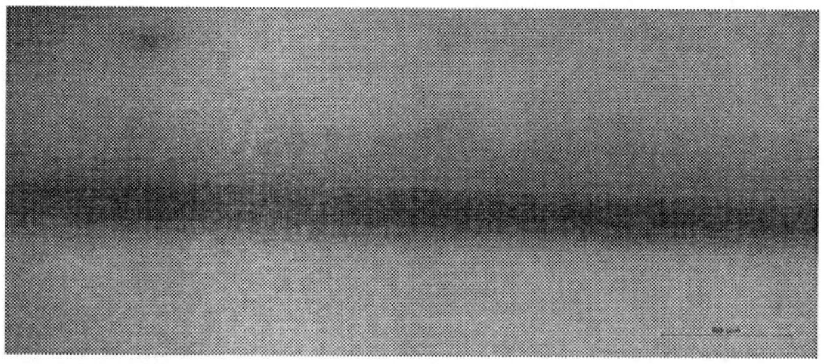

Figure. 2 Thickness of 40G:60Ch film reinforced with 10 mg

Despite that, the 40G:60Ch composite film with addition of different amount of zinc oxide nanoparticles had been determined to have an average thickness of 30 μm which measured by an optical microscope. The image for the thickness of the films have been captured from the microscope and illustrated in Figure 2. The rest of the films thickness was determined through same method. However, the addition of zinc oxide nanoparticles was found to have insignificant effect on the thickness of the film.

3.2 Film Morphology

The morphology of the film was observed under an optical microscope with magnification of 100×. The captured images for each of the film have been illustrated in Figure 3 and 4. There was no white spot being observed from the microscope. For the film with loaded with 1 mg and 10 mg of zinc oxide nanoparticles, white spots being observed and a relatively higher amount for the film loaded with 10 mg of nanoparticles. The zinc oxide nanoparticles which observed from the microscope was, most probably, caused by an unevenly distribution of the nanoparticles on the film and part of the nanoparticle have clumped together during the mixing of film forming solution with zinc oxide nanoparticles.

Figure 3: Surface of 40G:60Ch composite film with 1mg ZnO NPs

Figure 4: Surface of 40G:60Ch composite film with 10mg ZnO NPs

3.2.1 Light Transmission and Transparency

In this study, the light transmission and transparency of each film incorporated with ZnO NPs at different levels were conducted by using range of 200-800 nm wavelength. The result of light transmission with different wavelength were shown in Table 1. From the result obtained, the control film (without ZnO NPs) has a good barrier property in the UV-range of 200-300 nm. Regardless of zinc oxide nanoparticles, gelatine as protein based are generally reported to have an excellent barrier properties which contributed by the presence of high content of aromatic amino acids that absorb the UV light [3 p.1493]. When the visible wavelength ranges at 400-700 nm, the transmission for all the films were greater than 80 % and only control film (without ZnO NPs) have greater than 90 % of transmission in 800 nm wavelength.

Table 1: Light transmission and transparency of films obtained with different amount of ZnO NPs

Film	Light transmittance (%) at different wavelength (nm)							Transparency (A_{600}/mm)
	200	300	400	500	600	700	800	
40G:60Ch with 0mg ZnO NPs	0.10	76.17	87.53	89.69	89.25	89.94	90.09	1.65
40G:60Ch with 1mg ZnO NPs	0.09	70.17	84.45	86.97	87.27	87.89	88.49	1.97
40G:60Ch with 10mg ZnO NPs	0.10	68.04	82.55	84.79	85.82	86.63	87.45	2.21

From the graph that depicted in Figure 5 shows that the higher the levels of zinc oxide nanoparticles added into the 40G:60Ch composite film has lower the percent of light transmission through the film. The results indicated that zinc oxide nanoparticles have the ability to impede the light transmission through the film. According to Arfat *et al.* [7 p.106-107], the phenomena was mainly due to the obstacle of the light passage by the presence of the nanoparticles dispersed in the film matrix.

Despite that, the transparency of the film at wavelength of 600 nm were computed. The results show that the control film has higher transparency values than the film loaded with 10 mg zinc oxide nanoparticles indicating the control film are more transparent. The transparency values increase as the level of zinc oxide nanoparticles content increased indicating the decrease in the transparency of the resulting films. These results were in agreement with Arfat et al. [7 p.107] who reported that films based on fish skin gelatine added with zinc oxide nanoparticles was less transparent.

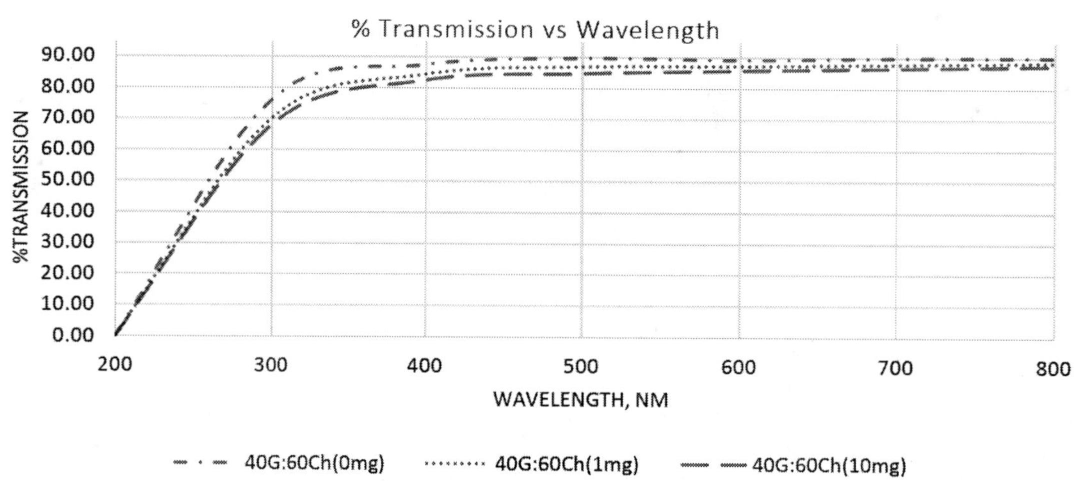

Figure 5: Percentage transmission against different wavelength

3.3 Solubility in Acid

The acid solubility is essential for the film because it allowed the film to withstand in an acidic medium while environment pH decreases. From the experiment, the acid solubility was found to be 100 %, all the dissolved completely after 24 hr immersion in 1 % acetic acid solution. Since the composite film was majorly made up by chitosan, the amine group of chitosan are protonated to form NH_3^+ when chitosan contact with the acid solution, thus chelating and polyelectrolyte properties of chitosan were mainly governed by the acidity of NH_3^+ [6 p.27].

Besides that, it was observed that the composite film loaded with 0 mg and 1 mg of zinc oxide nanoparticles have rolled up and became curled instantly when contact with 1% acetic acid solution. The composite film loaded with 1 mg of ZnO NPs was inadequate to hold and maintain the film in shape in an acidic medium. Only the composite film loaded with 10 mg of ZnO NPs have maintain in its original shape when contacted with the acidic solution. According to Baeza *et al.* [8], the nanoparticles 'entangled' in the film matrix acts as a reinforcement to the film that allow the film to reduce it segmental mobility and able to maintain its shape in an acidic medium. Hence, the addition of ZnO NPs has provided an advantage for the film to sustain in an acidic environment.

4. Conclusion

The properties of composite biopolymer (40G:60Ch) film can be improved by addition of zinc oxide nanoparticles. The film had increased the water resistance after incorporated with ZnO NPs. From the results, the higher the levels of ZnO NPs present in the film had significantly decreased their water solubility. Furthermore, the presence of nanoparticles played an important role when contact with acidic solution. It allowed the film to maintain in their original shape.

All the films were observed to have a smooth, flexible and shining surface. The transparency of the film which determined on wavelength 600 nm have showed that the film without zinc oxide nanoparticles have the most transparent property as compared to 1 mg and 10 mg nanoparticles incorporated film. Besides that, each of the film possesses a high UV absorption regardless of the presence of ZnO NPs.

Acknowledgement

This research was conducted as per the final year research project for Chemical Engineering discipline at the Swinburne University of Technology Sarawak Campus, with the help of the FYRP 1, 2016 research grant from the Swinburne University.

References

[1] Pavlath, AE & Orts, W 2009, 'Edible Films and Coatings: Why, What, and How?', in KC Huber and ME Embuscado (eds), *Edible Films and Coatings for Food Applications*, Springer New York, New York, NY, pp. 1-23.

[2] Sabir, S, Arshad, M & Chaudhari, SK 2014, 'Zinc Oxide Nanoparticles for Revolutionizing Agriculture: Synthesis and Applications', *The Scientific World Journal*, vol. **2014**, p. 1-8.

[3] Hosseini, SF, Rezaei, M, Zandi, M & Ghavi, FF 2013, 'Preparation and functional properties of fish gelatin–chitosan blend edible films', *Food Chemistry*, vol. **136**, no. 3–4, pp. 1490-1495.

[4] Benbettaïeb, N, Kurek, M, Bornaz, S & Debeaufort, F 2014, 'Barrier, structural and mechanical properties of bovine gelatin–chitosan blend films related to biopolymer interactions', *Journal of the Science of Food and Agriculture*, vol. **94**, no. 12, pp. 2409-2419.

[5] Cardoso, GP, Dutra, MP, Fontes, PR, Ramos, AdLS, Gomide, LAdM & Ramos, EM 2016, 'Selection of a chitosan gelatin-based edible coating for color preservation of beef in retail display', *Meat Science*, vol. **114**, 4, pp. 85-94.

[6] Sukkunta, S 2005, *Physical and mechanical properties of chitosan-gelatin based film*, thesis, Faculty of Graduate Studies, Mahidol University.

[7] Arfat, YA, Benjakul, S, Prodpran, T, Sumpavapol, P & Songtipya, P 2016, 'Physico-Mechanical Characterization and Antimicrobial Properties of Fish Protein Isolate/Fish Skin Gelatin-Zinc Oxide (ZnO) Nanocomposite Films', *Food and Bioprocess Technology*, vol. **9**, no. 1, pp. 101-112.

[8] Baeza, GP, Dessi, C, Costanzo, S, Zhao, D, Gong, S, Alegria, A, Colby, RH, Rubinstein, M, Vlassopoulos, D & Kumar, SK 2016, 'Network dynamics in nanofilled polymers', *Nature Communications*, vol. **7**, 04/25/online, p. 11368.

IOP Conf. Series: Materials Science and Engineering **429** (2018) 012040 doi:10.1088/1757-899X/429/1/012040

Comparative Study on the Extraction of Bioactive Secondary Metabolites from Pomelo and Pineapple Peels Extract

O Nurdalilah[1,a)], Y P Teoh[1,2,b)], Z X Ooi[3,c)] and S T Sam[4,d)]

[1]Department of Chemical Engineering Technology, Faculty of Engineering Technology, Universiti Malaysia Perlis (UniMAP), P.O Box 77, D/A Pejabat Pos Besar, 01000 Kangar, Perlis, Malaysia.
[2]Department of Petrochemical Engineering, Faculty of Engineering and Green Technology, Universiti Tunku Abdul Rahman, Jalan Universiti, Bandar Barat, 31900 Kampar, Perak, Malaysia.
[3]Department of Chemical Science, Faculty of Science, Universiti Tunku Abdul Rahman, Jalan Universiti, Bandar Barat, 31900 Kampar, Perak, Malaysia.
[4]School of Bioprocess Engineering, Universiti Malaysia Perlis (UniMAP), Kompleks Pusat Pengajian Jejawi 3, 02600 Arau, Perlis, Malaysia.

[a)]nurdalilahothman@gmail.com
Corresponding author: [b)]teohyp@utar.edu.my
[c)]zhongxian.ooi@gmail.com
[d)]stsam@unimap.edu.my

Abstract. In this study, a phytochemical analysis and in vitro analysis of antioxidant assay conducted for two different types of fruit peels extract which are pomelo peel (PP) and pineapple peel (PIP). For phytochemical analysis, total phenols content (TPC), total flavonoids content (TFC) and total tannins content (TTC) and for antioxidant assay, DPPH radical scavenging activity and ferric reducing ability power (FRAP) were determined using spectrophotometric methods. From the analysis, it was found that PP extract rich in TPC, TFC, TTC, DPPH antioxidant scavenging activity and FRAP assay compared to PIP with the values 6.3133 mg GAE/g of extract, 4.6675 mg QAE/g of extract and 0.1593 mg GAE/g of extract respectively. For antioxidant assays, PP extract showed higher percentage of antioxidant scavenging activity compared to PIP extract with 54.95% and also the increase in FRAP when the concentration of extract increased. Thus, PP was chosen to be conducted in a green synthesis of nanoparticles which can be applied in either pharmaceutical or environmental fields.

1. Introduction

A fast-growing economic in emerging-developing countries has encouraged an increased solid waste generation in that specific zone. Along with the economic development and business movement and consumption rate, daily generation and volume rate of municipal solid waste will accelerate. It has been proven that the higher advancement in economics, the higher the amount of municipal solid waste produced [1]. Approximately, 30000 tons of municipal solid wastes are generated daily, covering 83% of the country's waste generation, including agro wastes [2]. It is estimated that 15% of the total waste generated in Asia is agro waste, with agricultural waste generation in Malaysia at approximately 0.122 kg/cap/day in 2009 which is projected to reach 0.210 kg/cap/day by 2025 [3].

Global experienced a remarkable upsurge in production of fruit. Past decade, about 3% of the output has been growing at an annual rate. Malaysia besides Thailand, India and China recorded as one of the larger fruit production in Asia and predicted to increase in next few years ahead. Mohamed (2010) reported that 8.5% of Gross Domestic products contributed by agriculture came from industrial crops with 61% and agriculture food crops with (39%) respectively [4]. For the agriculture crops, they are divided into two categories which are crop residues and agro-industrial residues. Peels, husks hulls, bagasse are the example of agro-industrial residues. In Malaysia, the agricultural land covered only 20% for cultivated food crops while the fruit processing produced a great quantity of waste at remaining 80% land were poorly harvested or left to decay on the land.

Due to large amount of water content in agro waste, they are in wet and easily to be fermentable. These agro wastes may lead to the harbourage for insects and few types of pollutions such as odor, soil and serious environmental problems if they were no further processed. and also can contribute to serious environmental pollution if not processed further [5]. Besides, the statistical data provided from Fruit and Agricultural Statistics recorded Malaysia produced about 16% of pineapple and 2% of grapefruit include pomelo while the others are watermelon(13%), oranges(4%), banana (31%) and tropical fresh fruits include durian and rambutan (26%) and mangoes, mangosteens and guavas (8%) [6]. Non-edible plants produced are believed to be rich in bioactive secondary metabolites and they are cheap renewable feedstocks [7].

Bioactive secondary metabolites also known as bioactive compounds which other than primary metabolites are believed help to increase plants overall ability to survive and to overcome local challenges by allowing them to interact with surrounding [8]. Edible and non-edible plants such as vegetables, fruits, herbs and other plant materials are known to be rich in bioactive secondary metabolites. Seeds, peels, husks and among others are generated every year in the form of wastes and are poorly harvested or left to decay on the land. These bioactive secondary metabolites can be divided in three main categories which are terpenes and terpenoids (approximately 25 000 types, alkaloids (approximately 12 000 types and phenolic compounds (8000 types) and can be extracted using conventional or non-conventional extraction techniques [9].

Recently, researchers were found and able to develop high value added products such as cosmetics and medicines from these by-products and it seems that the recovery was to be economically attractive [10]. The intention of utilizing fruit wastes have slowly gaining popularity mainly the peels which some fruits represents almost 30% of the total weight. More recently, food enriched with fruit peels are been developed since the increasing interest natural sources of bioactive secondary metabolites as approach in functional food. However, the potential of the fruits peels are highly depended on their chemical composition. Hence in this study, the comparative study had been investigated on the pomelo peels (*Citrus maxima*) and pineapple peels (*Ananas comusus*) with the aim of exploiting the potential of high value products from these peels.

2. Procedure

2.1. Materials

Agricultural residues such as pomelo peels (PP) and pineapple peels (PIP) were obtained from a local food stall at Kangar, Perlis, Malaysia.

2.2. Chemicals

Methanol was purchased from Merck, n-hexane, Folin-Ciocatue reagents, sodium carbonate (Na_2CO_3), gallic acid (C_6 H_2 $(OH)_3$ COOH), potassium ferricyanide (K_3 [$Fe(CN)_6$]), tannic acid ($C_{76}H_{52}O_{46}$), sodium nitrite ($NaNO_2$), aluminium chloride hydrate ($AlCl_3.H_2O$), quercetin ($C_{15}H_{10}O_7$), 1-diphenyl-2-picrylhydrazyl (DPPH), sodium phosphate buffer (NaH_2PO_4), trichloroacetic acid ($C_2HCl_3O_2$) and ascorbic acid ($C_6H_8O_6$) were purchased from Bendosen.

IOP Conf. Series: Materials Science and Engineering **429** (2018) 012040 doi:10.1088/1757-899X/429/1/012040

2.3. Sample Pre-treatments

Agricultural residues collected and washed, dried in oven for 24 hours at 60 °C. The sample then were ground to 125 µm were collected and defatted using n-hexane (ratio of solid/liquid 1/10, w/w) at room temperature and then the solvent was evaporated.

2.4. Extraction of Fruit Peels

The extraction of phenolic compounds was carried out using solvent 80% methanol. The samples were added to a solvent 1:10 mixed well and kept at room temperature, for 3 days under constant stirring. The mixture was centrifuged at 6,000 rpm for 15 min and the supernatant was filtered through a filter paper. Then, the solvent was evaporated in a rotavapor. The extraction yield was expressed as dry matter percentage.

2.5. Total Flavonoids Content (TFC)

Total phenols content determined using method described by Stankovi (2011) [11]. The sample was re-dissolved in the extraction medium. To 0.5 ml methanolic extract, 2.5 ml 10% of Folin - Ciocalteu reactive, and 2.5 ml of 7.5 % aqueous solution of Na_2CO_3 were added. The mixture was kept for 30 min in the dark at room temperature. The absorbance was read at 765 nm using a UV/Vis spectrophotometer (UV-6450; Jenway, UK). Gallic acid (10 - 50 ppm) were used for constructing the standard curve, and the results were expressed as g of gallic acid equivalents (GAE)/100 g of extract.

2.6. Total Tannin Content (TTC)

Total tannins content determined by using method described in Tambe and Bhambar (2014) [12]. 0.1 ml methanolic extract, 0.5 ml of Folin - Ciocalteu reactive, 7.5 ml distilled water and 1 ml of 35 % aqueous solution of Na_2CO_3 were added in test tube. Then 0.9 ml of distilled water was added. The mixture kept for 30 min in the dark at room temperature. The absorbance read at 725 nm using a UV/Vis spectrophotometer (UV-6450; Jenway, UK). Gallic acid (20 - 100 ppm) was used for constructing the standard curve, and the results were expressed as g of gallic acid equivalents (GAE)/100 g of extract.

2.7. DPPH (1,1-diphenyl-2-picrylhydrazyl) Radical Scavenging Assay

DPPH radical-scavenging activity were performed by the method described by Atker (2010) [13]. For each determination, the stock solution (1mg/ml) was diluted with solvent. An aliquot of each dilution (0.5 mL) was mixed with methanolic solution of DPPH (5 mL, 0.06 mM). The mixtures were shaken vigorously and incubated at 37 °C in the dark for 30 min. At the same time, a control containing solvent (0.5 mL) and methanolic solution of DPPH (5 mL, 0.06 mM) were run. The absorbance measured at 517 nm against methanol as a blank. The percentage of DPPH scavenging was calculated as follows: DPPH radical scavenging activity (%) = $[(Abs_{control} - Abs_{sample}) / Abs_{control}] \times 100$.

2.8. Ferric Reducing Ability Power (FRAP) Assay

Ferric Reducing Ability Power determined by using method described Barku *et al.*, (2014) [14]. In 2.5 ml of methanolic extract at various concentration (31.25 -1000 ppm) were mixed to 2.5 ml of sodium phosphate buffer (pH 6.6) and 2.5 ml of 1 % potassium ferricyanide. The mixture were placed in water bath at 50 °C for 20 min. Then 2.5 ml of 10 % TCA (w/w) was added to mixture. The mixtures were centrifuged at 3000 rpm for 10 minutes. This will follow by the addition of 2.5 ml of distilled water and 0.5 ml of 0.1 % of ferric chloride. The absorbance recorded at 700 nm after 5 min. The antioxidant activity was calculated from the calibration curve of ascorbic acid (31.25 – 1000 ppm). Increased absorbance of the mixture indicates increased reducing power.

3. Results and Discussion

In this recent year, a great attention has been focused on the agro-industrial wastes, especially those containing residual phenols from used plant raw material. PP and PIP are example of the important dietary sources of antioxidant properties [12, 13].

Calorimetric analysis of phenols, flavonoids and tannins content indicated that the methanolic extract of PP had highly amounts of TPC, TFC and TTC compared to PIP and this in agreement with Romelle *et al.*, (2016) who studied the chemical composition of selected fruit peels [17]. The polyphenols, flavonoids and tannins content in PP extract was 6.3133 mg/g of dry weight of extract, expressed as gallic acid equivalents, 4.6675 mg/g of dry weight of extract, expressed as quercetin equivalents and 0.1593 mg/g of dry weight of extract, expressed as gallic acid equivalents respectively. A significant lower of the TPC, TFC and TTC in WP which the values were 1.6205 mg/g of dry weight of extract, expressed as gallic acid equivalents, 0.2675 mg/g of dry weight of extract, expressed as quercetin equivalents and 0.0164 mg/g of dry weight of extract, expressed as gallic acid equivalents respectively. The r^2 for standard curve of gallic acid in TPC is 0.9997, standard curve of quercetin in TFC is 0.9996 while for standard curve of gallic acid in TTC is 0.9918. The standard curve for TPC, TFC and TTC showed in Figure 1, Figure 2 and Figure 3 respectively. In biosynthesizing of nanoparticles, phenolic compounds play an important role for reduction and stabilization [18].

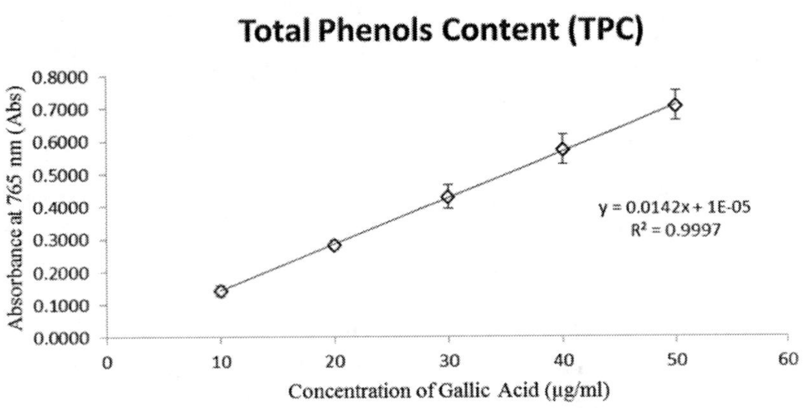

FIGURE 1. Graph Absorbance at 765 nm (Abs) against Concentration of Gallic Acid (µg/ml).

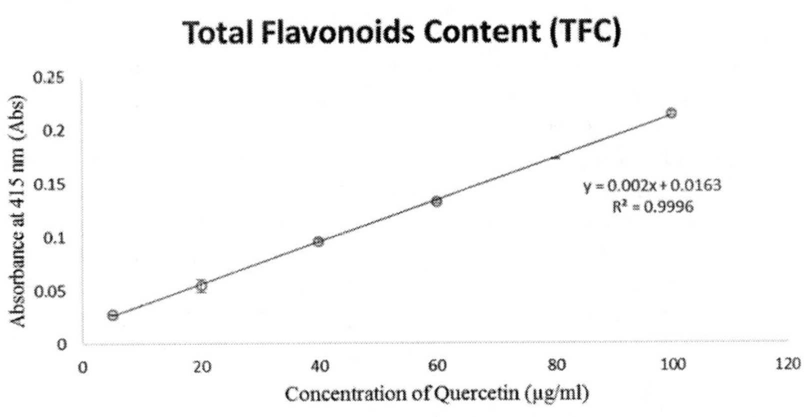

FIGURE 2. Graph Absorbance at 415 nm (Abs) against Concentration of Quercetin (µg/ml).

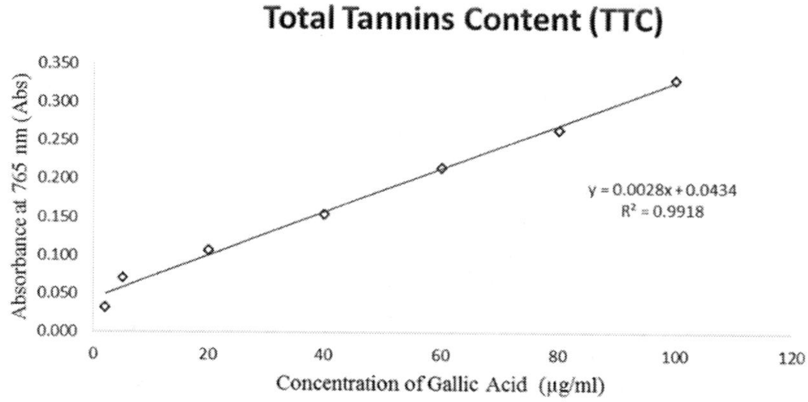

FIGURE 3. Graph Absorbance at 765 nm (Abs) against Concentration of Gallic Acid (μg/ml).

The antioxidant scavenging activity through the reaction between antioxidant molecules and radical results in the scavenging of the radical by hydrogen donation caused the decrease in absorbance of DPPH radical. In this research, the PP extract was found to be higher in DPPH scavenging activity with 54.95 % while in PIP was only 26.24 %. Table 1 showed there is a significant linear correlation was confirmed between the phenolic contents and the antioxidant scavenging activity of the fruit peels extract. The highest content of phenolic compounds indicated that these compounds contribute to the higher antioxidant scavenging activity [15, 16].

TABLE 1. TPC, TFC, TTC and DPPH Antioxidant Assay in Fruit Peels Extract.

Fruits	**TPC (g GAE/mg of extract)**	**TFC (g QAE/mg of extract)**	**TTC (g GAE/mg of extract)**	**DPPH antioxidant assay (%)**
Pomelo peels (PP)	6.3133	4.6675	0.1593	54.95
Pineapple peels (PIP)	1.6205	0.2675	0.0164	26.24

The PP extract showed the highest ferric reducing power ability compared to PIP extract at various concentration. For the iron reducing power assay, it was found that there was a tight relationship between the amount of phenolic contents and ferric reducing power ability. There is also a linear relationship between the phenolic contents and ferric reducing ability power [21]. Other than that, previous results showed higher in antioxidant scavenging activity also results in higher ferric reducing power ability power of bioactive secondary metabolites. These tight relationship of antioxidant properties have a good agreement with [18, 19]. Thus, higher in phenolic contents also results in antioxidant scavenging activity as well as ferric reducing ability power. However, in this studied, the ability of reducing power of the methanolic extracts for both PP and PIP were significantly lower than the synthetic antioxidant of standard ascorbic acid shown in Figure 4.

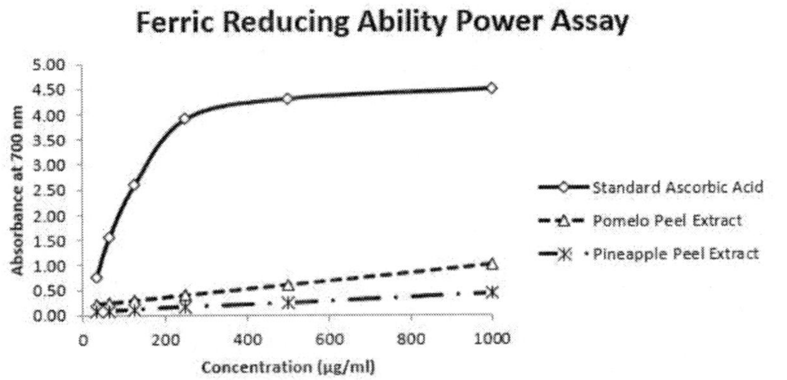

Figure 4. Iron Reducing Power Ability of PP and PIP at various concentration.

4. Conclusion

As the conclusion, it could be concluded that pomelo peels extract has higher TPC, TFC, TTC, DPPH scavenging activity and also ferric reducing power ability compared to pineapple peels extract. Besides the phenolic contents in peel itself, methanol act as solvent which has the highest polar solvent also played a vital role in extraction process. From the collected data for this analysis, it can be concluded that pomelo peel extract has potential to be conducted for further research mainly in synthesizing nanoparticles which these high value-added products can be applied either in pharmaceutical, food and also environmental scopes.

Acknowledgements

The authors would like to acknowledge the financial support from the Fundamental Research Grant Scheme (FRGS) under a grant number of FRGS/1/2015/TK02/UNIMAP/02/05 (Grant Account No.: 9003-00528) from the Ministry of Higher Education.

References

[1] C. Riber, "Waste to Energy in Malaysia," *Ramboll Waste-to-Energy*, 2012.
[2] S. H. Fauziah and P. Agamuthu, "Municipal Solid Waste Management in Malaysia: Strategies in Reducing The Dependency on Landfills," *J. Chem. Inf. Model.*, vol. **53**, p. 160, 2012.
[3] C. Bories, M. E. Borredon, E. Vedrenne, G. Vilarem, and P. Agamuthu, "Challenges and Opportunities in Agro-waste Management : An Asian Perspective What is AgroWaste ?," *J. Environ. Manage.*, vol. **143**, pp. 186–196, 2009.
[4] M. S. Mohamed and M. Y. Rokiah, "Tropical fruits and vegetables in Malaysia : Production and impact on health," no. August 2006, pp. 1–5, 2010.
[5] R. Shalini and D. K. Gupta, "Utilization of pomace from apple processing industries: A review," *J. Food Sci. Technol.*, vol. **47**, no. 4, pp. 365–371, 2010.
[6] E. Vladimir and E. K. Joseph, "Production Statistics - Crops, Crops Processed," 2011.
[7] S. M. Omar, G. C. Azucena, and S. V. Raul, "Agricultural residues as a source of bioactive natural products," *Phytochem. Rev.*, vol. **11**, no. 4, pp. 447–466, 2012.
[8] S. Vijayalaxmi, S. K. Jayalakshmi, and K. Sreeramulu, "Polyphenols from different agricultural residues: extraction, identification and their antioxidant properties," *J. Food Sci. Technol.*, vol. **52**, no. May, pp. 1–9, 2014.
[9] J. Azmir, I. S. M. Zaidul, M. M. Rahman, K. M. Sharif, A. Mohamed, F. Sahena, M. H. A. Jahurul, K. Ghafoor, N. A. N. Norulaini, and A. K. M. Omar, "Techniques for extraction of bioactive compounds from plant materials: A review," *J. Food Eng.*, vol. **117**, no. 4, pp. 426–436, 2013.

[10] I. S. Ashoush and M. G. E. Gadallah, "Utilization of Mango Peels and Seed Kernels Powders as Sources of Phytochemicals in Biscuit," *World J. Dairy Food Sci.*, vol. **6**, no. 1, pp. 35–42, 2011.

[11] M. S. Stankovi, "Total phenolic content, flavonoid concentration and antioxidant activity of Marrubium peregrinum L. extracts," *Kragujev. J. Sci.*, vol. **33**, no. July, pp. 63–72, 2011.

[12] V. D. Tambe and R. S. Bhambar, "Estimation of total phenol , tannin , alkaloid and flavonoid in hibiscus tiliaceus linn. wood extracts," vol. **2**, no. 4, pp. 41–47, 2014.

[13] M. S. Akter, M. Ahmed, and J. B. Eun, "Solvent effects on antioxidant properties of persimmon (Diospyros kaki L. cv. Daebong) seeds," *Int. J. Food Sci. Technol.*, vol. **45**, no. 11, pp. 2258–2264, 2010.

[14] V. Y. A. Barku, B. Y. Opoku, A. E. Owusu, and E. F. Mensah, "Antioxidant activity and the estimation of total phenolic and flavonoid contents of the root extract of Amaranthus spinosus," *Asian J. Plant Sci. Res.*, vol. **3**, no. 1, pp. 69–74, 2013.

[15] U. K. Ibrahim, N. Kamarrudin, M. U. H. Suzihaque, and S. A. Hashib, "Local Fruit Wastes as a Potential Source of Natural Antioxidant : An Overview," 2016.

[16] W. Suttirak and S. Manurakchinakorn, "In vitro antioxidant properties of mangosteen peel extract," *J. Food Sci. Technol.*, vol. **51**, no. 12, pp. 3546–3558, 2014.

[17] F. D. Romelle, A. Rani, and R. S. Manohar, "Chemical composition of some selected fruit peels," *Eur. J. Food Sci. Technol.*, vol. **4**, no. 4, pp. 12–21, 2016.

[18] A. Dzimitrowicz, P. Jamróz, G. C. diCenzo, I. Sergiel, T. Kozlecki, and P. Pohl, "Preparation and characterization of gold nanoparticles prepared with aqueous extracts of Lamiaceae plants and the effect of follow-up treatment with atmospheric pressure glow microdischarge," *Arab. J. Chem.*, 2016.

[19] G. Piluzza and S. Bullitta, "Correlations between phenolic content and antioxidant properties in twenty-four plant species of traditional ethnoveterinary use in the Mediterranean area.," *Pharm. Biol.*, vol. **49**, no. 3, pp. 240–7, 2011.

[20] M. M. Rashad, A. E. Mahmoud, M. M. Ali, M. U. Nooman, and A. S. Al-Kashef, "Antioxidant and anticancer agents produced from pineapple waste by solid state fermentation," *Int. J. Toxicol. Pharmacol. Res.*, vol. **7**, no. 6, pp. 287–296, 2015.

[21] N. Babbar, H. S. Oberoi, D. S. Uppal, and R. T. Patil, "Total phenolic content and antioxidant capacity of extracts obtained from six important fruit residues," *Food Res. Int.*, vol. **44**, no. 1, pp. 391–396, 2011.

[22] J. A. Larrauri, P. Rupérez, L. Bravo, and F. Saura-Calixto, "High dietary fibre powders from orange and lime peels: Associated polyphenols and antioxidant capacity," *Food Res. Int.*, vol. **29**, no. 8, pp. 757–762, 1996.

[23] Y. Q. Ma, J. C. Chen, D. H. Liu, and X. Q. Ye, "Effect of ultrasonic treatment on the total phenolic and antioxidant activity of extracts from citrus peel," *J. Food Sci.*, vol. **73**, no. 8, 2008.

IOP Conf. Series: Materials Science and Engineering **429** (2018) 012041 doi:10.1088/1757-899X/429/1/012041

Optimization of the product of nanocrystalline cellulose from coconut husks

O Nurdiana[1], S T Sam[1] and A M Faiq[1]

[1]School of Bioprocess Engineering, Universiti Malaysia Perlis (UniMAP), Kompleks Pusat Pengajian Jejawi 3, 02600 Arau, Perlis.

Email: nurdianothman@gmail.com

Abstract. Coconut husks can increase its value as agricultural wastes by utilizing these wastes to produce nanomaterials. Therefore, in this study prepared nanocrystalline cellulose (NCC) from coconut husks using sulphuric acid hydrolysis. The Box-Behnken Design (BBD) based on the response surface methodology was applied to study the effect of sulphuric acid concentration, reaction time and reaction temperature on the yield of NCc. The result showed that the yield of NCC was 44.94%, under optimum conditions of 64.6% concentration of sulphuric acid, 58.55 minutes and 44.56 °C. The X-ray diffraction (XRD) showed the crystallinity index of NCC was 68.3%.

1. Introduction

Cellulose, the most abundant and accessible organic compound in nature, has been increasingly as nanocrystalline cellulose (NCC) in the production of biodegradable materials. NCC are typically rod-shaped cellulose particles, with diameters of 1 to 100 nm and lengths of tens to hundreds nanometers [1]. Their main features include large specific surface area, very high modulus of elasticity and great aspect ratio [2], which determine their utilization as reinforcing agent in polymer composites [3]. For preparing NCC, one of the most commonly used methods is acid hydrolysis, based on the fact that the crystalline regions are insoluble under acid conditions [4]. Such a hydrolysis process can disrupt the disordered and amorphous parts of the cellulose and release ordered and definite crystals .

Over the past decade, several natural fibers such as flax, jute, sisal fibers and cellulosic component have been studied as resources for preparing NCC. However, the use of coconut husks has limited research for its utilization. Coconut husks as agriculture wastes are mostly used as floor mat, doormat and bushes. Coconut husks consists of 38-39% cellulose, 17.33% hemicellulose and 46-53% lignin [5]. Thus, coconut husks are suitable natural cellulose materials to be isolated as NCC because it has high performance to be reinforced with polymer in forming biodegradable film. Preparation of NCC from coconut husks also can greatly increase the economic value of these agricultural wastes.

Response surface methodology as an important experimental design and analysis methodology is to optimize the preparation conditions. The objective of this study is to prepare NCC from coconut husks by sulfuric acid hydrolysis and using Box-Behnken design in response surface methodology to improve the yield of NCC. The crystallinity index of NCC was optimize using X-Ray diffraction (XRD).

2. Materials and Methods

2.1. *Materials*

Coconut husks were purchased from the local market in Kangar, Perlis, Malaysia. Sodium hydroxide and sodium chlorite were purchased from Sigma-Aldrich and sulfuric acid was purchased from Vectec, PA. Glacial acetic acid was purchased from Fisher, Germany.

2.2. *Preparation of coconut husks fibres*

The coconut husks were cut and ground until it turned into powder. Then, the powder were sieved to 125 μm for the treatments. The coconut husks were then treated with 4wt% of sodium hydroxide (NaOH) for 2 hours at 80 °C with continuous stirring. Then, the fibres were repeatedly filtered with distilled water until the pH turned to neutral. After this alkaline treatment, the fibres were bleached using acetate buffer and 1.7wt% sodium chlorite ($NaClO_2$) at 80 °C for 2 hours. The fibres were undergone alkaline and bleaching treatment up to 4 times. The white fibres were then dried in air circulating oven for 24 hours at 50 °C.

2.3. *Preparation of nanocrystalline cellulose (NCC)*

Nanocrystalline cellulose (NCC) were prepared from the coconut husks fibres using acid hydrolysis. The coconut fibres were put into a 100 ml beaker, then the sulphuric acid concentration (60-70 wt%) was added into the beaker with the reaction temperature at (40-50 °C) and reaction time for (55-65 min) under continuous stirring. Then, 10-fold of cold water was added to the suspension to stop the hydrolysis process. The suspension was centrifuged for 10 min at 9000 rpm to remove the excessive sulphuric acid. After the suspension was centrifuged repeatedly, the suspension turned into milky suspension and were collected and dried for 24 hours to calculated the yield of coconut husks NCC.

2.4. *Experimental design*

To obtain the optimum preparation conditions of coconut husks NCC, the response surface method was used in this study. The sulphuric acid concentration (X_1), reaction temperature (X_2) and reaction time (X_3) were chosen as variables and yield (Y) was chosen as response value. The parameters of the test are shown in **Table 1.**

Table 1: The parameters that used for analyse NCC using the Design Expert 7.1.5

	Levels		
Factors	**-1**	**0**	**1**
X_1 (wt%)	60	65	70
X_2 (°C)	40	45	50
X_3 (min)	55	60	65

2.5. *X-ray Diffraction (XRD)*

The crystallinity of NCC were characterized using Bruker AXS D8 Advance diffractometer with a scanning rate of 5° per min. The range of diffraction angle were 2θ = 10 - 40°. The crystallinity index of the samples was calculated according to the Segal method as shown in equation below, where I_{002} for crystalline region and I_{am} for amorphous region.

$$CrI\ (\%) = \frac{I_{002} - I_{am}}{I_{002}}$$

3. Results and Analysis

3.1. Optimum preparation conditions of NCC from coconut husks

To obtain the optimum preparation conditions of NCC from coconut husks, sulphuric acid concentration (60-70 wt%), reaction temperature (40-50 °C) and reaction time (55-65 min) were used Box-Behnken design (BBD) and obtain the regression model. Design of Box-Behnken test and results of the yield for NCC obtained shows in **Table 2**.

Table 2: Design of BBD and percentage of yield for test

No	X_1 (wt%)	X_2 (°C)	X_3 (min)	Y (%)
1	60	40	60	43.67
2	70	40	60	43.26
3	60	50	60	43.28
4	70	50	60	43.37
5	60	45	55	44.08
6	70	45	55	44.06
7	60	45	65	44.11
8	70	45	65	43.54
9	65	40	55	44.22
10	65	50	55	43.42
11	65	40	65	42.91
12	65	50	65	43.58
13	65	45	60	44.76
14	65	45	60	44.84
15	65	45	60	44.93
16	65	45	60	44.92
17	65	45	60	44.89

Based on the design and results of Box-Behnken design in **Table 2**, the Design Expert software was used to fit the regression model and the obtained regression equation was expressed as follows:

$$Y = 44.81 - 0.11X_1 - 0.054X_2 - 0.21X_3 + 0.13X_1X_2 - 0.14X_1X_3 + 0.37X_2X_3 - 0.53X_1^2 - 0.95X_2^2 - 0.39X_3^2$$

The significant analysis was showed in **Table 3**. The significance level of the model was less than 0.0001, which showed that the model and method were reliable. The model determination coefficient R^2 was 0.9881 suggested the fitted model could explain 98.81% of the total variation, which further confirmed that the model could fit well the experimental results.

Table 3: Regression analysis of variance

Source	Sum of Squares	Degree of Freedom	Mean Square	F Value	P value
Model	7.28	9	0.81	64.31	< 0.0001
X_1	0.099	1	0.099	7.87	0.0263
X_2	0.023	1	0.023	1.84	0.2714
X_3	0.34	1	0.34	26.73	0.0013
X_1X_2	0.068	1	0.068	5.37	0.0535
X_1X_3	0.076	1	0.076	6.01	0.0440

Continue...

X_2X_3	0.54	1	0.54	42.95	0.0003
X_1^2	1.19	1	1.19	94.56	< 0.0001
X_2^2	3.77	1	3.77	299.86	< 0.0001
X_3^2	0.64	1	0.64	50.65	0.0002
Residual	0.088	7	0.013		
Lack of fit	0.069	3	0.023	4.69	0.0847
Pure Error	0.019	4	4.870E-003		
Cor Total	7.37	16			
R squared	0.9881				

The factors X_1, X_3, X_1X_3, X_2X_3, X_1^2, X_2^2 and X_3^2 significantly affected the yield of the of NCC ($p \leq 0.05$) and the factor that ($p \geq 0.1$) are not significant. So it was not linear relationship between the effects of different factors on the response value. **Figure 1 (a, b and c)** shows the result for response surface figure under these three factors. Taking the maximum yield as the objective, the 64.61 wt% of sulphuric acid, 44.55 °C of reaction temperature and 58.54 min of reaction time, the predicted yield of NCC was 44.84% under these conditions. The results showed that the response surface methodology can be used to optimize the preparation conditions of NCC in order to increase the yield.

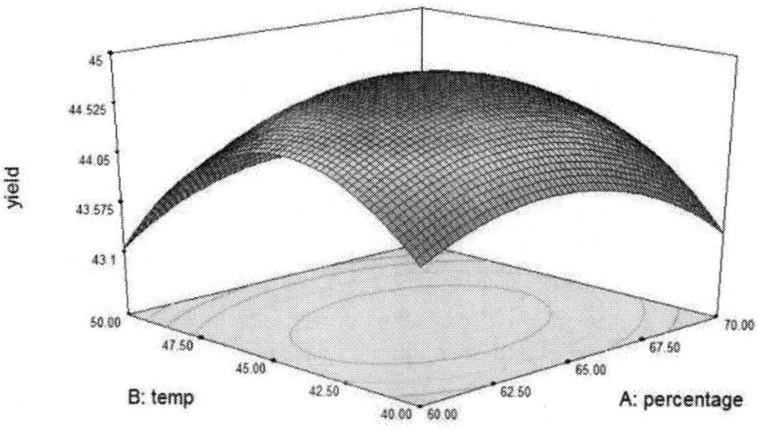

Figure 1(a): Response surface plot of yield of NCC (Factors X_1 and X_2)

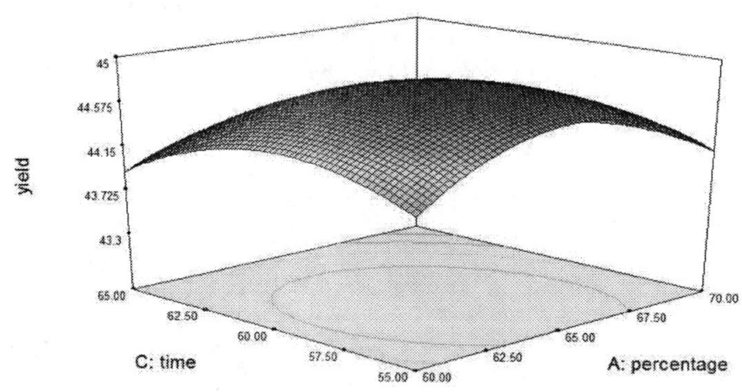

Figure 1(b): Response surface plot of yield of NCC (Factors X_1 and X_3)

IOP Conf. Series: Materials Science and Engineering **429** (2018) 012041 doi:10.1088/1757-899X/429/1/012041

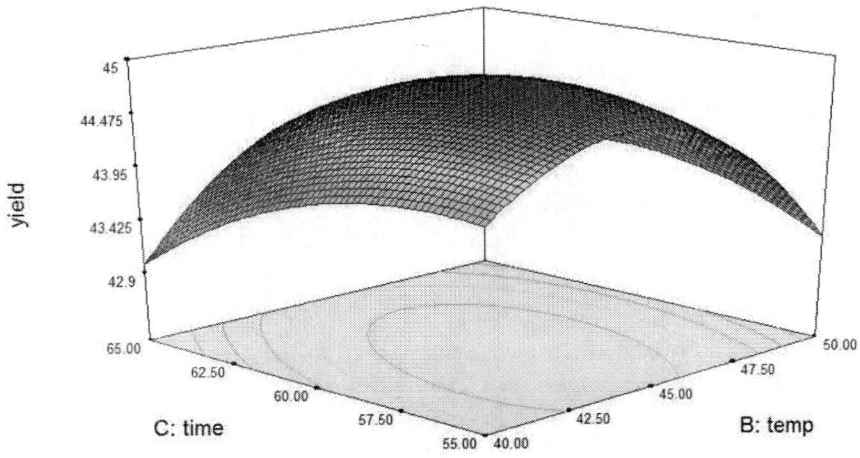

Figure 1(c): Response surface plot of yield of NCC (Factors X_2 and X_3)

3.2. XRD Analysis

Figure 2 shows the XRD pattern of NCC. In the diffraction pattern, there were crystalline peaks around 2θ= 16°, 22° and 35°, which presented the type of cellulose I pattern [6]. In this study, the crystallinity index of NCC for coconut husks was found about 68.3%, which indicating good crystallinity. This study also have a good agreement with Liu and his co-workers in 2015 stated that the angle of diffraction of NCC for peanut shells obtained were same with this study [7]. The intensity for NCC shows it is higher and indicates good crystallinity obtained for NCC.

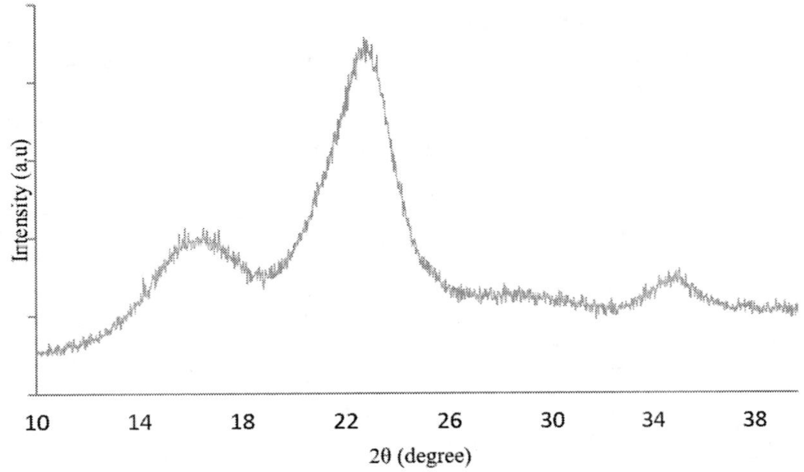

Figure 2: XRD pattern of NCC for coconut husks

4. Conclusion

This study investigated the optimisation of preparation of NCC from coconut husks using response surface methodology. It was found that the sulphuric acid concentration, reaction temperature and reaction time all affected the yield of NCC. The Box-Behnken design based on the response surface methodology can be used to establish the regression model of the yield of NCC. Based on the variance analysis, the regression equation can fit well the changes in the emulsifying properties. According to the regression equation, the optimum preparation conditions of NCC were 64.61 wt% of sulphuric acid, 44.55 °C of reaction temperature, 58.54 min of reaction time and the yield of NCC obtained was 44.84%. The crystallinity of NCC was 68.3% and it shows that the crystal type of NCC was cellulose I pattern.

Acknowledgments
The authors would like to acknowledge the support from the Fundamental Research Grant Scheme (FRGS) under a grant number of FRGS/1/2015/TK10/UNIMAP/03/5 from the Ministry of Higher Education of Malaysia.

References

[1] I Matos, A J Pereira, M Lince-Faria, L A Cameron, E D Salmon and H Maiato 2009 Synchronizing chromosome segregation by flux-dependent force equalization at kinetochores *J. Cell Biol.* **186** 11–26
[2] W P Flauzino Neto, H A Silvário, N O Dantas and D Pasquini 2013 Extraction and characterization of cellulose nanocrystals from agro-industrial residue - Soy hulls *Ind. Crops Prod.* **42** 480–488
[3] V Favier, H Chanzy and J Y Cavaille 1996 Polymer nanocomposites reinforced with cellulose whiskers 6365–6367
[4] Y Habibi, L A Lucia and O J Rojas 2010 Cellulose Nanocrystals : Chemistry, Self-Assembly and Applications 3479–3500
[5] R Li, J Fei, Y Cai, Y Li, J Feng and J Yao 2009 Cellulose whiskers extracted from mulberry: A novel biomass production *Carbohydr. Polym.* **76** 94–99
[6] D Klemm, B Heublein, H P Fink and A Bohn 2005 Cellulose: Fascinating biopolymer and sustainable raw material *Angew. Chemie - Int. Ed.* **44** 3358–3393
[7] X Liu, H Dong and H Hou 2015 Optimization of preparation of cellulose nanocrystals from peanut shells using response surface methodology *Adv. J. Food Sci. Technol.* **7** 466–473

IOP Conf. Series: Materials Science and Engineering **429** (2018) 012042 doi:10.1088/1757-899X/429/1/012042

Tensile Properties and Crystallinity of Crosslinked Nanocrystalline Cellulose/Chitosan Composite

P G Gan[1*], S T Sam[1], A M Faiq[1]

[1] School of Bioprocess Engineering, Universiti Malaysia Perlis (UniMAP), Kompleks Pusat Pengajian Jejawi 3, 02600 Arau, Perlis.

*Email: peigie023@gmail.com

Abstract. This study evaluated the properties of the polymeric composites obtained from nanocrystalline cellulose and chitosan using citric acid as the crosslinking agent prepared via solvent casting method. The tensile strength and Young's modulus of composites increased, but the elongation at break decreased with the nanocrystalline cellulose (NCC) loading. The addition of citric acid has further enhanced the tensile strength and Young's modulus. The morphology of the non-crosslinked and the citric acid crosslinked NCC/chitosan composites was investigated by the scanning electron microscopy (SEM) analysis. The X-ray diffraction (XRD) revealed that the relative crystallinity of composites was enhanced by the incorporation of NCC but the value declined by the addition of citric acid.

1. Introduction

Chitosan, a natural polymer consisting of 1,4-linked 2-amino-deoxy-β-ᴅ-glucan, can be obtained from the N-deacetylation of chitin. It is a promising and an attractive material for future applications due to its unique properties such as non-toxic, biodegradable, biocompatible and anti-microbial [1]. Chitosan can be applied in many areas such as packaging [2], biomedical purpose [3] and wastewater treatment [4], however, the chitosan film has poor barrier and mechanical properties which make it unsuitable for the aforementioned application [5]. Therefore, chitosan is often blended with reinforcing agent to improve its properties.

Nanocrystalline cellulose (NCC) is a highly crystalline nanometre rod-like cellulose which produced by acid hydrolysis with sulphuric acid being the most common-used acid. Acid hydrolysis involves the extraction of crystal from cellulose by hydrolysing the amorphous domains, resulting in highly crystalline nanoparticles [6]. The NCC synthesised (from plant cellulose) usually has a diameter of 5-20 nm and length of 100 -500 nm [7]. NCC reinforces the polymer matrix by forming the percolation network which interconnecting the NCC with hydrogen bonding. The presence of well-dispersed NCC as reinforcement in polymer matrices offers remarkable performances such as superior mechanical properties, biodegradable and highly crystalline leading to a new generation of biomaterial [10].

Besides adding reinforcing filler, polymers can also be chemically modified to improve its properties by using crosslinkers to alter the chemical structure of the matrix [10]. The crosslinking reaction occurs between the amino groups of chitosan and active groups of crosslinkers forms a three-dimensional network which limits the mobility of chitosan [11]. In this study, citric acid was used as the crosslinking agent for NCC/chitosan composites. Citric acid is a low cost and non-toxic poly(carboxylic) acid that widely used as a crosslinker for food and drug industry. Citric acid is categorised as safe for food usage due to its renewable and biodegradable properties [12]. Chitosan composites crosslinked with citric acid can undergo the amidation reaction at an elevated temperatures [13]. It was reported that addition of citric acid enhanced the mechanical properties of poly(vinyl alcohol)/starch significantly [14]. In the present study, the incorporation effect of NCC from empty fruit bunch (EFB) used as reinforcement and citric acid as crosslinking agent on mechanical properties of NCC/chitosan film was investigated. size. The morphology of the fracture surfaces of nanocomposites was also studied using field emission electron microscopy (FESEM).

2. Materials and Methods

2.1. Materials
EFB fibres was kindly provided by Taclico Company Sdn. Bhd. Chitosan with a degree of deacetylation 90.5% was provided by Cleo Shanghai Pharmaceutical. Sodium chlorite, acetic acid and sodium hydroxide were purchased from Sigma-Aldrich while citric acid and sulphuric acid were purchased from Merck. Potassium carbonate was provided by HmbG.

2.2. Preparation of Nanocrystalline Cellulose
NCC was isolated from EFB cellulose by acid hydrolysis. EFB fibres was grounded and sieved into a diameter of less than 125 μm. The fibres were pre-treated with deep eutectic solvent (DES) and bleaching treatments. DES was prepared by mixing and heating potassium carbonate and glycerol in a molar ratio of 1:7. The bleaching solution consisted of equal parts of acetate buffer, aqueous chlorite, and distilled water. Both DES and bleaching were performed at 80°C for 2 hours and repeated for three times. The treated cellulose was washed with distilled water and dried in oven at 60 °C for 2 hours.

Acid hydrolysis was carried out for 45 minutes at 45°C with a 64 wt% of sulphuric acid. After acid hydrolysis, the suspension was washed with distilled water and centrifuged at 9000 rpm for 10 minutes repeatedly until the suspension become turbid. 3 wt% of sodium hydroxide was added to neutralized the suspension. Ultrasonic treatment was applied for 10 minutes. The resulting suspension was kept in the refrigerator at 4°C for further usage.

2.3. Preparation of Chitosan Composites Reinforced with Nanocrystalline Cellulose
The chitosan solution was prepared by dissolving 1.5 wt% of chitosan powder into 2 v/v% acetic acid solutions at 60 °C for 4 hours. 10 wt% of glycerol and NCC were added into the chitosan solution and stirred for another hour. After removal of air bubbles, the NCC/chitosan solution was casted onto a 12 cm x 10cm petri dish and dried in an oven at 50 °C for 48 hours. The dried film has an average thickness of 0.09 mm.

For crosslinking process, 5wt% of citric acid was added into the NCC/chitosan solution. The solution was stirred vigorously in order to obtain homogeneous solution. The dried NCC/chitosan composite was cured in the oven at 105 °C for 10 minutes for the curing process.

2.4. Characterisation

2.4.1. Tensile Properties
Tensile properties of non-crosslinked and citric acid crosslinked NCC/chitosan composites were obtained using Instron universal tensile testing machine (UTM) according to ASTM standard method D882-A. The samples were prepared by cutting into rectangular specimens of 10 cm x 1 cm (length x width). A total of 10 measurements of were performed for each formulation and the average values were determined.

2.4.2. Scanning Electron Microscopy (SEM)
SEM was used to study the morphological fractured surface of non-crosslinked and citric acid crosslinked NCC/chitosan composites. To avert electrostatic charging, the external surface of samples

was coated with platinum using auto fine coater JFC 16000 before scanning. The samples were observed under the magnification of 5000 and the accelerating voltage of 10kV.

2.4.3. X-Ray Diffraction (XRD) Analysis

Crystallinity of non-crosslinked and citric acid crosslinked NCC/chitosan composites was determined using a Bruker D2 Phaser X-ray diffractometer at the operating voltage and current at 40 kV and 30 mA. Data were collected in a 2θ angle between 5 and 40°.

3. Results and Discussion

3.1. Tensile Properties

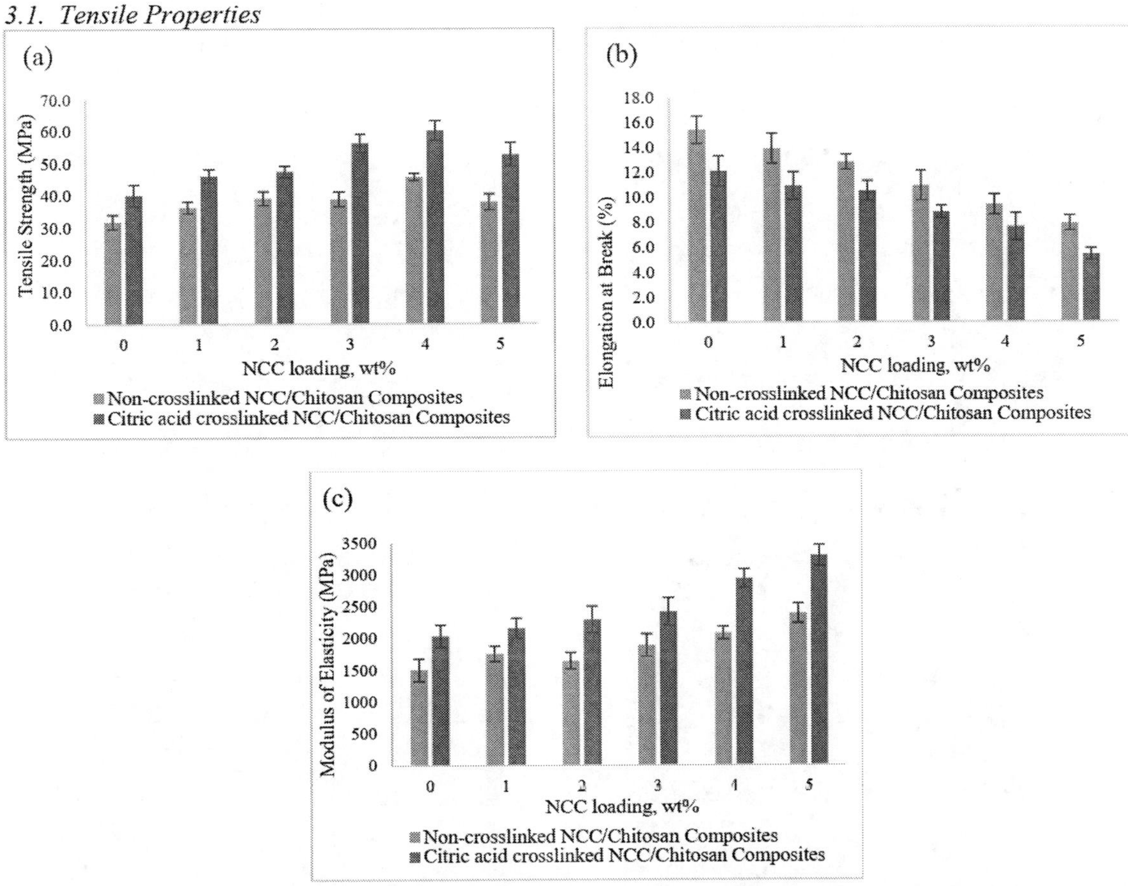

Figure 1: (a) Tensile strength, (b) elongation at break and (c) modulus of elasticity of non-crosslinked and citric acid crosslinked NCC/chitosan composites

Figure 1 shows the tensile properties of non-crosslinked and citric acid crosslinked NCC reinforced chitosan composites. The addition of NCC increased the tensile strength and modulus of elasticity values with respect to the neat chitosan film, proving the reinforcing effect induced by the NCC. Chitosan composite with the incorporation of 4 wt% NCC records the highest tensile strength of 45.7 MPa for non-crosslinked set. The enhancement could be attributed to the homogeneous dispersion between the NCC and chitosan matrix [15]. The good dispersion between filler and matrix also led to a better interfacial adhesion and enhanced the stress transfer to the filler which in turn to improve the tensile strength [16]. The crosslinking process has further increased the tensile strength by forming a more stable network between crosslinker and matrix. Similar result was also reported by Rivero et al.

[17] in which the addition of crosslinking agent has led to a formation of more rigid structure and hence, improved the tensile strength significantly.

However, the elongation at break of composites declined significantly with the incorporation of NCC in chitosan matrix as shown in Figure 1 (b). This observation may be attributed to the stiffening effect of the highly crystalline NCC which increased the rigidity of the composite. Furthermore, the addition of crosslinkers has limited the polymer chain mobility and flexibility of composites [1]. Besides, the structural reorganisation and moisture loss during curing process also led to the reduction of elongation at break [17].

The elasticity of modulus of NCC/chitosan composites increased linearly with the addition of NCC. This could be due to the formation of strong intramolecular or intermolecular hydrogen bonding of NCC with chitosan, as well as the stiffening effect, which resulted in the increment of elasticity of modulus [18]. Introduction of citric acid as crosslinker also interfered the NCC network formation in the chitosan matrix and restricted the mobility of chitosan segmental chain, consequently increased the modulus of elasticity dramatically [13]. The high performance of citric acid crosslinked NCC/chitosan composites proposed the favourable interactions between citric acid, NCC and chitosan matrix.

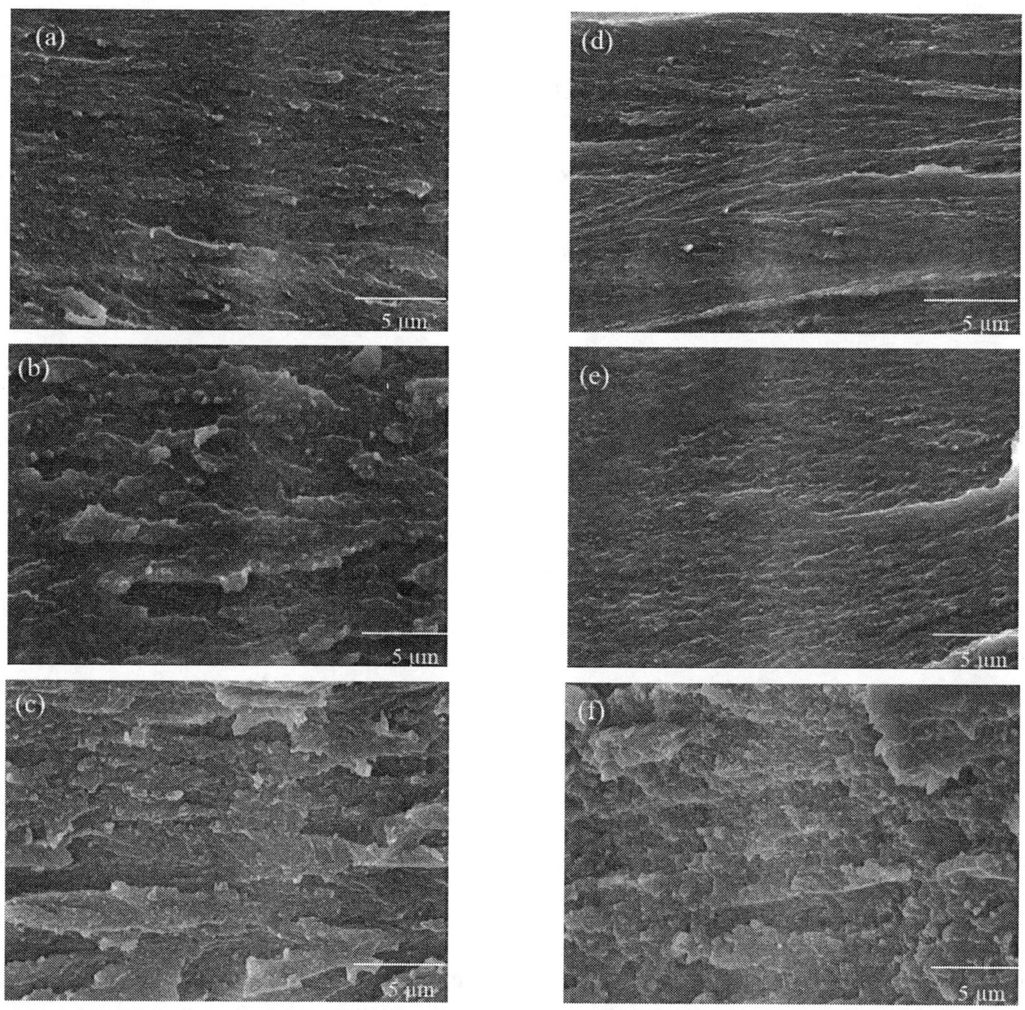

Figure 2: Fracture surface of the non-crosslinked chitosan composites with (a) 0% NCC, (b) 1% NCC & (c) 4% NCC and the citric acid crosslinked chitosan composites with (d) 0% NCC, (e) 1% NCC & (f) 4% NCC at 5000X magnification

The fracture surfaces of neat chitosan and chitosan composites were determined by SEM (Figure 2), in order to investigate the effect of NCC addition and the influence of crosslinker on the chitosan microstructure. The tensile fracture surface of non-crosslinked and citric acid crosslinked neat chitosan composites in Figure 2 (a) and (d) displayed a homogeneous and smooth surface with matrix tearing. The micrograph of non-crosslinked NCC/chitosan composites showed a rough surface. Furthermore, the fracture surface of 4 wt% NCC/chitosan composite showed a higher roughness due to the higher concentration of NCC with respect of 1 wt% NCC/chitosan composite. After the chemical modification using citric acid crosslinking agent, the film became more porous [5]. SEM analysis further verified the improved mechanical properties of chitosan composites due to the incorporation of NCC and citric acid.

3.2. X-Ray Diffraction (XRD) Analysis

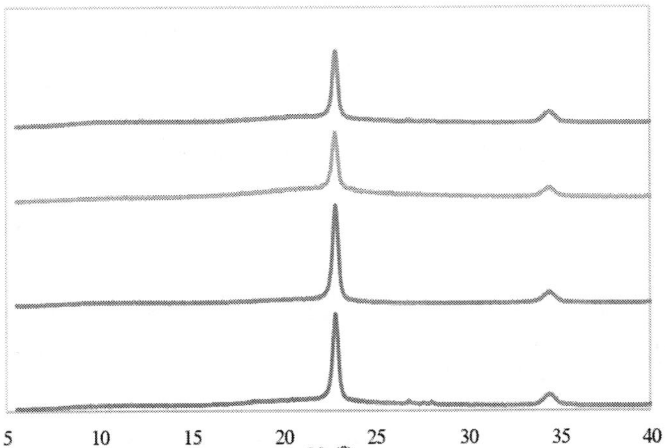

Figure 3: X-ray diffractograms of non-crosslinked and citric acid crosslinked NCC/chitosan composites

Table 1: Crystallinity index of non-crosslinked and citric acid crosslinked NCC/chitosan composites

Samples	Crystallinity index (%)
Neat chitosan	42.1
5 wt% NCC/chitosan	54.8
Citric acid crosslinked chitosan	31.4
Citric acid crosslinked 5 wt% NCC/chitosan	39.5

The X-ray diffraction patterns of non-crosslinked and citric acid crosslinked NCC/chitosan composites were exhibited in Figure 3. All composites showed major crystalline peaks at $2\theta = 12.5^0$, 22.6^0 and 34.7^0 , which corresponded to the hydrated "tendon" conformation of chitosan [19]. The chitosan composite with the addition of NCC has significantly elevated the peak at $2\theta = 22.6^0$. This indicated that NCC has a higher degree of crystallinity [20]. The introduction of NCC did not cause any significant shift in peaks in the NCC/chitosan composites. The addition of 5 wt% of NCC has increased the crystallinity index of composites from 42.1% to 54.8% as shown in Table 1.

In comparison with the non-crosslinked chitosan composites, the crosslinked chitosan composites have a significantly weakened peak at $2\theta = 22.6^0$ which indicated the effective crosslinking between chitosan and citric acid [21]. The average crystallinity index of crosslinked neat chitosan and

NCC/chitosan composites were reduced to 31.4% and 39.5% respectively. The crosslinking process hindered the close packing of the polymer chains by limiting the degree of freedom in the 3-D conformation, restricting or even inhibiting the formation of crystalline region. The introduction of crosslinking agent disturbed the intra and inters molecular hydrogen bonds of chitosan network which in turn lower the crystallinity index of chitosan composites [22].

4. Conclusion

NCC/chitosan composites were successfully produced and chemically crosslinked with citric acid. The tensile strength was improved notably by the addition of citric acid as crosslinking agent. The incorporation of NCC has increased the crystallinity of the chitosan composites, however, the crosslinking process has reduced the crystallinity index due to the breaking of crystalline domains of chitosan polymer.

Acknowledgments

The authors gratefully acknowledge the Ministry of Higher Education for the Fundamental Research Grant Scheme (FRGS/1/2015/TK10/UNIMAP/03/5).

References

[1] Yeng CM, Husseinsyah S, Ting SS 2013 Chitosan/corn cob biocomposite films by cross-linking with glutaraldehyde *BioResources* **8** 291023.
[2] Moraes Crizel T, de Oliveira Rios A, D. Alves V, Bandarra N, Moldão-Martins M, Hickmann Flôres S 2018 Active food packaging prepared with chitosan and olive pomace *Food Hydrocoll* **74** 13950
[3] Arce F, Pedrazzoli D, Manas-Zloczower I 2016 Thermoplastic Polyurethane Chitosan / Cellulose Nanocrystals Composites for Wound Healing Applications *Indianapolis* 175761.
[4] Nechita P 2017 *Applications of Chitosan in Wastewater Treatment* p 210
[5] Rubentheren V, Ward TA, Chee CY, Nair P 2015 Physical and chemical reinforcement of chitosan film using nanocrystalline cellulose and tannic acid *Cellulose* **22** 252941
[6] Lin N, Dufresne A 2014 Nanocellulose in biomedicine: Current status and future prospect *Eur Polym J* **59** 30225
[7] Abitbol T, Rivkin A, Cao Y, Nevo Y, Abraham E, Ben-Shalom T, et al 2016 Nanocellulose, a tiny fiber with huge applications *Curr Opin Biotechnol* **39** 076088
[8] Moriana R, Vilaplana F, Ek M 2016 Cellulose Nanocrystals from Forest Residues as Reinforcing Agents for Composites: A Study from Macro- to Nano-Dimensions *Carbohydr Polym* **139** 13949
[9] Gan PG, Sam ST, Abdullah MF, Zulkepli NN, Yeong YF 2017 Characterization of Nanocrystalline Cellulose Isolated from Empty Fruit Bunch Using Acid Hydrolysis. *Solid State Phenom* **264** 09012
[10] Rivero S, García MA, Pinotti A 2013 Physical and Chemical Treatments on Chitosan Matrix to Modify Film Properties and Kinetics of Biodegradation *J Mater Phys Chem* **1** 517.
[11] Yeng CM, Husseinsyah S, Ting SS 2015 A comparative study of different crosslinking agent-modified chitosan/corn cob biocomposite films *Polym Bull* **72** 791808
[12] Tieng W, Lin OH, Ting S, Rey A, C GN 2017 Tapioca Starch Based Green Nanocomposites with Environmental Friendly Cross-linker *Chem Eng Trans* **56** 4638
[13] Falamarzpour P, Behzad T, Zamani A 2017 Preparation of Nanocellulose Reinforced Chitosan Films , Cross-Linked by Adipic Acid *Int J Mol Sci* **18** 112
[14] Jose J, Al-Harthi MA 2017 Citric acid crosslinking of poly(vinyl alcohol)/starch/graphene nanocomposites for superior properties *Iran Polym J* **26** 579870.
[15] Rafieian F, Simonsen J 2014 Fabrication and characterization of carboxylated cellulose nanocrystals reinforced glutenin nanocomposite *Cellulose* **21** 416780

[16] Mohamad Haafiz MK, Hassan A, Abdul Khalil HPSS, Khan I, Inuwa IM, Saiful Islam M, et al 2015 Bionanocomposite based on cellulose nanowhisker from oil palm biomass-filled poly(lactic acid) *Polym Test* **48** 1339
[17] Rivero S, García MA, Pinotti A 2010 Crosslinking capacity of tannic acid in plasticized chitosan films *Carbohydr Polym* **82** 2706
[18] Haafiz MKM, Hassan A, Khalil HPSA, Fazita MRN, Islam MS, Inuwa IM, et al 2016 Exploring the effect of cellulose nanowhiskers isolated from oil palm biomass on polylactic acid properties *Int J Biol Macromol* **85** 3708
[19] Rivero S, García MA, Pinotti A 2012 Heat treatment to modify the structural and physical properties of chitosan-based films *J Agric Food Chem* **60** 4929
[20] Rubentheren V, Ward TA, Chee CY, Nair P, Salami E, Fearday C 2015 Effects of heat treatment on chitosan nanocomposite film reinforced with nanocrystalline cellulose and tannic acid *Carbohydr Polym* **140** 2028
[21] Fideles TB, Santos JL, Tomás H, Furtado GTFS, Lima DB, Borges SMP, et al 2018 Characterization of Chitosan Membranes Crosslinked by Sulfuric Acid *OALib* **05** 113 doi:10.4236/oalib.1104336.
[22] Julkapli NM, Ahmad Z, Akil HM 2009 X-ray diffraction studies of cross linked chitosan with different cross linking agents for waste water treatment application *AIP Conf. Proc.* **1202** 10611

IOP Conf. Series: Materials Science and Engineering **429** (2018) 012043 doi:10.1088/1757-899X/429/1/012043

Evaluation of the Biodegradation for Oil Absorbed Cassava Starch Film Filled with Kenaf Core Fiber

R. J. Walster[1, a], Rahman Rozyanty[1, b], A. W. M. Kahar[1, c]

[1]School of Materials Engineering, Universiti Malaysia Perlis, 02600, Kangar, Perlis, Malaysia

Email: [a]walsterjustin@yahoo.com, [b]rozyanty@unimap.edi.my, [c]kaharwahab@unimap.edu.my

Abstract. Cassava starch (CS) films filled with different loading of kenaf core fiber (KCF) were prepared and tested. Two tests were carried out; oil absorption test and biodegradtion test. The oil absorption test were carried out with different loadings of KCF (0%, 0.2, 0.4, 0.6%. 0.8% and 1.0%). The results showed an increase of oil intake as the kenaf loading increased. The biodegradation test was conducted via soil burial method. Two sets of film samples were tested; film samples with and without immersion of oil. Each set of the film samples recpectively compares between CS films with and without KCF loadings. The results of the biodegradation tests showed that CS films with KCF had lower amount of weight loss compared to CS without KCF. Furthermore, the result of biodegradation tests for film samples immersed in oil were lower compared to film samples with oil immersion.

1. Introduction

Nowadays, polymer based materials are used extensively in all kinds of fields due to their advantages of high formability, light weight and high strength. As a result, the consumption of polymer materials has globally increased since the 1950's [1]. However, most of these polymer materials are produced from petroleum based plastics. Petroleum based plastics have a large timeframe of working life. Their high durability results in the massive accumulation of plastic waste over the course of time, leading to various serious environment issues which requires collection and disposal, generally by burial in landfill, composting, or incineration [2].

As a result, the interest in the use of biopolymers has been increasing. According to Peterson et al. [3], this was due to the increase in concern regarding the price of crude oil and as well as global warming which had influenced the awareness of consumers significantly. Nowadays, biopolymers are obtainable without much difficulty, for instances; biopolymers can be obtained from waste produced by agricultural feedstock, microbial activities, and marine fauna.

Although biopolymers are incapable of completely replacing the conventional synthetic polymers, biopolymers are capable of other ways which synthetic polymers could not. Starch is an exampleof a promising material for biodegradable plastics due to its versatility as a biopolymer, which holds an enormous potential with low price for use in non-related food industries [4]. There are a variety of sources to obtain starch, for example; maize, potatoes, rice, corn and cassava [5] [6]. Based on Souza et al. [7], they described that films which are manufactured from starch are odorless, possessed neither taste nor colors, and are capable of biodegrading naturally. In another of their study [8], they stated the plasticization of cassava starch using glycerol. The results provided information on accurate formulation to be used to produce composite films with improved mechanical properties.

At present time, works regarding natural fiber polymer comppposites are increasing in number. Natural fibers can be incorporated as reinforcement filler as they are cheaper in cost and can be obtained with much ease compared to conventional mineral fillers. Natural fibers in general are polar and hydrophilic in nature, resulting in a poor mechanical properties when added into conventional plastics as they are more hydrophobic in nature [9]. In an attempt to overcome the polarity issue, Jutarat et al. [10] in one of their works studied the incorporation of starch and thermoplastic polymer blends. They reported natural fiber such as; cotton fibers, works well as reinforcing agents in their starch and polyethylene blends. Natural fibers and starch possesses good chemical compatibility since both originated from natural resources [11]. Hence, the mixing of natural fibers and thermoplastic starch will have reinforcing effects.

One of the most studied natural fibers today is the kenaf. Kenaf fibers have been used in many parts of the world since long time ago. In the past, the main usage of kenaf was to manufacture ropes, twines, coarse clothing, and paper. Even today there are some which continue such practices. Nevertheless, the kenaf are still in demand today however as reinforcing additives for polymer production [12]. The kenaf can be divided into two sections; the bast and the core. For this study, the kenaf core will be used as kenaf fibers since the core has better oil absorption properties compared to the bast section. In a report by Ridwan [13], he stated that kenaf core is a very high absorbent material and there are already plenty of commercialized products for toxic waste cleanup, remediation of chemically contaminated soils and oil spills on water. The main purpose of this study are to evaluate the effects different loading of KCF on oil absorption and natural degradation of cassava starch films via soil burial.

2. Experimental

2.1. *Materials*

Native cassava starch (brand: 'Kapal ABC' brand, imported from Thailand) was used without further purification. Analytical grade glycerol was purchased from Fisher Chemicals, UK. Kenaf core fibers were purchased from LKTN (Lembaga Kenaf Tembakau Negara). Cooking oil (brand: Knife, manufactured in Malaysia) were used to test for oil absorption of the CS films.

2.2. *Film Preparation*

Cassava starch films were prepared by mixing 6 g of CS, with 1.5 % (W/W) glycerol and different loading of KCF (0.2%, 0.4%, 0.6%, 0.8%, and 1% of W/W) using Isotemp stirring hot plate (Fisher Scientific) at 80 °C for 13-15 minutes or until a filmogenic solution was formed. The filmogenic solution was transferred on an polyacrylic mold (240 mm × 160 mm) via solution casting and left to dry in a ventilated oven at 40 °C for 24 hours. Two types of film samples were prepared, CS films without KCF (control) and CS films with KCF.

2.3. *Measurements of Film Thickness*

The thickness of each film was measured at three different positions to the nearest 0.0001 mm using dial thickness gauge (Precise brand, model: 330-101).

2.4. *Oil Absorption Test*

The oil absoprtion test were carried out by adapting the test methods from ASTM D570-98 [14]. The film samples were cut to bar shape (30 × 10 mm). The test samples were dried in a ventilated oven at 40 °C for 24 hours to remove any residual moisture and immersed into oil for 9 hours. The samples were weigh, and the weight of samples after immersed in oil were determined through the following equation:

$$O_o = \frac{(O_f - O_i)}{O_i} \times 100\% \quad (1)$$

where O_o represents the percentage of weight change, O_i represents the initial weight of samples before immersed in oil and O_f represents weight samples after immersed in oil. The CS films with and without KCF were tested.

2.5. Biodegradation Test

The biodegradation test was carried out via soil burial method by adaptation from SR EN ISO 846/2000. The sample specimens were cut into square shape (20 × 20 mm) were buried in a compost soil. The films samples were taken out of the soil, and cleaned carefully using a slightly damp tissue in order to remove adherent soil and dried on filter paper. The samples were stored in a desiccator for 24 h until they reached a constant weight. The weight of the samples was taken at regular interval of time (5 days) to record the weight loss. The weight loss degradation was determined through the following equation:

$$m_{loss} = \frac{m_i - m_f}{m_i} \times 100\% \quad (2)$$

where m_{loss} represents the weight loss, m_i represents the initial weight and m_f represents the final weight of the samples after each testing periods. The effects of the presence oil in biodegradation of CS films without KCF and CS films with the highest loading of KCF films were used.

3. Results and discussion

3.1. Oil Absorption Test

Fig. 1 shows the oil absorption of the CS/KCF with differing load of fibers for 9 hours. The results show that the oil uptake increased steadily up until the 6th hour and began to slowly form a plateau. All the reinforced films absorbed more than the un-filled CS films which was only 0.9% at the 9th hour of immersion in the oil. Films with 1% KCF loading absorb the most oil which was 3.98%. Moreover, the film with 0.2% loading absorbed the least oil with only 1.79%. It is clear that the KCF increased the oil absorption uptake of the CS film. From the results, it is shown that CS film does not absorb much oil. Most of the oil are probably absorbed by the KCF. Melissa et al. [15], stated in their report that polysaccharides possesses good film-forming properties, which provides efficient barriers against oils and lipids. Bazargan et al. [16], stated in his report that natural fibers possesses many free hydroxyl groups at the molecular level. Therefore, KCF are able to attract oil molecules.

IOP Conf. Series: Materials Science and Engineering **429** (2018) 012043 doi:10.1088/1757-899X/429/1/012043

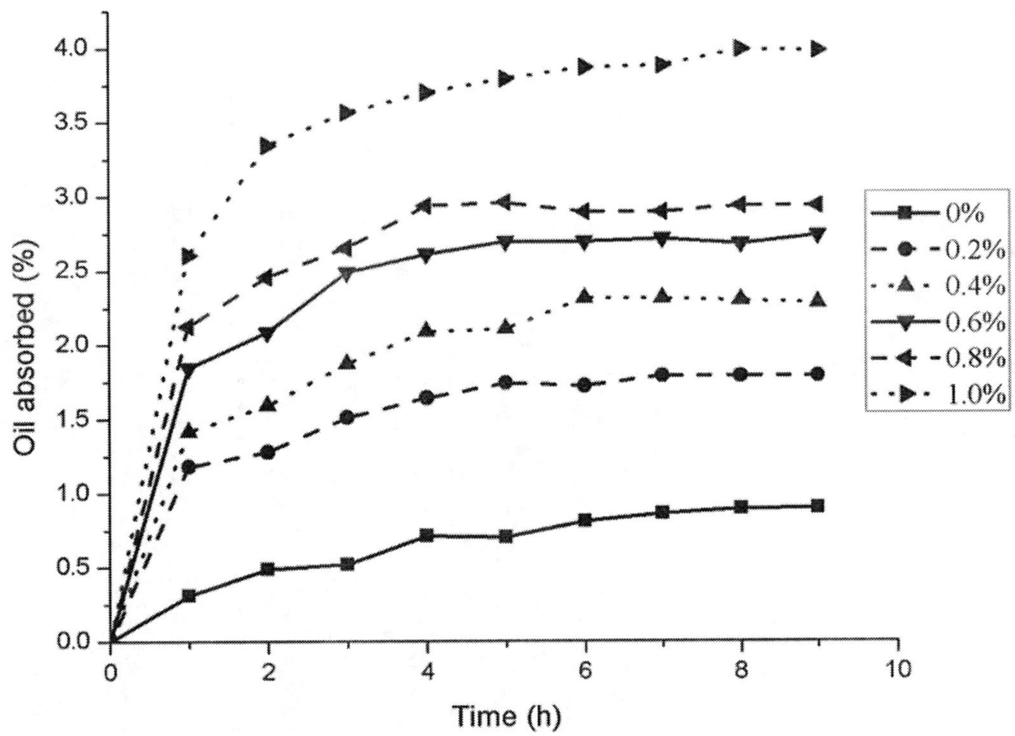

Fig. 1: Oil absorbed by CS/KCF films with different loading of KCF at different time(s).

3.2. Biodegradation Asssessment

The weight loss of film samples during 35 days of soil burial are shown in Fig 2. In all the films, the rapid increase in weight loss occured for the first 15 days, and gradually increased following after. Based on the graph, CS films without KCF loadings lost more weight compared to CS films with KCF. The film samples were all mainly made up of either starch or glycerol. Both CS and glycerol are hydrophilic in nature. Hence, the film samples were very susceptable to the water in the soil. According to Maran et al. [17], high concentration of starch and glycerol increases weight loss of starch films during soil degradation. They reported the starch and glycerol favors the absorption of water, increasing the water activity of the films and promotes the growth of microorganisms. Furthermore, the microorganissms which are present in the soil utilize the cassava starch as a source of nutrients which results in further degradation of the film samples. Lopez-Llorca et al. [18], reported the degradation of starch films and high microbial activity in the films which was buried in a forest soil. They stated that glycerol would eventually be assimilated into the soil or metabolized by microbes, increasing the degradation of the films. Based on the results, the incorporation of KCF had significantly reduced the weight loss of the CS films during the soil burial. In a study by Antonio et al. [19], they stated that strong interaction between fiber-matrix may resist the water from the soil from diffusing into the CS matrix. This will inhibit not only the susceptability of water towards the matrix but also mircoorganisms growth. The result showed a slight decrease in weight loss for both CS films with and without KCF after immersed in oil. This shows the oil absorbed are capable of influencing the degradation of the CS films. Oils are essentially non-polar, this makes the water in the soil which are opposite in polarity to repel each other. The repelling nature of the oils absorbed by KCF and water in soil makes it difficult for any interactions between CS and water. Thus, weight loss due to water of CS with KCF immersed in oil was less during the soil burial. However, CS films without KCF immersed in oil also experienced the same decrease in weight loss. This is probably due to some oil residue remaining on the surface of the CS films.

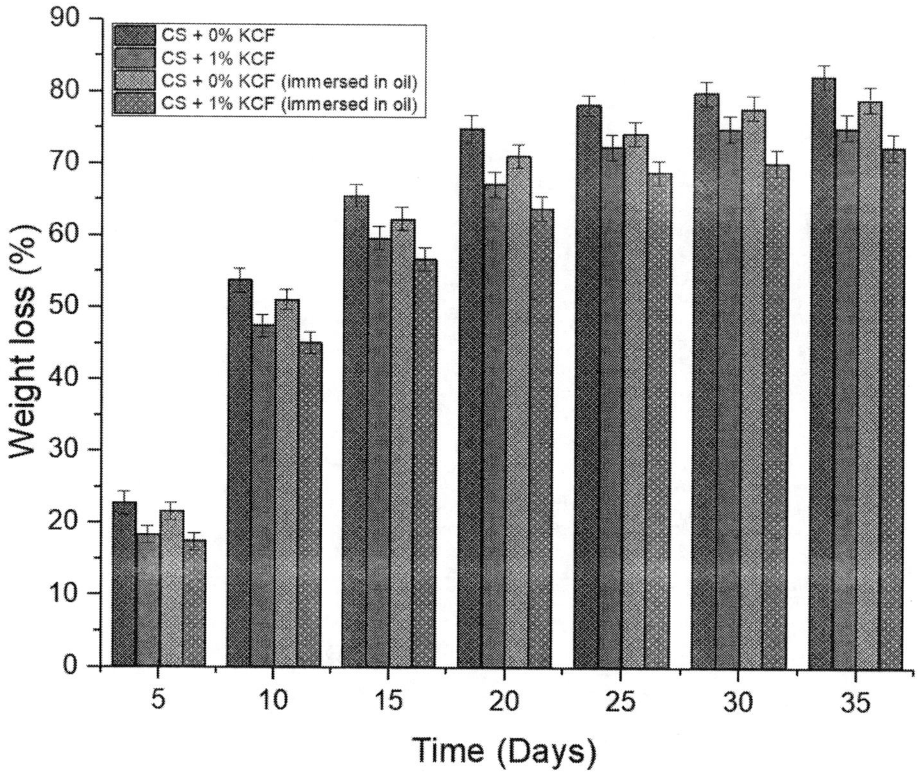

Fig 2. Weight loss (%) of CS film samples during soil burial

4. Conclusion

The effects of loading KCF on the oil absorption of cassava starch film was studied. The result of the oil absorbtion test showed an increase of oil absorption as the KCF loadings increased. As starch films are naturally resistant to oil, the incorporation of KCF are an improvement towards starch based plastic application for oil absorption purposses. However, incorporation of KCF in CS films showed a reduced in weight loss compared to CS films with no KCF loadings. Furthermore, the immersion of film samples in oil resulted in a slight decrease of weight loss in the soil burial test. Since the decrease in the weight loss of the films were only minor this does not impact the biodegradation properties in any major way.

References

[1] Deroiné M Et Al 2014 Accelerated Ageing Of Polylactide In Aqueous Environments: Comparative Study Between Distilled Water And Seawater Polym. Degrad. Stab. **108** Pp 319–329.

[2] Emilie G Melissa N Rowan T Bronwyn L And Peter H 2015 Effect Of Soil Environment On The Photo-Degradation Of Polyethylene Films J. Appl. Polym. Sci. **132** No 39

[3] Petersson L, Kvien I And Oksman K 2007 Structure And Thermal Properties Of Poly(Lactic Acid)/Cellulose Whiskers Nanocomposite Materials Compos. Sci. Technol. **67** No 11 Pp 2535–2544
[4] Yongshang L, Lihui W And Xiaodong C 2005 Biocomposites Of Plasticized Starch Reinforced With Cellulose Crystallites From Cottonseed Linter Macromol. Biosci. **5** No 11 Pp 1101–1107
[5] Chen Y, Liu C, Chang P R, Cao X And Anderson D P 2009 Bionanocomposites Based On Pea Starch And Cellulose Nanowhiskers Hydrolyzed From Pea Hull Fibre: Effect Of Hydrolysis Time Carbohydr. Polym. **76** No 4 Pp 607–615
[6] Galdeano M C, Mali S, Grossmann M V E, Yamashita F And García M A 2009 Effects Of Plasticizers On The Properties Of Oat Starch Films Mater. Sci. Eng. C **29** No 2 Pp 532–538
[7] Souza A, Ditchfield C And Tadini C Biodegradable Films Based On Biopolymers For Food Industries. 2009
[8] Souza A C, Benze R, Ferrão E S, Ditchfield C, Coelho A C V And Tadini C C 2012 Cassava Starch Biodegradable Films: Influence Of Glycerol And Clay Nanoparticles Content On Tensile And Barrier Properties And Glass Transition Temperature LWT - Food Sci. Technol. **46** No 1 Pp 110–117
[9] Mehta G, Mohanty A, Misra M And Drzal L Effect Of Novel Sizing On The Mechanical And Morphological Characteristics Of Natural Fiber Reinforced Unsaturated Polyester Resin Based Bio-Composites, **39**. 2004
[10] Jutarat P, Luxsana H, Darin P And Parichat C 2010 Property Improvement Of Thermoplastic Mung Bean Starch Using Cotton Fiber And Low-Density Polyethylene Starch - Stärke **62** No 8 Pp 435–443
[11] D. P A. T And M. K J 2004 Processing, Properties And Stability Of Biodegradable Composites Based On Mater-Bi® And Cellulose Fibres Polym. Adv. Technol. **14** No 11-12 Pp 749–756
[12] Nishino T, Hirao K, Kotera M, Nakamae K And Inagaki H 2003 Kenaf Reinforced Biodegradable Composite Compos. Sci. Technol. **63** No 9 Pp 1281–1286
[13] Ridzwan M And Ishak B 2007 A Comparison Of Mechanical Properties Between Kenaf Core Fiber And Kenaf Bast Fiber Reinforced Polyester Composites For. Res. P 27
[14] D570 1985 ASTM D 570 – 98 – Standard Test Method For Water Absorption Of Plastics ASTM Stand. **16** Pp 1–4
[15] Vieira M G A, Da Silva M A, Dos Santos L O And Beppu M M 2011 Natural-Based Plasticizers And Biopolymer Films: A Review Eur. Polym. J. **47** No 3 Pp 254–263
[16] Bazargan A, Tan J, Hui C W And Mckay G 2014 Utilization Of Rice Husks For The Production Of Oil Sorbent Materials Cellulose **21** No 3 Pp 1679–1688
[17] Maran J P, Sivakumar V, Thirugnanasambandham K And Sridhar R 2014 Degradation Behavior Of Biocomposites Based On Cassava Starch Buried Under Indoor Soil Conditions Carbohydr. Polym. **101** No 1 Pp 20–28
[18] Lopez-Llorca L V And Colom Valiente M F 1993 Study Of Biodegradation Of Starch-Plastic Films In Soil Using Scanning Electron Microscopy Micron **24** No 5 Pp 457–463
[19] Carvalho A, Curvelo A And A. M. Agnelli J Wood Pulp Reinforced Thermoplastic Starch Composites, **51**. 2002

Effect of Aluminium Filler Content on the Mechanical Properties of Polyurethane Foam

Sinar Arzuria Adnan[1], Firuz Zainuddin[1], Nur Hidayah Ahmad Zaidi[1], Yusrina Mat Daud[1], Heng Wei Soon[1], Nur Afikah Zulkeply[1] and Muhamad Fudhail Faiz Muhiyadin[1]

[1]School of Materials Engineering, Universiti Malaysia Perlis, 02600 Arau, Perlis, Malaysia.

sinar@unimap.edu.my

Abstract. Start In this work, the effects of Aluminium (Al) filler content in Polyurethane (PU) foam with different weight percentages (0.5 wt.% to 3.0 wt.%) were evaluated on morphology, density and compressive strength. The pure PU and PU/Al foam composites were produced from palm oil-based polyol (POP) and methylene diphenyl diisocyanate (MDI) with a ratio of 1:1. Scanning Electron Microscope (SEM) morphology showed that the cell structure of PU/Al foam composites is less homogeneous as compared to the pure PU. The PU/Al foam composites displayed a higher number of cells with a smaller cell size as the amount of Al content was increased. The addition of 1.5 wt.% Al on the PU foam showed the highest density and compressive strength in 0.073 g/cm^3 and 0.56 MPa, respectively. Meanwhile, the further addition of Al on PU foam has decreased the density and compressive strength value. The experimental compressive strength results have compared to the theoretical results that tested by using ANSYS MULTIPHYSICS (15.0) software.

1. Introduction

The Polyurethane (PU) foam has a special group of polymeric materials that are dissimilar from other plastic types. PU foam has a remarkably broad range of applications including insulators, automotive, refrigerators, mattress, construction, packaging, and several other industries because of its attractive characteristics in terms of lightweight, insulation materials, easy processing, and commercial potential availability [1, 2]. However, PU also presents some disadvantages, which prevents PU foam to be used in the structural industry due to its low mechanical strength [3]. To overcome this disadvantage, several types of fillers have been used to improve PU foam properties including inorganic materials such as dolomite, calcium carbonate, titanium dioxide, Aluminium (Al) silica, and talc, while some of the organic materials such as fillers are natural fibers and carbon black. Mechanical properties of PU foam also, basically, depends on their density and cell structure. PU foam with very fine cells and uniform cell distributions exhibits better mechanical properties [4].

Several work have been devoted in recent years to incorporate a metallic powders in polymer matrix with purpose to change its properties such thermal stability, conductivity or mechanical properties. Al fillers have high strength, low weight, easy machining, superior malleability, good corrosion resistance, and good thermal conductivity [5, 6]. The aim of this work is to investigate the effects of different Al contents on the morphology, density and compressive strength on the PU foam. The optimum compressive strength for PU/Al foam composites was studied by using ANSYS MULTIPHYSICS (15.0) software.

2. Experimental

2.1. Materials and Preparation of PU/Al Foam Composites

The commercial Aluminium (Al) powders were provided by the Acros Organics BVBA, Belgium. The palm oil-based polyol (POP) PolyGreen R3110 were supplied by the PolyGreen Chemicals Sdn. Bhd., Malaysia, and the isocyanate was the methylene diphenyl diisocyanate (MDI) Maskiminate 80, obtained from the Maskimi Polyol Sdn. Bhd. The PU foam was produced with a ratio of 1:1 to POP and MDI in presence of distilled water as a blowing agent. The POP was stirred at 2000 rpm until it turned milky yellow in color (for about 4 minutes). Then, the MDI was added and the mixtures were stirred for 15 minutes. Various Al filler contents were added into the mixtures as given in Table 1. The distilled water was added and continuously stirred for 4 minutes before placed in a mould. The PU/Al foam composites samples were cured at room temperature for 24 h and continued to be post cured for 2 h at 80°C.

Table 1: Formulation for pure PU and PU/Al foam composites (wt.%).

Sample	Palm oil-based polyol (POP)	Methylene diphenyl diisocyanate (MDI)	Aluminium (Al)	Distilled water
Pure PU	1	1.1	-	0.1
0.5 % PU/Al	1	1.1	0.5	0.1
1.0 % PU/ Al	1	1.1	1.0	0.1
1.5 % PU/ Al	1	1.1	1.5	0.1
2.0 % PU/ Al	1	1.1	2.0	0.1
2.5 % PU/ Al	1	1.1	2.5	0.1
3.0 % PU/ Al	1	1.1	3.0	0.1

2.2. Testing and Characterization

Scanning Electron Microscope (SEM) was used to observe the cell morphology of the PU/Al foam composites. From the SEM micrographs, the average cell size was obtained by using the Smile View software. The density was calculated from the ratio between the weight and volume of the foam samples. The compression tests were done by using the INSTRON Universal Testing Machine (UTM) according to ASTM D1621 at the crosshead speed of 10 mm/min with samples foam of 50 mm × 50 mm × 50 mm (width × length × thickness). The compression test results were compared to the theoretical results that were analyzed by using the ANSYS MULTIPHYSICS (15.0) software.

3. Results and Discussion

3.1. Morphology

The morphology of pure PU and PU/Al foam composites with different Al contents were observed using SEM, as can be seen in Figure 1(a-g). From these figures, the PU/Al foam composites morphology showed less homogeneity with the addition of Al fillers compared to the pure PU. The addition of Al fillers into the PU foam was found to have higher number of cells with smaller cell size as the amount of Al was increased. The cell size, density and compressive strength results are shown in Table 2.

Table 2: Density and compressive strength of pure PU and PU/Al foam composites.

Sample	Cell Size (μm)	Density (g/m^3)	Compressive strength (MPa)
Pure PU	1360	0.0680	0.077
0.5 % PU/Al	842	0.0690	0.085
1.0 % PU/Al	751	0.0710	0.087
1.5 % PU/Al	711	0.0730	0.094
2.0 % PU/Al	1142	0.0720	0.088
2.5 % PU/Al	1427	0.0705	0.083
3.0 % PU/Al	1681	0.0690	0.082

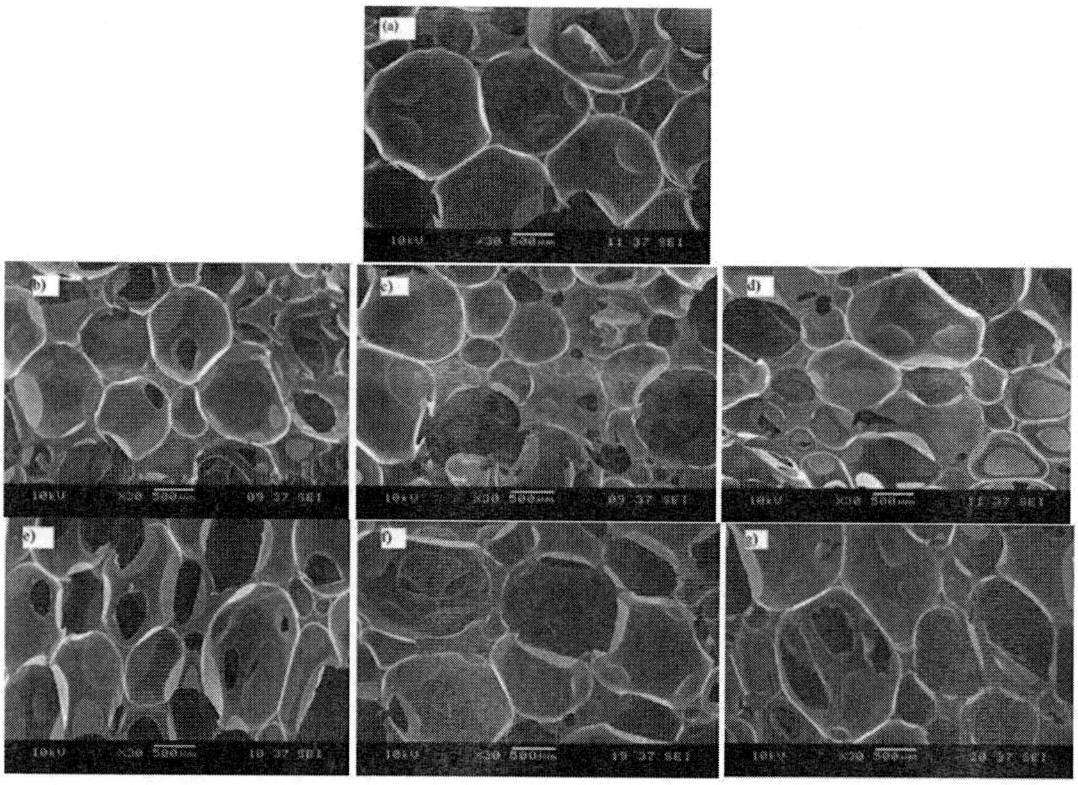

Figure 1: SEM morphology of: (a) pure PU, (b) 0.5 % PU/Al, (c) 1.0 % PU/Al, (d) 1.5 % PU/Al, (e) 2.0 % PU/Al, (f) 2.5 % PU/Al, and (g) 3.0 % PU/Al foam composites.

3.2. Density

Figure 2 displayed the effects of Al content of pure PU and PU/Al foam composites on density. The pure PU showed the lowest density as compared to the PU/Al foam composites. The addition of Al content to an optimum content of 1.5 wt.% was increased to the highest density of PU/Al foam composites with more number of cells and smaller cell size. Further addition of Al content after 1.5 wt.% showed that the decreasing density might be due to the agglomeration of Al filler. The Al cannot act as an effective nucleating agent to stabilize the foaming process; one of the consequences is that it can cause an inhomogeneous, larger cell size and alter the cell structure, which results in the decreasing PU/Al foam composites density [4].

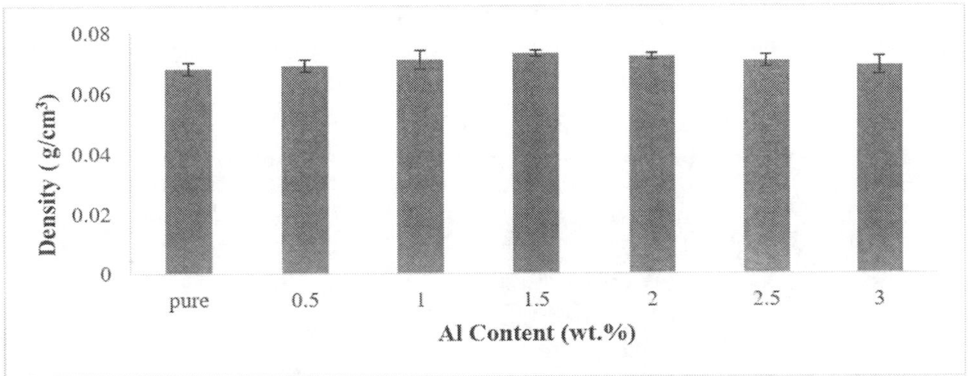

Figure 2: Density of: (a) pure PU, (b) 0.5 % PU/Al, (c) 1.0 % PU/Al, (d) 1.5 % PU/Al, (e) 2.0 % PU/Al, (f) 2.5 % PU/Al, and (g) 3.0 % PU/Al foam composites.

3.3. Compressive Strength

Figure 3 showed the changes of compressive strength from pure PU to 3 wt.% Al content on PU/Al foam composites that were divided into two stages, which takes on the trend of rising first, then, the dropping trend with increasing Al filler content. The addition of Al content until 1.5 wt.% was increased to the highest compressive strength due to the good adhesion between the Al filler and PU matrix. The stress transfers from PU matrix to the Al when the PU/Al foam composites were compressed had also contributed to the increasing compressive strength. However, the compressive strength had decreased, while the Al filler was added continuously after the optimum content of 1.5 wt.%. This decreasing trend can be explained by the agglomeration that tends to occur when the Al content is in excess and hence, cannot be dispersed uniformly [7]. The compressive strength results tallied with the density results as shown in Figure 2.

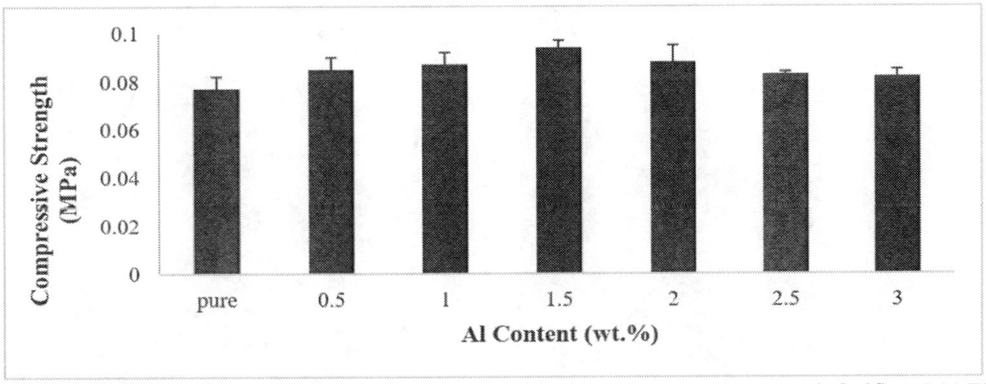

Figure 3: Compressive strength of: (a) pure PU, (b) 0.5 % PU/Al, (c) 1.0 % PU/Al, (d) 1.5 % PU/Al, (e) 2.0 % PU/Al, (f) 2.5 % PU/Al, and (g) 3.0 % PU/Al foam composites.

3.4. Simulation

The experimental results from the highest compression strength with 0.0924 MPa at 1.5 wt.% PU/Al foam composites were compared to the theoretical results that were tested by using the ANSYS MULTIPHYSICS (15.0) software. The modeling and photographs taken during the test work are shown in Figure 4(a-c). The modeling started with the blue color and the change of color indicated the different stress that passes through the 1.5 wt.% PU/Al foam composites. In the end, the red color revealed that the 1.5 wt% PU/Al foam composites have the highest stress with a theoretical result of 0.092 MPa. The correlation between the theoretical and experimental results displayed in small changes (2%) may be due to the PU foam process or temperature while conducting the compression test.

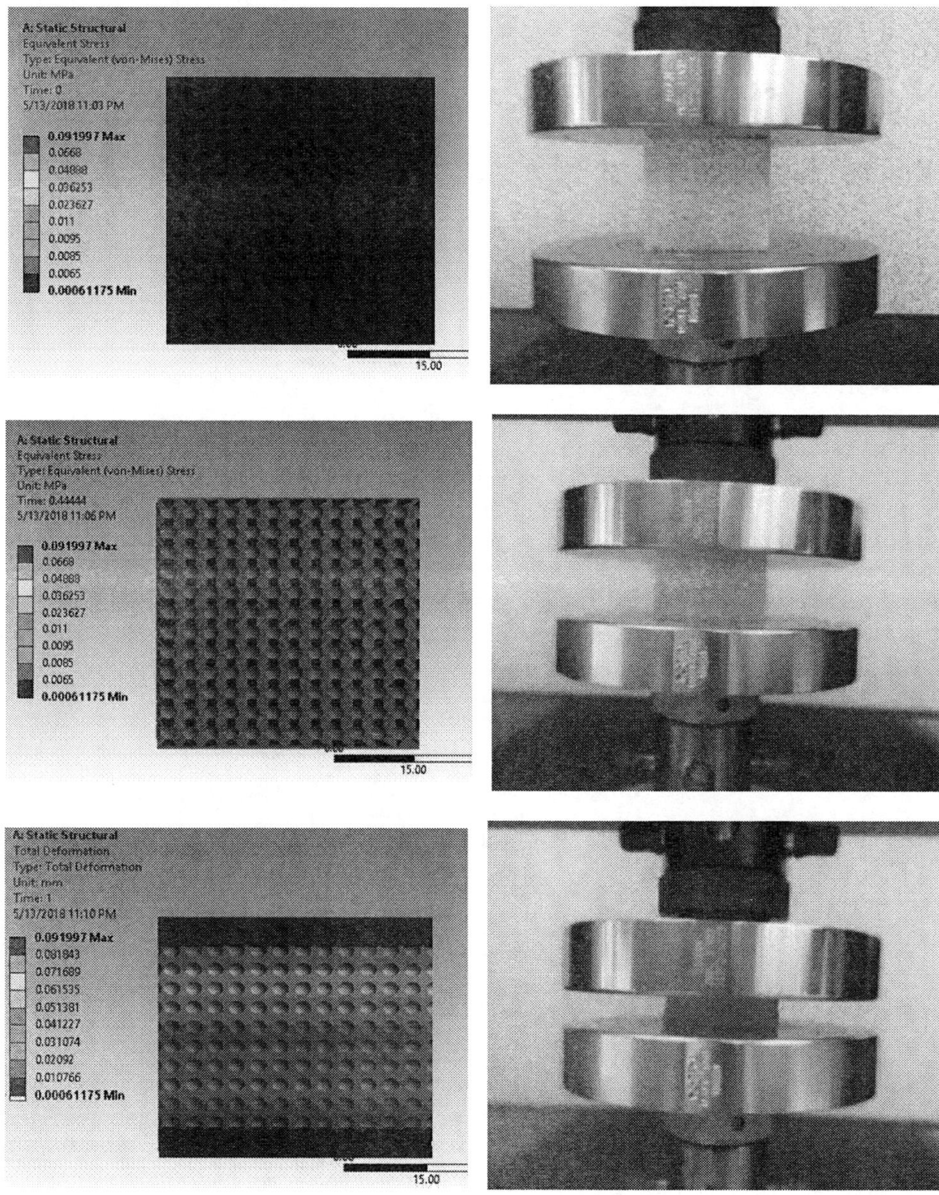

Figure 4: Modeling of compressive strength of 1.5 wt.% PU/Al foam composites: (a) before test, (b) during test, and (c) after test.

4. Results and Discussion

In this work, SEM micrographs have discovered that the addition of Al filler content in the PU foam had contributed to an inhomogeneous cell structure when compared to the pure PU. It was also found that the addition of Al content presented a higher number of cells with a smaller cell size. The incorporation of Al filler to 1.5 wt.% into PU foam had increased the density and compressive strength results. The correlation between the theoretical and experimental results displayed in small differences.

5. References

[1] A J. O. Akindoyo, M. D. H. Beg, S. Ghazali, M. R. Islam, N. Jeyaratnam and A. R. Yuvaraj 2016 *Royal Society of Chemistry Adv.* pp. 114453-114482.

[2] V. Dolomanova, J. C. M. Rauhe, L. R. Jensen, R. Pyrz and A. B. Timmons 2011 *Journal of Cellular Plastics 47(1)* pp. 81 -93.

[3] E. Ciecierska, M. J. Kowalska, P. Bazarnik, M. Kowalski, S. Krauze and M. Lewandowska 2016 *J Therm Anal Calorim., 123* pp. 283 – 291.

[4] L. Piszczyk, M. Strankowski, M. Danowska, J. T. Haponiuk and M. Gazda 2012 *European Polymer Journal 48* pp. 1726-1733.

[5] M. L. P. Luque, B. M. Simonet and M. Valcarcel 2013 *Trends in Analytical Chemistry, Vol. 47* pp. 99-110.

[6] M. Szycher 201. *Szycher's Handbook of Polyurethanes, Second Edition* 2nd ed. (CRC Press).

[7] J. M. Kim, Y. Lee, M. G. Jang, C. Han and W. N. Kim 2017 *Journal of Applied Polymer Science, 134(5)* pp. 1–9.

IOP Conf. Series: Materials Science and Engineering **429** (2018) 012045 doi:10.1088/1757-899X/429/1/012045

Dyeing Polyurethane Foam by Plant Based Pigment

F Zainuddin[1*], N S Ahmad Fauzi[1], S N F Syed Adam[1]

[1]School of Materials Engineering, Universiti Malaysia Perlis, 02600 Arau, Perlis Malaysia

*E-mail: firuz@unimap.edu.my

Abstract. Polyurethane rigid foam composites were produced by addition of plant-based filler as colour pigment. The filler used were turmeric powder, Kumkum powder, Telang's flower powder and dragon fruit peel's powder. These plant-based fillers were chosen due to their strong colour pigment that can enhanced the aesthetic value of the foam. In this study, these fillers were prepared by drying, blending and sieving, in order to obtain fine powder that can be added as particulate filler in polyurethane foams composite. The properties of the composite foam were compared with control PU foam with two variation of filler percentages (10wt% and 12wt%). The Fourier Transform Infrared (FTIR) spectra analysis show that the addition of filler in PU foam did not producing any new functional group, which proven that only physical mixing exists. On the other hand, compressive strength of 12wt% of filler is higher as compared to 10wt% and control PU foam. Morphology analysis by using Optical Microscope reveals that the addition of plant-based filler in PU foam has greatly enhanced the colour of the foam to strong yellow, red and blue respectively.

1. Introduction

Nowadays, it hard to live without coming across with some sort of polymer foams. This cellular material has been used in a wide variety of application such as disposable packaging of fast- food, cushioning of furniture and also in insulation application. Polymer foam started on 1940s when polystyrene was first invented as foam. Several years later, polyurethane (PU) foam was produced on Second World War by Dr. Otto Bayer [1]. PU foam has low thermal conductivity as compared to another polymer foamed that has been used in commercial world. Rigid PU foam were used in wide variety of application such as for sole insulation, coating, construction and many more, due to the unique and excellent combination between thermal insulation and mechanical properties.

Natural dyes have been used since ancient times for dyeing of body, food, walls of caves, textiles, leather and objects of daily use. Early of 16th century was the time when France, Holland and Germany began the cultivation of dye plants as an industry [2]. Natural dyes were used for coloring of textiles till 19th century when synthetic dyes pushed them out [3]. It is only since few decades ago that textile industries have turned to synthetic dyes, but they were so successful that natural dyes currently account only for about one percent of the total amount of dyes used worldwide. Although synthetic dyes are widely used, there are drawbacks that fears the consumers, especially on health effects to human body if exposes to high doses of dyes. Therefore, study on the use of natural dyes as an alternative for filler in polymer foam is needed.

2. Methodology

2.1. Materials
The primary materials to produce PU foam composite are isocyanate, palm oil-based polyol and natural dyes which are dragon fruit's peel, turmeric, Telang's flower and Kumkum powder. It is expected that these fillers will enhance the colour of the foam, from light yellow (control PU) to purple, bright yellow, blue and red, respectively.

2.1.1. Chemical
Methylene diphenyl diisocyanate (MDI) is a dark brown, fused solid. It is soluble in acetone, benzene, and kerosene. For this study, MDI was supplied by Maskimi Polyol Sdn. Bhd. The polyol with the series of R3110 Polygreen's bio-based was supplied by PolyGreen Chemicals. The bio content inside the polyol is 89%.

2.1.2. Plant Based Filler
The natural fillers that have been used are dragon fruit's peels, turmeric, Kumkum and Telang's flower. Generally, these fillers were purchased and taken from local sources. Turmeric and Kumkum powder were used as received, while dragon fruit's peels and Telang's flower were further processed to fine powder form.

2.2. Preparation of Natural Dyes
Natural dyes purchased were in different sized and shape. Therefore, the plant must be processed into powder foam with appropriate particle size. The plant in its original form, especially dragon fruit's peel and Telang's flower cannot be used as filler due to its size and need to be processed to required size. These plants must be cut into thin slices, dried under the sun for 24 hr and further dried in universal oven at 80°C for another 24 hr. Dried plants were then crushed by using blender, and sieved by using 63 microns siever.

2.3. Preparation of PU Foam Composite
The method for PU foam composite preparation was according to [4]. The polyol was first stirred by using overhead stirrer at 2000 rpm for 4 minutes. It must be stirred until its color changes into milky and creamy. Later, isocyanate was added, and the mixture were mixed at 2000 rpm for another 15 minutes. Then specific weight of plant-based powder was added and continues mixed at same speed for 4 minutes. Finally, water was added as blowing agent (5wt%) and continuously stirred for another 4 minutes until the mixture releasing heat that leads to initiation of chemical reaction.
The mixture was then poured into open mold with the dimension of 180 mm x 180 mm x 100 mm. The mold was placed in room temperature for 24 hours. After curing, the samples were taken out from mold for further characterizations.

2.4. Characterizations
There were several characterizations performed in order to analyze the effect of plant-based filler to the properties of the foam composites.

2.4.1. Fourier Transform Infrared Spectroscopy (FTIR)
Natural dyes were analyzed in the powdered form, with the amount of weight was approximately 10 mg. Nevertheless, PU foam composite was tested in bulk foam form, to ensure accurate data provided. The sample was scanned in the scan range of 4000 cm^{-1} to 250 cm^{-1}.

IOP Conf. Series: Materials Science and Engineering **429** (2018) 012045 doi:10.1088/1757-899X/429/1/012045

2.4.2. *Compression Test*

The compression test was conducted according to ASTM D1621-10 by using the Universal Testing Machine (UTM) Instron 5560 machine. The crosshead speed for compression test was 10 mm/min at 25°C with tolerance 3°C, and the load was applied until the foam was compressed to approximately 80% of its original thickness (height). The sample needed to be cut by using a band saw machine with dimension of 50 mm x 50 mm x 50 mm. The samples were compressed in the direction parallel to foam rise. Five sample of each composite composition were tested, and average value of compression strength were recorded.

2.4.3. *Optical Microscope Analysis*

Optical microscope analysis was conducted to observe the morphology, relative to the used of fillers in PU foams. It was also performed in order to determine the size of cells and their size distribution. The sample were cut into small size which was 10 mm x 10 mm x 10 mm.

3. Results and Discussions

The FTIR spectrum of PU foam composites shown in Figure 1 to Figure 5. The spectra exhibit the characteristic peaks of urethane bond and some functional group from dyes. As shown in Figure. 1, control PU foam exhibit the wavenumbers ranging from 2100 cm^{-1} to 2260 cm^{-1} attribute to-NCO groups (black box), 1709 cm^{-1} to 1712 cm^{-1} (-CO stretch, in red), 1517 cm^{-1} to1519 cm^{-1} (-NH bend, in green) and 1393 cm^{-1} to1395 cm^{-1} (-OCONH asymmetry stretching, in orange). All compound shows well defined hydroxyl band at range 3570 - 3200 cm^{-1}. According to Nur Azni et al, hydroxyl band is important in miscibility of filler and matrix since polyol and hydroxyl exhibit polar behavior to ease the formation.

Figure 2 shows PU foam composite with turmeric powder as filler. Turmeric is known producing curcumin radical, where the functional group is (-OCH_3) and was detected at 1050 cm^{-1} to 1150 cm^{-1}. As in Figure 3, PU/Telang's flower composite shows peak at 3010 cm^{-1} to 3100 cm^{-1} (=C-H), and for dragon fruit's peels (Figure4), the functional group is (-CO_2H), from ester 1000 cm^{-1} to 1300 cm^{-1} (-C-O). Moreover, Kumkum powder is producing crocin radical, and was detected at 2980 cm^{-1} - 3000 cm^{-1} (-CH stretch), and 1670 cm^{-1} - 1820 cm^{-1} (-C=O).

In general, there are no much differences in peaks from the results obtained due to the same aggregate of chemical and materials. It also can be concluded that the addition of plant filler did not affecting the chemical constituent of polyurethane foam.

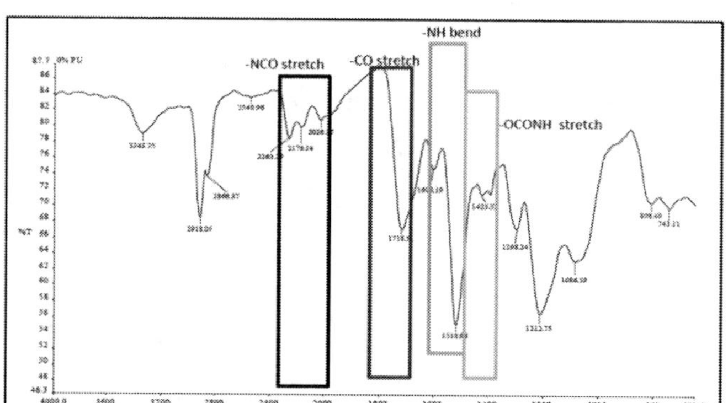

Figure 1: FTIR spectra of control PU foam

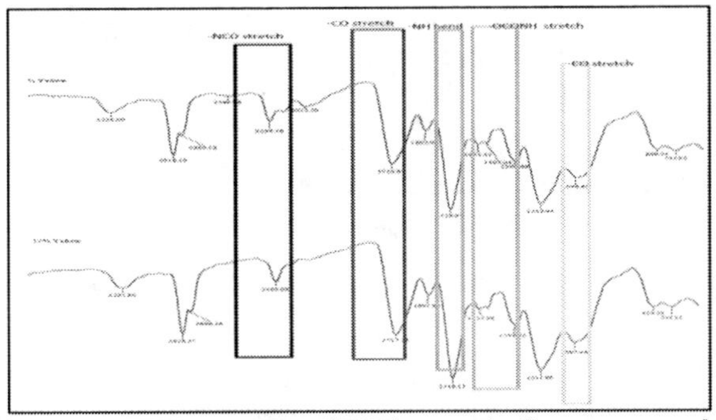

Figure 2: FTIR spectra of PU/turmeric foam composite

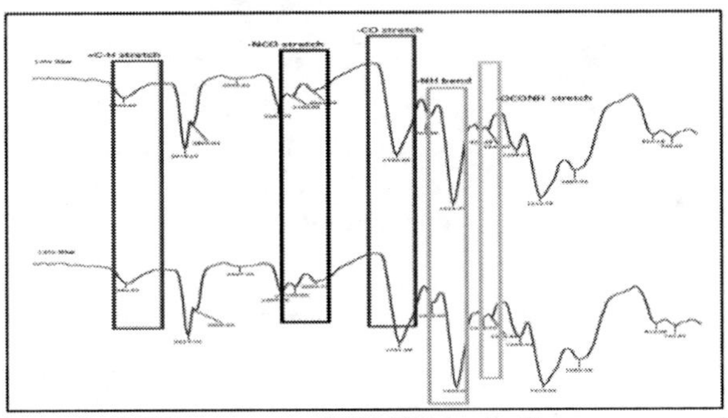

Figure 3: FTIR spectra of PU/Telang's flower composite

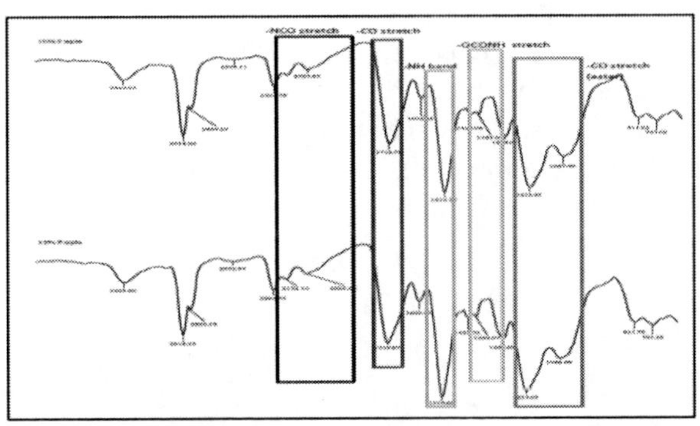

Figure 4: FTIR spectra of PU/dragon fruit's peel composite.

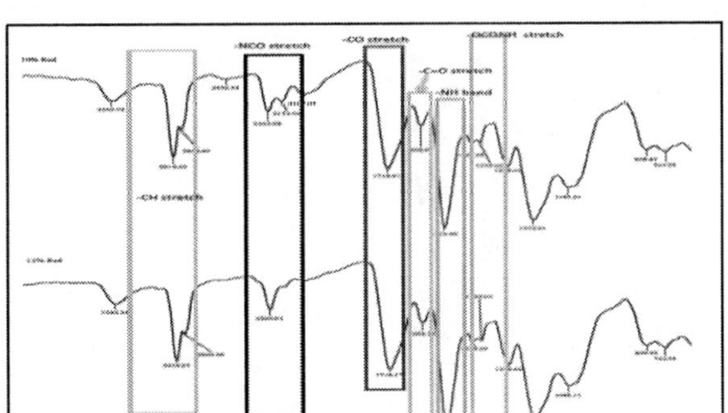

Figure 5: FTIR spectra of PU/Kumkum composite.

Figure 6 shows the compressive strength value of the composite with different percentages of natural dyes. It can be concluded that control PU has the lowest value which is 0.282MPa. As the percentages of natural dyes increases, the compressive strength was also increase. It is believed that plant-based filler was incorporated homogenously in PU foam, attributing to excellent load transfer, thus leads to good improvement of its compression strength property.

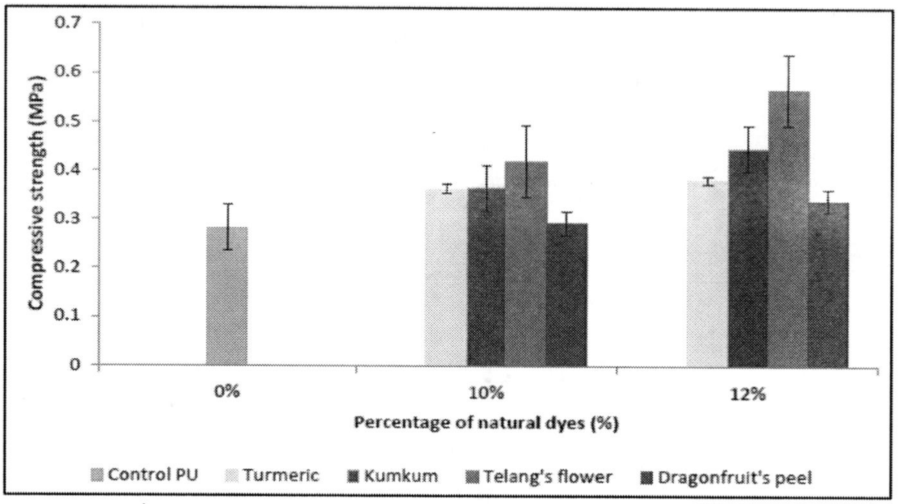

Figure 6: Compressive strength of PU foam composites.

PU foam composite with Telang's flower as filler has the highest compression strength value as compared to the other type of dyes. It is believed that the distribution of the filler inside the composite was at it best in PU/Telang's flower foam. As compared to PU/dragon fruit's peel foam, the compressive strength was the lowest. Dragon fruit's peel have a fiber like structure. Therefore, it might not be able to mixed homogenously with the PU foam.

Optical microscope (OM) result are shown in the Figure 7. It is noticed that almost all cells structure is in spherical shape with many cell wall, which proven that they are closed cells. It was found that the filler was dispersed homogenously inside the foam, with no discovery of agglomerate powder inside cell wall.

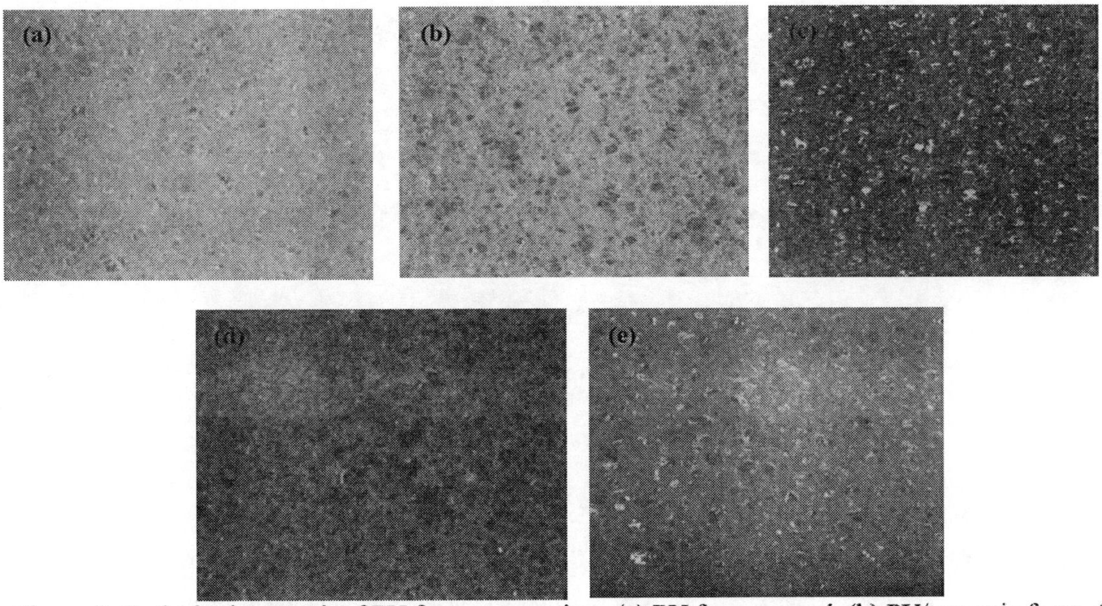

Figure 7: Optical micrograph of PU foam composites; (a) PU foam control, (b) PU/turmeric foam, (c) PU/Telang's flower foam, (d) PU/Kumkum foam, (e) PU/dragon fruit's peel foam

It was clearly shown that the addition of plant-based filler has enhancing the color of the foam. Other that dragon fruit's peel, these plant-based fillers was successfully highlighting the color of the foam, as compared to control PU foam (Figure 7(a)).

4. Conclusion

PU foam composite with variation of colour was successfully produced by incorporating natural dyes from plant-based filler. Compressive strength results show that by increasing the percentages of natural dyes, the strength of PU foam composites increases. The addition of this natural dyes was also found to enhance the colour of PU foam composite, making it suitable to be used in household and children appliances.

References

[1] D Klempner and K C Frisch 1991 *Handbook of Polymeric Foams and Foam Technology* Oxford University Press
[2] G Oertel 1993 *Polyurethane Handbook* Hanser Publisher
[3] M Desroches, M Escouvois, R Auvergne, S Caillol, B Boutevin 2012 *From Vegetable Oils to Polyurethanes: Synthetic Routes to Polyols and Main Industrial Products* Polymer Reviews **52** (1): 38–79.
[4] F Zainuddin, H M Akil, S Ahmad, R Rasid, & S N F S Adam 2014 *Drop weight impact analysis of laminated polyurethane foam/cloisite 30B clay*
[5] M A Nur Azni, A A Sinar, Z Firuz, & A G Supri 2014 *Deformation and energy absorption characteristics of biomass foam composites*

Effect of Heat Treatment on Microstructure and Mechanical Properties of AZ91 and AZ91 Reinforced Carbon Nanotube

A Ainol[1], A Z N Hidayah[1], F Najwa[1]

[1]School Of Materials Engineering, Universiti Malaysia Perlis (UniMAP), Kompleks Pusat Pengajian Jejawi 2, Taman Muhibbah, 02600 Jejawi,Arau, Perlis,Malaysia

*Email: mohdainol@gmail.com

Abstract. In present work the effect of heat treatment on microstructure and mechanical properties of AZ91 and AZ91 reinforced with carbon nanotube (CNT) were tested after undergoes Sintering,T1 at 450 °C , Solution Treatment,T4 at 415°C and Ageing Treatment,T6 at 715°C. Then, all samples will be analyze their microstructure morphology, phase composition, hardness and compression strength by using Scanning Electron Microscopy (SEM), X-ray Diffraction (XRD), Rockwell Hardness machine and Universal Tensile Machine (UTM). CNT were clustered homogeneously into the matrix of AZ91 due to successfully mechanical alloying using planetary mill for 20 hours. Precipitation of β-Phase ($Mg_{17}Al_{12}$) were reveal in XRD pattern usually highest intensity at 2θ=30°until 40°.Meanwhile, the ageing treatment,T6 nanocomposites found increase in hardness, compare to monolithic has lower value started form 4 hours until 8 hours of ageing.

1. Introduction

Powder metallurgy metal matrix composites based on AZ91 alloy matrix reinforced with 0.3 weight percent of multiwall carbon nanotube (MWCNT) were investigated from the point of view of their response to artificial aging as compared to the unreinforced AZ91 matrix alloy [1]. Magnesium alloys are utilized in engineering application for lightweight structural and functional parts in aircrafts, automotive applications, electronics, and other industrial fields, cause of their high specific strength, light automotive applications, electronics, and other industrial fields, due to the corresponding high specific strength, light specific weight, high damping capacity, recyclability, and electromagnetic shocks shielding ability [2]. In contrast, magnesium and its alloy were essentially deforming via its basal slip and twinning at room temperature, which limits their formability. Consequently, thermomechanical frequently carried out at high temperature [3]. It is generally considered that additional slip systems (prismatic and pyramidal) contribute significantly to deformation when the temperature is higher than 300°C, namely, when the associated critical resolved shear stresses are comparable [3].

CNT or multi-walled carbon nanotubes (MWCNT) have become popular for reinforcing metals and even polymers. These nanotubes show extremely high mechanical properties and flexibility as well as a high Young's modulus. Their small size reduces the possibility of thermal mismatch-induced dislocation generation at the matrix/tube interface, which is seen as a further advantage [4]. CNT was to be chosen to enhance magnesium composite properties. Recent researches producing Mg matrix composites reinforced with CNTs has been limited, looking to the problem of agglomeration of CNTs due to Wander wall forces but were largely focused on polymer matrix composites [5]. Therefore, mechanically alloying is one of green technique, where it's easier and cheapest way to overcome this problem. This technique was also approve as solution by (Q. Li et al., 2009) through Esawi et al. that tried to disperse carbon nanotubes in aluminium powder by mechanical alloying [6]. Jayakumar et al., (2013) reported that CNT in magnesium improved ductility [5] while Dieringa et.,(2011) addition of CNT increase the hardness and strength of AZ31 [4].

The aim of this research to to synthesis monolithic AZ91 and AZ91 reinforced with 0.3 weight fraction of CNT is to investigate the effect of physical properties of AZ91 with heat treatment is applied.

2. Experimental Procedure

Magnesium powder (Mg) supply by Strem Chemical, 99.8% purity with size of 75-100 μm, Aluminium powder (Al),95 % purity with size in range 50 nm, supply by Bendosen. Zinc powder (Zn) supply by Ensure, 95% purity with size < 45 μm. All the powder were used as matrix in composite. The multi-wall carbon nanotube (MWCNT) was supplied by ADV system with specification > 95% carbon purity was used as reinforcement materials. The specification of MWCNT is given in Table 1.

Table 1: Specification of multi-wall carbon nanotube

Materials	MWCNT
Diameter, nm	15-25
Length, μm	5-15
Purity	> 95%
Surface Area, m^2/g	150-210

The nominal composition (wt%) of AZ91 is 9% Al, 1% is Zn and others 90% is Mg. The 0.3% of CNT was added to monolithic AZ91 to investigate the influence of CNT toward composite. AZ91 and AZ91 + 0.3 wt% CNT was prepared by milling in planetary mill for 20 hours to ensure that all powder are fully homogeneous without fine the size of all powder.

The homogeneous blended powders monolithic and reinforced with CNT were compacted into cylinder palette with dimensions of samples is 12 mm D, 5.5 mm thickness. The pressure was used is 400 MPa. All samples pass through sintering process at temperature 450°C for 2 hours. After that, the composite undergo solution treatment,T4 at 415°C for a couple hours and lastly will continue with ageing treatment,T6 at 175°C for 2, 4 , 6 and 8 hours.

Specimen for microstructure was examined under SEM Hitachi TM 3000. The changes of phases was studied by using X-ray Diffraction (Bruker, D2 Phaser), using cuKα radiation (λ=1.5406). The XRD scans were acquired from 20° to 80°. Rockwell hardness testing was conducted on samples that were done undergoes T4 and T6 treatment. Force used is 147 N, 1/16 indentor with HrF mode. At least seven test was carry out in every sample and the mean value was reported. The specimen was polished using 60 ml Ehylene Glycol, 20 ml Acetic Acid, 1 ml Nitric Acid 19 ml distilled water.

3. Results and Discussion

3.1. Phase analysis Magnesium Composite

Figure 1 illustrate diffraction pattern for solution treatment while figure 2 until 4 illustrate diffraction pattern from the all sample after pass through ageing treatment, T6 at 175°C for 2, 4, 6 and 8 hours of ageing time. The X-ray diffraction is most powerful way to highlight the formation of intermetallic phase or $Mg_{17}Al_{12}$ in magnesium based alloy because it is very sensitive to any disturbance of periodicity of crystal lattice (Nur Hidayah et al., 2016). XRD pattern will show the intensity of α-Mg and β-$Mg_{17}Al_{12}$. Present of β-phase is important as hardening effect toward AZ91.

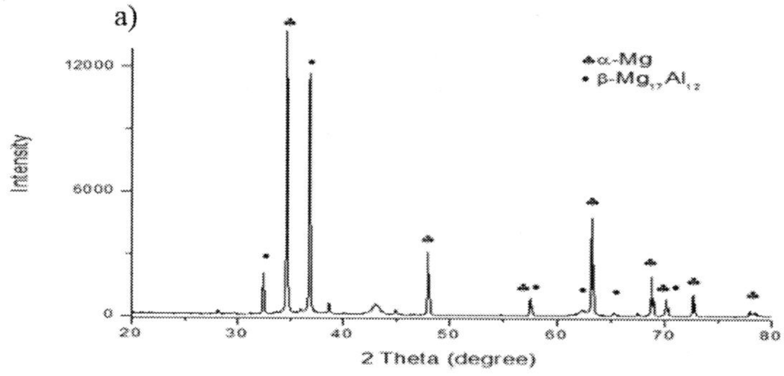

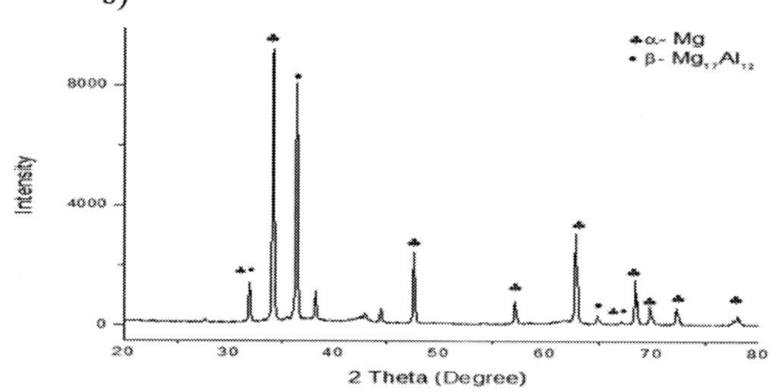

Figure 1: a) XRD pattern of solution treatment of AZ91 at 415°C for 2 hours b) XRD pattern of solution treatment of AZ91+0.3% CNT at 415°C for 2 hours.

On figure 1 a and b show the x-ray diffraction pattern for AZ91 and AZ91 + 0.3% CNT for solution treatment process at 415 °C temperature. AZ91 shows the decreasing number of α-phase compared to AZ91+0.3% CNT. At this process for AZ91, its shown six peak formation of β- phase where's 2 peaks sharing high intensity α and β phase on peak 57° and 70°. β-phase were form at 33°, 37°, 63°, and 65°.While the formation of β-phase on AZ91 + 0.3% CNT decrease in this process where it is formed at 37° and 65°. Two peak was shown formation of α and β-phase on 33° and 67°. This analysis can be conclude that solution treatment give positive effect on hardening of AZ91 not to AZ91+0.3% CNT.

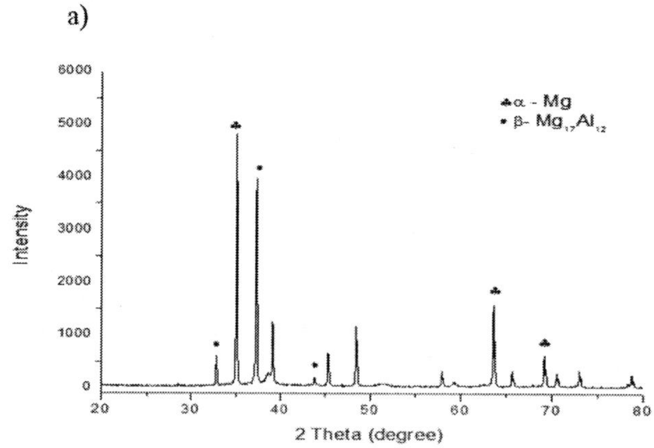

b)

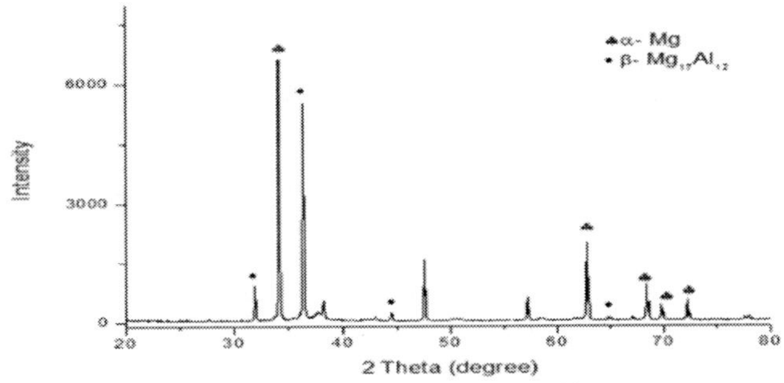

Figure 2: a) XRD of AZ91 after 2 hours ageing at 175°C b) XRD of AZ91+0.3% CNT after 2 hours ageing at 175°C.

c)

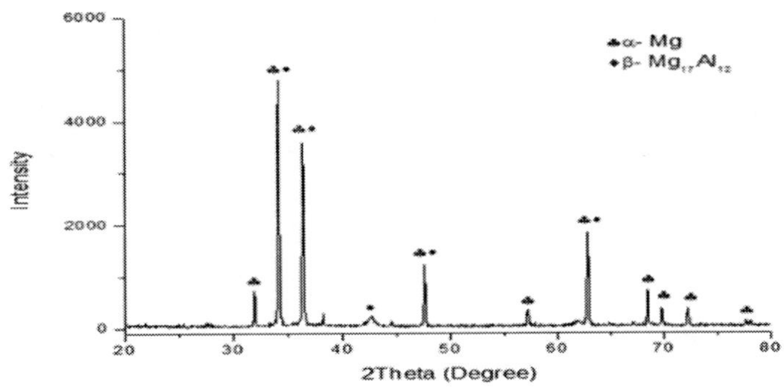

d)

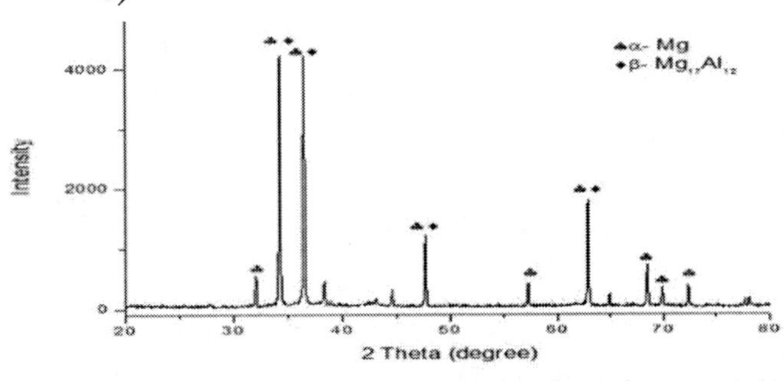

Figure 3: c) XRD of AZ91 after 4 hours ageing at 175°C d) XRD of AZ91+0.3% CNT after 4 hours ageing at 175°C.

e)

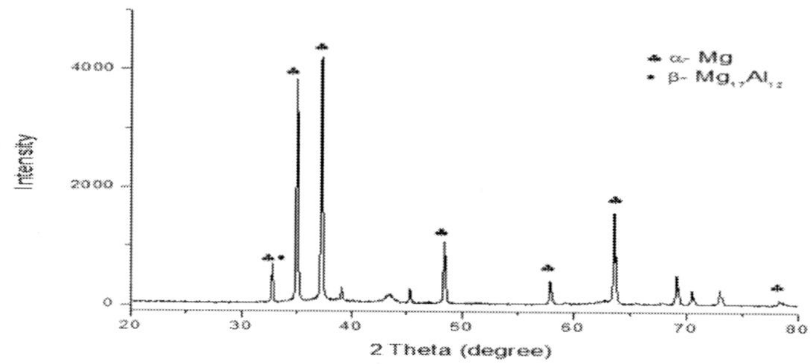

f)

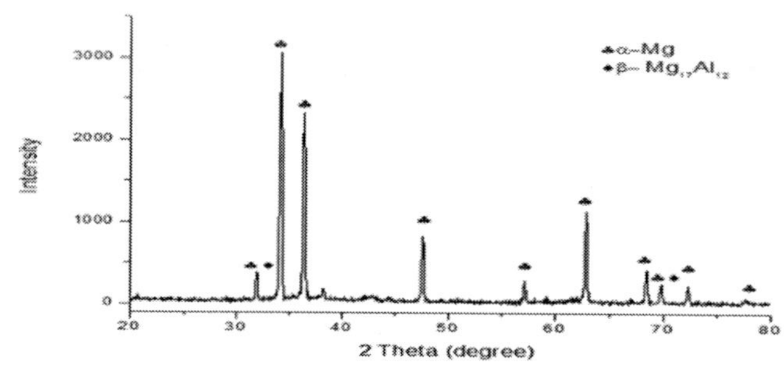

Figure 4: e) XRD of AZ91 after 6 hours ageing at 175°C f) XRD of AZ91+0.3% CNT after 6 hours ageing at 175°C.

g)

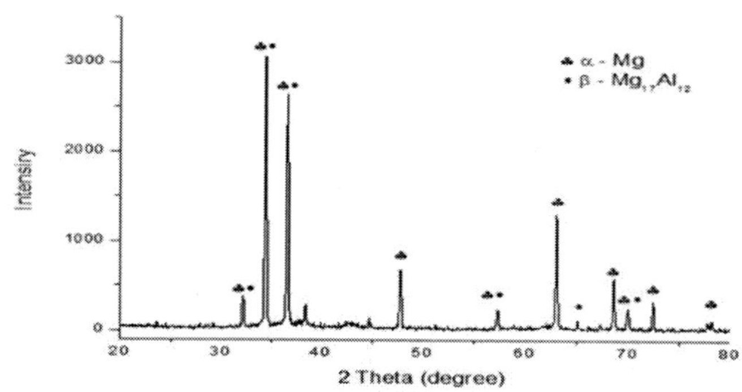

h)

h)

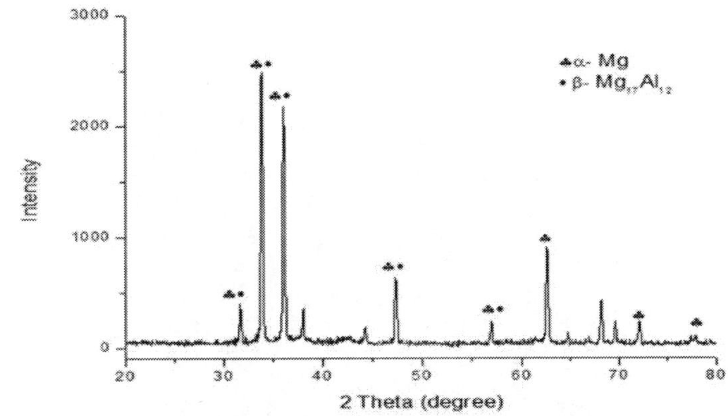

Figure 5: g) XRD of AZ91 after 8 hours ageing at 175°C h) XRD of AZ91+0.3% CNT after 8 hours ageing at 175°C.

Figure 2 a) shows X-ray diffraction for AZ91 while b) X-ray diffraction for AZ91+0.3% CNT for 2 hours of Ageing, T6 treatment. Figure shown that's, there are more β-phase of $Mg_{17}Al_{12}$ were form at peak 33°,37°, and 44°while α-phase of Mg on peak 35° and 64°. Although there are many of β-phase formation for AZ91 but it cannot compete with formation of β-phase of AZ91+0.3% CNT where's this phase have four peak on this samples at 32°, 36°, 44° and 65°. Others is α-phase of Mg. This result shown that 2 hours of Ageing give positive hardening effect toward AZ91+0.3% CNT compared to AZ91.

Figure 3 c) shows X-ray diffraction for AZ91 while d) X-ray diffraction for AZ91+0.3% CNT for 4 hours of Ageing, T6 treatment. As shown in figure, it shows there are many sharing peak between α and β-phase for both samples. For AZ91 formation of both sample at peak 34°,36°,47°and 64° while on AZ91+0.3% CNT at peak 34°,36°,47° and 63°. Only on peak 43° stand formation of β-phase, $Mg_{17}Al_{12}$ alone. This result shown positive hardening effect toward AZ91+0.3% CNT with have four peak of β-phase compared to AZ91 with have only three β-phase peaks.

Figure 4 e) shows X-ray diffraction for AZ91 while f) X-ray diffraction for AZ91+0.3% CNT for 6 hours of Ageing, T6 treatment. On this hours of ageing shown decreasing number formation of β-phase of Mg17Al12. AZ91 as shown in figure 4e) β-phase form on peak 33° only while on AZ91+0.3% CNT this phase form at 33° and 70° as shown in figure 4f). Others peak have full with formation of α-phase. This result shown that's, AZ91+0.3% CNT have good hardening effect compared with AZ91.

Figure 5 g) shows X-ray diffraction for AZ91 while h) X-ray diffraction for AZ91+0.3% CNT for 8 hours of Ageing,T6 treatment., there are more α-phase of Mg and β-phase of $Mg_{17}Al_{12}$ sharing their peak on 33°,34°,37°,57°and 70° while only single β-phase on peak 65°.But, for AZ91+0.3% CNT as shown in figure# h) shown that's α-phase and β-phase sharing their peak together at 33°,34°,36°,47° and 63°. There are no single β-phase stand alone. This result shown that 8 hours of Ageing give positive hardening effect toward AZ91 compared to AZ91+0.3% CNT.

3.2. Microstructure analysis

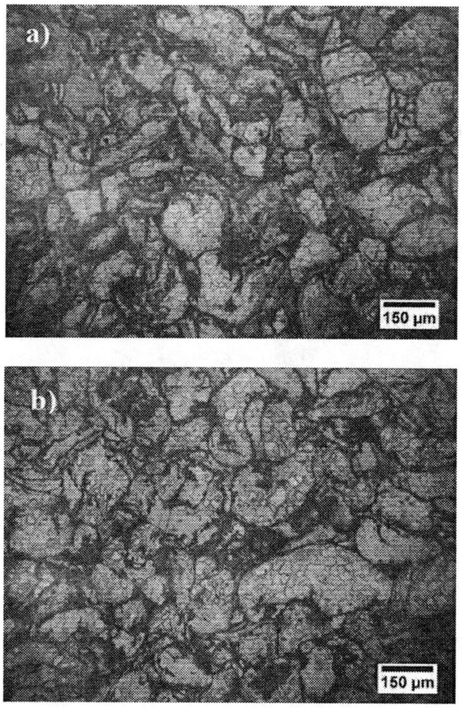

Figure 6: Optical microscope images solution treat sample at 120 minutes a) AZ91 b) AZ91+0.3% CNT.

Solution treatment of composite samples at 415 ◦C for 2 hours ensures that all the $Mg_{17}Al_{12}$ can be redissolved in to matrix. No massive $Mg_{17}Al_{12}$ has been seen with the solution treated samples. Figure 4 shows the microstructure of AZ91 and CNT added composite samples are solution treated for 120 min. The significant microstructural difference between these samples is the amount of discontinuous precipitates at the grain boundary. The black appearance at the grain boundary indicates the initiation of discontinuous precipitates. CNT added structure shows fewer amounts of discontinuous precipitates compared to the AZ91 composite [7].

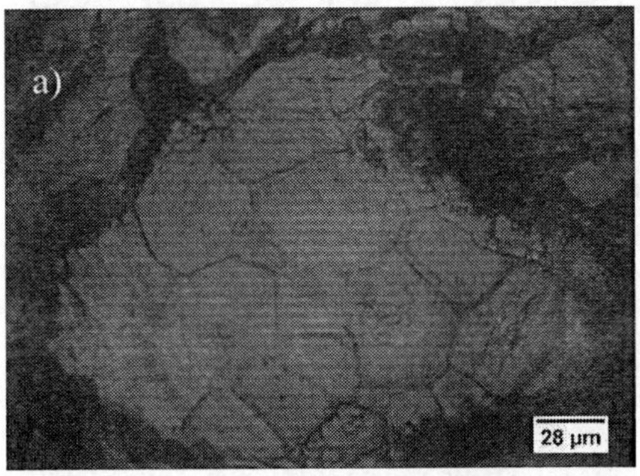

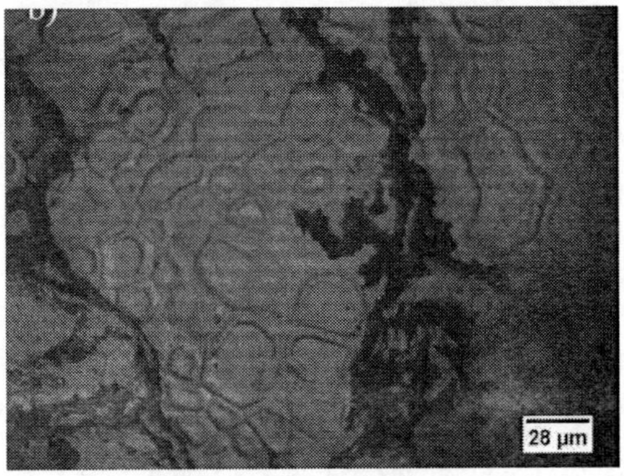

Figure 7: Optical microscope images ageing sample at 4 hours a) AZ91 b) AZ91+0.3% CNT.

The microstructural changes during the ageing treatment at 175 ◦C have been observed in all composite samples using optical microscopy. Figure 7 shows the microstructure of AZ91 and CNT added alloy samples aged for 4 hours. The black appearance at the grain boundary indicates the initiation of discontinuous precipitates. CNT added structure shows fewer amounts of discontinuous precipitates compared to the AZ91 composite. In fact, discontinuous precipitates are rarely seen with CNT added composite, which indicates that CNT effectively suppresses the nucleation of discontinuous precipitation. As the ageing time increased, the number of nucleation sites for discontinuous phase increases and the already formed discontinuous phases grow into the adjusted grains as evident from the microstructures of 6 hours aged samples (see Figure 8).

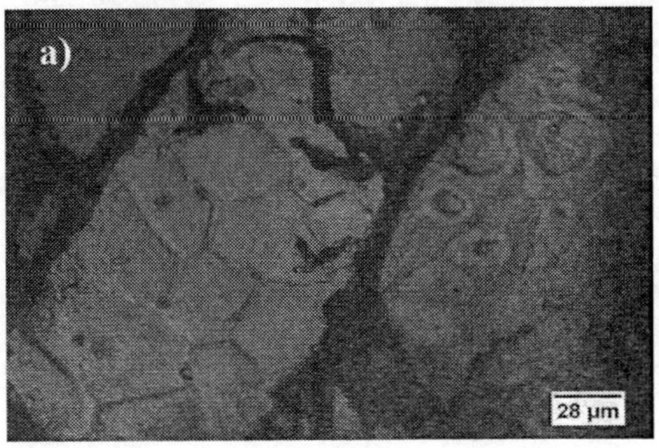

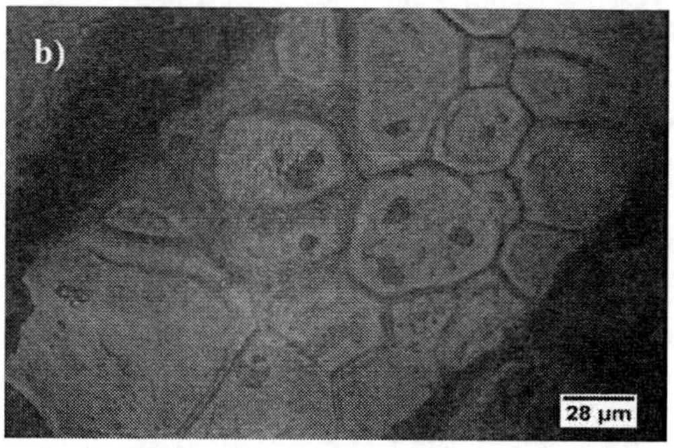

Figure 8: Optical microscope images ageing sample at 6 hours a) AZ91 b) AZ91+0.3% CNT.

3.3. Heat Treatment Behavior

The heat treatment behavior of hardness for AZ91 with and without CNT addition is shown in Figure 9.

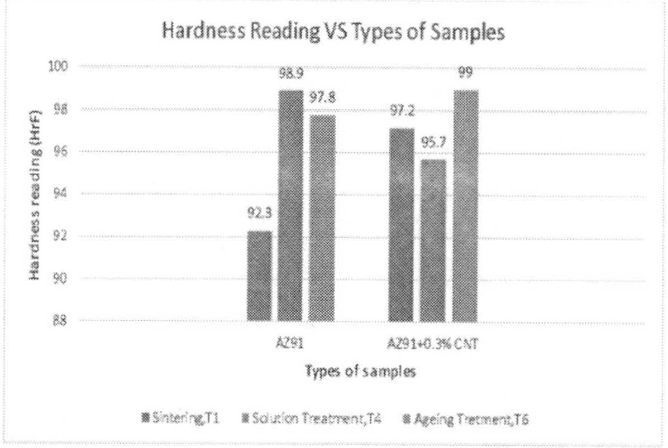

Figure 9: Rockwell Hardness (HrF) versus Type of Samples after pass through various type of heat treatment.

Rockwell hardness versus types of heat treatment was done by AZ91 composite is shown in figure 9. According to figure 9 the AZ91 reinforced CNT is highest hardness in sintering process compare to monolithic composite. But, after past through solution treatment, T4 it's shown that AZ91 increase drastically in hardness reading compared to AZ91 reinforced CNT drop into 95.7 N. Finally, AZ91 reinforced CNT slightly increase in hardness reading, 99.0 N after treated in T6 for 4 hours compared to AZ91 shown drop in hardness, 97.8 N after T4. This is contrast with result that be reported by Nur Hidayah et al.,(2015) the hardness for AZ91 + 0.3% CNT is lower than AZ91 [1]. In this research find that, 4 hours of ageing of AZ91 reinforced CNT give accelerate in hardness compared without CNT AZ91. But, same result was report by Nur Hidayah et al., state that hardness drop constantly until at 24 hours ageing. In this research, hardness reading of AZ91 reinforced CNT was drop when hours of ageing increase to 6 hours and 8 hours.

3.4. Compression Test Magnesium Composite

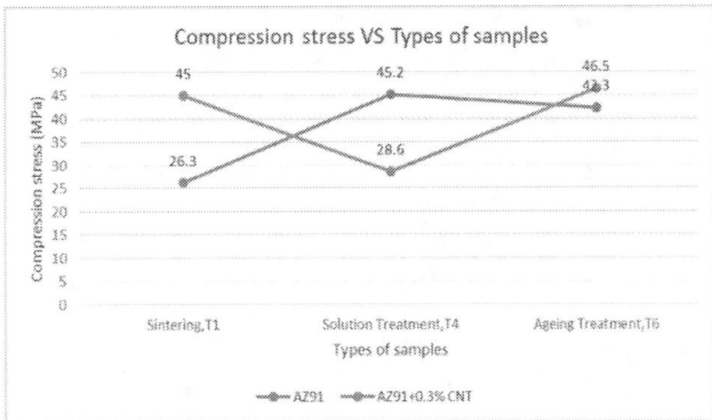

Figure 10: Compression test (MPa) versus Type of Samples after pass through various type of heat treatment.

Compression test was conducted for investigation of crack behavior in samples according to standard ASTM E9 at room temperature with L/D: 1.5 and strain rate of 1x10-4 s-1. The tests were continued till two points: 1- ultimate reduction of stress which was assumed as fracture point and 2- the point in which 8% strain was occurred. Figure 10 shown that sintering of AZ91 give low hardening. But, it was obviously slightly increased from T1 to T4 go through 45.2 MPa. This result shown that AZ91 give good hardening at Solution Treatment process compared to AZ91+0.3% CNT that show drastic slope of lowering in reading off compression stress on T4, 28.6 MPa. But, after undergoes T4, AZ91 are little drop in reading, 42.3 MPa compared to AZ91+0.3% CNT it was shown drastic positive slope of increasing in strength at Ageing treatment, T6. The reading is 46.5 MPa increase from 28.6 MPa on T4.

3.5. *Fracture Analysis under SEM*

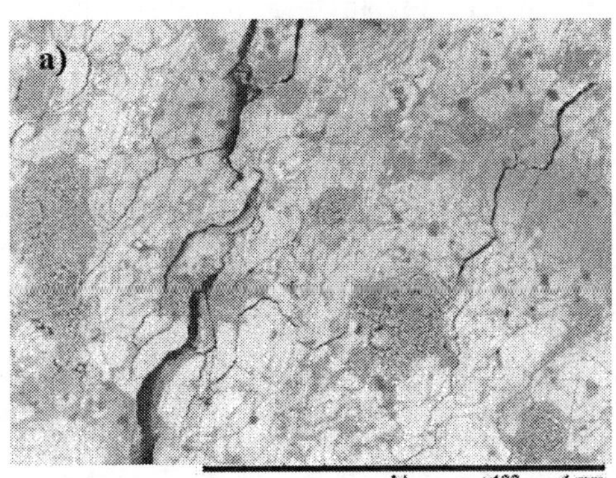

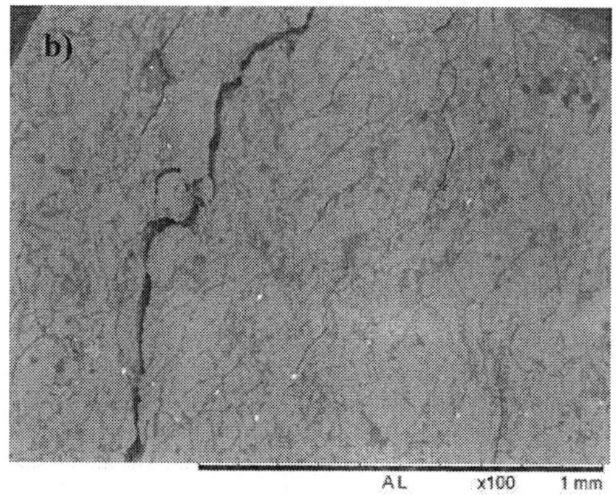

Figure 11: SEM image of crack propagation for AZ91+0.3% CNT a) Ageing for 2 hours b) Ageing for 8 hours.

The fracture surface shows on figure 11 elongated dimples indicating that significant plastic deformation takes place during compressive deformation [8]. Figure 11a) crack propagation aged sample of AZ91 + 0.3% CNT for 2 hours. It's shows that transgranular fracture. The fracture follows the edges of lattices in a granular material, ignoring the grains in the individual lattices. This results in a fairly smooth looking fracture with less sharp edges than one that follows the changing grains while in figure 11b) shown that intergranular fracture. The fracture that grows along the grain boundaries of the material.

4. Conclusion

AZ91 and AZ91 reinforced carbon nanotube were successfully fabricated using powder metallurgy method. The way to avoid clustering of CNT is by using mechanically alloying because its can dispersed CNT homogeneously into the matrix of AZ91. The current investigates effect of solution treatment, T4 and ageing treatment, T6 toward magnesium composite. T4 give positive hardening effect to AZ91 while T6 give positive hardening effect to AZ91 with addition of CNT. The CNT reinforced react as softening mechanism during initial ageing which was attributed to dislocation recovery before the formation of coherent precipitates which are responsible for strengthening the material in the early of ageing. Composite AZ91 + 0.3% CNT shows accelerated aging and achieve peak aged hardness at 4 hours of ageing time. The positive effect of hardening is expected due to the precipitation of $Mg_{17}Al_{12}$.

References

[1] Nur Hidayah, A. Z., Shamsul, J. B., Ahmad, K. R., Sinar, A. A., & Zaidi, A. M. A. (2016). Ageing Response of Powder Metallurgy AZ91 and AZ91 Reinforced with Multiwall Carbon Nanotube. *Materials Science Forum*, **857**, 261–265.

[2] Chen, Q., Li, K., Liu, Y., Zhao, Z., Tao, K., & Zhu, Q. (2017). Effects of heat treatment on the wear behavior of surfacing AZ91 magnesium alloy. *Journal of Materials Research*, **32**(11), 2161–2168.

[3] Li, J. Y., XIE, J. X., Jin, J. B., & Wang, Z. X. (2012). Microstructural evolution of AZ91 magnesium alloy during extrusion and heat treatment. *Transactions of Nonferrous Metals Society of China*, **22**(5), 1028–1034.

[4] Li, Q., Viereckl, A., Rottmair, C. A., & Singer, R. F. (2009). Improved processing of carbon nanotube/magnesium alloy composites. *Composites Science and Technology*, **69**(7–8), 1193–1199.

[5] Jayakumar, J., Raghunath, B. K., & Rao, T. H. (2013). Enhancing Microstructure and Mechanical Properties of AZ31-MWCNT Nanocomposites through Mechanical Alloying. *Advances in Materials Science and Engineering*, *2013*, 1–6.

[6] Dieringa, H. (2011). Properties of magnesium alloys reinforced with nanoparticles and carbon nanotubes: a review. *Journal of Materials Science*, **46**(2), 289–306.

[7] Srinivasan, A., Pillai, U. T. S., & Pai, B. C. (2007). Effect of Pb addition on ageing behavior of AZ91 magnesium alloy. *Materials Science and Engineering A*, **452–453**, 87–92.

[8] Chaubey, A. K., Scudino, S., Samadi Khoshkhoo, M., Prashanth, K. G., Mukhopadhyay, N. K., Mishra, B. K., & Eckert, J. (2014). High-strength ultrafine grain Mg-7.4%Al alloy synthesized by consolidation of mechanically.

IOP Conf. Series: Materials Science and Engineering **429** (2018) 012047 doi:10.1088/1757-899X/429/1/012047

Development of Driving Cycle for Passenger Car under Real World Driving Conditions in Kuala Lumpur, Malaysia

R Tharvin[1], N S Kamarrudin[1,2], Shahriman A B[1,2], I Zunaidi[3], Z M Razlan[1,2], W K Wan[1], A Harun[2], M S M Hashim[1,2], I Ibrahim[1,2], M K Faizi[1,2], M A M Saad[1,2], A R Mahayadin[2] and M F H Rani[2]

[1] School of Mechatronic Engineering, Universiti Malaysia Perlis, Kampus Tetap Pauh Putra, 02600 Arau, Perlis, Malaysia.

[2] Motorsport Technology Research Unit (MoTECH), Universiti Malaysia Perlis, Pauh Putra Campus, 02600 Arau, Perlis, Malaysia.

[3] Technopreneur at UniMAP Sdn Bhd, No 26, Block D, Bangunan PPPIPT, Pengkalan Jaya, Jalan Kangar-Alor Setar, 01000 Kangar, Perlis, Malaysia.

saifullah@unimap.edu.my

Abstract. The traffic condition becomes the main reason of pollutants in the developing countries. The exhaust emission and fuel consumption can be determined by constructing a driving cycle with its analyzed target parameters. The developed driving cycles are not same as other countries or cities due to its different traffic conditions. The methodology includes, route selection, data collection and cycle construction. The driving cycle was constructed from data collected from different routes of Kuala Lumpur during peak and off-peak hours. Micro-trip method was used to develop a complete KL cycle. A variety of parameters were studied and analyzed to determine and compare its driving pattern with other candidate cycles. The lowest percentage of overall error of candidate cycles was selected as KL cycle. The KL cycle was developed successfully with 17 numbers of micro-trips and a total time of 1219s. Then, the proposed driving cycle was compared with other existing cycles. From the comparison, it shows that, the KL driving cycle is similar to Beijing cycle with its driving parameters, except for the percentage of time taken in cruise mode.

1. Introduction

Driving cycle is a collection of data points that generate speed versus time graph which obtained from the real-world driving situation. It is usually developed from the data collected in a specific area or city, certain roads and routes [1]. The data collected mostly will represent the congested area to the free flow area during the peak hours, so that a typical driving pattern of the population of a place or city can be easily obtained. The definition of the driving cycle also can be referred to the operating conditions such as the idle state, acceleration, cruising, deceleration and steady state to show the type of pattern in an area of the city.

Subsequently, the operating conditions of a vehicle traveling in traffic are rarely available, its emissions and fuel consumption are often estimated from the design parameters, presumably operating under some selected average conditions [2]. This emission model is often represented by emission factors and are normally expressed in terms of grams of pollutants per unit of distance traveled. With the aid of driving cycle, we are able to estimate fuel consumption and emissions of vehicles within respective urban areas.

The driving cycle was categorized into legislative and non-legislative cycle. The higher authorities of road traffics will control the emissions from motor vehicles for legislative cycle so that it not exceed the statutory emission standard [3]. US 75 cycle, ECE cycle and Japan 10-15 mode cycle is the cycles that are implementing this legislative cycle to control the vehicle emissions. Non-legislative cycles are the cycle used to estimate the carbon emission and fuel consumption. The Melbourne peak cycle and Perth cycle, are some of the examples. Therefore, driving cycles has been developed for different cities to determine their driving characteristics as shown in Table 1.

The development of driving cycle includes route selection, data collection, and cycle construction. Route selection is usually done before data collection to ensure the targeted route are suitable to collect data according to the traffic conditions during peak hours. Data can be collected by using a chase-car method or on-board measurement to capture each second of travelling speed of a vehicle driven on selected routes [4]. The cycle can be constructed by extracting the collected speed data and separating it into micro-trips, and developing a driving cycle by generating a graph of speed versus time with the combined micro-trips and idle duration according to the assessment criteria. In this study, the objective is to develop a more sensible driving cycle to determine fuel consumption and exhaust emission for the Kuala Lumpur city. The use of real time data collection done to ensure the preciseness of developed driving cycle. Besides, the proposed methods can be a guideline for others to develop the driving cycle of other cities with other types of vehicle.

Table 1. Summary of real world driving cycles in the literature, including route selection, data collection and construction method to develop a drive cycle.

Driving cycle	Route selection	Data collection	Assessment criteria	Construction method
FTP72 FTP75	-Home-to-work trips	On-board measurement	Stop per distance Average speed Maximum speed Number of stops SAPD	Select the whole trip best fit the overall survey data
Sydney cycle	-Road classification -High emission -Traffic density	Chase car method	Speed / acceleration matrix Average speed Percentage of idle	Random selection of 2-min driving segment
Melbourne peak cycle	-Central business area -Arterials -Highways	Chase car method	Average speed Root mean square speed Root mean square acceleration Positive kinetic energy Percentage idle Speed / acceleration probability density function	Random selection of 100m driving segment
Perth cycle	-Geographic area	Chase car method	Average speed Stop per distance Positive kinetic energy Percentage of idle Speed distribution	Knight tour algorithm

2. Methodology

2.1. Route selection

In constructing of local driving cycle, the road route selection is crucial to identify and categorize the level of traffic flow condition from a congested zone to a free flow zone. The information can be

IOP Conf. Series: Materials Science and Engineering **429** (2018) 012047 doi:10.1088/1757-899X/429/1/012047

extracted from Road Traffic Volume Malaysia (RTVM) and referred as Level of Service (LOS). The traffic volume survey was done to determine the number of vehicles that passes through the area of interest in the region of Kuala Lumpur. Therefore, Road Traffic Volume Malaysia (RTVM) which is yearly statistic of transport report was referred to get the traffic volume of each route according to the percentage of the vehicles passed through the traffic census station. The traffic volume data were taken from 5 census station along the Kuala Lumpur region [5].

Since the number of census point in RTVM for KL was not adequate, pre-route selection was done to determine the congested area in the region of Kuala Lumpur. Pre-route selection was carried out by referring to the Goggle Map's live traffic where to determine the congested route and its peak hour helps in comparing the condition of the routes from time to time. The factors that need to be considered and influenced in route selection and data collection are construction activities that take place along the roadside, heavy rain and other factors such public holidays causes data variabilities when compared to the data collected on normal days.

2.2. On-road driving data collection

The speed-time data was collected using a chase-car technique. Chase-car method is a method where the drivers will randomly select a car and chase it along their route to collect the data. This technique should be repeated to obtain a huge number of data to construct a precise driving cycle [6]. Each equipment was calibrated before starting the data collection. The data were collected during the peak hours, off peak hours, weekends, and also at highway to categorize it according to its Level of Service (LOS) where the traffic flow can be determined from its travelling speed. Eight routes were selected to collect data during the peak hours, off peak hours and weekend while four routes were selected to collect data at highway which will differentiate the data from high speed to low speed data. To improve the efficiency and accuracy in data collection, each route was separated into two directions and the data was collected in each direction.

2.3. Generation of micro-trips

The driving cycle will be developed based on generated micro-trips which defined as driving activity between adjacent stop covers each "stop-go" condition. It may consist of acceleration, cruise and deceleration modes [7]. The micro-trips will be separated from the whole speed data that have been extracted from the recorded data.

2.4. Data analysis

The target parameters were defined and calculated based on the speed-time data obtained from on-roads data collection. The velocity, acceleration (or deceleration) and displacement in every second throughout the speed-time data are calculated. The assessment criteria used in developing the KL driving cycle to determine the driving pattern are as shown in Table 2 [8]. The driving modes of the vehicles were defined. Then, the micro-trips were arranged according to its specific speed range and the percentage number of micro-trips (%Nm) and percentage of time spent in micro-trips (%Tm) were calculated to be compare with the candidate cycles [9].

Table 2. Kuala Lumpur driving characteristics

No.	Parameters	Condition
1	Average speed, V_{avg} (km/h)	
2	Average running speed, $V1_{avg}$ (km/h)	
3	Average acceleration, Acc_{avg} (m/s^2)	
4	Average deceleration, Dec_{avg} (m/s^2)	
5	Percentage of idle, %Idle	Speed = 0, Acc = 0
6	Percentage of cruise, %Cruise	Speed > 0, Acc = 0
7	Percentage of acceleration, %Acc	Acc > 0
8	Percentage of deceleration, %Dec	Acc < 0
9	Number of stops per km, Stops/km	
10	Positive kinetic energy, PKE (m/s^2)	

2.5. Construction of driving cycle

A random selection of micro-trip was categorized depend on its speed range. The time taken to complete a driving cycle was set to 1200 seconds based on the time necessary to analyze the gas according to its samples [10]. A tolerance limit of 5% under-estimation and 15% over-estimation was considered in choosing the micro-trip to develop a better emission purpose driving cycle. Once the random selection of micro-trip was done, it's followed by calculating the driving parameters, percentage number of micro-trips (%Nm) and percentage of time spent in micro-trips (%Tm). A various number of candidate cycles were developed to compare between each other and to select the driving cycle with the lowest error as a KL driving cycle.

3. Result and discussion

The driving cycle construction with the procedure discussed above were applied to the data collected from the selected routes. Ten candidate cycle was proposed in order to get the best candidate cycle with a less overall percentage of error. The KL 10 candidate cycle was selected as a KL driving cycle with less error was recorded among the other developed candidate cycle as shown in Table 3. The developed KL driving cycle with its driving parameters are closely related to the target parameters is shown in Figure 1. Micro-trips become longer at the high speed range data against time taken. The developed KL driving cycle has completed a distance of 7.86 kilometers in 1219 seconds with a total of 17 intermediate stops between micro-trips.

Table 3. Sum of errors of constructed candidate cycles

	KL1	KL2	KL3	KL4	KL5	KL6	KL7	KL8	KL9	KL10
E1(%)	55.10	54.98	54.41	63.18	43.48	53.32	58.30	66.39	51.56	32.03
E2(%)	35.69	35.69	35.69	39.61	39.61	40.91	32.66	32.66	32.66	32.67
E2(%)	2.45	2.42	2.78	1.97	1.98	2.25	2.43	2.43	2.19	2.20
Sum of errors (%)	**37.08**	**37.02**	**36.82**	**41.99**	**32.14**	**37.45**	**37.92**	**41.97**	**34.49**	**24.73**

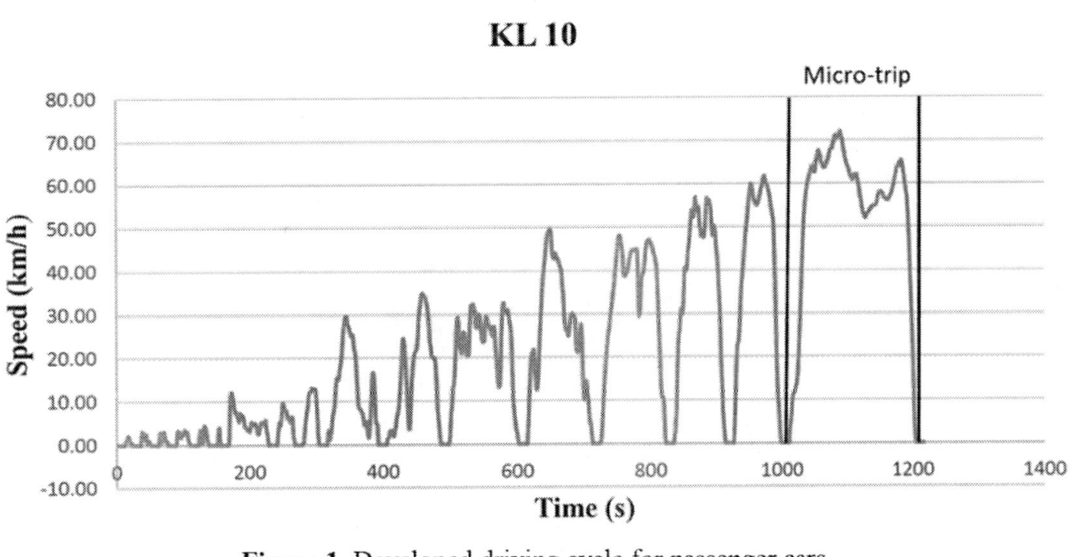

Figure 1. Developed driving cycle for passenger cars

The following observations are made from the developed driving cycle:

(1) Speed range below 10 km/h was dominant. This is due to congested road or heavy traffic condition which lead to have more stop-go condition.

(2) The micro-trips at the higher speed range are longer compared to the micro-trips at lower speed range. This is because the vehicle experiencing a free flow move at higher speed range with less frequent stop due to less traffic condition.

(3) A 23.21 km/h average speed was recorded in the developed KL driving cycle shows the vehicles are moving at lower speed and more micro-trips are found below the average speed. Therefore, more fuel consumption and emission takes place during that period due to frequent stop along the road.

To evaluate the developed KL cycle, a comparison was made between the existing international cycles as shown in Table 4. The driving cycle for passenger car was compared with US driving cycles such UDDS and LA92 which developed for emission purpose. Besides, it was also compared with the driving cycles of cities in China such Beijing driving cycle and Shanghai driving cycle and together with Hong Kong driving cycle.

For the average speed data, it shows that Beijing cycle is closely related to the KL driving cycle while the average speed for other driving cycles are a bit far from the developed KL cycle. The comparison for the average running speed is nearly correlated to the Beijing cycle. The average acceleration and average deceleration were the lowest than those mandatory cycles. The percentage of time spent in acceleration and deceleration is among the highest but closely related to the UDDS cycle. The percentage time spent in idle is similar to Hong Kong cycle, but the highest was Shanghai cycle. It has the lowest percentage of time spent in cruise mode compared to other existing cycle. Concerning the potential acceleration kinetic energy (PKE), the other cycles are generally larger than the developed KL cycle.

On overall consideration, it can be concluded that the traffic in Kuala Lumpur and Beijing is closely related to each other and the traffic condition in Kuala Lumpur is more congested compared to other existing driving cycles. Therefore, high fuel consumption and emission will be obtained from the tested vehicle using the KL driving cycle. Other than that, there are huge number of micro-trips found in the low range speed data than the high range speed data. It shows a high percentage of frequent stop of vehicles happen and causes traffic problems.

Table 4. Comparison of KL cycle with existing international cycles.

Driving Parameters	Chine driving cycle				US driving cycle (Emission purpose)	
	KL	Beijing	Shanghai	Hong Kong	UDDS	LA92
V_{avg} (km/h)	23.21	23.00	34.00	25.00	31.40	39.40
$V1_{avg}$ (km/h)	28.73	28.00	42.00	30.40	38.60	47.00
Acc_{avg} (m/s^2)	0.37	0.52	0.54	0.59	0.50	0.67
Dec_{avg} (m/s^2)	-0.38	-0.55	-0.68	-0.59	-0.57	-0.75
%Acc (%)	39.87	36.00	39.00	34.50	39.70	38.30
%Dec (%)	39.29	34.00	31.00	34.20	34.70	34.10
%Idle (%)	17.80	16.00	19.00	17.80	17.60	15.10
%Cruise (%)	3.04	10.00	11.00	12.00	8.00	12.50
PKE (m/s^2)	0.32	0.43	0.46	0.39	0.38	0.40

4. CONCLUSION

The objective of this paper is to develop a driving cycle to determine fuel consumption and emissions from vehicles. Due to data variation in constructing driving cycle, most of the developing countries' city routes having different driving cycle which are not same as one another.

A practical method was used to develop the driving cycle by using micro-trip method from real-world driving data. The driving parameters include average speed (Vavg), average running speed (V1avg), average acceleration (Accavg), average deceleration (Decavg), percentage of idle (%Idle), percentage of cruise (%Cruise), percentage of acceleration (%Acc), percentage of deceleration (%Dec), number of stops per kilometer travelled (Stops/km) and positive kinetic energy (PKE) were considered when constructing the driving cycle. The developed driving cycle was compared with other existing cycle, thus focuses the variety of traffic condition in Kuala Lumpur city.

There are some limitations found in this study. One of the limitation is, the high speed data from the highway data collection is less, which gives a less variety of choice to choose the micro-trip during the random selection of micro-trip to construct the driving cycle. Thus, it should be increased to improve the accuracy in selecting the microtrip. Besides, the inconsistency in driving behavior is not considered in this research. Other than that, the developed driving cycle mostly refers to Kuala Lumpur city and may not be suitable to use for other cities. This research can be used as a reference and further the studies to determine the fuel consumption and emission.

ACKNOWLEDGEMENT

Authors are thankful and acknowledge the financial support from Malaysia Automotive Institute (MAI) under research grant to MoTECH, UniMAP and to Perusahaan Otomobil Kedua Sendirian Berhad (PERODUA) for the technical and resources support.

REFERENCES

[1] Tamsanya S, Chungpaibulpatana S, Limmeechokchai B 2009 *International Journal of Automotive Technology* vol 10(2) p 251-264
[2] Ho S H, Wong Y D and Chang V W 2014 *Atmospheric environment* vol97 p 353-362
[3] Tong H Y, Hung W T and Cheung C S 1999 *Atmospheric Environment* vol 33(15) p 2323-2335.
[4] Galgamuwa U, Perera L, Bandara S. 2015 *Journal of Transportation Technologies* vol 5(04) p 191
[5] Highway Planning Unit 2016 *Road Traffic Volume Malaysia 2015* vol. M, p. 262
[6] Gamalath I, Fernando C, Galgamuwa U, Perera L and Bandara S. 2014 *Methodology to develop a driving cycle for a given mode and traffic corridor; case study for Galle Road, Colombo, Sri Lanka*
[7] Kamble S H, Mathew T V and Sharma G K. 2009 *Transportation Research Part D: Transport and Environment*. Vol 14(2) p 132-140
[8] Arun N H, Mahesh S, Ramadurai G and Nagendra S S 2017 *Sustainable cities and society* vol 32 p 508-512
[9] Tamsanya N U and Chungpaibulpatana S U 2009 *Journal of environmental sciences* vol 21(5) p 604-611.
[10] Xiao Z, Dui-Jia Z and Jun-Min S 20182 *Energy Procedia* vol 16 p 1868-1873

IOP Conf. Series: Materials Science and Engineering **429** (2018) 012048 doi:10.1088/1757-899X/429/1/012048

Crashworthiness Analysis of Bi-Tubular Aluminium Tubes with Varying Shapes in LS-Dyna

E H A Manaf, M T A Rahman*, A. Rahman, A S A Rahman, F Mat, K S Basaruddin

School of Mechatronic Engineering, Universiti Malaysia Perlis, Kampus Tetap Pauh Putra, 02600 Arau, Perlis, Malaysia.

*Corresponding author: tasyrif@unimap.edu.my

Abstract. The viability of thin-walled energy absorbers as impacts attenuators for Formula SAE racing cars was studied. The crashworthiness of bi-tubular tubes of varying geometries were explored. 9 designs grouped into 3 batches, were axially crush simulated in LS-Dyna. The tubes were set to 50mm long and made of aluminium 6061-T6. A convergence study was conducted, and the tubes were simulated using a meshing size of 1.5mm. The crushing energy applied is 490J. The LS-Dyna settings used were validated using experimental data from previous studies. It was found that the tube with an outer hexagon shape and an inner square shape had the best SEA. It was also observed that square-shaped geometries were the best in absorbing energy, with the most amount of energy being absorbed by a bi-tubular tube with outer and inner square geometries. Tubes with hexagonal geometries absorbed slightly less energy compared to tubes with square geometries, but they only differ by a small margin.

1. Introduction

The past century of human civilization has seen remarkable developments in the automotive industry generally in the engineering sector. Vehicles such as Koenigsegg Agera R and Bugatti Veyron are the pinnacles of speed machines and are exemplary of the extensive applications of engineering. However, with great power comes greater risk. In accordance with the laws of physics, the greater the velocity of a mass, the higher the kinetic energy the body has. In the event of a car crash, all of that energy would be transferred to the vehicle, and more critically, to the occupant of the vehicle, heightening the chances of severe injury. This is where impact attenuators (IA) come into play. They are devices that reduce the damage done to the vehicles and its occupants by absorbing the energy and deforming or redirecting the impact entirely thus safeguarding the occupants. This device is vital in racing competitions such as Formula SAE (FSAE). The impact attenuator acts as an energy absorber, absorbing the kinetic energy of the impact. The FSAE committee has given participating teams two options, whether to use a standard impact attenuator officially approved by the committee, or custom-make one. The custom-made one however must fulfil the requirements as stated in the rules [1].

Many parts of a FSAE racing car has been studied, including impact attenuators [2-7]. In a study of impact attenuator for racing vehicles, J. Wang et al experimented with the use of composite in impact attenuators for FSAE [8]. The basic structure chosen for the makeup of the IA was a half-corrugated beam. The structure was modelled in LS-DYNA and axial crushing of the model was studied. Material model 54 was chosen to simulate the fiber reinforced composite makeup of the IA. The rigid hammer responsible for the axial crushing was modelled using MAT_RIGID. Overall, the damage on the composite thin wall utilised shell 163 to reduce computation time. LS-OPT was then utilised to optimise the structure through Response Surface Methodology (RSM). They found that by increasing the diameter of the corrugated beams on the impact attenuator (IA), specific energy absorption (SEA) increased whilst increasing the height of the IA decreased SEA.

IOP Conf. Series: Materials Science and Engineering **429** (2018) 012048 doi:10.1088/1757-899X/429/1/012048

2. Cad Design

In the design of an impact attenuator, many different geometries and materials can be utilised to produce the desired part. Materials such as epoxy reinforced fibre and aluminium honeycomb sandwiches have been used in the makeup of an IA, proving that the possibilities are endless in the design and manufacture of IA's. In this study, the concept of thin-walled absorbers was applied to produce an IA that meets the FSAE requirements. Single and bi-tubular tubes of varying geometries were investigated. The different geometries of the tubes are presented in Figure 1, together with their respective codes. The outer and inner diameters of the tubes were set to 38.1mm and 25.4mm respectively. The wall thickness of all the tubes were set to 1mm and the lengths were set to 50mm. The geometries were drawn in Solidworks 2017 and were saved as IGES files to be further processed in LS-DYNA.

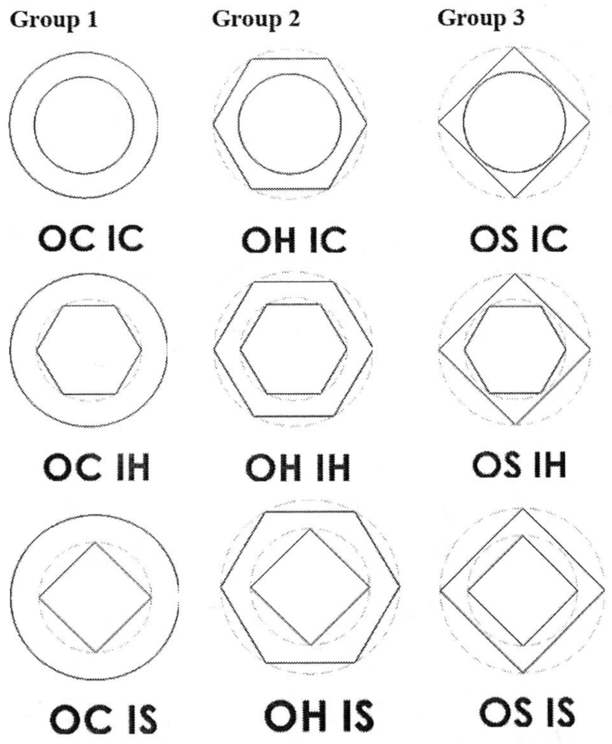

Figure 1: Varying tube geometries with their respective codes

The dimensions of the tubes were based off commonly available sizes on the market. All the tubes have the same outer diameter of 38.1mm or 1.5inches and inner diameter of 25.4mm or 1inch. The tubes have also been coded to easily identify them. The first two letters stand for the shape of the outer tube. OC for example stands for 'outer circle', meaning that the outer tube is circular. The next set of letters dictate the shape of the inner tube. IS for example means 'inner square' meaning the inner tube is square in shape. So, OH IH for example means the inner and outer tubes are both hexagonal in shape.

3. Simulation Validation

Before proceeding with the simulation on the different tube designs, the simulation configurations must first be validated to ensure that the results tally up with real life results. Based on research done in [9], the authors compared both experimental and simulation results on the axial crushing of empty aluminium tubes. They found that the major trends were well capture, but the difference is as high as 29%. Using that as a basis, a recreation of the setup was done, and the results were compared with the experimental results in [9]. The force versus displacement graph for both are shown in Figure 2.

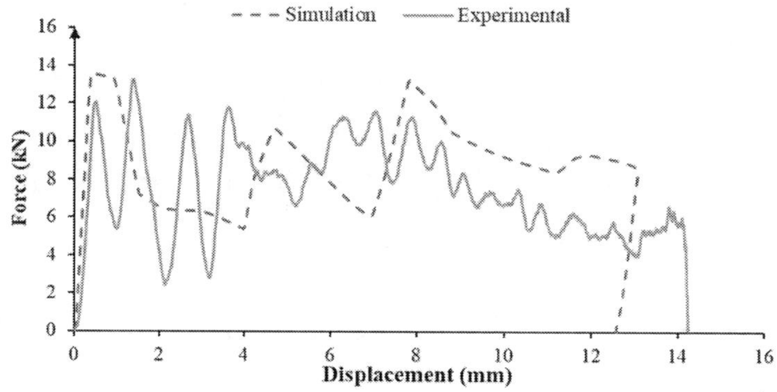

Figure 2: Force vs displacement of the experimental and simulation setup

Comparing the peak force for both the simulation and experiment, the percentage error is approximately 2.4%.

4. Finite Element Analysis In Ls-Dyna

The finite element model was set up as in Figure 6. The various tubes in IGES file format were imported into LS-PrePost and meshed using auto mesher with mesh size of 2mm. The tubes of 50mm length were placed in between two masses as in Figure 3. Both the top and bottom masses are rigid solid sections. The bottom mass acts as a supporting platform for the axial crushing whereas the top plate acts as the crusher. The set of nodes at the bottom of the tubes were fixed in place and the mass is set to approach the tubes at 7m/s. 490J of energy was applied. The dimensions of the masses were set to 80x80x2 mm.

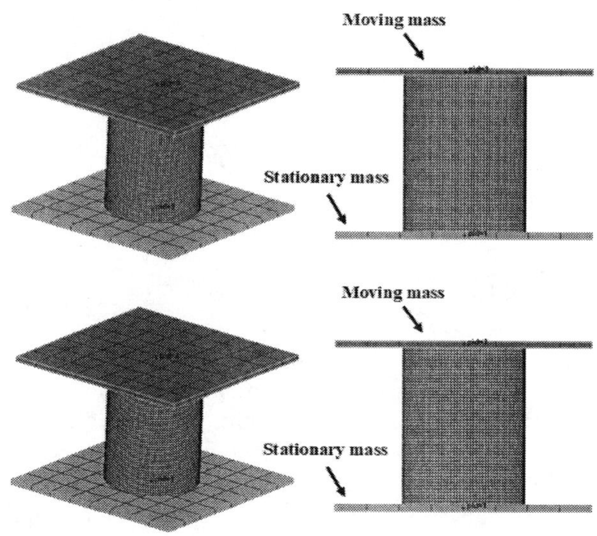

Figure 3: Simulation setup

In this setup, the tubes are made from aluminium 6061-T6. Material model 24, piecewise linear plasticity, was chosen to represent the material since it is an elasto-plastic material with strain-rate dependency. The properties of the 6061-T6 used were obtained from ASM International [10-11]. For the stationary and moving masses, MAT_RIGID was used since the masses are considered as rigid bodies in this simulation. In this study, the contact between the tubes and the moving mass is described by CONTACT AUTOMATIC NODE TO SURFACE. The moving mass is set as the master whereas the tubes were set as slaves. According to recommendations, this type of contact is suitable when the master part, in this case the moving mass is a rigid body. To prevent penetration of the bodies, automatic single surface was used.

5. Measuring Crashworthiness

Comparing how capable the tubes are in terms of absorbing energy without some sort of standard measurement would be like comparing apples to oranges. Thus, based on [12-15], a few parameters such as specific energy absorption (SEA), total energy absorption (TEA), initial peak force (Fpeak) and mean crush force (Fmean) can be used to compare the crashworthiness of the tubes. The parameters are expressed as the following equations

$$TEA = \int_0^{\delta} F\ (\delta)\ d\delta \quad (1.0)$$

$$SEA = TEA/m \quad (2.0)$$

$$F_{mean} = TEA/\delta \quad (3.0)$$

where δ is the displacement and m is the mass of the tube. For initial peak force Fpeak, it is the highest value on the force versus displacement graph.

6. Results And Discussions

6.1. Group 1 Results

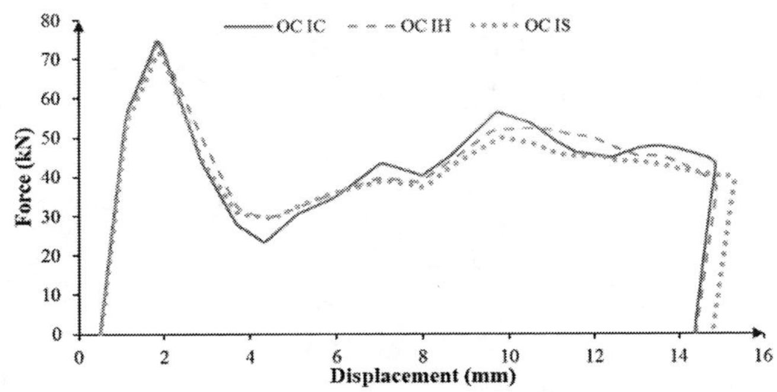

Figure 4: Force vs displacement graph for group 1 tubes

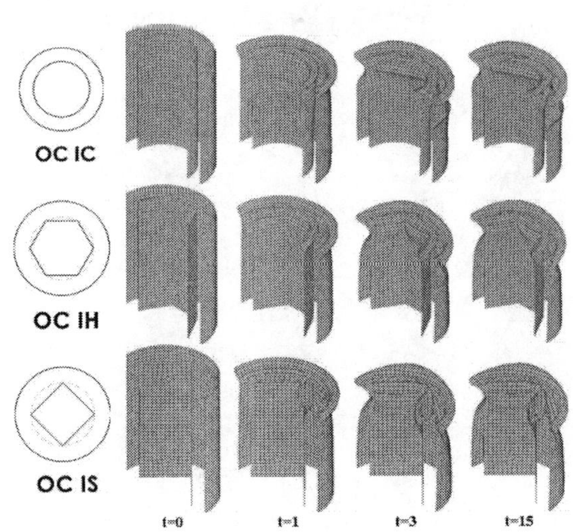

Figure 5: Group 1 tube collapse

6.2. Group 2 Results

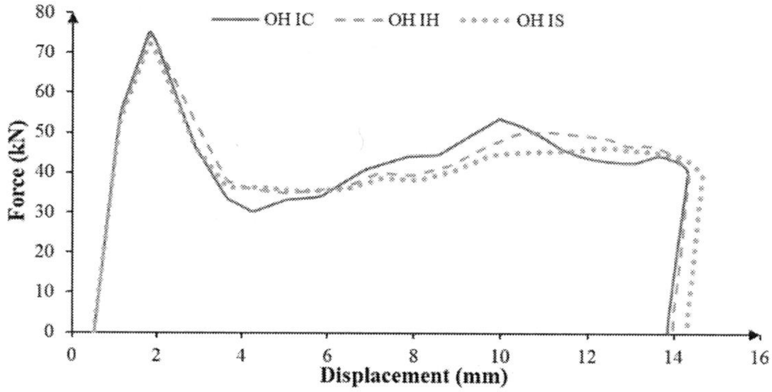

Figure 6: Force vs displacement graph for group 2 tubes

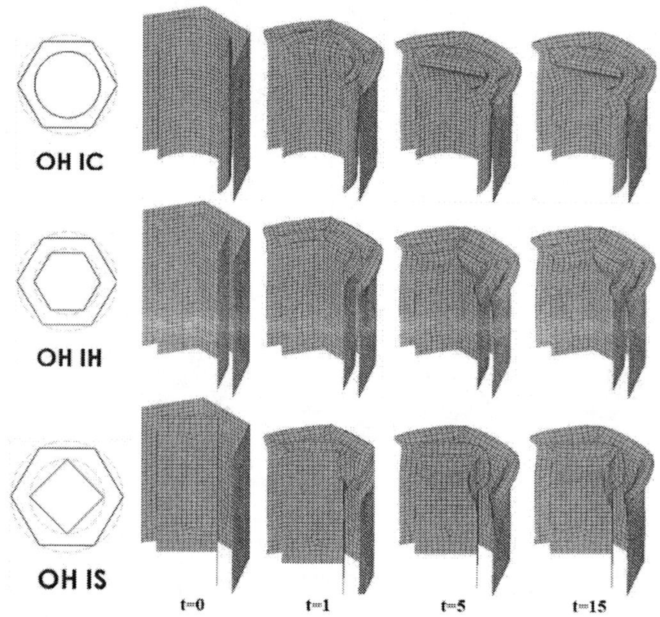

Figure 7: Group 2 tube collapse

6.3. Group 3 Results

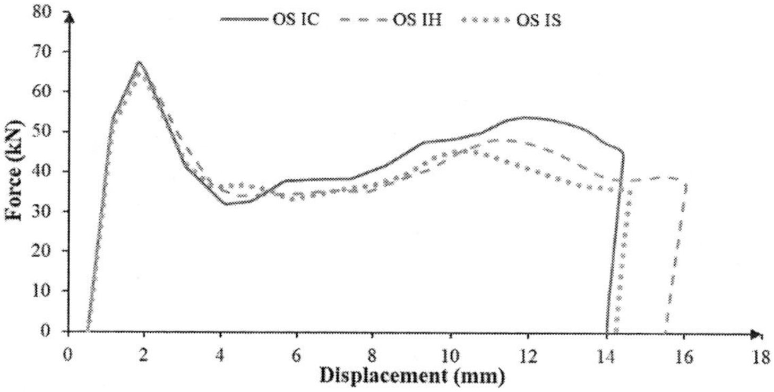

Figure 8: Force vs displacement graph for group 3 tubes

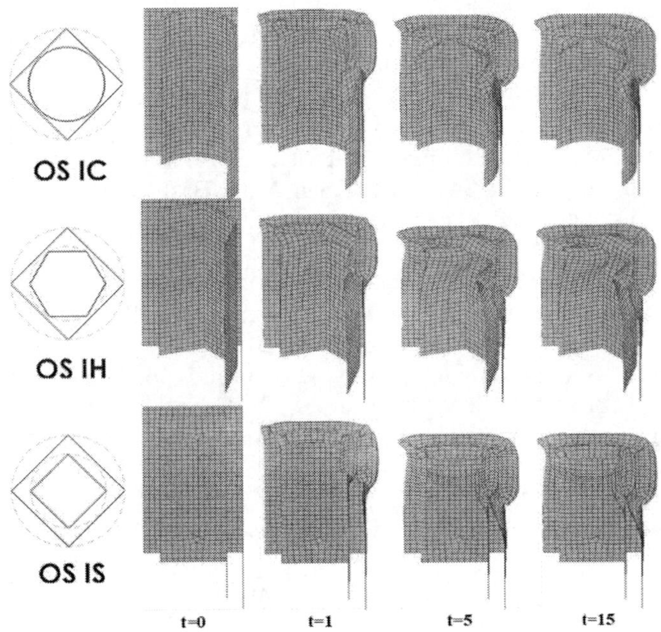

Figure 9: Group 3 tube collapse

The finite element analysis results for bi-tubular tubes with cylindrical outer tubes; OC IC, OC IH and OC IS are presented in Figure 4. The collapse progression of the tubes is shown in Figure 5. Looking at the force vs displacement graph, there seem to be little difference between OC IC, OC IH and OC IS. The same applies to the amount of energy absorbed for OC IC, OC IH and OC IC which are 478.5J, 480.23J and 483.46J respectively.

The finite element analysis results for bi-tubular tubes with hexagonal outer tubes are presented in Figure 6. Figure 7 shows the collapse of group 3 tubes with the progression of time. Similar to group 1 results, the force vs displacements graphs are almost identical. The same could be said for the total energy absorbed for group 2 designs which are 479.94J, 483.21J and 485.67J respectively. This may be due to the geometry of group 2 designs which have an outer hexagonal tube, behaving similar to circular tubes in terms of energy absorption.

The finite element analysis results for bi-tubular tubes with square outer tubes are presented in Figure 8. The collapse of the tubes with the progression of time is shown in Figure 9. An interesting observation to note in Figure 8 is the lower crushing force value for all group 3 designs compared to groups 1 and 2. The tubes in group 3 also has a larger average of total energy absorbed compared to the other groups. The energy absorbed for OS IC, OS IH and OS IS are 482.87J, 481.91J and 487.03J respectively.

7. Conclusions

This study has explored the viability of thin-walled energy absorbers as impact attenuators for Formula SAE racing cars. Bi-tubular tubes of varying geometries made of aluminium 6061-T6 were simulated in LS-Dyna. It was found that the bi-tubular tube with an outer hexagonal tube and an inner square tube was the best in terms of specific energy absorption. It was also observed that square-shaped geometries were the best in absorbing energy, with the most amount of energy being absorbed by a bi-tubular tube with outer and inner square geometries. Tubes with hexagonal geometries absorbed slightly less energy compared to tubes with square geometries, but they only differ by a small margin. It can be concluded that bi-tubular aluminium tubes could be used as impacts attenuators especially in vehicles such as Formula SAE racing cars.

References

[1] SAE International, "2017-18 Formula SAE® Rules," p. 180, 2017.
[2] A. F. A. Hamid, M. T. A. Rahman, S. F. Khan, A. H. Adom, M. A. Rahim, N. A. Rahim, M. H.

N. Ismail, and A. Norizan, "Connected car: Engines diagnostic via Internet of Things (IoT)," J. Phys. Conf. Ser., vol. 908, p. 12079, 2017.

[3] A. Norizan, M. T. A. Rahman, N. A. M. Amin, M. H. Basha, M. H. N. Ismail, and A. F. A. Hamid, "Study of intake manifold for Universiti Malaysia Perlis automotive racing team formula student race car," J. Phys. Conf. Ser., vol. 1, p. 12069, 2017.

[4] M. L. Mohamad, M. T. A. Rahman, S. F. Khan, M. H. Basha, A. H. Adom, and M. S. M. Hashim, "Design and static structural analysis of a race car chassis for Formula Society of Automotive Engineers (FSAE) event," J. Phys. Conf. Ser., vol. 908, p. 12042, 2017.

[5] M. Azmeer, M. H. Basha, M. F. Hamid, M. T. A. Rahman, and M. S. M. Hashim, "Design optimization of rear uprights for UniMAP Automotive Racing Team Formula SAE racing car," J. Phys. Conf. Ser., vol. 908, p. 12051, 2017.

[6] M N A Zaidie, M S M Hashim, M Tasyrif, M H Basha, I Ibrahim, N S Kamaruddin, Shahriman A B, "Analysis of a front suspension system for UniART FSAE car using FEA Analysis of a front suspension system for UniART FSAE car using FEA," J. Phys. Conf. Ser., vol. 908, p. 12058, 2017.

[7] Z. I. Zaman, K. S. Basaruddin, M. H. Basha, M. T. A. Rahman, and R. Daud, "Finite element prediction on the chassis design of UniART4 racing car," AIP Conf. Proc., vol. 1885, pp. 1–10, 2017.

[8] J. Wang *et al.*, "Design and experimental verification of composite impact attenuator for racing vehicles," *Composite Structures*, vol. 141. pp. 39–49, 2016.

[9] F. Mat, K. A. Ismail, M. Ahmad, and M. R. M. Rejab, "Foam filling effectiveness of conical aluminium tubes under dynamic axial and oblique loading," ARPN J. Eng. Appl. Sci., vol. 11, no. 12, pp. 7741–7747, 2016.

[10] L. Support, "Consistent units," LS-DYNA Support, 2012.

[11] V. 2 ASM International Handbook, Properties and selection: Nonferrous alloys and special-purpose materials, vol. 2. 1990.

[12] K. Vinayagar and A. Senthil Kumar, "Crashworthiness analysis of double section bi-tubular thin-walled structures," Thin-Walled Struct., vol. 112, no. June 2016, pp. 184–193, 2017.

[13] Q. Estrada, D. Szwedowicz, J. Silva-Aceves, T. Majewski, J. Vergara-Vazquez, and A. Rodriguez-Mendez, "Crashworthiness behavior of aluminum profiles with holes considering damage criteria and damage evolution," Int. J. Mech. Sci., vol. 131–132, no. August, pp. 776–791, 2017.

[14] L. Zhang, Z. Bai, and F. Bai, "Crashworthiness design for bio-inspired multi-cell tubes with quadrilateral, hexagonal and octagonal sections," Thin-Walled Struct., vol. 122, no. October 2017, pp. 42–51, 2018.

[15] X. Wu, "Crashworthiness analysis and optimization of fourier varying section tubes," Int. J. Non. Linear. Mech., vol. 92, no. February, pp. 41–58, 2017.

Acknowledgement

This research was supported by Geran Penyelidikan Khas Universiti Malaysia Perlis, No.9004-00063.

IOP Conf. Series: Materials Science and Engineering **429** (2018) 012049 doi:10.1088/1757-899X/429/1/012049

Design and Analysis of Impact Attenuator for a Formula Student Car: A Study between Singular and Bi-tubular Tubes of Varying Geometries

A. Rahman, M T A Rahman*, E H A Manaf, A S A Rahman

School of Mechatronic Engineering, Universiti Malaysia Perlis, Pauh Putra Main Campus, 02600 Arau, Perlis, Malaysia.

*Corresponding author: tasyrif@unimap.edu.my

Abstract. The viability of thin-walled energy absorbers as impact attenuators for Formula Student car was examined in this study, specifically the crashworthiness of singular and bi-tubular tubes of varying geometries. 12 designs were lumped into 4 groups, consisting of 3 singular tubes and 9 bi-tubular tubes of varying shapes and dimensions, and they were axially crush-simulated in LS-DYNA. The tubes were set to 50mm long and made of aluminium 6061-T6. The LS-DYNA settings were validated using published experimental data of similar studies. The singular tubes were found to have the highest values of specific energy absorption (SEA), with the singular hexagonal tube design recording the highest value of SEA compared to the other designs. It was also observed that square-shaped geometries were the best in absorbing energy, especially a bi-tubular tube with outer and inner square geometries.

1. Introduction

Many parts of a FSAE racing car has been studied, including impact attenuators [1-6]. The importance of impact attenuator (IA) in reducing the collision's impact during an accident has been frequently highlighted over the years [7,8], more so for racing/formula cars. A good design of IA is highly critical to absorb crash energy in a controlled manner and thus offer protection. In the Formula SAE (FSAE), this device is one of the most important aspect that will be judged during the tournament. While the FSAE committee has given participating teams options whether to use a standard IA that was officially approved by the committee, or a custom-built one, most teams tend to come up with their novel IA design to reduce cost. Additionally, any customized IA must fulfil the requirements as stated in the FSAE rules [9].

2. CAD Design

In this study, the concept of thin-walled absorbers was applied to produce an IA that meets the FSAE requirements. Figure 1 displays single and bi-tubular tubes of varying geometries investigated in this study, along with their respective codes. The outer and inner diameters of the tubes were set to 38.1mm and 25.4mm respectively. The wall thickness of all the tubes were set to 1mm and the lengths were set to 50mm. The geometries were drawn in SolidWorks 2017 and were saved as IGES files to be further processed in LS-DYNA.

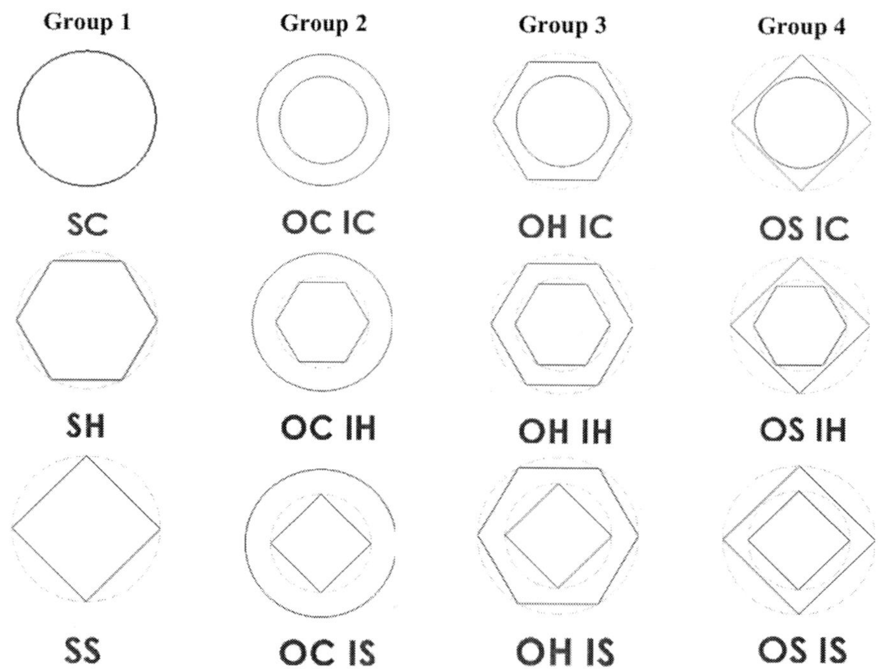

Figure 1: Varying tube geometries with their respective codes. The four groups used to classify the tubes geometries are also clearly indicated.

The dimensions of the tubes were based off commonly available sizes on the market. All the tubes have the same outer diameter of 38.1mm and inner diameter of 25.4mm. The tubes have also been coded to easily identify them. For Group 1 tubes, the first 'S' stands for single, meaning singular tubes whereas the second alphabet stands for the shape of the singular tubes. SH for example means a 'Singular Hexagonal' tube. For Groups 2 to 4, the first two letter stand for the shape of the outer tube. OC for example stands for 'Outer Circle', meaning that the outer tube is circular. The next set of letters dictate the shape of the inner tube. IS for example means 'Inner Square' meaning the inner tube is square in shape. OH IH for example means the inner and outer tubes are both hexagonal in shape.

3. Finite Element Analysis in LS-DYNA

The finite element model was set-up as demonstrated in Figure 2. The various tubes in IGES file format were imported into LS-PrePost and the elements (mesh size of 2mm) were generated using *auto-mesher function*. The tubes of 50mm length were placed in between two masses as illustrated in Figure 2. Both the top and bottom masses are rigid solid sections. The bottom mass acts as a supporting platform for the axial crushing whereas the top plate acts as the crusher. The set of nodes at the bottom of the tubes were fixed in place, while the mass approaching the tubes was set to be at 7m/s. Next, a total of 490J of impact energy was applied to the model. The dimensions of the masses (i.e. top and bottom plates) were set to 80x80x2 mm.

IOP Conf. Series: Materials Science and Engineering **429** (2018) 012049 doi:10.1088/1757-899X/429/1/012049

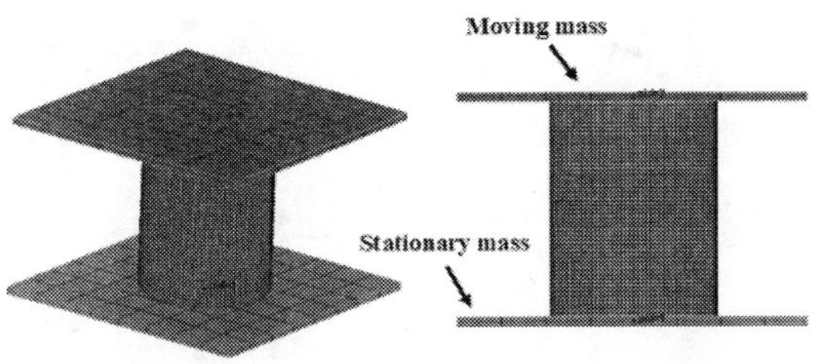

Figure 2: Model set-up

In this set-up, the tubes are made from aluminium 6061-T6. Material model 24, piecewise linear plasticity, was chosen to represent the material since it is an elasto-plastic material with strain-rate dependency. The properties of the 6061-T6 used were obtained from ASM International [10,11]. For the stationary and moving masses, MAT_RIGID was used since the masses are considered as rigid bodies in this simulation. In this study, the contact between the tubes and the moving mass is described by CONTACT AUTOMATIC NODE TO SURFACE. The moving mass was set as the master whereas the tubes were set as slaves. This type of configuration is commonly used when the moving mass is a rigid body. To prevent penetration of the bodies, automatic single surface was applied.

4. Measuring Crashworthiness

Equations 1-3, which were based on [12-15], were used to compare the crashworthiness of the tubes. Specific energy absorption (SEA), total energy absorption (TEA), initial peak force (F_{peak}) and mean crush force (F_{mean}) can be expressed using the following equations:

$$TEA = \int_0^{\delta} F\ (\delta)\ d\delta \tag{1}$$

$$SEA = TEA/m \tag{2}$$

$$F_{mean} = TEA/\delta \tag{3}$$

where δ is the displacement and m is the mass of the tube. For initial peak force F_{peak}, it is the highest value on the force versus displacement graph.

5. Results and Discussions

The results from the finite element models for the singular tubes are displayed in Figure 3, while Figures 4-6 illustrate the simulation outputs for the bi-tubular tubes. Comparing the initial peak force for all three tubes in Group 1, SC and SH seem to have the same value for peak force whereas SS has a lower value of approximately 34kN. Lower crushing force means lower deceleration for the rigid body, making it more suitable for the application of impact attenuators in vehicles since humans cannot handle high G-forces [16]. In terms of energy absorption, SC, SH and SS absorbed 479.67J, 483.18J and 482.79J respectively. When looking at the displacement of the models, SS displaced the most at approximately 27mm. As for Group 2 results, there seem to be very minor differences between OC IC, OC IH and OC IS. Likewise, this behaviour was also repeated for the amount of energy absorbed for OC IC, OC IH and OC IC which were 478.5J, 480.23J and 483.46J respectively

IOP Conf. Series: Materials Science and Engineering **429** (2018) 012049 doi:10.1088/1757-899X/429/1/012049

Results	Force vs displacement graph	Tube collapse with respect to time for respective group
Figure 3: Results for Group 1 (singular tubes)	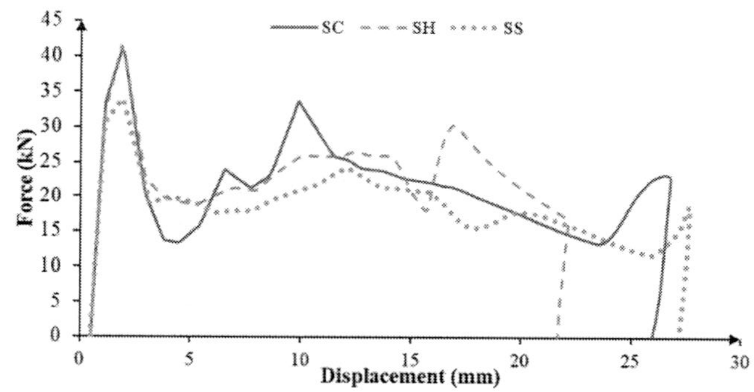	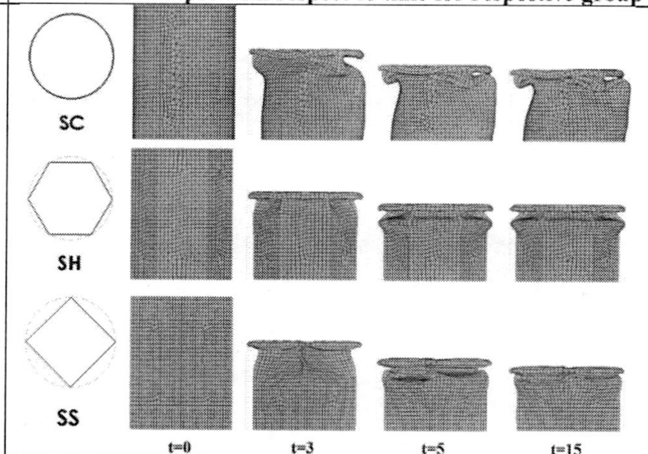
Figure 4: Results for Group 2 (bi-tubular tubes)	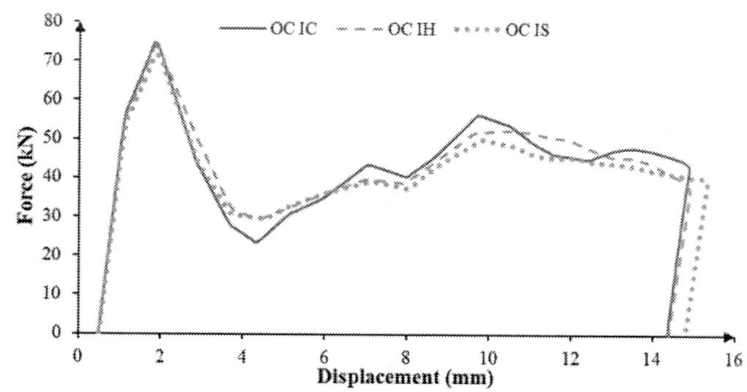	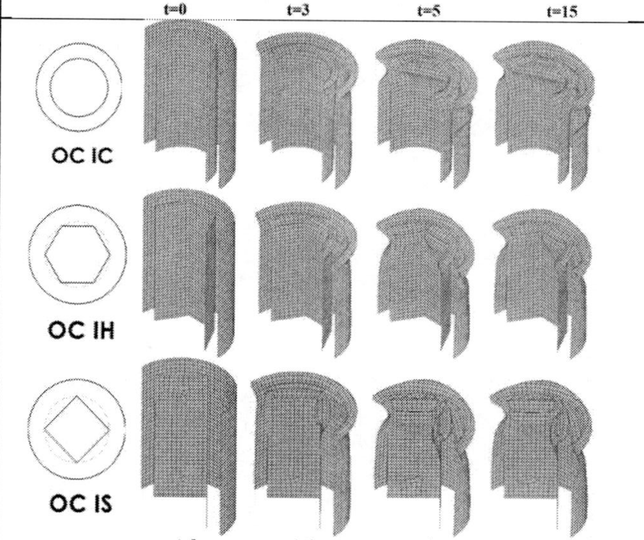

IOP Conf. Series: Materials Science and Engineering **429** (2018) 012049 doi:10.1088/1757-899X/429/1/012049

Results	Force vs displacement graph	Tube collapse with respect to time for respective group
Figure 5: Results for Group 3 (bi-tubular tubes)	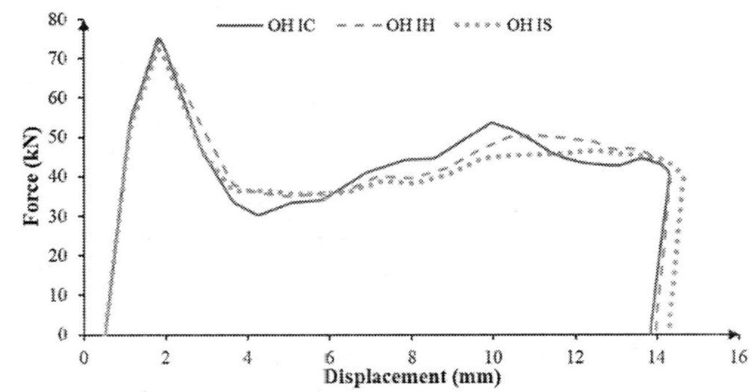	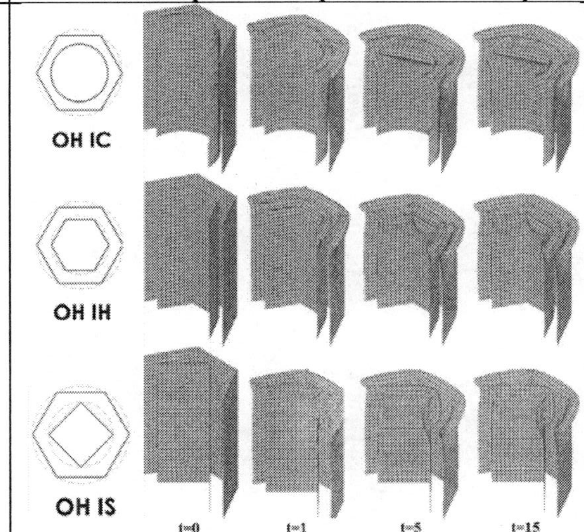
Figure 6: Results for Group 4 (bi-tubular tubes)	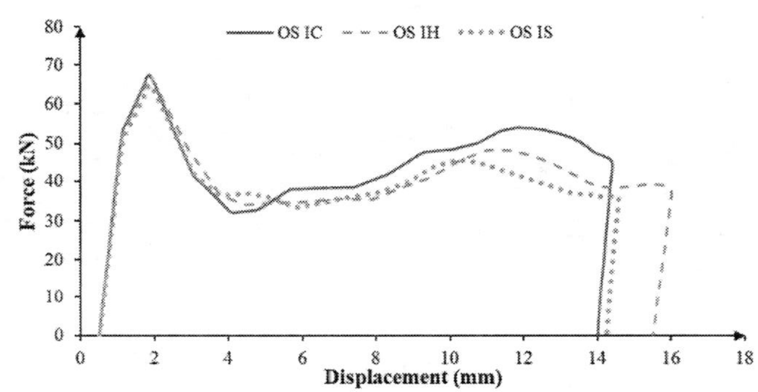	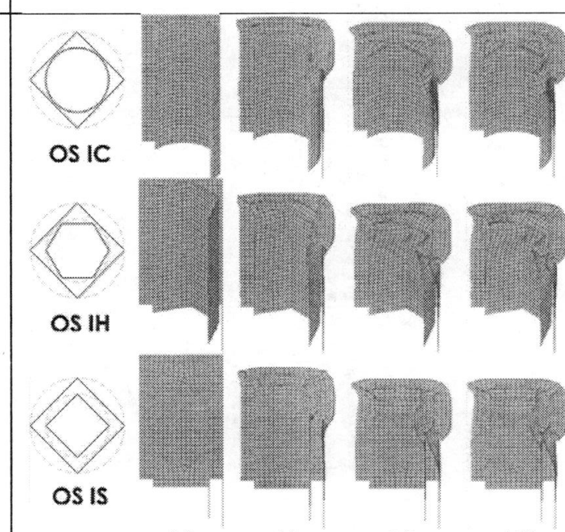

The force vs displacements graph for Group 3 however, shows an almost identical results to the one produced by Group 2. The same can also be said for the total energy absorbed for the Group 3 designs which were 479.94J, 483.21J and 485.67J respectively. This may be due to the geometry of group 3 designs which have an outer hexagonal tube, hence resembling the circular tubes in terms of energy absorption.

Finally, Figure 6 demonstrates that all Group 4 designs exhibit lower crushing force value when compared with Groups 2 and 3. Additionally, it is also apparent from Figure 3 that having an outer square tube helps to reduce the initial peak force. Nonetheless, it is interesting to note that tubes in Group 4 also have a larger average of total energy absorbed when compared to the other groups. For instance, the energy absorbed for OS IC, OS IH and OS IS are 482.87J, 481.91J and 487.03J respectively.

6. Conclusions

This study has explored the viability of thin-walled energy absorbers as impact attenuators for Formula SAE racing cars. Singular and bi-tubular tubes of varying geometries made of aluminium 6061-T6 were simulated in LS-Dyna and then compared in terms of the amount of energy absorbed, peak crushing force and specific energy absorption. It was found that the singular tubes were the best in terms of specific energy absorption, with the singular hexagonal tube recording the highest value of SEA out of all the designs.

It was also observed that square-shaped geometries were the best in absorbing energy, with the most amount of energy being absorbed by a bi-tubular tube with outer and inner square geometries. Tubes with hexagonal geometries absorbed slightly less energy compared to tubes with square geometries, except for the singular hexagonal tube, but they only differ by a small margin.

Overall, this study has demonstrated the capability and suitability of both singular and bi-tubular aluminium tubes as impact attenuators that can be applied in Formula SAE racing cars.

References

[1] A. F. A. Hamid, M. T. A. Rahman, S. F. Khan, A. H. Adom, M. A. Rahim, N. A. Rahim, M. H. N. Ismail, and A. Norizan, "Connected car: Engines diagnostic via Internet of Things (IoT)," J. Phys. Conf. Ser., vol. 908, p. 12079, 2017.

[2] A. Norizan, M. T. A. Rahman, N. A. M. Amin, M. H. Basha, M. H. N. Ismail, and A. F. A. Hamid, "Study of intake manifold for Universiti Malaysia Perlis automotive racing team formula student race car," J. Phys. Conf. Ser., vol. 1, p. 12069, 2017.

[3] M. L. Mohamad, M. T. A. Rahman, S. F. Khan, M. H. Basha, A. H. Adom, and M. S. M. Hashim, "Design and static structural analysis of a race car chassis for Formula Society of Automotive Engineers (FSAE) event," J. Phys. Conf. Ser., vol. 908, p. 12042, 2017.

[4] M. Azmeer, M. H. Basha, M. F. Hamid, M. T. A. Rahman, and M. S. M. Hashim, "Design optimization of rear uprights for UniMAP Automotive Racing Team Formula SAE racing car," J. Phys. Conf. Ser., vol. 908, p. 12051, 2017.

[5] M N A Zaidie, M S M Hashim, M Tasyrif, M H Basha, I Ibrahim, N S Kamaruddin, Shahriman A B, "Analysis of a front suspension system for UniART FSAE car using FEA Analysis of a front suspension system for UniART FSAE car using FEA," J. Phys. Conf. Ser., vol. 908, p. 12058, 2017.

[6] Z. I. Zaman, K. S. Basaruddin, M. H. Basha, M. T. A. Rahman, and R. Daud, "Finite element prediction on the chassis design of UniART4 racing car," AIP Conf. Proc., vol. 1885, pp. 1–10, 2017.

[7] Boria, S., Obradovic, J., Belinggardi G., Composites Part B: Engineering, 79, 20-29 (2015)

[8] Ubaidillah, Mazlan, S.A., Sutrisno, J., Zamzuri, Mechanics and Materials, 663, 695-699 (2014).

[9] SAE International, "2017-18 Formula SAE® Rules," p. 180, 2017.

[10] L. Support, "Consistent units," LS-DYNA Support, 2012.

[11] V. 2 ASM International Handbook, Properties and selection: Nonferrous alloys and special-purpose materials, vol. 2. 1990.

[12] K. Vinayagar and A. Senthil Kumar, "Crashworthiness analysis of double section bi-tubular thin-walled structures," Thin-Walled Struct., vol. 112, no. June 2016, pp. 184–193, 2017.

[13] Q. Estrada, D. Szwedowicz, J. Silva-Aceves, T. Majewski, J. Vergara-Vazquez, and A.

Rodriguez-Mendez, "Crashworthiness behavior of aluminum profiles with holes considering damage criteria and damage evolution," Int. J. Mech. Sci., vol. 131–132, no. August, pp. 776–791, 2017.
[14] L. Zhang, Z. Bai, and F. Bai, "Crashworthiness design for bio-inspired multi-cell tubes with quadrilateral, hexagonal and octagonal sections," Thin-Walled Struct., vol. 122, no. October 2017, pp. 42–51, 2018.
[15] X. Wu, "Crashworthiness analysis and optimization of fourier varying section tubes," Int. J. Non. Linear. Mech., vol. 92, no. February, pp. 41–58, 2017.
[16] G. Lu and T. Yu, Energy Absorption of Structures and Materials. 2003.

Acknowledgement
The authors would like to extend their sincere gratitude for the funding provided by the Universiti Malaysia Perlis (UniMAP) (*Geran Penyelidikan Khas* - No.9004-00063) which enable this research to be carried out.

Terengganu routes representation for development of Malaysia Driving Cycle: Route selection methodology

M F H Rani[1], A R Mahayadin[1], Shahriman A B[1], Z M Razlan[1], I Zunaidi[4], W K Wan[1], A Harun[1], M S M Hashim[1], I Ibrahim[1], N S Kamarrudin[1], M K Faizi[1], M A M Saad[1], M Sahari[2], M S Sarip[2], Z M Isa[3] and Y Nagaya[3]

[1] Motorsports Technology Research Unit, Universiti Malaysia Perlis, 02600 Arau, Perlis, Malaysia.
[2] Malaysia Automotive Institute (MAI), 63000 Cyberjaya, Selangor, Malaysia.
[3] Perodua Manufacturing Sdn Bhd, 48009 Rawang, Selangor, Malaysia.
[4] Technopreneur at UniMAP Sdn Bhd, 01000 Kangar, Perlis, Malaysia.

Corresponding author: shahriman@unimap.edu.my

Abstract. A program to develop a harmonized light duty test cycle that will represent typical conditions for Malaysia is going to be developed, named Malaysia Driving Cycle (MDC). The development of MDC will adapt the methodology from the worldwide harmonized light duty driving test cycle (WLTC). Terengganu has been chosen as one of the regions for in-use driving data collection. In developing MDC, the route selection phase must be carried out first before proceeding to the route data collection. This paper describes briefly on the methodology used for the route selection, the selected routes and the driving conditions for route data collection in Terengganu region. Road Traffic Volume Malaysia (RTVM) version 2015 and Google Maps are used to study the route selection in Terengganu region. By adapting the methodology of WLTC, 11 routes in the Terengganu region have been selected to contribute the speed-time data for development of Malaysia Driving Cycle (MDC). The selected routes consist of urban, rural, and motorway road. For the route data collection, the driving conditions (peak, off-peak, weekend) are considered to collect the samples for each route and direction.

Keywords. Route selection, driving cycle, Terengganu region, Malaysia driving cycle, driving condition, light duty driving test cycle.

1. Introduction

Transport carbon emissions are the major contributory of air pollution [1]. Emission standards play a major role in controlling the vehicular emissions for a cleaner air. Many countries use driving cycles to determine regional vehicular emissions due to traffic patterns [2]. By using measured emission inventories as baseline emission, standards are set up or emission zones are established to control emissions. Through the working party on pollution and energy transport program (GRPE), a worldwide harmonized light duty driving test cycle (WLTC) was introduced to represent typical driving characteristics around the world [3]. A driving cycle is a series of data points that represents speed versus time [4], speed and gear selection as a function of time [2], speed versus distance [5], or time versus gradient [6] in a specific region or part of a road segment. Driving cycles have been used throughout the world for traffic engineering [7], emission purposes [8] and to estimate the fuel

IOP Conf. Series: Materials Science and Engineering **429** (2018) 012050 doi:10.1088/1757-899X/429/1/012050

consumption [9]. A test vehicle will run on a chassis dynamometer according to developed driving cycles to estimate the fuel consumption and to develop the emission inventories by collecting the total emission. Then, the collected emissions are analysed by using a gas analyser [10]. Malaysia Automotive Institute (MAI), an agency under the Ministry of International Trade and Industry (MITI), initiated a program to develop a harmonized light duty test cycle for Malaysia, known as Malaysia Driving Cycle (MDC), which will represent typical conditions for Malaysia. Terengganu, which is located in the east coast of Peninsular Malaysia, has been chosen as one of the regions for in-use driving data collection. In the process of developing MDC, the route selection phase must be carried out first before proceeding to the route data collection, the cycle construction and the cycle assessment. This paper describes briefly on the methodology used for the route selection, the selected routes and the driving conditions for route data collection in Terengganu region.

2. Methodology of Terengganu route selection

Figure 1 presents the flowchart of Terengganu route selection methodology. All routes in Terengganu region were studied by referring to Road Traffic Volume Malaysia (RTVM) version 2015 published by the Ministry of Works, Malaysia [11]. The traffic flow and quality of traffic service for each route is categorized based on level of service (LOS). Raw data of traffic volume is extracted first from RTVM 2015 then ranked up based on the LOS and peak hour traffic flow as shown in Figure 2. All routes are selected based on consideration of LOS and traffic volume. If the selected route is in the same networks or chains, the route is discarded as described in [12]. The route selection was also verified with Google Maps to confirm the traffic conditions of selected routes. On the other hand, Google Maps will not only be utilized to assess the road geography of selected routes, but also to determine the start/end point. Self-assessment and route validation are obligatory to ensure the selected routes are free from major road construction and any road diversion before route data collection is started.

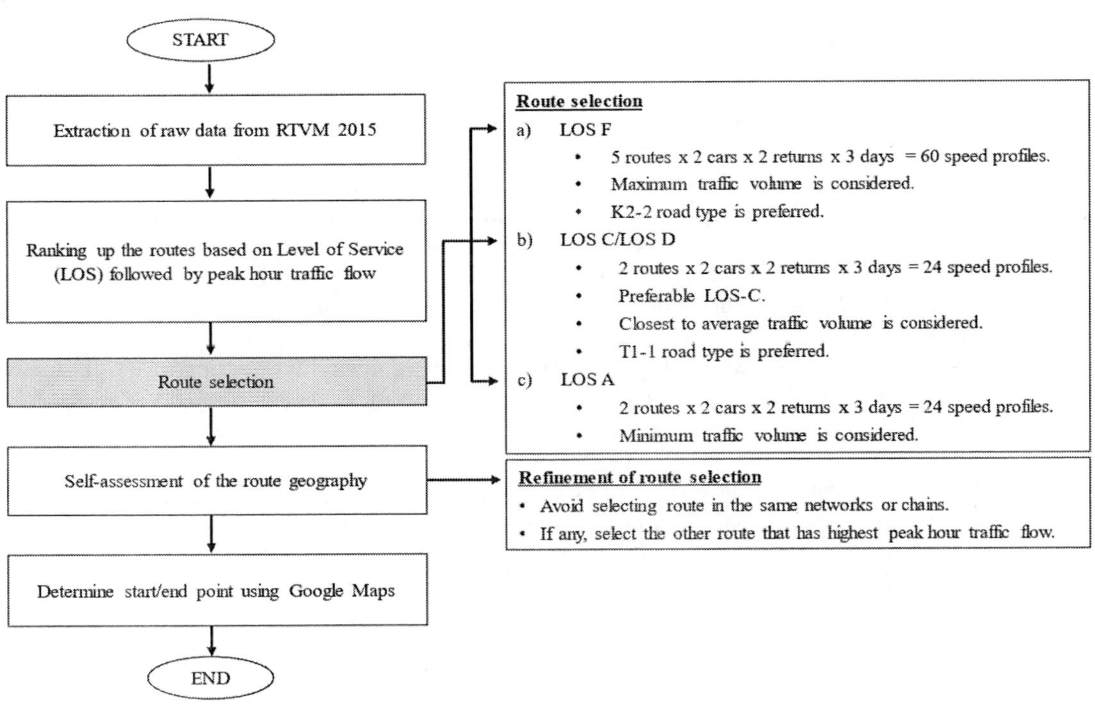

Figure 1. Flowchart of Terengganu route selection methodology.

Census station number	Level of Service (LOS)	Traffic flow (Peak hour)	Traffic flow (16H)	Location of census station (km)	Type of carriageway
TR405	F	7443	61598	15.2	K 2-2
TR406	F	2073	20738	15.5	T 1-1
TR103	F	1969	23584	62.0	T 1-1
TR205	E	5233	35009	102.2	K 2-2
TR202	D	3901	28171	80.2	K 2-2
TR302	D	1649	16258	124.2	T 1-1
TR101	D	1492	15204	3.0	T 1-1
TR104	D	1344	15569	70.8	T 1-1
TR404	D	1186	11455	33.1	T 1-1
TR401	C	3006	25586	16.8	K 2-2
TR102	C	1306	15598	4.0	T 1-1
TR204	B	1163	10166	4.1	T 1-1
TR105	B	1135	11940	1.6	T 1-1
TR403	B	1101	1186	26.0	T 1-1
TR502	B	986	9561	146.0	T 1-1
TR402	A	2194	3006	17.0	K 2-2
TR407	A	1798	7443	17.0	K 2-2
TR203	A	1773	14526	5.6	K 2-2
TR503	A	1663	17601	5.0	K 2-2
TR408	A	1100	2073	1.6	T 1-1
TR504	A	937	8092	1.0	K 2-2
TR201	A	628	8068	67.2	T 1-1
TR304	A	467	3878	7.0	T 1-1
TR206	A	401	4758	79.3	T 1-1
TR305	A	367	4105	48.0	T 1-1
TR207	A	301	3423	111.2	T 1-1
TR501	A	298	2884	11.0	T 1-1
TR301	A	153	1860	1.0	T 1-1
TR303	A	105	755	7.0	T 1-1

Figure 2. Ranked data based on LOS and peak hour traffic flow for Terengganu region.

3. Terengganu routes representation

In our case of study, 11 routes in Terengganu region have been selected to contribute to the speed-time data for development of Malaysia Driving Cycle (MDC), which consists of urban, rural, and motorway road based on WLTC [3]. Figure 3 illustrates the locations of all route selection in the Terengganu region. The routes for urban and rural roads are determined based on LOS F, C and A, while the East Coast Expressway (ECE) represents the motorway road at Terengganu. The data samples are collected in both directions for each selected route by passing through the census station. For example, R1D1 represents route 1 with direction 1, has to pass through the census station TR405. For motorway, the routes are denoted by "H" that represents highway. For example, H1D1 represents highway 1 with direction 1. The start and end points are important to be determined before proceeding to route data collection and they are necessary especially for the routes without the census station.

Table 1 summarizes the selected urban roads that represent LOS F at Terengganu, which there are 3 routes with census stations and 2 routes without census stations. Route 1 (R1), covers the location from Gong Badak to Kuala Terengganu, route 2 (R2), covers the location from Kuala Terengganu to Chendering through Surau Panjang, and route 3 (R3), covers the location from Kuala Terengganu to Jerteh. Another 2 routes are selected to cover the urban roads in the inner city of Kuala Terengganu (KT). Meanwhile, route 4 (R4), covers the north inner city and route 5 (R5) covers the south inner city as shown in Figure 4.

Table 1. Selected urban roads.

Route	Census station no.	LOS	Start point	End point	Peak hour	Distance covered (km)
R1D1	TR405	F	Petron Gong Badak	Shell Bukit Tunggal	0700 - 0800	6.8
R1D2	TR405	F	Shell Bukit Tunggal	Petron Gong Badak	1700 - 1800	8.2
R2D1	TR406	F	SMK Seri Payong	Masjid Kg. Kubang Ikan	1700 - 1800	7.4
R2D2	TR406	F	Masjid Kg. Kubang Ikan	SMK Seri Payong	1700 - 1800	7.4
R3D1	TR103	F	Masjid Al-Falah	Shell Kg. Padang Luas	1200 - 1300	6.3
R3D2	TR103	F	Shell Kg. Padang Luas	Masjid Al-Falah	1200 - 1300	7.5
R4D1	KT North Inner City	F	Shell Cabang Tiga	Food Court Batu Burok	1700 - 1830	5.5
R4D2	KT North Inner City	F	Food Court Batu Burok	Shell Cabang Tiga	1700 - 1830	7.2
R5D1	KT South Inner City	F	Plaza Perabot Perdana	Shell Batu Burok	1700 - 1830	5.6
R5D2	KT South Inner City	F	Shell Batu Burok	Plaza Perabot Perdana	1700 - 1830	6.4

Table 2 summarizes the selected rural roads at Terengganu identifying 2 routes that represent LOS C and 2 routes that represent LOS A. For LOS C, route 6 (R6), covers the location from Kuala Terengganu to Jerteh through Tepoh, while route 7 (R7), covers the location from Kuala Terengganu to Jerteh through Mengabang Panjang and Batu Rakit. Meanwhile, for LOS A, route 8 (R8), covers the location from Kuala Terengganu to Kuala Berang and route 9 (R9), covers the location from Kuala Berang to Tanggol.

Table 2. Selected rural roads.

Route	Census station no.	LOS	Start point	End point	Peak hour	Distance covered (km)
R6D1	TR401	C	Caltex Bukit Jong	Shell Kuala Nerus	0700 - 0800	9.3
R6D2	TR401	C	Shell Kuala Nerus	Caltex Bukit Jong	0700 - 0800	7.7
R7D1	TR404	C	Masjid Al-Ehsan	Petronas Mukim Merang	1700 - 1800	8.5
R7D2	TR404	C	Petronas Mukim Merang	Masjid Al-Ehsan	1700 - 1800	8.5
R8D1	TR407	A	Caltex Bukit Payong	Shell Wakaf Tapai 2	0700 - 0800	13.0
R8D2	TR407	A	Shell Wakaf Tapai 2	Caltex Bukit Payong	1700 - 1800	14.0
R9D1	TR501	A	Masjid Kg. Pengkalan Berangan	SK Tanggol	0700 - 0800	12.5
R9D2	TR501	A	SK Tanggol	Masjid Kg. Pengkalan Berangan	1700 - 1800	12.5

Table 3 summarizes the selected motorway roads at Terengganu that cover the East Coast Expressway (ECE), which consists of ECE1 (LPT1) and ECE2 (LPT2). Highway 1 (H1), covers the location from Karak to Jabor (LPT1), while highway 2 (H2), covers the location from Jabor to Kuala Nerus (LPT2).

Table 3. Selected motorway roads.

Route	Census station no.	LOS	Start point	End point	Distance covered (km)
H1D1	-	-	Plaza Tol Jabor Outbound	Plaza Tol Gambang Outbound	36.1
H1D2	-	-	Plaza Tol Gambang Outbound	Plaza Tol Jabor Outbound	36.7
H2D1	-	-	Plaza Tol Kuala Terengganu	Plaza Tol Ajil	31.7
H2D2	-	-	Plaza Tol Ajil	Plaza Tol Kuala Terengganu	32.7

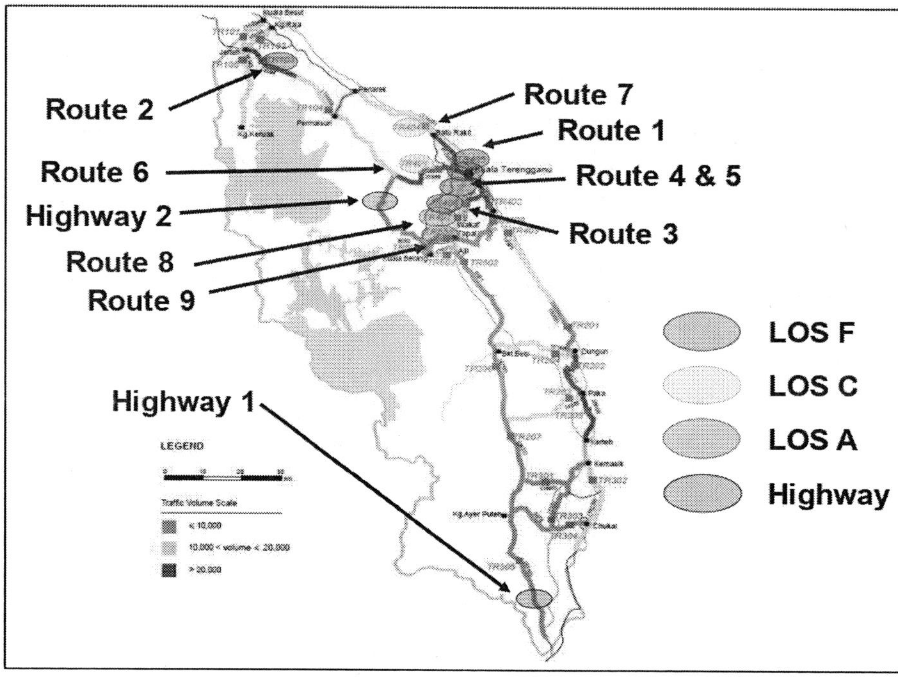

Figure 3. Locations of route selection at Terengganu.

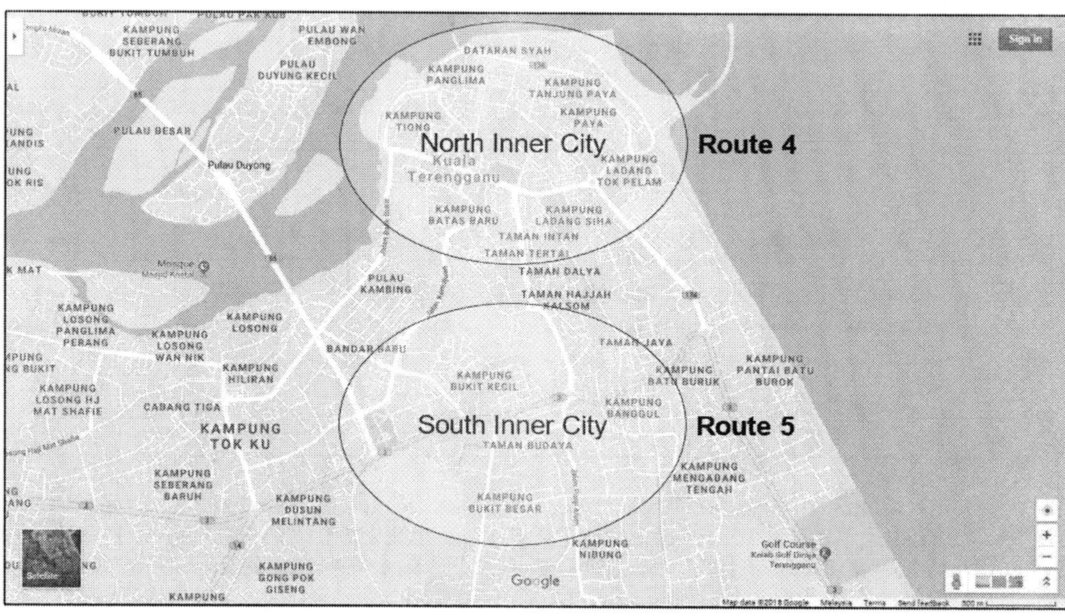

Figure 4. Location of north (Route 4) and south (Route 5) inner city in Kuala Terengganu.

4. Self-assessment of the road geography and traffic conditions

Figure 5 demonstrates a Google Maps assessment and verification of traffic conditions for the selected routes in Terengganu region. Google Maps is not only capable to view the live traffic conditions, but also the typical traffic conditions in an area for a specific day and time. Verification of traffic conditions by Google Maps is necessary rather than relying solely on RTVM data since the traffic conditions can be displayed more effectively.

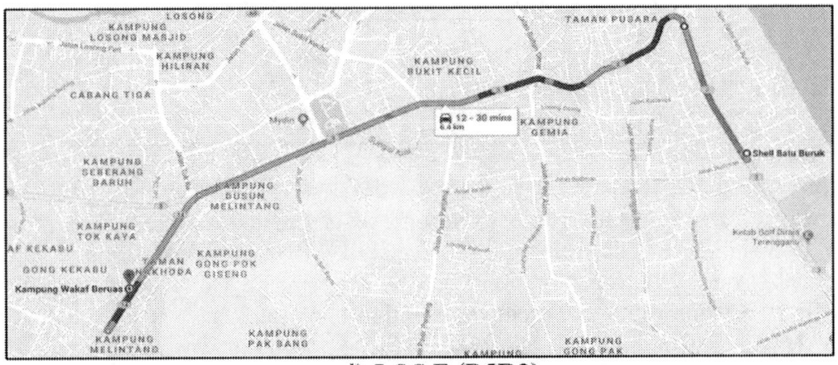

i) LOS F (R5D2)

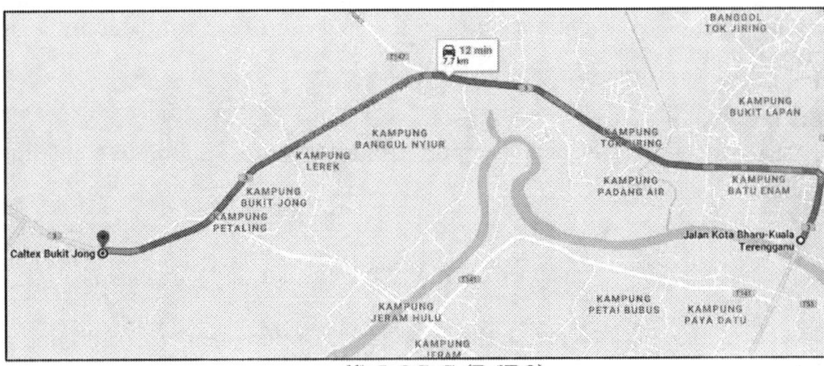

ii) LOS C (R6D2)

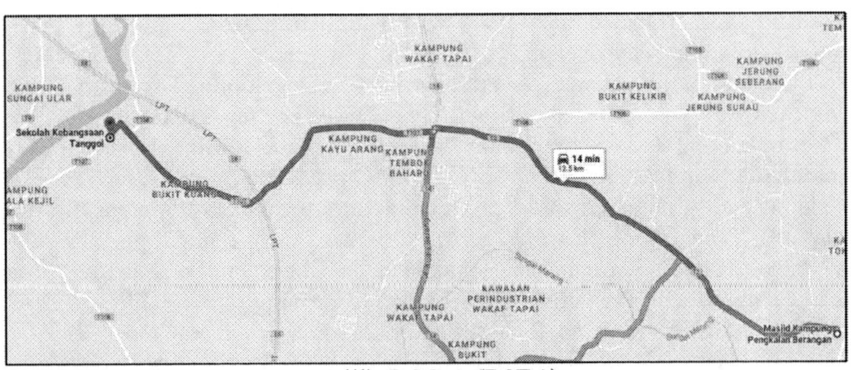

iii) LOS A (R9D1)

Figure 5. Examples of traffic condition assessment by Google Maps.

5. Driving conditions for route data collection

Apart from selecting routes based on the road types (urban, rural and motorway), the driving conditions (peak, off-peak and weekend) were also taken into account in the development of WLTC [3]. The peak hours for each route have been stated in Table 1, Table 2 and Table 3. There is no peak hour for motorway roads. Therefore, for Terengganu region, the data samples for peak and off-peak conditions are collected on Monday, Tuesday and Wednesday (weekday). In the meantime, the weekend data samples are collected on Saturday. Table 4 summarizes the number of samples to be collected for each route and direction based on the driving conditions.

Table 4. Driving conditions for route data collection

Driving conditions	Samples for each route and direction
Peak	3
Off-peak	2
Weekend	2
Highway	2

6. Conclusion

This paper mainly focuses on the route selection at Terengganu as one of the regions in Malaysia contributing to speed-time data collection for overall MDC data collection campaign. The route selection are based on RTVM data version 2015 by referring to the level of service (LOS) and peak hour traffic flow for each route in Terengganu region. 11 routes have been selected to represent the Terengganu region, which consists of urban, rural, and motorway road based on WLTC. It is recommended to verify the traffic conditions by Google Maps and to perform self-assessment to ensure the selected routes are free from major road construction and any route diversion. The speed-time data collection can be executed once all the selected routes are validated to be suitable for route data collection.

Acknowledgments

The authors acknowledge the financial support from Malaysia Automotive Institute (MAI) under research grant of Malaysia Driving Cycle.

References

[1] F. J. Kelly and T. Zhu, "Transport solutions for cleaner air," *Science*, vol. 352, no. 6288. pp. 934–936, 2016.

[2] T. Barlow, S. Latham, I. Mccrae, and P. Boulter, "A reference book of driving cycles for use in the measurement of road vehicle emissions," *TRL Publ. Proj. Rep.*, vol. 12, no. 1, p. 280, Jan. 2009.

[3] M. Tutuianu *et al.*, "Development of a Worldwide harmonized Light duty driving Test Cycle," *Tech. Rep.*, vol. 3, no. January, pp. 7–10, 2014.

[4] H. Y. Tong and W. T. Hung, "A Framework for Developing Driving Cycles with On-Road Driving Data," *Transp. Rev.*, vol. 30, no. 5, pp. 589–615, Sep. 2010.

[5] K. S. Nesamani and K. P. Subramanian, "Development of a driving cycle for intra-city buses in Chennai, India," *Atmos. Environ.*, vol. 45, no. 31, pp. 5469–5476, Oct. 2011.

[6] D. S. Han *et al.*, "Characterization of driving patterns and development of a driving cycle in a military area," *Transp. Res. Part D Transp. Environ.*, vol. 17, no. 7, pp. 519–524, Oct. 2012.

[7] M. A. Samad Kamal, J. Imura, T. Hayakawa, A. Ohata, and K. Aihara, "Smart Driving of a Vehicle Using Model Predictive Control for Improving Traffic Flow," *IEEE Trans. Intell. Transp. Syst.*, vol. 15, no. 2, pp. 878–888, Apr. 2014.

[8] G. O. Duarte, G. A. Gonçalves, and T. L. Farias, "Analysis of fuel consumption and pollutant emissions of regulated and alternative driving cycles based on real-world measurements," *Transp. Res. Part D Transp. Environ.*, vol. 44, pp. 43–54, May 2016.

[9] S. Tamsanya, S. Chungpaibulpatana, and B. Limmeechokchai, "Development of a driving cycle for the measurement of fuel consumption and exhaust emissions of automobiles in Bangkok during peak periods," *Int. J. Automot. Technol.*, vol. 10, no. 2, pp. 251–264, Apr. 2009.

[10] H.-L. Chiang, P.-H. Huang, Y.-M. Lai, and T.-Y. Lee, "Comparison of the regulated air pollutant emission characteristics of real-world driving cycle and ECE cycle for motorcycles," *Atmos. Environ.*, vol. 87, pp. 1–9, Apr. 2014.

[11] MOW, "Road Traffic Volume Malaysia 2015," 2016.

[12] A. R. Mahayadin *et al.*, "Efficient methodology of route selection for driving cycle development," *J. Phys. Conf. Ser.*, vol. 908, no. 1, 2017.

MODENAS CT115s intake port performance enhancement through analyzing volumetric and geometrical design with valve lift adjustment by using CFD Simulation

Efi N E[1], Shahriman A B[1,*], Rojan M A[1], I Zunaidi[1], Z M Razlan[1], W K Wan[1], M S M Hashim[1], N S Kamarrudin[1], A. Harun[2], I Ibrahim[1] and Azizul Aziz I[3]

[1] School of Mechatronic Engineering, Universiti Malaysia Perlis, Pauh Putra Campus, 02600 Arau, Perlis, Malaysia.
[2]School of Microelectronic Engineering, Universiti Malaysia Perlis, Pauh Putra Campus, 02600 Arau, Perlis, Malaysia.
[3] Modenas Research and Development Division, Kawasan Perusahaan, 08300, Gurun Kedah.

[*]Corresponding author: shahriman@unimap.edu.my

Abstract This paper reviews improvement of intake port for Modenas CT115s by using CATIA software for the design and ANSYS software for the simulation result. Intake port is referring to the passage in the cylinder head which connect the intake manifold to the intake valve through which the fuel-air mixture proceeds on its way to the cylinders. The results obtained has been analysed and interpreted in formed of plotted graph. From the analysis, increasing the inlet of the intake port size can increasing the volume flow rate. The main impact is high angle of intake port that can give high torque and power. The decision to choose the proper angle for the each design is important because it will affect the major result for the simulation. Thus, it can be concluded that the increase of the inlet size and using the best angle of the intake port can increase the engine performance which allow higher fuel-air mixture entrance during the combustion process.

Keywords: intake port, geometrical design, air flow pattern, flow characteristic, engine performance

1. Introduction

The basic elements of an internal combustion engine requires three main thing which is air, fuel and a source of ignition [1]. Development of internal combustion engine involves resolving several problem issues. Focusing on MODENAS CT115s intake port which powered using 115cc four-stroke engine is the main project purpose. MODENAS or the specific name is National Motorcycle and Engine Company is a national Malaysian motorcycle company. MODENAS research and development (R&D) team now have put a great effort to improve the engine performance of MODENAS CT115s in order to be competitive among to its categories. Design of the cylinder head for automotive spark ignition (SI) engine is important because it affects the performance, fuel economy and emissions [2]. Accordingly, the manufacturer company has focus on the plan to improve the geometry of intake port to produce high scale fluid motion, which is can generate better tumble in the engine cylinder [3]. Intake port flow is measured by the increasing of horsepower produced and coefficient of discharge (CD).

IOP Conf. Series: Materials Science and Engineering **429** (2018) 012051 doi:10.1088/1757-899X/429/1/012051

Several of intake port geometry patterns will affected the amount of air flow that entered the port [4]. In this study focuses on the angle between the entrance plane and the outlet plane (AC), angle between the upper wall and the outlet plane (AV) and the diameter of inlet as shown in figure 2.1[8]. This investigation studied were strictly to know the effect of intake port cylinder head modification to improve the engine performance. The tumble motion that had been naturally produced by the geometry of intake port contributed proper and quick air-fuel mixing that produced higher efficiency does increasing the engine performance [5,6]. The simulation result between the existing and improvement intake port testing will be compared with three different condition of valve lift 0.5mm(low), 3mm(medium) and 6mm(high). These condition can directly the different of flow entrance[7].Therefore, it is critical to determine intake port to achieve flow coefficient target.

2. Methodology

2.1. A. Design using CATIA software

The intake port design in this project has been divided into two categories which is the existing design and the improvement design. CATIA is multi-platform software. By using CATIA, drafting the dimension intake port design can be made to determine the exactly parameter of the existing and improvement design [6]. Table 1 shows the parameters for existing and improvement design.

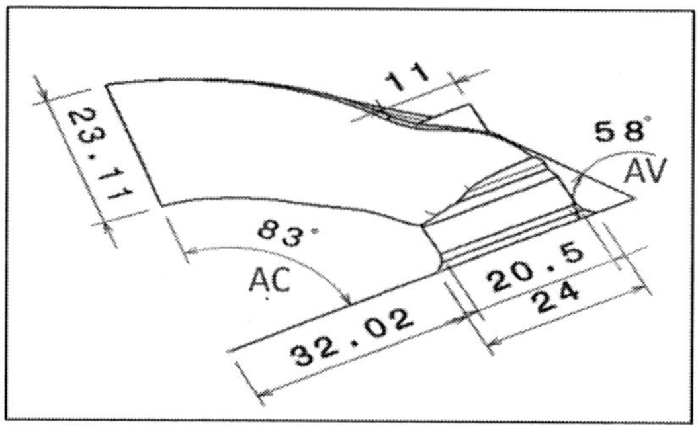

Figure 2.1: Parameter list for existing design

Table 1: Improvement intake port parameter

DESIGN	ANGLE AC (°)	ANGLE AV (°)	INLET DIAMETER (MM)
EXISTING	83	58	23.1
A	37.5	60	25
B	42.5	60	25
C	47.5	60	25
D	52.5	60	25
E	57.5	60	25
F	62.5	60	25

2.2. Simulation using ANSYS software.

In order to study the intake port of MODENAS CT115s the simulation analysis was modules using computational fluid dynamic. ANSYS is an analysis tools that are used to improve the intake passage design by simulation air flow in the design. ANSYS simulation provides the pressure and the velocity of the air flow throughout the solution domain with the boundary condition and complex geometries.

IOP Conf. Series: Materials Science and Engineering **429** (2018) 012051 doi:10.1088/1757-899X/429/1/012051

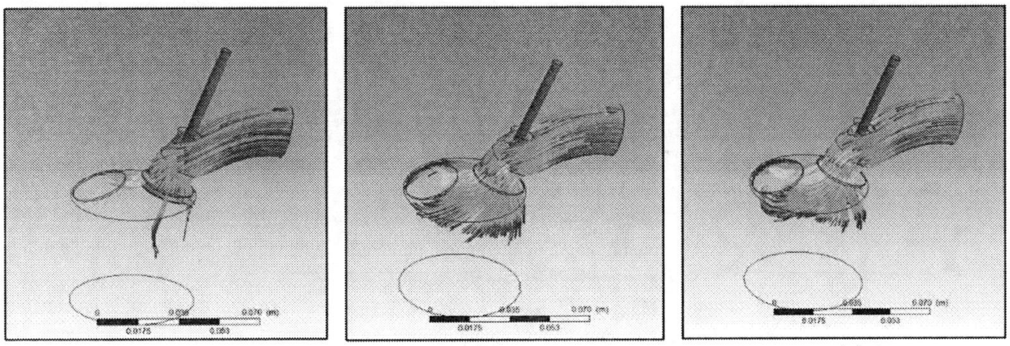

(a) (b) (c)

Figure 2.2: Existing design streamline flow pattern, (a) 0.5mm, (b) 3mm, (c) 6mm.

3. Result and Discussion

The first condition is low valve lift which is 0.5mm, second is 3mm which is medium valve lift and the last is high valve lift which is 6mm. However, for the existing intake port design the mass flow rate result show that the peak flow rate is in 3mm valve lift compared to the 6mm valve lift. This is because the flow rate that enters the intake port is already in the maximum flow.

Increasing the inlet size and the angle of intake port make a huge different in simulation result. For the mass flow rate result at high valve lift (6mm), improvement design D show that the mass flow is 6.87 ×10-5 kg/s which is the highest compared to the others design. Different result shows at the original intake port result which is 0.16 ×10-5 kg/s much lower than improvement design D. However in low valve lift (3mm), mass flow rate 4.65×10-5kg/s is increasing about 60% from the existing intake port which is 3.17×10-5 kg/s. Moreover, for the existing intake port volume flow rate the result also show that the maximum flow rate is in 3mm valve lift compared to the 6mm valve lift.

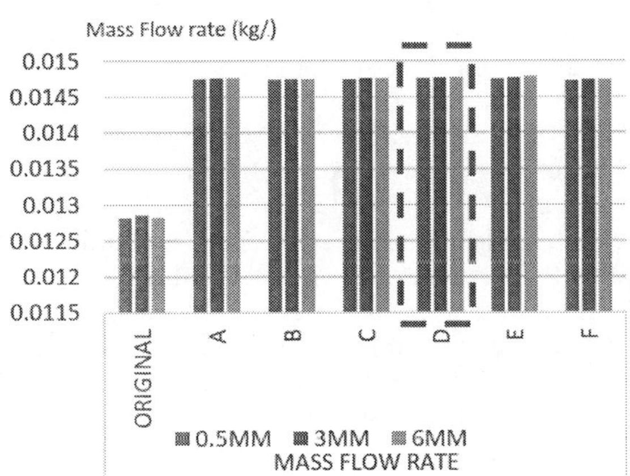

Figure 3.1: Comparison for mass flow rate

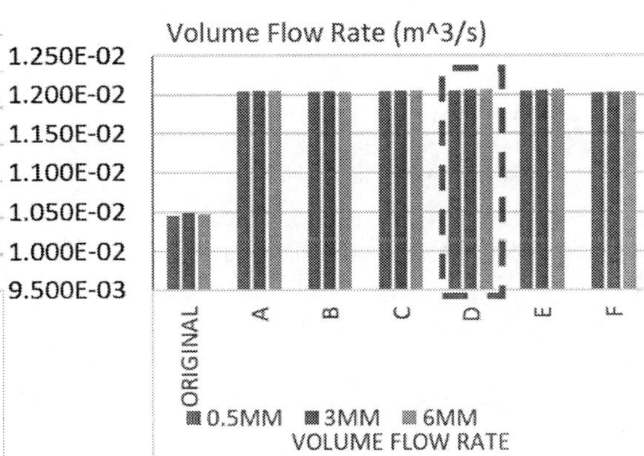

Figure 3.2: Comparison for volume flow rate

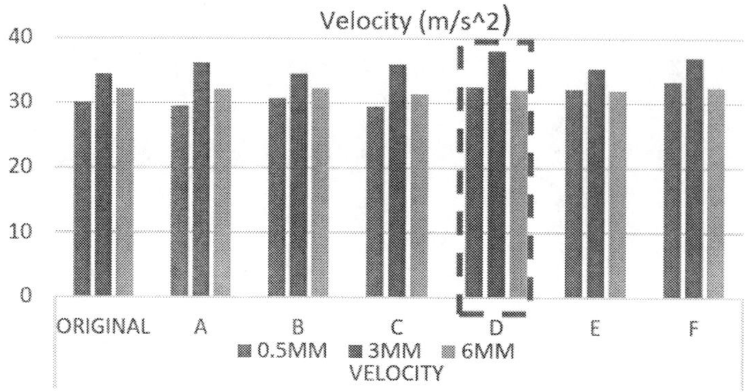

Figure 3.3: Comparison for velocity

To validate this study of MODENAS CT115s intake port by optimizing engine performance through analyzing volumetric and geometrical design at intake port system, the performance of the original intake port was measured on the flow bench test and simulation test by using ANSYS software. From comparison between the experimental data and the simulated results for the existing intake port, it can be found that the different of the flow coefficient is less than 3.5%.The result obtain from the CFD analysis shows the differences of mass flow rate and tumble ratio between the existing design and the improvement design. The variation of flow for the original design create an improvement value of the mass flow rate as the valve lift increased from lower to medium but decrease from medium to high valve lift. However for the improvement design A, from the result, the optimal value of mass flow rate at the high valve lift was achieved compared to the original intake port.

The valve lift from low to medium and medium to high not only affect the movement of flow in the cylinder but also lead the change of the cross section. The variation of flow cross section, which is cause by the reciprocating motion of valve may result separation of flow field and the backflow of the outlet of the intake port further decrease the flow capacity. The improvement design A is the best design for producing greater percentage of mass flow rate and volume flow rate in the intake port compare to the actual design. Since the result for the improvement design is exceed 10% as required from MODENAS Research and Development Team, means the target and the main objective for this research is achieved.

4. Conclusion

Based on the research and analysis that had been done for intake port of MODENAS CT115s, the experimental and simulation results show that design approach can accurately build the intake port and achieve high performance. Result that been compare for the improvement design and the existing design is mass flow rate, volume flow rate, pressure and velocity. Parameters Av is to determine the direction of the jet flow in the outlet. Parameter Ac controls the direction of the entrance flow and the overall direction of the flow in the intake port. The increase of parameter Ac enhances the intensity of the large-scale vortex in the cylinder. Meanwhile, it causes the formation of the large-scale vortex in advance. In conclusion by increasing the inlet of the intake port will increase the volume flow rate at the beginning of the flow and increasing the velocity result. Other main impact is high angle of intake port that can give high torque and power. The decision to choose the proper angle for the each design is important because it will affect the major result for the simulation and it give some advantages and disadvantages. This is because if the research does exceed the proper angle of the intake port it will give bad result due to the limit that near to the ideal curve condition and hard to install to the motorcycle engine thus it will make the engine performance as low than the ideal performance.

Acknowledgements

I would like to express my deepest appreciation to all those who provided me the possibility to complete this paper. A special gratitude I give to my supervisor and MODENAS Research and Development team whose contribution in stimulating suggestions and encouragement, helped me to coordinate my project students especially in writing this paper.

References

[1] C. G. et al. LUO Maji, 2003 "Three-dimensional transient numerical simulation for intake process in the engine intake port-valve-cylinder system Crank angle (degree) Intake valve moving schedules," Three-dimensional transient Numer. Simul. intake Process engine intake port-valve-cylinder Syst., vol. 1, no. 309–316,.

[2] P. Mandloi and G. Verma, "Design Optimization of an in-Cylinder Engine Intake Port," Nafems World Congr. 2009, 2009.

[3] S. K. Sabale and S. B. Sanap, "Design and Analysis of Intake Port of Diesel Engine for Target Value of Swirl," Am. J. Mech. Eng., vol. 1, no. 5, pp. 138–142, 2013.

[4] A. A. A. M. F. Samuri, M. F. Muhamad Said, , N. A. Mohamad Shafie, Z. Abdul Latiff, "Intake Port Flow Study on Cylinder Head Using Flowbench," J. Transp. Syst. Eng., vol. 2, pp. 29–35, 2015.

[5] Y. L. Qi, L. C. Dong, H. Liu, P. V. Puzinauskas, and K. C. Midkiff, "Optimization of intake port design for SI engine," Int. J. Automot. Technol., vol. 13, no. 6, pp. 861–872, 2012.

[6] S. Falfari, F. Brusiani, and G. M. Bianchi, "Numerical analysis of in-cylinder tumble flow structures - Parametric 0D model development," Energy Procedia, vol. 45, pp. 987–996, 2014.

[7] D. Wentworth-linton and S. Gao, "A Computational Study of the Effect of Intake Design on Volumetric Efficiency for Best Performance in Motorsport," vol. 11, no. 5, pp. 1103–1111, 2017.

[8] T. Wang, "Full-Parameter Approach for the Intake Port Design of a Four-Valve Direct-Injection Gasoline Engine Full-Parameter Approach for the Intake Port Design of a Four-Valve Direct-Injection Gasoline Engine," no. September, pp. 0–11, 2015.

A review on effect of intake system flow by various designs to engine performance

Efi N E[1], Shahriman A B[1], Rojan M A[1], I Zunaidi[1], Z M Razlan[1], W K Wan[1], M S M Hashim[1], N S Kamarrudin[1] , A Harun[2], I Ibrahim[1] and Azizul Aziz I[3]

[1] School of Mechatronic Engineering, Universiti Malaysia Perlis, Pauh Putra Campus, 02600 Arau, Perlis, Malaysia.

[2]School of Microelectronic Engineering, Universiti Malaysia Perlis, Pauh Putra Campus, 02600 Arau, Perlis, Malaysia.

[3] Modenas Research and Development Division, Kawasan Perusahaan, 08300, Gurun, Kedah

Corresponding author: shahriman@unimap.edu.my

Abstract This paper presents a critical review of effect of intake manifold flow to four stroke single cylinder engine performances. Intake manifold and its components play critical role in deciding the performance of engine. Engine can be easily breathing related to flow pattern by best design of intake manifold that can contribute to performance. Flow pattern of intake manifold affect the volumetric efficiency, mass flow rate and velocity which ultimately makes impact up on the engine power & torque. This paper reviews work done by various prominent researchers in the field of intake tuning and variable intake manifold. It can be easily concluded that for getting best flow pattern of intake manifold throughout the engine operating speed the intake geometry need to be changed.

Keywords: intake port, geometrical design, flow pattern, flow characteristics, power & torque.

1. Introduction

Intake manifold is the part of an engine and piping system which deliver incoming air to the cylinders, usually made of cast metal, plastic or composite material. In most SI engines, fuel is added to the air in the intake manifold system either by fuel injectors or with a carburettor. Some intake manifolds are heated to enhance fuel evaporation. The individual pipe to a single cylinder is called runner. Other parts in intake system are air filter, intake pipe and plenum.
Various injection timing, fuel delivery pressure and flow rate are the several ways in enhance the engine performance. Improving the air flow rate through intake port is one of the ways which can enhance the turbulence intensity at spark time and it will increase the stability of combustion and reducing emission. The technologies for high power include optimized combustion system. An intake flow which is generation of flow structures are related closely to the design process and performance of the 4-stroke engines. The tumble motion is necessary to increase the turbulence level which contributed proper and quick mixing of fresh charge which leads to effective combustion with reduced emission. The intake port design plays a critical role in the development of modern internal

combustion engines (ICE). Geometrical properties are importance to improve the gasoline engine performance in terms of efficiency.

2. iterature Re iew

M.F. Samuri et.al done experiment on 16-Valve DOHC engine which more focuses on air flow at intake port. They record and analyzed the existing data by using flow bench machine. For improvement design the polish the intake port as maximum possible and test again by using flow bench machine. Result for both design are comparing. The increase in flow from improvement design is due to the occurrence of the flow in the intake port. The larger section area of port together with small cross section near intake valve causes the velocity increase and quality of flow.

M. Eng. M Gandyk et.al in their research of combustion engine intake port design on 2015 prove that phenomena effect of geometry changing in theory and in simulated engine are similar. Shortening and increasing are of ducts makes high torque on high engine speed. Final improvement gives higher power, up to 1.5hp and about 1Nm higher torque in wide range of engine speed. The maximum power could be higher but the engine torque would be worst.

Ujwal D. Patil studied on cylinder head intake port design in his paper more focus on developing intake port geometry to meet swirl ratio required to meet emission.

Research Methodology Steps

- Provides the literature to know about flow patterns, designing parameters, types of ports to be used and methodology for design of Intake ports.
- Development efforts are focused while designing intake ports with improvement to obtain the required swirl and flow patterns
- Defining the full valve lift 3D model and in cylinder air flow analysis through CFD package to know swirl, velocity vectors streamlines formation for full lift model.
- Validation of results by prototyping testing.

Dominic Wentworth Linton and Shian Gao in their research work were studied on the effect of intake design on volumetric efficiency which focus to selecting the best stack length for the given discipline of motorsport. Pressure and velocity contours around the valve guide show the effect of porting on smoothing the airflow into cylinder. Opening the throat of the valve port helps to smooth the airflow into the cylinder and improve the power delivery. From the result prove shorter stack length is more suited for high rpm applications while longer stack shift the bulk of the power lower down the rpm range.

The key findings from the study were:

- The valve lift dimension 3mm for low, 7mm for medium and 10mm for high valve lift.
- The stack length use 40mm, 90mm and 120mm.
- The rpm tested are range between 2500-8000rpm.

Y. bifeng Jia Hekun Liu Shengji research on “Notice of Retraction intake port of the non-road small spark-ignition engine,” state that maintaining the flow of cross-sectional area at the middle of the intake port can prevent the formation of dead air and handle the changes of structural in the corner. The tumble and swirl motion that had been naturally produced by the geometry of intake port contributed proper and quick air-fuel mixing that produced higher efficiency does increasing the engine performance.In this research, the coefficient discharge value will indicates the flow efficiency in the intake port area. When the engine is in operation, the intake valve will lift and the discharge coefficient of inlet poppet valve will occur.

The geometry of intake port for a gasoline engine has an important effect on tumble and swirl motion in cylinder. The tumble motion can generate the air and fuel to form a well mixture while Swirl can be generated by geometry of the intake port that provide a tangent components to the intake flow as it enters the cylinder. J. Rabault study on “International Journal of Heat and Fluid Flow A study using PIV of the intake flow in a diesel engine cylinder,” found that function of tumble motion is to increase the turbulence level which give good mixing of fresh charge which leads to effective combustion does increase the engine performance

M. F. Hushim prove that the cross-section of the flow is greater at the entry point of the intake port and the smaller at the exit of the intake port. This will cause of increasing of the velocity does improving the engine performance. B. Y. Hong Han-Chi study on "Optimization of Intake and Exhaust System for FSAE Car Based on Orthogonal Array Testing," found one of the most important subsystems of the engine is the intake port. Intake port is the entrance point for air flow into the combustion chamber. In order to increase the power output of the engine, the quantity of the air flowing into the engine, intake port must to be maximized as possible.

3. Principle of intake manifold and intake port

Air Flow Motions

The air-flow motions in the intake passage are in turbulent motions. These flows help to improve the rate of transfer and mixing the fuel and air in the combustion chamber. B. M. Krishna and J. M. Mallikarjuna in their research, the rates of momentum and mass transfer also can be increased when there is the presence turbulent flow. From C.W. Hong and S.D Tarng, the term of turbulent flow will be referred as swirl when the flow enters and rotates in the combustion chamber. Organized rotations that are produced by air-flow in the intake passage are the definition of the swirl motion. From K. Lee et.al swirl also can be referred as the directional flow of air and fuel mixture that are affected by the shape of the intake port or its angle of entry into the cylinder block.

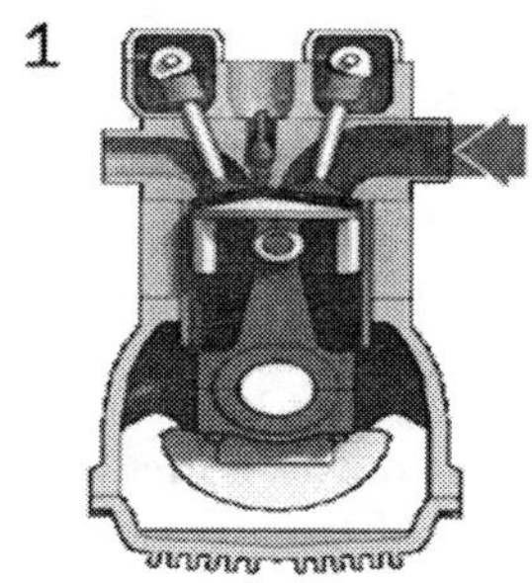

Figure 3.1: Intake air flow motion

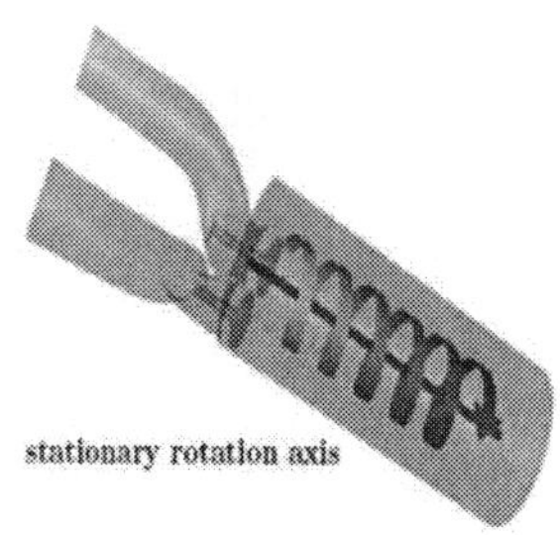

Figure 3.2: Intake Air for stationary rotation axis

The flow of swirl in the combustion chamber. The mixing of air and fuel will be more homogeneous where there is the presence of swirl motion. The swirl motion enters the combustion chamber until the ignition and the combustion process started. The swirl flow is very important where it can help to increase the air-fuel mixing and improve the combustion process.

Besides, swirl also helps to minimize engine knocking, and performed the complete combustion. An optimal swirl ratio is not only good for optimum combustion, but also for an optimal emission reduction. In racing tournament, the swirl value in the cylinder block is very important in order to optimize the engine performances by A. Shamsudeen et.al. The effective combustion occurs when there is high swirl near the TDC when the ignition occurs.

Intake Port Geometry

"Full-Parameter Approach for the Intake Port Design of a Four-Valve Direct-Injection Gasoline" by T. Wang study focus on tangential port template for a four valve gasoline engine with symmetrical arrangement. Inlet intake port must larger size to the outlet intake port.In this research also clearly classified in figure 3.3 and named each angle and area for intake port geometry in table 1.

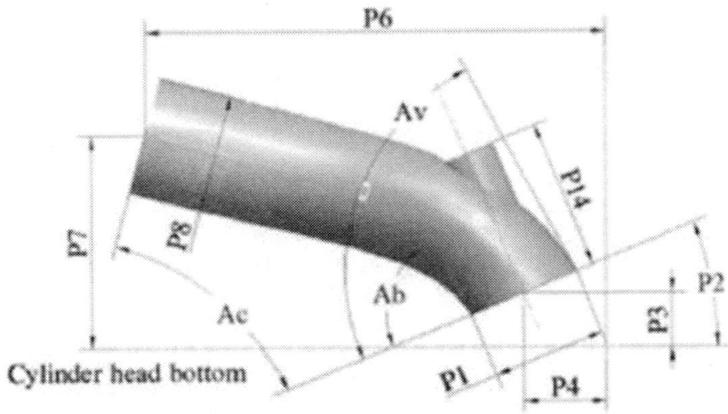

Figure 3.3: Intake Port Geometry

Parameter	escription	nit
P1	Outlet diameter of intake port	mm
P2	Angle between the outlet plane and the cylinder head bottom	Deg
P3	Distance between the outlet center and the cylinder head bottom	mm
P4	Distance between the outlet center and the axis of the cylinder	mm
P5	Distance between the outlet center and the symmetry plane of the intake port	mm
P6	Distance between the entrance center and the axis of the cylinder	mm
P7	Distance between the entrance center and the cylinder of the bottom	mm
P8	Eight of the entrance	mm
P9	Width of the entrance	mm
P10	Fillet of the entrance (not shown)	mm
P11	Bifurcation of the intake port	mm
P12	Pattern draft of the part model along the parting line (not shown)	mm
P13	Diameter of the boss	mm
P14	Distance between the outlet plane and the boss plane	mm
Av	Angle between the upper wall and the outlet plane	Deg
Ab	Angle between the lower wall and the outlet plane	Deg
Ar	Ratio between the entrance area of one leg and the outlet area	
Ac	Angle between the entrance plane and the outlet plane	Deg

Table 1: Parameter List

. Conclusion

After reviewing work of different authors as above it can be easily concluded that intake geometry make considerable impact on engine performance. The main impact is on the volumetric efficiency of the engine which ultimately affects the torque & power produced at different speed. The intake port size is bigger give high torque and power to engine. Other main impact is high angle of intake port that can give high torque and power. For future study, is toward studying for geometry and angle of intake port by theoretical and experimental.

. Acknowledgment

The author acknowledge the institute authorities for supporting the present work to carry out in the institute. Authors also acknowledge the various researchers whose work has been reviewed and reported in this paper.

6. References

1. A. M. F. Samuri, M. F. Muhamad Said, , N. A. Mohamad Shafie, Z. Abdul Latiff, "Intake Port Flow Study on Cylinder Head Using Flowbench," J. Transp. Syst. Eng., vol. 2, pp. 29–35, 2015.
2. M. E. M. Gandyk, M. Tkaczyk, P. D. Dsc, P. L. J. Sitnik, and P. T. Uzuntonev, "COMBUSTION ENGINE INTAKE PORT DESIGN ANALYSIS," vol. 1, no. 4, pp. 4–6, 2015.
3. U. D. Patil, "Cylinder Head Intake Port Design & In-Cylinder Air-flow Patterns, Streamlines formations, Swirl Generation Analysis to Evaluate Performance & Emissions," Int. J. Eng. Res. Technol., vol. 2, no. 9, pp. 2789–2793, 2013.
4. D. Wentworth-linton and S. Gao, "A Computational Study of the Effect of Intake Design on Volumetric Efficiency for Best Performance in Motorsport," vol. 11, no. 5, pp. 1103–1111, 2017.
5. Y. bifeng Jia Hekun Liu Shengji,Wang Jian, "Notice of Retraction intake port of the non-road small spark-ignition engine," Anal. flow F. Param. Optim. intake port non-road small spark-ignition engine, vol. 1, pp. 328–330, 2010.
6. J. Rabault, J. A. Vernet, B. Lindgren, and P. H. Alfredsson, "International Journal of Heat and Fluid Flow A study using PIV of the intake flow in a diesel engine cylinder," Int. J. Heat Fluid Flow, vol. 62, pp. 56–67, 2016.
7. M. F. Hushim, A. J. Alimin, M. A. Razali, A. N. Mohammed, A. Sapit, and J. C. M. Carvajal, "Air flow behaviour on different intake manifold angles for small 4-stroke PFI retrofit KIT system," ARPN J. Eng. Appl. Sci., vol. 11, no. 12, pp. 7565–7571, 2016.
8. B. Y. Hong Han-Chi, Huang Hong-wu, "Optimization of Intake and Exhaust System for FSAE Car Based on Orthogonal Array Testing," vol. 2, no. 3, pp. 392–396, 2012.
9. K. Lee, C. Bae, and K. Kang, "The effects of tumble and swirl flows on flame propagation in a four-valve S.I. engine," Appl. Therm. Eng., vol. 27, no. 11–12, pp. 2122–2130, 2007.
10. B. M. Krishna and J. M. Mallikarjuna, "Tumble Flow Analysis in an Unfired Engine Using Particle Image Velocimetry," vol. 3, no. 6, pp. 430–435, 2009.
11. C.-W. Hong and S.-D. Tarng, "In-Cylinder Tumble Flow Field Measurements and Predictions," J. Eng. Gas Turbines Power, vol. 123, no. 1, p. 139, 2001.
12. A. Shamsudeen, S. Abdullah, and A. K. Ariffin, "Finite Element Simulation of a Steady-State Stress Distribution in a Four Stroke Compressed Natural Gas-Direct Injection Engine Cylinder Head."
13. T. Wang, "Full-Parameter Approach for the Intake Port Design of a Four-Valve Direct-Injection Gasoline Engine Full-Parameter Approach for the Intake Port Design of a Four-Valve Direct-Injection Gasoline Engine," no. September, pp. 0–11, 2015.

IOP Conf. Series: Materials Science and Engineering **429** (2018) 012053 doi:10.1088/1757-899X/429/1/012053

Modenas CT115s performance enhancement by redesign internal geometry to improve the flow pattern of air box

Efi N E[1], Shahriman A B[1*], Rojan M A[1], I Zunaidi[1] , Z M Razlan[1], W K Wan[1], M S M Hashim[1], N S Kamarrudin[1], A Harun[2], I Ibrahim[1] and Azizul Aziz I[3]

[1] School of Mechatronic Engineering, Universiti Malaysia Perlis, Pauh Putra Campus, 02600 Arau, Perlis, Malaysia.
[2]School of Microelectronic Engineering, Universiti Malaysia Perlis, Pauh Putra Campus, 02600 Arau, Perlis, Malaysia
[3] Modenas Research and Development Division, Kawasan Perusahaan, 08300, Gurun, Kedah.

*Corresponding author: shahriman@unimap.edu.my

Abstract. This paper reviews a study of engine performance MODENAS CT15s by redesign internal geometry to improve flow pattern of air box. Air box is an empty chamber on the inlet of most combustion engines combustion process. It collects air from outside and feeds it to the intake hoses of each cylinder. The entire air box collects the fresh air and later gets it filtered through the air filter present inside it. Engine can be easily breathing related to flow pattern by best design of air box that can contribute to performance. Flow pattern of air box affect the volumetric efficiency, mass flow rate and velocity which ultimately makes impact up on the engine power & torque. 3D Computational Fluid Dynamics (CFD) analysis was carried out for an existing model and several improvement models to make comparison of air flow behaviour through the air box geometry. Based on the results, geometrical changes like sigle blade placement in inlet plenum showed good improvement in flow behaviour compare to existing model. By using 3D CFD analysis, the best design of the air box for a motorcycle engine is achieved with considerable reduction in development time and cost.

Keywords: Air box, internal geometry, air flow pattern, CFD simulation, performance enhancement.

1. Introduction

The iteration of engine performance is the major concern in the automotive industry [1]. The aim for this project is to study engine performance MODENAS CT115s by redesign internal geometry to improve flow pattern of air box. Here, focuses on the CT115s air box model as the benchmark. However, this latest model still needs some improvement so that it can produce superior performance. The improvement that can be done is by redesign internal geometry of air box system in order to improve engine performance. The design of the internal air box geometry including the place of the air filter will effect on mass flow rate, volume flow rate and pressure in inlet and outlet of air box, thus will influences the engine performance[2]. Besides, this project has allowed to study about different air box internal geometry and to understand what gives the best engine performance with all parameter that will be deliberated in relations of graphical design methods before select the most appropriate design for further study

Air box is an empty air chamber that collects air from the outside and feed it to the intake hoses of each cylinder for the combustion engine [3]. Generally, the old model engine takes air directly from surrounding into each carburettor different with now days engines takes air into an air box that linked by horses to every carburettor, thus avoiding an extra intake [4]. The research possible solution for optimization of air box especially on the internal geometrical design that gives a big impact in combustion process in addition to determine the performance of the engine can be. The analysis of the air box will be done by using computerized engineering software such as CATIA and CFD ANSYS. This analysis is designed so that optimization of the air box can be done, thus the best design of air box can be determined.

2. Methodology

2.1. Air Box Model in CAD Software

The software that has been used to model this air flow is CATIA V5. CATIA V5 is the drawing tools software that enables us to draw the exact dimension of MODENAS CT115s air box model. The air box modelling in this project can divided into two types which are the actual design and the proposed designs as shown in figure below which are Figure 1 until 3 shows overall view of CAD model of proposed design with blade.

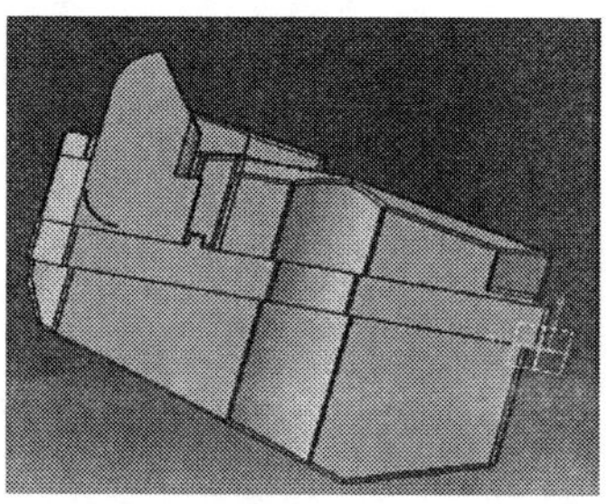

Figure 1: Design 1 with single blade

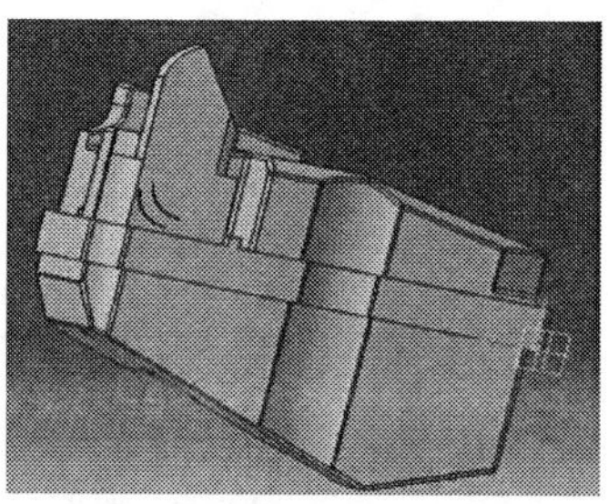

Figure 2: Design 2 with double blade

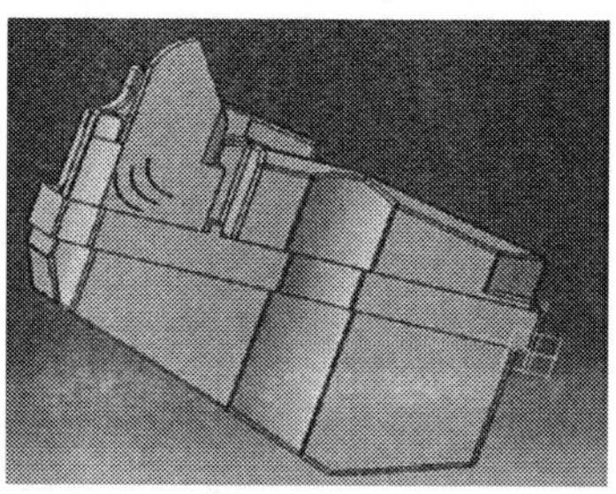

Figure 3: Design 3 with three blade

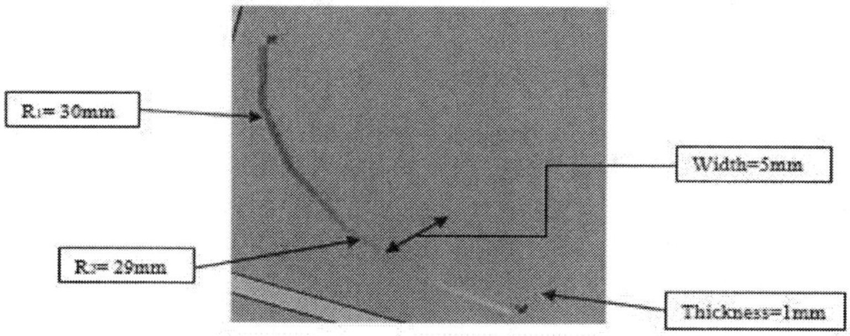

Figure 4: Dimension of blade

For proposed designs, blade are created in the inlet duct of the air box. It is because there have a statement said that adding more blade on the critical region, may improve the design of air box system [3]. In addition, all dimensions for each design of blade are the same including their thickness and width as shown in Figure 4. For thickness, 1 mm has been taken as a boundary because the best design of blade thickness is as thin as possible in order to improve the air flow [7], and its width, 5mm

2.2. Computational Fluid Dynamics Analysis

IOP Conf. Series: Materials Science and Engineering **429** (2018) 012053 doi:10.1088/1757-899X/429/1/012053

The analysis of this project is performed using the Computational Fluid Dynamic (CFD) fluid flow fluent to determine the efficiency of the air flow. Besides, CFD is an analysis tools that are used to improve the design of air box system by simulating the air flow in it. The specific name of the CFD tool that has been used in this project is ANSYS software. Based on figure 5 shows the velocity streamline and different result of every design.

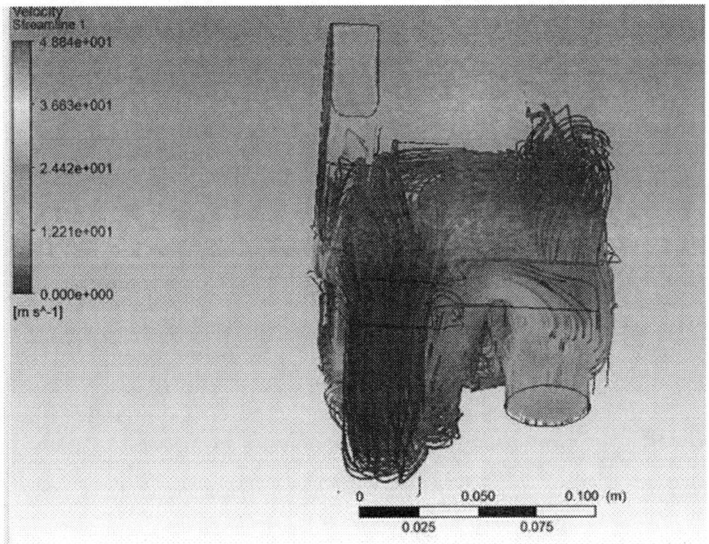

Figure 5: Air Flow of Existing Design of air box

3. Result and Discussion

The result will divided into two categories which are result for actual design MODENAS CT115s and proposed designs which are Design 1, Design 2 and Design 3. Table 1 below shows the comparison parameter of analysis data for all the designs.

Table 1: Comparison parameter of analysis data for existing design and proposed designs

Type of design	*Mass Flow Rate (kg/s)*	*Volumetric Flow Rate (m^3/s)*	*Reynolds Number*	*Velocity Magnitude (m/s)*	*Total pressure (Pascal)*
Existing Design	$3.20\,x10^{-5}$	$5.06\,x10^{-5}$	*Min = 0.00203* *Max = 5200*	*20.1545*	*256.8430*
Design 1	$4.82\,x10^{-5}$	$6.38\,x10^{-5}$	*Min = 0.00312* *Max = 5360*	*20.1682*	*257.2849*
Design 2	$3.66\,x10^{-5}$	$5.44\,x10^{-5}$	*Min = 0.00217* *Max= 5090*	*20.1609*	*256.4939*
Design 3	$3.02\,x10^{-5}$	$4.92\,x10^{-5}$	*Min = 0.00160* *Max= 4740*	*20.1626*	*257.4713*

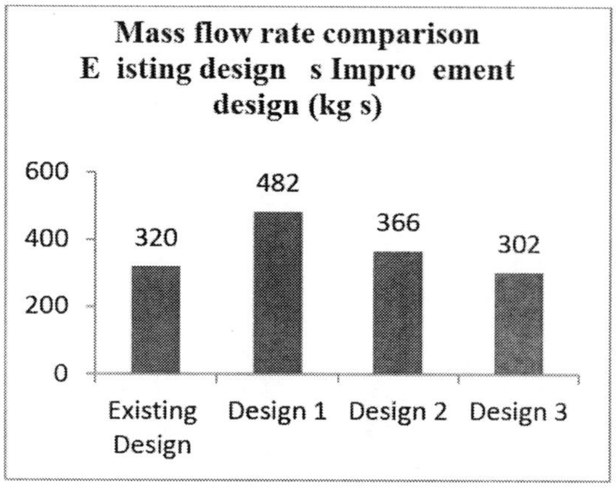

Figure 6: Mass flow rate comparison

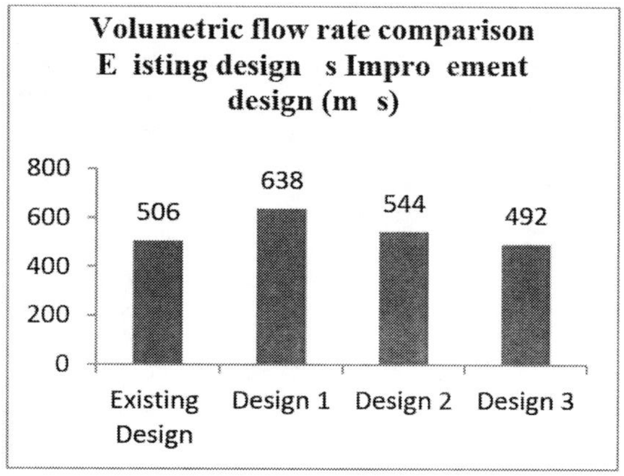

Figure 7: Volumetric flow rate comparison

Air box play main part in getting good quality air into vehicle engine. It advances the burning competence and also decreases air pollution. 3D Computational Fluid Dynamics (CFD) analysis carried out for an existing design and several improvement design to understand the flow actions through the air box system and geometry. Results got from CFD analysis of the existing design showed real situation for the air flow. Based on existing design CFD results, geometrical changes like sigle blade placement in inlet plenum showed good improvement in flow behavior compare to existing model. By expending 3D CFD analysis, ideal design of the air box system for vehicle is completed with substantial reduction in progress time and cost [3]. The ideal result to reach determined mass flow rate for air in the air box system it is need to reduce the pressure difference over the flow constraint in the air box [11]. When the air flow increases, the pressure drop in the air box might also increases [12]. The finest overall plan for this improvement design is to redesign the internal geometry of the air box so that can launch the air flow in the air box, thus can optimize the engine performance. From the data collected through the many simulations and analysis, it can be detected that a single design of blade has capability to achieve the highest value of mass flow rate and volumetric flow rate compare to existing design and other improvement design.

4. Conclusion

Conclusion that can be made from the analysis from Figure 6 and Figure 7, it shows that the improvement design of air box model which is also known as the Improvement 1 produced the best result in terms of mass flow rate and volumetric flow rate compared with the actual design MODENAS CT115s and other design. The CFD analysis studies that are performed on the air flow coefficient have provide important information that the reason why the air flow coefficient become difference are because it affected by the geometry design. From this study, it show that some new changes that can make for improvement on the air box so it can be function better for the system. This study also helps to develop the standard CFD work procedure for future air box geometric design optimization.

Acknowledgements

I would like to express my deepest appreciation to all those who provided me the possibility to complete this paper. A special gratitude I give to my supervisor and MODENAS Research and Development team whose contribution in stimulating suggestions and encouragement, helped me to coordinate my project students especially in writing this paper

References

[1] M. Rosli, "The Automobile Industry and Performance of Malaysian Auto Production," Journal of Economic Cooperation, vol. 27, no. 1, pp. 89-114, 2006.

[2] Makgata, K. W., " Computational Analysis and Optimisation of The Inlet System of A High-Performance Rally Engine," Reserch Master thesis, Faculty of Eng. Built Env. and Information

Tech., Pretoria Univ, South Africa, Pretoria, 2005

[3] T. C. Khindir, " Design Special Types of Blades in Air Box of Car Engines," International Journal of Enhanced Research in Sc. Tech. & Eng., vol. 4, pp. 121-127, March 2015.

[4] A.S Phulpagar and N.S Gohel, "CFD Analysis of Air Intake Systems," International Journal on Theoretical and Applied Research in Mechanical Eng. (IJTARME), vol. 4, pp. 2319-3182, 2015.

[5] A. J. Balinga and Dr. L. L. R. Rodrigues, " Business Plan – The Secret to Success," The International Journal of Business & Manangement, vol.3, issue 4, pp. 210-214, Apr. 2015.

[6] Cyr, L., "Creating a Business Plan", Boston, Massachusetts: Harvard Business School Press, 2007.

[7] Y. Zhang and Y. Jia, "2D Automatic Body-fitted Structured Mesh Generation using Advancing Extraction Method," Journal of Computational Physics, vol. 353, pp. 316-335, 2018.

[8] Aman Mohd. Ihsan Mamata, N. N., Salmiah Kasolanga, Aminuddin Zulkiflia and Rizalman Mamat. "Effects of Air Intake Pressure to the Fuel Economy and Exhaust Emissions on a Small SI Engine," The Malaysian International Tribology Conference (MITC), Kuala Lumpur, Malaysia, 2013.

[9] A. Z. Tayfik, N. Rashid, Mazlan, M. Faruq, and M. Zahir, "Electric car chassis design and analysis by using CATIA V5 R19," Journal of Mechanical and Civil Eng., vol. II, issue 4, Jul.-Aug. 2014.

[10] S. Pal, H. Reddy Tiyyagura, I. Drstvensek and C. S. Kumar, " The effect of post-processing and machining process parameters on properties of Stainless Steel PHI product produced by Direct Metal Laser Sintering," International Conference on Manufacturing Engineering and Materials (ICMEM), Novy Smokovec, Slovakia, June 2016.

[11] Sundar Raj, C., & Sendilvelan, S. Effect of oxygenated hydrocarbon additives on exhaust emission of a diesel engine. International Journal of Automotive and Mechanical Engineering, 2, pp 144-156. 2010.

[12] W. "Optimization of Air Intake System for Inline-4 Diesel Engine." Bachelor of Mechanical Engineering with Automotive Engineering, June 2012.

Valve Lift Overlapping Effect by Analyzing Engine Charge Flow in Various Cam Profiles for on Modenas CT115s using CFD Simulation

Efi N E[1], Shahriman A B[1], Rojan M A[1], I Zunaidi[1], Z M Razlan[1], W K Wan[1], M S M Hashim[1], N S Kamarrudin[1], A Harun[2], I Ibrahim[1] and Azizul Aziz I[3]

[1] School of Mechatronic Engineering, Universiti Malaysia Perlis, Pauh Putra Campus, 02600 Arau, Perlis, Malaysia.

[2]School of Microelectronic Engineering, Universiti Malaysia Perlis, Pauh Putra Campus, 02600 Arau, Perlis, Malaysia

[3] Modenas Research and Development Division, Kawasan Perusahaan 08300 Gurun,Kedah

Corresponding author: shahriman@unimap.edu.my

Abstract Cam profile plays a critical role which adjusts the valve lift to allow the air fuel mixture flow in into cylinder in order to obtain the optimum performance of engine. This paper reviews the effect of valve lift overlap by analyzing charge flow in various cam profiles for and determine the optimum overlapping valve lift to enhancement of cam performance. By using Ansys software, the parameters like volumetric efficiency, mass flow rate and velocity will be measured which ultimately make impact up on the engine power and torque. Therefore, for this study 0.5mm to 3mm are tested to figure out the effects in combustion process. Overall, by considering the discussion, the results shows that valve lift adjustment of 2mm for both intake and exhaust valve has the best performance among others.

Keywords: cam profile, overlap valve lift, air flow pattern, cfd simulation.

1. Introduction

In general, cam is a device, formulated with three basic components of kinematics cam, which acts as the driving part a follower, serves as a pioneer component and fastened frame [2]. Cam is a mechanical component that handles other segments called followers, throughout a certain movement by higher connections. A cam can be customized, mostly in the form of an engine component that is always associated with followers. The main function of the camshaft in the engine is used to move a motion to the intake and exhaust valves to control the closing and opening of the valves in an internal combustion engine [1, 5]. Valve train operation carried out by the action of the cam and cam geometry. If transporting the movement is not perfect, the engine's performance will be also affected [4]. As the cam pivots, the point contact moves, and hence the operative force is transferred to the follower [7]. Besides, cam is a component that works to transmit a power by motion of mechanical from a speed constant to variable of speed and widely used in automotive industry due to the superior properties for high speed in operation, accuracy in motion, rigidity of structure and low cost production. The location and shape of the lobe will determined the timing of the valve opening-closing motion, its lifting height and valve movement rate for one complete cycle in which could affect the engine performances. The simultaneous open time of the inlet and exhaust valves is referred to as valve overlap [3, 4]. The inlet valve is usually opened well before the TDC so that the overlap period is located on either side of the TDC. In addition to reducing the residual gas fraction, overlap aids inducting the fresh charge from the inlet port provided that the inlet pressure is higher than the exhaust pressure [6]. Other than that, valve overlap is problematic in engines where the fuel is premixed with the inlet air prior to entering the cylinder. A small part of the fresh charge may slip from the inlet to the exhaust port during the overlap period, particularly if the overlap period is long. Long overlap promotes high-speed operation as the scavenging improves but at low speed the combustion quality deteriorates and cyclic variation increases [5]. In this project, optimizing the engine performance by increasing the performance of cam by improve the cam profile curve using mathematical method is main objective of MODENAS R&D team and this research. Cam curve performances analysis will be conducted based on its several radius of base circle, rb from Polynomial method and all possible engineering parameters in order to determine the best cam design that will produce high performances throughout. The analysis of the cam will be done by using computerized engineering software such as CATIA V5 for designing and, ANSYS for simulation. This analysis is designed so that optimization of the cam profile can be done and thus, the factor affecting the optimization of the cam can be determined. Then, the results from the analysis could contribute in enhancing the engine performance for the MODENAS motorcycles.

2. Methodology

A. Design using CATIA software

For this study, several parts of Modenas Ct115s engine such as intake port, exhaust port, intake valve, exhaust valve and combustion cylinder are designed by using CATIA software. CATIA is multi-platform software. By using CATIA, drafting the dimension intake port design can be made to determine the exactly parameter of the existing design as shown in figure 1.[6] Figure 2 illustrates the adjustment of valve lift for intake assembly.

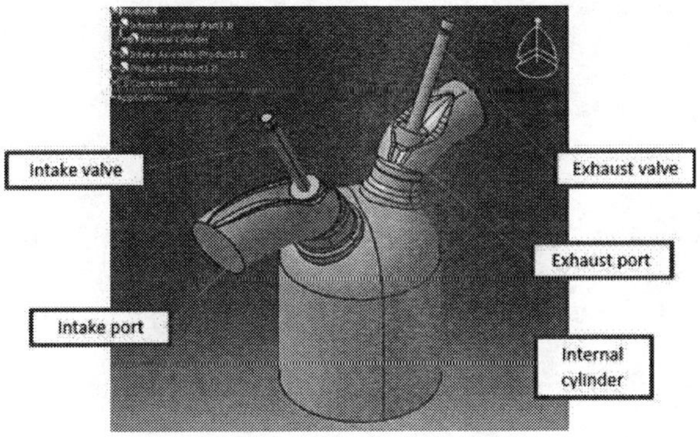

Figure 1: Isometric view of assembly drawing

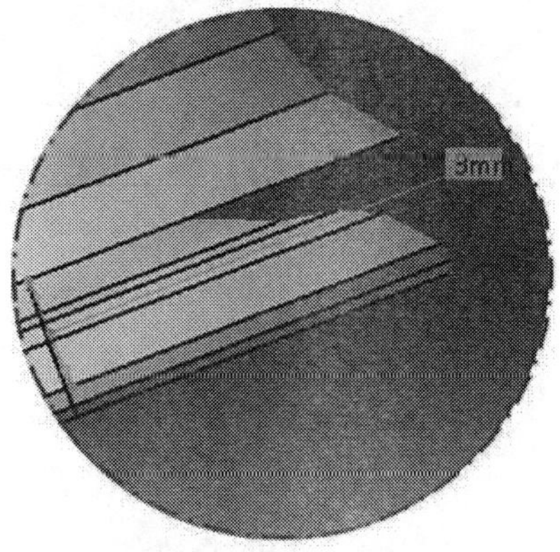

Figure 2: The adjusment point of valve lift between port and valve

B. ANSYS Simulation

Since the parameter of the research is considering the air flow to determine the mass flow rate and volumetric flow rate, the suitable analysis to be used is FLUENT Analysis. Therefore, for this study 0.5mm to 3mm are tested to figure out the effects in combustion process. In this research, they have been transferred to the ANSYS software to be analyzed and simulated. The current simulation of ANSYS has partially done somehow need to be analyzed in detail. It is because there are some parameters that need to be considered, such as mass flow rate, volume flow rate, inlet velocity, pressure and others. Thus, more researches need to be discovered in order to obtain the best solution from it.

3. Result and iscussion

In this research, valve lift overlap is considered as the variables in order to obtain the value of mass flow rate and volumetric efficiency. Figure 3 below show the assembly design that undergoes the analysis and simulation by ANSYS which explained the results gained by differentiation of the valve lift which are 0.5 to 3.0 mm.

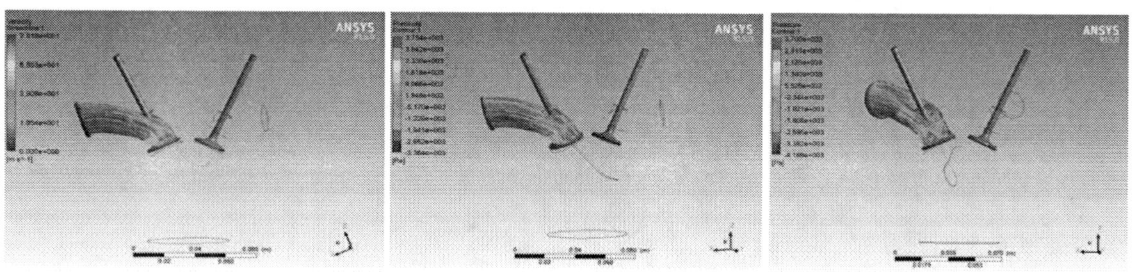

a) Valve lift 0.5mm b) valve lift 1.0mm c) valve lift 1.5mm

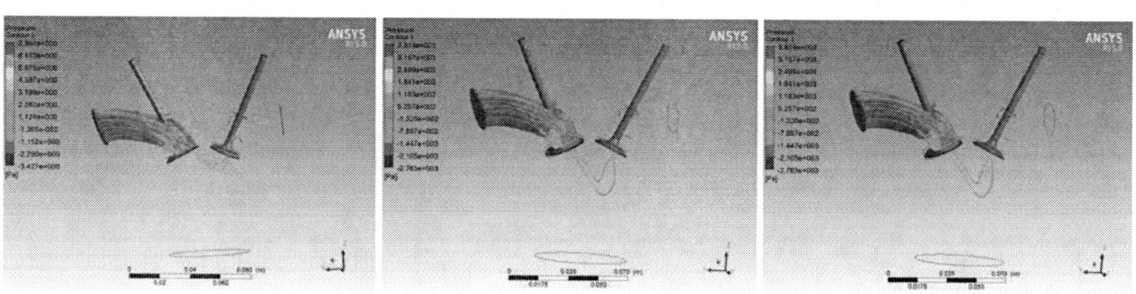

d) Valve lift 2.0mm e) Valve lift 2.5mm f) Valve lift 3.0mm

Figure 3: Simulation for 0.5mm to 3.0mm valve lift overlapping

By simulation analysis, the data results obtained are collected whereby mass flow rate and volume flow rate are considered in table 1.

Val e lift O erlap (net)	**Mass flow rate, (kg s)**	**Volume flow rate, (m^3 s)**
0.5 mm	4.2805517E-05	3.4942645E-05
1.0 mm	3.5963294E-05	2.9357330E-05
1.5 mm	5.0495768E-05	4.1221065E-05
2. mm	**. 2E**	**. 3 E**
2.5 mm	1.1988420E-05	9.7870779E-06
3.0 mm	3.3470984E-05	2.7323056E-05

Table 1:Mass flow rate and volume flow rate for each valve lift overlap

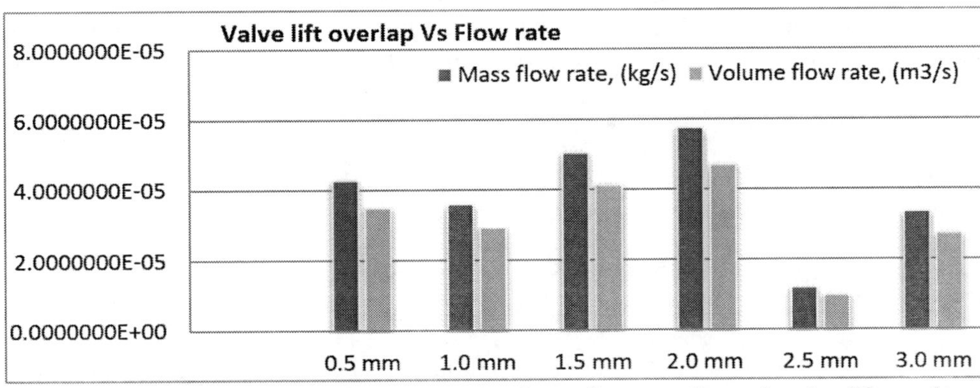

Figure : Comparison for mass flow rate and volume flow rate for each valve lift overlapping

The purpose of this chapter is to show how the valve lift parameters affect the mass flow rate, and volumetric efficiency. Results from the ANSYS Workbench calculations are illustrated as shown in figure 4.
Theoretically, smaller valve overlap will improve the engine performance because when the air fuel in cylinder that enter is compressed more tight, the combustion of fuel is actually almost to completely burnt which perform a better fuel efficiency. However, if the valve lift overlap is too low, it would causes an insufficient air fuel to enter through the cylinder at the intake stroke while at the exhaust stroke, burnt compressed gases are not be able to expelled from the cylinder as a whole, and caused the remains of the gases left inside. Consequence, there will be a performance loss and fuel efficiency reduction. Mass flow rate through intake valve with different valve lift speed for peak volumetric efficiency increases while the peak volumetric efficiency itself is slightly decreased. The peak volumetric efficiency occurs at high engine speed as the valve lift increases while the volumetric efficiency itself increases. A lower valve lift will result in a better volumetric efficiency at low engine speed, and higher valve lift has a greater volumetric efficiency at high engine speed. At the high engine speed, a low valve lift result in a smaller mass flow per crank angle during most of the open duration of the intake stroke. Since volumetric efficiency is calculated based on the integral of mass flow into the cylinder through the intake valve, a smaller mass flow per crank angle will result in a lower volumetric efficiency.

. Conclusion

The research and analysis for the valve lift overlap is carried out by using the simulation of ANSYS Workbench software in order to determine the mass flow rate and volumetric efficiency. As shown in graph result, valve lift of 2mm has the highest mass flow rate and volume flow rate; velocity magnitude; and static pressure. Based on the results of the simulation, the following conclusions can be drawn:

(1) By increasing intake valve lift, the engine speed for peak volumetric efficiency will increase.
(2) IVC timing has much more dominant influence on volumetric efficiency than intake valve opening (IVO) timing.
(3) Volumetric efficiency will decrease with a retarded intake valve closing (IVC) time at low to normal engine speed, due to reverse flow. It will increase at higher engine speeds.

Overall, by considering the discussion, the results from ANSYS simulation analysis shows that valve lift adjustment of 2mm for both intake and exhaust valve has the best performance among others. Hence, 2mm is selected as the optimum valve lift overlap.

. Acknowledgements

I would like to express my deepest appreciation to all those who provided me the possibility to complete this paper. A special gratitude I give to my supervisor and MODENAS Research and Development team whose contribution in stimulating suggestions and encouragement, helped me to coordinate my project students especially in writing this paper.

6. References

[1] Zhang Y, Zhao H, Ojapah M, Cairns A. CAI combustion of gasoline and its mixture with ethanol in a 2-stroke poppet valve DI gasoline engine. Fuel 2013; 109: 661–8

[2] Zhao H, editor. HCCI and CAI engines for the automotive industry. Cambridge: Woodhead Publishing; 2007.

[3] Heywood JB, Sher E. The two-stroke cycle engine: its development, operation and design. Warrendale: SAE International–Taylor and Francis; 1999.

[4] De Bellis V, Gimelli A, Muccillo M. Effects of pre-lift intake valve strategies on the performance of a DISI VVA turbocharged engine at part and full load operation. Energy Procedia 2015; 81:874–82.

[5] Zhang D, Hill PG. Effect of swirl on combustion in a short cylindrical chamber. Combust Flame 1996; 106:318-32.

[6] Kim J, Kim H, Yoon S, Sa S. Effect of intake valve swirl on fuel-gas mixing and subsequent combustion in a CAI engine. Int J Automot Technol 2008; 9: 649e57.

[7] Clenci AC, Iorga-Siman V, Deligant M, Podevin P, Descombes G, Niculescu R. A CFD (computational fluid dynamics) study on the effects of operating an engine with low intake valve lift at idle corresponding speed. Energy 2014; 71: 202-17.

IOP Conf. Series: Materials Science and Engineering **429** (2018) 012055 doi:10.1088/1757-899X/429/1/012055

Enhancement of Engine Performance through analyzing internal part design on Modenas CT115s exhaust using CFD Simulation

Efi N E[1], Shahriman A B[1], Rojan M A[1], I Zunaidi[1], Z M Razlan[1], W K Wan[1], M S M Hashim[1], N S Kamarrudin[1], A Harun[2], I Ibrahim[1] and Azizul Aziz I[3]

[1] School of Mechatronic Engineering, Universiti Malaysia Perlis, Pauh Putra Campus, 02600 Arau, Perlis, Malaysia.
[2]School of Microelectronic Engineering, Universiti Malaysia Perlis, Pauh Putra Campus, 02600 Arau, Perlis, Malaysia.
[3]Modenas Research and Development Divison, Kawasan Perusahaan, 08300 Gurun, Kedah.

Corresponding author: shahriman@unimap.edu.my

Abstract Exhaust is made to make a way that guide and expel exhaust gasses far from the engine. Inside exhaust system, there are four noteworthy parts which are muffler, perforation, resonator chamber, inlet and outlet. For mechanical performance of perforated resonator can be controlled by size of holes. The aim of this paper is to analyzing internal part geometry of exhaust muffler in order to increase Modenas CT115s performance by using Ansys software as a programmed to calculate the velocity magnitude and the flow rate of exhaust muffler. Consequently, high proficiency inside the exhaust system will create a brilliant execution and high quality performance. From the simulation shows that decreasing the size of holes give ultimately impact to performance of exhaust.

Keywords: exhaust, mass flow rate, velocity magnitude, perforated tube holes, engine performance.

1. Introduction

Focusing on exhaust system for the four stroke engine, inside exhaust system, there are four noteworthy parts which are muffler, perforation, resonator chamber, inlet and outlet. Basically, exhaust system also known as muffler is made to route exhaust gases away from the engine. When the fuel is burn inside the engine, the fume and gases produced need to be taken away to avoid performance engine loss. The muffler also captures harmful toxins in gases before they are released to the atmosphere. It is made to make a way that guide and expel exhaust gasses far from the engine. [1]. consequently, high proficiency inside the exhaust system will create a brilliant execution and high quality performance. Each part conveys a function that suit with its purpose. To start with designer of suppressor design was granted to Milton O. Reeves and Marshall T. Reeves [2]. Essentially mufflers are characterized into two classes which are dissipative and reflective. In dissipative muffler, pressure waves is change over into heat energy by using the principle of converting exhaust noise energy. Next, reflective muffler utilizes complex muffler parts to create pressure waves that will counteract the dynamic pressure wave with approaching pressure [2, 3]. Selection of suitable muffler is based on transmission loss and it is vary with changes in geometry parameters such as number of holes, diameter of pipe and number of pipes

[3]. A lot of literature review had been made to enhance the performance of engine by designing diverse mufflers with various diameter of the perforated hole, distinctive length of the perforated tube and different perforated holes design [7]. Perforated tubes is a tube that generally made up with stainless steel and have gaps or hole punched around its outline. These tube are given to control the stream of the gasses all through the exhaust muffler [3, 4]. One of its significant capacity is to expand the transmission loss thus will increase the performance of the engine. From the study, it has been observed that the number of holes and the size of the hole in the perforated tube affects the amount of transmission loss in the mufflers [5, 6, 8]. The bigger the diameter of the hole, lower will be the backpressure and hence lower will be the transmission loss. Length of perforated tube likewise considerably affects the transmission loss and it increase with the increasing in the length of the perforated tube [4, 6].

2. Methodology

2.1. CATIA V5 Software

To calculate the exhaust profile the parameters that influence the motion of the exhaust mufflers need to be identify first by using CATIA V5. The exhaust dimensions need to be analyze and identified before continuing to the experiment. In this research, the method had been used are by CATIA V5 software to identified the part. The software that has been utilized to display the air flow is CATIA V5. Perforated tube can be redesign by changing its holes diameter whether to increase the diameter or decrease the diameter. The air flow modeling in this project is been divided into two groups which is the existing design and the improvement design.

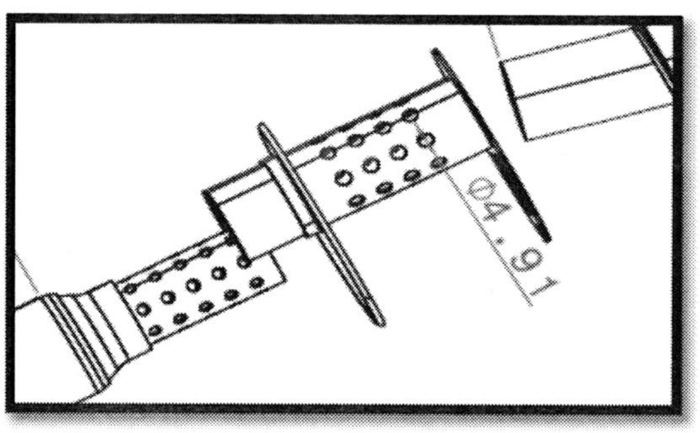

Figure 2.1: Dimension for existing design of perforated tube holes

Table 1: Existing and improvement diameter

esign	E isting iameter (mm)	Impro ement iameter (mm)
A	4.91	4.41
B	4.91	3.91

2.2. ANSYS Workbench 15.0 Software

The actual design of the air flow model is created based on exact dimension of the CT115s engine's exhaust. Next, it is import into ANSYS Workbench 15.0 software. In the mesh generation, proper steps should be conducted properly to avoid failure in analysis. If there is improper mesh define, the fluid zone will cause failure. The critical area need to be choose carefully especially for the inlet, outlet and wall part body. For the existing design, the approximate meshing cell are shown in figure 2.2.

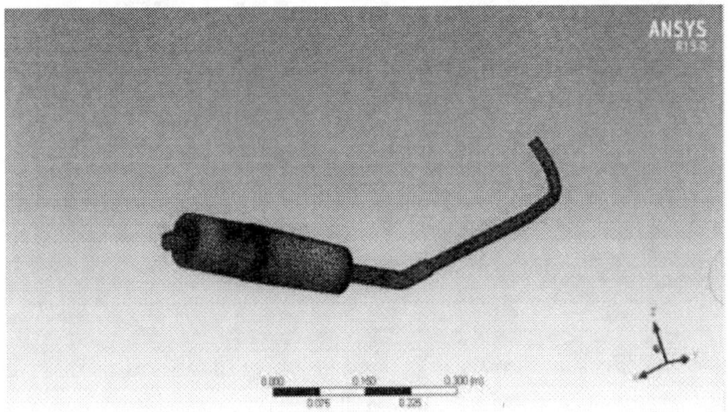

Figure 2.2: Existing MODENAS CT115s exhaust design Meshing cell analysis

3. Result and Discussion

Exhaust Modenas CT115s and proposed design has been tested by using ANSYS Workbench 15.0 software to determine the velocity air flow and the pressure. All the design then undergo different type of analysis test. From the figure 3.1 until 3.3, shown the velocity streamline and pressure contour for the actual and improvement design part of exhaust muffler. The red colour in both figures indicates the maximum values while the blue colour indicates the minimum values.

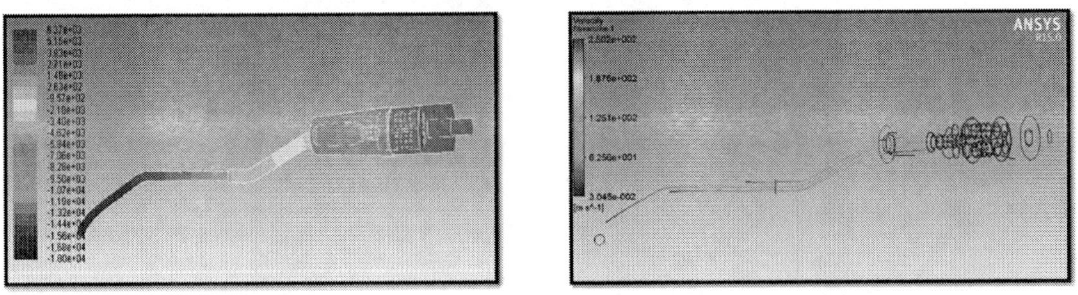

(a) (b)

Figure 3.1: Pressure contour (a) and velocity magnitude (b) for existing design

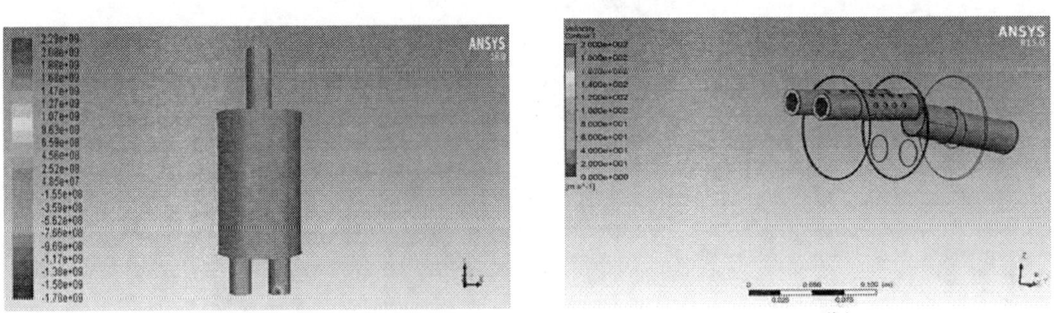

(a) (b)

Figure 3.2: Pressure contour (a) and velocity magnitude (b) for Design A

IOP Conf. Series: Materials Science and Engineering **429** (2018) 012055 doi:10.1088/1757-899X/429/1/012055

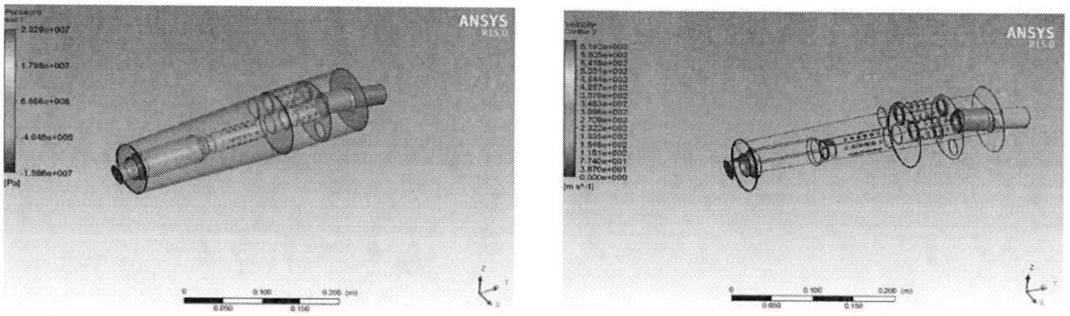

(a) (b)

Figure 3.3: Pressure contour (a) and velocity magnitude (b) for Design B

Increasing the size of hole in exhaust make a huge different in simulation result. For the mass flow rate result at improvement design B show that the mass flow is 0.15644 kg/s which is the highest compared to the others design. Different result shows at the existing design result which is 0.03256 kg/s much lower than improvement design B. For velocity magnitude design B also improve from 1.876e2 m/s for existing design to 6.192e2m/s. Table 2 shows all the result for all design.

Table 2: All the result for all design

	EXISTING DESIGN	**DESIGN A**	**DESIGN B**
MASS FLOW RATE (KG/S)	0.03256	0.10345	0.15644
VELOCITY MAGNITUDE (M/S)	1.876e^2	2.0×10^2	$6.192e\times10^2$
PRESSURE (PA)	6367.66	-1.7838×10^9	2.929×10^7

4. Conclusion

Based on the research and analysis that had been done for internal part of exhaust which focusing on redesign the perforated tube in the muffler, the size of hole give ultimately impact to the performance of gases flow out to the surrounding. In this research, diameter of hole is decreasing 0.5mm from existing value. From the simulation result clearly shows that smaller the size of holes give higher performance in term of mass flow rate and velocity. For this research, conclusion can be make as lower the value of diameter for perforated tube will contribute to higher performance of exhaust. Therefore, for the future work, the optimum value for size of hole and number of holes are suggested to be investigate in detail for general small engine.

Acknowledgements

I would like to express my deepest appreciation to all those who provided me the possibility to complete this paper. A special gratitude I give to my supervisor and MODENAS Research and Development team whose contribution in stimulating suggestions and encouragement, helped me to coordinate my project students especially in writing this paper.

References

[1] Suvro Sen (2013, Jan. 27) How Does Motorcycle Exhaust System Work [Online]. Available: https://www.bikebd.com/how-does-motorcycle-exhaust-systems-work

[2] Wu T.W Boundary element acoustics fundamentals and Computers codes, WIT Press, Southampton. 2000

[3] V. M. Mundhe and E. R. Deore, "Design and Optimization of Perforated Muffler in an Automobile Exhaust System," vol. 1, no. 8, pp. 390–395, 2015.

[4] A. Pratap, U. Kalita, and S. Kumar, "Optimization of Transmission Loss of Perforated Tube Muffler by Using CAE Tool ANSYS," vol. 2, no. 5, pp. 1–5, 2015.

[5] A.R Mohanty and S.P Pattnaik. "An Optimal Design Methodology for a Family of Perforated Mufflers". SAE Paper no. 2005-26-053, 2005.

[6] Taylor W. Lee Ray.: "Muffler Characterization with Implementation of the Finite Element Method and Experimental Techniques", Master's Thesis, Paper 381, Michigan Technological University, 2011.

[7] Haluk Erol and Ozcan Ahmetoglu. "Acoustic Attenuation in Fully Filled Perforated Dissipative Mufflers with Extended Inlet/Outlet", 13th International Congress on Sound and Vibration, Vienna, Austria, 2006.[8] Fangsen Cui, Ying Wang and Richard Chao CAI. "Improving Muffler Performance Using Simulation-Based Design". Inter. Noise, Melbourne, Australia, 16th-19th November 2014.

IOP Conf. Series: Materials Science and Engineering **429** (2018) 012056 doi:10.1088/1757-899X/429/1/012056

Development of an Automatic Vehicular Heatstroke Detection System

***S N David Chua[1], W J Goh[1], S F Lim[1], A Joseph[1], Oon Y B[2] and C V Sia[3]**

[1] Faculty of Engineering, Universiti Malaysia Sarawak, 94300, Sarawak, Malaysia
[2] Faculty of Cognitive Sciences and Human Development, University Malaysia Sarawak, 94300, Sarawak, Malaysia
[3] University of Technology (Sarawak Campus), 93350, Kuching, Sarawak, Malaysia

*csndavid@unimas.my; davidchua2@mail.dcu.ie

Abstract. The cases of children being left in a car and died of heatstroke are on the rise every year. Car manufacturers have not implemented any sort of alert system that could prevent these unfortunate events from happening. There are many baby car seat alert systems available in the market, however, they are mostly inconsistent and unreliable. There is a need for more effective and efficient alert system to remind the users of their forgotten child. This paper aims to provide solutions to caretakers through application of multiple sensors for integration of a child detection system and also to discover the best placement of the sensors for a more effective system. The selection method is based on experimental setup with a test subject which is a human being. A prototype for the detection system is fabricated and results show a combination of ultrasonic and motion sensors produces the best results. The best position to place the ultrasonic and motion sensor is determined to be above the child seat and in the middle of the car respectively. The prototype of the detection and alert system is validated via a set of assessments to determine its accuracy, adaptability and reliability.

1. Introduction

Heatstroke is a condition where body overheats to a certain degree from a prolonged exposure in high temperature surroundings which might ultimately lead to permanent damage and even death. The hot weather in countries like Malaysia makes it even easier to get a heatstroke or also called hyperthermia. There have been many cases over the years where a baby or a child left in the car under the scotching weather has died of heatstroke. Based on a fact sheet by a non-profit organisation, there were 793 children that died in vehicles due to heatstroke since 1998 in the U.S. [1]. From a research done to a total of 171 fatalities in the U.S. of heat related deaths to young children in parked cars from 1995-2002, 73% (125) were children who were left unattended by adults [2]. One of the reasons that children are left unattended is because of the carelessness of parents who forget or unaware of the present of their child in the car. The parents fail to realise how hot a car would become if it is exposed to direct sunlight for a period of time. A normal human being body temperature is around 37 Celsius. Heatstroke would occur when the body temperature reaches around 40.5 Celsius. It could take as little as fifteen minutes being stuck in an overheated car for a child to suffer heatstroke [3]. There is no guarantee it can be prevented from just opening the car windows by a small gap. Children die of heatstroke in cars when parents get distracted from their daily routines and does not realise that they left their child unattended. The busy schedule of parents combined with the tendency to seat their sleeping child in the rear seat is one of the major reasons why these unwanted cases are happening. Moreover, tinting of car windows prevents the child to be noticed by passers-by.

2. Related work and products

There are a few studies and research done on the development of a device to alert the parents of an unattended child left in a car in recent years. However, all the inventions are yet to provide an effective and comprehensive warning and feedback system for the unattended child in the car [4]. There are also some products being sold in the market that parents can buy to prevent such a death-defying situation from occurring, but they are mostly unreliable and inconsistent [5]. Types of product that are currently available in the market can be divided into three categories i.e. seat based reminder system, buckle based reminder system, and reminder system installed in car.

The first kind of heatstroke detection system is through attaching the device to the car seat. The device is placed at the base of the seat which usually makes use of pressure sensor to determine the presence of a child. It comes with an alarm system that can be hooked as a keychain or installed on a phone. The alarm device is connected to the sensor on the seat via a communication medium such as radio frequency, Wi-Fi or Bluetooth. The alarm device alerts the driver when the device is set apart and a weight is detected on the seat. A study that reported the problem with pressure sensing products was due to positioning of the pressure sensors [6].

There are also products that are attached to a baby seatbelt buckle to determine the presence of the child. It uses a sensor to sense if the seatbelt tongue is connected to the buckle. The alarm system functions similarly to the pressure type detection system where the alarm system will go off if the child is left buckled to the seat and the parents has gone off to a certain distance away from the car. There are also some problems with this type of system as it works with the condition that the child is buckled up to the seat. The weakness of this type of detection system is that if the child is unbuckled or the child unbuckles by himself from the seat, it will fail to trigger.

There is also a type of heatstroke detection system that is installed within the car. This kind of device usually uses a human presence sensor to determine if a child is being left unattended in the car. The examples of sensors that can sense the presence of a human being are thermal sensor, carbon dioxide sensor, motion sensor, facial recognition, radar and sound sensor. This kind of system can be installed during the manufacturing process of the car but the downside is that it is dearer compared to seat based type and buckle type.

2.1. Types of sensors for heatstroke detection system

A human presence sensor is required for the system to detect a child that is left unattended in a car. The sensor must accurately determine if there is a child left in the car and able to differentiate between a non-living object and a child that is unattended in the car. The presence of a human being can be detected using thermal, motion, carbon dioxide and distance. Table 1 shows the comparisons between the human presence sensors available in the market.

2.1.1. Thermal human presence sensor. A thermal sensor captures the heat condition in the monitored area and determines the range of the surrounding temperature. Then, a dynamic threshold is obtained to constantly update the range of the surrounding temperature. The presence of a human can be detected if there is a temperature higher than the threshold value. It only takes less than two seconds for the system integrated with the thermal sensor to sense a human being [7]. A research was done using heat based human sensing method to determine the location of the user for an intelligent fan [8]. Thermal and visual information integration was introduced to detect and identify a human [9]. The proposed system has main modules that works with visual images and thermal images. Thermopile sensors can also be used to detect human presence in a supervised environment [10]. Thermopile sensor was chosen because of its low cost and intrusive qualities. The sensor reacts to the level of infrared signals generated by the temperature of human subject.

2.1.2. Carbon dioxide sensor. There are two types of carbon dioxide (CO2) sensor. The first one uses a chemical reaction to determine the level of CO2 in the area. The second simpler CO2 sensor is the Non-dispersive Infrared (NDIR) sensor. This sensor uses infrared light and a detector to analyse the drop in light through space. Carbon dioxide sensors needs design algorithms for estimating the presence of a human being. There are models to determine occupancy using a carbon dioxide sensor [11].

2.1.3. Motion sensor. A typical motion sensor to detect human presence is Passive Infrared (PIR) sensor. It uses the change in infrared light reflected onto it from objects to determine if there is motion in its field of view. Based on a research on PIR sensor to monitor occupancy patterns, it is found that if the sensor is placed in a room smaller than the sensor's maximum detection range, it is even possible to discriminate between different movements and provide a rough estimate of energy expenditure [12]. The disadvantage of PIR sensor is if there is no movement, it results in a false negative as the occupant may not be moving [13]. In order to address this problem, researchers came up with a solution which utilises particle filtering method [14]. Particle filtering method which moves particles and eliminates incorrect state transition particles.

2.1.4. Ultrasonic sensor. Ultrasonic array sensor transmits short bursts of sinusoidal pulses periodically. Human presence is determined processing the received echoes at the receiver array. A combination of ultrasonic array sensor and tracking algorithm was proposed for the detection of a human in a confined space as it can detect minor motion that are usually missed by another detecting device [15]. An ultrasonic sensor can even detect muscle activity by designing a complex algorithm that estimates joint torque of knee [16].

2.1.5. Pressure sensor. A pressure sensor is used to measure the changes in applied pressure or force and converting them into electric charge. When place in the car seat, it can be used to detect the presence of a human being by arranging sensor logic to differentiate between a person, child or belonging. A research proposed to have four pressure pads on each seat of the car to distinguish between person, child and luggage [17]. Based on a study, it is found that by using a movement detection algorithm, the pressure sensor can detect small movements such as breathing [18].

Table 1. Overview of human presence sensors.

Sensor	Advantages	Disadvantages
Thermal	Able to detect human being and animals. Simple algorithm. Non-contact temperature measurement. Able to determine the position of an object. High sensitivity. Low noise.	Detection of temperature depends on the proximity of the object. High cost.
Carbon dioxide	Able to detect multiple people at a time.	Inaccurate if car windows are opened. Calibration drift. Sophisticated method required to estimate occupancy. Low sensitivity if CO_2 level is too low. Preheated is required.
Motion	Fast response time. Can detect small movement. Low power consumption. Low cost.	Unable to detect stationary human being. False detection if range is not set properly.
Ultrasonic	Free from external interference (e.g. temperature, humidity, noise and dust). Emitted waves are harmless to human body.	Requires complex algorithm to detect human being.
Pressure	Can differentiate between an object, adult and child with the use of multiple sensors. Simple mechanism. Low cost.	Troublesome to install. Difficult to maintain or replace.

IOP Conf. Series: Materials Science and Engineering **429** (2018) 012056 doi:10.1088/1757-899X/429/1/012056

3. Materials and methodology

3.1. Main system
Figure 1 shows the system block diagram of the product constructed to fulfil the aims of the product design. Power is supplied to the whole system and the sensor unit detects signals that represent a child is being left unattended and in danger. This information is transferred to the processor that determines if the child is left unattended and in danger. The processor will then relay the decision to the response unit which will convey the processor decision to users.

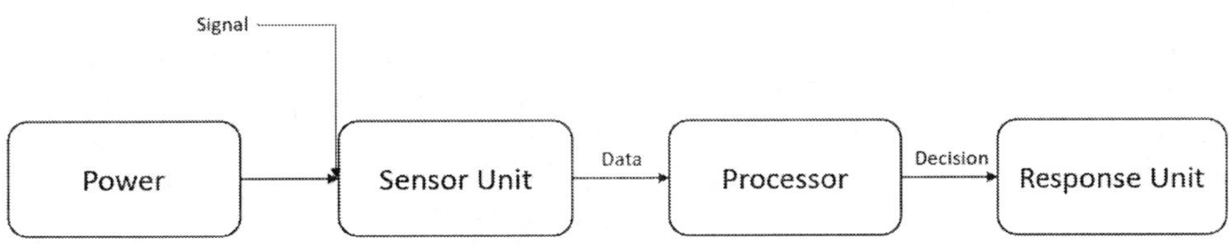

Figure 1. Main system block diagram.

3.2. Components
Basically, the detection system consists of sensors, processors, response system, transmitter and receiver. Vital signs to indicate life are the means to determine if a child is present within the car. Amongst sensors selected for experiment are Passive Infrared (PIR) motion sensor, HC-SR04 Ultrasonic Sensor, Analog Infrared CO_2 sensor, and Melexis MLX90614 Infrared Temperature sensor. Arduino UNO is used for constructing and programming electronic devices and it is based on ATmega329P as its brain. The nRF24L01 is used to transfer simple data from one Arduino UNO to Arduino Pro mini in the keychain device. When an unattended child is detected in the hot car, the Arduino Uno sends a signal via the transmitter to the keychain device. The receiver receives the signal and alerts the user of the unattended child in the car.

3.3. Experimental setup

3.3.1. Sensor selection criteria. The selection of the sensors to detect the presence of life in the car is based on the four selection criteria which are range, response time, false alarm and cost. The range of the sensor must be at least more than two meters so that the location of the sensor is not limited, and the number of required sensors can be reduced. The experiment is done with a sensor placed in line of sight of a human being, then the distance that is able to stimulate the sensor is recorded. Response time is the most critical criteria in the selection of the sensors as users must be alerted as soon as the unattended child is detected. The experiment starts with a test subject (human being) sitting in the car. The time taken to detect the test subject is recorded from the time the system is turn on. False alarm is the largest factor that causes unreliability of existing vehicular heatstroke detection systems. Therefore, sensors are tested if it reacts to surrounding movements and movements of the car itself. Cost is a huge determining factor in the overall justification of the system which might influence its marketability. The cost of the sensors is justified through market survey and benchmarking with the existing available products.

3.3.2. Receiver device selection. A receiver device using an Arduino Nano with a radio frequency transmitter is used for this research. The receiver device is assessed based on its range and accuracy through obstacles. The range is tested by having a test subject (human being) walking in line of sight from the vehicle holding the receiver device with the transmitter inside the vehicle. For the accuracy through obstacles, the test subject with the receiver device stands behind a wall between the vehicle and the receiver device. The distance of the user from the vehicle is measured to determine the range of the device. The experiment is repeated for 5 times.

3.3.3. Placement locations. The selected sensors are subjected to finding the best placement locations once they are finalised. The selection criteria for the best placement locations is determined by the reaction time to detect life and range coverage of the sensors. The sensors are placed on the front, middle and back ceiling, above the front and rear seats of the vehicle. The test object is placed on the front and rear seats in separate experiments to determine the most suitable sensors placement.

3.4. Validation of detection system

The validation of the whole device which consists of the Arduino UNO integrated with the sensors chosen are done based on three criteria which are accuracy, resistance and reliability. The device must be able to accurately determine the presence of an unattended child. The assessment covers the detection of stationary human and non-living objects. Moving human beings and a moving object must be differentiated accurately by the sensor. The device must be able to resist and adapt to extreme conditions such as hot weather. The device is placed under hot sun for an hour and systems functionality is tested after subjected to the harsh conditions. The reliability of the system is for the system to operate and response to many different situations. Situations such as a child moving around in the car, a sleeping motionless child, the range of connection and obstacles between systems are tested to prove its reliability.

. Results and discussion

Table 2 shows the outcomes of the testing results based on the selection criteria mentioned. Both the ultrasonic sensor and motion sensor met the minimum requirement of two meters range (sufficient to cover the entire vehicle) despite less than three meters of detection. The Melexis temperature sensor provided the fastest response time but the range of detection were very limited. In terms of the false alarm testing, all the four types of sensor responded without giving false alarm. The CO2 sensor provided the best range, but its response time was extremely slow as it needed preheating time and the cost of the sensor was high when compared to others. The non-contact temperature sensor had very small range of detection and it was more expensive than the ultrasonic and motion sensors.

Table 2. Summary result of sensors evaluation.

Criteria	PIR	Ultrasonic	CO2	Melexis
Range	> 2 m	> 2 m	> 3 m	< 0.1 m
A erage response time	5.0 s	2.9 s	> 5 min	1.8 s
False alarm	No	No	No	No
Cost	RM5.30	RM25	RM285	RM137.50

Table 3 shows the scores for each type of sensors based on its weightage. Obviously, the total scores for ultrasonic and motion sensors stood out from the evaluations. The ultrasonic sensor and motion sensor were the most suitable sensors for developing the detection system based on the criteria chosen. A detection system based on the ultrasonic sensor and motion sensor was therefore developed for this research.

Table 3. Decision matrix.

Criteria	Weightage	PIR	Ultrasonic	CO2	Melexis
Range	3	4	4	5	1
A erage reaction time	5	4	4	1	5
False alarm	4	4	5	1	2
Cost	4	5	5	3	4

Total	68	72	36	52

The receiver device was validated based on its range and its capability transmitting the signal through obstacles. The device could receive the signal between 3 and 4 meters around the vehicle and this value diminished when there was a wall in between the receiver device and the vehicle.

The best placement locations of the sensors were carried out for the ultrasonic sensor and motion sensor since it was chosen as the most suitable sensors for integration of the child detection system. The best placement of the ultrasonic sensor to detect the child was determined to be above the child seat as it used distance to detect the child in a straight line and gave the best result. The best placement for the motion sensor was on the ceiling in the middle of the vehicle. The sensor was able to pick-up movements anywhere in the vehicle at that position.

Figure 2 and 3 showed the prototyping of the detection system. The whole device was validated based on the assessment criteria mentioned. The validation showed the detection system passed the accuracy test, high resistance and reliable. The device was able to detect a stationary or motionless sleeping child and differentiate between a living and non-living object. The device also could withstand the harsh conditions as in the hot vehicle and function as required. The device was able to pick-up the movements and alert the user of the unattended child when the child got out of the baby seat and moved around in the vehicle.

Figure 2. Main device (Prototype).

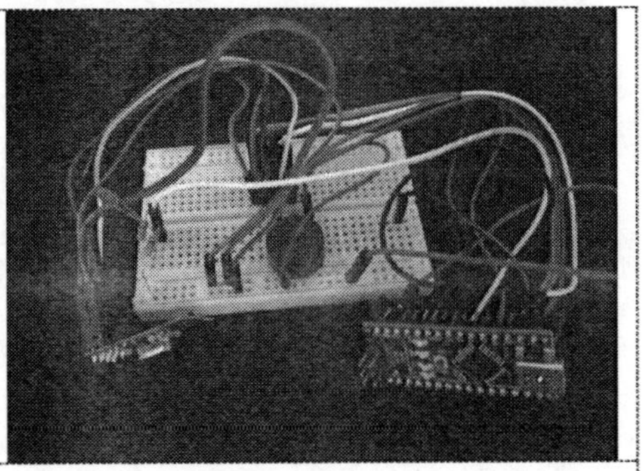

Figure 3. Receiver device (Prototype)

. Conclusions

The aim to develop an automatic vehicular heatstroke detection system to reduce the unfortunate cases of a child death due to heatstroke after being left unattended in the car was achieved. The most suitable sensors which were the ultrasonic and motion sensor were selected for integration of the detection system. The sensors were used for fabricating the prototype in this research. The best placement locations of the sensors were determined. The detection system was also validated based on criteria set which were accuracy, resistance and reliability. The prototype was able to achieve the results required to function as an automatic vehicular heatstroke detection system without false alarms.

. References

[1] Kidsandcars.org U.S. *Child Vehicular Heatstroke Deaths Fact Sheet* 2017

[2] Guard A and Gallagher S S 2005 *Heat related deaths to young children in parked cars: an analysis of 171 fatalities in the United States, 1995-2002* Inj. Prev. **11** 1 p 33–37

[3] Grundstein A, Duzinski S and Null J 2017 *Impact of dangerous microclimate conditions within enclosed vehicle on pediatric thermoregulation* Theor. Appl. Climatol. **12** 1–2 p 103–110

[4] Sulaiman N, Ghazali K H, Jadin M S, Hadi A A, Najib M S, Zain M S M, Halim F A, Daud S M, Zahed N and Abdullah A A 2017 *Development of comprehensive unattended child warning and feedback system in vehicle* MATEC Web Conf. 90 vol 01008 p 1–9

[5] Rudd R, Prasad A, Weston D and Wietholter K 2015 *Functional assessment of unattended child*

reminder systems Report No. DOT HS 812187 (Washington DC: National Highway Traffic Safety Administration) p 76

[6] Karaman A, Augustine S Z T, Bofeng L, Lawrence G H T, Ruo X Y, Sheng Q and Yao Y 2013 *ThermSafe: Child Heat Injury Prevention in Heated Locked Cars* Imperial College London p 1–9

[7] Tan T F, Teoh S S, Fow J E and Yen K S 2016 *Embedded Human Detection System Based on Thermal and Infrared Sensors for Anti-Poaching Application*. IEEE Conference on System, Process and Control (Melaka: Malaysia) p 37–42

[8] Parnin S and Rahman M M 2016 *Human Location Detection System Using MicroElectromechanical Sensor for Intelligent Fan* International Conference on Mechanical, Automotive and Aerospace Engineering vol 184 012042 p 1–8

[9] Correa M, Hermosilla G, Verschae R and Ruiz-del-Solar J 2012 *Human Detection and Identification by Robots Using Thermal and Visual Information in Domestic Environments* Journal of Intelligent & Robotic Systems 1–2 p 223–243

[10] Guettari T, Boudy J, Benkelfat B, Chollet G, Baldinger J, Dore P and Istrate D 2014 *Thermal Signal Analysis in Smart Home Environment for Detecting a Human Presence* 1st International Conference on Advanced Technologies, (Sousse: Tunisia) p 334–339

[11] Wang S, Burnett J and Chong H 1999 *Experimental Validation of CO2-Based Occupancy Detection for Demand-Controlled Ventilation* Indoor and Built Environment 6 p 377–391

[12] Kaushik A R and Celler B G 2006 *Characterization of passive infrared sensors for monitoring occupancy pattern* Annual International Conference of the IEEE Engineering in Medicine and Biology – Proceedings (New York City: USA) p 5257–5260

[13] Luppe C and Shabani A 2017 *Towards Reliable Intelligent Occupancy Detection for Smart Building Applications* IEEE 30th Canadian Conference on Electrical and Computer Engineering (CCECE) p 1120–1123

[14] Yokoishi T, Mitsugi J, Nakamura O and Murai J 2012 *Room occupancy determination with particle filtering of networked pyroelectric infrared (PIR) sensor data* Proceedings of IEEE Sensors p 3–6

[15] Caicedo D and Pandharipande A 2012 *Ultrasonic Array Sensor for Indooer Presence Detection* In 20th European Signal Processing Conference (EUSIPCO 2012) p 175–179

[16] Koyama T, Tanaka T, Kaneko S, Moromugi S and Feng M Q 2005 *Integral ultrasonic muscle activity sensor for detecting human motion* IEEE International Conference on Systems, Man and Cybernetics vol 2 (Waikoloa: USA) p 1–6

[17] Garethiya S, Agrawal H, Gite S, Suresh V, Kudale A, Wable G and Yendargaye G R 2015 *Affordable system for alerting, monitoring and controlling heat stroke inside vehicles* International Conference on Industrial Instrumentation and Control (ICIC) p 1506–1511

[18] Holtzman M, Townsend D, Goubran R and Knoefel F 2011 *Breathing sensor selection during movement* Proceedings of the Annual International Conference of the IEEE Engineering in Medicine and Biology Society (EMBS) p 381–384

Acknowledgments

Authors wishing to acknowledge assistance from colleagues especially for providing the physical vehicle for experimental tests in this research. This research was also funded by the UNIMAS Research Grant Scheme SGS Grant: F02/SGS/1600/2017.

IOP Conf. Series: Materials Science and Engineering **429** (2018) 012057 doi:10.1088/1757-899X/429/1/012057

Comparative Study on Fuel Consumption and Different Driving Cycles for a Passenger Car in Malaysia via 1-D Simulation

M S Mustaqim[2], M S M Hashim[1,2]*, Shahriman A B[1,2], Z M Razlan[1,2], I Zunaidi[3], W K Wan[2], A Harun[1,2], N S Kamarrudin[1,2], I Ibrahim[1,2], M K Faizi[1,2], M A M Saad[1,2], M F H Rani[1,2], A R Mahayadin[1,2], Efi N E[1,2], M S Sarip[4], Z M Isa[5] and Y Nagaya[5]

[1]Motorsport Technology Research Unit (MoTECH), School of Mechatronic Engineering, Universiti Malaysia Perlis (UniMAP), 02600 Arau, Perlis, Malaysia.

[2]School of Mechatronic Engineering, Universiti Malaysia Perlis, Pauh Putra Campus, 02600 Arau, Perlis, Malaysia.

[3]Technopreneur at UniMAP Sdn Bhd, No 26, Blok D, Bangunan PPPIT, Pengkalan Jaya, Jalan Kangar – Alor Setar, 01000 Kangar, Perlis, Malaysia.

[4]Malaysia Automotive Institute (MAI), Block 2280, Jalan Usahawan 2, Cyber 6, 63000 Cyberjaya, Selangor, Malaysia.

[5]Perusahaan Otomobil Kedua Sdn Bhd (PERODUA), Sungai Choh, Locked Bag 226, 48009 Rawang, Selangor Darul Ehsan, Malaysia.

* Corresponding author: sanihashim@unimap.edu.my

Abstract. The purpose of this study is to analyze the fuel consumption of passenger cars in Malaysia by using 1D-simulation. The analysis of fuel consumption will focus on common model of cars in Malaysia and will be carried out by using existing driving cycle. New European Driving Cycle (NEDC), Federal Test Procedure (FTP75) Driving Cycle, and Worldwide Harmonized Light Vehicle Test Procedure (WLTC) Driving Cycle were used as the established driving profiles. The new local driving cycle which was developed by UniMAP Motorsport Technology Research Unit (MoTECH) was used as the main driving profile for the simulation. A 1.3L Perodua Myvi 2012 with manual transmission was been selected as the vehicle model for the project. The simulation results were compared with the established driving cycles. The results show the average fuel consumption and emissions of passenger cars is critically affected by the driving behaviors' and vehicle engines condition. The aggressive driving gives the higher rate of fuel consumption and also the cold start-up engine affecting the fuel devours and emissions rate. In addition, comparison of the obtained simulation results with manufacturer's published data showed a good agreement thus indicating suitability of the model to be used in further research.

1. Introduction

Fuel consumption is the measure of fuel that vehicle utilized it to travel a particular range at a specific velocity. It is rate at which an engine uses fuel, which is expressed in units as miles per gallon or liters per kilometer. Malaysia is a quickly creating nation in Asia, where the interest for traveler autos expands each year. The increasing of demand for passenger cars in Malaysia was caused the rate of air pollution becomes higher since the rate of exhaust emissions from the cars was increased indirectly by large uses of fuel consumption. The larger uses of fuel consumption will increased the rate of exhaust emissions that indirectly can pollute the air quality. Hence, it is important to carry out the study on average fuel consumption and exhaust emission from the cars as a measure of energy optimization and environmental quality. Fuel consumption is the amount of fuel a vehicle utilizes to travel in a particular space at a particular velocity. It is rate at which an engine uses fuel, expressed in units such as miles per gallon or liters per kilometer. Fuel consumption or fuel utilization is an essential engineering degree which straightforwardly leads towards to fuel devoured per 100 kilometres. It is valuable since it capable as a coordinate degree of fuel optimization [1]. The fuel consumption has utilized as a standard to calculate the fleet normal fuel utilization execution for the urban and interstate cycles.

The exhaust emissions and fuel utilization were critically affected by the vehicle start-up conditions factor [2]. The cold start-up and hot start-up are depends on the catalytic converter. The hot start-up prefers a better fuel proficiency and lower emission rate compared to the cold start-up. Speeding up mode is critical factors that influencing the fuel utilization. A study has been conducted by Iftishah [3] which is to evaluate the fuel utilization from the graphic data obtained precisely. The study is divided into two parts, simulation and experiment.

In this project, fuel consumption of passenger cars in Malaysia was analyzed by using the GT-SUITE simulation software. A 1.3L Perodua Myvi model has been selected as the projects' model and the driving profile are based on the local driving cycles. The simulations will give the outcomes about the average of fuel consumption. Then, the simulation results will be validated by made a comparison with the real world data and also with the specification data from the manufacturer.

2. Development of a car modelling

The car modelling was developed based on 1.3L Perodua Myvi model using GT-Suite software. The specifications of the car are given by Table 1.

Table 1. Specifications of 1.3L Perodua Myvi 2012 Products [4].

Engine Type		1NR-VE, 4-cylinders
Engine Displacement		1329
Maximum Output,	kW/rpm	70/6000
Maximum Torque,	Nm/rpm	121/4000
Kerb Weight	kg	975
Wheelbase	mm	2200

A complete car was modelled in GT-Suite consists of nine parts of component which are engine, clutch, transmission, vehicle body, driveshaft, differential, axles, tires, and environment as shown in Figure 1. The attributes for each parts of components is filled by using the approximately real car specifications. There were several attributes that set to the default since certain data cannot retrived from the manufacturer. However, the input parameter still in standard ranges and approximately to the real data. Among the parts of components for car modelling, there were several critical parts that highly affected the result which is engine, vehicle body, transmission and environment. Table 2 shows the input parameter for several parts of components of car modelling.

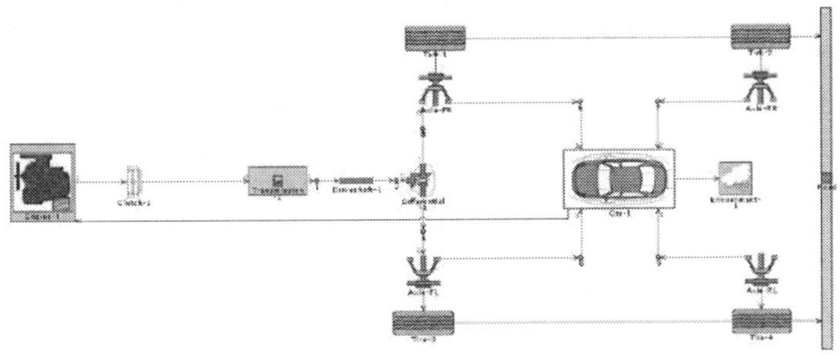

Figure 1. Complete Block Diagram for Car Modelling

Table 2. Parameters of 1.3L Perodua Myvi 2012 model. [3]

Parameter			Value
Maximum engine torque	T_{engine}		116 Nm
Total gear ratio	Gear ratio	1st gear	11.01
		2nd gear	6.15
		3rd gear	4.03
		4th gear	2.81
Efficiency of gear transmission	H_t		0.85
Area of frontal	A		2306 m^2
Coefficient of air drag	C_d		0.32
Density of air	ρ		1.225 kg/m3
Radius of tire	r		0.288 m
Mass of driver + mass of vehicle	m		1055 kg
Coefficient of rolling resistance			0.01

For the driving profile, there were three types of existing driving cycle is used for the simulation process which is New European Driving Cycle (NEDC), Environmental Protection Agency Federal Test Procedure (FTP75), and Worldwide Harmonized Light Vehicle Test Procedure (WLTP). All this type of driving cycles were selected to make a comparison with the simulation analysis by using local driving cycle which named as Urban Driving Cycles (UDC). Hence, all four type of driving cycles were used during the simulation analysis and some comparisons between them would be made.

3. Results and Discussions

According to the fuel performance factory claims made by Zal [6], the 1.3L Perodua Myvi 2012-2014 made products has used the NEDC for their fuel economy and performances. The Perodua claims that their 2012's 1.3L Myvi had average fuel consumption at 11.9 km/L. The simulation has been conducted as 1D simulation in GT-Suite for each of the driving cycle. The results were tabulated and represented in Table 3.

Table 3: Comparison of Simulation Results between Different Type of Driving Cycle

		Type of Driving Cycle			
Criterion	Unit	WLTP	FTP75	NEDC	UDC
Engine Total Fuel Consumption	g	675.90	1427.29	1358.88	928.33
Total Distance Travelled	m	11001.91	17717.06	22613.55	13259.05
Average Vehicle Speed	km/h	33.37	34.03	43.23	31.82

Maximum Vehicle Speed	km/h	120.03	91.20	123.09	87.35
Average Fuel Consumption	km/L	12.30	9.38	12.64	10.80
Average CO Emissions Rate	g/h	0.17	0.13	0.15	0.15
Total CO Emissions	g	203.76	237.48	266.47	220.19
Average NOx Emissions Rate	g/h	0.03	0.04	0.03	0.03
Total NOx Emissions	g	38.00	74.29	62.03	49.56
Average HC Emissions Rate	g/h	0.01	0.01	0.01	0.01
Total HC Emissions	g	11.22	14.58	14.45	13.38

Thus, the simulation result that used the NEDC and UDC as the driving cycle profile is been critically compared since the claims made by Zal [6] also state the factory data used the NEDC as their parameter input data. By referring the engine total fuel consumption criteria, the NEDC simulation result gave 1358.88g based on its profile transient while the UDC simulation result give 928.33g of engine total fuel consumption based on its own profile transient. To calculate the average mass of fuel consume per kilometre for each driving cycles simulation results, the Equation 1 is been used.

$$Average\ Mass\ of\ Fuel\ Consumption = \frac{Engine\ Total\ Fuel\ Consumption}{Total\ Distance\ Travelled} \quad (1)$$

The average mass of fuel consumption for the NEDC is 60 g/km while the average mass of fuel consume for the UDC was about 70g/km approximately. It was about 14% percentage errors between both simulations data. Table 4 shows the comparison for average fuel consumption between NEDC, UDC and factory data claim by Perodua.

Table 4: Comparison of Average Fuel Consumption between NEDC, KLDC and Factory Data Claims by Perodua

CRITERIA	**FACTORY DATA (1.3L PERODUA MYVI)**	**NEDC SIMULATION RESULTS**	**UDC SIMULATION RESULTS**
Average Fuel Consumption	11.9 km/L	12.3 km/L	10.8 km/L

To verify the vehicle car modelling developed from this project, some comparison are been made by referring the factory data claims and compared it with the NEDC simulation results. Equation 2 was used to calculate the percentage error between the simulation data and factory claimed data.

$$Percentage\ Error(\%) = \frac{|Original\ data - Simulation\ data|}{Original\ Data} X\ 100\% \quad (2)$$

By applying the Equation 2, the percentage error between NEDC simulations data and factory data was about 3.36% approximately. The percentage error between both data is very acceptable and improve the accuracy of the results. Thus, it can validate the vehicle car model that has been developed during the car modelling stage of the project and make the car modelling by using the GT-SUITE software was able to used for simulation analysis.

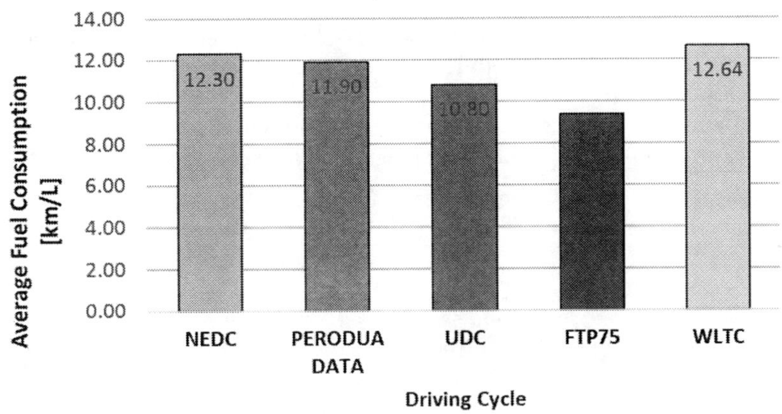

Figure 2: Comparison of average fuel consumption for different driving cycles and Perodua Data.

Figure 2 shows the comparison of average fuel consumption between different driving cycles and compared to the factory data claims by the manufacturer. From the results, the average fuel consumption of NEDC is the nearest with the factory data claims. The second nearest with the factory data claim was theUDC simulations data which is 10.8km/L. The percentage error between the UDC simulation data and factory data claim was about 9.24%. This shows that Malaysian driving behaviour are more aggressive, thus the fuel consumption is higher. Moreover, the developed driving cycle was based on data from urban area, which consist much of decelaration-acceleration conditions. Nutramon [7] had claims that the rough driving behaviour has used the brake for multiple times during sudden decceleration and acceleration, may affect the condition of engine states which may used larger power for the braking mechanism. FTP75 gave the highest average fuel consumption rate while WLTC gave the lowest average fuel consumption rate, which are 9.38 km/L and 12.64 km/L, respectively.

4. Conclusion

This study had compared the fuel consumption for a passenger car in Malaysia through four different driving cycles by using GT-SUITE. The tested car modeling has been developed based on 1.3L Perodua Myvi Model 2012. Data specifications from the previous study and Perodua factory data had been used for modelling. One of the driving cycles was developed for urban area in Malaysia. This driving cycle will reflect the driving behaviour of Malaysian, namely UDC. From the results, Malaysian driving behaviour were more aggressive, thus the fuel consumption is higher compared to NEDC and factory claims fuel consumption. For validation of car modelling, the simulation results of NEDC was compared to the factory data that also used the European drive cycle as its driving profile and give a percentage error about 3.36%, which is acceptable. This study is the preliminary study to the fuel consumption and exhaust emission investigation in Malaysia using driving cycle that reflects driving behaviour of Malaysian.

5. References

[1] C. Fischer, Assessment of Fuel Economy Technologies for Light-duty Vehicles. Washington, D.C.: National Academies Press, 2011.

[2] K. Den Braven, A. Abdel-Rahim, A. Battles and K. Henrickson, "Integrating Field-Collected Data Into Microscopic Simulation", Modeling Vehicle Fuel Consumption and Emissions at Signalized Intersection Approaches, 2012.

[3] M. Iftishah Ramdan and C. Piew Lim, "Perodua Myvi Engine Fuel Consumption Map and Fuel Economy Vehicle Simulation on The Drive Cycles Based on Malaysian Roads", MATEC Web of Conferences, pp. 1-7, 2017.

[4] "Myvi Specification", Perodua. [Online]. Available: http://www.perodua.com.my/specification/myvi. [Accessed: 11- Feb- 2018].
[5] EPA (U.S. Environmental Protection Agency). 2008. Light-Duty Automotive Technology and Fuel Economy Trends: 1975 Through 2008. EPA420-R-08-015. September. Washington, D.C.
[6] P. Zal, "Perodua Myvi 1.3 SX (man. 5) (2012) Detail performance and fuel economy data factory claim and ProfessCars™ estimation", Automobile Catalog, 2012. [Online]. Available: - http://www.automobile-catalog.com/auta_perf1.php. [Accessed: 11- Apr-2018]
[7] T. Nutramon and C. Supachart, "Influence of Driving Cycles on Exhaust Emissions and Fuel Consumption of Gasoline Passenger Car in Bangkok", Journal of Environmental Sciences, vol. 21, no. 5, pp. 604-611, 2009.

Acknowledgments
The authors would like to acknowledge Malaysia Automotive Institute (MAI) for funding this study and PERODUA for technical supports.

IOP Conf. Series: Materials Science and Engineering **429** (2018) 012058 doi:10.1088/1757-899X/429/1/012058

Potential of Nano-Fluid Application in Deep Well Drilling Operation Challenges

A R Ismail[1*], A R A Hakim[2] and M N A M Norddin[1]

[1]Faculty of Chemical & Energy Engineering, Universiti Teknologi Malaysia, Malaysia
[2]Petronas Carigali Sdn. Bhd., Malaysia

*Email: razak@utm.my

Abstract. Nanotechnology have vast potentials for applications in oil & gas industry. Nanoparticles are ultra-fine in nature, usually larger than an atom cluster but smaller than ordinary micro particles and thus have very high specific surface area with enormous potential interactions with its surroundings. In fact, some nanoparticles exhibit high thermal conductivity ability and affinity to acid gases. These traits and features possessed by nanomaterials intrigues the industry players and makes them zoomed on its potential application in the industry. In this study, the potentials of nanomaterials application in formulating "smart" drilling fluids for deep well drilling operations are discussed. Major deep well drilling fluid related challenges such as narrow drilling tolerance window, HPHT environment, excessive fluid loss and extensive formation damage, wellbore stability issues, differential sticking and low rate of penetration could be addressed through the addition of specific functionally tailored nanoparticles in the fluid formulation. It has been proven at least experimentally in laboratories and some minor field applications that addition of nanomaterials to the formulation would bring positive impacts on the plastic viscosity, yield strength, gel strength, filtrate loss volume and the lubricating effects of drilling fluids, making them feasible to be used in addressing the challenges in deep well drilling operations as discussed in this project paper. Due to the rheological enhancements, nano-fluids application is deemed technically feasible, cost effective and more environmental friendly due to the low concentrations requirement, in formulating smart fluids that could help to combat and address the major and prominent challenges in deep well drilling operations.

1. Introduction

Hydrocarbons are finite resources and hence the surging demand made by the global community will need to be addressed via expansion of the reserves through continuous exploration activities and maximum recovery. The depleting amount of reserves in the low hanging fruits category has forced the oil and gas players to venture further into the sea; offshore exploration and even deep water and ultra-deep water exploration.

However, as the hydrocarbons exploration and extraction activities moves further deep down the Earth's crust, the challenges and risks involved prevails. Deep well drilling activities are commonly associated with complex drilling hazards usually not encountered in onshore and shallow drilling operations. In fact, the recent improvements in drilling techniques have imposed a new set of requirements on the drilling fluid; e.g. horizontal and extended reach drilling, multilateral drilling, mono-bore drilling, etc. Therefore, the formulation of a drilling fluid that is able to live up to the

IOP Conf. Series: Materials Science and Engineering **429** (2018) 012058 doi:10.1088/1757-899X/429/1/012058

expectation from technical, commercial and environmental aspects, is paramount in the current drilling environment.

In the efforts to formulate such wholesome drilling fluid, the industry has called for the mapping of some of the technological breakthroughs in fields like nanotechnology, with the hope that they will help in curbing the challenges that exist in the deep water plays aforementioned. This paper will hence entail the analysis and the discussion of the potential and feasibility of nanomaterials in addressing the challenges in deep water drilling focusing on the fluid system.

2. Deep Well Drilling Challenges for Conventional Fluid

As we explore farther and deeper into the Earth's crust for new hydrocarbon prospects, the drilling challenges gets tougher and intensified. The problems in many cases can be attributed to the performance of the fluids component used in drilling operations. Such problems may include well control issues, poor wellbore cleaning, loss circulation zones, fluid gelation, reservoir fluid invasions etc. Thus, the industry would need to figure out a robust solution to overcome these challenges [1]. Hence, the need to formulate a new comprehensive drilling fluid which is capable of good continuous performance under the adverse conditions presented by the deep water environment. Some of the prominent challenges in deep water drilling operations are discussed below.

2.1. Narrow Drilling Tolerance Window (DTW)

DTW is the range or difference between pore pressure and fracture gradient where the drilling operations can safely operate in without the risk of kicks or blowout. It is preferable to operate in big DTW as the range would provide a certain extent of safety and flexibility for the operators to formulate the drilling fluid to cater for different pressure zones. However, should the DTW becomes narrower, especially in deep well environment, the formulation of drilling fluid to cater for the different pressure zones becomes trickier. Figure 1 explains the DTW comparison between shallow well environment and deep well environment. DTW is indicated by the area between the blue dotted line (Pore Pressure Gradient) and the red dotted line (Fracture Pressure Gradient).

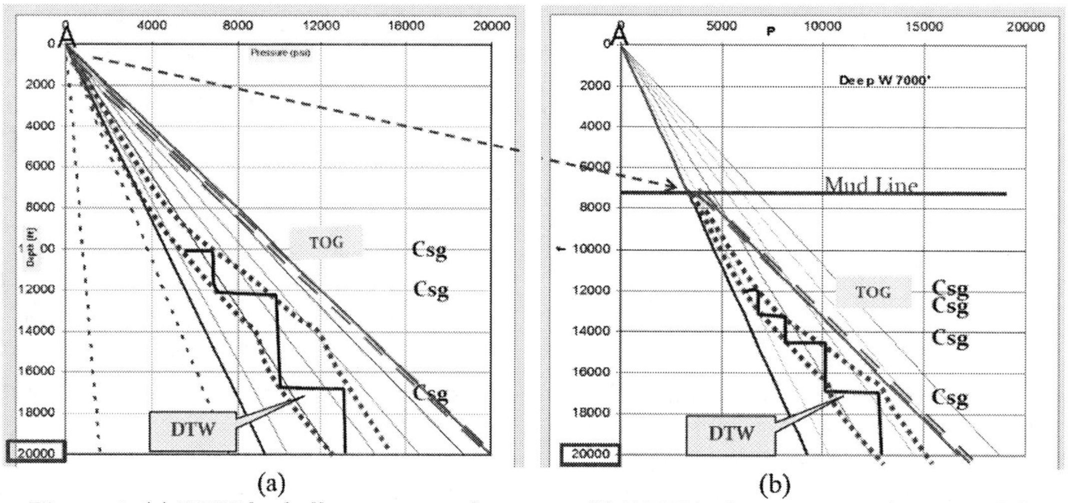

(a) (b)

Figure 1. (a) DTW in shallow water environment, (b) DTW in deep water environment [2].

The narrow DTW in deepest water environment has increased the complexity of the drilling activities in these areas; especially in maintaining the pressure balance inside the wellbore to prevent undesired events such as kicks or blowout. Hence, it is vital for the drilling fluid formulated to be able to function in such environment.

2.2. High Pressure - High Temperature (HPHT) Environment

Deep wells are mostly associated with HPHT environment. As the depth of drilling increased, more severe drilling conditions can be expected which may exceed temperature of 600°F (315°C) and pressure as high as 40,000 psi [3]. Drilling in a HPHT environment may affect the efficiency of the drilling operations when the drilling equipment, tools and materials ability to perform optimally is jeopardized due to the HPHT environment. More critically, the efficiency of the drilling fluid properties will be severely impacted due to the extreme temperature and pressure. The HPHT environment may change the rheological properties of the drilling fluid especially the polymer-based additives which will degrade under high temperature. When drilling fluids destabilize, it will cause various drilling, logging and completion issues which may eventually lead to kicks or even blowout.

With conventional invert emulsion fluids, the major challenges encountered under HPHT environment are related to the thermal degradation of emulsifiers and fluid-loss reducer that can lead to gelation and loss of rheological properties that can cause weighting material sagging, and other associated well control problems. An HTHP rheology plot of a conventional invert emulsion fluid is shown in Figure 2.

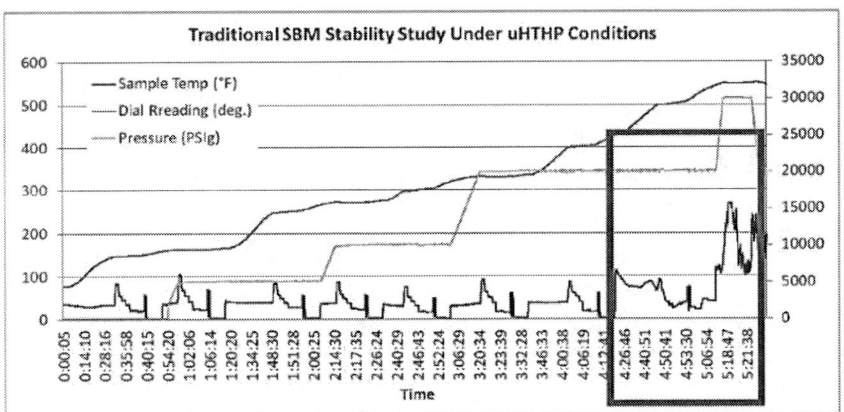

Figure 2. Effect of temperature and pressure on the rheology of a conventional invert fluid. The boxed area shows the impact on rheology after 60 minutes at temperatures above 400°F (204°C) [4]

2.3. Well Formation Instability

The instability of the well formation is another major challenge need to be faced in deep well drilling operations. Formation instability such as caving, sloughing and collapsing are some of the examples frequently faced especially while drilling through shale formations. They happen mostly due to the hydration of the shale formations or by formation geo-mechanical failure due to the in-situ formation pressure anomalies.

In many deep well cases, both factors co-exist and effective simultaneously, hence, may cause the shale formation to swell and cause geo-mechanical failures. These may result in the slough of the shale formation which may lead to other drilling problems (stuck pipe and etc). The hydration of the formation caused by fluid invasion, can generally be minimized by the formulation of a proper fluid chemistry – adding appropriate additives (shale hydration inhibitors) such as bentonite into the drilling fluid. Hence, it is important to have a robust drilling fluid with the ability to prevent excessive fluid invasion into the formation while strengthening the mechanical strength of the formation.

2.4. Fluid Circulation Loss

During a typical overbalanced drilling, drilling fluids are used to prevent influx of formation fluids into the wellbore. However, in doing so, the drilling fluids will in turn invade the formation to a particular extent especially in the highly porous and permeable zones and naturally fractured zones (Fig. 3).

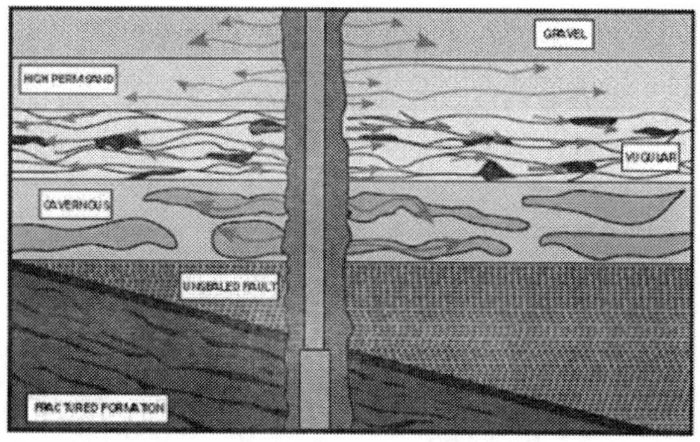

Figure 3. Example of naturally occurring loss zones [5]

Lost circulation may lead to well control issues and may even results in potential kicks and blowouts which will bring adverse impacts to the environment and personnel. Hence, in the effort to minimize the fluid loss through the invasion, drilling fluids are often added with plugging materials such as bentonite to aid in plugging the pores on the porous and permeable zones, forming mud cakes in doing so. Mud cakes presence is very critical as it defines the effectiveness of the plugging in avoiding fluid loss. The mud cakes however, could become too thick, it impedes the movement of the drill strings; making it stuck. This is often called differential sticking and this will be discussed in the following section.

2.5. *Differential Sticking*

Differential sticking occurs when the drill string is "sucked" up against the formation due to the pressure difference between the wellbore and the formation. This normally occurs at permeable zones where the mud cake is formed. Thick mud-cake may increase the likelihood of differential sticking during drilling operations. Hence, it is vital that the mud cake formed by the drilling fluid is kept at a minimum thickness while maintaining its effectiveness in preventing or minimizing fluid loss. This is another area that nanotechnology might come into the picture to aid.

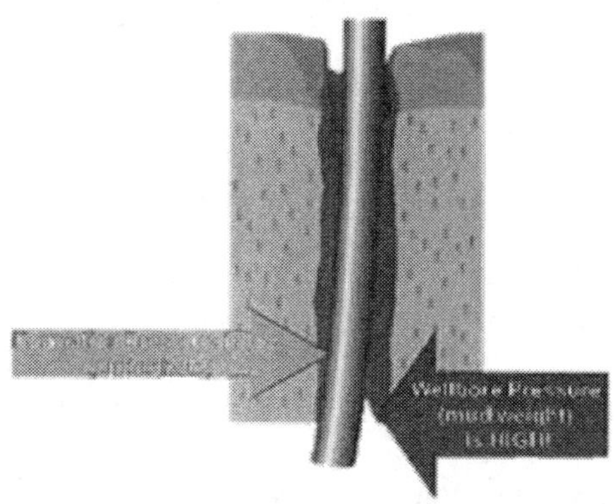

Figure 4. Differential sticking of drill string in wellbore [5]

3. Nanotechnology and Nanomaterials

"Nano" denotes a thousand millionths (10–9), hence, a nanometer is a millionth of a meter. That equals to the width of 10 hydrogen atoms. Indicatively, a human red blood cell is approximately 7,000 nanometers wide and a water molecule is 0.1 – 0.3 nanometers across. At nanoscale, the laws of classical dynamics begin to deteriorate and nanoparticles traveling at high velocities are subject to Einstein's law of relativistic-quantum mechanics. In other words, Quantum Mechanics becomes extremely important at this atomic scale [6]. Amazing phenomena begins to occur on the particles in this domain and this is what intrigues the industries to take a deep interest in nanotechnology.

4. Addressing the Challenges in Deep Well Drilling Operation: Nanotech Way

Due to the enhanced physio-chemical, mechanical, electrical, thermal, hydrodynamic properties and interaction potential, nanomaterials are deemed to be the most promising option for smart fluid design in oil and gas field application [7]. Moreover, with the possibility of manufacturing tailored made nanomaterials with custom made functional behaviour, nanotechnology widened the door to the development of a new generation of drilling fluids.

These ultra-fine particles are usually larger than an atom but smaller than the micro particles. They have a high specific surface area to volume ratio, with enormous area for interactions with its surroundings. Hence, due to the nanoscale particle dimension, nanoparticles are thus expected to deliver superior fluid properties at a significantly reduced additive concentration [8]. In highlighting the potential of nanoparticles in addressing the technical challenges in deep water arena, the author has gathered and analyzed papers and reports that highlights the impacts of nanoparticles application in the field as well as those works in experimental stage in the laboratories.

4.1. HPHT Environment

The stability of additives in HPHT conditions is often the determinant for the drilling fluids to be able to fulfill the expected functional tasks as temperature above 200°C and downhole pressure of 20,000 psi and above are very common in deep water wells. In fact, when temperature exceeds 400°F (204°C), the stable additive options are almost nonexistent for both water based and non-aqueous based fluids [9]. The extremely high surface area to volume ratio of nanoparticles leads to a more effective interaction between the nanoparticles surfaces with external and internal rock surfaces and enhance the thermal conductivity of nano-fluids, which in turn enhances the efficiency of the cooling system. Hence, the nanoparticles, with their excellent thermal conductivity are deemed to be the most suitable materials of choice for application in such environment [8].

Experimental study on OBM with nanomaterials added as replacement of the conventional polymer-based additives and the fluid formulation was subjected to high temperature (175°C) aging for up to 96 hours. The rheological properties measured showed that the nano-fluid formulation maintained the desirable properties after aging and exposure to high temperature [10].

4.2. Loss Control & Formation Damage

Mud filtrates is believed to be the major factor contributing towards formation damage while drilling. Hence, formulation of fluids with no or negligible filtrate loss potential in downhole conditions is essential to avoid or minimize excessive filtrate loss and most of the mud induced formation damage. The prevention of circulation and filtrate loss by current drilling fluid with conventional lost circulation materials showed limited success in these environments. However, due to the extremely small size of the nanoparticles, they are able to provide effective sealing effect in the porous and permeable zones, fractured and cavernous formations. Due to the extremely small size, nanoparticles are able to access the smallest pores and pore throats in the matrix and act as sealing agents in all lithology types including even shales.

4.3. Wellbore Stability & Narrow Drilling Tolerance Window

One of the causes of wellbore instability in deep water environment is because the formations below the deep sea bed are weakly bonded with little cementation and inter-particle cohesion [11]. This is due to the presence of a big water column during the deposition instead of a long rock column as compared to onshore and shallow water formations of similar vertical depth. In other words, the

IOP Conf. Series: Materials Science and Engineering **429** (2018) 012058 doi:10.1088/1757-899X/429/1/012058

formations in the deep environment are often unconsolidated and loose as compared to that of onshore and shallow environment.

Previous experience and drilling data reveals that various drilling problems arise while drilling unconsolidated formations using the conventional macro or even micro-material-based drilling fluid. These fluid formulations are often not able to prevent sanding, hole collapse, washout, fracturing problems due to their inability to generate effective inter-particle cohesion and cementation to strengthen near wellbore formation. Due to the ultra-fine size, nanomaterials are able to access the pores and the inter-granular contact surfaces of the unconsolidated sand particles. Hence, nanomaterials with gluing and cementing properties are expected to form bonded networks of particles within the formation matrix creating an integrated ring of rock mass around the wellbore wall [8].

4.4. Differential Sticking

Since the presence of thick mud cake may increase the likelihood of differential sticking during drilling operations, it is important to produce a thin and low sticking mud cake on the wellbore wall in preventing pipe sticking problems [8]. As established in the previous sections of this paper, nano-fluids has proven to possess the ability to produce thin, non-erodible, impermeable mud cake that are effective in preventing fluid loss. In addition, nano-fluids would be able to reduce the adhesive tendency of mud cake by forming another thin non-adhesive nano-film on the drill pipe surface. This would help to decrease the likelihood of stuck pipe to occur.

Due to the extremely small particle size, the nanoparticles added into the formulation of nano-fluids can easily enter the pipe-mud cake interface to form an effective lubricating film, isolate the mud cakes or even destroy the mud cake for an effective de-bonding of the pipe from the mud cake surface. This demonstrates the potential of nanomaterials application in the formulation of both drilling and spotting fluids to address the pipe sticking problems. Therefore, in curbing the stuck pipe problems, the researchers have decided to use nanoparticles (carbon black particles) in their drilling fluid formulation in their experimental study [12]. The specific gravity of carbon black is typically in the range of 1.9 – 2.1 and its thermal stability is up to 3,000°F (1,649°C). The estimated initial diameter of carbon black particles is about 30 nanometers in which after aggregation this value will be about 150 to 500 nanometers.

4.5. Rate of Penetration and Bits Balling

Solid contents of in the drilling fluid is one of the factors that reduces the ROP. Hence, it is important to keep the amount of solid contents in the fluid at a minimum required level [7]. It has been established in the previous section that nanomaterials application in the drilling fluid would be able to form a continuous layer of a very thin film on the surface of wellbore surfaces, mud cake and even on the surface of the drill pipes. This characteristic of nano-fluids would be able to provide sufficient lubricating features, minimize the frictional resistance between the drill pipes and the wellbore surfaces, minimize the torque and drag problems and hence improve the overall ROP.

Apart from that, the progressive sticking and accumulation of this gumbo shale on the bit tooth gaps and the tooth surfaces can turn the drill bit into a ball-like structure which would drastically reduce the ROP and the drilling efficiencies [8]. This could lead to a significant increase in the total drilling cost should the operations need to be stopped to clean or change the drill bit. Hence, nano-fluid formulation containing a combination of inhibitive and hydrophobic film forming nanomaterials can eliminate and minimize the bit balling problems and could potentially lead to cost saving measures. In conclusion, the enhanced lubricating effect provided by nano-fluids application would bring positive impacts on the overall drilling performance, increases the ROP, minimizes tools and equipment break down related problems and hence, brings in a significant potential cost saving and value creation for the drilling operation.

5. Health, Safety and Environment

Technically, from environmental compliance point of view, the application of nanoparticles in drilling fluid may be deemed a more environmental friendly measures as the concentration requirement is low (typically less than 1% concentration) in the fluid formulation. Hence, nano-fluids could be the fluid of choice in conducting drilling operations in sensitive environments [7].

However, at the moment, the eco-toxicological and toxicological data library of nanomaterials applications in oilfield environments is barely minimum and thus insufficient to make any broad statements regarding the health and ecological effects of these nanoparticles. Hence, it is highly recommending for the related government and non-governmental bodies to establish specific local work instructions and enforce mandatory use of appropriate personnel protective equipment (PPE) to minimize or prevent any unwanted health and environmental risks and exposure [9].

6. Conclusions

The reviewed literature suggests that the current drilling operation is facing many fluid and materials based challenges, some of which may be able to be addressed through application of nanomaterials. Nanotechnology offers many potential solutions to drilling fluid related problems especially in deep well environment that cannot be effectively solved with the current conventional fluid formulation.

The laboratory tests and a few field application results reviewed here shows that although most of the nano-fluids and solutions are still in the laboratory testing stage, they could be extensively applied in the field once the economic challenges have been resolved. The tremendous improvement it brought unto the rheological properties of the drilling fluid has made it the suitable option (if not the only option), to be used in technically challenging environment of deep well drilling operations.

Hence, through the discussion made, it is concluded that though there has been limited application of nanomaterials in the drilling fluid formulation to date, the numerous benefits of its application in addressing some of the major problems and issues in drilling operations especially deep well drilling operations cannot be ignored.

Acknowledgments

The authors wish to thank Universiti Teknologi Malaysia for funding this project under Research University Grant vote Q. J130000.2546.17H77.

References

[1] Kelessidis V C 2009 Drilling Fluid Challenges for Oil-Well Deep Drilling *9th International Multidisciplinary Scientific Geoconference & Expo, Modern Management* Albena, Bulgaria

[2] Shaker SS 2011 The Deep Water Drilling Tolerance Window: Walking a Fine Line, Case Histories from Gulf of Mexico. Search and Discovery *AAPG Annual Convention and Exhibition* Houston, Texas, USA

[3] Agarwal S, Tran P, Soong Y, Martello D and Gupta R K 2011 Flow Behavior of Nanoparticle Stabilized Drilling Fluids and Effect of High Temperature Aging *AADE National Technical Conference and Exhibition* Houston, Texas

[4] Stefano G D, Stamatakis E, Young S 2013 Development and Application of Ultra-HPHT Drilling Fluids *11th Offshore Mediterranean Conference and Exhibition* Ravenna, Italy

[5] Fereday K 2012 Downhole Drilling Problems: Drilling Mysteries Revealed! *BP's 58th Annual Meeting* San Francisco

[6] Esmaeili A 2009 Applications of Nanotechnology in Oil and Gas Industry *Petrotech* pp 9-76 New Delhi, India

[7] Amanullah M, Al-Arfaj M K and Al-Abdullatif Z 2011 Preliminary Test Results of Nano-based Drilling Fluids for Oil and Gas Field Application *SPE/IADC Drilling Conference and Exhibition* Amsterdam, The Netherlands

[8] Amanullah M and Al-Tahini A M 2009 Nano-Technology- Its Significance in Smart Fluid Development for Oil and Gas Field Application *SPE Saudi Arabia Section Technical Symposium and Exhibition* AlKhobar, Saudi Arabia

[9] Hoelscher K P, Young S, Friedheim J and Stefano G D 2013 Nanotechnology Application in Drilling Fluid *11th Offshore Mediterranean Conference and Exhibition* Ravenna, Italy

[10] Agarwal S, Walker L M, Prieve D C, Soong, P Y and Gupta R K 2009 Using Nanoparticles and Nanofluids to Tailor Transport Properties of Drilling Fluids for HTHP Operations *AADE National Technical Conference & Exhibition* New Orleans, Louisiana

[11] Amanullah M and Richard B 2006 Experimental Evaluation of the Formation Strengthening Potential of a Novel Gel System *IADC/SPE Asia Pacific Drilling Technology Conference* Bangkok, Thailand

[12] Paiaman A M and Al-Anazi B D 2008 Using Nanoparticles to Decrease Differential Pipe Sticking and Its Feasibility in Iranian Oil Fields *Oil & Gas Business* Sharif University of Technology

IOP Conf. Series: Materials Science and Engineering **429** (2018) 012059 doi:10.1088/1757-899X/429/1/012059

Hydrolysis Empty Fruit Bunch (EFB) Using Green Solvent

A Ibrahim[1], M F Abdullah[1,2*], S T Sam[1*]

[1]School of Bioprocess Engineering, Universiti Malaysia Perlis, Kompleks Pengajian Jejawi 3, Kawasan Perindustrian Jejawi, 02600 Arau, Perlis, Malaysia.
[2]Centre of Excellence Biomass Utilization (CoEBU), Universiti Malaysia Perlis (UniMAP), Kompleks Pusat Pengajian Jejawi 3, 02600 Arau, Perlis, Malaysia.

*Email: faiq@unimap.edu.my

Abstract. Empty fruit bunch (EFB) contained high amount of cellulose and can be further utilized for nanocrystalline cellulose (NCC) production. Increasing awareness on environment has attracted attention on green solvent compared to the usage of conventional strong acid in nanocellulose isolation. In this study, NCC has been isolated from raw EFB using environmental friendly solvent, natural deep eutectic solvent (NADES). The purpose of this study is to investigate the effect of different types of NADES on NCC isolation. Three different types of NADES based on three different organic acids (citric acid, malic acid and lactic acid) have been used in this study. The solubility of NCC in these NADES was measured while their physical characterization was determined using Fourier transform infrared (FTIR), scanning electron microscopy (SEM) and field emission scanning electron microscopy (FESEM). From the results, EFB dissolution in NADES made up of citric acid shows higher solubility with a value of 2.74%. FTIR spectra shows similar spectra between treated EFB and NCC isolated using NADES. From FESEM micrograph, citric acid based NADES is able to produce NCC in the range of 25 to 37 nm. Based on the results, NCC has been successfully isolated using NADES and further improvement can be applied to enhance its production.

1. Introduction

Over the last 25 years, Malaysia has been one of the global oil palm producer contributing about 40-60% of the world oil palm [1]. Consequently, this industry generates lots of lignocellulosic waste such as empty fruit bunch (EFB). EFB from oil palm consist of 44.4% of cellulose, 30.9% of hemicellulose and 14.2% of lignin [2]. Inexpensive, readily available and high cellulose content are the factors that have raised the interest in utilizing them in energy production, medical and pharmaceutical products, and industrial biodegradable products [3]. Besides, the high cellulose content in EFB can be used to produce nanocrystalline cellulose (NCC).

NCC is also known as cellulose nanowhiskers and cellulose microcrystals [4]. NCC are rod like particles and highly crystalline with a diameter of 1-100nm while its length varies between tens to hundreds of nanometers[5]. NCC are generated when polysaccharides bound at the fibril surface are removed which result in the cleavage and destruction of the vulnerable amorphous region [6]. NCC possesses remarkable properties such as large surface area, superior mechanical strength, high specific surface area, non-toxic and biodegradability which make it as one of the most promising new class of nanomaterial [7]. NCC is usually produced by conventional method using strong acid hydrolysis to remove amorphous parts so that highly crystalline NCC can be obtained [7]. However, concerns are raised due to the toxicity of the NCC produced through conventional method as the utilization of concentrated acid and alkali are hazardous and detrimental to environment [8]. Thus, a new solvent named natural deep eutectic solvent (NADES) is seen as an alternative due to its biodegradability, low toxicity and low cost [9].

NADES is a sub-class of deep eutectic solvent (DES) and it is a mixtures between hydrogen bond acceptor (HBA) and hydrogen bond donor (HBD) components, which comprised of two or three natural compounds [10]. NADES is synthesized from low-cost and naturally available chemicals which is sustainable, biodegradable and has a good physicochemical properties as solvent [9]. Besides, it has low toxicity as it is usually made of compounds that are naturally occurred in the environment[11]. Therefore, to fulfill the increasing interest on green technology, NADES is the most promising solvent to be studied for its potential in the hydrolysis of EFB to obtain NCC. However, the potential of NADES in isolating NCC has yet to be investigated. Thus, the main aim of this study is to discover the potential of NADES in NCC production.

2. Materials and method

EFB was collected from Taclico Company Sdn. Bhd. Chemicals used in this study including sodium hydroxide (NaOH), sodium chlorite ($NaClO_2$), acetic acid (>95%), choline chloride (99%), citric acid 1-hydrate (99.5-100%), DL-malic acid (99+%), DL-lactic acid (85%) and ethanol (95%) were acquired from A.R. Alatan Sains (K) Sdn. Bhd., Malaysia and Modern-Lab Chemicals Sdn. Bhd., Malaysia.

2.1. EFB Pre-treatment

Prior usage, a miniature grinder was used to grind EFB into smaller size and then sieved to get 250μm particle size. Lignin was removed using alkaline treatment by heating 10 g of EFB in 200 mL of 3% (w/v) NaOH solution at 85°C for 1 to 2 hours [12,13]. The mixture was filtered and washed with distilled water until the pH of the wash reached pH 7. Alkaline-treated EFB was mixed with bleaching solution at a solid to liquid ratio of 1:30 and heated at 70°C for 1 hour. 1% (w/v) of $NaClO_2$ was used as bleaching solution and was prepared by putting few drops of glacial acetic acid in the solution to obtain pH 4 [14]. The mixture was filtered and washed with distilled water repeatedly to achieve pH 7.

2.2. NADES Preparation

Three types of choline chloride based NADES was prepared by mixing all components and heated at 80°C for 1 to 2 hours until homogenous liquid NADES was formed [15] 10 wt% of water was added to ease the homogenization of the mixture [16]. Table 1 shows the ratio of NADES prepared in this study.

Table 1. NADES composition and ratio.

HBA	HBD	Ratio	NADES
	Citric acid	2:1	1
Choline chloride	Malic acid	2:1	2
	Lactic acid	2:1	3

2.3. EFB dissolution

EFB was dissolved in NADES to obtain NCC by referring to the method used by previous study [5,17]. Treated EFB was dissolved in NADES at a ratio of 1:40 and heated at 85°C. The mixture was heated and stirred vigorously for 6 hours. Three different NADES as shown in Table 1 before was used to dissolve the treated EFB. Cold distilled water was added into the mixture after 6 hours of heating to stop the reaction. The mixture was washed with ethanol first and then distilled water using centrifuge at 8000 rpm for 10 minutes. During centrifugation step, NCC suspension was collected until there was no more cloudy suspension obtained. NCC suspension was then dried to obtain NCC powder for characterization and solubility measurement. Solubility of NCC in NADES was calculated as the equation (1) shown below.

$$\text{Solubility (\%)} = \frac{\text{Weight of dried NCC}}{\text{Initial weight of treated EFB}} \times 100 \quad (1)$$

2.4. Fourier Transform Infrared Spectroscopy (FTIR)
NCC obtained was analyzed to observe changes in their functional group using Shimadzu IR spectrophotometer model Prestige-21. NCC was mixed with potassium bromide (KBr) to form a pellet and then viewed in the spectrometer using the range of 500 cm^{-1} to 4000 cm^{-1}[18].

2.5. Field Emission Scanning Electron Microscope (FESEM)
The morphology of NCC obtained was determined using field emission scanning electron microscope (Carl Zeiss Leo Supra 50 VP). Scanning electron microscope (Zeis Supra 35VP) was used to evaluate the morphology of raw EFB, alkaline treated EFB and bleached EFB. Platinum coating was used to coat all samples prior analysis.

3. Results and discussion

3.1. Solubility
The ability of NADES to dissolve EFB to produce NCC has been measured in this study. Figure 1 shows the solubility of EFB obtained using different types of NADES. From the results obtained, it is found that NADES 1 that is made up of citric acid is able to dissolve EFB more followed by NADES 2 and NADES 3.

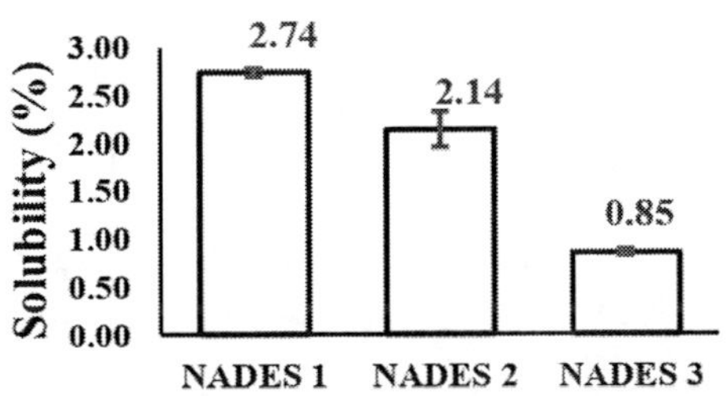

Figure 1. Nanocellulose solubility based on different types of NADES.

It is known that NADES is able to donate or accept electron or proton, making it as an excellent dissolution agent [19]. Besides, NADES also able to promote the hydrolysis of inulin to fructose and followed by the dehydration of fructose to hydroxymethylfurfural (HMF) [20]. Based on Figure 1, NADES 1 is able to dissolve 2.74% of cellulose and consequently produce more nanocellulose compared to the other NADES. In previous study, fructose has been successfully converted to HMF when NADES made up of choline chloride and citric acid is used.[20]. This shows that NADES is able to hydrolyze carbohydrates such as cellulose into NCC. NADES is formed by hydrogen bonding between HBA and HBD. This creates a competition between NADES and cellulose in terms of hydrogen bonding. From the point of view of molecular structure of organic acid used, citric acid contains higher numbers of carbonyl groups and hydroxyl group as compared to malic acid and lactic acid. The amount of carbonyl group and hydroxyl groups in NADES is imperative for the formation of hydrogen bonding with cellulose and consequently increased cellulose solubility [21].

3.2. FTIR Analysis

FTIR spectra of raw EFB, alkaline treated EFB and bleached EFB are portrayed in Figure 2. From the results obtained, EFB undergoes some changes on its chemical structures and functional groups. A broad band between 3000-3500 cm^{-1} appears in raw, alkaline treated and bleached EFB indicating the stretching vibration on free –OH group resulting from the hydrogen bonding in cellulose molecules [5]. The peak that appeared at 1635-1646 cm^{-1} shows the –OH bending vibration of the absorbed water molecules. Meanwhile, the stretching vibration of saturated C-H in cellulose and hemicellulose can be identified by the peak that occurred in the range of 2903 cm^{-1} to 2918 cm^{-1} [22]. Besides, the peak at 1508 cm^{-1} in raw EFB indicates the presence of C=C stretching vibration in the aromatic lignin. This particular peak is not observed in the bleached sample proving that lignin has been removed successfully [23]. The removal of lignin also can be confirmed by the disappearance of peak 1246 cm^{-1} which signify the C-O-C stretching of aromatic ether linkages in raw EFB [24]. The deformation of the vibration of C-H and C-O-C pyranose can be found prominently in bleached EFB at 1105 cm^{-1} suggesting the removal of hemicellulose and lignin [25].

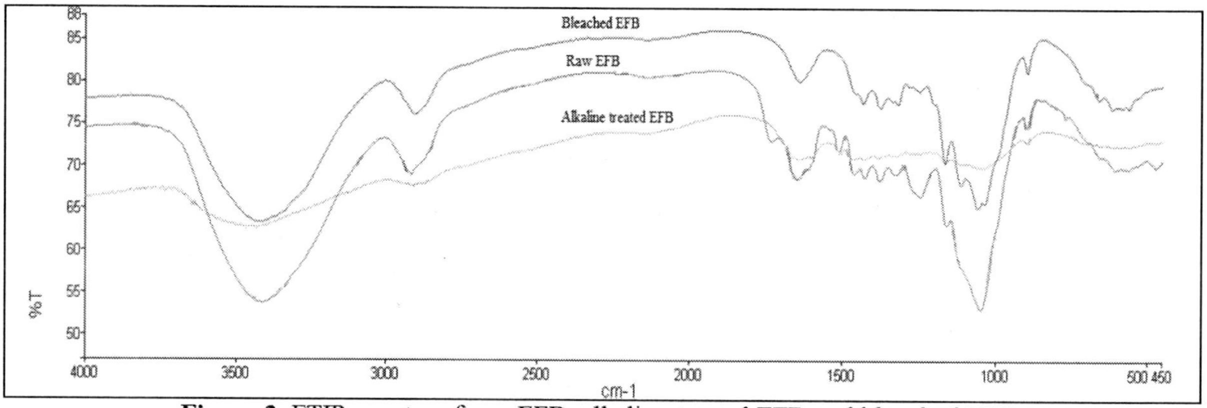

Figure 2. FTIR spectra of raw EFB, alkaline treated EFB and bleached EFB.

FTIR spectra of NCC shows similar cellulose fingerprint region as FTIR spectra of treated EFB indicating that NADES is able to maintain cellulose structure (Figure 3). The peak found at 897 cm^{-1} indicates the deformation of C-H glycosidic bond with a ring vibration contribution and the O-H bending. These shows the presence of β-glycosidic linkage between the anhydroglucose units of cellulose which can also be known as the amorphous region [26]. Based on the samples, NCC obtained from NADES 1 shows the highest transmittance value suggesting that more β-glycosidic bonds have been destroyed. Cellulose structure is further confirmed by the appearance of peak near 1044 cm^{-1} which represents the C-O and C-H stretching vibration [27]. From the spectra obtained, NADES 1 shows higher intensity of this peak which signifies higher crystallinity of the as-produced NCC. The spectra obtained from all treatments suggest that nanocellulose obtained is in cellulose I form due to the presence of peak at 1424 cm^{-1} and 1163 cm^{-1} [18].

IOP Conf. Series: Materials Science and Engineering **429** (2018) 012059 doi:10.1088/1757-899X/429/1/012059

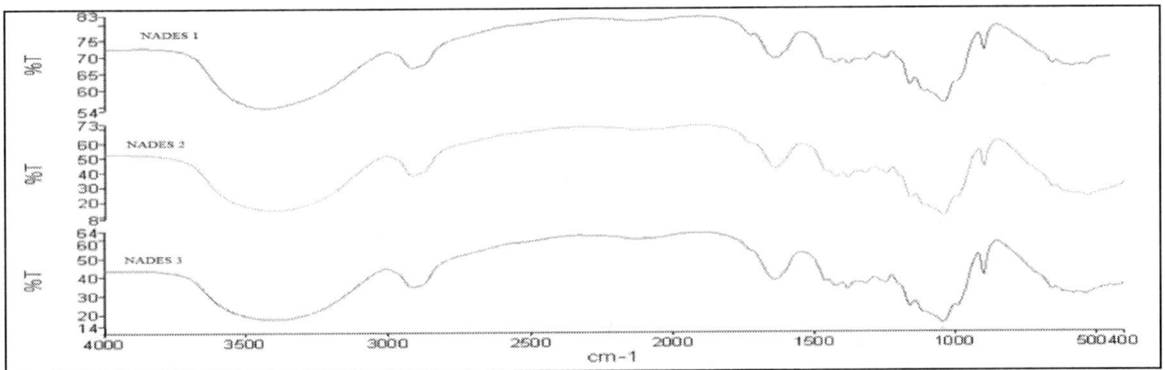

Figure 3. FTIR spectra of NCC obtained from the usage of different NADES.

3.3. SEM and FESEM Analysis

The SEM and FESEM micrographs of raw EFB, treated EFB and NCC obtained from NADES are shown in Figure 4. In Figure 4(a), SEM micrograph of raw EFB shows a smooth, compact structure and non-fibrous component due to the presence of lignin and hemicellulose [28]. These materials hold together bundles of individual fibrils tightly like a cement [29]. Silica can also be found on the surface of raw EFB and has been removed during alkaline treatment as shown in Figure 4(b). This is indicated by the formation of empty holes which is previously occupied by silica. It is important to remove silica during pre-treatment process to ease the penetration of hemicellulose [30]. Further treatment with bleaching agent reduces the diameter of the fiber by defibrillation as shown in Figure 4(c). Defibrillation process is initiates by oxidizing the lignin and promote the formation of hydroxyl carbonyl and carboxylic group which help to enhance lignin solubilisation [23].

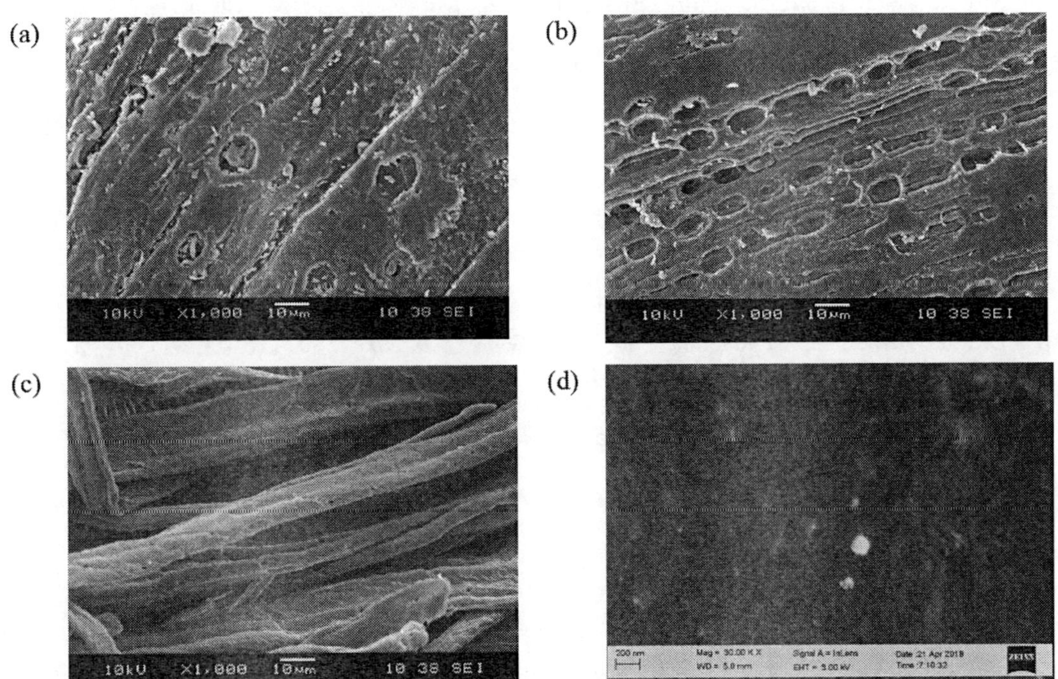

Figure 4. SEM micrograph of a) raw EFB, b) alkaline treated EFB, c) bleached EFB and d) FESEM micrograph of NCC obtained using NADES 1.

Figure 4(d) shows the FESEM micrograph of NCC obtained using NADES 1. The size of NCC obtained ranged of 25 to 37 nm resulting from the breakdown of cellulose fibril into nano-sized cellulose by disruption of hydrogen bonding. The inter-cellulose hydrogen bonding competes with hydrogen bonding formed between carbonyl group of NADES 1 and hydroxyl group of cellulose [21]. Thus, the ability of NADES 1 to form more hydrogen bonding with cellulose has shown its capability to hydrolyse cellulose into NCC.

4. Conclusion

A green solvent synthesized from natural components is able to hydrolyze raw EFB into NCC based on the results obtained in this study. NADES 1 synthesized from the mixture of citric acid and choline chloride has shown better results as compared to other NADES types. NADES 1 able to dissolve more treated EFB into NCC with a solubility value of 2.74%. Moreover, the usage of NADES in producing NCC from EFB does not change the cellulose structure, as proved by similar FTIR spectra of NCC to those of treated EFB. Besides, FESEM micrograph of NCC obtained using NADES 1 shows that NADES 1 able to hydrolyze cellulose into NCC with a size ranging from 25 to 37 nm. The increasing awareness on environment and attention on biodegradable as well as biocompatible NCC has seen NADES as a promising green solvent. The potential of NADES in isolating NCC from lignocellulosic materials should be further investigated to improve its efficiency.

Acknowledgement

The authors would like to acknowledge the financial support given by the Ministry of Higher Education Malaysia under Fundamental Research Grant Scheme (FRGS) with a grant number of FRGS/1/2015/TK10/UNIMAP/03/05.

References

[1] Gan, S, Zakaria, S, Ng, P, Chia, C H and Chen, R S. 2015. *BioResources*, **11**, 126–39.
[2] Fahma, F, Iwamoto, S, Hori, N, Iwata, T and Takemura, A. 2010. *Cellulose*, **17**, 977–85.
[3] Mohtar, S S, Tengku Malim Busu, T N Z, Md Noor, A M, Shaari, N, Yusoff, N A, Bustam@Khalil, M A et al. 2015. Bioresource Technology, **192**, 212–8.
[4] Moon, R J, Martini, A, Nairn, J, Simonsen, J and Youngblood, J. 2011. *Chemical Society Reviews*, **40**, 3941.
[5] Xiao, Y T, Chin, W L and Abd Hamid, S B. 2015. *Advanced Materials Research*, **1087**, 106–10.
[6] Klemm, D, Kramer, F, Moritz, S, Lindström, T, Ankerfors, M, Gray, D et al. 2011. *Angewandte Chemie - International Edition*, **50**, 5438–66.
[7] Oun, A A and Rhim, J-W. 2016. *Carbohydrate Polymers*, **150**, 187–200.
[8] Ni, H, Zeng, S, Wu, J, Cheng, X, Luo, T, Wang, W et al. 2012. *Bio-Medical Materials and Engineering*, **22**, 121–7.
[9] Dai, Y, Verpoorte, R and Choi, Y H. 2014. *Food Chemistry*, **159**, 116–21.
[10] Radošević, K, Čanak, I, Panić, M, Markov, K, Bubalo, M C, Frece, J et al. 2018. *Environmental Science and Pollution Research*, **25**, 14188–96.
[11] Kudłak, B, Owczarek, K and Namieśnik, J. 2015. *Environmental Science and Pollution Research*, **22**, 11975–92.
[12] Nomanbhay, S M, Hussain, R and Palanisamy, K. 2013. *Journal of Sustainable Bioenergy Systems*, **03**, 7–17.
[13] Suvachittanont, S and Ratanapan, P. 2013. *J Chem Chem Eng*, **7**, 1136–41.
[14] Amiralian, N, Annamalai, P K, Memmott, P and Martin, D J. 2015. *Cellulose*, **22**, 2483–98.
[15] Duan, L, Dou, L-L, Guo, L, Li, P and Liu, E-H. 2016. *ACS Sustainable Chemistry & Engineering*, **4**, 2405–11.
[16] Dai, Y, van Spronsen, J, Witkamp, G-J, Verpoorte, R and Choi, Y H. 2013. *Analytica Chimica Acta*, Elsevier B.V. **766**, 61–8.

[17] Jablonský, M, Andrea, Š and Kamenská, L. 2015. **10**, 8039–47.
[18] Lamaming, J, Hashim, R, Leh, C P, Sulaiman, O, Sugimoto, T and Nasir, M. 2015. *Carbohydrate Polymers*, **134**, 534–40.
[19] Owczarek, K, Szczepanska, N, Plotka - Wasylka, J, Rutkowska, M, Shyshchak, O, Bratychak, M et al. 2016. *Chemistry & Chemical Technology*, **10**, 601–6.
[20] Vigier, K D O, Chatel, G and Jérôme, F. 2015. *ChemCatChem*, **7**, 1250–60.
[21] Zhou, E and Liu, H. 2014. *Asian Journal of Chemistry*, **26**, 3626–30.
[22] Li, M, Cheng, Y L, Fu, N, Li, D, Adhikari, B and Chen, X D. 2014. *International Journal of Food Engineering*, **10**, 427–36.
[23] Goh, K Y, Ching, Y C, Chuah, C H, Abdullah, L C and Liou, N-S. 2016. *Cellulose*, **23**, 379–90.
[24] Shanmugarajah, B, Loo, P, Mei, I, Chew, L, Yaw, S and Tan, K W. 2015. *Chemical Engineering Transactions*, **45**, 1705–10.
[25] Zulnazri, Z, Anjana, F and Roesyadi, A. 2017. *The Journal of Pure and Applied Chemistry Research*, **6**, 14–21.
[26] Fatah, I Y A, Abdul Khalil, H P S, Hossain, M S, Aziz, A A, Davoudpour, Y, Dungani, R et al. 2014. *Polymers*, **6**, 2611–24.
[27] Zain, N F M, Yusop, S M and Ahmad, I. 2014. *Journal of Nutrition & Food Sciences*, **05**, 10–3.
[28] Nascimento, P, Marim, R, Carvalho, G and Mali, S. 2016. *Materials Research*, **19**, 167–74.
[29] Adel, A M, Abd El-Wahab, Z H, Ibrahim, A A and Al-Shemy, M T. 2011. *Carbohydrate Polymers*, **83**, 676–87.
[30] Nor, N A M, Mustapha, W A W and Hassan, O. 2016. *Procedia Chemistry*, **18**, 147–54.

IOP Conf. Series: Materials Science and Engineering **429** (2018) 012060 doi:10.1088/1757-899X/429/1/012060

Optimizing Yield of Microcrystalline Cellulose from Empty Fruit Bunch Via Hydrolysis Using Ionic Liquid

J J M Fong [1*], A Ibrahim [1], M F Abdulah [1,2*], S T Sam[1]

[1] School of Bioprocess Engineering, Universiti Malaysia Perlis (UniMAP), Kompleks Pusat Pengajian Jejawi 3, 02600, Arau, Perlis, Malaysia
[2] Centre of Excellence Biomass Utilization (CoEBU), Universiti Malaysia Perlis (UniMAP), Kompleks Pusat Pengajian Jejawi 3, 02600, Arau, Perlis, Malaysia

*Email: faiq@unimap.edu.my

Abstract. Microcrystalline cellulose (MCC) has been widely recognized as an important material with high potential in contributing to wide range of field including nanocomposite, biomedical applications, electroactive polymers, and many others. MCC has been isolated extensively using conventional method such as acid hydrolysis. However, utilization of hazardous and volatile solvent in acid hydrolysis has raised concern on the impact to the environment and the toxicity level of the as-produced MCC for potential application in biomedical field. In this study, new generation of environmental benign solvent named ionic liquid (IL) was utilized to extract MCC from empty fruit bunch (EFB). The yield of MCC has been optimized by varying three factors, namely temperature, reaction time and solvent concentration. It has been observed that the highest yield of MCC produced using IL is 31.44%, achieved at optimum conditions of 95°C, 129 minutes and 95% solvent concentration. The Fourier transform infrared (FTIR) spectra have confirmed the absence of absorption bands corresponding to lignin and hemicellulose, suggesting that almost pure MCC has been obtained. Scanning electron microscopy (SEM) micrographs have further revealed that the MCC had been agglomerated which resulted in micro-size cellulose ranging from 10 – 20 μm, highly probably due to inappropriate drying method.

1. Introduction

Cellulose, in conjunction with hemicellulose and lignin, is found abundantly in all plant materials. These biopolymers combine and are made up into the main constituent of the biomass termed as lignocellulosic. Cellulose, in particular acts as the utmost plenteous sustainable biopolymer material with high value and availability fits undeniably with the current growing demand in advantage of its biocompatible and environmental friendly properties [1]. Lignocellulosic materials can be isolated from fruit mesocarp, frond, trunk and empty fruit bunch (EFB). Among these varieties of fiber supply available in the tree's part, EFB is more favourable in terms of its high availability, low price and its potential to yield high percentage of fibers making it possible to substitute the natural fiber resources [2–5]. Being the most abundant biomass materials which are constantly discharged as waste from the palm oil mill, efforts to extract the high value lignocellulosic materials as value-added products have gained increasing interest [5]. Being the most significant lignocellulosic material in EFB, cellulose is made up of adjoining glucose units bonded to each other by β-1,4-glycosidic bonds [6]. Microfibrils which are the linear cellulose chains bundled together are composed of amorphous and crystalline regions. Cellulose is recognized as an important renewable resource with wide applications. However, since cellulose comprised of a straight glucose polymer alignment which forms a highly strong supramolecular arrangement due to hydrogen bonding, it is highly resistant to enzymatic degradation. From a greener perspective, designing an effective and environmental friendly cellulose hydrolysis method is highly desired [6–8].

Although cellulose is the most accessible polymer in nature, it is only rather recently that cellulose has attained notability as micro and nanomaterial, in form of microcrystalline cellulose (MCC) and nanocrystalline cellulose (NCC) [9]. MCC and NCC are commonly prepared via conventional acid hydrolysis. However, the use of conventional acids poses several problems such as producing potential hazardous organic compounds which led to environmental pollution. Hence, ionic liquid (IL) is proposed as an environmental benign and highly efficient solvent for the preparation of crystalline cellulose, particularly MCC [6–11]. In this study, one type of IL named 1-butyl-methylimidazolium chloride (BmimCl) is used to investigate the potential of IL in the preparation of MCC from EFB. In addition, the optimization study to determine the optimum conditions for the preparation of MCC is also conducted in this work. The effects of three process parameters namely temperature, reaction time and solvent concentration on the yield of MCC are investigated with the aid of Design of Experiment (DOE) software. Subsequently, the physical properties of raw EFB, treated EFB and the as-produced MCC are determined using Fourier transform infrared (FTIR) spectroscopy and scanning electron microscopy (SEM).

2. Materials and methods

EFB was collected from Felda Palm Industries Sdn. Bhd., Kulai, Malaysia. BmimCl used in this study was acquired from Shanghai Chong Jie Chemical Co. Ltd., China while sodium hydroxide (NaOH) was obtained from Sigma-Aldrich, Malaysia. All chemicals were used as received.

2.1. Pre-treatment of EFB

EFB was first sun-dried for 5 days and then was cut into smaller pieces. The EFB was further dried in an oven at 80°C overnight before it was ground into smaller size using a grinder. Next, 50 g of the ground EFB was pretreated using 500 mL of 2% (w/v) NaOH solution. The pretreatment process was carried out at 80°C for 2 hours while continuously stirred. Then, the mixture was left to cool down before it was washed with distilled water to remove the NaOH solution [12]. The pretreated EFB was subsequently dried at 60°C for 8 hours and kept for further used.

2.2. Hydrolysis of EFB using IL

Hydrolysis of EFB was carried out by treated the EFB at different operating conditions using BmimCl. Three process parameters (i.e. temperature, reaction time and solvent concentration) were investigated to optimize the yield of MCC from EFB. For each designated run, 60 mL of desired concentration of BmimCl was first heated to its desired temperature using a heating plate (Model 11-102-50SH, Fisher Scientific, USA). 3 g of pretreated EFB were then added to the BmimCl. The mixture was stirred continuously at a designated temperature for a specific reaction time. Afterwards, the cold water was added to the mixture to stop the reaction. Next, the mixture was diluted with distilled water and then was centrifuged at 8000 rpm for 10 minutes to extract the MCC. MCC suspension was collected during the centrifugation step until no more cloudy suspension was obtained. The MCC suspension was dried in an oven at 80°C overnight to obtain MCC powder. The as-produced MCC powder was then weighed for yield measurement and kept in a desiccator for further characterization. The yield of MCC was calculated using equation (1) as shown below.

$$\text{Yield (\%)} = \frac{\text{Weight of dried MCC}}{\text{Initial weight of treated EFB}} \times 100 \qquad (1)$$

2.3. Optimization of Yield of MCC

2.3.1. Preliminary Tests
Preliminary test was conducted to determine the range of each process parameter to be used for optimization study. Ranges selected for preliminary test for the effect of temperature, reaction time and solvent concentration were 60°C – 130°C, 40 min – 130 min and 40% - 100%, respectively. In this test, one-factor-at-a-time (OFAT) method was used.

2.3.2. Design of Experiment (DOE)
The smaller ranges of the process parameters were determined from the preliminary tests and subsequently selected to be used in the optimization study using Design of Experiment (DOE). In this study, Central Composite Design (CCD) under Response Surface Methodology (RSM) was used to investigate the optimum process parameters for the preparation of higher yield of MCC.

2.4. Fourier Transform Infrared (FTIR) Spectroscopy
The change of functional groups of raw EFB, treated EFB and the as-produced MCC was recorded using a FTIR spectrometer (Model Spectrum 65, Perkin Elmer). All samples were mixed with potassium bromide (KBr) powder with weight ratio of 1:100 and subsequently compressed into a disc. The spectra for each sample were recorded as an average of 64 scans at a resolution of 4 cm^{-1} ranging from 500 cm^{-1} to 4000 cm^{-1} [13,14].

2.5. Scanning Electron Microscopy (SEM)
The surface morphology of raw EFB, treated EFB and MCC was analysed using a scanning electron microscope (Model JSM-6100, Jeol). All samples were gold coated and then mounted in the microscope using conductive carbon tape. All samples were ensured to be fully dried before scanned at accelerating voltage of 10kV.

3. Results and discussion

3.1. Preliminary Tests
The yield of MCC from preliminary tests using OFAT for different process parameters (i.e. temperature, reaction time and solvent concentration) were shown in Figure 1. Preliminary tests were carried out to determine the smaller range of process parameter prior to the optimization study. From Figure 1(a) – (c), the selected range for optimization study for temperature, reaction time and solvent concentration are 80°C – 120°C, 90 min - 130 min and 55% - 95%, respectively. For all three process parameters, it can be observed that the increase of temperature, reaction time and solvent concentration led to the increasing yield of MCC until it reached a maximum value. This is highly probably due to more amorphous region have been removed by the reaction between IL and cellulose fibres. However, as all three parameters were further increased, the MCC yield started to decline as the fibres have further broken down into soluble sugar.

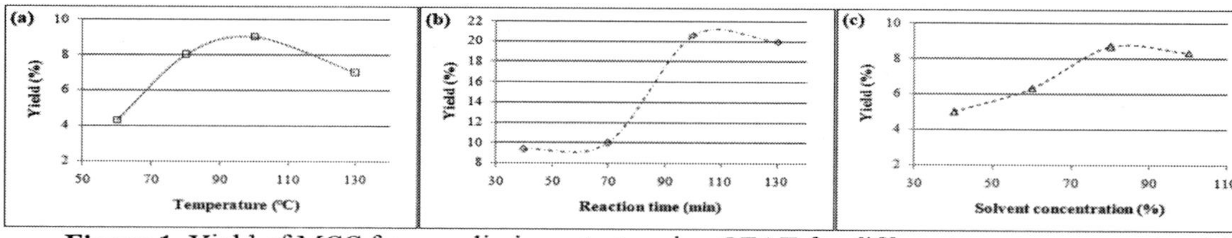

Figure 1. Yield of MCC from preliminary tests using OFAT for different process parameter, (a) temperature, (b) reaction time, and (c) solvent concentration.

3.2. Optimization of Yield of MCC

The optimization study was conducted using DOE by applying the results from the preliminary tests. For simplicity reason, the three process parameters (i.e. temperature, reaction time and solvent concentration) investigated in the optimization study were denoted as A, B and C, respectively. The yield of MCC prepared at different designated conditions is shown in Table 1. The highest MCC yield (31.33%) was obtained at conditions of 100°C, 110 min and solvent concentration of 95% (run 2). Meanwhile, the lowest MCC yield (3.33%) was recorded from run 3, where the MCC was prepared at conditions of 120°C, 90 min and solvent concentration of 55%.

Table 1. Yield of MCC prepared at different designated conditions.

Run	Process Parameter			Response
	Temperature, A (°C)	**Reaction time, B (min)**	**Solvent concentration, C (%)**	**Yield (%)**
1	80	90	95	18.33
2	**100**	**110**	**95**	**31.33**
3	120	90	55	3.33
4	100	90	75	9.33
5	100	110	75	14.00
6	100	110	75	13.00
7	120	90	95	14.33
8	120	110	75	7.67
9	80	110	75	7.00
10	80	130	95	30.67
11	80	90	55	3.67
12	100	130	75	10.00
13	100	110	75	11.00
14	80	130	55	4.00
15	100	110	75	9.67
16	120	130	55	5.33
17	120	130	95	22.33
18	100	110	75	11.33
19	100	110	55	6.33
20	100	110	75	9.00

3.2.1. Analysis of Variance (ANOVA)

The analysis of variance (ANOVA) was applied in DOE to measure the precision of the model obtained from the results of the experiments. The statistical data of the variation of mean square ratio and mean square residual error was shown in Table 2. R-squared represents the determination coefficient which denotes the square of correlation coefficient for the response. In this study, the R-squared obtained is 0.9504. The closer the R squared value to 1, the better the correlation between the experimental and the predicted values. Hence, the model will predict the response and fits the data better. The adjusted R-squared value is 0.9058 while the predicted R-squared value is 0.6538. Although the predicted R-squared value is not as close to the adjusted R-squared value, the predicted R-squared value is considered to be in reasonable agreement with the aforementioned adjusted R-squared value.

Table 2. Statistical data for ANOVA.

Parameter	Value	Parameter	Value
Std. Dev.	2.47	**R-Squared**	0.9504
Mean	12.08	**Adj R-Squared**	0.9058
C.V. %	20.48	**Pred R-Squared**	0.6538
PRESS	427.58	**Adeq Precision**	15.7360

Meanwhile, the adequate precision measures the signal to noise ratio. An adequate precision value of more than 4 is desirable so that the effective determination of the model could be obtained [15]. The adequate precision was determined to be 15.7360, which implies an adequate signal and validating the model precision. The standard deviation value of 2.47 which is relatively low value shows that the model has a good precision and lower dispersion. Whilst the coefficient of variation (C.V.) value obtained is 20.48%. This value represents the ratio of standard deviation to mean and expresses precision and repeatability. The low value of C.V. shows that the factors are highly reliable during the optimization.

The result of ANOVA for response surface quadratic mode is shown in Table 3. The ANOVA result indicates that the investigated process parameters are significantly affecting the yield of MCC, as implied by the model F-value of 21.30. There is only 0.01% (p-value) possibility that the F-value of a model this large could occur due to noise. The result also revealed that the model terms were significant. The individual effect of solvent concentration (C) is influencing the yield of MCC the most, as shown by high F-value of 145.33. In addition, B, C, BC, A2 and C2 were also concluded as significant model terms, in which all their p-values are less than 0.1000. The F-value of lack of fit of 2.36 implies that the lack of fit is not significant relative to the pure error value of 18.21. There is an 18.36% possibility that the F-value of lack of fit this large could occur due to noise. Non significant lack of fit is desirable as it shows that the model fits well.

Table 3. ANOVA for response surface quadratic mode.

Source	Sum of Square	df	Mean Square	F-Value	p-value (Prob>F)	
Model	1174.01	9	130.45	21.30	< 0.0001	significant
A-Temperature	11.41	1	11.41	1.86	0.2022	
B-Reaction time	54.48	1	54.48	8.90	0.0137	
C-Solvent concentration	889.81	1	889.81	145.33	< 0.0001	
AB	0.89	1	0.89	0.15	0.7108	
AC	22.21	1	22.21	3.63	0.0860	
BC	40.55	1	40.55	6.62	0.0277	
A^2	47.34	1	47.34	7.73	0.0194	
B^2	9.10	1	9.10	1.49	0.2508	
C^2	148.40	1	148.40	24.24	0.0006	
Residual	61.23	10	6.12			
Lack of Fit	43.02	5	8.60	2.36	0.1836	Not significant
Pure Error	18.21	5	3.64			
Cor Total	1235.24	19				

3.2.2. Response Surface Analysis

The interaction and relationship between process parameters and the response (MCC yield) can be indicated by three dimensional (3D) contour plots. Figure 2 portrays 3D contour view for the effects of different combination of process parameters on MCC yield. From Figure 2(a), the combined effect of temperature and process time is less significant on the MCC yield, as represented by blue and green region showing less significant response. Meanwhile, a saddle shape was observed in Figure 2(b) which represented the combined effect of temperature and solvent concentration on MCC yield. The aforementioned shape was observed as the solvent concentration is more significantly affected MCC yield as compared to the temperature. Similar saddle shape also has been observed in Figure 2(c), in which solvent concentration also shown superior effect on MCC yield as compared to the reaction time. The result indicates that reasonably high MCC yield could be prepared by employing high solvent concentration coupled with moderate values of temperature and reaction time.

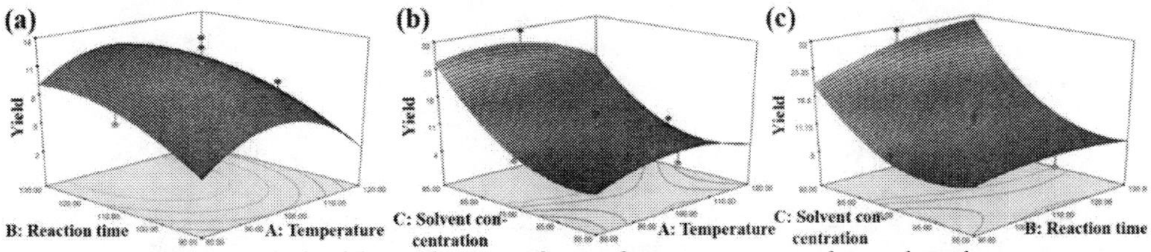

Figure 2. 3D contour view for the combined effects of (a) temperature and reaction time, (b) temperature and solvent concentration, and (c) reaction time and solvent concentration on MCC yield.

3.2.3. Optimization Validation

In order to validate the model, the DOE proposed a solution run with the predicted maximum yield of MCC (Table 4). The suggested optimum run was carried out for three times and the average percentage error between actual and predicted yield was calculated and recorded. The average percentage error (3.15%) was revealed to be less than 5%, thus the predicted model was realistically accepted.

Table 4. Experimental validation for the optimization of the yield of MCC.

Solution no.	Temperature, A (°C)	Reaction time, B (min)	Solvent concentration, C (%)	Predicted yield (%)	Actual yield (%)	Percentage error (%)
1	95	129	95	31.37	32.00	2.02
2	95	129	95	31.37	32.33	3.07
3	95	129	95	31.37	30.00	4.36
Average					**31.44**	**3.15**

3.3. FTIR Analysis

FTIR spectra of raw EFB, treated EFB and the as-produced MCC are shown in Figure 3. All three spectra show two main absorbance regions which are the low wavelength region at 500 cm^{-1} to 1750 cm^{-1} and the high wavelength region ranging from 2800 cm^{-1} to 3500 cm^{-1}. The absorption peaks found in all spectra in the range of 3400cm^{-1} to 3450 cm^{-1} are attributed to the hydroxyl groups which are one of the functional group found in lignocellulosic materials [16]. On the other hand, the peaks found in the range of 2850 cm^{-1} to 2930 cm^{-1} are attributed to the aliphatic saturated C-H stretching vibrations in cellulose, hemicellulose and lignin [17]. In comparing the raw EFB spectrum with both treated EFB and MCC spectra, the most noticeable difference is the absence of the vibration peak at 1737.39 cm^{-1}. This peak is associated with the C=O stretching of the methyl ester and carboxylate groups in pectin and the acetyl and uronic ester groups in hemicelluloses. It may also be caused by the ester linkage of carboxylic group of the ferulic and p-coumaric acids of lignin and/or hemicelluloses [18]. Hence, the disappearance of this peak indicates that the hemicellulose and lignin components have been effectively removed post alkaline treatment.

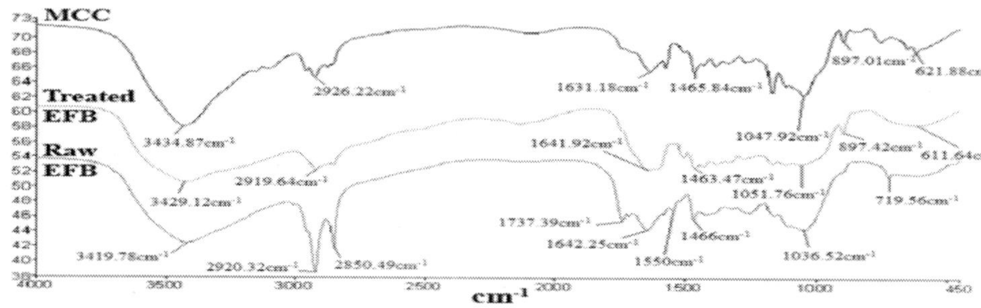

Figure 3. FTIR spectra of raw EFB, treated EFB and the as-produced MCC.

On the other hand, the peaks present at 1630 cm^{-1} to 1645 cm^{-1} correspond to the O-H bending due to absorption of water by cellulose [19]. This peak is found in both treated fibres probably due to reaction of sodium hydroxide with the hydroxyl groups of celluloses and hence subsequently lead to formation of water molecules. The same vibration peak is also found at 1642.25 cm^{-1} of untreated fibres. This shows that even though the fibres were subjected to drying process, the water adsorbed in the cellulose molecules was very difficult to be extracted due to the cellulose-water interaction [20]. The peaks at 1500 cm^{-1} to 1600 cm^{-1} are due to C=C aromatic skeletal vibration of lignin [21]. This can be seen in the spectra of raw EFB at 1550 cm^{-1} indicating the presence of lignin. Peaks in this range however do not appear in both treated EFB and MCC spectra. This indicates that lignin has been removed post alkaline and IL treatment. The vibration band in the range of 1450 cm^{-1} to 1475 cm^{-1} could be identified as the bending vibration of CH_2 group in aliphatic compounds and is found in all spectra. The peak observed in all of the spectra at 1030 cm^{-1} to 1060 cm^{-1} is due to the C-O stretching assignment of primary alcohol [21]. The increase in the intensity of this band indicates an increase in crystallinity of the samples. Another significant absorption band, which increases continuously with alkaline and IL treatment is the peak at 897 cm^{-1} which corresponds to glycosidic –C-H- deformation, with a ring vibration contribution and –O-H bending. This peak features the characteristic of the β-glycosidic linkage between the anhydroglucose units in cellulose [22]. The deformation, wagging and twisting modes of anhydro-glucopyranose vibration is shown from 600 cm^{-1} to 750 cm^{-1} [23].

3.4. SEM Analysis

SEM micrographs of raw EFB, treated EFB and MCC at different magnification are shown in Figure 4. Raw and rough surface morphology has been observed for the raw EFB sample while the treated EFB sample shows smooth and holes-like surface. The holes-like surface of the treated EFB was observed highly probably due to successful removal of lignin and hemicellulose post alkaline treatment. Whereas the SEM micrographs of the as-produced MCC at 1000 and 5000 magnifications are portrayed in Figure 4(c) and (d), respectively. The MCC produced are made up of the crystalline regions from the cellulosic fibres after the removal of amorphous regions at the cellulose polymer. The MCC sample displays a rigid and conglomerate texture with size ranging from 10 – 20 μm, in which the fibers are fused together giving a relatively homogeneous macrostructure. It is anticipated that smaller particles in nanosize could be prepared by applying more appropriate drying approach such as freeze drying method.

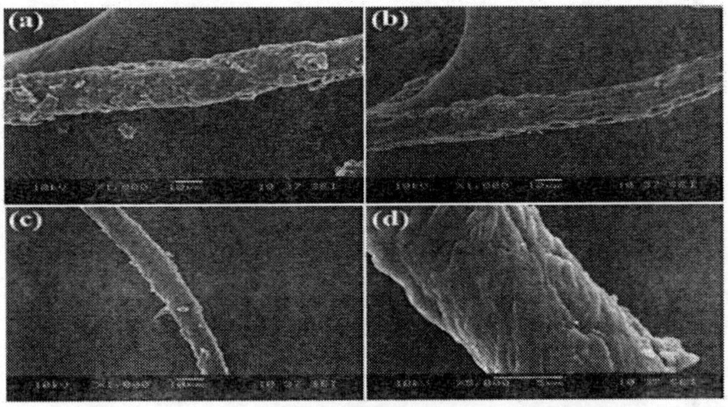

Figure 4. SEM micrographs of (a) raw EFB at 1000x, (b) treated EFB at 1000x, and MCC at (c) 1000x and (d) 5000x.

4. Conclusion

Successful production of MCC from EFB could be achieved via hydrolysis using IL. The highest yield of MCC (31.44%) can be prepared at optimized conditions of 95°C, 129 minutes and at solvent concentration of 95%. Almost pure MCC is prepared as lignin, hemicellulose and other non-cellulosic components have been successfully removed from the raw EFB, as shown by FTIR spectra. SEM analysis also have revealed that the as-produced MCC shown rigid and conglomerate structure with size ranging from 10 – 20 µm. As more environmental-friendly approach attracts ever-increasing attention in the field of nanomaterial production, IL has shown remarkable potential to be used as environmental benign solvent for the preparation of micro- and nano-size cellulose.

Acknowledgment

The authors would like to acknowledge the support from the Fundamental Research Grant Scheme (FRGS) under a grant number of FRGS/1/2015/TK10/UNIMAP/03/5 from the Ministry of Higher Education Malaysia.

References

[1] Brinchi L Cotana F Fortunati E and Kenny J M 2013 *Carbohydrate Polymers*, **94**, 154–69.
[2] Chang S H 2014 *Biomass and Bioenergy*, **62**, 174–81.
[3] Palamae S Dechatiwongse P Choorit W Chisti Y and Prasertsan P 2017 *Carbohydrate Polymers*, **155** 491–7.
[4] Wanrosli W D Rohaizu R Ghazali A 2011 *Carbohydrate Polymers* **84**, 262–7.
[5] Zianor A Z A Beg M D H Rosli M Y Ramli R Junadi N Alam A K M M 2017 *Carbohydrate Polymers* **162** 115–20.
[6] Mohtar S S Tengku Malim Busu, T N Z Md Noor A M Shaari N and Mat H 2017 *Carbohydrate Polymers*, **166**, 291–9.
[7] Ling, Z, Chen, S, Zhang, X, Takabe, K and Xu, F. 2017. Scientific Reports, **7**, 10230.
[8] Song, H, Niu, Y, Yu, J, Zhang, J, Wang, Z and He, J. 2013. *Soft Matter, The Royal Society of Chemistry*. **9**, 3013–20.
[9] Isik, M, Sardon, H and Mecerreyes, D. 2014. *International Journal of Molecular Sciences*, **15**, 11922–40.
[10] Abdulkhani, A, Hojati Marvast, E, Ashori, A and Karimi, A N. 2013. *Carbohydrate Polymers*, **95**, 57–63.
[11] Parveen, F, Patra, T and Upadhyayula, S. 2016. *Carbohydrate Polymers*, **135**, 280–4.
[12] Rosa, M F, Medeiros, E S, Malmonge, J A, Gregorski, K S, Wood, D F, Mattoso, L H C et al. 2010. *Carbohydrate Polymers*, **81**, 83–92.
[13] Fahma, F, Iwamoto, S, Hori, N, Iwata, T and Takemura, A. 2010. *Cellulose*, **17**, 977–85.

[14] Jiang, F and Hsieh, Y Lo. 2013. *Carbohydrate Polymers*, **95**, 32–40.
[15] Abd Hamid, S B, Chowdhury, Z and Ziaul Karim, M. 2014. *Bioresources*, **9**, 7403-7426.
[16] Mandal, A and Chakrabarty, D. 2011. *Carbohydrate Polymers*, **86**, 1291–9.
[17] Shanmugarajah, B, Loo, P, Mei, I, Chew, L, Yaw, S and Tan, K W. 2015. *Chemical Engineering Transactions*, **45**, 1705–10.
[18] Han, J. 2013. Preparation and characterization of cellulose nanoparticles and their application in biopolymeric nanocomposites.
[19] Xiao, Y T, Chin, W L and Abd Hamid, S B. 2015. *Advanced Materials Research*, **1087**, 106–10.
[20] S. Lani, N, Ngadi, N, Johari, A and Jusoh, M. 2014. *J. Nanomater*, **2014**, 702538.
[21] Rehman, N, de Miranda, M I G, Rosa, S M L, Pimentel, D M, Nachtigall, S M B and Bica, C I D. 2014. *Journal of Polymers and the Environment*, **22**, 252–9.
[22] Rosli, N A, Ahmad, I and Abdullah, I. 2013. *BioResources*, 8, 1893–908.
[23] Man, Z, Muhammad, N, Sarwono, A, Bustam, M A, Vignesh Kumar, M and Rafiq, S. 2011. *Journal of Polymers and the Environment*, **19**, 726–31.

IOP Conf. Series: Materials Science and Engineering **429** (2018) 012061 doi:10.1088/1757-899X/429/1/012061

Effect of Emulsion Temperature on Properties of Conductive Epoxy Porous Prepared by Single Emulsion Technique

L Anusha[1], Z Zakaria[1], A W M Kahar[1], D N U Lan[1]*

[1]School of Materials Engineering, Universiti Malaysia Perlis, Kompleks Pusat Pengajian Jejawi II, 02600 Arau, Perlis, Malaysia.

*Email: uylan@unimap.edu.my

Abstract. Currently, conductive porous is used in composite to reduce the conductive filler loading and improve the conductive property. In this study, conductive epoxy porous (CEP) had been fabricated by single emulsion technique from the mixture of epoxy, polyamide hardener, carbon black filler and sodium bicarbonate. Epoxy mixture was dropped into a heated corn oil at different temperatures (140, 160 and 180 °C) under stirring of 1000 rpm. Because of the immiscibility of epoxy and corn oil, epoxy droplets were formed. Initial epoxy droplets were broken into smaller droplets due to applied stirring shear. In addition, the decomposed gas (carbonic gas and water vapour) from sodium bicarbonate also caused an internal expanding pressure to further break the epoxy droplet into many fine droplets. Curing reaction of epoxy and polyamide occur simultaneously formed CEP. It was found that all the CEPs exhibited higher conductivity compared to control epoxy and epoxy composite filled equivalent content of carbon black. Furthermore, emulsion temperature of 160 °C produced the smallest porous CEP, which performed the highest electrical property.

1. Introduction

Conductive microspheres have been paid excessive attention by researchers and industries due to many advantageous properties such as similar density to polymer matrix, high specific surface area, excellent thermal property and reasonable electrical conductive property [1,2]. Conductive microspheres can be obtained by coating conductive filler on the surface of non-conductive microspheres. The common microspheres used widely are glass microspheres, cenospheres, carbon microspheres and polymer microspheres [3, 4, 5]. Hence, conductive microspheres have been used to replace bare conductive filler in conductive polymer composites and applied in electronic, biomedical and sensor devices [6].

Numerous occasions researcher used emulsion technique to produce conductive microspheres. Emulsions are systems consisting of two or more liquid immiscible phases wherein one liquid is dispersed in another liquid phase [7]. Previous study reported fabrication of conductive epoxy porous by single emulsion method. The epoxy mixture was dropped and stirred in heated corn oil. Due to immiscible between epoxy mixture and corn oil, the epoxy mixture droplets form a spheres shape in heated corn oil. These epoxy mixture droplets contain blowing agent, thus, this epoxy droplets blown and simultaneously cured after receiving heat from corn oil [8]. To obtain good porous materials, the blowing agent has to be carefully selected depending on processing condition sort of temperature [9]. Elzbieta & Pawel stated that temperature has influence in liberation gas once blowing agent decompose resulted different expansion formed in polymer matrix [10]. In this study, conductive

IOP Conf. Series: Materials Science and Engineering **429** (2018) 012061 doi:10.1088/1757-899X/429/1/012061

epoxy porous prepared by single emulsion technique and the aim of this research was to discover the influence of emulsion temperatures of corn oil has an effect on the properties of conductive epoxy porous.

2. Experimental

2.1. Materials

Epoxy resin DER 331 and polyamide A062 was purchased from Euro Chemo-Pharma Sdn. Bhd. Epoxy resin DER 331 has epoxide equivalent weight of 182–192, density and viscosity at 25 °C of 1.16 g/cm^3 and 11–14 Pas. Polyamide A062 has equivalent weight per H active of 110, density and viscosity at 25 °C of 0.96 g/cm^3 and 35–45 Pas, respectively. Polyamine was purchased from Euro Chemo-Pharma Sdn. Bhd. White crystalline powder of sodium bicarbonate was purchased from HmbG Chemicals having density of 2.22 g/cm^3 was used as the blowing agent. Corn oil was purchased from Euro Chemo-Pharma Sdn. Bhd and light yellow in apperance. Corn oil has density and viscosity at 25 °C of 0.91 g/ml and 50 Poise. Carbon black N330 (HAF) was purchased from Mega Makmur Saintifik Sdn. Bhd. Carbon black has iodine absorption number of 82 ± 5 g/kg and CTAB surface area of 79~87 10^3 m^2/kg.

2.2. Preparation of Conductive Epoxy Porous (CEP)

The porous were prepared based on polyamide dominant which is polyamide-epoxy adducts. The weight ratio of epoxy and polyamide was 1E:2P. Epoxy was mixed with sodium bicarbonate by using over-head stirrer with the speed of 300 rpm for 3 minutes. Carbon black was added continuously and mixed at the speed of 300 rpm for another 2 minutes. Lastly, polyamide was added into the mixture and stirred for 3 minutes at the same speed. Epoxy mixture was continuously dropped into corn oil at 160 °C (weight ratio of epoxy mixture to corn oil of 1:9) and stirred at 1000 rpm for another 1 hour. The laboratory syringe was used as dropper with a nozzle diameter of 0.5 cm to drop epoxy mixture into corn oil. The porous was produced by allow it to cool down at 80 °C before filtered and washed five times with detergent solution at 60 °C. The ratio of detergent to water used is 1:20. The detergent solution was changed every 15 minutes. Drying and post-cure process were carried out at 80 °C for 4 hours in air-circulating oven.

2.3. Preparation of Epoxy filled CEP for Conductive Testing

The weight ratio of epoxy and polyamine was 2:1. 20 phr of CEP was mixed with epoxy by using over-head stirrer with the speed of 300 rpm for 3 minutes. Polyamine was added into the mixture and stirred for another 3 minutes at the same speed. The mixture was casted into PP mold with dimension of 24 x 20 x 4 mm and allowed to harden at room temperature for 24 hours.

2.4. Characterization

The morphology of CEP was carried out using a scanning electron microscope (SEM), model JSM 6260 LE JOEL. CEP was coated with gold/platinum layer to avoid electrostatic charging during examination. Bulk density of CBEP was determined according to ASTM D7481-09. Bulk density was carried out by filling 100 ml of CEP in 100 ml cylinder. The mass of 100 ml CEP was weight. Bulk density was calculated by using mass diving to volume. Pycnometer density of CEP was measured by using gas pycnometer density analyser, AccuPyc II 1340 V1.0.5. The 10 cm³ of size chamber and Helium gas was used. Differential scanning calorimetry (DSC) of CEP was measured in temperature range from 25–200 °C with a heating rate 10 °C/min under nitrogen atmosphere. Thermo-gravimetric analyzer (TGA) of CEP was determined in temperature range from 25–600 °C with a heating rate at 20 °C/min under a nitrogen flow of 50 ml/min by using Pyris Diamond TGA from PerkinElmer. Electrical conductivity test of CBEP was obtained by using Keithley 4200 Semiconductor Characterization System.

IOP Conf. Series: Materials Science and Engineering **429** (2018) 012061 doi:10.1088/1757-899X/429/1/012061

3. Results and iscussion

3.1. Morphology Analysis

Epoxy droplets receiving heat from corn oil would induce two reactions such of curing reaction and decomposed reaction of sodium bicarbonate. These both reactions depend greatly on the heating temperature. Figure 1 displayed the morphology of CEP prepared in different emulsion temperatures. T140 showed bigger porous particles compared to these of T160 and T180. Furthermore, epoxy particles in T140 agglomerated to form large lumps. In expected, T180 possessed the smallest epoxy particles and the most separated epoxy groups compared to others (T140 and T160). It is because the fast decomposition of sodium bicarbonate at 180 °C produced carbonic and vapour too vigorously that low viscosity epoxy droplets could be burst into many small droplets.

Figure 1: Scanning electron microscopy images of CEP for different emulsion temperatures at 500X magnification: (a) T140, (b) T160 and (c) T180.

3.2. Density Analysis

Figure 2 demonstrated the effect of different emulsion temperatures towards bulk density and pycnometer density of conductive epoxy porous. Figure 3 showed volume of CEPs in a same mass. Bulk density described as a direct measure of the packing closeness for particles in volume [11]. Surprisingly that T180 showed the highest bulk density while T160 possessed the lowest bulk density. It could be T180 had more small groups of epoxy particles, and then they can pack easier and occupy less spacing compared to others. T160 could have more bulky groups which composed from attaching epoxy particles and large pores. Therefore, same weight of T160 could occupy the highest volume compared to other hence exhibit the lowest bulk density as reported in Figure 3. This bulk density promised the potential of CEP dispersing well in epoxy composite and lead to better conductivity, which will be discussed in the next section. In contrast, pycnometer density decreases with increasing in emulsion temperatures. It could be some closed pore presenting among the packing epoxy particles in T160 and T180 so the Helium gas during pycnometer measurement could not able to enter. Hence,

IOP Conf. Series: Materials Science and Engineering **429** (2018) 012061 doi:10.1088/1757-899X/429/1/012061

the density of T160 and T180 were lower than T140. Especially in T180 which had more epoxy packing from small epoxy particles.

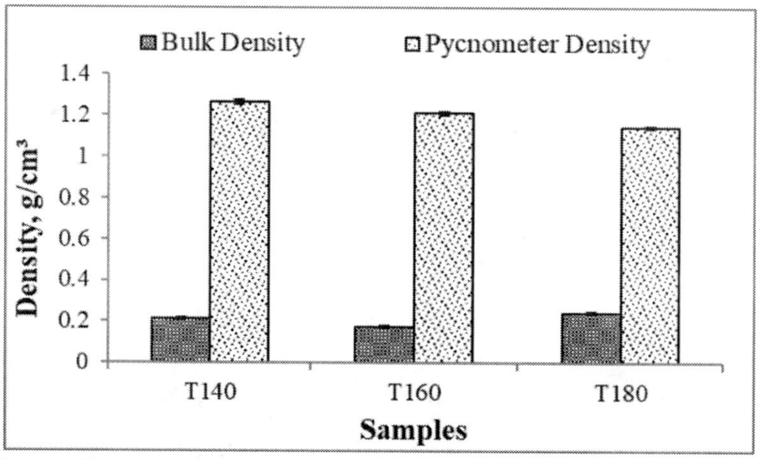

Figure 2: The effect of different emulsion temperatures of conductive epoxy porous towards bulk density and pycnometer density.

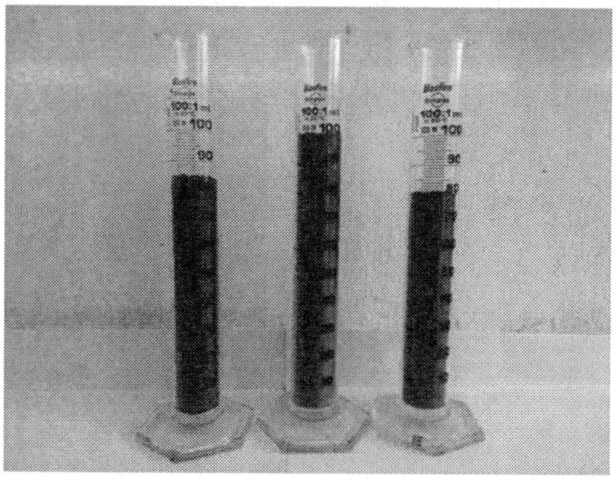

Figure 3: Volume of CEPs in a same mass: (a) T140, (b) T160 and (c) T180.

3.3 Thermo-gravimetric Analyzer (TGA)

The results of TGA weight loss and derivative thermograms of CEPs was reported in Figure 4.10(a) and 4(b). According to Figure 4(a), there is no decomposed peak occurred at before 160 °C so all sodium bicarbonate was decomposed completely. It was interested to find that the onset degradation was in well agreement to the bulk density results, where T160 started the earliest degradation following by T140 and lastly T180. Onset degradation depends strongly on the heat circulation in the sample. T160 had lowest bulk density, more volume occupying, larger open pores, hence more heat circulation and easier to be burnt. The best onset degradation of T180 was thanks to its closed pore, which was also proven from pycnometer density result. However, once the degradation start T180 exhibited the fastest rate of degradation because it had the smallest epoxy particles. After starting degradation, T140 had better heat conductivity hence it could be degraded faster than T160. These facts explained the results of temperature at 50% weight loss.

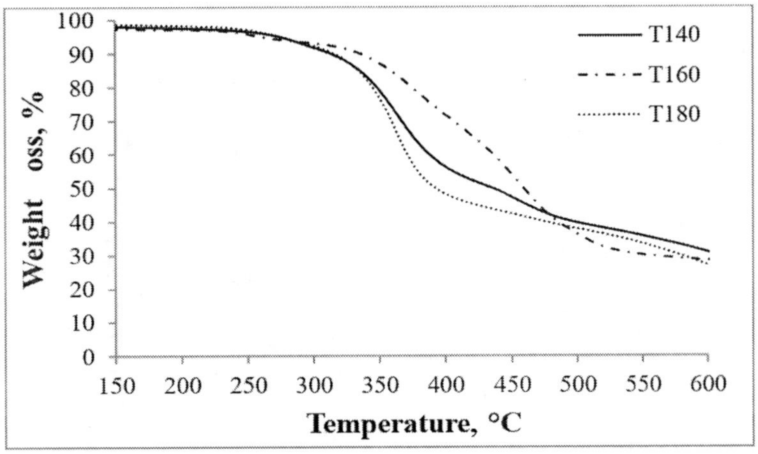

Figure 4(a): Weight loss of CEPs for different emulsion temperatures as a function of temperature.

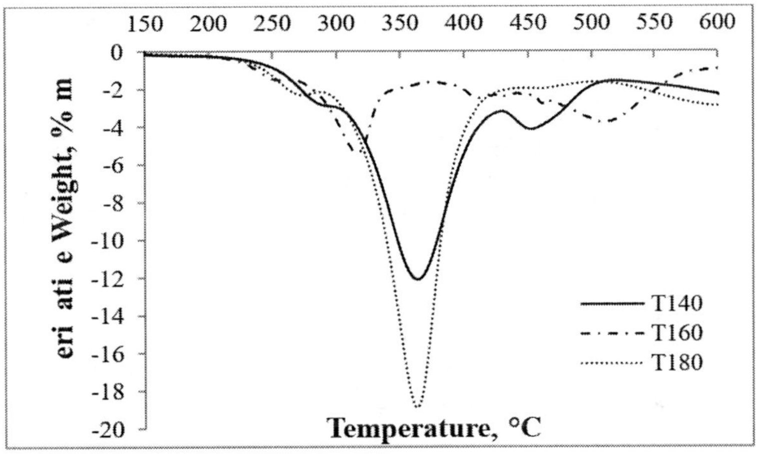

Figure 4(b): Derivative weight loss of CEPs for different emulsion temperatures as a function of temperature.

3.4 Conductive Behaviour

Figure 5 illustrated comparison of conductivity behaviour toward different epoxy composites. It was found that electrical conductivity of epoxy filled 20 wt% of CEP fabricated from different emulsion temperatures possessed highest electrical conductivity compared to control epoxy (control) and epoxy composite filled equivalent content of carbon black (CB). It can be implied that addition of 20 wt% of CEP into epoxy-polyamine composites exhibited better electrical conductivity compared to direct equivalent carbon black filler when same amount of carbon black (1.8 %) was used. This conductive behaviour proved the advantage of the proposed CEP, where the conduction was occurred via the connection of carbon black in composite. According to Figure 5, epoxy filled T160 indicates the highest electrical conductivity, while T140 exhibited the lowest electrical conductivity. This property reflected well the morphology of T160 and the bulk density. T160 exposed larger open pores and lowest bulk density, it means T160 could spread more throughout in epoxy composite. Therefore, the better carbon black dispersion in epoxy composite, and the higher conductivity is. The lower electrical conductivity of T180 could be due to its closed pore in the packing of small epoxy particles.

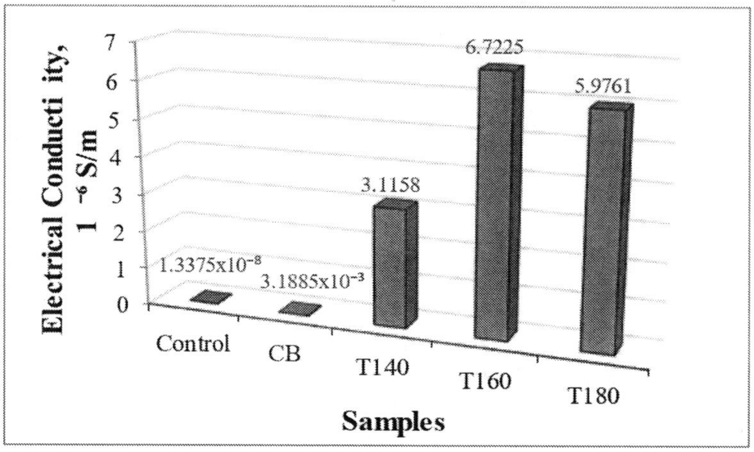

Figure . Electrical conductivity of epoxy composites.

4. Conclusion

Conductive epoxy porous was prepared by single emulsion technique. Emulsion temperatures are plays important role to determine properties of conductive epoxy porous. Epoxy mixture was dropped into a heated corn oil at different temperatures (140, 160 and 180 °C) under stirring. It was found that T140 showed bigger porous particles compared to T160 and T180. But, at higher emulsion temperatures of 180 °C, sodium bicarbonate decomposed produced carbonic and vapour too vigorously that low viscosity epoxy droplets could be burst into many small droplets were proved through the morphology. The use of 160 °C (T160) was optimum emulsion temperature offers better expansion behaviour. Besides, epoxy filled T160 exhibited the highest electrical conductivity compared to others because of T160 could spread more in which supported by bulk density result.

Acknowledgements

The authors would like to acknowledge Ministry of Higher Education Malaysia for providing the ERGS grant (grant no: 9010-00030) that has made this work possible.

References

[1] Xuemei H and Hao Y 2013 *J. Nanomater.* **2013** 1–8.

[2] Jain S, Whalley D C, Cottrill M, Helland T, Kristiansen H, Redford K, and Liu C 2013 *Conf. on Electronic Components & Technology* 796–802.

[3] Ferreira J, Capela C and Costa J 2010 *Compos Part A Appl Sci Manuf* **41** 345–352.

[4] Zeng A, Xiong W and Xu J 2005 *Surf. Coat. Technol.* **197** 142–147.

[5] Koopman M, Gouadec G, Carlisle K, Chawla K K and Gladyz G 2004 *Scripta Mater.* **50** 593–596.

[6] Qian H, Lin G, Zhang Y, Gunawan P, and Xu R 2007 *Nanotechnol.* **18** 355602.

[7] Adheed U A S, Premkumar J and Ranganathan T V 2014 *Int. J. Eng. Res. Appl.* **4** 242–48.

[8] Anusha L, Nurul A M N, Zakaria Z and Lan D N U 2017 *Macromol. Symp.* **371** 44–49.

[9] Tomasz G, Tomasz J, Ivan G and Grzegorz K 2015 *Adv. Sci. Technol. Res. J.* **9** 81–88.

[10] Elzbieta and Pawel 2013 *Cell. Polym.* **32** 257–278.

[11] Davies C, Tallon S and Brown N 2005 *Chem. Eng. Res. Des.* **83** 782–787.

IOP Conf. Series: Materials Science and Engineering **429** (2018) 012062 doi:10.1088/1757-899X/429/1/012062

Effect of Oregano Essential on the Properties of Mango Kernel Starch Films

N Sarifuddin[1*], N A A Shahrim[1], K S Azman[1]

[1]Department of Manufacturing & Materials Engineering, International Islamic University Malaysia (IIUM), Jalan Gombak, 53100 Kuala Lumpur, Malaysia

*Email: norshahida@iium.edu.my

Abstract. The mango kernel starch (MKS) - glycerol (GLY) films were prepared via solution casting method. The blend ratio of MKS and GLY was fixed at 1:1 wt. %, while the composition of oregano essential oil (OEO) was varied from 0, 0.3, 0.4, 0.5, and 0.6 wt. %. In this study, the effect of OEO on mechanical, morphological and biodegradable properties of MKS - GLY films were evaluated. From the analysis, the optimum tensile strength of MKS – GLY film was obtained at 1.10 MPa, while the addition of 0.4 wt. % OEO to MKS – GLY film resulted in optimum tensile properties, which give a smooth surface with less voids, as presented in the micrographs by scanning electron microscope (SEM). Moreover, the addition of 0.4 wt. % OEO to MKS – GLY film exhibited minimum weight loss during soil burial test, thus confirmed the function of OEO as an antimicrobial agent.

1. Introduction

In recent years, the usage of starch-based film as a potential alternative to petroleum-based plastics has been raising an interest among researchers to promotes sustainability and focuses the negative impact of non-biodegradable plastics pose on the environment [1]. Specifically, starch has been extensively utilized for packaging application due to its availability, affordability, renewability, and biodegradability. Mango kernel starch (MKS) was found to be one of potential biopolymer material for the preparation of biodegradable films because this type of starch is usually ignored and wasted after eaten or processed [2].

Mango (*Mangifera indica* L.) is a typical tropical fruit that are normally found in Asia, African and Central America [3]. Generally, mango seed is a single flat oval shape seed that are fibrous or hairy on the surface depending on the harvest. The mango seed consists of strong layer bound to the kernel and seed content of mango weight are range from 9% to 23% depend on the different mixture of mangoes while kernel content of seed range around 45.7% to 72.8%. On top of that, mango seed kernel was high in potassium, magnesium, calcium, phosphorus and sodium also consists of crude protein, oil, ash, crude fiber, and carbohydrate [3].

However, native starch usually leads to the brittleness of the fabricated films because of the strong intermolecular forces between the starch chains. Hence, the incorporation of glycerol as a plasticizer with starch reduces the internal hydrogen bonding along polymer chains, thus increasing the intermolecular spacing, enhancing the film flexibility [1,2]. Yet, the plasticized starch-based films possessed some drawbacks, such as short shelf life and safety of contained products for food packaging industry. For that reason, natural antimicrobial and antioxidant like oregano essential oil (OEO) is a promising alternative on replacing synthetic additives [4].

Therefore, the present study aimed to utilize the mango kernel starch for the preparation of biodegradable film. The effects of different concentration of oregano essential oil as both antimicrobial and antioxidant agent on mechanical, morphological and biodegradable properties of MKS films were investigated.

2. Materials and methods

2.1. Materials

Assorted mango seeds were obtained from the local market. Sodium hydrosulfide hydrate (NaHS) flakes (≥ 67% purity) were used to extract starch from mango seeds via centrifugation method, ACROS Organics™. Glycerol (plasticizer) (99% purity) purchased from Merck Chemicals (Malaysia) was used as received to produce thermoplastic starch. Oregano essential oil (natural extract) was purchased from Soap Cart Penang (Malaysia) was used as received.

2.2. Extraction of Starch from Mango Kernel

Dried mango kernel was cut into 2 cm^2 and soaked in 0.16% NaHS solution for 24 hours at 40 °C. The slurry was filtered using tea strainer. The filtrate was ground in kitchen blender and wash thoroughly with distilled water. The settle starch was centrifuge at 5000 rpm for 30 minutes and dried at 40°C overnight [5].

2.3. Preparation of Mango Kernel Starch Films

The mango kernel starch (MKS) films were prepared by solution casting method proposed by Nawab et al. [2] with some modification. The ratio of starch to glycerol was fixed at 1:1 wt. %, while the amount of oregano essential oil (OEO) was varied from 0.3 to 0.6 wt. % with the total mass of the system is 100 g. The materials' abbreviations and corresponding sample compositions are listed in Table 1. The mixture was heated up to 90°C on a hot plate and held at the particular temperature for 20 minutes with constant stirring speed at 250 rpm. Then, the OEO was added to the solution, continuously stirred for 10 minutes, and cooled to 65°C and poured onto acrylic plates. The cast films were dried for about 24 hours in the oven at 40°C and later peeled from the plates. Then, the peeled films were stored in the desiccator maintained at 23°C and 30% relative humidity prior to characterization.

Table 1. Used symbols and corresponding compositions of each sample

Materials	Abbreviations	Samples (weight proportions, gram)			
		Starch	Glycerol	Oregano Essential Oil	Distilled Water
MKS-GLY	MF-T1	1.0	1.0	-	98.0
MKS-GLY-OEO1	MF-ORE0.3	1.0	1.0	0.3	97.7
MKS-GLY-OEO2	MF-ORE0.4	1.0	1.0	0.4	97.6
MKS-GLY-OEO3	MF-ORE0.5	1.0	1.0	0.5	97.5
MKS-GLY-OEO4	MF-ORE0.6	1.0	1.0	0.6	97.4

2.4. Tensile Test

Tensile test was performed by using MTS tensile tester (Shimadzu; Material Testing System) according to ASTM D882-02 with 5 mm/min speed tensile and 5 kN of load. Testing was done on five samples each measuring 70 mm x 10 mm that were cut from the cast films into strips according to standard ISO–527.

2.5. Scanning Electron Microscopy (SEM) Test

Microstructure of the MKS powder and films were observed using a SEM machine (JEOL, Model JSM 5600 SEM; JEOL Ltd., Tokyo, Japan). Before analysis, the sample was mounted on the aluminum holder using carbon tape and sputter coated with palladium (Polaron SC515) with 8 kV accelerating potential.

2.6. Soil Burial Test

The films measuring 20 mm x 20 mm were cut, weighed, and wrapped in the plastic (synthetic) net. Then, the samples of each composition were buried at depth of 50 mm in different pots filled with composted soil and placed at the open area so that they were exposed to the natural weather. The films were removed at regular time intervals (2 days), dried and weighed on the electronic mass balance. The weight loss of the sample over time was evaluated during the soil burial test using the equation:

$$Weight\ loss\ (\%) = \frac{w_i - w_d}{w_i} \times 100$$

where w_i is the initial weight of the sample and w_d is the weight of the sample after seven days.

3. Results and Discussion

3.1. Tensile Properties

The addition of OEO greatly affected the tensile strength (TS) and elongation at break (E%) of MKS-GLY films. Fig. 1 and Fig. 2 shows the effect of increasing OEO content of MKS-GLY films on tensile properties.

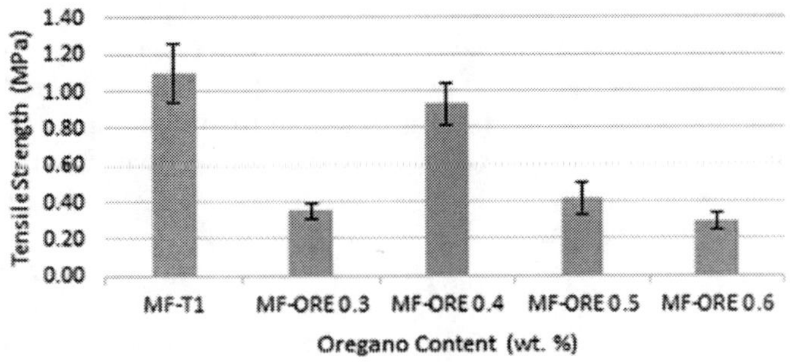

Fig. 1. Effect of increasing oregano content on the tensile strength of mango kernel starch films.

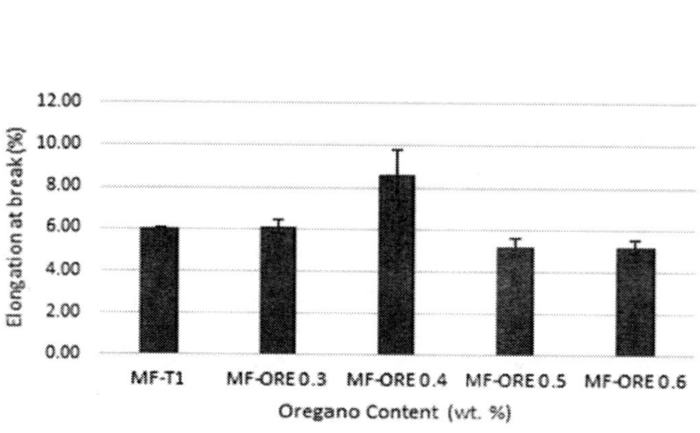

Fig. 2. Effect of increasing oregano content on the elongation at break of mango kernel starch films.

The tensile strength (TS) of 0 wt. % OEO to the MK starch film has contributed to the highest strength of 1.10 MPa, followed by 0.4 wt. % OEO with 0.93 MPa. Initially, addition of OEO to the MK film shows an increment and once it reaches optimum tensile strength, it shows a gradual decrease. On top of that, MKS films that contained excess oil showed poor mechanical performance than the control films especially when higher oil concentration (above 0.4 wt. % OEO) was added, this will entrap the oil content in the polymer network and affecting the network weakening due to lack of miscibility of components, hence will reduce overall cohesion forces of the polymer networks [6]. However, elongation at break for 0.4 wt. % OEO shows the high flexibility compared to the control film (MF-T1) due to the interactions between oil's component and polymer chains, thus provide strong plasticization effect in the matrix that restrain polymer chain aggregation caused by excess oil [6].

3.2. Morphological properties

The morphology of mango kernel starch and casted MKS-GLY films incorporated with OEO (magnification of 1000×) were presented in Fig. 3. Fig. 3(a) revealed the MKS has a fibrous shape, non-uniform particle size and irregular surface. As a result from plasticization of MKS with 1 wt. % glycerol, less coarse granules were observed in Figure 3(b), probably due to starch granule fusing into homogeneous phase and has smooth surface without gross defects. However, the addition of 0.3 wt. % OEO shows a disrupt homogeneity of the blend on the surface as it has rough surface and some void and starch granule pull out from the matrix (Fig. 3(c)). Further addition of OEO at 0.4 wt. % shows a homogenous, clearer and smoother surface (Fig. 3(d)). It shows that the starch molecules have been well dispersed as not many starch granules were detected on the surface film due to all the starch-glycerol and OEO blend uniformly in the matrix, thus showed the integrity of structure and improve the mechanical properties such as tensile strength. Fig. 3(e) and 3(f) illustrates that the addition of OEO more than 0.4 wt. % show the presence of a few starch granules but the blend shows a less homogenous as the surface was not smooth compared to 0.4% concentration due to oregano oil could not bind properly to the starch matrix caused by destabilization of the matrix. Less homogenous and void cracks on the surface morphology of the film led to lower tensile properties.

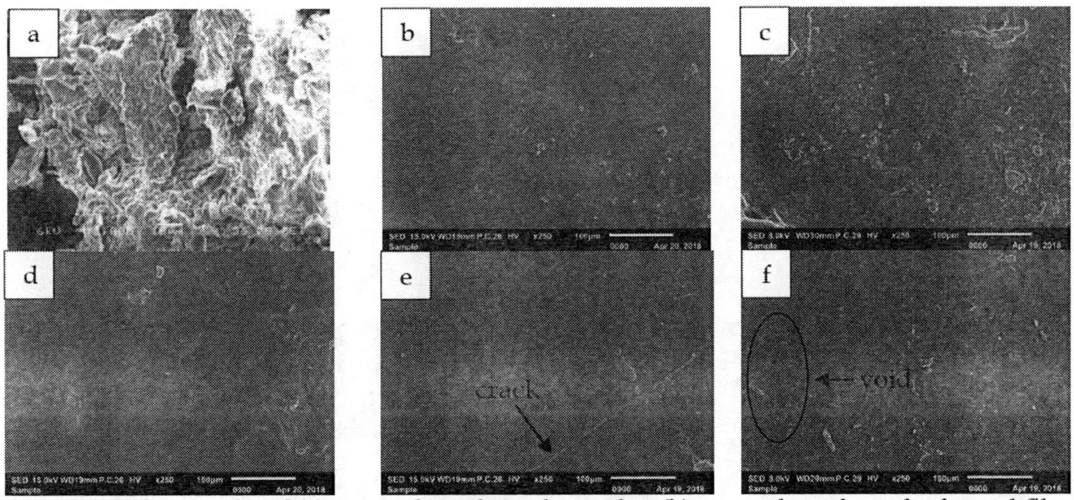

Fig. 3. SEM micrographs of (a) mango kernel starch powder, (b) mango kernel starch-glycerol film, and mango kernel starch-glycerol film incorporated with oregano essential oil at (c) 0.3, (d) 0.4, (e) 0.5, and (f) 0.6 wt. %

3.3. Biodegradation of films in soil

The biodegradability behavior of mango kernel starch-glycerol films with variations of oregano essential oil content are determined based on their respective weight loss and macroscopic appearance shown in Fig. 4 and Fig. 5.

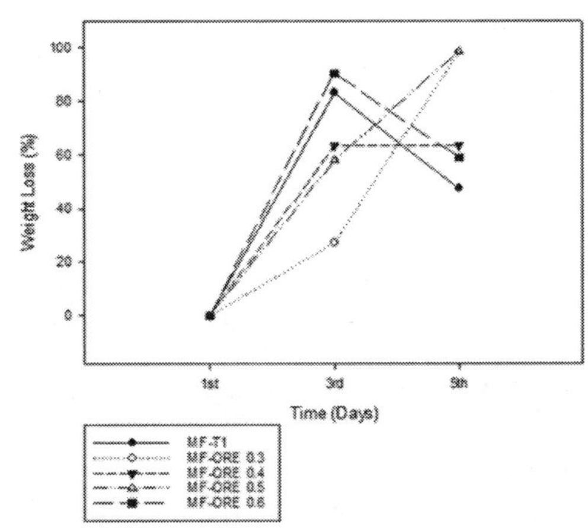

Fig. 4. Weight loss of mango kernel starch-glycerol films incorporated with oregano essential oil after buried in soil for 5 days

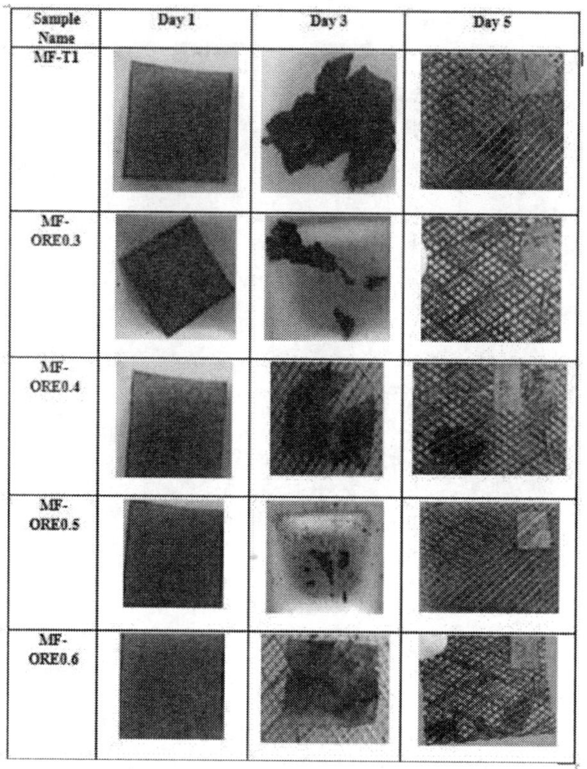

Fig. 5. Macroscopic appearance of mango kernel starch-glycerol films incorporated with oregano essential oil after being buried for 5 days

From Fig. 4, it can be observed that the percent weight loss shows a continuous increase, however after day 3, samples that contain 0 wt. %, and 0.6 wt. % concentration of OEO shows a sharp drop as in Fig. 5 shows the appearance of starch-glycerol film incorporated with OEO after 5 days. This is due to the presence of water in the films permeates through soil and diffused into the films causing it to become soft and swelling at day 3 and at the same time rapid the degradation rate, thus after day 3 the film slowly degrade due to intermolecular bond is weakening

Next, MK films contain 0.3 wt. % and 0.5 wt. % OEO, the percent weight loss shows a constant escalation in percent weight loss and at day 5, both samples show an almost 100% percent weight loss. This is further evidenced by image in Fig. 5 whereby, no films residue left in the wire trap and this represent a fully degradation after 5 days buried in compost soil. Unlike other results in Fig. 4, 0.4 wt. % OEO shows a minimum weight loss after day 3. This is related to the antimicrobial properties that helps to reduce the growth rate of microorganism at the films compared to control film. This could be supported SEM morphology in Fig. 6. From the figure, sample with 0 wt. % OEO shows more formation of microorganism which is known as spores on the surface of the film compare to 0.4 wt. % concentration that has less spores.

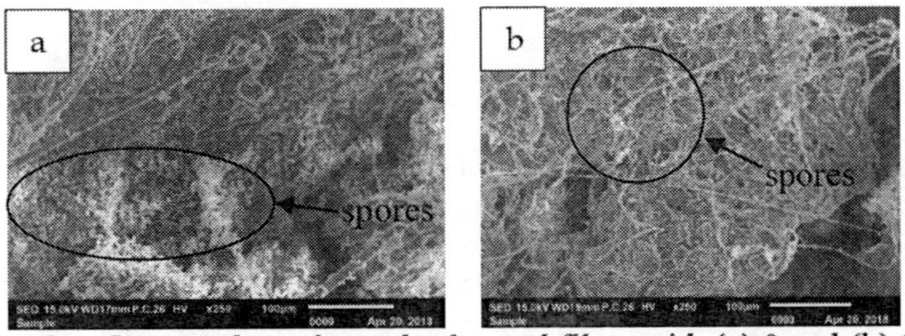

Fig. 6. SEM micrographs of mango kernel starch-glycerol films with (a) 0 and (b) 0.4 wt. % OEO after soil burial for 5 days

4. Conclusion

This study has shown that oregano essential oil as the natural extract effectively incorporated with starch and glycerol blend via solution casting method. Mango kernel starch-glycerol films with 0.4 wt. % oregano essential oil has achieved the optimum tensile properties with smooth film surface. In addition, mango kernel starch-glycerol films showed their degradability under natural weather, but considerably decreased the weight loss and reduced the size of films.

References

[1] Sanyang M L, Sapuan S M, Jawaid M, Ishak M R and Sahari J 2015 *Polymers* **7** 1106–1124
[2] Nawab A, Alam F, Haq M A and Hasnain A 2016 *Starch/Staerke* **68**
[3] Kittiphoom S 2012 *Int. Food Res. J.* **19** 1325–1335
[4] Aziz M and Karboune S 2016 *Crit. Rev. Food Sci. Nutr.* **8398**
[5] Nawab A, Alam F, Haq M A, Lutfi Z and Hasnain A 2017 *Int. J. Biol. Macromol.* **98** 869-876
[6] Cano A, Cháfer M, Chiralt A and González-Martínez C 2015 *Foods* **5** 3

IOP Conf. Series: Materials Science and Engineering **429** (2018) 012063 doi:10.1088/1757-899X/429/1/012063

Wear Prediction via Accelerated Test on Chromium Based Hard Coatings for Gas Turbine Interfaces Applications Up to 370 °C

S M Yunus[1*], **A A Pauzi**[1], **S Husin**[2]

[1]Researcher, TNB Research Sdn Bhd, Selangor, Malaysia
[2]Principal Researcher, TNB Research Sdn Bhd, Selangor, Malaysia

*Email: salmi.yunus@tnb.com.my

Abstract. Chromium (Cr) based hard coatings offer excellent wear resistance and thermal conductivity properties. Applying additives, such as carbon (C) and cobalt (Co) may improve hardness thus delay wear. Therefore, Cr-C and Cr-Co based coating systems were selected to investigate their potential on the mating surfaces of Ni-based superalloy combustor liners of commercial gas turbines. One surface contact on the commercial combustor liner was selected, in which was exposed to temperatures of 370 °C during its operation. In this study, the selected hard coatings were deposited onto nickel-based substrates via air plasma. An accelerated wear test was conducted at two different operating temperatures; at 100 °C and 200 °C to predict the wear characteristics at elevated temperatures for longer operation, from 8,000 Operating Hours (OH) to 12,000 OH. Both coatings exhibited a decreased hardness with increasing temperatures. It was found that high severity in volume loss suffered by Cr-Co coatings. In summary, it was quantitatively predicted that Cr-C coated Ni alloy was found to be much better than the Cr-Co based coating to delay the wear in gas turbines thus prolong the operation from 8,000 to 12,000 OH at 370 °C.

1. Introduction

Suitable material selection of the hard coatings will intensify the wear delaying process especially in high temperature applications like gas turbine. To suit the gas turbine applications, the chromium-based hard coating is a preferable choice due to its ability to provide protection for the substrate in a high-temperature environment [1]. Chromium-cobalt (Cr-Co) based and chromium-carbide (Cr-C) based hard coatings are amongst the popular choices in resisting wear. Not limited to wear resistance applications, plasma-sprayed chromium-based hard coating has also been effectively applied for the erosion-corrosion resistance of boiler steels in the coal-fired boiler environment [2]. The wear behaviours of Cr-Co based coating are almost similar at elevated temperatures up to 600 °C whereas Cr-C based coating are up to 500 °C [3,4]. Hardness is one of the mechanical characteristics evaluated to describe the wear behaviour of a material. In general, good coatings must be harder than the protected surface to reduce wear damage under the same operational conditions [5]. Pin-on-disc test is chosen due to its ability to simulate metal-to-metal contact under laboratory conditions [6,7]. Neither a standard nor customised wear test could completely imitate actual operating conditions since the mode of relative movement between the mating surfaces varies with different components. However, they can be used for coating development or in a coating comparative study [8]. An accelerated wear test via pin-on-disc is selected by researchers and applied in many applications, such as in the coating industry and surgical orthopaedics, to vary the choice of materials and operating conditions [9,10]. The accelerated test that has been conducted was aimed to reproduce the wear process in a much shorter time than in the field, without changing the degradation mechanism [11]. This accelerated test has speed up the wear process and obtain reliability volume loss for both Cr-Co based and Cr-C based coating in a timely manner. This accelerated-time approach then is used to predict the volume losses by both at 12,000 OH which based on the accelerated test via pin-on-disc.

2. Methodology

Cr-C based and Cr-Co based hard coatings were deposited on the nickel-based superalloy substrate using the air plasma spray technique. A field test was carried out on both hard coatings to obtain any relevant wear correlation between the two coatings. Exposure of Cr-C based and Cr-Co based coatings was conducted on two gas turbine units, which made of similar brand and model, for 8,000 OH. An accelerated test was carried out via pin-on-disc. The parameter of the accelerated test used was tabulated in Table 2-1. Both deposited coatings were measured using Vickers hardness tester. Surface hardness changes are measured after both tests are completed. Any correlation that is possibly established between surface hardness and wear behaviour of both hard coatings was determined. The volumetric loss was calculated for each contact temperature; 100 °C and 200 °C and for each sliding time of 3600 s, 10,800 s and 18,000 s. The height, h of the pin is measured before and after each test condition. Each test is repeated three times with 25 readings recorded, (for repeatability). The best-fit curve via logarithmic trend line was established on the volume loss results of the accelerated test, at both temperatures of 100 °C and 200 °C. The established equation of the curve was practically useful to link the accelerated test results to the field test and determine the accelerated time. The volume loss by both Cr-C and Cr-Co coatings at 12,000 OH was determined via similar equation. The maximum volume loss was used as acceptance limit [12]. Once the dimensions exceeded the maximum volume loss limit, the component will not be reliable for further operation.

Table 2.1: Parameter used in the accelerated test (pin-on-disc).

Variable	
Independent	**Dependent**
	Coating material:
	i. Cr-C based coating
	ii. Cr-Co based coating
Normal load (100 N)	Contact temperature:
Rotational speed (200 rpm)	i. 100 °C
Pin height (35 mm)	ii. 200 °C
Pin diameter (12 mm)	Sliding time:
Track diameter (100 mm)	i. 3,600 s (1 hour)
	ii. 10,800 s (3 hours)
	iii. 18,000 s (5 hours)

3. Results and Discussion

3.1. *Hardness Changes*

Referring to Figure 1, for the Cr-Co coated mating surfaces, the hardness of the worn surfaces that were exposed to 370 °C decreased by 7 % and it was also as expected due to the remaining hard coating on the surface. In contrast, the substantial decrement in the hardness has been observed on the worn Cr-C coated surfaces reaching the hardness of the underneath surface material. There were many studies by other researchers found that exposures of carbide coatings to elevated temperatures up to

400 – 500 °C had decreased its hardness. Many discussions were reported to explain this such as the variations of fracture toughness during the measurement, carbide formation and distribution throughout the worn surface and density of the oxides contents within the coated surface [13, 14, 15].

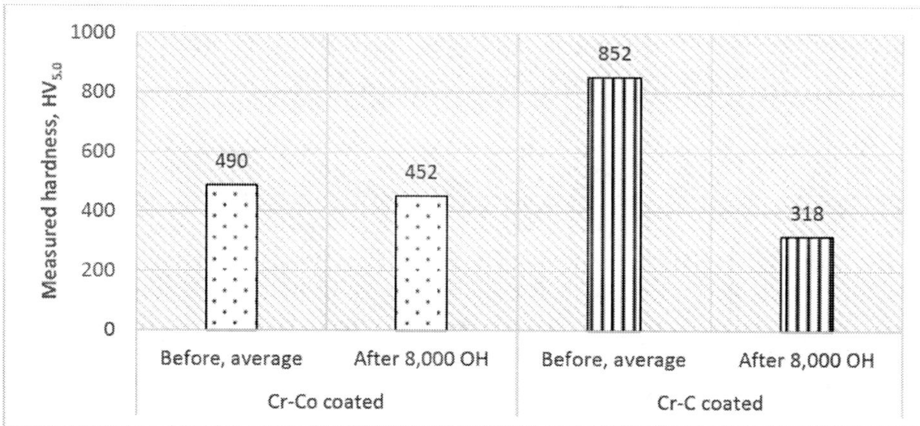

Figure 1: Hardness of Cr-Co coated and Cr-C coated mating surfaces before and after 8,000 OH exposure during field test.

Figure 2 illustrates the changes in hardness of the coatings with increasing temperature. The initial hardness of Cr-C coated pins was observed to be 17 % higher than Cr-Co coated pins. Each coating exhibited similar trends, where both coatings tended to decrease linearly by applying higher temperatures. The higher hardness of Cr-C coatings before and after the accelerated test hypothesised that they will exhibit better wear resistance than Cr-Co based coatings.

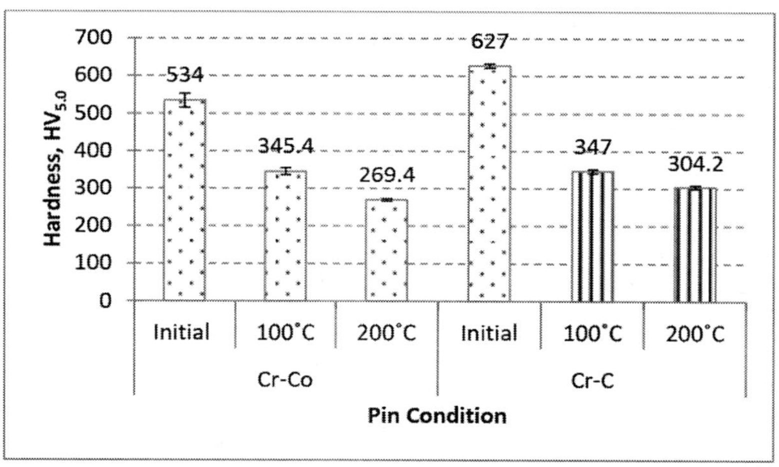

Figure 2: Average hardness measured on Cr-Co and Cr-C coated pins, before and after the test.

3.2. Volume Losses

Figure 3 illustrates the volume loss experienced by the gas turbine interfaces at 370 °C in Cr-Co coated and Cr-C coated conditions after 8,000 OH of exposure. Among these two hard coatings, Cr-Co based was slightly better due to its slightly lower volume loss compared to Cr-C based.

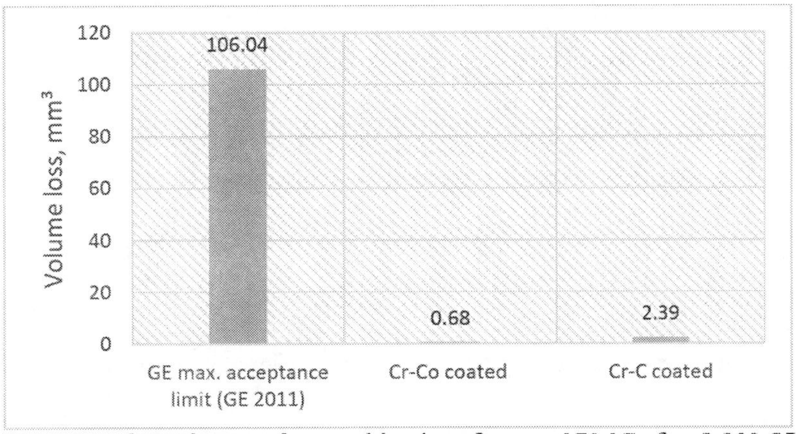

Figure 3: Volume losses of gas turbine interfaces at 370 °C after 8,000 OH.

Figure 4 illustrates volume losses of Cr-Co coated pins and Cr-C coated pins from the accelerated test. It was observed that by applying an increment in the operating temperature, this induced a slight change in the volume. The finding agrees with researchers observing that contact temperatures would trigger only slight changes in the wear rate [16]. The Cr-Co coating exhibited a higher amount of volume loss than Cr-C coating. The higher volume losses by Cr-Co coated pins met the hypothesis due to its lower initial hardness.

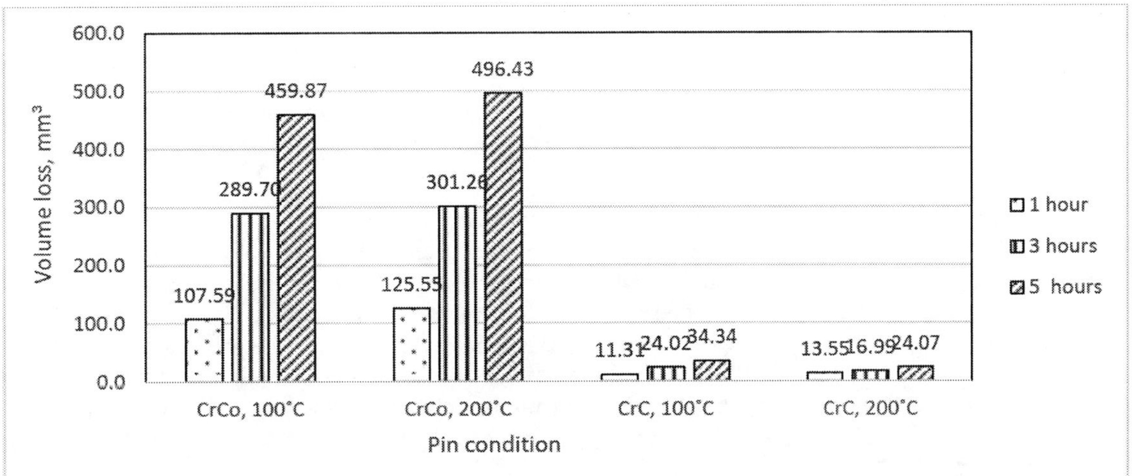

Figure 4: Volume loss of both coatings through the accelerated test.

3.3. Prediction of Volume Loss at 12,000 OH

Figure 5 and Figure 3-6 show the best-fit curve trend line generated on the accelerated test results for Cr-Co and Cr-C coating, respectively at 100 °C and 200 °C. R^2 value which almost reached one (R^2~1) on the graphs indicated that the best-fit curve has been developed.

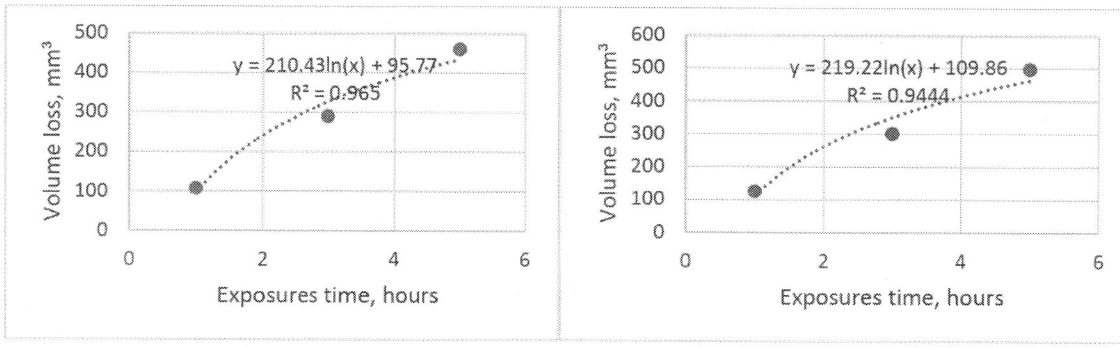

(a) (b)

Figure 5: The best-fit curve trend line generated on the accelerated test results for Cr-Co coating at (a) 100 °C and (b) 200 °C.

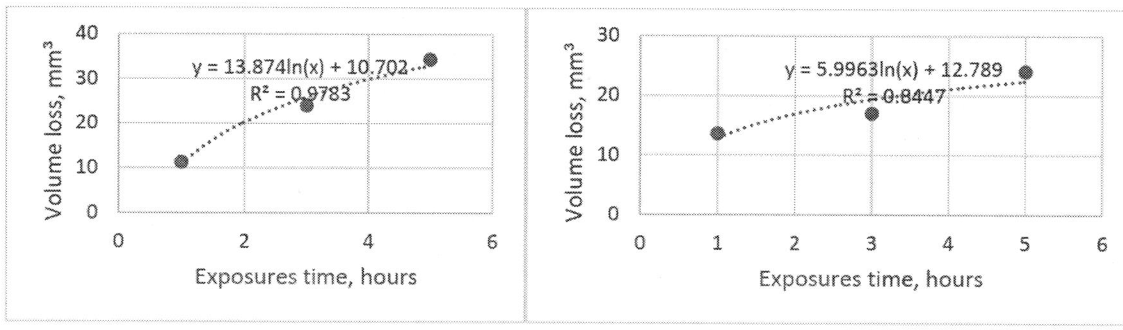

(a) (b)

Figure 6: The best-fit curve trend line generated on the accelerated test results for Cr-C coating at (a) 100 °C and (b) 200 °C.

Figure 7 illustrates the predicted volume loss for the fuel nozzle collar at 12,000 OH, which were calculated from the generated equation. The maximum volume loss at 8,000 OH recommended by OEM is used as the acceptance criteria, which 106.04 mm³ [12]. It can be seen that the predicted volume losses shown at 12,000 OH were in the acceptance range for both Cr-C coated and Cr-Co coated interfaces. Cr-C coating displayed a very convincing choice where the predicted volume loss at 12,000 OH is of 7.55 % for 100 °C-curve and 4.55 % for 200 °C-curve, from the maximum acceptance limit. In contrast, Cr-Co coating displayed as unsuitable choice when the predicted volume loss at 12,000 OH is of 80.99 % for 100 °C-curve and 84.46 % for 200 °C-curve, from the maximum acceptance limit.

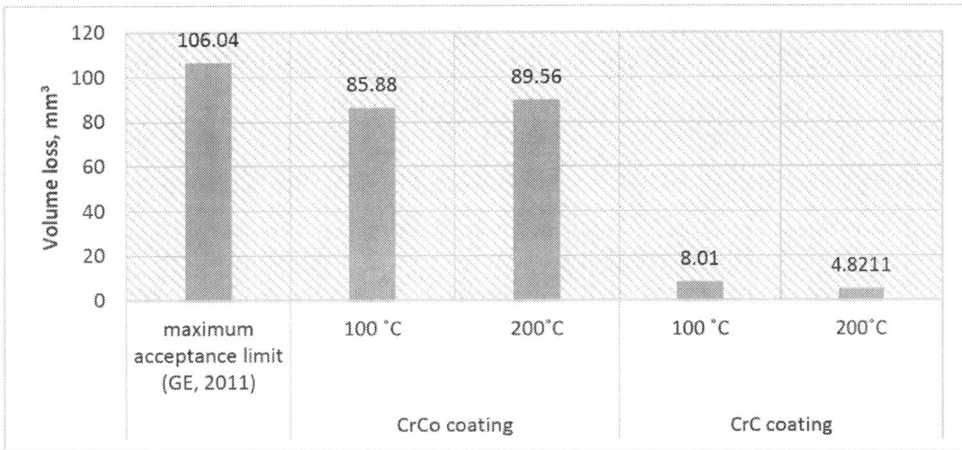

Figure 7: Prediction of the volume loss for the fuel nozzle collar at 12,000 OH exposed at 370 °C obtained from the accelerated test curve.

4. Conclusions

The application of Cr-Co and Cr-C coatings has tremendously improved the volume loss of interfaces in gas turbine. From field test at contact temperature of 370 °C, Cr-Co coating exhibited only 0.68 mm^3 which 0.64 % from the maximum acceptance limit whereas Cr-C coating exhibited only 2.39 mm^3 which 2.25 % from the maximum acceptance limit. From the best-fit curve which generated from the accelerated test results, it was found that the volume losses by Cr-Co coating at 12,000 OH are 89.56 mm^3 (from 200 °C-curve) and 85.88 mm^3 (from 100 °C-curve). This was about 84.5 % and 81.0 % from the maximum acceptance limit by OEM, respectively. Meanwhile, the volume losses by Cr-C coating at 12,000 OH are 4.82 mm^3 (from 200 °C-curve) and 8.01 mm^3 (from 100 °C-curve). This was only about 4.5 % and 7.6 % from the maximum acceptance limit by OEM, respectively. Considering 200 °C-curve as the nearest contact temperature to 370 °C, it can be seen that Cr-C coating is 94.6 % better than Cr-Co coating, of the predicted volume loss at 12,000 OH, which is good indication in delaying wear. It can be concluded that Cr-C coating is recommended to prolong for further 4,000 OH.

Acknowledgement

The authors would like to thank Tenaga Nasional Berhad (TNB) for the financial aid and special thanks to the Materials Engineering and Testing Group of TNB Research Sdn Bhd, and National University of Malaysia for their full cooperation in this work.

References

[1] Singh H *et al* 2007 *Materials and Corrosion* **58(2)** 92
[2] Jazi H S 2012 *Advanced Plasma Spray Applications* (InTech)
[3] Hou G *et al* 2015 *Acta Materialia* **95** 164
[4] Beckman B 2013 *Wear of a Gas Turbine Friction Damper* (MANE-6960 Friction, Wear, and Lubrication of Materials)
[5] Kato K 2000 *Wear* **241** 151
[6] Talluri S 2000 *Pretreatment of Small Four-Stroke Engine Components for No-Oil Hot Tests* (Virginia Polytechnic Institute and State University)
[7] Gobind, Grover N and Parshad J 2015 *Journal of Mechanical and Civil Engineering* **12(2)** 101
[8] Ghabchi A 2011 *Wear-resistant carbide-based thermal sprayed coatings: process, properties, mechanical degradation and wear* (Stony Brook University, New York)
[9] Kuforiji C and Nganbe M 2017 *Tribology International* **111** 100

IOP Conf. Series: Materials Science and Engineering **429** (2018) 012063 doi:10.1088/1757-899X/429/1/012063

[10] Yamaleyeva D N, Baker P A and Catledge A A 2016 *J. of Young Investigators* **31(3)** 21
[11] Gu X *et al* 2009 *Service Life Prediction of PoLymeric Materials* (Springer)
[12] Anon 2011 *GEK 99607 – Inspection and Maintenance* (General Electric)
[13] Zia A W *et al* 2017 *Surface & Coatings Technology* **320** 118
[14] Tobi A L M *et al* 2017 *IOP Conf. Series: Materials Science and Engineering* **165** 012020
[15] Mojena M A *et al* 2017 *Wear* **376-377** 557
[16] Falqueto L E *et al* 2017 *Wear* **376-377(B)** 1739

IOP Conf. Series: Materials Science and Engineering **429** (2018) 012064 doi:10.1088/1757-899X/429/1/012064

Optimization of *Diplazium Esculentum* Extract Using Pressurized Hot Water Extractor by Box-Behnken Design of Experiments and Its Antioxidative Behavior

A J Nur[1], F K Khairul[1,2*], M A Nuradibah[1], S S S Noor [3]

[1]School of Bioprocess Engineering, Universiti Malaysia Perlis, 02600 Arau, Perlis, Malaysia.
[2]Centre of Excellence for Biomass Utilization, School of Bioprocess Engineering, Universiti Malaysia Perlis
[3]School of Chemical Sciences and Food Technology, Faculty of Science and Technology, Universiti Kebangsaan Malaysia, 43600 Bangi, Selangor.

*Email: khairulfarihan@unimap.edu.my

Abstract. Pressurized hot water extraction (PHWE) was implemented in attempt to reduce the use of toxic organic solvent in extracting bioactive compounds from *Diplazium esculentum*. The antioxidant assays, which were the DPPH scavenging activity, total phenolic contents and total flavonoid contents were analyzed on PHW extracts. By applying Box-Behnken design under response surface methodology (RSM), the optimized condition for the best antioxidant activity of PHWE was at 175 °C, 21 min extraction time and 50 mL water volume added to 2 g of dried *D. esculentum*. The DPPH antiradical efficiency of *D. esculentum* give EC_{50} values of 1241.14 µg/mL and time to reach steady state (T_{EC50}) was 79.83 min. Thus, the extract was categorized as slow in scavenging the DPPH free radicals.

1. Introduction

Diplazium esculentum (Retz.) Swartz also known as *Athyrium esculentum* (Retz.) Copel is in the family of *Athyriaceae*. This vegetable ferns (known as pucuk paku in Malaysia) grows wildly at moist or wet areas such as partially shaded river banks or homestead ground. The diploid plant occurs throughout Asia and Oceania (Indonesia, Philippines, Thailand, Papua New Guinea, Bangladesh, Japan and Fiji [1,2]. The fronds are use in traditional remedy to treat jaundice, constipation, earache, fever, measles and dermatitis [3,4].

D. esculentum were freshly used, boiled with water or soaked with organic solvents (hexane, acetone and alcohols) prior to analysis of the bioactive compounds in several studies [5,6,7]. However, there is no recent extraction method has been applied on *D. esculentum* up to this date. Thus, pressurized hot water extraction (PHWE) was applied in this recent study on *D. esculentum* due to the advantages of non-toxic, non-flammable and environmental friendly criteria of water as solvents.

PHWE refers to the extraction with water under pressure at the temperature between the boiling and critical temperature (100 and 374 °C respectively) [8]. The extraction of the bioactive compounds in plants could be affected by the plant species, extraction temperature, time, water-sample ratio and/or pressure4, [9, 10, 11].

This research focuses on the extraction of *D. esculentum* active compounds which contribute to DPPH antiradical activity. Optimization of the extraction condition of PHWE for the DPPH, total pehnolic contents (TPC) and total flavonoid contents (TFC) will be analyzed using Box-Behnken design of experiment.

2. Materials and Methods

2.1. Plant Material
D. esculentum were bought from local market in Arau, Perlis. The fronds and stems were dried overnight in the oven (Binder, Malaysia) at 60 °C and ground to a fine powder (< 65 μm) for the extraction processes.

2.2. Optimization Using Design of Experiment
Optimization of extraction condition was done by using three-level three factor Box-Behnken design with total of 17 runs including five centre points under RSM. The analysis was carried out using Stat-Ease software (Design-Expert 7.1.5, 2008). The range of parameters (Table 1) used was based on the earlier screening results of the antioxidant assays at various PHWE condition.

Table 1: Range of parameters used for the optimization of antioxidant activity.

Factors	Level	
	Minimum limit (-1)	Maximum limit (+1)
Extraction temperature (°C)	125	175
Extraction time (minutes)	10	30
Volume of distilled water (mL)	30	50

2.3. DPPH Scavenging Activity
Antioxidant activity was determined using DPPH scavenging method [12].

2.4. Total Phenolic Content (TPC)
The Folin-Coiocalteu colorimetric method with minor modification was used to determine TPC in *D. linearis* extracts [13].

2.5. Total Flavonoid Content (TFC)
The total of flavonoids in D. linearis extracts were determined using aluminum chloride colorimetric method [13].

2.6. DPPH Antiradical Efficiency
The DPPH antiradical efficiency of the optimized PHW extracts of *D. esculentum* and reference standard quercetin were detrmined using method reported by Sánchez-Moreno et al. [14].

2.7. Statistical Analysis

The results were presented as mean ± SEM of 6 independent experiments. Statistical values ($P<0.05$) were determine by Student's t-test using GraphPad Prism 5.02, GraphPad Software Inc.

3. Results and Discussion

3.1. Optimization of Three Responses Using Box-Behnken Design

3.1.1. Effects of Parameters on DPPH Scavenging Activity

The parameters units in the design model (Table 2) were coded as A, B and C for extraction temperature, extraction time and volume of distilled water respectively. Analysis of variance shows the reduced two factors interaction design model was significant for DPPH scavenging activity response ($p < 0.001$) with no lack of fit. The Adj-R^2 value for the respond was 0.8089 much higher than the Pred-R^2 (0.4609) could be due to large block effects.

The Adj-R^2 is a statistic that is adjusted for the size or number of factors in the experiment and can decrease if insignificant variables are added to the design model. Reducing insignificant term in the model would produce a more precise data predictor [15]. The interaction between A and B was insignificant, therefore was eliminated from the design model.

The main effect of scavenging activity was extraction temperature (A) ($P < 0.001$). While extraction time (B) and water volume (C) on its own has no significant effects on the DPPH test but they were significantly involved in the interaction of AC and BC. Thus, terms B and C were not eliminated from the model design.

The equation of the reduced model term was:

$$DPPH = 70.18235 + 3.23750A + 0.38750B - 0.90000C + 1.75000AC + 2.10000BC$$

Referring to the interaction of A and C at three level of B (-1, 0 and +1) on DPPH radical scavenging activity, not obvious increment on DPPH response at low level of C (-1.00 C) from low to high level of A, but significant increment can be seen at high level of C (+1.00 C). Both level of water volume (C), require high temperature for good DPPH assay result.

In AC interaction, at high level of B (+1.00 B), the highest DPPH test response was at high level of C with prediction of 76.76% scavenging activity (Figure 1c). Lowering B level to 0.00 B (Figure 1b), DPPH test response was increased at low level of C, but still the highest point was at +1.00 C with reduced DPPH test prediction to that of 74.27%. At lowest level of B (-1.00 B) (Figure 1a), the DPPH response was highest at -1.00 C with 74.28% and while +1.00 C showed further reducing DPPH response. Thus, good DPPH scavenger compounds can best be extracted at high temperature, and the extraction time was highly influenced by water volume. This is supported by the DPPH antioxidant activity of kaffir lime fruit peel. The DPPH scavenging activity was increasing with extraction temperature up to 200 °C but no significant effect of extraction time on PHW extraction condition [16].

Table 2: The Box-Behnken design model with three responses

Std	Run	Factor 1 A: Temperature (°C)	Factor 2 B: Time (min)	Factor 3 C: Water volume (mL)	Response 1 DPPH[a]	Response 2 TPC[b]	Response 3 TFC[c]
4	1	1	1	0	73.9	79.6	14.5
9	2	0	-1	-1	71.1	80.1	14.5
5	3	-1	0	-1	70.7	90	13.2
14	4	0	0	0	70.1	85.3	14.6
6	5	1	0	-1	74.5	86.7	13.3
2	6	1	-1	0	72.7	78.3	14.7
10	7	0	1	-1	68.2	85.4	16.6
17	8	0	0	0	69.6	84.5	14.9
16	9	0	0	0	70.9	85.1	14.1
15	10	0	0	0	70.1	85.3	14.6
8	11	1	0	1	73.8	94.5	16.1
1	12	-1	-1	0	68	79.6	10.9
7	13	-1	0	1	63	78.5	10.3
11	14	0	-1	1	67.5	88.8	14.8
12	15	0	1	1	73	96.8	14.4
3	16	-1	1	0	67.3	79.7	13.9
13	17	0	0	0	68.7	83.9	14.2

[a] scavenging activity (%)
[b] μg catechin equivalent/mg dry extracts
[c] μg rutin equivalent/mg dry extracts

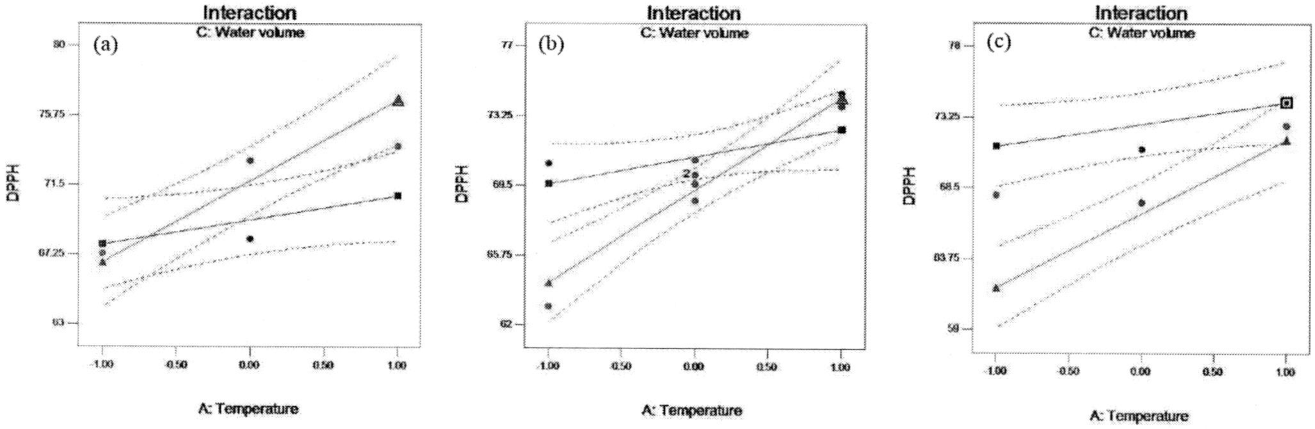

Figure 1: The interaction of extraction temperature and volume of water effects on DPPH scavenging activity response at: a; 10 min, b; 20 min, c; 30 min. The terms -1.000 and +1.000 codes for low and high level of: A; temperature (125 °C and 175 °C), B; extraction time (10 and 30 min), C; water volume (30 and 50mL).

3.1.2. Effects of Parameters on TPC

Quadratic model was suggested by Design Expert for the TPC response. Analysis of variance shows the model was significant to the response with significant lack of fit. Statistically, the three factors (A, B and C) seems to be insignificant model terms ($P > 0.05$), therefore reducing the factors resulting a significant model design with R^2 value of 0.6342. Low value of R^2 does not mean the regression model cannot be acceptable. This is because, the R^2 value will keep increasing by the addition of another model terms including insignificant terms which would results in poor prediction on new data[22]. In TPC response, AC and C^2 are significant model terms with the following equation:

$$TPC = 82.367 + 1.413A + 2.05C + 4.825AC + 5.233C^2$$

The interaction of factor B with the factor A and C respectively shows no significant effects on model design, thus B was eliminated from the model term. However, A and C were involved in an interaction (Figure 2a). At low level of C (-1.000 C), TPC value will decrease with increasing of A whereas, at high level of C (+1.000 C), TPC value will increase along with A.

From Figure 2b, TPC will increase with elevated temperature but no effects on extraction time variation. This finding was supported by the result of parallel increment of TPC and extraction temperature in bitter melon subcritical water extraction. The suitable temperature was in the range between 150-200 °C. Higher extraction temperature may cause degradation of phenolic compounds [17].

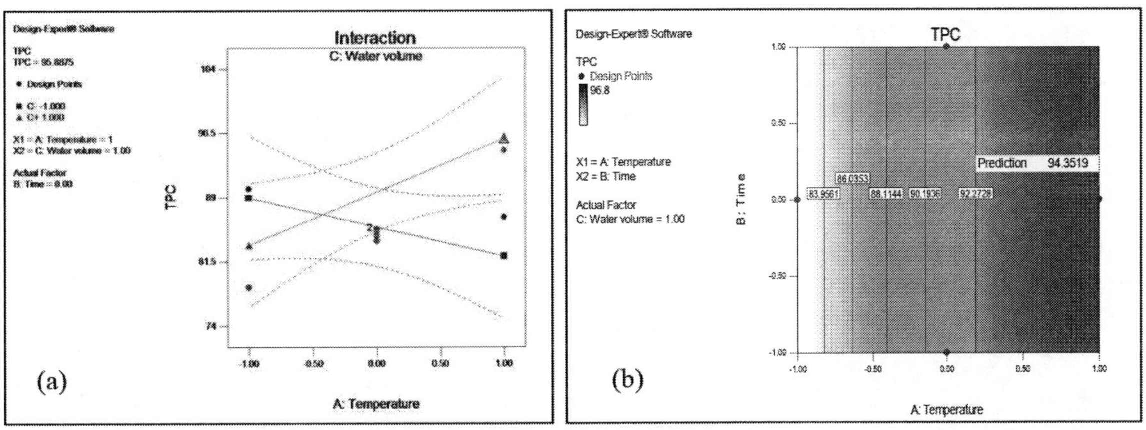

Figure 2: a) The interaction of extraction time and temperature at 50 mL water volume. The terms -1.000 and +1.000 codes for low and high level of: A; temperature (125 °C and 175 °C), B; extraction time (10 and 30min), C; water volume (30 and 50 mL). b) The interaction of extraction time and volume of water effects on TFC at: a; 125 °C, b; 150 °C, c; 175 °C. The terms -1.000 and +1.000 codes for low and high level of: A; temperature (125 °C and 175 °C), B; extraction time (10 and 30 min), C; water volume (30 and 50 mL).

3.1.3. Effects of Parameters on TFC

Analysis of variance on quadratic model applied for TFC response shows the parameters were significant ($P < 0.001$) for the model design with no lack of fit except for water volume (C). However, C was not eliminated in the model design because there were interaction between A (temperature) and B (time) respectively with significant model terms with the following equation:

$$TFC = 14.48 + 1.29A + 0.56B - 0.25C - 0.80AB + 1.42AC - 0.63BC - 1.41A^2 + 0.44B^2 + 0.16C^2$$

IOP Conf. Series: Materials Science and Engineering **429** (2018) 012064 doi:10.1088/1757-899X/429/1/012064

In BC interaction at low extraction temperature (-1.000 A) (Figure 3a), more time is required (+1.000 B) for the lowest water volume (-1.000 C) to get high TFC with prediction of 15.96 µg rutin equivalent/mg dry extract. By increasing the temperature to 0.000 A, the TFC was increasing at both level of C factors (Figure 3b). The highest prediction was 16.43 µg rutin equivalent/mg dry extract at high level of both B and C. The results was further increased at higher temperature (+1.000 A) for high level of C with design point of 16.91 µg rutin equivalent/mg dry extract (Figure 3c). While at low level of C, the TFC was reducing. Thus, temperature highly influenced TFC result with interaction on extraction time and water volume.

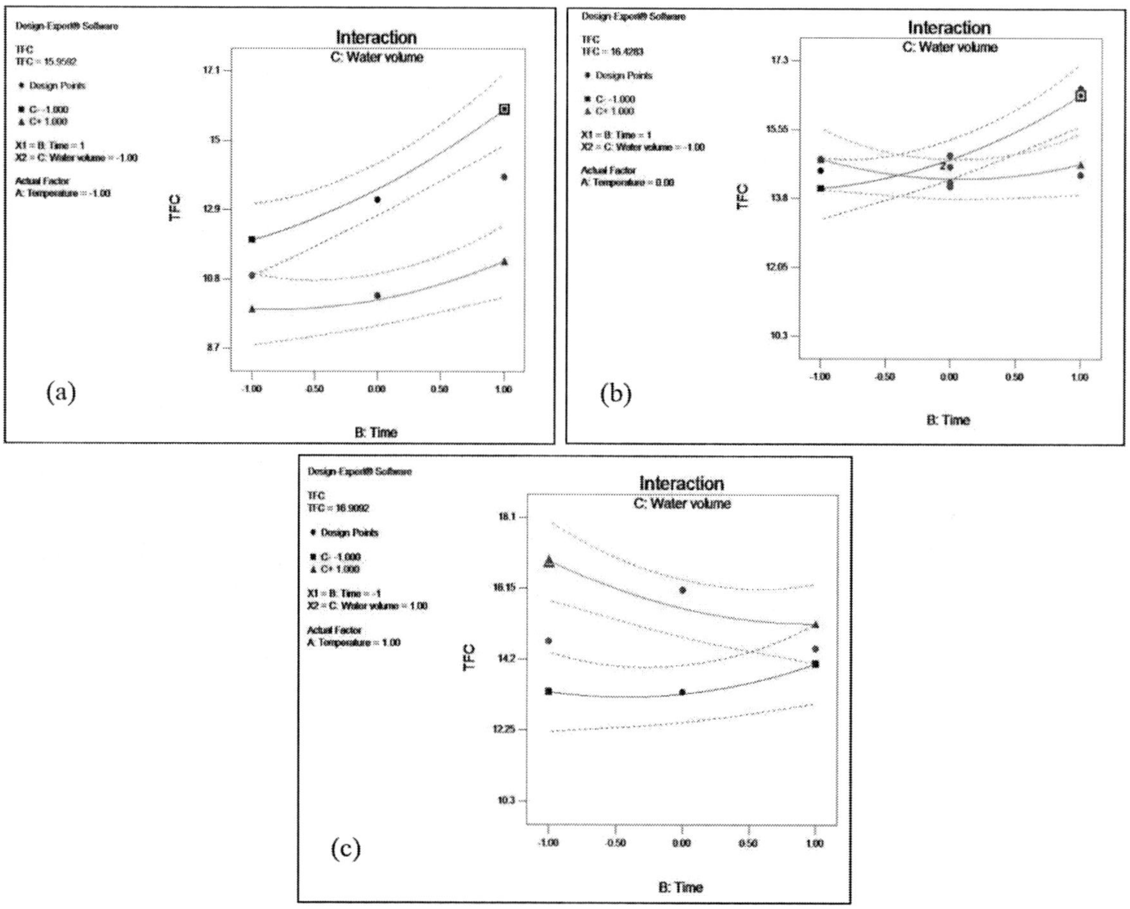

Figure 3: The interaction of extraction time and volume of water effects on TFC at: a; 125 °C, b; 150 °C, c; 175 °C. The terms -1.000 and +1.000 codes for low and high level of: A; temperature (125 °C and 175 °C), B; extraction time (10 and 30 min), C; water volume (30 and 50 mL).

3.1.4. Validation of the Optimized Condition

Suggested condition for optimum antioxidant activity of *D. esculentum* PHW extraction was 175 °C extraction temperature at 21 min and 50 mL of distilled water with desirability of 0.828. Conformation run was done on optimized extraction condition. The three analyses were done on the optimized extract and it was shown that DPPH scavenging activity of actual results was lower than predicted (67.65 ± 1.2 and 74.5% respectively). The TPC of real experiment is much higher than predicted (125.7 ± 16.4 µg and 95.89 µg catechin equivalent/mg dry weight respectively) while TFC of actual result was slightly lower than the predicted (11.12 ± 0.58 µg and 15.53 µg rutin equivalent/mg dry weight respectively). The design model of TPC has low pred-R^2, which could explain the cause of huge differences between the predicted and actual result. It can be concluded that, longer extraction time gave better results for DPPH scavenging activity and total phenolic compounds but less time needed for high total flavonoid contents. Increasing the extraction time could increase the contacts time between solvents and samples.

3.2. DPPH Antiradical Efficiency

The DPPH scavenging activity of standard quercetin at four different concentrations (i.e. 5, 10, 20 and 50 µg/mL) from 1 to 120 min were observed (Figure 4A). The scavenging activities were increasing with standard concentration. At low concentrations, the scavenging activities were less than 30% activity, while at higher concentration reaching almost 80% activities. At low concentration, the time needed to reach the steady states was longer (120 min) and at 20 µg/mL, took lesser time (80 min). At 50 µg/mL, only 9 min was required to reach the steady state. Similar trends were seen in *D. esculentum* extracts at various and higher concentrations than pure quercetin (500 – 2500 µg/mL) (Figure 4B). At 1000 µg/mL and lower concentrations, the DPPH scavenging activities were lower than 55% activities. The highest activities were at the highest concentration (2500 µg/mL) with 77% activities. At low concentrations, the time required to reach steady states were longer (100-120 min), while at 1500, 2000, and 2500 µg/mL needed 80, 50 and 20 min respectively to reach steady states.

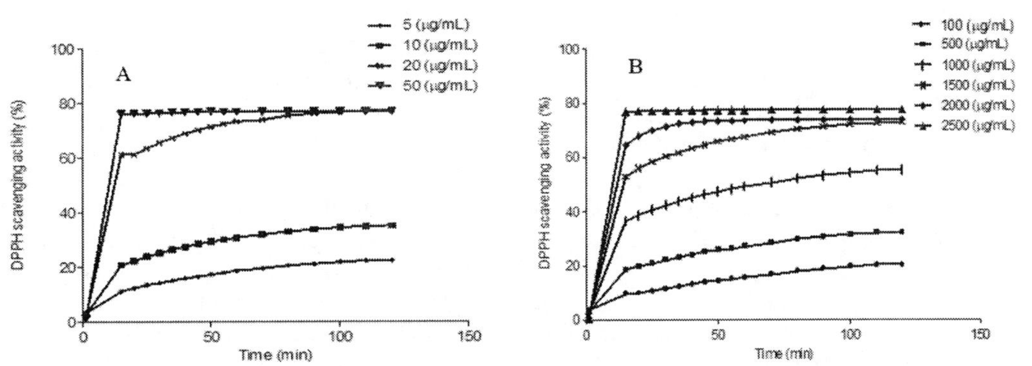

Figure 4: A) Kinetics of quercetin at different concentrations. B) Kinetics of *D. esculentum* at different concentrations.

Thus, the kinetic behaviour of antioxidant depends on the concentration or the amount of the compounds to react with free radicals. Rapid kinetic behaviour occur when the percentage of remaining DPPH reaching a steady state in less than 5 min as of ascorbic acid. Intermediate kinetic behaviour occur in within 5 to 30 min, while longer time (> 30 min) was categorized as slow behaviour [18]. From the results, it can be concluded that the kinetic behaviour of quercetin and *D. esculentum* were intermediate at high concentration and slow at low concentration.
The kinetic behaviours of quercetin and *D. esculentum* extracts follow a general multiplicative model:

$$\ln [\%\ \text{DPPH remaining}] = b \ln \text{time} + \ln a$$

where b is the slope and a is the intercept. The slope determined the rate of percentage of remaining DPPH (% DPPH remaining/min) after scavenged by the antiradicals (quercetin and *D. esculentum* extracts).

The slope is steeper at higher concentration and the remaining DPPH radicals are lower [14]. From the results in Table 3, the slope is steeper with increasing sample concentrations, which is the rate of scavenging antiradical is higher. However, at 50 μg/mL (quercetin) and 2000 and 2500 μg/mL (*D. esculentum*) the slope is less steep because at these concentrations, the antiradical activities have reached pleateu at early time (9, 50 and 20 min respectively). The remaining DPPH radicals were reducing at higher concentration at the time to reach steady state because there are more antiradicals compounds available to scavenge the DPPH radicals at a time.

Table 3: Kinetic behaviour of quercetin and *D. esculentum*

Sample	Concentration (μg/mL)	Slope (% DPPH/min)	Correlation coefficients	Remaining % DPPH at steady state
Quercetin	5	-0.07	-0.991	77.34
	10	-0.102	-0.995	64.99
	20	-0.298	-0.983	23.75
	50	-0.02	-0.892	23.82
D. esculentum	100	-0.067	-0.967	79.87
	500	-0.095	-0.986	68.01
	1000	-0.178	-0.994	45.74
	1500	-0.275	-0.997	28.76
	2000	-0.123	-0.782	26.92
	2500	-0.014	-0.996	23.26

[a] Value mean is ± standard deviation.
[b] Time needed to reach the steady state to EC_{50}.

The amount of antioxidant required to decrease the initial DPPH radical concentration by 50% is known as EC_{50}. Since the DPPH antiradical activity is dose dependent process, the antiradical efficiency was determined by EC_{50} values of each sample (Table 4). The EC_{50} value was determined by plotting the concentrations of antioxidants against the percentages of remaining DPPH radicals [19]. The lower the EC_{50} value, the higher the antioxidant power of the compound.

Phenolic compounds have diverse EC_{50} values depends on the structure and the mechanism involves to scavenge the DPPH radicals [20,21]. The efficiency of antiradicals are not solely dependent on the EC_{50} of the compounds because the rate of reaction depends on the concentration. Thus, TEC_{50} were determined (Table 4) which is the time for the sample at EC_{50} to reach equilibrium [22].

Table 4: The concentration needed to reduce DPPH concentration by 50% and their kinetic classification.

Sample	$EC_{50}{}^{a}$ (μg/mL)	Ranges of times at the steady state (min) for the sample concentrations	$T_{EC50}{}^{a,b}$ (min)	Classification
Quercetin	15.98 ± 9	9.0-120.0	95.29 ± 0.1	Slow
D. esculentum	1241.14 ± 7	20.0-120.0	79.83 ± 0.5	Slow

[a] Value mean is ± standard deviation.
[b] Time needed to reach the steady state to EC_{50}.

The lower the EC_{50}, the lower the TEC_{50} but in this study, TEC_{50} value for both samples were almost similiar alhough the EC_{50} value were significantly different. Similar condition had been observed in gallic acid and (-)-epicatechin,where EC_{50} for both compound were quite similar but the TEC_{50} value for gallic acid was far lower than (-)-epicatechin[14]. This shows that there was uncertainty in the determination of the antioxidant activity using EC_{50} and TEC_{50} values. Therefore, the efficiency of phenolic compounds to scavenge the DPPH were further determined by antiradical efficiency equation :

$$\text{Antiradical efficiency} = 1/\ EC_{50}.TEC_{50}$$

The antiradical efficiency of quercetin ($0.65\text{x}10^{3}$) was higher compared to *D. esculentum* ($0.01\text{x}10^{-3}$) and both were classified as low antiradical efficiency (Table 5). The percentage of remaining DPPH radicals against the concentration of these two samples were expressed by exponential model.

$$\ln\ [\%\ \text{DPPH remaining}] = b\ [\text{antioxidant}] + \ln a$$

From this model, the slope of linear regression of quercetin was higher than *D. esculentum* extracts. The steeper the slope, the lower EC_{50} would give higher antiradical efficiency [14]. Therefore, the antiradical efficiency of *D. esculentum* was lower than quercetin.

Table 5: Antiradical efficiencies of quercetin and *D. esculentum*.

Sample	Slope ($\text{x } 10^{-3}$)	Correlation coefficients	Antiradical efficiency ($\text{x } 10^{-3}$)[a]	Classification
Quercetin	-0.083	-0.960	0.65	Low
D. esculentum	-0.001	-0.958	0.01	Low

[a] Antiradical efficiency = 1/ $EC_{50}.TEC_{50}$

Acknowledgements
The author would like to acknowledge the support from the Fundamental Research Grant Scheme (FRGS) under a grant number of FRGS/1/2016/TK02/UNIMAP/03/2 from the Ministry of Higher Education Malaysia.

References
[1] Noweg, T.; Abdullah, A.R.; Nidang, D. Forest plants as vegetables for communities bordering the Crocker Range National Park. *ASEAN Review of Biodiversity and Environmental Conservation (ARBEC).* 2003, 18.
[2] Sarker, S.K.; Hossain, A.B.M. Pteridophytes of greater Mymensingh district of Bangladesh used as vegetables and medicines. *Bangladesh Journal of Plant Taxonomy*. 2009, 16(1), 47-56.
[3] Baas, P.; Kalkman, K.; Gessink, R. *The plant diversity of Malesia.* Springer Netherlands. 1990, **1**, 420
[4] Bernama, *Promoting local herbal industry to local market*, in *Borneo Post*. 2009, Malaysia. p. B13.
[5] Roy, S.; Hazra, B.; Mandal, N.; Chaudhuri, T.P. Assessment of the Antioxidant and Free Radical Scavenging Activities of Methanolic Extract of Diplazium esculentum. *Int. J. Food Prod. 2013,* **16**(6), 1351-1370
[6] Tongco, J.V.; Villaber, R.A.; Aguda, R., Razal, Ramon. Nutritional and Phytochemical Screening, and Total Phenolic and Flavonoid Content of Diplazium esculentum (Retz.) Sw. from the Philippines. *J Chem Pharm Res*. 2014, **6**(8):238-242

[7] Akter, S.; Hossain, M.M.; Ara, I.; Akhtar, P. Investigation of in vitro antioxidant, antimicrobial and cytotoxic activity of Diplazium esculentum (Retz.) Sw. *Int. J. Adv. Pharm. Biol. Chem.* 2014, **3**(3), 723-733

[8] Teo, C. C.; Tan, S. N.; Yong, J. W. H.; Hew, C. S.; Ong, E. S. Pressurized hot water extraction (PHWE). *J. Chromatogr. A*. 2010, 2484-2494.

[9] Rangsriwong, P.; Rangkadilok, N.; Satayavivad, J.; Goto, M.; Shotipruk, A. *Subcritical water extraction of polyphenolic compounds from Terminalia chebula Retz. fruits.* Sep. Purif. Technol. 2009, **66**, 51-56.

[10] Gbashi, S.; Njobeh, P.; Steenkamp, P.; Tutu, H.; Madala, N. The effect of temperature and methanol-water mixture on pressurized hot water extraction (PHWE) of anti-HIV analogoues from Bidens pilosa. *Chem Cent J.* 2016, **10**(1), 1752-1789.

[11] Baharuddin, N. A. F.; Nordin, M. F. M.; Morad, N. A.; Rasidek, N. A. Pressurized hot water extraction of phenolic and antioxidant activity of Clinacanthus nutans leaves using accelerated solvent extractor. *Aust. J. Basic & Appl. Sci.* 2017, **11**(3), 56-63.

[12] Rajesh, K.D.; Vasantha, S.; Panneerselvam, A.; Valsala Rajesh, N.; Jeyathilakan, N. Phytochemical Analysis, *in vitro* Antioxidant Potential and Gas Chromatography Mass Spectrometry Studies of Dicranopteris Linearis. *Asian J. Pharm. Clin. Res.* 2016, **9**, 220.

[13] Vijayalaxmi, S.; Jayalakshmi, S. K.; Sreeramulu, K. Polyphenols from different agricultural residues: extraction identification and their antioxidant properties. *J. Food* Sci. *Technol.* 2015, **52**(5), 2761-2769.

[14] Sánchez-Moreno, C.; Larrauri, J.A.; Saura-Calixto, F. A procedure to measure the antiradical efficiency of polyphenols. *J.* Sci. *Food* Agric. 1998. **76**(2): p. 270-276.

[15] Montgomery, D.C. *Design and analysis of experiments*. 5th ed. 1997, United States of America: John Wiley & Sons, Inc. 699.

[16] Khuwijitjaru, P.; Chalooddong, K.; Adachi, S. Phenolic content and radical scavenging capacity of kaffir lime fruit peel extracts obtained by pressurized hot water extraction. *Food* Sci. *Technol. Res.* 2008. **14**(1): p. 1-4.

[17] Budrat, P.; Shotipruk, A. Enhanced recovery of phenolic compounds from bitter melon (Momordica charantia) by subcritical water extraction. *Sep. Purif. Technol.* 2009, 66(1), 125-129.

[18] Brand-Williams, W.; Cuvelier, M.E.; Berset, C. Use of a free radical method to evaluate antioxidant activity. *LWT-Food Sci Technol.* 1995. **28**(1): p. 25-30.

[19] Villaño, D.; Fernández-Pachón, M. S.; Moyá, M. L.; Troncoso, A. M.; García-Parrilla, M.C. Radical scavenging ability of polyphenolic compounds towards DPPH free radical. *Talanta*, 2007. **71**(1), 230-235.

[20] Bondet, V.; Brand-Williams, W.; Berset, C. *Kinetics and Mechanisms of Antioxidant Activity using the DPPH Free Radical Method.* Lebensmittel-Wissenschaft und-Technologie. 1997, **30**(6), 609-615.

[21] Tsimogiannis, D.I.; Oreopoulou, V. The contribution of flavonoid C-ring on the DPPH free radical scavenging efficiency. A kinetic approach for the 3',4'-hydroxy substituted members. *Innov. Food* Sci. *Emerg. Technol.* 2006, **7**(1-2), 140-146.

[22] Tsao, R.; Deng, Z. Separation procedures for naturally occurring antioxidant phytochemicals. *J. Chromatogr. B*. 2004, **812**(1-2), 85-99.

IOP Conf. Series: Materials Science and Engineering **429** (2018) 012065 doi:10.1088/1757-899X/429/1/012065

Antioxidative Properties and Sensory Evaluation of Perlis Sunshine Mango Seed Flour Body Scrub

Rohazita Bahari[1,2], Khairul Farihan Kasim[1,2], Kho Sy Xuan[1], Noor Shazliana Aizee Abidin[1]

[1] School of Bioprocess Engineering, Universiti Malaysia Perlis, Perlis, Malaysia.
[2]Centre of Excellence for Biomass Utilization, School of Bioprocess Engineering, Universiti Malaysia Perlis.

*Email: abc@unimap.edu.my

Abstract. The aim of this work was to utilize Perlis Sunshine Mango Seed Flour (MSF) as natural active ingredient in body scrub for skin care cosmetology. The body scrubs were prepared in two forms i.e. rough salt and oil-in-water (O/W) emulsion. Both types of body scrub were formulated as based and containing Perlis Sunshine MSF. The determination of total phenolic content (FC-reagent method), ascorbic acid content (HPLC) and total antioxidant activity (DPPH method) of the body scrubs were evaluated for the possible application in dermo-cosmetic field. The O/W MSF body scrub extract contains 6.65 ± 0.16 GAE mg/g of total phenolic content (TPC) with 32.29 ± 2.60% of DPPH activity inhibition. The rough salt MSF body scrub extract had shown highest ascorbic acid content compared to other body scrubs. For the sensory analysis, most of the body scrubs testers satisfied with the O/W body scrubs compared to rough salt body scrubs in all the attributed tested.

1. Introduction

Mango (*Mangifera indica* L) is a tropical seasonal fruits which grown over tropical countries and sub-tropical regions. Perlis Sunshine Mango can be obtained in Perlis and the production of mangoes is two times per season [1]. Mango fruit is the second largest in tropical crop behind fruit bananas [2]. Ripe mangoes are used to make frozen mango products, canned fruits products, pickles and beverages in food industries. Industrially, the edible parts of mango fruits are used to process mango products while peels and seeds are dispensed as waste. The vast amount of these industrial wastes is a pollution source to the environment [3].

Mango by-products, especially seeds and peels are recognized as natural organic wastes which contain high level of nutrients that can be utilized [4]. Economic losses is a factor come to the food industries if the huge amount of industrial waste did not utilized effectively. Economic losses occur due to high expenses cost for solving the industrial waste and low recovery obtained to the industry. Therefore, it is necessary and important to reuse the plant by-products from environmental and economic perspective [3].

Mango seed and mango peel has been studied and utilized to produce biscuit flour products. Mango seed powder contain high protein, total phenolic and ash contents, whereas mango peel powder contain high level of crude fiber and antioxidant activity. Both mango seed and peel powder can help in prevention of diseases due to their natural and unharmed bioactive substances [5]. However, mango

IOP Conf. Series: Materials Science and Engineering **429** (2018) 012065 doi:10.1088/1757-899X/429/1/012065

seed flour has been reported to have extremely bitter taste. Hence, the utilization of mango seed flour in pharmaceutical and cosmetic aspects are much more recommended compared to food industry.

Antioxidant compounds are an essential formulate recipe for cosmetic products such as body scrub due to it provides a protection against reactive oxygen species (ROS) deleterious effect to the human skin. Thus, the mango seeds flour have a potential as a value added to the body scrub formulation by enhancing the function value of the body scrub products. Therefore, this study is focus on the utilization of mango seed flour in the formulation of body scrub and to evaluate the consumer acceptance of the products.

2. Materials

2.1. Chemicals
2,2-diphenyl-1-picrylhydrazyl free radical (DPPH) was purchased from Merck. Follin-Ciocalteu reagent was purchased from Sigma-aldrich Co. Glycerol, gallic acid, ascorbic acid (99%) and methanol were purchased from HmBG.

2.2. Plant Materials
Perlis Sunshine *Mangifera indica* seed powder was purchased from local farmers around Perlis.

3. Procedures

3.1. Body Scrub Formulation
There are two types of formulations had been used to prepare the body scrub i.e. rough salt and oil in water (O/W) emulsion. Composition for the rough salt and O/W emulsion with and without MSF are tabulated in Table 1 and 2 respectively.

Table 1: Composition of rough salt body scrubs.

Composition (INCI)	Function	Base	MSF
		Amount (g)	
Sodium Chloride	Abrasive	50.00	49.00
Helianthus annuus (Sunflower) oil	Emollient	5.00	5.00
Olea europea (olive) fruit oil	Emollient	5.00	5.00
Jasminum officinale (Jasmine) oil	Emollient/Fragrance	2.00	2.00
Mangifera indica seed powder	Active principle	-	1.00

Table 2: Composition of O/W emulsion body scrubs.

Composition (INCI)	Function	Base	MSF
		Amount (g)	
	Phase A		
Glyceryl stearate SE, Glyceryl stearate, Cetearyl alcohol	Base self emulsifying	13.00	13.00
Helianthus annuus (Sunflower) oil	Emollient	5.00	5.00
Olea europea (olive) fruit oil	Emollient	5.00	5.00
Jasminum officinale (Jasmine)oil	Emollient/Fragrance	2.00	2.00
Phenoxyethanol	Preservative	0.50	0.50
	Phase B		
Aqua	Vehicle	66.40	65.40
Disodium EDTA	Chelating	0.10	0.10
Glycerin	Moist	3.00	3.00
	Phase C		
Sodium chloride	Abrasive	5.00	5.00
Mangifera indica seed powder	Active principle	-	1.00

3.2. Body Scrub Extration

Body scrub extracts were prepared by dissolving 5 g of the body scrubs in 100 mL of methanol in the beakers for 30 min and followed by sonication for 20 min. The mixtures then were filtered by using No. 1 Whatman filter paper. The collected filtrates were evaporated at 67 °C (boiling point of methanol) under vacuum by rotary evaporator for 5 to 6 min until the extracts were fully concentrated. The extracts were kept at 4 °C until further used [6].

3.3. Total Phenolic Compounds

The Folin-Coiocalteu colorimetric method was used to determine TPC in body scrubs extracts [7]. 0.1 mL of sample extracts was added into 0.5 mL of Folin-Ciocalteu and 7.4 mL of distilled water. The mixtures were mixed thoroughly and incubate for 3 min. Then, 2 mL of sodium carbonate solution (15%) was added and it was kept in dark place for 30 min. The mixture was measured at 765 nm in a UV-VIS. The TPC was expressed as mg gallic acid equivalents (GAE)/g extract.

3.4. Ascorbic Acid Content

The ascorbic acid (vitamin C) content in the body scrubs were measured by using High Performance Liquid Chromatography (HPLC). 0.1 mL of each of the body scrub extracts was dissolved in 9.9 mL distilled water in a test tube. The mixtures were mixed well by using the vortex and centrifuged at 3000 rpm for 5 min at room temperature. The supernatants are collected and aliquots of the samples were filtered through a μm PFTE syringe filter before analysed with HPLC [5].

3.5. Total Antioxidant Activity

DPPH method was used to evaluate the total antioxidant activity of the body scrubs. A methanol solution with DPPH was used as control (2 mL DPPH solution and 1 mL of methanol). The test samples were prepared by mixing 0.1 mL of each body scrub extracts with a methanol solution of 0.1 mM DPPH under a 200 rpm speed in the dark [8]. Next, the mixtures were left to incubate for 30 minutes in the dark and the absorbance were measured at 517 nm. The percentage of free radical activities in the tested samples were calculates as following equation 1.

$$\% \, inhibition = \left[\frac{(Ac-As)}{Ac} \times 100\%\right] \quad (1)$$

Where,
A_c = absorbance value for control
A_s = absorbance value for tested sample

3.6. Sensory Evaluation

Twenty panelist, female, aged between 20-2 years old were randomly recruited from School of Bioprocess Engineering's students. The panelists were asked to evaluate spreading, aroma, greasiness, allergy and smoothness of the body scrubs using a three-scale assessment (good, normal and worse). The body scrubs were blind-labeled and coded with using a random three-digit number. Samples were applied individually, one after another, under normal lighting conditions at room temperature and a balanced complete block design was used to assign the presentation order. Water and tissue were provided to clean and rinse the skin between samples.

4. Results and Discussion

4.1. Total Phenolic Content (TPC) of the Formulated Body Scrubs

The total phenolic contents (TPC) of different type of body scrubs are shown in Table 3. As expected, the body scrubs containing MSF had higher total phenolic content compared to the bases body scrubs.

Table 3: The total phenolic content (GAE mg/g) of the body scrubs.

Samples	TPC(GAE mg/g)
Rough salt base scrub	0.32 ± 0.02
Rough salt MSF scrub	5.92 ± 0.12
O/W base scrub	0.48 ± 0.01
O/W MSF scrub	6.65 ±0.16a

4.2. Determination of Ascorbic Acid in the Formulated Body Scrubs

Ascorbic acid in the samples was determined based on the area under the peak of the HPLC chromatogram. Height of the peak was referred to high or low selectivity of the method to separate interested compound. Thus, the high and sharp peak means the interested compound had been separated efficiently from other compound [9]. The obtained results of HPLC analysis were summarized in Table 4.

Table 4: Ascorbic acid peak in the body scrubs from HPLC.

Samples	Retention time (min)	Area under the peak	Height of the peak
Standard ascorbic acid	3.299	3718149	177323
Rough salt base body scrub	Not detected	Not detected	Not detected
Rough salt MSF body scrub	2.582	294268	5437
O/W base body scrub	Not detected	Not detected	Not detected
O/W MSF body scrub	3.427	3565	392

For the rough salt base body scrub and O/W base body scrub, the occurrence of ascorbic acid peak in the report were absence and thus the retention time of ascorbic acid is not obtained. Hence, it can be said that both type body scrub did not containing ascorbic acid. For the rough salt MSF body scrub and O/W MSF body scrub, the retention time of ascorbic acid were 2.582 minutes and 3.427 minutes. The increasing or decreasing of retention time of the samples to the retention time of standard ascorbic acid sample can be caused by the presence of contaminants [9]. Besides, both body scrub had shown more than one peak in the chromatograms and this means there were presence of impurities in the sample and these impurities may have low level difference to diode array spectrum of ascorbic acid [10]. The content of the ascorbic acid content in both scrubs can be said very low due to the area under the peak of both body scrubs were relatively low to the area under peak of standard ascorbic acid sample. By comparing the area under the peak of both body scrubs, since the area of rough salt MSF body scrub is greater than the area O/W MSF body scrub. Thus, the rough salt MSF body scrub had containing higher ascorbic acid content than O/W MSF body scrub.

4.3. Determination of Total Antioxidant Activity of the Formulated Body Scrubs

This DDPH method is depended on the reduction of free radical DPPH by antioxidant to DPPHH as in resulting absorbance decreased from DPPH [11]. The lower absorbance in the reaction mixture has indicated higher free radical activity. This means alcoholic extract of the body scrub with O/W MSF body scrub form had captured more free radicals formed by DPPH and this resulting into decrease in absorbance with increase in inhibition value [11]. As expected, all bases body scrubs had displayed a much lower potency as compared to those body scrubs containing Perlis Sunshine Mango Seed Flour (Table 5). The O/W MSF body scrub had better antioxidant potential and followed by rough salt type body scrub with Perlis Sunshine MSF.

Table 5: DPPH activity (%) of the body scrubs.

Samples	DPPH Activity (%)
Rough salt base body scrub	4.17 ±1.04
Rough salt MSF body scrub	28.65 ± 2.08
O/W base body scrub	6.52 ± 0.52
O/W MSF body scrub	32.29 ± 2.60

Data given are mean ± standard deviation for three different value of each sample.

4.4. Evaluation of Sensory Evaluation

The aim of the sensory evaluation was to evaluate the personal response of preference to the formulated body scrubs. The sensory evaluation is importance for the evaluation and optimization to the effects of cosmetic recipes. Twenty of body scrubs evaluation questionnaire had been collected and the data were analyzed in figures below. Most of participants considered O/W body scrubs had good performance compare to rough salt body scrubs in spreading, aroma, greasing and smoothness attributes. Major participants comment that the strong oily smell in rough salt body scrubs had covered the aroma of fragrance oil and thus most of them cannot accept the smell of the rough salt type body scrub. Besides, most of the participants comment that the greasing effect from rough salt body scrubs were too oily and sticky for them. Lastly, all participants (100 %) had stated that they were not allergy to all type of body scrubs that had been prepared. However, there were still have some comments about the O/W type body scrub from the participants. Most of participants think that the abrasive effect of O/W body scrubs to their skin were not strong enough for them and the texture of the body scrubs were a bit similar to the texture of lotion. Based on the evaluation of body scrubs above, it can be conclude that most participants were more prefer to O/W body scrubs compared to rough salt body scrub in spreading, aroma, greasing, and smoothness attributes.

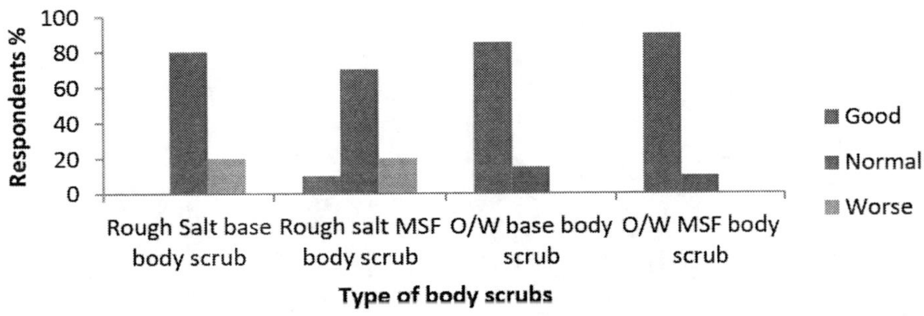

Figure 1: Spreading assessment to the body scrubs by participants

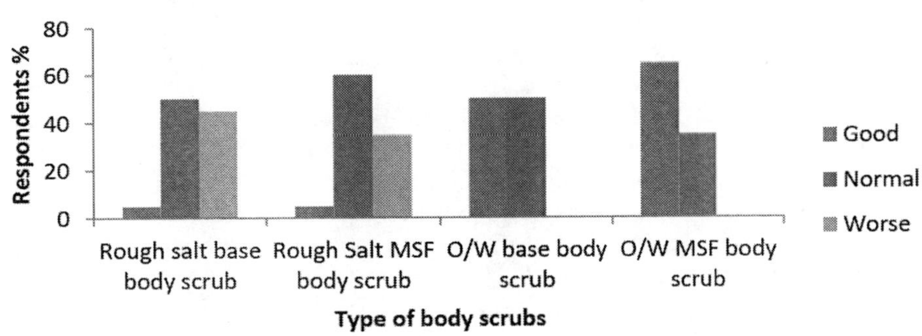

Figure 2: Aroma assessment to the body scrubs by participants.

IOP Conf. Series: Materials Science and Engineering **429** (2018) 012065 doi:10.1088/1757-899X/429/1/012065

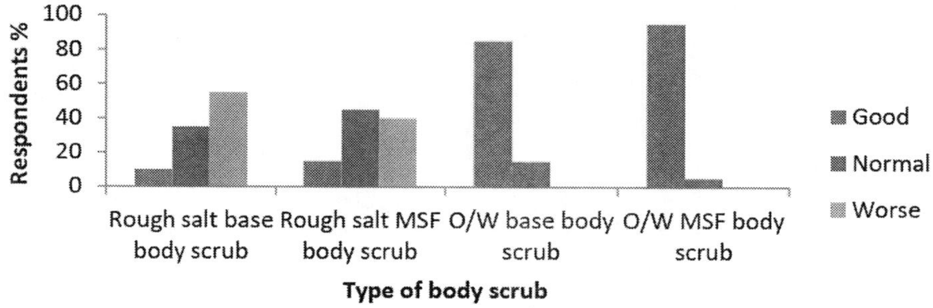

Figure 3: Greasing assessment to the body scrubs by participants.

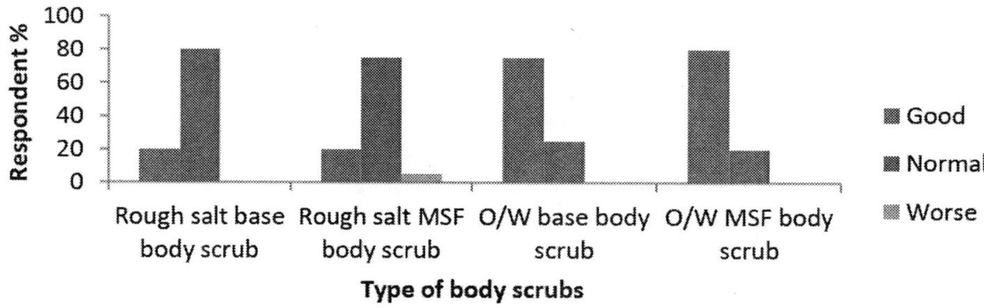

Figure 4: Smoothness assessment to the body scrubs by participants.

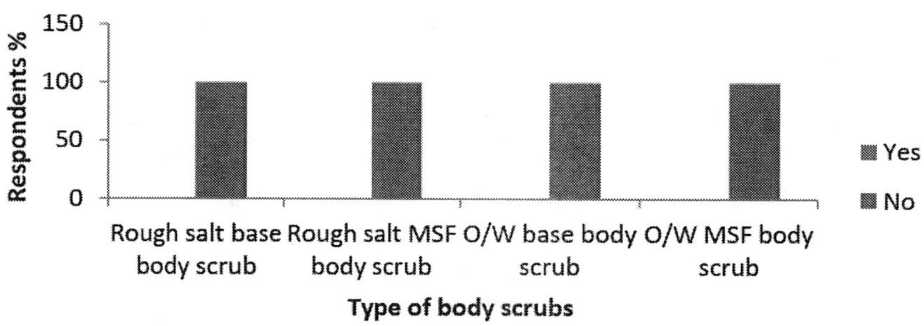

Figure 5: Allergic assessment to the body scrubs by participants.

5. Conclusion

The O/W MSF body scrub content the highest total phenolic content (6.65 ± 0.16 mg/g) and better antioxidant potential ($32.29 \pm 2.60\%$) compared to the other body scrubs formulation. From the HPLC peaks obtained, it shows that rough salt MSF body scrub and O/W MSF body scrub had shown tiny peak in the chromatograms; this means both body scrub contains ascorbic acid but with low content. However, in the consumer acceptance test, O/W body scrubs were more preferred compared to rough salt body scrubs in overall attributes tested.

Acknowledgements
The author would like to acknowledge the support from the Short Term Grant (STG) under a grant number of STG 9001-00555 from University Malaysia Perlis.

References

[1] Abdelhady, M. I. S., Motaal, A. A., & Beerhues, L. (2011). Total Phenolic Content and Antioxidant Activity of Standardized Extracts from Leaves and Cell Cultures of Three Callistemon Species. American Journal of Plant Science, 847–850.
[2] Jahurul, M. H. A., Zaidul, I. S. M., Ghafoor, K., Al-Juhaimi, F. Y., Nyam, K. L., Norulaini, N. A. N., & Mohd Omar, A. K. (2015). Mango (Mangifera indica L.) by-products and their valuable components: A review. Food Chemistry, 183, 173–180.
[3] Abidin, N., Nor, M., & Jaafar, B. (2015). Chemical Composition , Antioxidant Activity and Functional Properties of Mango (*Mangifera indica L. var* Perlis Sunshine) Peel Flour. *Applied Mechanics and Materials*, 754-755.
[4] Aslam, H. K. W., Ianam, M., Raheem, U., Ramzan, R., Shakeel, A., Shoaib, M., & Sakandar, H. A. (2014). Utilization of Mango Waste Material (Peel , Kernel) To Enhance Dietary Fiber Content and Antioxidant Properties of Biscuit. J*ournal Global Innovation Agriculture Social Science*, 2(2), 76–81.
[5] Castro, R. N., Azeredo, L. C., Azeredo, M. A. A., & de Sampaio, C. S. T. (2001). HPLC assay for the determination of ascorbic acid in honey samples. *Journal of Liquid Chromatography & Related Technologies*, 24(7), 1015–1020.
[6] Khan, H. M. S., Iqbal, M., Rasul, A., & Bhatti, N. S. (2010). Formulation and Characterization of a Cream Containing Extract of Fenugreek Seeds. Acta Poloniae Pharmaceutica-Drug Research, 67(2), 173–178.
[7] Sinica, D. P., Jadhav, A. P., Kareparamban, J. A., Nikam, P. H., & Kadam, V. J. (2012). Pelagia Research Library Spectrophotometric Estimation of Ferulic Acid from Ferula asafoetida by Folin-Ciocalteu's Reagent. Der Pharmacia Sinica, 3(6), 680–684.
[8] Sprada, G., Da Mota, T., Arantes, A. B., Sacchetti, G., Spagnoletti, A., Ziosi, P. & Manfredini, S. (2014). Antioxidant Activity of Cosmetic Formulations Based on Novel Extracts from Seeds of Brazilian Araucaria angustifolia (Bertoll) Kuntze Characterization, HPTLC, ORAC, PCL. Journal of Cosmetics Dermatological Sciences and Applications Journal of Cosmetics, 4(4), 190–202.
[9] Vallat, A. (2007). Troubleshooting guide – HPLC.
[10] Kaushal, C. K., & Srivastava, B. (2010). A process of method development : A chromatographic approach. Journal of Chemical and Pharmaceutical Research,2(2), 519–545.
[11] Leaves, L. (2014). Antioxidant Activity by DPPH Radical Scavenging Method of Ageratum conyzoides. *American Journal of Ethnomedicine,* 1(4), 244–249.

IOP Conf. Series: Materials Science and Engineering **429** (2018) 012066 doi:10.1088/1757-899X/429/1/012066

Development of Antioxidant Film Based on Blends of *Stenochlaena palustris* Flour and Poly(Vinyl Alcohol)

AH Ma'Radzi [1*], KF Kasim[1], N Mahmud[2], NAAA Rahman[1] and N Jullok[1]

[1]School of Bioprocess Engineering, University Malaysia Perlis, Perlis, Malaysia.
[2]Faculty of Industrial Science and Technology, University Malaysia Pahang, Pahang, Malaysia.

*Email: akmalhadi@unimap.edu.my

Abstract. The aim of this work is to develop antioxidant films based on blends of *Stenochlaena palustris* flour and poly(vinyl alcohol) (PVA). Four different PVA types were tested. All films produced were analyzed for their antioxidant activity by diphenyl-picrylhydrazyl radical scavenging (DPPH) method. This study found that S. palustris flour-88% HD PVA blends film contains the highest antioxidant activity. The efficiency study of the film shows the EC_{50}, T_{EC50} and Antioxidant Efficiency value of 20 mg/ml, 55 min and 0.91×10^{-3} respectively. The use of low concentration PVA and addition of glycerol made the film more flexible with greater elongation at break but low tensile strength. The formation of hydrogen bonds between the blend components was confirmed by the FTIR spectra analysis. The thermogravimetric analysis shows the thermal degradation of the film produced.

1. Introduction

In recent years, antioxidants have gained a lot of importance because of their potential as prophylactic and therapeutic agents in many diseases. Antioxidants are substances which counteract free radicals and prevent the damage caused by them [1]. Plant sourced food antioxidants like vitamin C, vitamin E, carotenes, phenolic acids, phytate and phytoestrogens have been recognized as having the potential to reduce disease risk. Antioxidants which have been used in diet or as formulated agents have some common problems in their efficacy as therapeutic agents which are related to their physicochemical and biopharmaceutical properties. Many of the antioxidants have poor oral bioavailability, due to low solubility, permeability, stability and/or drug biotransformation before they reach systemic circulation [1]. Development of antioxidant films based on blends of *Stenochlaena palustris* (Burm) flour and poly(vinyl alcohol) (PVA) will somehow eliminate the used through oral route. This antioxidant films can be used mainly in cosmetic applications and a potential material in production of biodegradable packaging or plastic.

S. palustris a scrambling fern, is distributed in a large part of the tropical areas from southern and northern India through Malaysia to Polynesia and Australia [2]. The red young fronds are used as spinach and are said to be particularly beneficial when eaten at breakfast-time during the fasting month. The juice is used to cure fevers, the long rhizomes to bind baskets and fish traps, and a very durable rope can be made from the stems [3].

2. Experimental Section

2.1. Materials

Poly(vinyl alcohol) (99-100% HD, 95% HD, 88% HD, & 75% HD) were purchased from Acros Organics. 2,2-diphenilpicrylhydrazyl free radical (DPPH) was purchased from Merck. Kaempferol, myricetin, quercetin dihydrate, and rutin hydrate were purchased from Sigma-aldrich Co. Glycerol and methanol were purchased from HmBG. S. palustris were collected from open area at Wang Tepus, Jitra, Kedah. Fresh S. palustris leaveswere dried in the oven at 60°C for 24 h. The dried samples then ground and the resulting flour were sieved at 45 μm to produce fine flour.

2.2. Sample Preparation

S. palustris flour/PVA blended films were made by the casting technique. The two solutions were first prepared separately. The *S. palustris* flour solution was prepared with 4 g of *S. palustris* flour/100 g solution. The *S. palustris* flour and distilled water were mixed for 2 hours with a magnetic stirrer. The *S. palustris* flour dispersion was then heated to 75°C and glycerol (20 g glycerol/100 g polymers) was added. The temperature was increased to 85°C for 60 min with constant magnetic stirring [4]. The PVA solution was prepared with PVA (4 g/100 g solution) and distilled water. The PVA was stirred for at least 4 h at room temperature, and then heated at 85 to 95°C depending of the PVA type, until the PVA was completely dissolved [4]. Films were produced by varying the PVA content from 10 to 100% (w/w) of the dry matter. Blends of *S. palustris* flour and PVA solutions were prepared by mixing these two solutions. The blends were heated at 85°C for 1 h with constant stirring but avoid frothing to increase the interaction between the components [4]. Water was added to maintain the volume. The film forming solutions were cast onto a glass plate with thickness of about 0.5 - 1 mm, after which air bubbles were removed by flaming and each solution was dried for 24 h at 55 °C to form the desired films. Complete drying was avoided as some moisture was required for films to remain flexible and not to crack. The films were finally removed (by peeling) from the glass.

2.3. Antioxidant Assay

Antioxidant properties of the films were carried out by using DPPH method as described in [5] with slight modification. An aliquot of 22 μL of the sample was added to 200 μL of DPPH (60 μM) in a 96-well flat-bottom microplate. The absorbance was recorded in a VersaMax microplate reader at 517 nm after 30 min incubation time at 37°C [5]. The radical scavenging activity was calculated by the following equation:

$$\text{DPPH activity}(\%) = \left[1 - \frac{\text{Absorbance of sample at 517nm}}{\text{Absorbance of DPPH at 517nm}}\right] x100 \quad (1)$$

2.4. Efficiency Study

The efficiency of the antioxidant activity was carried out only on blended film of 88% HDPVA with *S. palustris* flour at 55% PVA concentration as this film show the best antioxidant activities. An aliquot of 22 μL of the sample at various concentration (10 – 100 mg/mL) was added to 200 μL of DPPH (60 μM) in a 96-well flat-bottom microplate. The absorbance was recorded in a VersaMax microplate reader at 517 nm after 30 min incubation time until plateau at 37°C [6]. The remaining level of DPPH (% $DPPH_{REM}$) in the reaction medium was calculated using the following relation:

$$\% \text{ of remaining DPPH} = 100.\left(A_{s517nm(t=30)}/A_{c517nm}\right) \quad (2)$$

where $A_{s\,517\,nm}$ is the absorbance of the sample measured at t = 30 min and $A_{c\,517\,nm}$ is the absorbance of the control. The percentage of remaining DPPH against the standard concentration was then plotted to obtain the amount of antioxidant necessary to decrease the initial DPPH concentration by 50 %. The time needed to reach the steady state to EC_{50} concentration (T_{EC50}) was calculated graphically. Taking into account that both, EC_{50} and T_{EC50}, affect the antiradical capacity, a new parameter: antioxidant efficiency (AE) [7], which combines these two factors, was defined:

$$AE = 1/EC_{50}T_{EC_{50}} \quad (3)$$

2.5. Film Characterization

For Fourier transform infrared (FTIR) spectroscopy analysis, the infrared spectra were recorded between 4000 and 600 cm^{-1} at 4 cm^{-1} of resolution with a Fourier transform IR (FT-IR) spectrometer RX1 (Perkin Elmer). The films were dried at 100 °C for 48 h and ground into powder before the infrared (IR) measurement. Thermogravimetric analysis was carried our using a Perkin Elmer (TGA7) under nitrogen atmosphere. Sample (about 10 mg) were heated from room temperature to 500 °C at a heating rate of 10 °C/min. The tensile strength (σ_b) and elongation at break (ε_b) of the films in both dry and wet states were measured using a universal testing machine Haunsfield (H10KS), at a tensile rate of 10 mm/min. The wet films were measured immediately after immersed in water for 30 min. The σ_b and ε_b values are the averages of three measurements.

3. Results and discussion

3.1. PVA Selection

Four different types of PVA were tested in the formulation of blended films with equal proportions of *S. palustris* flour-PVA. All films were homogeneous and easily peelable from the support except for 75% HD PVA. The characteristics of the film produced were simplified in Table 1. Even though it had been dried for 72 h, the *S. palustris* flour-75% HD PVA film was unable to peel smoothly. This may due to the low compatibility between this partially hydrolyzed and low molecular weight PVA and the *S. palustris* flour. Several factors associated with PVA may explain the effects of molecular weight and concentration.

At a high molecular weight polymer with a high degree of hydrolysis, the solution viscosity increases significantly and hence reduced rate of solvent evaporation [8]. When using low molecular weight PVA, the viscosity of solution was lower thus increased the rate of solvent evaporation. This conditions cause less formation of hydrogen bond between the PVA chains and water molecules. Low molecular weight PVA comparatively has low ability of film producing.

Table 1 Characteristics of film produced

PVA Types	Film produced
99-100% HD	Easily peeled at high concentration and at low concentration
95% HD	Easily peeled at high concentration and at low concentration
88% HD	Easily peeled at high concentration and at low concentration
75% HD	Not peeled at high and low concentration, tend to rupture when separate from support

3.2. Antioxidant Activity

The antioxidant activity of the *S. palustris* flour-PVA films were carried out after 1 h of incubation time at 37 °C with three different concentration of the film (25, 50 & 100 mg/mL) and five different PVA concentration (10, 32.5, 55, 77.5 & 100%). Percentage of antioxidant activity was calculated and presented as graph percentage of antioxidant activity versus PVA concentration. Fig 1, 2 and 3, shows the antioxidant activity of *S. palustris* flour-99-100% HD PVA blends film, *S. palustris* flour-95% HD

PVA blends film & *S. palustris* flour-88% HD PVA blends film respectively. Antioxidant activities were lowest at 100% PVA concentration for all type of PVA. Comparing all three types of PVA, 88% HD PVA shows highest antioxidant content for all concentration used. While 95% HD PVA shows lowest antioxidant content compare to the other two types of PVA, might be due to the higher molecular weight of 95% HD PVA.

In the higher molecular weight PVA, in which the polymer component was much more concentrated and viscous, there were very high tendency of the antioxidant from the *S. palustris* flour react more on the concentrated PVA. Therefore, the amount of hydrogen or electron that will react to DPPH much more lower resulting low antioxidant activity. The highest antioxidant activities were recorded from film developed from 88% HD PVA at 55% PVA concentration.

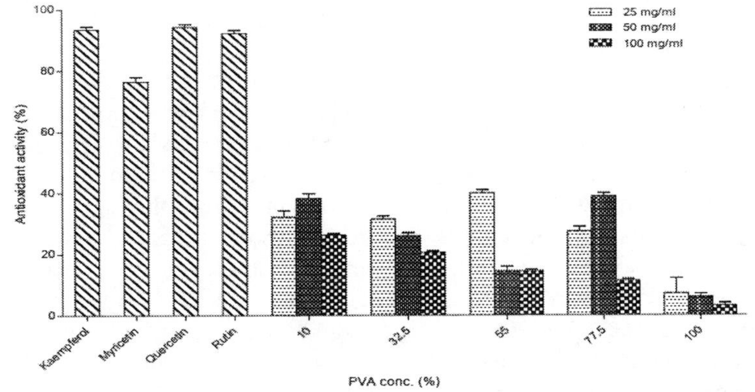

Fig. 1 Antioxidant activity of *S. palustris* flour- 99-100% HD PVA blends film at different concentrations of the PVA and different dilution rate of the film. Values are the means ± SEM (n = 4).

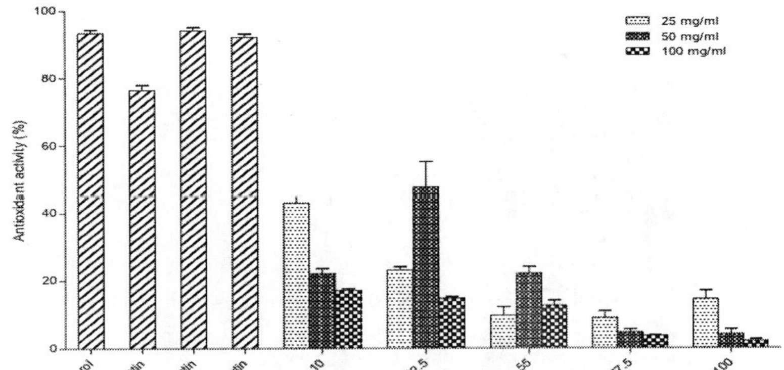

Fig. 2 Antioxidant activity of *S. palustris* flour- 95% HD PVA blends film at different concentrations of the PVA and different dilution rate of the film. Values are the means ± SEM (n = 4).

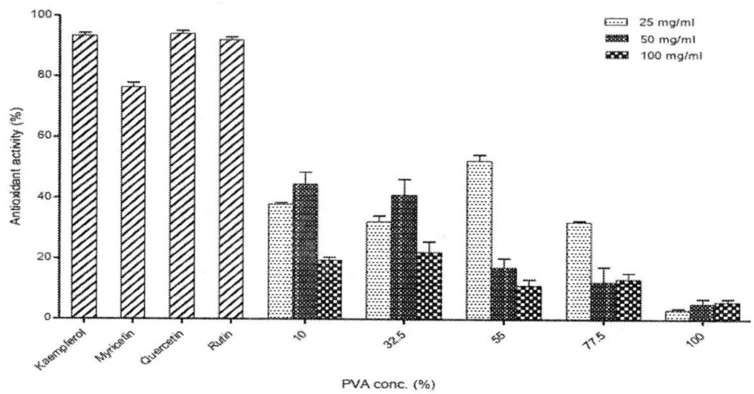

Fig. 3 Antioxidant activity of *S. palustris* flour- 88% HD PVA blends film at different concentrations of the PVA and different dilution rate of the film. Values are the means ± SEM (n = 4).

3.3. Efficiency Study

The efficiency of the antioxidant activity was conducted on blends film of 88% HD PVA with *S. palustris* flour at 55% PVA concentration of varies concentration (10 - 100 mg/mL) (Fig. 4). Reacting DPPH solution with each concentration of films sample for every 30 min incubation time until plateau at 37 °C facilitates the extraction of antioxidant compounds from the sample thereby increasing the measured antioxidant activity of the sample. The results showed that the absorbance decreased as the radical was scavenged by antiradicals, through donation of hydrogen, to give the reduced form DPPH-H. Basically, the antioxidant activities of the blends films were increased with concentration up to 20 mg/mL and further increased up to 100 mg/mL resulted in significant decrease of antioxidant activity. This might be explained by the fact that at higher concentrations, films act as oxygen carrying agent and serves as pro-oxidant in the oxidation of lipid [9].

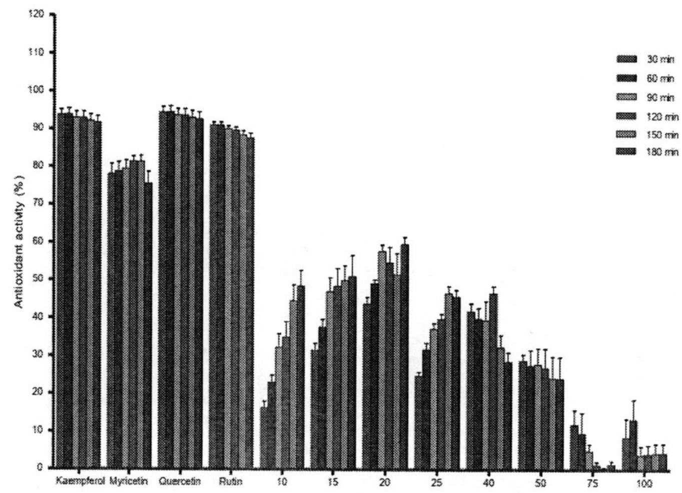

Fig. 4 Antioxidant activity of various concentration of *S. palustris*-88% HD PVA blends films. Values show means ± SEM (n = 4).

Fig. 5 illustrates antioxidant behavior of *S. palustris*-PVA blend film, the values of % DPPH remaining as a function of time are presented at a concentration of sample in the reaction mixture varies from 10 to 100 mg/mL. In the presence of film solutions a rapid initial decrease of DPPH content was followed by slow subsequent disappearance of DPPH. There were significant differences between the slopes after the end of the initial fast step. These differences were related to the role of secondary slow reactions (dimerization or disproportionation) of the radicals formed.

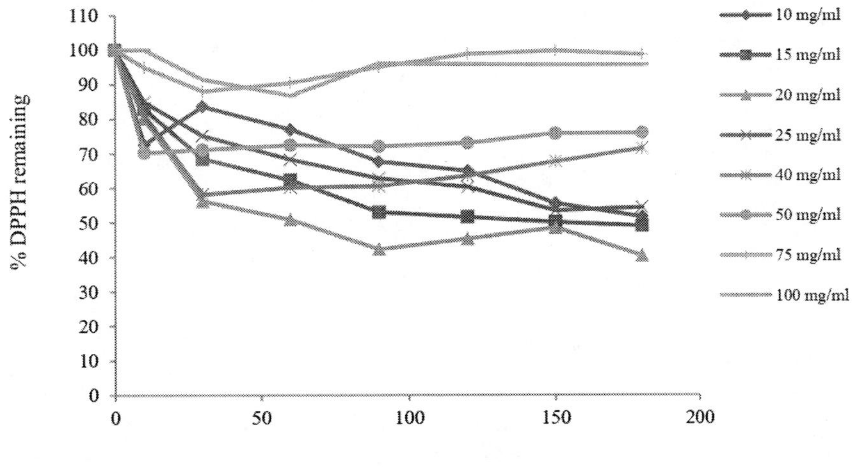

Fig. 5 Antioxidant behaviour of different film concentration

Antioxidants in blends film may not necessarily freely available to react with DPPH, hence they react at different rates and the reaction will often not go to completion in a reasonable assay time. Therefore, the sample size that can lower the initial absorbance of DPPH solution by 50% has been chosen as the endpoint for measuring the antioxidant activity [10]. This sample size was called EC_{50}. The lower EC_{50}, the higher the antioxidant activity of a compound is [11]. Fig. 6 shows the EC_{50} value was 20 mg/mL.

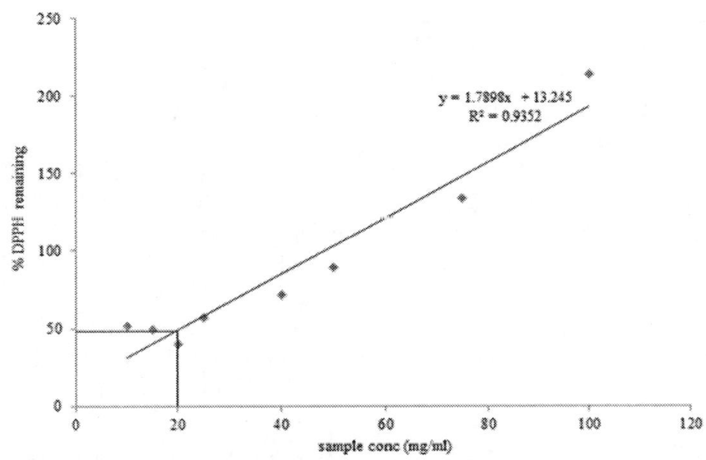

Fig. 6 Determination of EC_{50}

T_{EC50} was the time needed to reach a steady state at the concentration corresponding to EC_{50}. This parameter was obtained by plotting the times at steady state against the concentration for each antioxidant compound [7], as illustrated in Fig. 7. Based on T_{EC50} values determined, the kinetic behavior of the antioxidant compound can be classified as follows: < 5 min (intermediate); 5 – 100 min (rapid) and > 100 min (slow). The T_{EC50} value of *S. palustris*/PVA blend film was about 55 min, therefore it was classified as rapid.

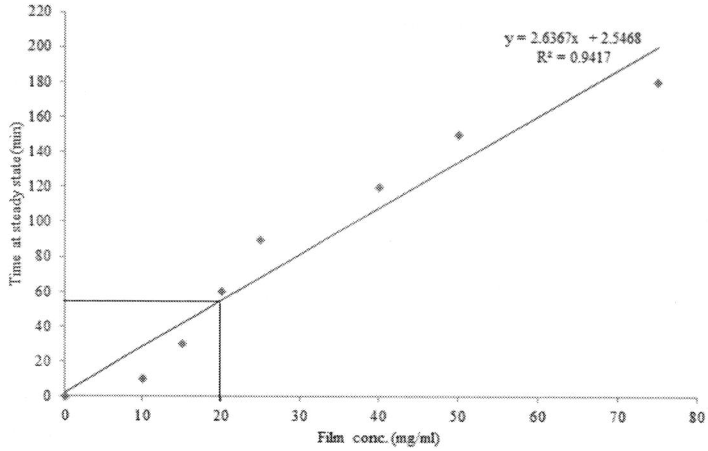

Fig. 7 Determination of the time needed to reach the steady state to EC_{50} concentration.

DPPH method permits to evaluate not only the electron or hydrogen atom donating properties of antioxidants, but also the rate of their reaction towards the free radicals. The AE, was calculated in order to easily characterize the behaviour of a substance as antioxidant. It was found that AE value of the studied film was 0.91×10^{-3}. The AE values were classified as follows: $AE \leq 1 \times 10^{-3}$ = low; $1 \times 10^{-3} < AE \leq 5 \times 10^{-3}$ = medium; $5 \times 10^{-3} < AE \leq 10 \times 10^{-3}$ = high and $AE > 10 \times 10^{-3}$ = very high [12]. Therefore, the AE of *S. palustris*/PVA blend film was suggested as low rate of reaction. AE was more adequate parameter than widely used EC_{50}, which is not completely discriminatory to select the antioxidant compound.

3.4. Film Characterization

The main IR bands observed in *S. palustris* flour-PVA 88% HD (55% PVA concentration) film spectra were then investigated. The *S. palustris* flour-88% HD PVA blended film spectrum displayed the typical profile of a polysaccharide in the 1600 to 600 cm^{-1} range (characteristic peaks attributed to COC bond stretching), due to starch content of *S. palustris* flour [4]. No strong peak was observed indicate that low amount of anhydroglucose ring presence in *S. palustris* flour. The band at 3899.73 cm^{-1} was due to hydrogen-bonded hydroxyl group (O – H) [13].

Assessing the thermogravimetric data of a polymer was means to asses the phase transformation (degradation) of the polymer. When studying the phase transformation of a polymer, one of the most important point to determine was the glass transition temperature; where the polymer structure turns rubbery upon heating and glassy upon cooling. The thermal degradation of blends film is known to take place in various steps such as; a small weight loss of about 1-7 % below 100 °C is assigned to the release of moisture from the sample. The sample show one stage degradation within the range of 105 - 300 °C. The thermal stability also depends on molecular weight and crystallinity of the polymer. Generally, lower the molecular weight or lower the cristallinity, the easier the degradation of the polymer.

Fig. 8 shows the variations of tensile strength and elongation at break of the film. The values of tensile strength and elongation at break after performed the experiment at triplet were 0.0977 ± 0.0144 MPa and $266 \pm 19.95\%$ using 0.5500 ± 0.3606 N load at break. Considering this result, it shows that film produced were actually low in elasticity due to effect of low concentration of PVA and plasticization. The use of glycerol as plasticizer caused a reduction in the resistance (decrease in TS) and an increase in the flexibility (increase in EB) of the films [14]. This behaviour, typical of the plasticization phenomenon, is a consequence of the increase in molecular mobility with the addition of plasticizer and may be due to the absorbed moisture in the protein films causing protein chains to slip over one another.

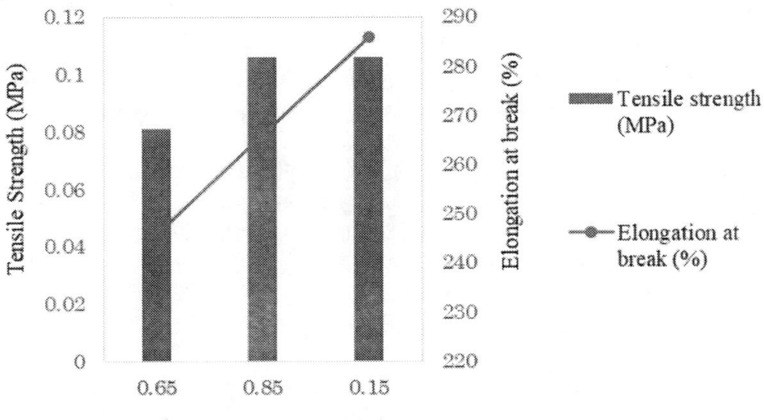

Fig. 8 The tensile strength and elongation at break of *S. palustris*-88% HD PVA (55 % PVA concentration) blends film for different load at break.

4. Conclusions

S. palustris flour-PVA blend film had been successfully developed by using three types of PVA; 88% HD, 95% HD, and 99-100% HD, however unsuccessfully by using 75% HD PVA. The highest antioxidant activity was show by the film formulated with 88% HD PVA at 55% concentration. The efficiency study conducted shows that 20 mg/mL of the film concentration gives the highest antioxidant activity. The EC_{50} value also shows the concentration of 20 mg/mL was needed to reduce radical concentration by 50%. Antioxidant efficiency (AE) of the film, which combines with the T_{EC50} value of time at steady state need to achieve EC_{50} gives the value of 0.91×10^{-3}. The use of glycerol plasticizer gives high elongation at break but low tensile strength of films produced. FTIR analysis gives identical polysaccharide profile and presence of hydrogen bonded hydroxyl group (O – H). The TGA analysis shows the thermal degradation of polymer. The glass transition temperature of blends film was at 305 °C and the mass residue was 4.48%.

Acknowledgements

This work is supported by Ministry of Higher Education (MOHE) Malaysia under Fundamental Research Grant Scheme (FRGS/1/2015/TK05/UNIMAP/02/8).

References

[1] Rein M J, Renouf M, Cruz-Hernandez C, Actis-Goretta L, Thakkar S K, da Silva Pinto M 2013 Bioavailability of bioactive food compounds: a challenging journey to bioefficacy British J. Clinical Pharm. **75** 588

[2] Chai TT, Panirchellvum E, Ong HC, Wong FC (2012) Phenolic contents and antioxidant properties of Stenochlaena palustris, an edible medicinal fern. Botanical Studies **53**: 439-446.

[3] Chambers, T.C. (2013) A review of the genus Stenochlaena (Blechnaceae, subfamily Stenochlaenoideae). Telopea **15**: 13–36.

[4] Nadiarid JE, Paulo JAS and Florencia CM 2009 *Carbohydrate Polym.* **75** 592

[5] Moradi M Tajik H Razavi Rohani S M Oromiehie A R Malekinejad H Aliakbarlu J 2012 Characterization of antioxidant chitosan film incorporated with Zataria Multiflora Boiss essential oil and grape seed extract *Food Sci.Tech.* **46** 477

[6] Khairul FK, Nur Azirah J, Ku Syahidah KI and Zulkali MMD 2009 *Proc. Mal. Tech. Uni. Conf. Eng. and Tech.*.

IOP Conf. Series: Materials Science and Engineering **429** (2018) 012066 doi:10.1088/1757-899X/429/1/012066

[7] Concepcion SM, Jose AL and Fulgencio SC 1998 A procedure to measure the antiradical efficiency of polyphenols *J. Sci. Food Agri.* **76** 270
[8] Sousa AMM *et al* 2015 Electrospinning of agar/PVA aqueous solutions and its relation with rheological properties *Carbohydrate Polym.* **115** 348
[9] Sampath S *et al* 2016 *RSC Adv.* **6** 12382
[10] Apak R, Özyürek M, Güçlü K and Çapanoılu E 2016 Antioxidant Activity/Capacity Measurement. 1. Classification, Physicochemical Principles, Mechanisms, and Electron Transfer (ET)-Based Assays *J. Agri. Food Chem.* **64** 997
[11] Tan J B L and Lim Y Y 2015 Critical analysis of current methods for assessing the in vitro antioxidant and antibacterial activity of plant extracts *Food Chem.* **172** 814
[12] Petra T, Miran B and Helena A 2009 A kinetic model for evaluation of the antioxidant activity of several rosemary extracts J. of Food Chem. **115** 740
[13] Popescu MC, Froidevaux J, Navi P and Popescu CM 2013 Structural modifications of Tilia cordata wood during heat treatment investigated by FT-IR and 2D IR correlation spectroscopy *J Mol. Struc.* **1033** 176
[14] Ahmad M *et al* 2015 Optical and thermo-mechanical properties of composite films based on fish gelatin/rice flour fabricated by casting technique *Prog. Org. Coat.* **84**:115

IOP Conf. Series: Materials Science and Engineering **429** (2018) 012067 doi:10.1088/1757-899X/429/1/012067

Synthesis and Preparation of Asymmetric PVDF-co-PTFE/DES Supported Membrane for CO_2/N_2 Separation

M N Amira[1*], M D Irfan Hatim[1], N Jullok[1], M Syahmie Rasidi[2], Hameed R Alamery[1]

[1]School of Bioprocess Engineering, Universiti Malaysia Perlis, 02600 Arau, Perlis, Malaysia
[2]School of Material Engineering, Universiti Malaysia Perlis, 02600 Arau, Perlis, Malaysia

*Email: amira.m.nasib@gmail.my

Abstract. Incorporation liquid into membrane polymer pores structure would overcome challenges in developing a membrane with increasing gas permeability while maintaining the excellent selectivity. Asymmetric Polyvinylidene fluoride-co-polytetrafluoroethylene (PVDF-co-PTFE) and PVDF-co-PTFE/Deep Eutectic Solvent (PVDF-co-PTFE/DES) supported membrane was prepared using phase inversion method. Membranes prepared were coated with 3 wt. % of polydimethylsiloxane (PDMS) in n-hexane and were dried in a room temperature. While, deep eutectic solvent (DES) was prepared by mixing choline chloride and ethylene glycol with a ratio of 1:3. The membranes was characterized using scanning electron microscopy (SEM), fourier transform infrared (FTIR) and gravimetric methodology to determine the membrane porosity. The membrane performance was carried out using gas permeation test at 3.5 psi. FTIR spectra revealed that existence of DES in the membrane pore structures. Empty PVDF-co-PTFE membrane coated with PDMS shows CO_2 permeance at 12 x 10^3 GPU with CO_2/N_2 selectivity at 0.72. Interestingly, incorporation of DES showing an improvement for both CO_2 permeance and CO_2/N_2 selectivity up to 49% and 36% respectively. Thus, employing DES as a liquid filler into PVDF-co-PTFE matrix shows a promising membrane for CO_2/N_2 separation.

1. Introduction

Carbon dioxide (CO_2) separation via polymeric membrane shows as one of the good potential technologies to replace conventional techniques such as chemical absorption, physical adsorption and cryogenic distillation. Gas separation involving membrane technology has certain advantages which are compactness and light in weight, low labor intensity, modular design allows easy expansion or operation at partial capacity, low maintenance due to no moving parts, low energy requirements and also low in cost [1]. However, polymeric membranes performance is limit by Robeson's trade-off limit between selectivity and permeability [2]. Thus combination of liquid with membranes for CO_2 separation is a one of the approach to surpass Robeson's trade-off limit, called supported (or immobilized) liquid membrane (SLM). SLM is a non-dispersive-type liquid membrane, where the liquid-phase selective material is immobilized in the membrane pores by capillary forces. Traditional SLMs have been develop by using amine-based aqueous solution such as diethylenetriamine (DETA), diaminoethane (DAE), diglycolamine (DGA) and triethylene glycol (TEG) [5].

First SILM was reported by Scovazzo et al. in a year of 2002, where ionic liquid (IL) was impregnated into the membrane pores for CO_2/N_2 separation. They found that supported ionic liquid membrane shows high in selectivity with high in flux in comparison to polymer membranes itself [3]. Even though ILs have been reported widely as a green solvent for the past two decades due to their tuneable physiochemical properties, there are still some challenged need to be encounter such as poor biodegradability, biocompatibility and sustainability. Thus, deep eutectic solvents (DES) has become a new substitute to ILs, where DES also possess an ionic character where it is a mixture of organic compounds having a melting point significantly lower than that of either individual component. DES have similar characteristics to ILs but are cheaper to produce due to lower cost of raw materials, less toxic and biodegradable [4].

The application of SLM is still not widely been used in industrial level due to the mechanical stability of liquid to retain in the membrane pore support. Changes of the operating conditions especially on pressure variations is considered as the most challenging issue. Yet, having high viscosity of DES would help to prevent the liquid loss. However, this makes filling up the pores is quite challenging, hence a vacuum method impregnation needs to be applied. Apart from that, having a very narrow pore size distribution of membrane pores support also could possibly reduce the chances of liquid loss [5].

In order to develop high performance of SLM for CO_2 gas separation, good interaction between liquid medium with the polymer membrane is compulsory. The right choice of the liquid medium is necessary to avoid carrier degradation and the loss of membrane lifetime. Thus, this study aims to develop an asymmetric PVDF-co-PTFE membrane as a support and the pores to be filled up with DES via vacuum method. Polyethylene glycol (PEG) will be added as a pore additive. While, DES is synthesized using choline chloride and ethylene glycol. The influence of DES in the polymer pores is investigated.

2. Experimental

2.1. Materials
Poly(vinylidene fluoride)-co-Poly(tetrafluoroethylene) PVDF-co-PTFE was obtained from Arkema Pte Ltd, Singapore. N,N-dimethylacetamide (DMAc) solvent was purchased from Merck Milipore Corporation with $\geq$ 99.0% purity. Polyethylene glycol, PEG with a reagent grade and average MW 10,000 was purchased from Sigma Aldrich with $\geq$ 99.0% purity. In this study, ethylene glycol ($\geq$ 99.0% purity) was bought from Merck Milipore Corporation. For coating solution, polydimethylsiloxane (PDMS) was purchased from Sigma Aldrich and n-hexane was purchased from Merck.

2.2. Preparations of Polymer Solution
Homogenous polymer solutions were prepared by dissolving 20 wt. % of PVDF-co-PTFE polymer in 77 wt. % of DMAc solvent. Then 3 wt. % of PEG additives was added and continuously stirred at 200 rpm until a clear and homogenous solutions were obtained. Throughout the stirring process, the temperature of the mixture was maintained at 50 °C. A homogenous polymer solution was then degassed in a room temperature for 24 hours.

2.3. Phase Inversion Method
The non-solvent induced phase separation (NIPS) technique was chosen to cast a flat sheet membrane. A polymer solution was casted onto a glass plate using a hand-casting knife with a knife gap set at 400±10 μm and followed by exposed to the surrounding air for 30 seconds. The cast film together with the glass plate was immersed into a coagulation bath for 24 hours. The coagulation bath medium consists of 75 wt. % of distilled water and 25 wt. % of ethanol. The formed membrane was taken out from the coagulation bath and dried at room temperature for 24 hours [6].

2.4. Synthesis of Deep Eutectic Solvent

The deep eutectic solvent (DES) used was a mixture of ChCl and ethylene glycol with a ratio of 1:3. DES was prepared by adding choline chloride in the ethylene glycol at room temperature. The mixture was continuously agitated at 200 rpm until a clear and homogenous solution was obtained.

2.5. Preparation of Supported Deep Eutectic Solvent Membranes

All supported membranes throughout this study were prepared by using vacuum immobilization techniques. The prepared membrane was submerged into the deep eutectic solvent in a petri dish and both were placed in the desiccator. A vacuum pump was applied and the pressure was set at -10 mbar for 3 hours at room temperature. The desiccator was aerated for every hour to ensure that the membrane pores were filled with DES. The excess DES on the membrane surface was wiped off with a filter paper. The weight of membrane before and after impregnation were recorded to determine the amount of DES immobilized in the membrane pores.

2.6. Membrane Coating

The membrane was then brought into contact with 3 wt. % PDMS/n-hexane solution for 10 minutes to seal pinholes before it was cured at room temperature for 24 hours.

2.7. Membrane Characterization

The prepared membrane top surface and the cross section morphological studies was carried out by using a scanning electron microscope (SEM) model JSM 6260 LE JEOL. The pore sizes were observed through the SEM images and measured by using SEMAFORE software.

FTIR analysis was used to study the effect of addition DES into the membrane by observing at their functional group presence before and after functionalized membrane. The FTIR analysis was carried out by using a FTIR spectroscopy, Perkin Elmer, Model L128044. The attenuated total reflectance (ATR) method was used. The 4 scans were recorded per sample in the frequency range 4000 – 650 cm^{-1}, with a resolution of 4 cm^{-1}.

The porosity of the prepared membrane was analysed by using an octanol. The sample was immersed in octanol for 15s to improve the hydrophilicity. The sample was later immersed in distilled water for 2 minutes before dried using a filter paper. Mass of samples before and after soaked in both octanol and distilled water were recorded. Equation (1) was used to calculate the membrane porosity.

$$\omega = \frac{m_n/\rho_n}{m_n/\rho_n + m_p/\rho_p} \tag{1}$$

Where ω is the porosity of the membrane, m_p is the mass of dry membrane, m_n is the mass of absorbed octanol, ρ_p is the density of the membrane, and ρ_n is the density of octanol. This method is used to estimate the porosity by determined the weight of liquid contained in the membrane pores.

The membrane was characterized using pure gas permeation tests. The pure gas permeance was determined by a constant volume. The membranes were placed into the permeation cell with an effective area of 11.34 cm^2 and exposed to pure CO_2 or N_2 respectively. The transmembrane pressure was constant at 3.5 psi and the temperature was 27 °C. The permeance (pressure-normalized flux) of gas *i* was calculated as follows;

$$\left(\frac{P_i}{l}\right) = \frac{1}{A\Delta p} \times \frac{dV_i}{dt} \tag{2}$$

Where *i* represents the gas penetrant, V_i is the volume of the gas permeated through the membrane (cm^3, STP), *A* is the effective membrane area (cm^2), *t* is the permeation time (s) and Δp is the transmembrane pressure drop (cmHg). The permeance is expressed in gas permeation unit, GPU, as

$$GPU = 1 \times 10^{-6} \frac{cm^3(STP)}{cm^2.S.cmHg}$$

Meanwhile, the selectivity was obtained by

$$\alpha_{i/j} = \frac{(P/l)_i}{(P/l)_j} \quad (3)$$

Where $\alpha_{i/j}$ is the selectivity of gas i over gas j. Both $(P/l)_i$ and $(P/l)_j$ are the permeance of gas i and j, respectively.

3. Results and Discussion

3.1. Characterization

The SEM images of empty PVDF-co-PTFE membrane cross sectional area and top surface are presented in Figure 1. Existence of pores on top of membrane surface is believe due to addition PEG as a pore additives. PEG as a polymer additives will increase the polymer solution viscosity, improve pore formation and suppress macro-void formation. This is the reason existence of combination both micro voids and sponge like structure as can be seen in the cross sectional area of the membrane [7]. The formation of sponge like structure also can be said due to the formulation of bath medium. The coagulation bath medium plays an important role in the formation of membranes via the phase inversion method. This phenomenon is related to the coagulation ability of the non-solvent. The presence of 25 wt. % of ethanol in the bath medium was believe to reduce the polymer precipitation rate in phase inversion process. Similar phenomena can be seen from a research conducted by Deshmukh and Li in a year 1998 [8]. Table 1 listed the empty PVDF-co-PTFE membrane configuration used for both DES impregnation and gas permeation test. Both membrane pore size and thickness were measured by using SEMAFORE software while the membrane porosity was obtained via gravimetric method.

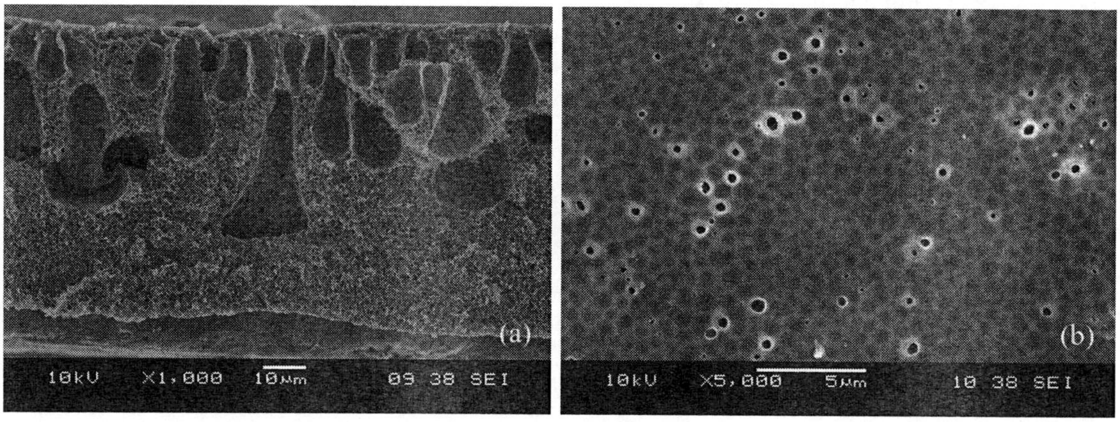

Figure 1. Micrographs of empty PVDF-co-PTFE membrane: (a) cross sectional view and (b) the membrane top surface.

Table 1. Characteristics of empty PVDF-co-PTFE membrane

Pore size (μm)	Minimum	0.6 ± 0.2
	Maximum	10.5 ± 6.0
Thickness (μm)	73.8	
Porosity, ε (%)	73.5 ± 2.0	

The FTIR characterization of empty PVDF-co-PTFE and PVDF-co-PTFE/DES supported membranes are shown in Figure 2. The characteristic peaks of PVDF are at 830 cm^{-1} and 882 cm^{-1}. The first peak is assigned to CH_2 rocking vibration and the peak at 882 cm^{-1} is the CH out-of plane deformations [9]. While, three peaks at 1396 cm^{-1}, 1166 cm^{-1} and 1072 cm^{-1} correspond to PTFE structure. The first two bands are assigned to the C-F stretching vibrations of PTFE [10]. Compared with empty PVDF-co-PTFE membrane, the additional spectra of PVDF-co-PTFE/DES supported membrane at 3370 cm^{-1}, which corresponds to N-H stretching, indicates the successful impregnation of DES in PVDF-co-PTFE membrane pores [11].

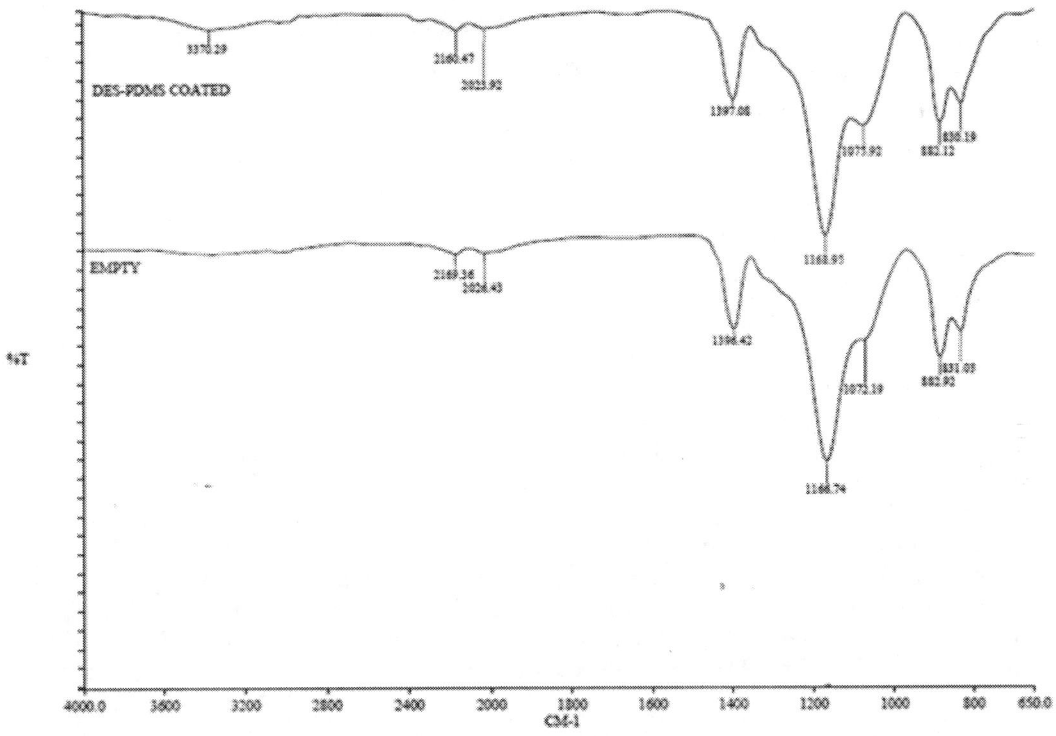

Figure 2. FTIR spectra of membrane samples.

3.2. Gas Permeation

The gas permeation was conducted to evaluate ideal separation properties of both membranes presented in Table 2. The gas permeation of PVDF-co-PTFE/DES supported membrane shows better in both CO_2 permeance and CO_2/N_2 selectivity. In SLM, the main transport mechanism is solution diffusion. A gas molecule dissolves into the membrane, diffuses through the membrane and finally release to the permeate side of the membrane. The carbon dioxide permeability is greatly dependent on its solubility coefficient within the membrane [12]. According to Leron and Li (2013), solubility of CO_2 in the DES was comparable with those in the ionic liquids such as 1-butyl-3-methylimidazolium tetrafluoroborate and 1-butyl-3-methylimidazolium dicyanamide. Their findings also indicates the CO_2

solubility in the DES are better than CO_2 solubility in those found in 1-ethyl-3-methylimidazolium based ionic liquids [13]. Table 3 shows relative comparison between this works to literature data using different polymer-liquid membrane combination. Although DES is a new green solvent incorporated into the polymer pores, significant improvement of both CO_2 permeance and CO_2/N_2 selectivity somehow is comparable with literature data.

Table 2. CO_2 and N_2 for empty PVDF-co-PTFE and PVDF-co-PTFE/DES supported membrane.

Membrane	Permeability of N_2 (GPU)	Permeability of CO_2 (GPU)	Selectivity, α (CO_2/N_2)
Empty PVDF-co-PTFE membrane	17×10^3	12×10^3	0.72
PVDF-co-PTFE/DES supported membrane	12.7×10^3	25.5×10^3	2.00

*The gas permeation was conducted at 3.5 psi, 27 °C

Table 3. Comparison between literatures

Membrane support	Liquid	Permeability of CO_2 (GPU)	Selectivity, α (CO_2/N_2)	Reference
Porous alumina	Ionic liquid 1-(2-(2-(2-Methoxyethoxyl)ethoxy)ethy)-1-methylpyrrolidinium p-toluenesulfonate	1.12	15.42	[14]
Polyethersulfone	1-n-butyl-3-methylimidazolium tetrafluoroborate	13.8	60	[15]
PVDF-co-PTFE	Deep eutectic solvent	25.5×10^3	2.0	In this work

4. Conclusion

Deep eutectic solvent from choline chloride and ethylene glycol has been successfully synthesized and incorporated into PVDF-co-PTFE membrane pores. The membrane was characterized via SEM and FTIR have shown the existence of DES in the membrane. As expected, PVDF-co-PTFE/DES supported membrane show an improvement in both CO_2 permeance and CO_2/N_2 selectivity as compared to empty PVDF-co-PTFE membrane.

Acknowledgement

The authors gratefully acknowledge the Ministry of Higher Education (MOHE) for the scholarship and support from the Short Term Grant under a grant number of 9001-00547, Universiti Malaysia Perlis (UniMAP).

References

[1] Ravanchi MT, Kaghazchi T, Kargari A 2009 Application of Membrane Separation Processes in Petrochemical Industry: A Review. *Desalination.* **235**(1) 199-244.
[2] Robeson LM 2008 The upper bound revisited. *Journal of Membrane Science.* **320**(1-2) 390-400.
[3] Scovazzo P, Visser AE, Davis Jr JH, Rogers RD, Koval CA, DuBois DL, et al. Supported Ionic Liquid Membranes and Facilitated Ionic Liquid Membranes. ACS Publications; 2002.

[4] Paiva A, Craveiro R, Aroso I, Martins M, Reis RL, Duarte ARC 2014 Natural deep eutectic solvents–solvents for the 21st century. *ACS Sustainable Chemistry & Engineering.* **2**(5) 1063-71.
[5] Grünauer J, Shishatskiy S, Abetz C, Abetz V, Filiz V 2015 Ionic Liquids Supported by Isoporous Membranes for CO 2/N 2 Gas Separation Applications. *Journal of Membrane Science.* **494** 224-33.
[6] Feng C, Shi B, Li G, Wu Y 2004 Preparation and Properties Of Microporous Membrane from Poly (Vinylidene Fluoride-Co-Tetrafluoroethylene)(F2. 4) for Membrane Distillation. *Journal of Membrane Science.* **237**(1) 15-24.
[7] Aroon M, Ismail A, Montazer-Rahmati M, Matsuura T 2010 Morphology And Permeation Properties of Polysulfone Membranes for Gas Separation: Effects of Non-Solvent Additives and Co-Solvent. *Separation and Purification Technology.* **72**(2) 194-202.
[8] Deshmukh S, Li K 1998 Effect of Ethanol Composition in Water Coagulation Bath on Morphology of PVDF Hollow Fibre Membranes. *Journal of Membrane Science.* **150**(1) 75-85.
[9] Ma J, Zhao J, Ren Z, Li L 2012 Preparation and Characterization of PVDF-PFSA Flat Sheet Ultrafiltration Membranes. *Frontiers of Chemical Science and Engineering.* **6**(3) 301-10.
[10] Huang Q-l, Xiao C-f, Hu X-y, Li X-f 2011 Study on the Effects and Properties of Hydrophobic Poly (Tetrafluoroethylene) Membrane. *Desalination.* **277**(1-3) 187-92.
[11] AlOmar MK, Hayyan M, Alsaadi MA, Akib S, Hayyan A, Hashim MA 2016 Glycerol-Based Deep Eutectic Solvents: Physical Properties. *Journal of Molecular Liquids.* **215** 98-103.
[12] Luebke D, Ilconich JB, Myers C, Pennline HW. Carbon Dioxide Separation with Supported Ionic Liquid Membranes. National Energy Technology Laboratory (NETL), Pittsburgh, PA, and Morgantown, WV; 2007.
[13] Leron RB, Li M-H 2013 Solubility of Carbon Dioxide in A Eutectic Mixture of Choline Chloride and Glycerol at Moderate Pressures. *The Journal of Chemical Thermodynamics.* **57** 131.
[14] Hojniak SD, Khan AL, Hollóczki O, Kirchner B, Vankelecom IF, Dehaen W, et al. 2013 Separation of Carbon Dioxide from Nitrogen or Methane by Supported Ionic Liquid Membranes (Silms): Influence of the Cation Charge of the Ionic Liquid. *The Journal of Physical Chemistry B.* **117**(48) 15131-40.
[15] Zhao W, He G, Zhang L, Ju J, Dou H, Nie F, et al. 2010 Effect of Water in Ionic Liquid on the Separation Performance of Supported Ionic Liquid Membrane for CO_2/N2. *Journal of Membrane Science.* **350**(1-2) 279-85.

IOP Conf. Series: Materials Science and Engineering **429** (2018) 012068 doi:10.1088/1757-899X/429/1/012068

Thermal Curing of Benzoxazine Blends from Cardanol-Furfurylamine and Diphenolic Acid-Aniline

Viet C X[1*], Lan D N U[2], Hung T K[1]

[1] Department of Polymer Materials, Ho Chi Minh University of Technology, Vietnam
[2] School of Materials Engineering, University Malaysia Perlis, Malaysia

*Email: caoxuanviet@hcmut.edu.vn

Abstract. Cardanol-furfurylamine benzoxazine (CFB) is a bio-based resin thanks to cardanol as a by-product of cashew nut shell liquid. Polybenzoxanize obtained from CFB exhibited high thermal stability, however, it requires a high thermal curing temperature. This research was aimed to reduce the curing temperature of CFB by blending it and diphenolic acid-aniline benzoxazine (DAB). The blend molar ratio of CFB:DAB was 4:1. Under thermal curing, carboxyl group in DAB was degraded and accelerated ring-opening polymerization of CFB/DAB blend so that no catalyst was used. Different curing temperatures of CFB/DAB blend was investigated. Results from fourier transform infrared (FTIR) and differential scanning calorimetry (DSC) proved that the ring-opening initiated at 115.2 °C for DAB, 141.6 °C for blend and 206.2 °C for CFB. It proved that CFB/DAB blend could start thermal curing at an obvious lower temperature. Moreover, polybenzoxazine synthesized from CFB/DAB blend exhibited good thermal stability compared to DAB-polybenzoxazine and obtained higher char yields compared to CFB-polybenzoxazine.

1. Introduction

Polybenzoxazines (PBzs) appeared as attractive candidates over traditional phenolic polymers due to their superior properties including high thermal stability, superior modulus properties, low water absorption, near zero shrinkage, and no release of by-products [1]. Recently, polybenzoxazine technology is now finding a breakthrough through the preparation of their monomers from renewable resources in the same way as other petroleum-based polymers. Among these renewable resource materials, cardanol is considered as an important starting material bio-based benzoxazine synthesis purposes due to its unique structural features, abundant availability and low cost. Cardanol is a major constituent of the cashew nut shell liquid (CNSL) which is an agricultural by-product abundantly available in many parts of the world, particularly in ASEAN countries. In other hand, the bio-based polybenzoxazines containing furan units has demonstrated excellent properties due to the aromatic characters of furan ring, like the high performance furan-based epoxy systems [2]. The furanic ring of furfurylamine is highly reactive and promotes an increase of the crosslinking density. When comparing cardanol-based monofunctional benzoxazines to other amine derivatives, such as aniline or benzylamine (T_g around 55°C), was observed for the furfurylamine-based PBz material. Moreover, furfurylamine is synthesized from furfural which is a green platform product derived from a variety of agricultural by-products such as corncobs and wheat bran like other fatty amines containing a long hydrocarbon chain [3].

However, cardanol-furfurylamine benzoxazines (CFB) still requires a relatively high polymerization temperature. In the present research, an attempt was done to lower curing temperature of CFB by blending with benzoxazine derived from diphenolic acid (DPA) which is also commercially produced from cellulose-rich biomass. The presence of a carboxylic acid in DPA monomer should lower the high temperature needed to complete curing of these benzoxazines [4].

2. Experimental

2.1. Materials

All reagents and solvents were used as received from commercial suppliers. Cardanol was procured from Son Chau Co., Ltd (Viet Nam). Cardanol ($C_{21}H_{36-n}O$, n = 0, 2, 4, 6) was distilled at 230–240 °C under 2–4 mm Hg. Diphenolic acid (95%), furfuryl amine (99%), sodium sulfate (99%) and sodium hydroxide (97%) were purchased from Sigma-Aldrich. Paraformaldehyde (95%), aniline (99.5%) and chloroform (99%) were obtained from Merck. Toluene (99%) was provided by Prolabo. All reactions were carried out in oven-dried flask.

2.2. Synthesis of Cardanol – Furfurylamine Benzoxazine (CFB)

Cardanol-furfurylamine benzoxazine was synthesized by a method adopted from earlier report [4]. Cardanol (0.033 mol) was heated to 50 °C in a 100 ml three-necked round flask equipped with magnetic stirrer and thermometer. Then paraformaldehyde (0.066 mol) and aniline (0.033 mol) were added drop wise under vacuum environment. The temperature was gradually raised to 80-90 °C and kept at this temperature for half an hour until the mixture becomes homogenous and turns dark brownish red color.

Then, the temperature was further raised to 120–130 °C and refluxed for another 2h. The reaction mixture was then cooled to room temperature. The product was washed with 2N NaOH followed by washing with distilled water in a separating funnel. The organic phase was dried over sodium sulphate (Na_2SO_4) and filtered to give a red brow oil. The solvent was removed under reduced pressure and then dried at 60 °C for 12h under vacuum to obtain the final product which is CFB with a yield of 88%.

2.3. Synthesis of Diphenolic Acid-Anilne Benzoxazine (DAB)

Diphenolic acid-aniline (DAB) were synthesized according to the reported procedure [5]. 1,3,5-Triphenylhexahydro-1,3,5-triazine and (0.02 mol), paraformaldehyde (0.06 mol), 4,4'-bis(4-hydroxyphenyl) pentanoic acid (0.03 mol) and 100 ml of toluene were placed into 250 ml two-necked round-bottom flask. The reaction mixture was heated at 110 °C for 6 h. The resulting light orange solution was filtered and concentrated under vacuum to obtain a syrup that was subsequently dried under high vacuum giving a yellowish solid with a yield of 89% of DAB.

2.4. Preparation of CFB/DAB Blend – Thermal Curing

CFB/DAB blend was prepared in a 4:1 molar ratio. CFB and DAB were taken and dissolved in 10 ml of chloroform and stirred at room temperature for 1 h to obtain a homogenous solution form. Afterwards, the mixture was dropped and casted on a glass plates. The solution was dried under vacuum for 2h at 80 °C. Thermal curing of blends was carried out in an oven at 150 °C, 180 °C and 220 °C for 1 hour and further characterized by DSC and FTIR. Similar process was used to prepare CFB and DAB polybenzoxazine.

2.5. Measurements

2.5.1. Structural Characterization

IOP Conf. Series: Materials Science and Engineering **429** (2018) 012068 doi:10.1088/1757-899X/429/1/012068

^{1}H (500 MHz) and ^{13}C (125.8 MHz) nuclear magnetic resonance (NMR) spectra were obtained using a Bruker Avance AM500 FT-NMR spectrometer with Fourier transform and $CDCl_3$ as solvent. The chemical shift are given relative to tetra methyl silane (TMS). The NMR measurements and analysis were performed at Center for Apply Spectroscopy, Institute of Chemistry, Vietnam Academy of Science and Technology (VAST), Hanoi, Vietnam. Fourier transform infrared (FTIR) spectra of the samples were recorded on a Bruker Tensor 37 spectrophotometer with a resolution from 4000 – 400 cm^{-1} in the absorbance and transmittance modes.

2.5.2. Thermal Characterization

Differential scanning calorimetric (DSC) studies were carried out on a NETZSCH DSC 204F1 Phoenix thermal analyzer using N_2 as a purge gas, heated from room temperature to 300 °C at scanning rate of 10 °C/min. The thermal stability of polymer was studied by thermogravimetric analysis (TGA) using a TGA Q500 instrument. Cured samples at 230 °C were analyzed at a temperature rate of 10 °C /min from 25 °C to 800 °C under nitrogen environment.

3. Results and Discussion

3.1. NMR Spectroscopy of the Synthesized Benzoxazines

The 1H NMR spectra shown in Figure 1 confirm the structure of CFB benzoxazine. The oxazine ring protons (Ar-CH2-N and O-CH2-N) give resonance peaks at 3.9 and 4.8 ppm. The furan ring was characterized with the peaks at 6.3 to 7.3 ppm. The peaks from 0.8 to 2.7 ppm, at 5.0 ppm, and 5.4 ppm, were attributed to the unsaturated alkyl side chain of the cardanol [4].

Figure 2 shows ^{1}H NMR and ^{13}C NMR spectra of DAB. It can be seen that the nuclear magnetic resonance spectroscopy (^{1}H NMR) spectrum of the DPA-b exhibits not only the specific signals of the benzoxazine ring, but also chemical shifts that belong to the alkyl chain and the aromatic signals. Notably, while the two signals at 4.56 and 5.32 ppm correspond to -CH_2 protons of benzoxazine ring, methyl protons of the pentanoic acid appears at 1.51 ppm (singlet, 3H). Alkyl protons of the propionic acid moiety resonate at 2.13 ppm (triplet, C-CH_2, 2H) and 2.35 ppm (triplet, -CH_2-COOH, 2H). The COOH proton is not observed in ^{1}H NMR when $CDCl_3$ is used as solvent [6].

The corresponding ^{13}C NMR spectrum also supports the structure of the monomer. The two singlets at 79.23 ppm and 50.6 ppm are typical of the carbon resonances of –N-CH_2-O- and –N-CH_2-Ph of the oxazine ring, respectively. Most importantly, the signal appears at 179.1 ppm confirmed the presence of COOH group in the monomer structure [5].

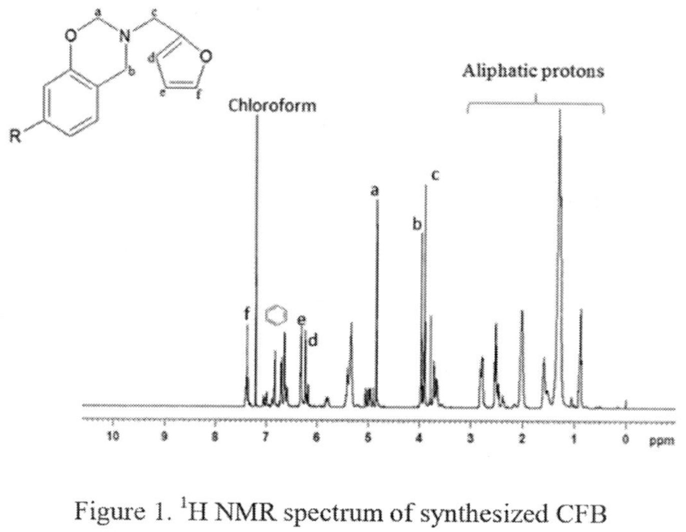

Figure 1. ^{1}H NMR spectrum of synthesized CFB

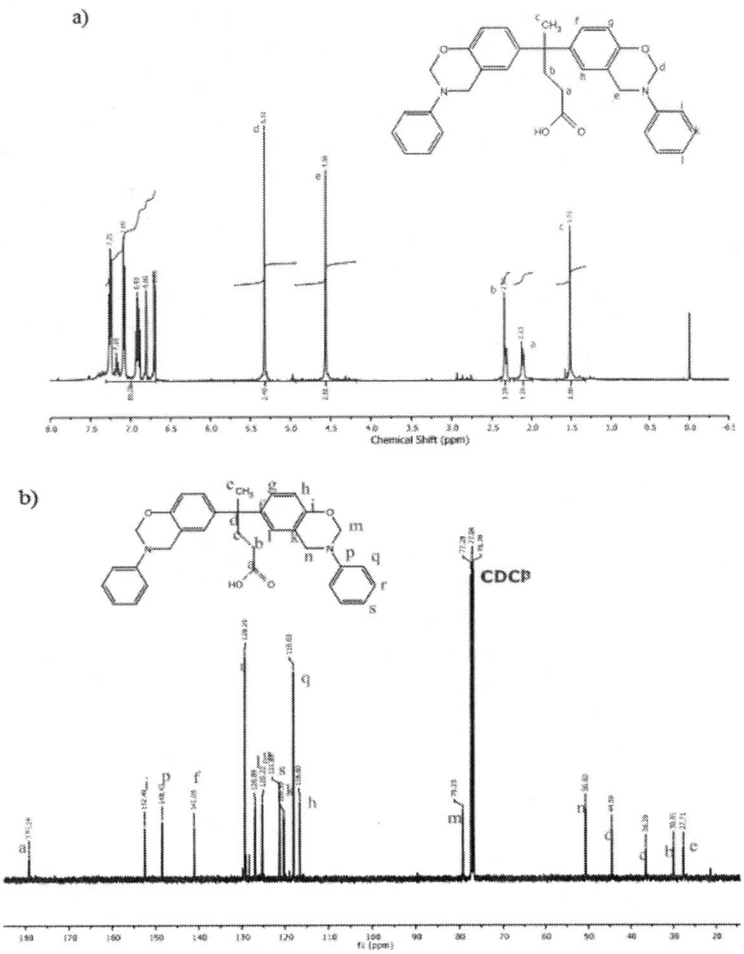

Figure 2 (a): ^{1}H NMR, (b) ^{13}C NMR spectra of synthesized DAB.

Figure 3 shows the distinguished curing behaviour of CFB, DAB and their blend. DAB could have the earliest curing temperature due to its carbonyl group, while CFB has the highest one. However, the curing energy of DAB was more than 10 times higher than that of CFB. As expected, the blend could have lower ROP initiation and endset temperature. It means carbonyl group of DAB could also accelerate the ROP reaction of CFB, even the high molar ratio of 4 CFB: 1 DAB.

CFB/DAB blend was cured at 150 °C, 180 °C, and 220 °C; and its curing energy and residue cure of were recorded seen in Figure 4. The curing energy of CFB/DAB blend was 93.8 J/g (Table 1), so its residue cure was calculated as 52.24% at 150°C and 3.2% at 180°C. In the other words, CFB/DAB blend was cured 47.76% at 150 °C, 96.8% at 180 °C and fully cured at 220 °C. Glass transition was detected from the DSC curve of CFB/DAB blend cured at 220°C. CFB/DAB blend was able fully cured at 220°C which is significantly lower than the curing temperature of CFB. Suggestion of polymerization of CFB and DFB are shown in Figure 5(a) and (b). It is noted that not all the COOH group will be esterification.

IOP Conf. Series: Materials Science and Engineering **429** (2018) 012068 doi:10.1088/1757-899X/429/1/012068

Table 1: Thermal curing behaviour of CFB, DAB and their blend using DSC

Monomer	$T_{initiation}$ (°C)	T_{peak}(°C)	T_{endset} (°C)	ΔH(J/g)
CFB	206.2	251.6	281.2	25.49
CFB/DAB blend	141.6	210.0	243.0	93.88
DAB	115.2	177.2	242.7	272.6

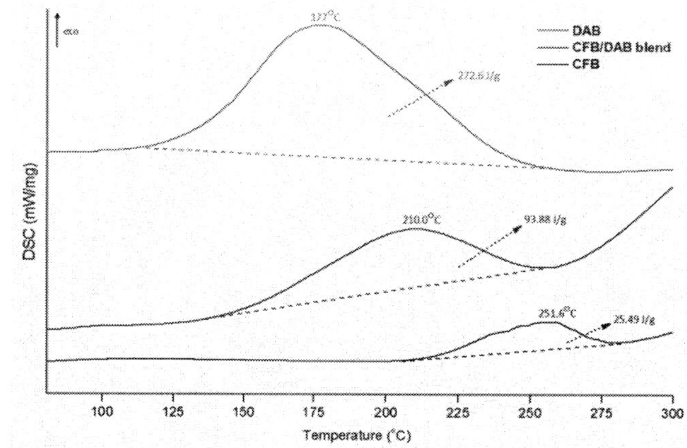

Figure 3. Thermal curing of DAB, CFB and their blend by using DSC

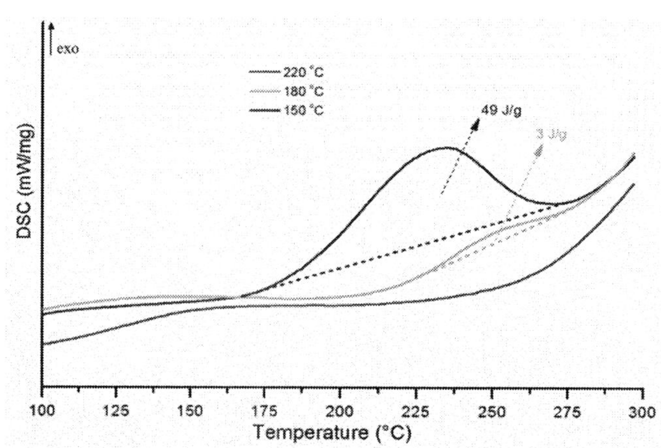

Figure 4. DSC of CFB/DAB blend after cured at 150 °C, 180 °C, and 220 °C

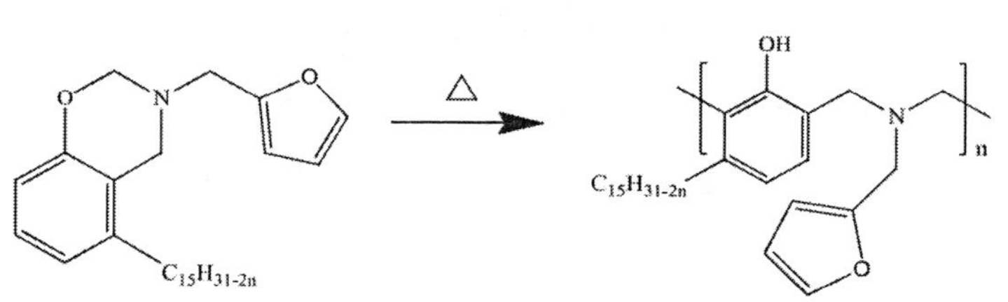

Figure 5(a). Thermal curing reaction of CFB

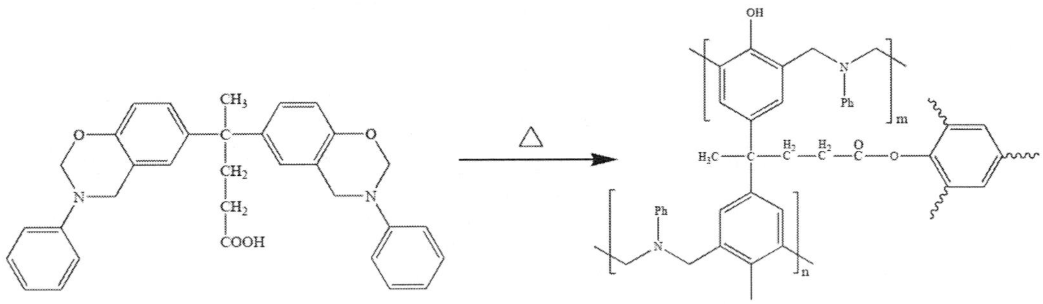

Figure 5(b). Thermal curing reaction of DAB

FTIR spectra of monomers blend and poly-benzoxazine after cured at 150 °C, 180 °C and 220 °C. The peaks at 1721cm^{-1} and 1410 cm^{-1} of COOH group in DAB disappeared in all poly-benzoxazines. The carboxyl group could react with hydroxyl group to form ester (Figure 5(b)). Therefore the peak 1410 cm^{-1}of OH belonged to COOH group reduced. Similar finding was also reported by Zuniga et. al [4]. In the other side, three main groups of benzoxazine are 1360 cm^{-1} (CH_2 wagging), ~1248 cm^{-1} (Ar-CO-asymmetric stretching) and 960 cm^{-1} (benzene attached to oxazine ring), which also decreased in all poly-benzoxazines proving the occurrence of ROP reaction. The peak 690 cm^{-1} of tri-substituted aromatic ring of benzoxazine also reduces.

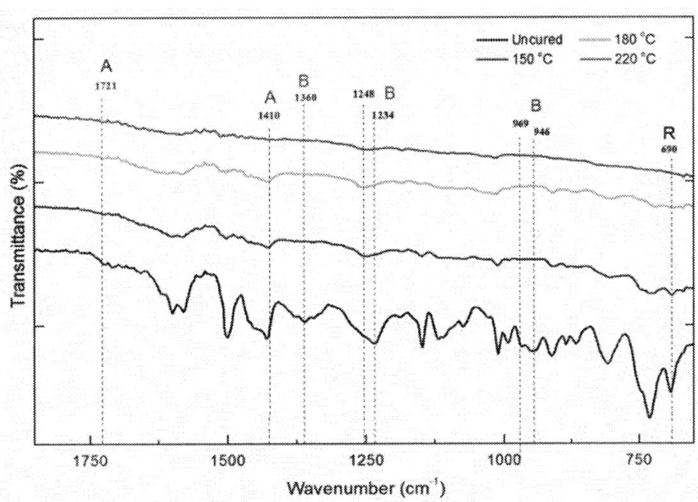

Figure 6. FTIR of CFB/DAB blend before and after cured, A: COOH, B: peaks of benzoxazine, R: tri-substituted.

Thermal stability of poly-benzoxazines synthesized from CFB, DAB and their blend is reported in Table 2 and Figure 7. DAB -benzoxazine exhibited the early degradation at 221.1 °C because of the degradation of remained carboxyl group or new ester groups. CFB poly-benzoxazine had a significantly high thermal stability compared to DAB. Therefore, the poly-benzoxazine of CFB/DAB blend also performed well thermal stability.

Table 2: Thermal properties of Poly-benzoxazine from CFB, DAB and their blend

	T_{onset} (°C)	$T_{5\%}$ (°C)	$T_{10\%}$ (°C)	Char yield (%)
CFB	436.9	402.5	428.5	14.7
Blend	427.8	360.5	399.5	18.3
DAB	221.1	255.5	297.5	30.3

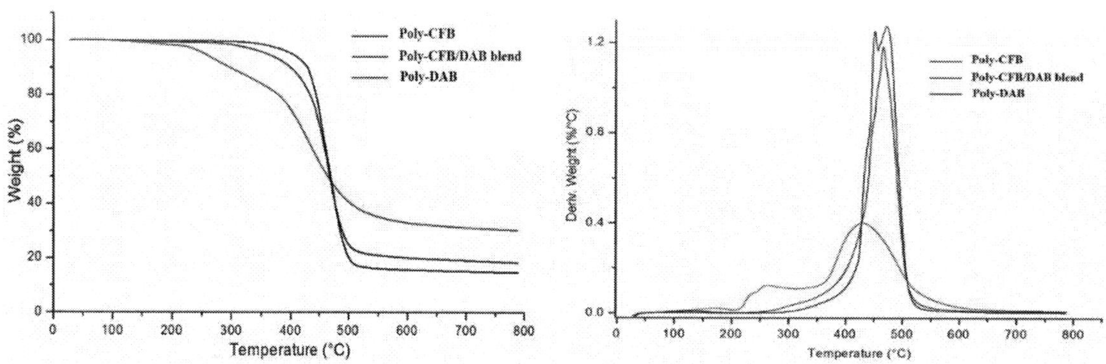

Figure 7. TGA and DTG curves of polybenzoxazine from CFB, DAB and their blend

4. Conclusion

Cardanol-furfurrylamine benzoxanize as a bio-benzoxazine was synthesized. Its poly-benzoxazine exhibited a distinguished high thermal stability. However, cardanol-furfurrylamine benzoxanize needed a high temperature for thermal curing. Addition of dibisphenol acid – aniline benzoxazine with the blend ratio of 4 CFB/1 DAB was successfully reduced the peak temperature of thermal curing from 251.6 °C of CFB to 210.0 °C of CFB/DAB blend. Further investigation on the ability to form interpenetrating polymer network of CFB and DFB could be carried out, as well as the reaction of CFB and DFB during thermal curing can be examined.

References

[1] T Takeichi T Kawauchi and T Agag 2008 High Performance Polybenzoxazines as a Novel Type of Phenolic Resin, Polymer Journal, **40** (12) 1121–1131.

[2] F Hu J J La Scala J M Sadler G R Palmese 2014 Synthesis and characterization of thermosetting furan-based epoxy systems. Macromolecules **47**:3332–3342

[3] P Thirukumaran A Shakila Parveen M Sarojadevi 2014 Synthesis and characterization of novel bio based benzoxazines from eugenol, RSC Adv. **4** 7959–7966.

[4] P Thirukumaran R Sathiyamoorthi A Shakila Parveen M Sarojadevi 2014 New benzoxazines from renewable resources for green composite applications, **37**(2), 573-582.

[5] C Zúñiga M S Larrechi G Lligadas J C Ronda M Galià and V Cádiz 2011 Polybenzoxazines from renewable diphenolic acid, Journal of Polymer Science Part A: Polymer Chemistry, **49** (5), 1219–1227.

[6] C Zúñiga M S Larrechi G Lligadas J C Ronda M Galià and V Cádiz 2012 Self-foaming diphenolic acid benzoxazine, Polymer, **53**(15), 3089-3095.

[7] Pretsch, E., Bühlmann, P., Badertscher, M. - Structure Determination of Organic Compounds, Springer, 2009, pp. 69-242.

IOP Conf. Series: Materials Science and Engineering **429** (2018) 012069 doi:10.1088/1757-899X/429/1/012069

Computational Fluid Dynamic Analysis of Coolant Flow in Milling of Titanium grade-2 under Multiple Nozzle

LDK Catherine*, Darulihshan Bin Abdul Hamid, M.S.N. Ibrahim, A. Zakaria

Universiti Kuala Lumpur (IPROM)

*louis.denis06 @s.unikl.edu.my

Abstract. The application and behaviour of coolant in end milling of hard-to-cut material such as titanium alloy is very critical due to the high cutting temperature generated at the tool tip that can be more than 600°C. Titanium is generally well known to have a very low thermal conductivity that prevent a rapid dissipation of the heat generated during machining due to the friction that occur in the metal removal process. In addition, hard-to-cut material have a very high hardness that can range from 198 to 1300 HV at Vickers hardness. For all this reasons, the optimization of the coolant that includes positioning, inclination and distance of the nozzle to the tool tip is very vital to provide an efficient cooling during the machining operation. In this paper, a three dimensional static analysis of the application of coolant through two nozzles was studied by using computational fluid dynamics (CFD). It was found that the velocity of the coolant was more uniformly distributed and effective when two nozzles were applied at an angle of 12.5 ° and a distance of 25 mm from the tool tip.

1. Introduction

The milling operation of hard-to-cut material such as titanium has been subject to numerous investigation on the various ways of controlling the high heat generated during the cutting operation. The low thermal conductivity of titanium that can range from 14W/m.K for titanium grade 1 to 6.7 W/m.K for grade 5 contribute in the building up of high heat in the cutting zone between the tool tip and the workpiece interface due to a poor dissipation of the heat [1]. In addition, this issue trigger a rapid tool wear, poor surface finish, high cutting forces and high power consumption among others. In the machining of titanium, coolant is also necessary to prevent any types of oxidation due to the environment [2]. Without a proper cooling system in the machining of titanium, the build-up of the cutting heat within the material contribute in its embrittlement couple with a reduction in the overall strength of the material.Next, another works also suggest that coolant can directly influence the surface quality [3]. Furthermore, the most common types of coolant that are widely use in the industry are synthetic, semi-synthetic and mineral oil based [4]. These types of coolant offer a good stability in terms of chemical reaction during the machining operation. On the other hand, the high content of toxic substance and quick contamination of this coolant are consider as a hazard to human health and the environment [5]. The best alternate type of coolant that is slowly being accepted by the manufacturing community is the vegetable oils (VO) base cutting fluids because of its high biodegradability. The molecular structure of the vegetable oil that is thick, tenacious and vigorous film layer is very useful in machining precisely for hard-to-cut material such as titanium because it can absorb the contact pressure at the tool-workpiece interface [6]. Furthermore, VO have two other attractive qualities namely a low thermal conductivity and coefficient of friction that are 0.172 W/m.K and 0.03 whereas mineral oil has a thermal conductivity of 0.125 W/m.K and a coefficient of friction of 0.07. The most common coolant supply system that is use in machining is flood cooling due to its low installation and maintenance cost, the flow rate in this system can range between 10 L/min to 225 L/min depending on the applied parameters [7]. It was demonstrated that supplying the cutting zone with plenty of fluid contribute in cooling down the heat generated and flush away the chips more efficiently [8]. However, the high fluid consumption added to a relatively low penetration to the cutting zone, especially at high-speed cutting, were considered as some major disadvantages in applying flood-cooling system. In addition, in the last decade, two other cooling system were introduce for machining purpose namely high pressure cooling (HPC) and cryogenic cooling [9]. HPC system can supply the cutting fluid to the cutting zone at pressure of up to 200 bars. This system was also compare against other system, and some researchers found out that it was very efficient in improving the tool life and the chips breaking [10]. Moreover, cryogenic cooling is another system where liquid nitrogen (LN2) is use in machining and it was reveal that it reduces tool wear and extend the tool life but the high investment cost of this system limit it to some selected application only [11].

The minimum quantity lubrication (MQL) is yet another effective cooling system in terms of sustainability, environment friendly and economical [12,13]. The effectiveness of a cooling system like MQL can also be enhance by positioning the nozzle at the most appropriate location in a machining setup [14]. In a study on the influence of nozzle positioning, two specific position namely at 45° and 135° were analysed. The two nozzles were position in relation to the feed direction during the machining of AL5083 using MQL system. The authors concluded that when the nozzle is position at 135°, a lower tool flank wear of 0.098 and 0.095 mm is achievable with a flow rate of 0.04 and 0.06 ml/min [15]. In a further study, two different orientations were evaluated, one was position at 12.5° and the other at 45°. It was reveal that at an inclination of 12.5°, 50 % extension in the tool life was achieve with average flank wear of 0.3 mm compared to that of 45° [16]. In another experiment, micro-milling of Ti-6Al-4V using two different cooling system namely MQL and jet-cooling was investigated. It was concluded that the tool wear was 1.6 % less with feed direction under MQL compared to 6.15 % with jet cooling [17]. The spraying distance for the coolant is also a very important factor to consider in milling operation. In a study related to MQL, two scenarios of the positioning and distances of the spraying nozzle was investigated. In the first situation, the spraying nozzle was position at 135 ° with a distance of 25 mm to the cutting zone and in the second situation, it was position at 90° and 45 mm. It was found that the first scenario was more effective as the cutting temperature decreases by 2.5 and 20°C compared to 90° nozzle angle and 45 mm spraying distance [18]. MQL is mostly considered as a lubricating rather than a cooling method and this properties is very critical when apply in the machining of hard-to-cut material such as Inconel and Titanium due to the high heat generated couple with a low thermal conductivity [19].

The application of computational fluid dynamics (CFD) simulation is not well-establish in machining operation due to many challenges. Among the various challenges that arise in the modelling process of the fluid are geometrical boundary conditions, resistance coefficients, driving forces and interactions between different processes [20]. Moreover, the proper selection of the turbulence models and the types of meshing are very important to produce a more realistic situation with good accuracy [21,22]. In machining operation, the situation is more complex due to the various variables such as coefficient of friction that differ for each material due to their hardness. In addition, the meshing of smaller dimensions around the boundary layers is also another critical step in the CFD. Two major types of CFD simulation are currently available namely 2D and 3D simulation. The 3D simulation is more complex due to the high amount of mesh elements but better results can be generated compare to a 2D. Vortices is another element that are formed when a rotation is involve and it is define as a flow with circular and rotary route, dissipating energy. It is also form in different size and react according to the model of the boundary conditions. In a three dimensional simulation, the developed turbulence due to the rotation can be observe. Many methods had been define to identify vortices and the most popular for CFD simulation are the q-criterion and the $\lambda 2$-method [23,24]. The work presented uses 3D CFD methods to investigate the velocity and flow direction of the coolant in a milling operation of titanium grade 2. The results presented could be used to improve the setup of the coolant nozzle especially during machining of hard-to-cut material such as titanium. In addition, a more efficient way of applying the coolant can contribute directly in lowering the energy consumption due to an excessive cutting forces from the machine tools.

2. Methodology of simulation.

The overall simulation was carried out base on a one-tooth milling cutter with a diameter of 25 mm. In addition, the cutting insert are coated and especially design for machining of hard-to-cut material such as titanium. The spindle speed that was applied in this current study was set at 1000 rpm couple with water as cutting fluid. Figure 1 shows the steps in developing the meshing of the milling cutter. At the initial stage a CAD model is develop base on the actual dimension of the milling cutter by using Solidworks software. Next, Altair Hyperworks software, precisely Acusolve was used to construct the meshing model using tetrahedron elements for the cutting tool, with a maximum meshed of 5 mm. The region of fine meshing are located at all high turbulence region so that a more accurate prediction for the flow field can be generated. The complex features with very small dimensions are all fitted with small mesh. The CFD domain was set in an area of 100 x 100 mm and the simulated workpiece was

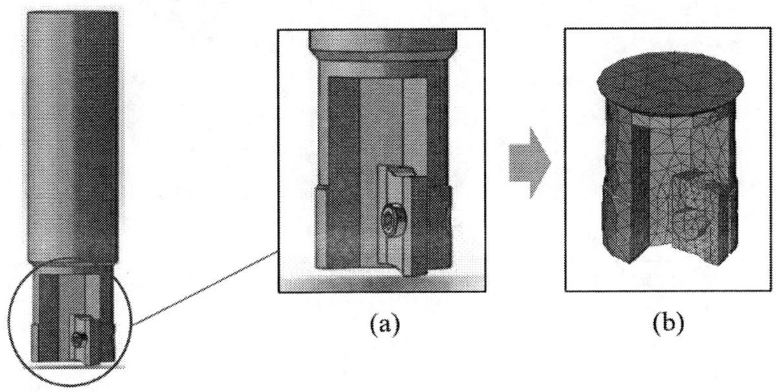

Figure 1*. Building up process of (a) CAD model and (b) meshing of model*

represented by a block with a dimension of 50 x 50 mm as shown in Figure 3 (b) under the cutting tool. In this current paper, the effect of multiple amount of nozzles on the fluid velocity was investigated. In the first scenario, one nozzle was position against the clockwise rotation of the milling cutter and in the second scenario, a second nozzle was position with the rotation of the milling cutter as shown in Figure 2. The nozzles are both incline at an angle of 12.5° with an impinging distance of 25 mm between the nozzle tip and the tool tip. Next, in both scenario, a constant flowrate of 3.1 L/min was applied through a nozzle diameter of 1.75 mm. This particular diameter was based on the contraction ratio (inlet to exit diameter ratio) to achieve high jet stream quality [25].

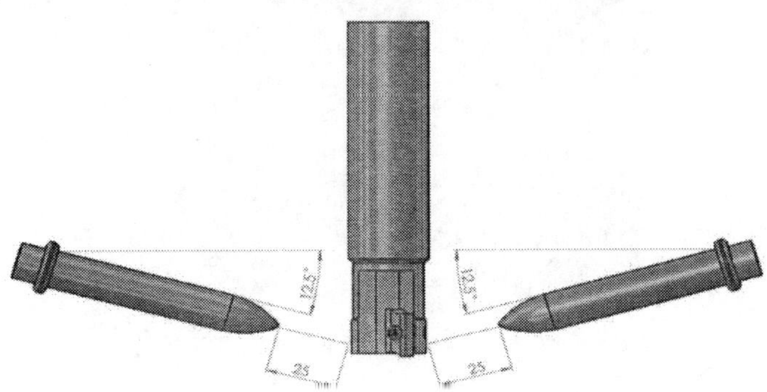

Figure 2*. Schematic view for the nozzle setup*

IOP Conf. Series: Materials Science and Engineering **429** (2018) 012069 doi:10.1088/1757-899X/429/1/012069

3. Results and discussion

The first scenario was the application of cutting fluid through nozzle "B" only as shown in Figure 3 (b). The simulated model generate the pathway created by the flow of the coolant around the milling cutter and the colors represent the variation in the velocity recorded when the cutter is rotating at a speed of 1000 rpm. The values of the velocity have a range of 0.031-11.592 m/s and the highest velocity was recorded along the edge of the cutter. It was also observe that an average velocity of 8.289 m/s was located in the section of the insert that suggest a proper penetration of the coolant towards the cutter edge. In addition, the velocity contours also indicate that the distribution of the coolant near the cutting zone is efficient as the velocity contour is not scattered but instead it creates a normal vortex that is triggered by the rotation of the milling cutter.

Fluid : Water Tool diameter: 25 mm (one insert)	Number of nozzle: 1 Angle of inclination : 12.5° Impinging distance : 25 mm

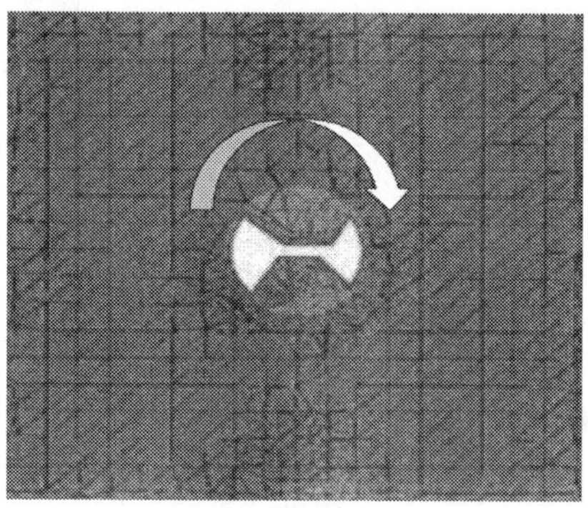

(a)

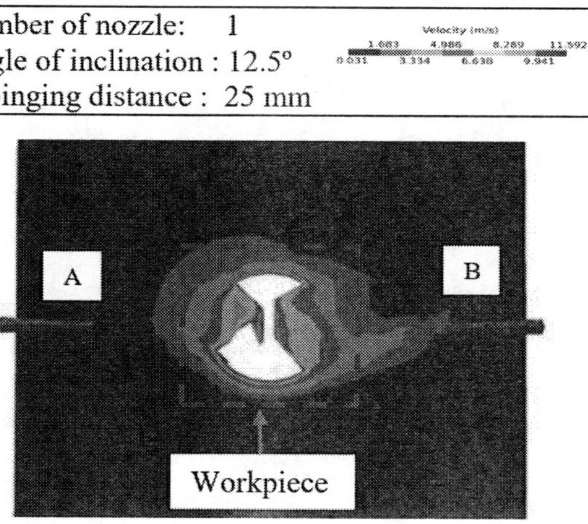

(b)

Figure 3. *The simulated output (a) meshing of boundary area (b) velocity contour*

Furthermore, for the second scenario, the water jet was applied through both nozzles A and B simultaneously. The flow rate was maintained at 3.1 L/min. Then the simulated results were analysed. The generated velocity contour for the second scenario suggest that the application of two nozzles offer a wider cooling area around the cutting zone. More importantly, the inner section of the insert experience a higher velocity as shown in Figure 4. Biermann et al. state that a higher velocity of the coolant at the cutting edge prevent high thermal load on the cutting edges which most of the time results in higher tool wear [24].

Fluid : Water Tool diameter: 25 mm (one insert)	Number of nozzle: 2 Angle of inclination : 12.5° Impinging distance: 25 mm	Number of nodes: 6316 Number of elements: 30971 Number of surfaces: 5570

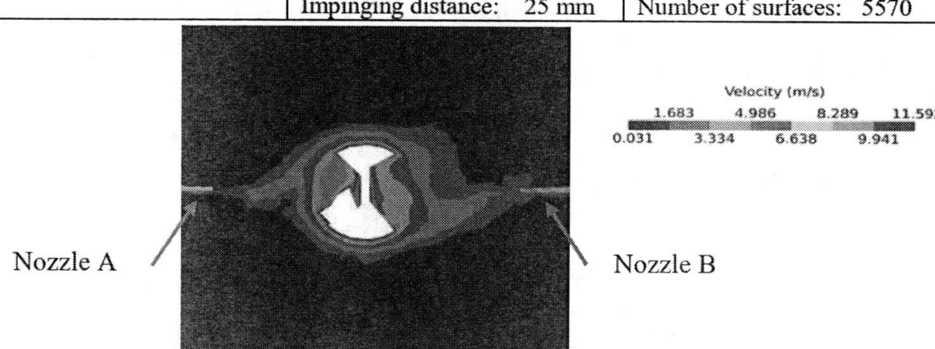

Figure 4. *Velocity contour for two nozzles*

Furthermore, the application of the two nozzles setup provides a more efficient supply of coolant to the cutting zone that is vital in machining especially when dealing with hard-to-cut material such as Inconel and Titanium among others that generate high intensity of heat due to their unique properties such as low thermal conductivity and low elastic modulus. Figure 5 shows the velocity field developed when the water jet is apply through both nozzles towards the rotating cutting tool that was set at a spindle speed of 1000 rpm. A high velocity of water jet in the range of 9.287 to 11.592 m/s were recorded around the edge of the cutting tool. In addition, a higher velocity in the range of 6.981 to 9.287 m/s was also noted nearby the cutting insert that is much higher than in the first scenario with the one nozzle setup. The high velocity of the coolant at the cutting edge allow it to be more wetted.

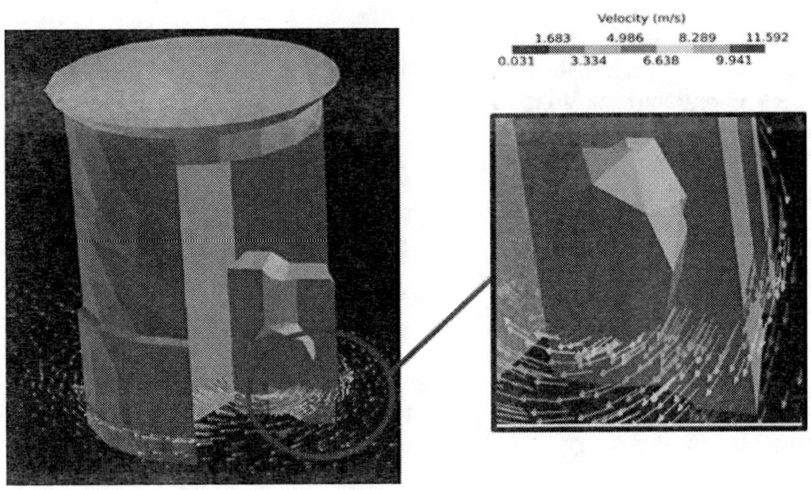

Figure 5*. Velocity contour at the insert cutting edge*

4. Conclusion

In this paper, CFD simulation was used to evaluate the efficiency of cooling a cutter with single and two nozzles setup together with a specific angle and distance. It was possible to define the path line of the coolant from the nozzle tip towards the cutting tool and its surrounding. It was also noted that in both cases whether with one or two nozzles, a zone of turbulence was created at the edge of the cutter and a high velocity was recorded. The turbulence also show that the stream that are closer to the cutting tool tend to have a high velocity compare to those away from the tool. The parameters for both setup was quite efficient in supplying the coolant at the cutter because the cutting fluid was not totally thrown away from the cutting zone. Next, when using one nozzle only, the coolant tend to scatter outside the cutting zone whereas when both nozzles were apply, the cooling zone remain more compact around the cutting zone that allow the milling cutter temperature to remain at a reasonable intensity and also contribute to extend the tool life and delay the tool wear. According to the stream flow, it also suggest that the region of high cutting temperature that is at the cutting edge of the insert is well wetted when two nozzle setup are used compare to a single one. Additionally the results will be examined by temperature and cutting test with the insert.

5. References

[1] Ezugwu EO, Da Silva RB, Sales WF, Machado AR. Overview of the machining of titanium alloys. In reference module in earth systems and environmental sciences: Elsevier: Amsterdam, The Netherlands, 2017.

[2] Camara MA, Rubio JC, Abrão AM, Davim JP. State of the art on micromilling of materials, a review. Journal of Materials Science & Technology. 2012 Aug 1;28 (8):673-85.

[3] Catherine LDK, Ma'arof RAR, Suresh S. A Study on the Impact of the Milling Parameters on the Surface Roughness when Using Polyurethane Board as a Base Material in Manufacturing Automotive Checking Fixtures. InMaterials Science Forum 2015 Jun 1 (Vol. 819, p. 449). Trans Tech Publications Ltd.

[4] Samuel J, Rafiee J, Dhiman P, Yu ZZ, Koratkar N. Graphene colloidal suspensions as high performance semi-synthetic metal-working fluids. The Journal of Physical Chemistry C. 2011 Feb 10; 115 (8):3410-5.

[5] Srikant RR, Ramana VS. Performance evaluation of vegetable emulsifier based green cutting fluid in turning of American Iron and Steel Institute (AISI) 1040 steel–an initiative towards sustainable manufacturing. Journal of Cleaner Production. 2015 Dec 1; 108: 104- 9.

[6] Lawal SA, Choudhury IA, Sadiq IO, Oyewole A. Vegetable-oil based metalworking fluids research developments for machining processes: survey, applications and challenges. Manufacturing Review. 2014; 1:22.

[7] Kalpakjian S, Schmid SR. Manufacturing engineering and technology. Upper Saddle River, NJ, USA: Pearson; 2014.

[8] Pathan A, Kadam M. Experimental study on effect of cutting parameters on chip-tool interface temperature and chip formation in turning en-31 hardened steel under flooded and mql conditions. Int. J. Innov. Sci. Eng. Technol. 2015; 2: 724-9.

[9] Vagnorius Z, Sørby K. Effect of high-pressure cooling on life of SiAlON tools in machining of Inconel 718. The International Journal of Advanced Manufacturing Technology. 2011 Apr 1; 54(1-4):83-92.

[10] Busch K, Hochmuth C, Pause B, Stoll A, Wertheim R. Investigation of cooling and lubrication strategies for machining high-temperature alloys. Procedia CIRP. 2016 Jan 1; 41: 835-40.

[11] Benedicto E, Carou D, Rubio EM. Technical, economic and environmental review of the lubrication/cooling systems used in machining processes. Procedia Engineering. 2017 Jan 1; 184: 99-116.

[12] Marksberry PW, Jawahir IS. A comprehensive tool-wear/tool-life performance model in the evaluation of NDM (near dry machining) for sustainable manufacturing. International Journal of Machine Tools and Manufacture. 2008 Jun 1; 48(7-8):878-86.

[13] Shen CH. The importance of diamond coated tools for agile manufacturing and dry machining. Surface and Coatings Technology. 1996 Dec 15; 86: 672-7.

[14] Iturbe A, Hormaetxe E, Garay A, Arrazola PJ. Surface integrity analysis when machining inconel 718 with conventional and cryogenic cooling. Procedia CIRP. 2016 Jan 1;45:67-70.

[15] De Lacalle LL, Angulo C, Lamikiz A, Sanchez JA. Experimental and numerical investigation of the effect of spray cutting fluids in high speed milling. Journal of Materials Processing Technology. 2006 Feb 20; 172(1):11-5.

[16] Mulyadi I. Improving the performance of minimum quantity lubrication in high-speed milling and environmental performance analysis.2013.

[17] Vazquez E, Gomar J, Ciurana J, Rodríguez CA. Analyzing effects of cooling and lubrication conditions in micromilling of Ti6Al4V. Journal of Cleaner Production. 2015 Jan 15;87:906-13.

[18] Liu ZQ, Cai XJ, Chen M, An QL. Investigation of cutting force and temperature of end-milling Ti–6Al–4V with different minimum quantity lubrication (MQL) parameters. Proceedings of the Institution of Mechanical Engineers, Part B: Journal of Engineering Manufacture. 2011 Aug; 225(8):1273-9.

[19] Shokrani A, Dhokia V, Newman ST. Environmentally conscious machining of difficult-to-machine materials with regard to cutting fluids. International Journal of Machine Tools and Manufacture. 2012 Jun 1; 57:83-101.

[20] Oezkaya E, Beer N, Biermann D. Experimental studies and CFD simulation of the internal cooling conditions when drilling Inconel 718. International Journal of Machine Tools and Manufacture. 2016 Sep 1; 108:52-65.

[21] Catherine LDK, Darulihsan Bin Abdul Hamid, Koshy GS, Zakaria A. Review on finite element modeling for prediction of cutting forces and temperature pattern behaviour during the milling machining of commercially pure Titanium grade 2. (In progress to be published in International journal of nanoelectronics and materials)

[22] Jeong J, Hussain F. On the identification of a vortex. Journal of fluid mechanics. 1995 Feb;285: 69-94.

[23] Hunt JC. Vorticity and vortex dynamics in complex turbulent flows. In Canadian Society for Mechanical Engineering, Transactions (ISSN 0315-8977), vol. 11, no. 1, 1987, p. 21-35. 1987 (Vol. 11, pp. 21-35).

[24] Morgan MN, Baines-Jones V. On the coherent length of fluid nozzles in grinding. In Key Engineering Materials 2009 (Vol. 404, pp. 61-67). Trans Tech Publications.

[25] Biermann D, Oezkaya E. CFD simulation for internal coolant channel design of tapping tools to reduce tool wear. CIRP Annals. 2017 Jan 1; 66(1):109-12.

Acknowledgments
The authors would like to thank Unikl NuSim lab and their team members, for supporting this research.

Cooling capacity of 1 HP room air conditioner – comparison between indoor room testing condition of ISO 5151 standard requirement and average Malaysian preferable condition

M F H Rani[1,2], Z M Razlan[1,2], Shahriman A B[1,2], I Zunaidi[3], W K Wan[2], M S M Hashim[1,2], I Ibrahim[1,2], N S Kamarrudin[1,2], R Ayob[2] and N N A Noramzil[2]

[1] Motorsports Technology Research Unit (MoTECH), Universiti Malaysia Perlis, Pauh Putra Campus, 02600 Arau, Perlis, Malaysia.
[2] School of Mechatronic Engineering, Universiti Malaysia Perlis, Pauh Putra Campus, 02600 Arau, Perlis, Malaysia.
[3] Technopreneur at UniMAP Sdn Bhd, No 26, Blok D, Bangunan PPPIT, Pengkalan Jaya, Jalan Kangar – Alor Setar, 01000 Kangar, Perlis, Malaysia.

Corresponding author: zuradzman@unimap.edu.my

Abstract. Energy efficiency rating and labelling program encourages the manufacturers in ranking the performance of air conditioners by specifying the energy consumption and energy saving for all the tested models. The star rating benefits the consumers to buy air conditioners that work more efficient and save more money after a long period of time. The international standard ISO 5151 regulates the testing and rating for performance of non-ducted air conditioners and heat pumps. However, Malaysia experiences hotter climate since it is near to the equator. Thus, Malaysians have other preferable condition, which is different with the comfort zone specified in the standard test condition. An Annual Performance Factor (APF) test of a room air conditioner was conducted to compare the average Malaysian preferable condition and the standard test condition. Results showed that the indoor dry bulb/wet bulb temperature, 27 °C/19 °C of ISO 5151 standard requirement shall be applied for all room air conditioners in Malaysia. Although Malaysians favour the comfort zone of indoor dry bulb/wet bulb temperature, 25 °C/18 °C, the cooling capacity, the energy efficiency ratio and the coefficient of performance at standard requirement is higher, which are more than 2.559 %, 2.051 % and 2.049 % respectively after comparison. The results confirm that the room air conditioners at Malaysia should comply with the standard test condition based on ISO 5151 for better energy and cost efficient.

Keywords. Room air conditioner, standard test condition, dry bulb temperature, wet bulb temperature, cooling capacity, average Malaysian preferable condition and energy efficiency ratio.

1. Introduction

The growing standards of living and urbanization in Malaysia has increased the usage of room air conditioners (RACs) in the residences, stores and office buildings [1, 2, 3]. The split-type air conditioners are the common choice by Malaysian due to their aesthetic appeal and smaller setup space requirement [4]. This type consist of two parts, which are the outdoor unit and the indoor unit. By fitting outside the house or building, the outdoor unit comprises components like the compressor, the condenser and the expansion valve. While the indoor unit includes the evaporator and the cooling fan [5]. The economic growth is not only drives the usage demand of air conditioners (ACs), but also increases the electrical energy consumption in Malaysia [6]. Since 2005 [7], Energy Commission of Malaysia has enforced an energy efficiency rating and labelling program that offers good benefits for the consumers, governments as well as the environment. As one of the electrical appliances mentioned in the program, all RACs at Malaysia should comply with the Malaysia Standard (MS) ISO 5151 before they can be sold to the customers. The standard test condition specifies that the human comfort zone to be 27 °C/19 °C, dry bulb/wet bulb temperature [8]. However, Malaysia is located near to the equator and experiences hot and humid climates throughout the year. Based on MS ISO 1525, for the thermal comfort zone at Malaysia, the dry bulb temperature is recommended between 23 °C to 26 °C with relative humidity of 55 % to 70 % [9]. Hence, an Annual Performance Factor (APF) test of a room air conditioner should be performed to compare the performance between the standard test condition and the preferable condition at Malaysia. The average Malaysian (MY) preferable condition will be tested and compared with the standard test condition based on the cooling capacity, the energy efficiency ratio (EER) and the coefficient of performance (COP). Table 1 shows the EER's classification of air conditioner based on the energy star rating at Malaysia.

Table 1. Classification of EER based on the energy star rating of air conditioner at Malaysia [10].

Energy star rating	Cooling capacity < 4.5 kW	4.5 kW < Cooling capacity < 7.1 kW
5	> 11.94	> 10.71
4	11.16 – 11.93	9.83 – 10.70
3	10.37 – 11.15	8.94 – 9.82
2	9.56 – 10.36	8.03 – 8.93
1	9.00 – 9.55	7.50 – 8.02

$$Q_{cool}(kJ/s) = \left\{ \frac{\left(h_{sucked\ air} - h_{discharged\ air}\right)(kJ/kg)}{v_{discharged\ air}(m^3/kg)} \right\} \times \dot{m}_{air\ at\ nozzle}(m^3/s) \quad (1)$$

$$EER\ (Btu/hW) = \frac{Q_{cool}(Btu/h)}{P_{input\ power}(W)} \quad (2)$$

$$COP = \frac{P_{output\ power}(W)}{P_{input\ power}(W)} \quad (3)$$

2. Annual Performance Factor (APF) test methodology

An Annual Performance Factor (APF) test facility was used to conduct the cooling capacity and COP test by setting up an air-enthalpy method apparatus as shown in Figure 1. The facility consists of two rooms – an outdoor-unit room and indoor unit-room, which the temperature and humidity can be controlled independently. A 1 horsepower (HP) room air conditioner with R22 refrigerant was used throughout the standard cooling mode testing. Results from both of the test conditions will be computed to analyse the energy efficiency ratio of room air conditioner.

2.1. Experimental conditions

Figure 1 illustrates the APF test facility, while Figure 2 illustrates the flowchart of APF test for a room air conditioner. In this study, two different indoor dry bulb/wet bulb temperatures were tested based on the standard test condition and the comfort zone at Malaysia. For the standard test condition, the dry bulb/wet bulb temperatures of air entering indoor side were determined based on the ISO 5151. While the average Malaysian preferable condition is considered for the comfort zone at Malaysia. During the standard cooling mode testing, a 100 % load was set based on the rated capacity of the RAC, which was 2.81 kW. The APF test was experimented for three times for each condition to obtain reliable and accurate results. The experimental conditions are summarized in Table 2.

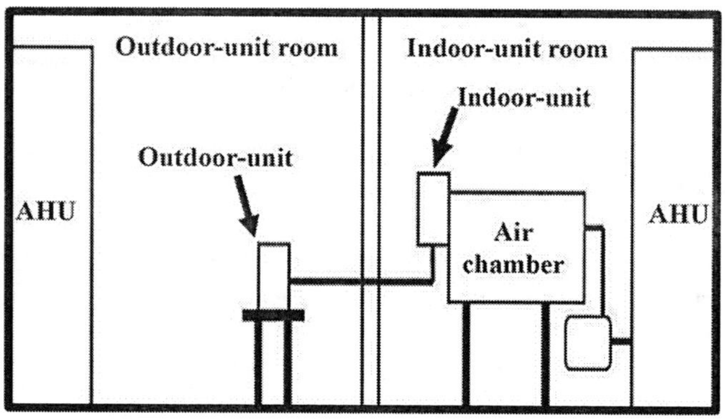

Figure 1. APF test facility as illustrated by [11].

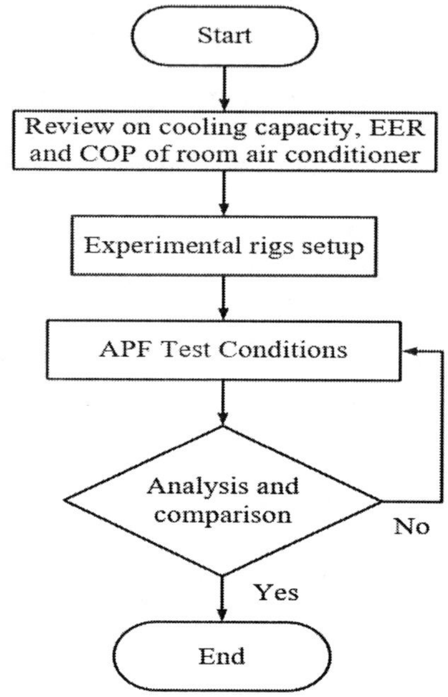

Figure 2. Flowchart of APF test for a room air conditioner.

Table 2. Experimental conditions.

Setting parameters in APF test facility	Test conditions	
	Standard test condition	Average MY preferable condition
Indoor dry bulb temperature, °C	27	25
Indoor wet bulb temperature, °C	19	18
Outdoor dry bulb temperature, °C	35	35
Outdoor wet bulb temperature, °C	24	24
Power source frequency, Hz	50	50
Source voltage, V	220.5	220.5

3. Result and discussion

Several parameters were collected to compare the performance of room air conditioner at standard test condition and at average Malaysian preferable condition. Equations (1), (2) and (3) were used to analyse the results as summarized in Table 3. The test conditions were compared based on the cooling capacity, the energy efficiency ratio (EER) and the coefficient of performance (COP). Based on Figure 3 (a), the standard test condition provided higher cooling capacity, which is good for heat exchange although the power input is higher in comparison with the average Malaysian (MY) preferable condition. The same trends were obtained when comparing the test conditions based on the EER, Figure 3 (b) and the COP, Figure 3 (c), which the standard test condition overshadowed the average MY preferable condition.

Table 3. Results of indoor standard test (IS0 5151) condition and indoor average MY preferable condition.

Analysed parameters during the APF test of room air conditioner	Indoor dry bulb/wet bulb temperature	
	27°C/19°C	25°C/18°C
Enthalpy of sucked air, kJ/kg	53.771	50.665
Enthalpy of discharged air, kJ/kg	40.156	37.317
Specific volume of air at nozzle, m^3/kg	0.839	0.835
Air flow rate, m^3/min	10.223	10.113
Cooling capacity, J/s	2766.111	2695.327
Cooling capacity, BTU/h	9438.362	9196.840
Power input, W	868.5	864.0
Energy efficiency ratio (EER), BTU/hW	10.867	10.644
Power output, W	2798.133	2726.600
Coefficient of performance (COP)	3.222	3.156

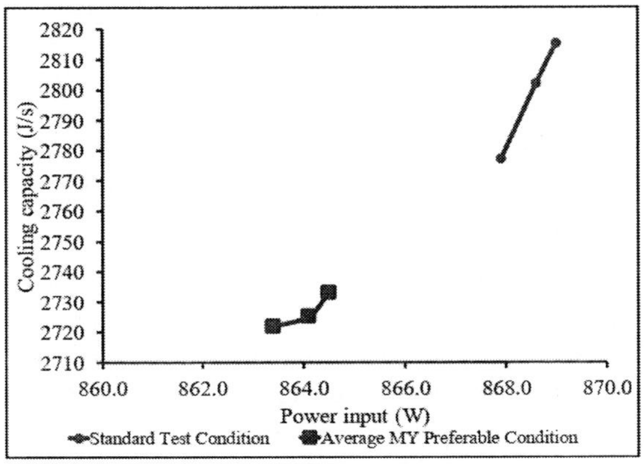

(a)

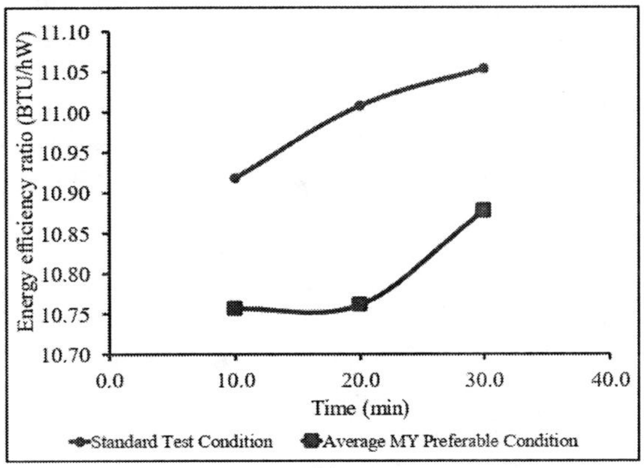

(b)

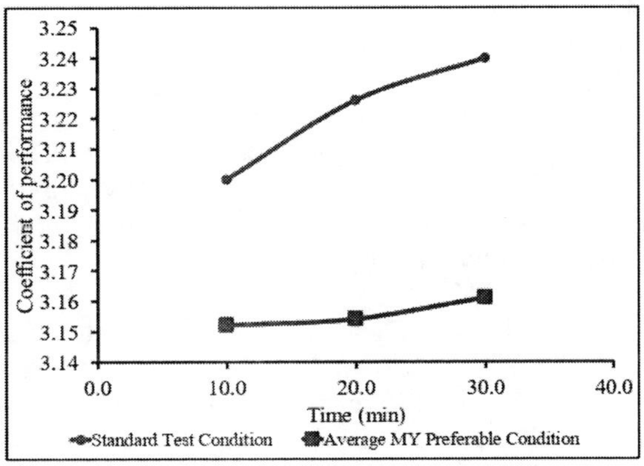

(c)

Figure 3. Comparison of test conditions based on (a) the cooling capacity, (b) the EER and (c) the COP.

As shown in Table 3 and Figure 3, the cooling capacity, the energy efficiency ratio (EER) and the coefficient of performance (COP) of the standard test condition showed higher values than the average Malaysian preferable condition. Based on the energy star rating in Table 1, a room conditioner with a higher EER will have a lesser energy consumption and promote better energy saving. The average Malaysian preferable condition showed poorer value of EER, indicating the room air conditioner will consume more electricity. Thus, this condition is not suitable to be applied in the testing and rating for performance of room air conditioner.

4. Conclusion

The introduction of energy efficiency standards mitigate the future demand of energy and reduce the environmental degradation. This study is helpful in validating the cooling capacity, the EER and the COP of a room air conditioner for human comfort zone in Malaysia. The indoor room test condition of ISO 5151 standard requirement was compared with the average Malaysian preferable indoor room condition. The standard test condition indoor dry/wet bulb temperature, 27 °C/19 °C is suitable to be used in Malaysia for testing and rating performance since the standard test condition results in higher EER value rather than referring to the indoor dry/wet bulb temperature, 25 °C/18 °C. The results conclude that the room air conditioners at Malaysia shall comply with the ISO 5151 standard requirement, which will help the consumers to save more money in the long run.

Acknowledgments

The authors acknowledge School of Mechatronic Engineering, Universiti Malaysia Perlis and Hitachi Air Conditioning Products (M) Sdn. Bhd. for the lab facilities. Special thanks to those who contributed to this project directly or indirectly.

References

[1] C. Siwar, F. Ahmed, A. Bashawir, and M. S. Mia, "Urbanization and Urban Poverty in Malaysia: Consequences and Vulnerability," *J. Appl. Sci.*, vol. 16, no. 4, pp. 154–160, Apr. 2016.
[2] D. R. Kubota, Tetsu; Jeong, Sangwoo; Toe, Doris Hooi Chyee; Ossen, "Energy Consumption and Air-Conditioning Usage in Residential Buildings of Malaysia," *J. Int. Dev. Coop.*, vol. 17, no. 3, pp. 61–69, 2011.
[3] Information on https://www.nst.com.my/business/2018/02/335371/malaysia-post-59pct-gdp-growth-2017
[4] Information on https://www.recommend.my/blog/6-air-conditioner-tips/?hlid=3fhYhK
[5] ASHRAE, *ASHRAE Handbook - HVAC Applications*, SI Unit. Atlanta, GA, 2015.
[6] C. Chong, W. Ni, L. Ma, P. Liu, and Z. Li, "The use of energy in Malaysia: Tracing energy flows from primary source to end use," *Energies*, vol. 8, no. 4, pp. 2828–2866, 2015.
[7] Information on http://www.st.gov.my/index.php
[8] DSM, "MS ISO 5151:2004," 2004.
[9] Y. H. Yau, L. C. Ding, and B. T. Chew, "Thermal comfort and Indoor Air Quality at Green Building in Malaysia," *Int. J. Low Carbon Technol.*, vol. 1, no. 2, pp. 167–176, 2006.
[10] K. Mardappan, "Minimum Energy Perfomance Standards In Malaysia," 2016.
[11] H. Horie and E. Hihara, "Study on Annual Performance of Room Air Conditioners under Partial Load Condition," no. 1982, pp. 1–10, 2012.

A Study on Possibility of CFD Simulation on Air Simulation in Minor Operation Theatre

M N Rahman Y*, Z M Razlan, M Izham, M I Omar, N A A Zambri, Shahriman A B, I Zunaidi and W K Wan

School of Mechatronic Engineering, Universiti Malaysia Perlis, Pauh Putra Campus, 02600 Arau, Perlis, Malaysia.

*Corresponding author: nurrahman@unimap.edu.my

Abstract. Inadequate air flow distribution in room is undesired as it will cause discomfort to the occupant. Fair air flow distribution is important as it will affect the performance of the occupant as well. The objective of this research work is to develop a realistic Computational Fluid Dynamics (CFD) model of Minor Operation Theatre for investigating the air flow pattern in Minor Operation Theatre. One unit of Air Conditioner in Minor Operation Theatre has been selected for this study. The distribution of air flow pattern is analysed. The real measurement is conducted using a digital air flow anemometer with USB interface and the channels digital handheld temperature thermometer meter data loggers for validation purpose. CFD simulation is conducted by applying ANSYS Fluent software. The result of measurement and simulation are compared and validated. Based on the simulation and experiment findings, it proofs that ANSYS Fluent can be used for air simulation and relocation recommendation is proposed with the purpose for better distribution of air flow in Minor Operation Theatre.

1. Introduction

A research work for better Minor Operation Theatre (OT) concurrently is running as development for better quality of air. The surgeons and patients are expecting comfortable environmental of operation room as they spend most of their time in that room [1][2][3][4]. This research work execution based on a need to develop a realistic CFD model of Minor Operation Theatre room and pursue by investigating the pattern air flow distribution in the Minor Operation Theatre of UniMAP [2][5]. The pattern of air flow in the Minor Operation Theatre in UniMAP was assumed is not uniform for current location of AC unit in the Minor Operation Theatre. It was also recognized that the location of AC unit in the Minor Operation Theatre will affect the distribution of air flow in the Minor Operation Theatre and then influences the change of thermal sensation among occupants. Therefore, an identification on the locations where air distribution is inadequate needs a Computational Fluid Dynamics (CFD) method as a tool to investigate the air flow pattern and distribution as well. This has been supported by previous research works [6][7][8][9].

2. Methodology

2.1. Minor Operation Theatre Data Collection

One of Minor Operation Theatre at Universiti Malaysia Perlis Health Centre (Figure 1) with Spilt Unit of Air Conditoning has been selected for this study. The details of the room are given in Table 1 as below:

Table 1. The details of the room.

Items	Minor Operation Theatre
Dimension (m)	2.89 (H) x 3.21(W) x 3.4(L)
Air-conditioning Configuration	Single Unit

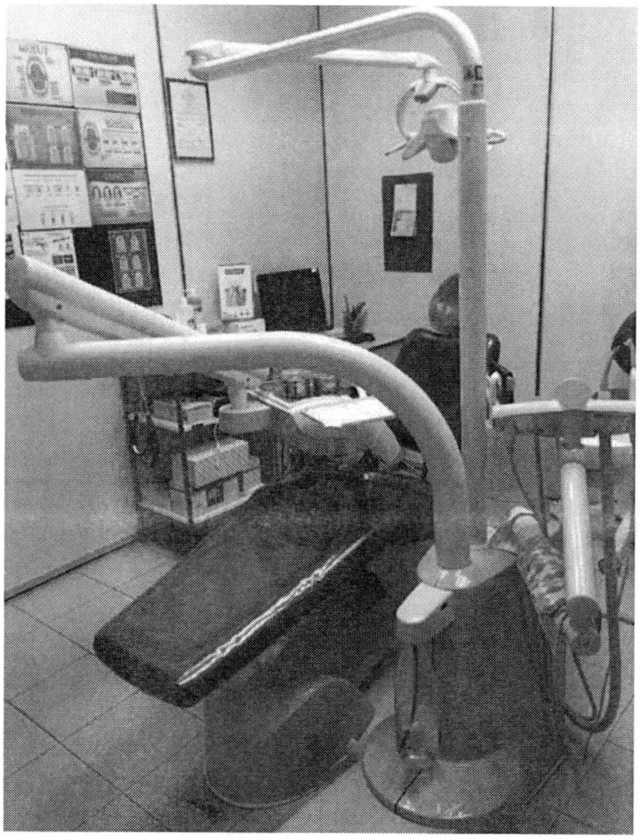

Figure 1. Minor Operation Theatre

2.2. Collection Data of Air Characters in Minor Operation Theatre

The velocity and temperature of air were chosen as the air characters that need to be measured. The digital air flow anemometer with USB interface was applied for air flow velocity data collection and that device were directly located at AC unit nozzle part. Meanwhile, the channels digital handheld temperature thermometer meter data loggers were applied for the air temperature data collection. The Minor Operation Theatre was divided into 9 zones vertically and each zone was divided into 4 points horizontally as shown Figure 2. The height of each sensors was adjusted to 0.5 m above from the floor level.

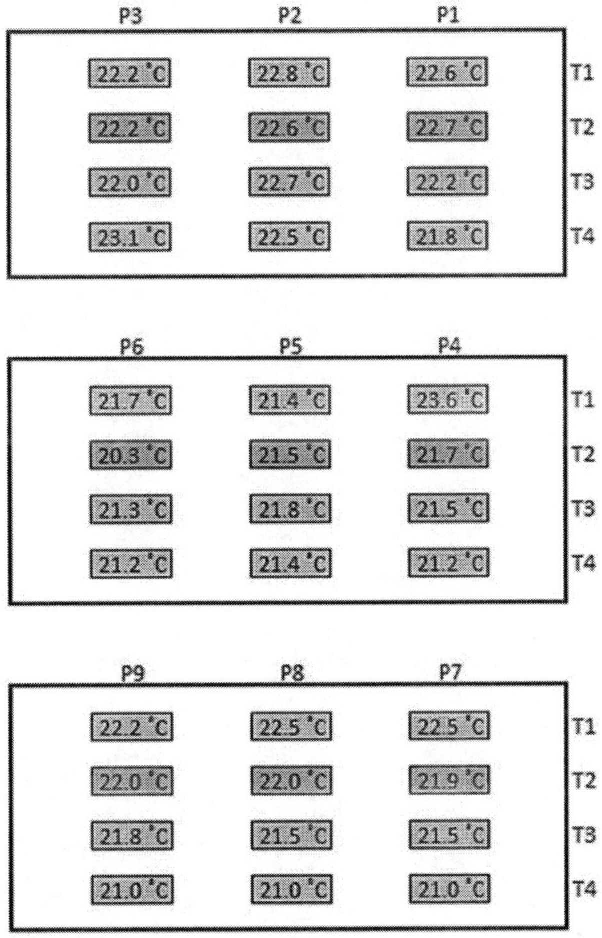

Figure 2. Cross section of the room with temperature distribution

2.3. Computational Fluid Dynamics Simulation Methodology

The commercial software named ANSYS Fluent 17.1 was utilized for simulation work in this study. In this study, the Minor Operation Theatre room (Figure 3) modeling must go through Pre-Processing, FLUENT-Solver and finally FLUENT Post-Processing for Minor Operation Theatre as shown in figure 4. The assumptions must be done while doing this simulation work due to commercial software limit and to reduce calculation time. The assumptions for this study are given as follows:

- The AC unit in Minor Operation Theatre rooms is well maintained, fully functioning and running well.
- The measurements and simulation are done in a completely sealed OT space, with all the doors and windows closed.
- The outdoor temperature on the surface of the room is constant
- Internal heat source emitted from the desktop, digital devices and lamps will be neglected due to minimal effect on the temperature.
- The effects of physical condition of the room (type of floor construction, difference in location of glass windows and doors) are neglected in this study.
- A chair, table, and bed are included in this simulation.

IOP Conf. Series: Materials Science and Engineering **429** (2018) 012071 doi:10.1088/1757-899X/429/1/012071

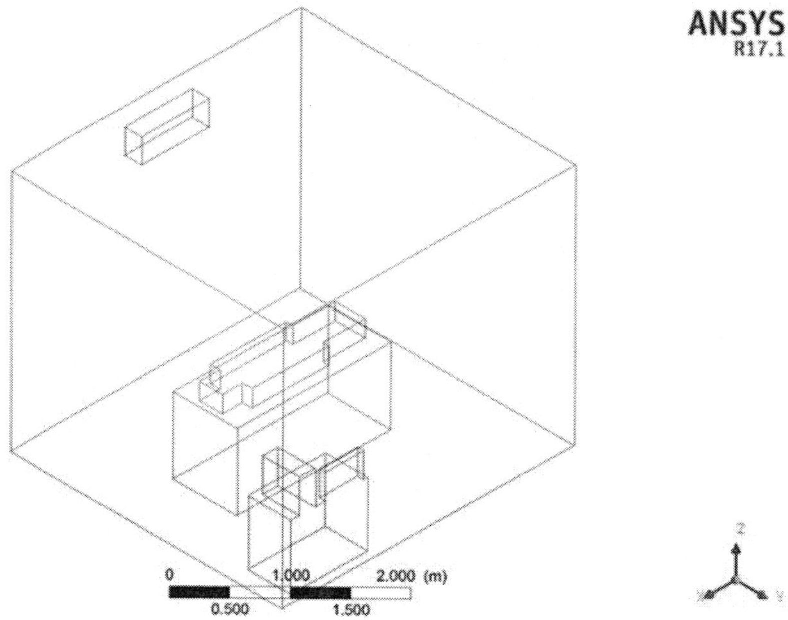

Figure 3. Geometry Model of Minor Operation Theatre.

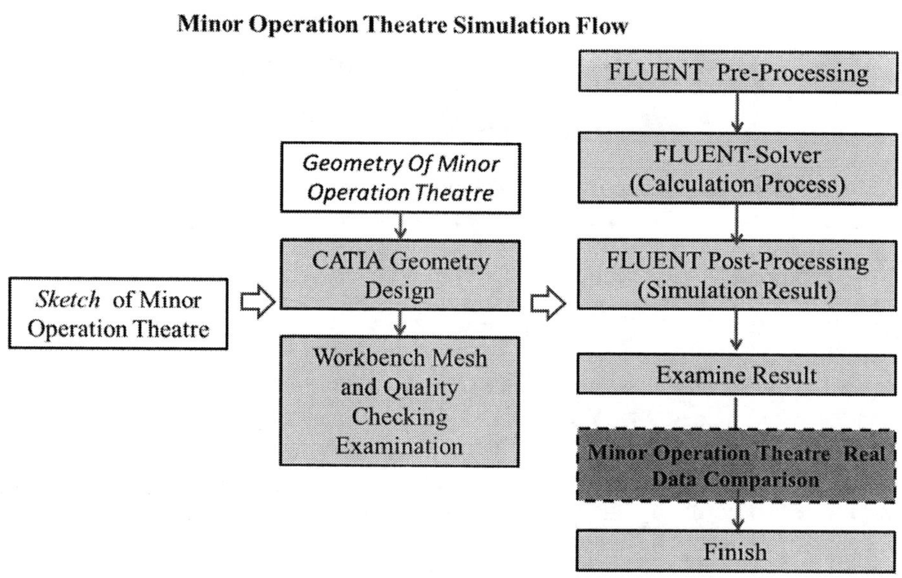

2

Figure 4. Flow chart of Minor Operation Theatre Simulation Work.

The simulation details are summarized in table 2.

Table 2. Summary of simulation work details.

etails	Minor Operation Theatre	
General	Gravity Force	On
	Energy	On
Models	Viscous model	K-Epsilon (2 Equation)
	k-Epsilon Model	RNG
	Near-Wall Treatment	Standard Wall Functions
Boundary Conditions	AC Nozzle Velocity	2.8 m/s
Mesh Size	Number of Nodes	96469
	Number of Cells	112933
	Gradient	Least Squares Cell Based
	Pressure	Second Order
Solution Method	Momentum	Second Order Upwind
	Turbulent Kinetic Energy	Second Order Upwind
	Turbulent Dissipation Rate	Second Order Upwind
	Energy	Second Order Upwind

3. Result and Discussions

Based on the real measurement result as shown in table 3, the average airflow temperature for Minor Operation Theatre was 21.0 °C and the average air velocity for Minor Operation Theatre was 2.52 m/s. Meanwhile the average airflow temperature and the average air velocity for CFD Simulation of Minor Operation Theatre are 17.0 °C and 2.62 m/s respectively. The CFD approach was successfully performed to obtain the velocity and temperature of air flow in Minor Operation Theatre.

The simulation results were related to the standard methods which are redefining the geometry of the Minor Operation Theatre room, optimization in mesh density, boundary conditions setting, solver setting, right solution methods and acceptable results were obtained from CFD post.

Figure 5 and 6 shows the CFD stream plot of the air velocity inside Minor Operation Theatre and the slices of temperature contour of Minor Operation Theatre. Based on the both figures (Figure 5 and 6), single unit AC shows inadequate distribution in air flow pattern in Minor Operation Theatre cause inadequate distribution in cold air.

Table 3. Summary of results of Minor Operation Theatre

Details	Real Measurement of Minor Operation Theatre	CFD Simulation of Minor Operation Theatre
Average Air Temperature (°C)	21.0	17.0
Average Air Velocity (m/s)	2.52	2.62

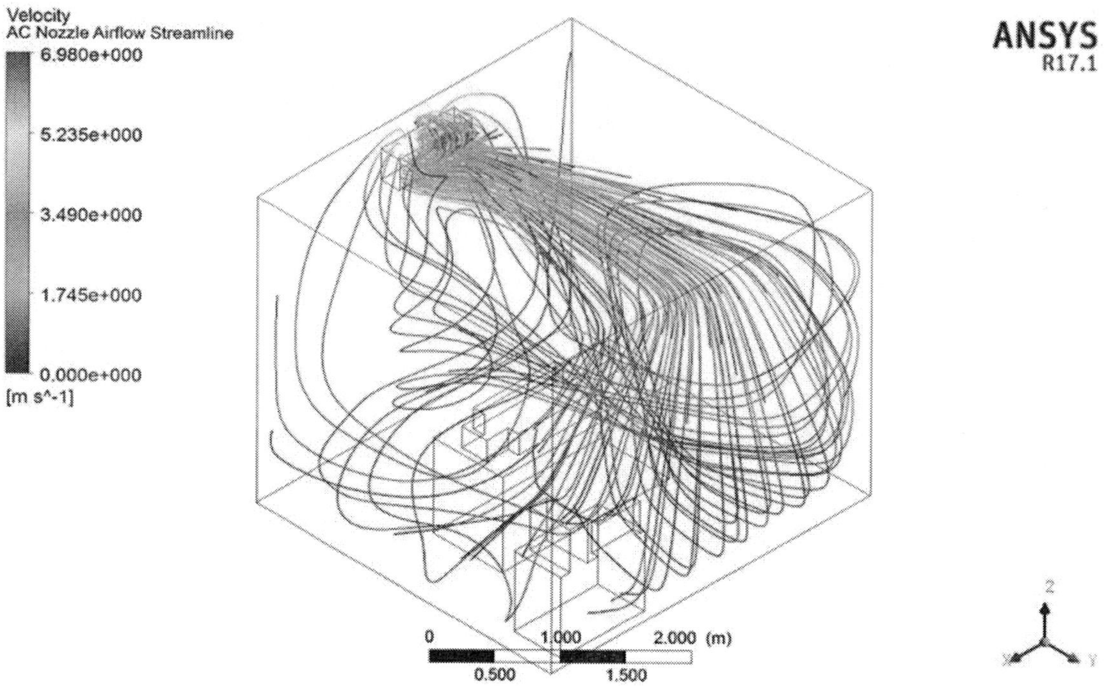

Figure 5. AC Nozzle Airflow Streamline for Minor Operation Theatre.

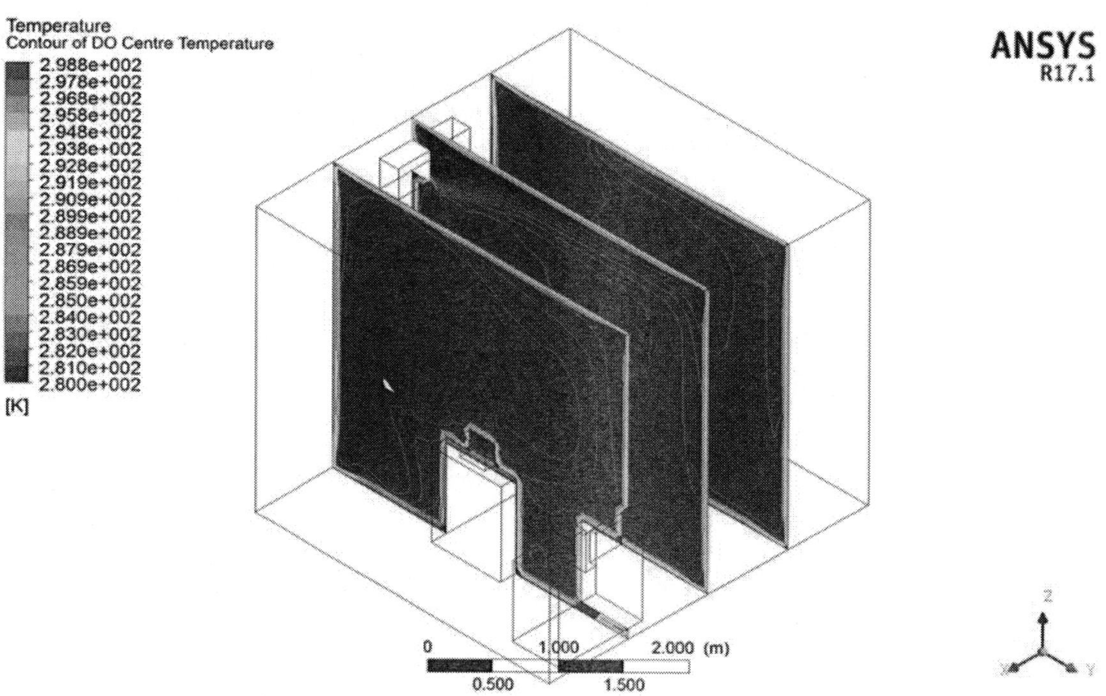

Figure 6. Slices of Temperature Contour of Minor Operation Theatre based on location in figure 2.

4. Conclusion

The CFD simulation result for Minor Operation Theatre demonstrated differences in real measurement about 19% for air flow velocity and for the 0.04% temperature. The errors that occurred due to the uncontrolled mesh density and wall of Minor Operation Theatre in actual Minor Operation Theatre which hard to be quantified for CFD simulation. Based on the observation, better location of AC unit must be proposed for better distribution of air flow in Minor Operation Theatre and it reveals that ANSYS Fluent can be utilized for air simulation in Minor Operation Theatre.

References

[1] R. Khatri, A. P. Singh, and V. R. Khare, "Identification of Ideal Air Temperature Distribution Using Different Location for Air Conditioner in a Room Integrated with EATHE - A CFD Based Approach," *Energy Procedia*, vol. 109, no. November 2016, pp. 11–17, 2017.

[2] A. M. Paul Danca, Florin Bode, Ilinca Nastase, "On the possibility of CFD modeling of the indoor environment in a vehicle," in *Sustainable Solutions for Energy and Environment, EENVIRO 2016, 26-28 October 2016, Bucharest, Romania*, 2017, pp. 656–663.

[3] S. Sadrizadeh, *Design of Hospital Operating Room Ventilation using Computational Fluid Dynamics (PhD thesis)*, no. February. 2016.

[4] S.A. Nada, H.M. El-Batsh, H.F. Elattar,N.M. Ali, "CFD investigation of airflow pattern, temperature distribution and thermal comfort of UFAD system for theater buildings applications," *J. Builiding Eng.*, vol. 6, pp. 274–300, 2016.

[5] T. C. Stevenson, "Experimental investigation of hospital operating room air distribution (MSc Thesis)," no. May, p. 97, 2008.

[6] S. Lin, B. T. Tee, and C. F. Tan, "Indoor airflow simulation inside lecture room: A CFD approach," *IOP Conf. Ser. Mater. Sci. Eng.*, vol. 88, no. 1, pp. 1–7, 2015.

[7] A. Abduladheem, K. S. M. Sahari, H. Hasini, W. Ahmed, and R. A. Mahdi, "Ventilation Air Distribution in Hospital Operating," vol. 2, no. 11, pp. 81–85, 2013.

[8] Yuanda Cheng, Jianlei Niu, Naiping Gao, "Stratified air distribution systems in a large lecture theater : A numerical method to optimize thermal comfort and maximize energy saving," *Energy Build.*, vol. 55, pp. 515–525, 2012.

[9] J. Lim, S. Chong, A. Husain, and T. B. Tuan, "Simulation of Airflow in Lecture Rooms."

IOP Conf. Series: Materials Science and Engineering **429** (2018) 012072 doi:10.1088/1757-899X/429/1/012072

Analysis of Suitable Thermodynamic Properties and Conditions for Dew Point Cooling system

N N A Noramzil[1,*], Z M Razlan[1], I Zunaidi[1], M Izham[1], Shahriman A B[1], A Harun[2], M S M Hashim[1], M K Faizi[1], I Ibrahim[1], N S Kamarrudin[1], W K Wan[1] and M F H Rani[1]

[1]School of Mechatronic Engineering, Universiti Malaysia Perlis, Pauh Putra Campus, 02600, Arau, Perlis, Malaysia
[2]School of Microelectronic Engineering, Universiti Malaysia Perlis, Pauh Putra Campus, 02600, Arau, Perlis, Malaysia

*Corresponding author: zuradzman@unimap.edu.my

Abstract. This study introduced dew point cooling system based on Maisotsenko Cycle Air-Conditioning (MAC/M-Cycle) for hot and humid climate. The purpose of this study to analyze the air thermodynamic properties and conditions supplied to the conditioned space. The room model has been developed to investigate the dry bulb temperature, specific humidity, and velocity distribution in the room. The simulation results showed a significance difference cooling capacity between these two systems. Consequently, this integrated system can realize significant energy saving and indoor air quality.

1. Introduction

Nowadays, clean and environment friendly energy is one of the fundamental human needs. Currently, usage of existing energy are mostly inefficient and/or involved in technology that harmful to environment. Such as, heating, ventilation, and air-conditioning (HVAC) are the biggest energy consumer today [1] due most air-conditioning systems are based on the vapor compression cycles which are consuming primary energy [2] and causing ozone layer depletion, greenhouse effect and global warming [3].

Besides, fossil fuels are diminishing rapidly and these reserves (oil, coal, and natural gas) contributing almost 80% of the world's primary energy demand [4]. In addition, the existing renewable energy such as hydro, geothermal, solar, and the wind are contributing 13.1% only while nuclear power is contributing 6.5% of the total energy supply [4,5].

So, make environmentally friendly cooling techniques and low-cost air-conditioning will reduce the energy consumption. Direct and/or indirect evaporative cooling systems are environmental friendly and low-cost but due to thermodynamic limitation, it couldn't maximize the utilization for this system in some conditions or situations such as at Oman, Turkey and Saud Arabia [2].

The dew point cooling system utilized the psychrometric renewable energy in the indirect evaporative cooling process by combining it with the thermodynamic processes of heat transfer with a much different airflow [6]. These system are based on Maisotsenko Cycle (M-Cycle) which has been discovered and patterned by Valeriy Maisotsenko at 1976 with patent number SU979796 and SU620745 [7]. This system combines the heat exchanger and advance indirect evaporative cooling (IEC) process to overcome the cooling limit which can reject heat of product air to the ambient dew point temperature theoretically compared to IEC process which can cool the air only to the incoming wet-bulb temperature [7].

Different from conventional systems which are used chlorofluorocarbon-based refrigerants (CFCs), dew point cooling system just uses water by utilizing water evaporation and latent heat [8], so it is more an environmental friendly. Besides, just using some fans and without using any compressor or pump, electricity demand is very low [8]. But, due to thermodynamic limitations, it is unfeasible air conditioning system in many situations such as in humid environment.

2. Methodology

The mathematical description and the cooling system configuration will be discussed in this section.

2.1. *Mathematical description of the system*

The governing equation of this study can be described as follow. The conservation of mass or also known as the continuity equation is expressed as:

$$\frac{\partial u}{\partial x}+\frac{\partial v}{\partial y}=0 \tag{1}$$

The conservation of momentum equations is given as:

$$\rho\left(u\frac{\partial u}{\partial x}+v\frac{\partial u}{\partial y}\right)=\mu\frac{\partial^2 u}{\partial y^2}-\frac{\partial P}{\partial x} \tag{2}$$

Mass flow rate of a fluid, need most on this analysis where can be described as below:

$$\dot{m}=\rho VA \tag{3}$$

where ρ is the density of the fluid, V is the average fluid velocity, and A is the cross-sectional area.

2.1.1. Cooling load calculation. Based on ASHRAE, there are many methods available to calculate the heat load in the conditioned space such as total equivalent temperature differential/time-averaging method, a method of the transfer function and the heat balance method [9]. In this study, the cooling load temperature difference (CLTD)/cooling load factor (CLF) method employed to calculate the heat gain and cooling load of the conditioned space.

2.1.2. Performance evaluation. Since this M-Cycle system could be cooled without dehumidification, there is only sensible cooling capacity described as:

$$\dot{Q}_s=\dot{m}_1 c_p \Delta T_1=\dot{m}_1 c_p\left(T_{1,in}-T_{1,out}\right) \tag{4}$$

where $\dot{m}_1$ is the mass flow rate of the product air, c_p is the specific heat of air, ($\Delta T_1=T_{1,in}-T_{1,out}$) is the temperature difference between the inlet and outlet of product air, and ($\Delta h_1=h_{1,in}-h_{1,out}$) is the enthalpy difference of product air.

$$\dot{Q}_l-\dot{m}_1 h_{fg}\Delta\omega_1-\dot{m}_1 h_{fg}\left(\omega_{1,in}-\omega_{1,out}\right) \tag{5}$$

$$\dot{Q}_{tot,\mathrm{cooling}}=\dot{Q}_s=\dot{m}_1\left(h_{1,in}-h_{1,out}\right) \tag{6}$$

2.2. *Cooling system configuration*

The conditions of the system both system illustrated as in Fig. 1. As in Fig. 1(a), where M is a mixture air of O the outside air and R the return air. While S is the supply air after the air goes through the cooling coil and may reach the dew point temperature of cooling coil.

Vapor Compression Air Conditioning (VAC) system analyzed in general with recirculation air and using a cooling coil to cool the supply air. While dew point cooling system based on M-Cycle (MAC) analyzed integrated with desiccant wheel where the sensible cooling provided by dew point cooling system and the latent cooling is provided by desiccant wheel.

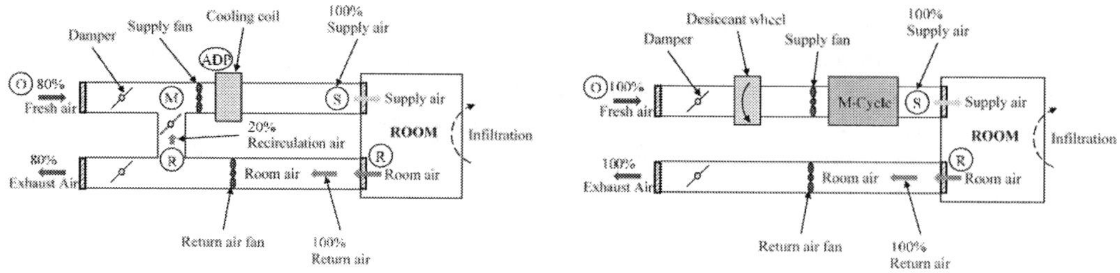

(a) (b)

Figure 1 Schematic diagram for: (a) VAC system, (b) Dew point cooling system

3. Results

The total heat gain of the conditioned space is 2.0763kW with 1.9851kW and 0.0912kW for sensible and latent heat gain, respectively with room condition 27 °C dry bulb temperature and 19 °C wet bulb temperature. The temperature, specific humidity and velocity distribution discussed as follow section.

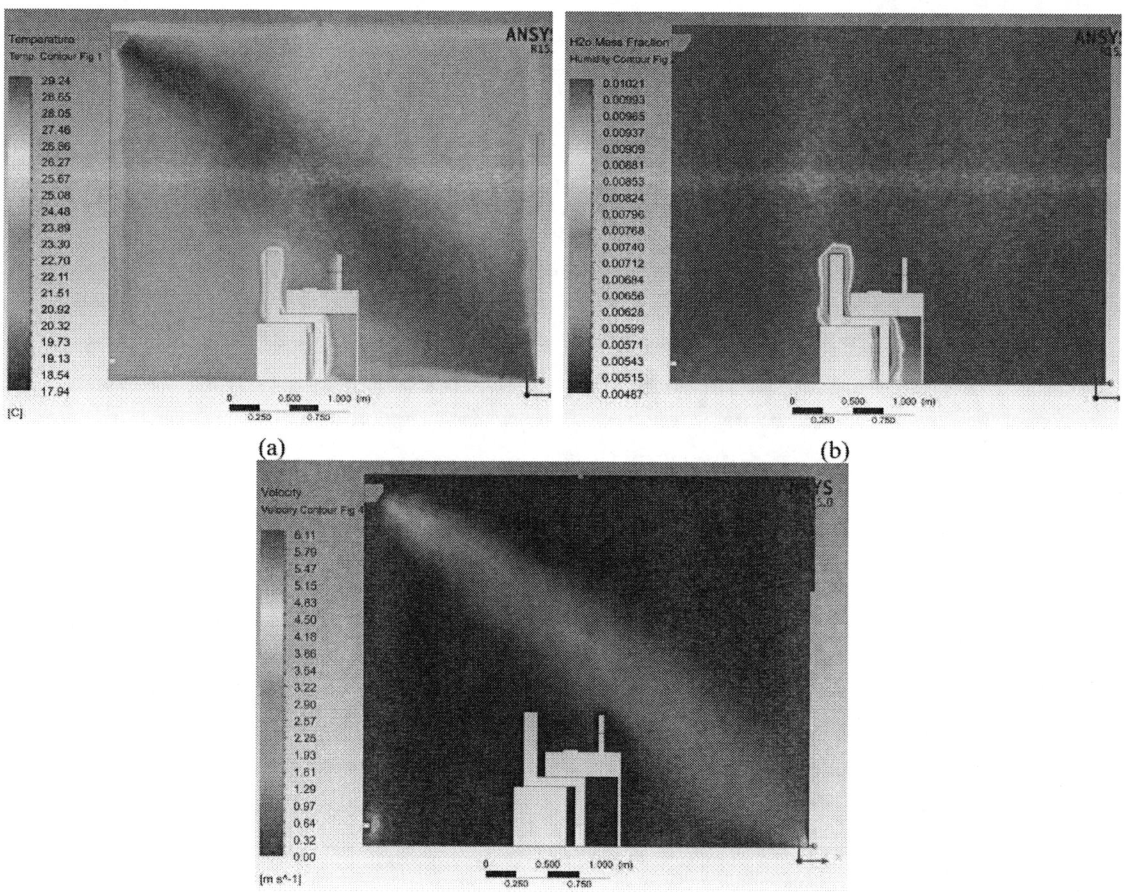

(c)

Figure 2 Contour distribution for: (a) temperature, (b) specific humidity, (c) velocity.

3.1. Temperature distribution
Initial inlet temperature range 16-17 °C, it is shown that the difference temperature distribution for VAC and dew point cooling system where the supply air temperature is in the range from 15.27 °C to 17.94 °C. Where in general the VAC system can provide lower temperature for both cases. This can explain due to the temperature blow to the exhaust fan and fast infiltration and 100% return air which is cool air will become exhausted air while the VAC system recirculated that cool air.

3.2. Specific humidity distribution
With initial inlet range of specific humidity is 10.2-10.4kg/kg, result shows that the humidity increased in small range from 0.1 to 0.3 g/kg only. This is due to have continuously gain humidity from other factor which is from occupant and infiltration.

3.3. Velocity distribution
With initial inlet velocity range 1.5-2.5 m/s, it's shown that the velocity supply air for M-Cycle case 1 not distributed much due to high velocity from the blower. The infiltration of VAC system shows contribution in velocity but not for M-Cycle system.

3.4. Cooling capacity and thermal comfort
Based on the cooling capacity as shown in Table 1, generally, the type 2 supply air is higher than type 1 supply air with discrepancy range from 0.38 to 0.68 kW which is M-Cycle case 2 can provide the higher cooling capacity than the others. The cooling capacity does not reach the heat load due to heat loss to surrounding and latent load does not remove totally in the room while latent heat gain is high 0.0912kW.

Based on thermal comfort [10], the supply air after 1-hour operation still in the range of the thermal comfort. The mean velocity suggested from [11] which is below than 0.82m/s for light sedentary activity while the mean velocity provided in the room is in range 0.12-0.14m/s.

The best type and case is Type 2 with condition 2, it is shown in the Fig. 2 with contour of distribution for temperature, specific humidity and velocity. The location of contour located effectively in the middle of the room where occupant will sit most of the time.

Table 1. Cooling capacity of the room

Type	Case	T_{su}, (°C)	ω_{su}, (kg/kg)	V_{su}, (m/s)	T_{ro}, (°C)	ω_{ro}, (kg/kg)	V_{ro}, (m/s)	Δh, (kJ/kg)	$\dot{Q}_{room}$, (kW)
	VAC	16	0.0104	1.5	19	0.0103	0.1246	8.48	0.9779
1	MAC 1	16	0.0104	1.5	19.5	0.01025	0.1358	8.12	0.9364
	MAC 2	16	0.0104	1.5	19.7	0.01035	0.1333	7.74	0.8926
	VAC	16.98	0.0102	2.436	19.8	0.0101	0.1246	7.70	1.4421
2	MAC 1	16.98	0.0102	2.436	19.5	0.0101	0.1358	6.50	1.2173
	MAC 2	16.98	0.0102	2.436	20.5	0.0101	0.1333	7.50	1.4046

4. Conclusion and recommendation
Based on this study, dew point cooling system can provide less cooling capacity than VAC system with discrepancy 0.2kW which is not much different with the conventional cooling system in term of the cooling load. The system also needs an additional system to implement the system in a humid climate, where in this study the dew point cooling system integrated with the solid desiccant system. The suitable thermodynamic properties for dew point cooling system based on this study is 16.98 °C dry bulb temperature, 0.0102kg/kg and the velocity of the inlet is 2.436 m/s.

Future study can compare the dew point cooling system integrated with other system such as Vapor-AC, Desiccant-AC, or Ejector-AC. Besides, M-Cycle alone have many configurations for future study to increase the efficiency such as perforated working channel and a different configuration of working

IOP Conf. Series: Materials Science and Engineering **429** (2018) 012072 doi:10.1088/1757-899X/429/1/012072

and product channel. Based on this study also, the return air has wasted to exhaust air where the air is already cooled. The future project can study the effectiveness of using return air as working air.

Acknowledgments

Thousands The authors acknowledge School of Mechatronic Engineering, Universiti Malaysia Perlis for the facilities. Special thanks to those who contributed to this project directly or indirectly.

References

[1] Caliskan H, I Dincer, and A Hepbasli 2011 Exergetic and sustainability performance comparison of novel and conventional air cooling systems for building applications *Energy Build.* **43** 1461–72

[2] Sultan M, I I El-Sharkawy, T Miyazaki, B B Saha, and S Koyama 2015 An overview of solid desiccant dehumidification and air conditioning systems *Renew. Sustain. Energy Rev.* **46** 16–29

[3] Heidarinejad G, M Heidarinejad, S Delfani, and J Esmaeelian 2008 Feasibility of using various kinds of cooling systems in a multi-climates country *Energy Build.* **40** 1946–53

[4] Asif M 2009 Sustainable energy options for Pakistan *Renew. Sustain. Energy Rev.* **13** 903–9

[5] International Energy Agency (IEA) 2007 *Renewables in Global Energy Supply : An IEA Fact Sheet* (France: International Energy Agency (IEA) Publications)

[6] Caliskan H, A Hepbasli, I Dincer, and V Maisotsenko 2011 Thermodynamic performance assessment of a novel air cooling cycle: Maisotsenko cycle *Int. J. Refrig.* **34** 980–90

[7] Mahmood M H, M Sultan, T Miyazaki, S Koyama, and V S Maisotsenko 2016 Overview of the Maisotsenko cycle – A way towards dew point evaporative cooling *Renew. Sustain. Energy Rev.* **66** 537–55

[8] Tertipis D and E Rogdakis 2015 Maisotsenko cycle: technology overview and energy-saving potential in cooling systems *Energy Emiss. Control Technol.* **3** 15–22

[9] Gadalla M and M Saghafifar 2016 Performance assessment and transient optimization of air precooling in multi-stage solid desiccant air conditioning systems *Energy Convers. Manag.* **119** 187–202

[10] Shaharon M N and J Jalaludin 2012 Thermal Comfort Assessment-A Study Toward Workers ' Satisfaction in a Low Energy Office Building Mohd Nafiz Shaharon and Juliana Jalaludin Department of Environmental and Occupational Health , Faculty of Medicine and Health Sciences , *Am. J. Appl. Sci.* **9** 1037–45

[11] International Organization for Standardization (ISO) 2005 *ISO 7730:2005 (E) Ergonomics of the thermal environment Analytical determination and interpretation of thermal comfort using calculation of the PMV and PPD indices and local thermal comfort criteria* vol 3 (Switzerland: International Organization for Standardization (ISO))

IOP Conf. Series: Materials Science and Engineering **429** (2018) 012073 doi:10.1088/1757-899X/429/1/012073

The Comparison between the Performance of Water and CuO/Water nanofluid in Improving the Heat Transfer of Small Spaces

Nazih A Bin-Abdun[1,2,a], Z M Razlan[1], Shahriman A B[1], C H Voon[3], W K Wan[1], I Zunaidi[1], M K Albzeirat[4] and M F H Rani[1]

[1]School of Mechatronic Engineering, Universiti Malaysia Perlis, Pauh Putra Campus, 02600 Arau, Perlis, Malaysia

[2]Seiyun Community College, Department Air-Conditioning, Hadhramoot, Yemen

[3]Institute of Nano Electronic Engineering, University Malaysia Perlis, Seriab, 01000, Kangar, Perlis, Malaysia

[4]School of Manufacturing Engineering, University Malaysia Perlis, Pauh Putra Campus, 02600 Arau, Perlis, Malaysia

[a]Corresponding author: khsallehnaka82y@gmail.com

Abstract. This paper examines the design of heat exchanger made of copper pipe for cooling small space system and the effects of water and CuO/distilled water like coolants channeling inside the heat exchanger to increase the transportation of heat between the compartment and the heat exchanger and comparison between them. The nano-fluid was prepared by dispersing a nanoparticle (CuO) in base fluid (distilled water). Nano-fluid with a nominal diameter of 50 nm at volume frictions (0.27 vol. %) at compartment temperature was used for these investigations. The analysis showed that secondary cooling system by means of nano-fluid (CuO/distilled water) has advantages in enhancing the flow rate heat transfer, best from base fluid (distilled water) also in Nusselt number.

1. Introduction

The nano-fluid process is a recent development in heat transfer area. A considerable amount of past research and experiments have been published on nano-fluid; covering the aspect of improving heat transfer performance [1] and miniaturization designs [2]. Recently, most industries have become aware of its significance. Researchers have identified many different ways to determine the abilities of heat transfer for nano-fluid through experimental data. In relation to this, the availability of nano-fluid in liquid phase provides more reliable application [3]. The usage of forced convective heat transfer application varies in its utilization in automotive, industrial cooling, solar devices and chemical industry [4]. In light of these researchers and experiments, the work of an analytical study on the operations of the heat transfer in cooling small spaces system by using conventional fluid (water) and also nano-fluid (CuO/water) to improve heat transfer is concluded.

2. Research design

A cooling system in this paper is a closed system, and the purpose of secondary cooling is to remove heat generated from small spaces to the surrounding space, for the rate heat transfer improvement, as shown in Figure 1. Using distilled water and CuO/distilled water, such as coolant for this system.

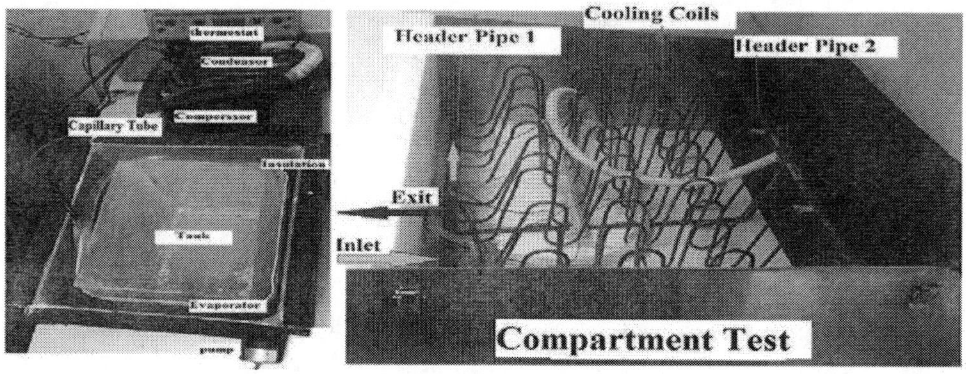

Figure 1. Design of Cooling System

The values for thermophysical properties of the base fluid (distilled water) and nano-fluid (CuO/distilled water) are taken at the average temperature 22 °C and 23.5°C respectively of cooling coils.

Table 1. Elements constant in this study

Elements	**Results**
The inner surfaces of the copper tube	smooth
Volume flow rate of the pump (Q)	9 Lit/min
The surface area of copper tube (one coil) ($\mathbf{A_{si}}$)	$0.022\ m^2$
The cross-sectional area of copper tube (one coil) ($\mathbf{A_{ci}}$)	$1.9635 \times 10^{-5} m^2$
length of one coil	1.4 m
Number of coils	12
Gravity acceleration (g)	$9.81\ m/s^2$
Diameter of (CuO) nanoparticles	50nm
Volume concentration of (CuO) nanoparticles	0.27 %
The ambient temperature in the compartment	31 °C
Dimensional of compartment	(82.5*49*19) cm^3

Correlations given in literature are used to find the thermophysical properties of distilled water and CuO/ distilled water.

Table 2. Thermophysical properties relationships of distilled water [5]

Thermophysical properties	Relationships
Thermal Conductivity (W/m.°C)	$k_{bf} = 0.56502 + 0.002636T - 0.0001251T^{1.5} - 1.51549 \times 10^{-6}T^2 - 0.00094129\ T^{0.5}$ (1)
Dynamic Viscosity (Pa.s)	$\mu_{bf} = \frac{1}{557.8248 + 19.4084\ T + 0.13604\ T^2 - 3.11608 \times 10^{-4}T^3}$ (2)
Density (kg/m³)	$\rho_{bf} = 999.7968 + 0.0683\ T - 0.01074\ T^2 + 0.0008214\ T^{2.5} - 2.30309 \times 10^{-5}T^3$ (3)

Specific Heat (kJ/kg K)	$Cp_{bf} = 4.2174 - 0.005618\,T + 0.001299\,T^{1.5} - 0.0001153\,T^2 + 4.14964 \times 10^{-6} T^{2.5}$	(4)

Table 3. Thermophysical properties relationships of CuO/ distilled water

Thermophysical properties	Relationships	
Thermal Conductivity (W/m.°C) [6]	$K_{nf} = K_{bf} + K_{bf}[\,3.761\,ø + 0.0179\,T - 307]$	(5)
Dynamic Viscosity (Pa.s) [7]	$\mu_{nf} = (1+2.5\varnothing_p)\,\mu_{bf}$	(6)
Density (kg/m³) [8]	$\rho_{nf} = \varnothing\,\rho_p + (1-\varnothing)\,\rho_{bf}$	(7)
Specific Heat (kJ/kg K) [8]	$Cp_{nf} = \dfrac{\varnothing\,\rho_p\,Cp_p + (1-\varnothing)(\rho_{bf} \times Cp_{bf})}{\rho_{nf}}$	(8)

3. Results

As shown in Figure 2 (b), a relationship between thermal conductivity and the increment of temperature is studied. From the graph, it shows the thermal conductivity of distilled water, and the x-axis shows the increment of temperature from 0 °C to 40 °C.

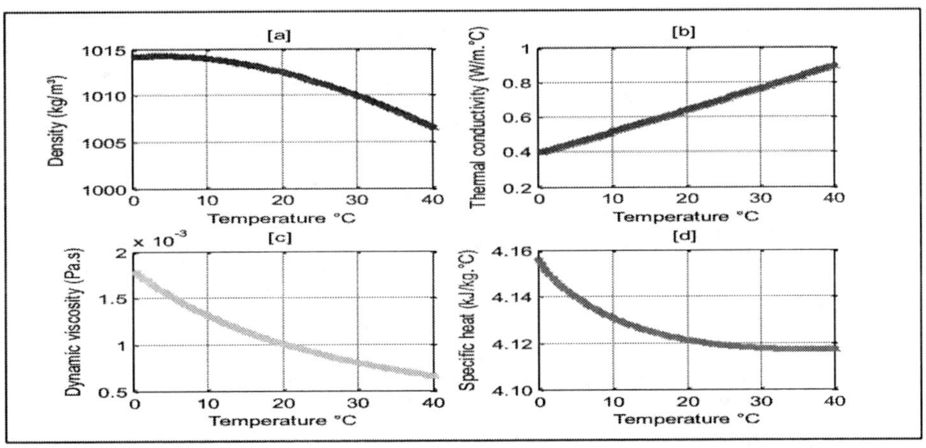

Figure 2. Relationship of Thermophysical Properties of the Distilled Water and Temperature

What we can explain from this graph is that as the temperature increased, the thermal conductivity is also gradually increased. This increment is because of the increment of the temperature of the distilled water it will change the property characteristic of the substance as from the empirical formula studied from Popiel and Wojtkowiak relations [5].

Figures 2 (a), 2 (c) and 2 (d), show the density, dynamic viscosity, specific heat of distilled water variation of as the temperature increases from 0 °C to 40 °C. This data is important to the study of the heat transfer coefficient for base fluid (distilled water).

Tables 4 shows the calculation result of heat transfer coefficient and mass flow rate for base fluid. The bulk mean temperature was 22 °C. Noted that, while conducting the experiment, all piping surfaces were insulated, excluding the cooling coils to prevent heat exchange processes at the unnecessary area.

Table 4. Summary of heat transfer coefficient, mass flow rate and heat transfer rate for water when bulk mean temperature (t_b) = 22 °C

Base fluid	Coefficient of heat transfer (h) W/m².°C	Mass flow rate ($\dot{m}$)	Heat transfer rate ($\dot{Q}$)

Distilled water	704.8	0.008 Kg/sec	1632 W

As shown in Figure 3 (b), a relationship between thermal conductivity and the increment of temperature is studied. From the graph, it shows the thermal conductivity of nano-fluid (CuO/ water), and the x-axis shows the increment of temperature from 0 °C to 40 °C.

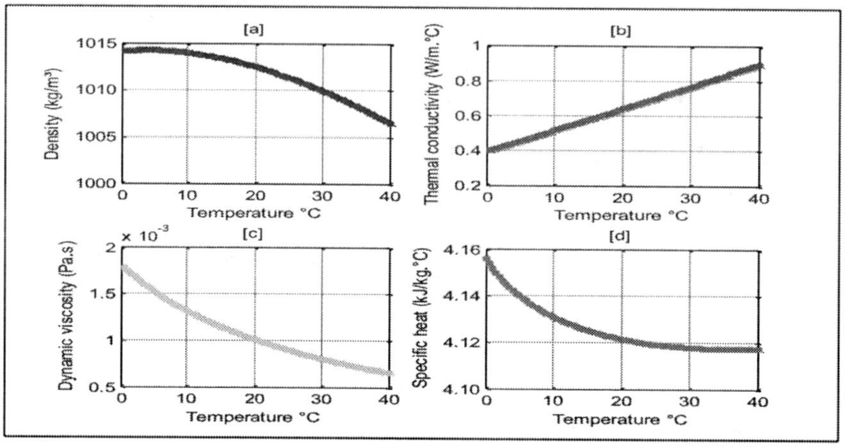

Figure 3. Variation in the Thermophysical Properties of Nano-fluids (CuO /Distilled Water) with Temperature for Constant Volume Concentrations

What we can explain from this graph is that as the temperature increased this lead to a significant increase in thermal conductivity. This increment is because of the increment of the temperature of nano-fluid it will change the property characteristic of the substance as from the empirical formula studied, from Li and Peterson formula [5].

Figures 3(a), 3(c) and 3(d), show the density, dynamic viscosity, specific heat of (CuO/distilled water) variation of as the temperature increases from 0 °C to 40 °C. This data is important to the study of the heat transfer coefficient for base fluid (CuO/distilled water).

Table 5 shows the calculation result of heat transfer coefficient and mass flow rate for nano-fluid. The bulk mean temperature was 23.5 °C. Noted that, while conducting the experiment, all piping surfaces were insulated, excluding the heat exchanger to prevent heat exchange processes at the unnecessary area.

Table 5. Summary of heat transfer coefficient, mass flow rate and heat transfer rate for nano-fluid (CuO/ distilled water) when bulk mean temperature (t_b) = 23.5°C

Nano-fluid	Coefficient of heat transfer (h) W/m^2.°C	Mass flow rate ($\dot{m}$)	Heat transfer rate ($\dot{Q}$)
CuO/distilled water	3088.2	0.0081 kg/sec	4794 W

4. Discussions

Comparison of thermophysical properties for water and CuO/distilled water:

Figure 4 (a) shows the variation of densities of water and CuO/distilled water for different temperatures. This graph indicates that as the temperature increases, there is a decrease in density of nano-fluid (CuO/distilled water); also there is a decrease in density of water. The graph also shows that the density of nano-fluid (CuO/distilled water) is higher than the density of water.

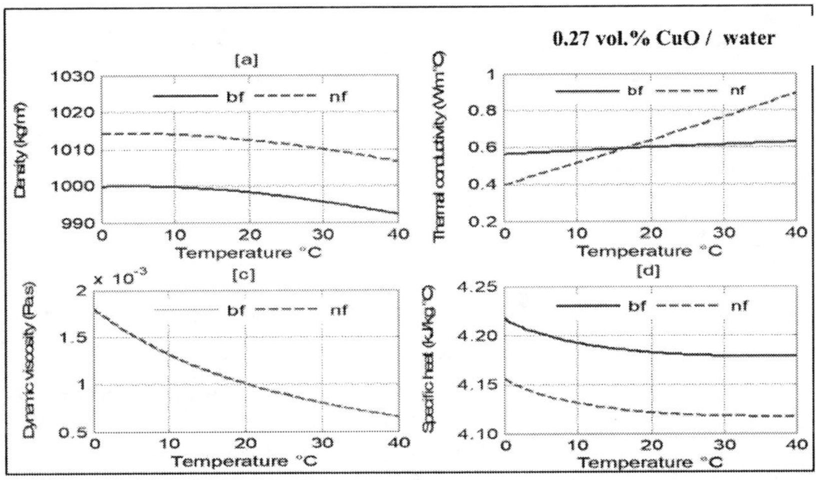

Figure 4: Comparison of Thermophysical Properties of (Distilled water) and (CuO/Distilled Water) with Different Temperature

Figure 4(b) shows the increment comparison thermal conductivities of water and CuO/distilled water (0.27 vol.% concentration at different temperatures). This graph indicates that as the temperature increases, there is a significant increase in thermal conductivity of nano-fluid (CuO/distilled water), while there is a slight increase in thermal conductivity of water. This significant increase in thermal conductivity of nano-fluid was due to the presence of nanoparticles (CuO) dispersed in water.

Figure 4(c) shows the dynamic viscosity variation of water and nano-fluid. This graph indicates that as the temperature increases, there is a decrease in dynamic viscosity of nano-fluid; also there is a decrease in dynamic viscosity of water. The graph also shows that dynamic viscosities of nano-fluid and water are close to each other.

Figure 4(d) shows the variation of specific heats of distilled water and nano-fluid at different temperatures. It is noted that as the temperature increases, there is a decrease in specific heat of nano-fluid; also there is a decrease in specific heat of distilled water. The graph also shows that the specific heat of distilled water is higher than the specific heat of nano-fluid.

Comparison of calculations for water and CuO/distilled water:
Table 6 shows the calculation results of head loss, pressure drop, heat transfer coefficient, mass flow rate and heat transfer rate for nano-fluid (CuO/distilled water) and water. When the average temperature was 22 °C of water and the bulk mean temperature was 23.5 °C of nano-fluid (CuO/distilled water). It shows that there are improvements in heat transfer coefficient and heat transfer rate of (CuO/distilled water), while there is a slight increment in pressure drop, compared to that for (distilled water).

TABLE 6. Results of calculations for distilled water and (CuO/distilled water)

Types	(distilled water) t_b= 22 °C	(CuO/distilled water) t_b= 23.5 °C
Total head loss (m)	5.15	4.96
Total pressure drop (kPa)	50.40921	50.77806
Heat transfer coefficient (W/m^2.°C)	704.8	3088.2
Mass flow rate (kg/s)	0.008	0.0081
Heat transfer rate (kW)	1.632	4.794

Comparison of non-dimensional values for distilled water and CuO/distilled water:

Table 7 shows the calculation results of non-dimensional for nano-fluid and base fluid. When the bulk mean temperature was 22 °C of base fluid (distilled water) and the average temperature was 23.5 °C of nano-fluid (CuO/distilled water). We note that there is an increase in Reynolds number and Nusselt number of (CuO/distilled water), while the Prandtl number and Peclet number are decreasing comparatively with (distilled water).

Table 7. Results of non-dimensional values for water and (CuO / water)

Types	(distilled water) t_b= 22°C	(CuO/distilled water) t_b= 23.5°C
Reynolds number	2121.3	2213
Prandtl number	6.6	5.6
Peclet number	14000	12378

5. Conclusions

The heat transfer rate for base fluid (distilled water) and nano-fluid (CuO/distilled water) flowing in cooling coils was investigated. It is found that the heat transfer rate improvement is attained by using 50 nm diameters of nanoparticles dispersed in distilled water with volume concentrations of 0.27 % of nano-fluid (CuO/distilled water) used in cooling coils when processed under the laminar flow. Results showed the following:

1- The thermal conductivity of CuO/distilled water nano-fluid improved 5.83 % compared to distilled water.

2- The heat transfer coefficient of CuO/distilled water nano-fluid improved 61.888 % compared to distilled water.

3- The heat transfer rate of CuO/distilled water nanofluid improved 49 % compared to distilled water.

4-The Nusselt number of CuO/water nano-fluid increased 25.7 % compared to distilled water).

References

[1] Y. Li, J. Zhou, S. Tung, E. Schneider and SH. Xi. A review on development of nanofluid preparation and characterization. Powder Technology, 196, 89. (2009).

[2] S.U.S. Choi, D.A. Singer and H.P. Wang, Development and application of non-Newtonian flows. ASME, New York, vol. FED 231, p.99. (1995).

[3] N. A. Bin-Abdun., Z. M. Razlan, A. B. Shahriman, K. Wan, D. Hazry, S. F. Ahmed "Base fluid in improving heat transfer for EV car battery" (AIP Conference, 2015), 090036. http://doi.org/10.1063/1.4915880

[4] K.V. Wang and O. Leon, Applications of nanofluids: current and future. Adv. Mech. Eng., 1-11. (2010).

[5] C. O. Popiel, and J. Wojtkowiak, Simple Formulas for Thermophysical Properties of Liquid Water for Heat Transfer Calculations (from 0°C to 150°C). Heat Transfer Engineering, 19(3), 87-101. (1998, January).

[6] C. H. Li, G. P. Peterson. Experimental investigation of temperature and volume fraction variations on the effective thermal conductivity of nanoparticle suspensions (nanofluids). Journal of Applied Physics, 99(08), 084314. (2006).

[7] J. F. Brady, The Einstein viscosity correction in n dimensions. Int. Journal multiphase flow, 10(I), 113–114. (1984).

[8] B. C. Pak and Y. I. Cho. Hydrodynamic and Heat Transfer Study of Dispersed Fluids with Submicron Metallic Oxide Particles. Experimental Heat Transfer, 11(2), 151-170. (1998, April).

Acknowledgments
Thanks to all the authors and those who contributed and helped in completing this paper.

The Effects of Nanofluid (Al_2O_3/De-ionized Water) with Different Concentrations in Improving the Transportation of heat in Small Space

Nazih A Bin-Abdun[1,2,a], Z M Razlan[1], Shahriman A B[1], C H Voon[3], W K Wan[1], I Zunaidi[1], M K Albzeirat[4] and M Isa A[3]

[1]School of Mechatronic Engineering, Universiti Malaysia Perlis, Pauh Putra Campus, 02600 Arau, Perlis, Malaysia

[2]Seiyun Community College, Department Air-Conditioning, Hadhramoot, Yemen

[3] Institute of Nano Electronic Engineering, University Malaysia Perlis, Seriab, 01000, Kangar, Perlis, Malaysia

[4]School of Manufacturing Engineering, University Malaysia Perlis, Pauh Putra Campus, 02600 Arau, Perlis, Malaysia

[a]Corresponding author: khsallehnaka82y@gmail.com

Abstract. An experimental study on forced convective heat transfer and flow thermophysical properties of a nanofluid consisting of three various concentrations of Al_2O_3 nanofluids (0.13, 0.25 and 0.64%) vol.% and deionized water which flows in a cooling coil made of copper pipe (heat exchanger) under laminar flow conditions are investigated. The Al_2O_3 nanoparticles of 90 nm diameter are used in this study. Working fluid flow rate of 0.5 L/min to have the fully laminar regime (1320.4< Re< 1169.5).The results demonstrate that the heat transfer coefficient of Al_2O_3/ deionized water nanofluid when concentration 0.64 vol.% is slightly higher than that of the deionized water alone at same conditions (inlet temperature and mass flow).

1. Introduction

Transportation of heat is one of the significant manufacturing processes. Traditional heat transfer working fluids like oil, ethylene glycol, water, etc, have not given enough performance for cooling systems because of their comparatively low thermal conductivity. It was noted that most recent research areas are interested in transport of heat improvement by forced convection, it has been already demonstrated that adding a nanoparticle to these traditional working fluids could improve their thermal performance, Wang et al (1999). Thence, the development of highly efficient working fluids is disbanding the disadvantages of traditional fluids has become one of the most significant priorities in the cooling systems.

2. Experimental Apparatus

As shown in figure 1, the experimental devices used includes working fluid flow lines (copper pipes), a centrifugal pump, a flowmeter, a tank of working fluids, two-water bath (one to control the working fluid temperature and another one is for controlling the heat sources temperature, i.e., the water tank that simulates the battery), two cooling coils (heat exchanger), data logger and twelve-thermocouples (K-type) were used for temperature measurement. The working fluid was pumped to the copper tube coils (heat exchanger) in the test compartment from a tank that the temperature was controlled by the first temperature bath. A water that the temperature was being controlled from the second temperature bath was pumped toward tank (simulated battery) in the test compartment as the heat load of the battery. A heat exchange occurs in the test compartment and the temperature differences between the inlet and outlet of the working fluid flows were measured. The details of tested conditions were shown in table 1. Al_2O_3/deionized water nanofluids had been prepared by two-step method so that dispersing Al_2O_3 nano-powder in deionized water and stirring the mixture for approximately 3-5 hrs, then the mixture was placed in an ultrasonic bath about than 8-12 h to complete the mixing processes.

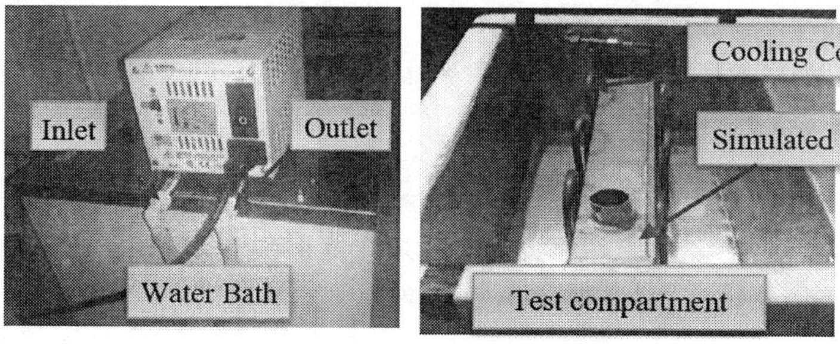

Figure 1. Design of Cooling System

The purpose of the conducted experiment is to study the heat exchange or heat removal rate from the simulated battery model to the working fluid inside the cooling coil for heat exchange rate enhancement. Utilizing de-ionized water and Al_2O_3/deionized water as the working fluid for cooling system design.

Table 1. Specifications and test conditions

Elements	Specifications
Volume flow rate of the pump (Q)	0.5 L/min.
The surface area of copper tube (one coil) ($\mathbf{A_{Si}}$)	$0.0346\ m^2$
Cross-sectional area of copper tube (one coil) ($\mathbf{A_{ci}}$)	$1.9635 \times 10^{-5} m^2$
length of one coil (L)	2.2 m
Number of coils (N)	2
The ambient temperature in the compartment	39 °C
Working fluid temperature in the tank	29 °C
Test compartment size	(0.435×0.27×0.07) m^3

3. Shape and Size Nanoparticles

It is necessary to investigate the nanoparticles size and shape, as spherical shape particles give higher thermal conductivity improvement from cylindrical particles Xie et al (2003). Wu et al (2010) proposed the Scanning Electron Microscope (SEM) was the used to study the shape, and size of nanoparticles. Deionized water was used as the base fluid. The nanoparticles (Al_2O_3, 99.8% metals oxide) was purchased from US-Research Nanomaterial.

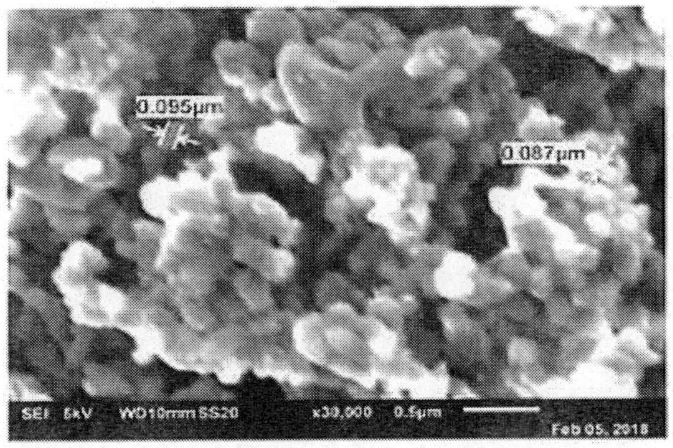

Figure 2. SEM image of dry Al_2O_3 nanoparticles

Figure 2, the SEM image of dry Al_2O_3 nanoparticles which shows the nanometre-sized (87-95nm), and most of the nanoparticles are spherical in shape.

4. Thermophysical Properties of Working Fluids

The values for thermophysical properties of the nano-fluid (Al_2O_3/De-ionized Water) at different concentrations are taken at the average temperature of working fluid in cooling coils.
The density of Al_2O_3/de-ionized water is the calculated values by using analytical models suggested by Pak and Cho (1998), while thermal conductivity, viscosity, and specific heat have been measured by using a KD2 Pro, viscometer (model LVDV- Pro, Brookfield Instruments) and differential scanning calorimetry(DSC) respectively.

4.1. Influence of particle concentrations and temperature on thermal conductivity

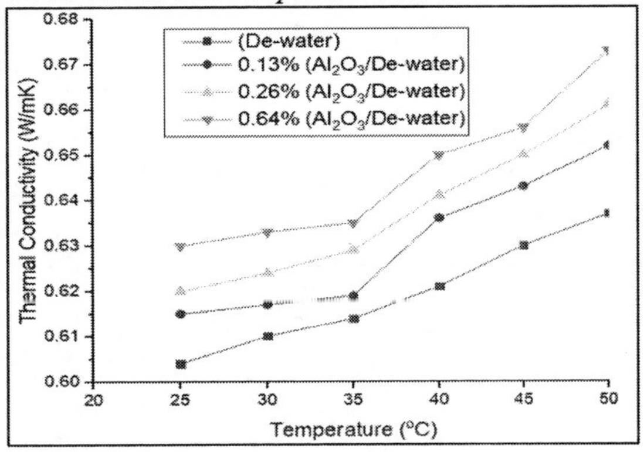

Figure 3. Influence of fraction concentration and temperature on Al_2O_3/de-ionized water nanofluid thermal conductivity

Figure 3, it is noted that the thermal conductivity increments with increasing volume concentration and temperature. The measured thermal conductivity of 0.13%, 0.26% and 0.64% nanofluid at 50°C is 6.0%, 6.6% and 6.8% respectively higher than the nanofluid at 25°C. This is due to the high temperature, which resulting in the viscosity decreases.

4.2. Influence of particle concentrations and temperature on dynamic viscosity

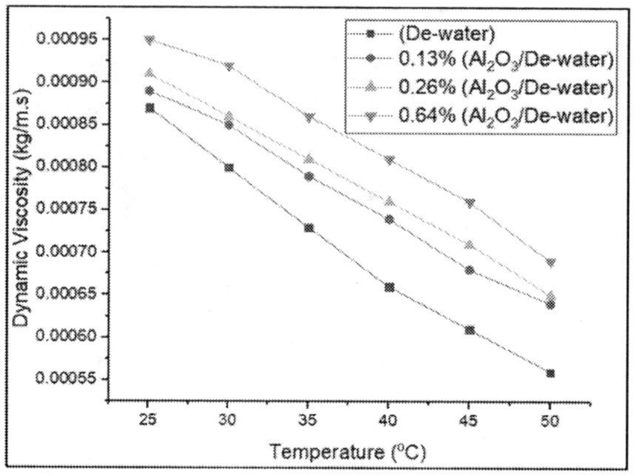

Figure 4. Influence of fraction concentration and temperature on Al_2O_3/de-ionized water nanofluid viscosity

Figure 4, shows the nanofluid (Al_2O_3/de-ionized water) viscosity decreases with increasing temperature. The 0.13% nanofluid at 50 °C is 39% lower than those at 25°C. The 0.26% and 0.64% nanofluid also show similar observation with 40% and 42% decrement respectively. This is because at elevated temperature the shear stress decreases which leads to Brownian motion and intense the nanoparticle. And it is observed that the dynamic viscosity of nanofluid (Al_2O_3/de-ionized water) increases with increasing nanoparticle volume concentration in de-ionized water.

4.3. Influence of particle concentrations and temperature on specific heat

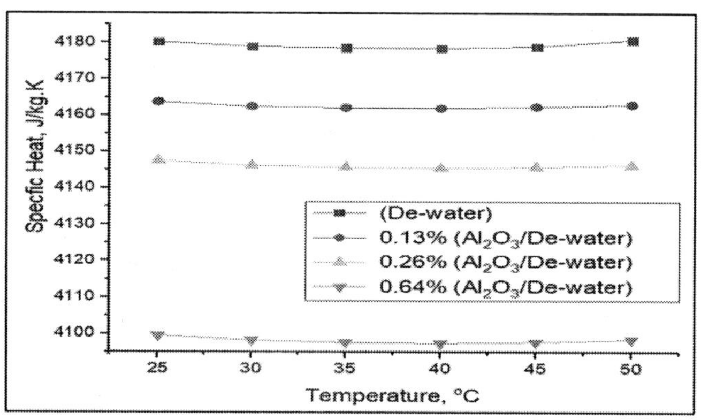

Figure 5. Influence of fraction concentration and temperature on Al_2O_3/de-ionized water nanofluid specific heat

Figure 5, it is noted that the specific heat capacity of Al_2O_3/ de-ionized water decreases moderately with increasing temperature, and decreases substantially as the particle volume concentration. This data indicates that less heat input is required to increase the temperature of Al_2O_3/de-ionized water at higher particle volume concentration. Thence, lower the specific heat of Al_2O_3/de-ionized water can lead to higher convective heat transfer.

4.4. Influence of particle concentrations and temperature on density

The density of nanofluid and base fluid are calculated at the average bulk temperature, which proposed by Pak and Cho (1998) and Abbasian and Amani (2012) respectively.

$$\rho_{nf} = \phi\, \rho_{np} + (1-\phi)\, \rho_{bf} \quad (1)$$

$$\rho_{bf} = -9.339158\times10^{-8}\, T_b^{\,4} + 1.364893\times10^{-4}\, T_b^{\,3} - 0.07714568\, T_b^{\,2} + 19.251515\, T_b - 764.475639 \quad (2)$$

In Equations (1,2), ρ is density of working fluids, 'T_b' indicates bulk temperature which is the average values of inlet and outlet temperature of the deionized water moving through the heat exchanger, ' ϕ ' indicates volume concentration, 'np' indicates (Al_2O_3) nanoparticle, 'nf' indicates (Al_2O_3/deionized water) nanofluid and 'bf' indicates deionized water (base fluid).

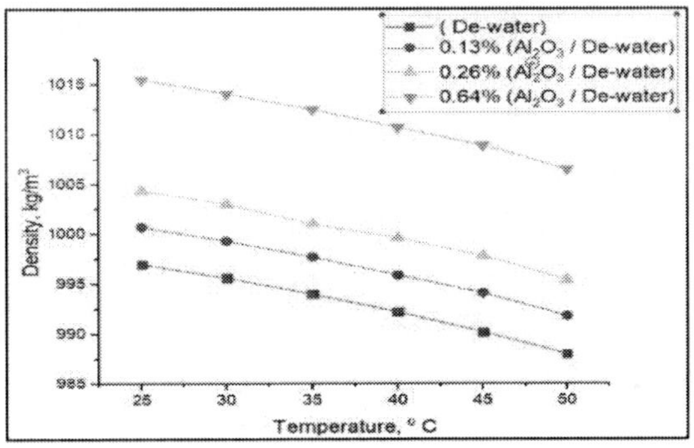

Figure 6. Influence of fraction concentration and temperature on Al_2O_3/de-ionized water nanofluid density

Figure 6, it is noted that the density of Al_2O_3/de-ionized water increases substantially with increasing particle volume concentration and decreases with increasing temperature. The thermophysical properties of the base fluid and nanofluid as the experimental working fluid, at a volume concentration of 0.13 %, 0.26 % and 0.64% of Al_2O_3 nanofluids.

Table 2. Summary of thermophysical properties of working fluids

Volume concentration	0 %	0.13 %	0.26 %	0.64 %
Thermal conductivity, W/m.K	0.610	0.618	0.624	0.633
Dynamic viscosity, kg/m.s	0.00080	0.00085	0.00086	0.00092
Specific heat, J/kg K	4179.1	4162.5	4146.2	4098.2
Density, kg/m^3	996	999	1003	1014

5. Calculation of Heat Transfer Procedure

To get the coefficient of heat transfer (h) and Nusselt number (Nu), the following method has been performed. According to Newton's cooling law:

$$Q = h\,A\,\Delta T = h\,A\,(\Delta T_b - \Delta T_w) \Rightarrow \frac{Q}{A} = q = h\,(\Delta T_b - \Delta T_w) \quad (3)$$

Heat transfer rate can be defined as:

$$Q = m^{\bullet}\, Cp\, \Delta T = m^{\bullet}\, Cp\, (T_{in} - T_{out}) \quad (4)$$

Regarding the equation of Q in the above equations:

$$h = \frac{Nu_{av}\, k}{d_i} \quad (5)$$

Nusselt number was determined from Kay's correlation which is referred than Hausen.

$$Nu_{av} = \frac{h\, d_i}{k} = 3.66 + \frac{0.0668\, Re\, Pr\, (d_i / L)}{1 + 0.04\left[(d_i / L)\, Re\, Pr\right]^{2/3}} \quad (6)$$

$$Pr = \frac{\text{Molecular diffusivity of momentum}}{\text{Molecular diffusivity of heat}} = \frac{\nu}{\alpha} = \frac{\mu \times C_P}{k} \quad (7)$$

Where Q is the heat transfer rate, $\dot{m}$ is the mass flow rate, A is the area of cooling coils pipes, ΔT is the temperature variation of the cooling working fluids, T_{in} and T_{out} are inlet and outlet temperatures, T_w is tube wall temperature, T_b is bulk temperature, d_i is inside diameter of the pipe, Cp is the specific heat of working fluids, μ is the dynamic viscosity of working fluid, k is the thermal conductivity of working fluids and Nu_{av} is average Nusselt number for the total heat exchanger.

6. Results and Discussions

Firstly, before conducting methodical experiments on the application of nanofluids as working fluid in the heat exchanger, some experiments were carried out on deionized water for verification the heat exchanger and accuracy of the experimental setup. Noting that the Nusselt number increase with increasing the Reynolds number.

Table 3, shows the calculation results of three important non-dimensional quantities for nanofluid and base fluid. As expected, the Nusselt number is seen to rise with increasing the Reynolds number, while the Prandtl number is increasing with increases of volume fraction.

Table 3. Results of non-dimensional quantities for (Al_2O_3/ De-ionized water) at bulk mean temperature with different concentrations

Types	0 %	0.13%	0.26%	0.64%
Reynolds number	1320.42	1254.74	1237.42	1169.52
Prandtl number	5.480	5.695	5.714	5.956
Nusselt number	4.5330	4.5234	4.5160	4.5045

Table 4 shows the results of heat transfer coefficient and heat transfer rate for base fluid and nano-fluid. Noted that, while conducting the experiment, all piping surfaces were insulated, excluding from the heat exchange zone.

Table 4. Summary of heat transfer coefficient, mass flow rate and heat transfer rate for nano-fluid (Al_2O_3/ De-ionized water) at bulk mean temperature with different concentrations

Volume Concentrations (%)	The coefficient of heat transfer (h) $W/m^2.°C$	Heat transfer rate ($\dot{Q}$) W
0	553.03	30.90
0.13	558.73	38.62
0.26	563.60	50.64
0.64	570.30	78.84

7. Conclusions

In this paper, the heat transfer rate of two working fluid in the simulated battery cooling system (heat exchanger) has been measured experimentally. The working fluids are base fluid (deionized water) and nanofluid (Al_2O_3/deionized water) at various volume concentrations without surfactants. The results are concluded as follows:

- After preparation of the nanofluid and then based on the photo capturing method to check the stability of nanoparticles inside the deionized water, it was noted that Al_2O_3/deionized water nanofluid can be made fairly stable by using magnetic stirrer and ultra-sonication.

- It was observed that the nanofluid at the volumetric concentration of 0.13%, had stabilized for one week after preparation.
- Dispersion of the Al_2O_3 nanoparticles stable into the deionized water increases the viscosity and thermal conductivity of the Al_2O_3/deionized water nanofluid, this augmentation increases with the increase in nanoparticle volume concentrations.
- Experimentally, noted that the dispersion of the Al_2O_3 nanoparticles in deionized water gives significantly improve the heat transfer rate of the heat exchanger. The degree of the heat transfer improvement depends on the amount of Al_2O_3 nanoparticle added to deionized water. When the volume concentration of 0.64 vol. %, the heat transfer rate improvement of 60 % compared to deionized water was recorded.

References

[1] Wang X, Xu X, and S Choi, S U 1999 Thermal conductivity of nanoparticle-fluid mixture. Journal of Thermophysics and Heat Transfer, 13(4), 474-480.
[2] Xie H, Lee H, Youn W, and Choi M 2003, Nanofluids Containing Multiwalled Carbon Nanotubes and Their Enhanced Thermal Conductivities. Journal of Applied Physics 94: 4967–4971.
[3] Wu S, Zhu D, Zhang X, and Huang J 2010 "Preparation and melting/freezing characteristics of Cu/paraffin nanofluid as phase-change material (PCM)," Energy and Fuels, vol. 24, no. 3, pp. 1894–1898.
[4] Pak B C and Cho Y 1998 "Hydrodynamic and Heat Transfer Study of Dispersed Fluids with Submicron Metallic Oxide Particles," Exp. Heat Transfer, 11, pp. 151–170.
[5] Abbasian Arani A and Amani J 2012 Experimental study on the effect of TiO2–water nanofluid on heat transfer and pressure drop. Experimental Thermal and Fluid Science. 42107-115.
[6] Prasad P V D and Gupta A V S S K S 2016 Experimental investigation on enhancement of heat transfer using Al2O3/water nanofluid in a u-tube with twisted tape inserts. International Communications in Heat and Mass Transfer, 75, 154–161.

Acknowledgments
Thanks to all the authors and those who contributed and helped in completing this paper.

Design and Analysis of an Operative Inlet

S Aziz, NAM Amin , MTA Rahman, A Rahman, ARA Syayuthi, MSAMajid and S Suhaimi

School of Mechatronic Engineering, Universiti Malaysia Perlis, Perlis, Malaysia
*nasrulamri.mohdamin@unimap.edu.my

Abstract. This paper describes the effect of intake manifold with the application of intake runner as the operative inlet used in racing car that participate in Formula Student (FSAE) competition. The design analysis of the runner is carried out to give uniform distribution of the air flow getting into engine. Air flow behavior plays an important role in order to provide better air fuel mixture inside the intake manifold before it is drawn to the combustion chamber. The results from this study indicate that longer runner gives low percentage of pressure loss but cause the air velocity to drop. Changing the outlet diameter of runner gives higher velocity at the smaller diameter and highest angle resulted higher air mass flow rate inside the runner. The new design of runner is built of 60mm length, 55mm outlet diameter and 55° angle and the results show 0.05% improvement occur in air flow distribution at the end of each runner outlet.

Keywords—Intake manifold runner geometry, Air flow behaviour, Pressure loss, Air velocity, Mass flow rate

1. Introduction

Formula SAE is an event organized by Society of Automotive Engineers International (SAE) to enhanced student in designing automobile racing car. The concept of Formula SAE is fully designed and build vehicle independently by a team of student from worldwide universities. The finished product of the racing cars are evaluated based on the potential as production car. The car is tested on the two main event which is static and dynamic event including technical inspection, cost report, presentation, engineering characteristic design, time attack performance and high-performance endurance challenge.

The engine and vehicle design in Formula SAE competition need to comply with a strict regulation provided by the organizer. Regarding the engine intake line an air restrictor of circular cross-section must not greater than 20 mm and should be fitted between the throttle valve and the engine inlet [1]. The function of restrictor is to limit the engine air flow rate as it is strongly affect the volumetric efficiency and the maximum power of the engine. All parts of the engine air and fuel control systems (including the throttle or carburetor, and the complete air intake system, including the air cleaner and any air boxes) must lie within the surface defined by the top of the roll bar and the outside edge of the four tires [1].

Excellent engine performance requires the simultaneous combination of good airflow in intake manifold and good combustion characteristic. To maximize the mass of air inducted into the cylinder during the suction stroke, the intake manifold design, which plays an important role, has to be optimized. The air has to be distributed equally in all four cylinders for multi-cylinder engine. Intake runner is a series of tubes or channel in the manifold distribute the air/fuel mixture from the plenum chamber to the individual cylinder. The geometry of the intake runner plays an important role as the high pressure of air can be produced for the air/fuel mixture distributes into the cylinder.

UniMAP Automotive Racing Team (UniART) has participating FSAE-ASEAN since 2016. The formula car was developed using Suzuki GSX-R 600 2005 engine model. All the participant team need to follow the rule and regulation provided by the FSAE-ASEAN for passing all the inspection before they can run their car on track. The main rule for air intake system is that all gasoline internal combustion engines must have a restrictor on intake manifold. Due to this application of this restrictor, the performance of engine has decreased 10% to 30% of its original performance. Besides, an insufficient air flow in intake manifold entering engine cylinder. The air flow has the biggest effect in internal combustion engine. Incorrect air fuel mixture can cause inconsistence combustion in cylinder. Due to this problem, the engine was unable to achieve its idle RPM. The RPM for idle speed on previous UniART racing car is 4000 RPM. The standard idle RPM for this engine is about 1000 RPM. The engine needs to breathe in more air to keep the engine run. This is why the idle RPM is high due to the engine needs more pumping air into the combustion chamber.

Numerous research in optimising the performance of UniART racing car have been carried out [2-7]. With the same aim, this paper provided a study on the geometry of the intake runner to stabilize the air flow in the intake manifold into the combustion chamber. Thus, with the better air distribution at each outlet of intake manifold, it could help in improving the volumetric efficiency of the engine.

2. Computational fluid dynamic (CFD) analysis inside the intake manifold runner

Air flow analysis inside the intake manifold runner was performed using CFD software – Ansys CFX. The objective of this study is to investigate the air flow behavior and the effects on the flow inside the runner with different geometry to the engine performance. The geometry parameters studied are the length, diameter and bending angle of the intake runner, presented in the following subsections.

Three results characteristic are discussed, which are; percentage of pressure loss, air velocity and mass flow rate. Each geometry is tested under the steady state condition. The parameters used in the simulation are tabulated in Table 1.

Table 1. Boundary conditions specification for simulation study

Parameters	Conditions
Inlet Pressure	101325 Pa
Outlet Pressure	0 Pa
Intake Runner Wall	No slip wall and smooth wall roughness
Air Velocity	15 m/s
Air Temperature	300 K

2.1 Analysis on the intake runner length

Runner length has big effect over torque curve in RPM range. Longer runner length would give higher torque at lower engine RPM while shorter length tend to give peak torque at high engine RPM [8]. The runner are tested for the length of 20 mm, 40 mm, 60 mm, 80 mm, and 100 mm (figure 1). The diameter of the inlet and outlet of the runner is fixed at 55 mm to suit the outlet of the intake manifold and the inlet of the engine cylinder. The runner length is selected based on less value in pressure loss, higher air velocity flow into engine and highest mass flow rate into the combustion chamber. The results of the runner length effect on the pressure loss, air velocity and mass flow rate are presented in figures 2(i), (ii) and (iii) respectively. The pressure loss is the pressure different between inlet and outlet of the runner. The lowest amount of pressure loss is to be found. The air velocity is the speed of air flow through the runner. Highest air velocity is to be select in the runner design process. The mass flow rate is the amount of air entering the runner. The more mass air flow get into the cylinder, the better engine performance of the engine. The intake runner must be kept as short as possible to minimize fuel delivery lag [8]

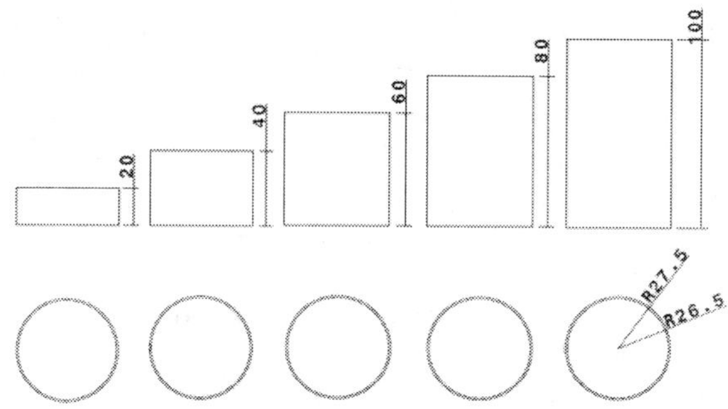

Figure 1. Intake Runner Length

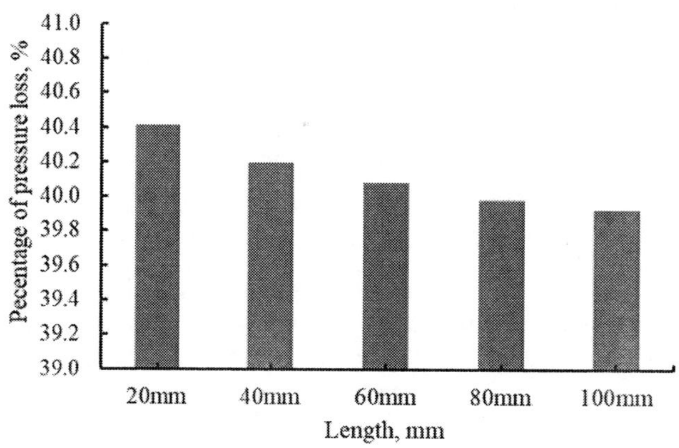

Figure 2(i) Graphical result for pressure loss at different length

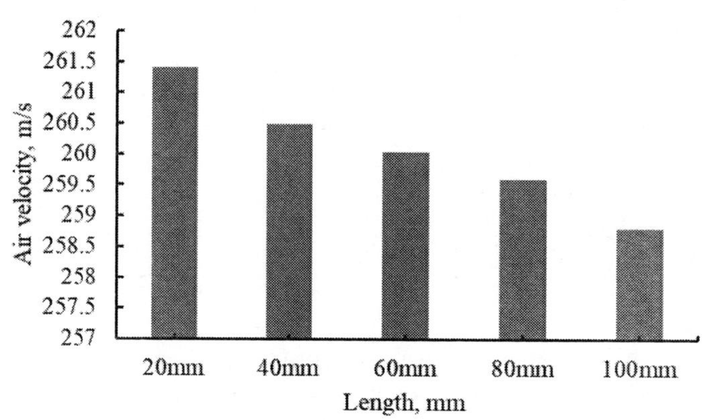

Figure 2(ii) Graphical result for air velocity at different length

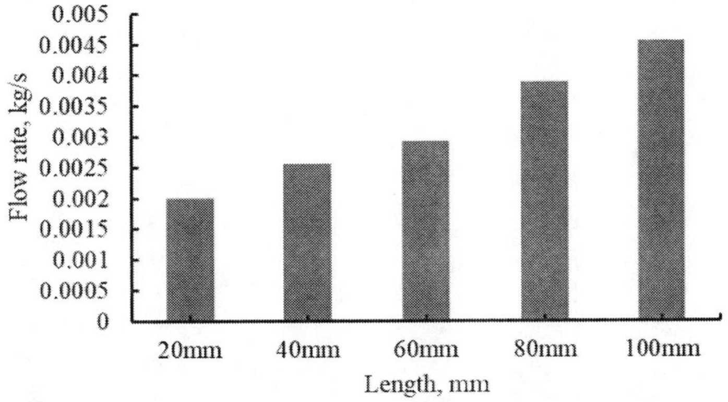

Figure 2(iii) Graphical result for mass flow rate at different length

From figure 2(i), it can be seen that the highest pressure loss occurs at 20 mm runner length. The reason why the value is higher is because air flow stream does not have enough time to build back the pressure flow in the runner. The 60 mm length of runner have resulted 40.1% of pressure loss. All these value shows only a slightly difference since the tested runner length is perform on small increment of the dimension. Meanwhile, figure 2(ii) shows that air velocity are the highest (261.4 m/s) at 20 mm runner length. This is because the air flow travel small distance in short runner length. Thus, it can be concluded that he runner needs to be kept as short as possible so high air flow speed can enter the engine [9]. For the analysis of the effect of runner length towards the mass flow rate, it can be seen that the speed is lowest at 100 mm runner length. This high amount of mass flow occurs is because of the surface area in the pipe is larger than other tested pipe. The lowest mass flow rate shows at length 20 mm which the surface area is smaller inside the runner pipe.

2.2 Analysis on the runner diameter

A second approach in increasing the volumetric efficiency of the engine is to vary the diameter of intake runner. Basically, larger cross sectional of the air passage will cause low air velocity flow through it. This is because more flow resistance occurs during suction stroke. By changing the cross sectional area at the outlet of the runner, it creates nozzle either converging or diverging depending on the change made [10]. The 60 mm length is selected to be fixed. The diameter varies in range of 40 mm to 50 mm as shown in figure 3. The dimension cannot be more than the size of the inlet of the engine since it cannot be fit with the inlet.

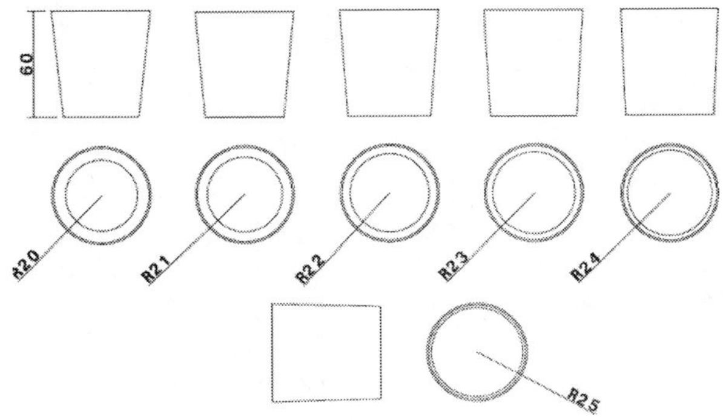

Figure 3. Intake Runner Diameter

Generally, larger diameter of air passage will cause low air velocity flow through it. This is because more flow resistance occurs during suction stroke. By changing the cross sectional area at the outlet of the runner, it creates nozzle either converging or diverging depending on the change made [11]. The results of the analysis are shown in Figure 4(i), (ii) and (iii).

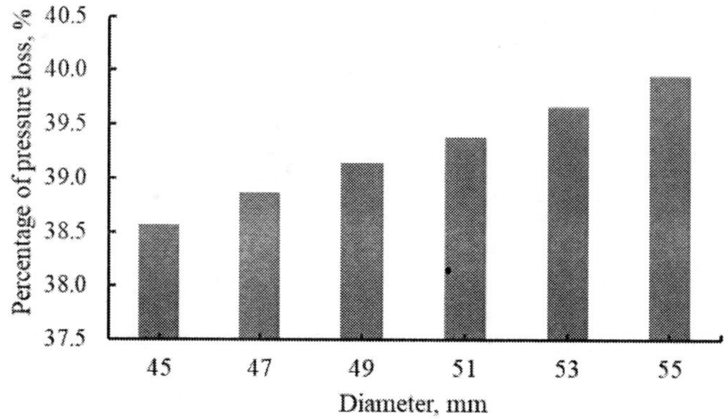

Figure (i) Graphical result for pressure loss at different diameter

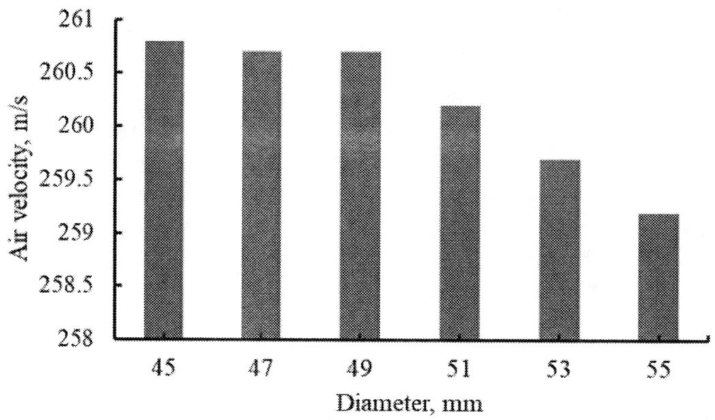

Figure (ii) Graphical result for air velocity at different diameter

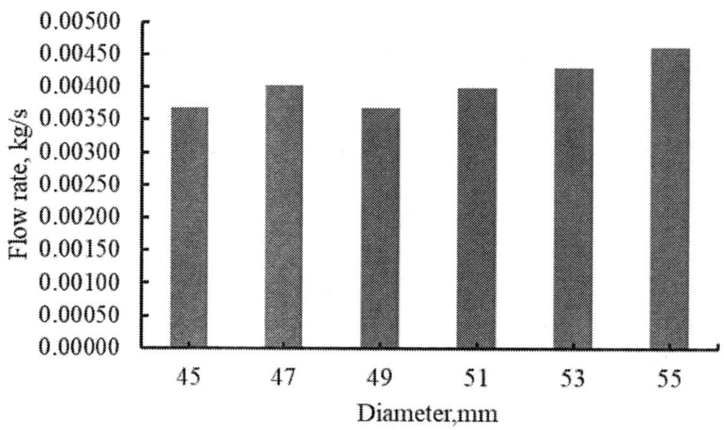

Figure (iii) Graphical result for mass flow rate at different diameter

In figure 4(i), it can be seen that the changes in diameter is obviously have caused the pressure loss in the system. Theoretically, the high pressure loss should occur at diameter of 45 mm but since the analysis result only varies at the end of the outlet and no sudden changes of the diameter occur that's why the highest pressure loss shows at 55 mm diameter.

Meanwhile, as depicted in figure 4(ii), air velocity is the highest and lowest at 45 mm and 55 mmm respectively. The change in cross sectional area at the outlet of the runner created a nozzle either converging or diverging depending on the change made. This will affect either increasing or decreasing in air speed at the entrance into engine cylinder. Decreasing the diameter value would theoretically cause the air velocity to reach a point where it is no longer affected by the restrictor, but by the runner diameter itself.

Figure 4(iii) showed the trends of mass flow rate with the different diameter. From the graph, the data indicated that the smallest (investigated value) diameter results lowest mass flow rate, and vice versa. The amount of mass flow rate is high because there is no obstacle in the air flow. The surface are also play roles since the larger surface area gives advantage for mass flow rate in the flow. The lowest mass flow occur at diameter 45mm because the less surface area traveled by the air.

2.3 Analysis on the intake runner bending angle

Air flow behavior plays an important role to the system as it to provide better air fuel mixture getting into combustion chamber [9]. High angle of intake manifold results better engine performance in terms of high brake torque, brake power, brake mean effective pressure, and volumetric efficiency [12]. Bends and junctions for example will create turbulent air flow characteristics which will choke the flow in the pipe [9]. Pressure differences also will be created on either side of the pipe in bends geometry, because as the air goes through the bend it is subject to a centrifugal effect where the air is pushed to the outer radius resulting in higher pressure and velocity on one side and lower on the other. The test angle are varies from 40° to 60°. The diameter is fixed at 55 mm and length to be at 60 mm.

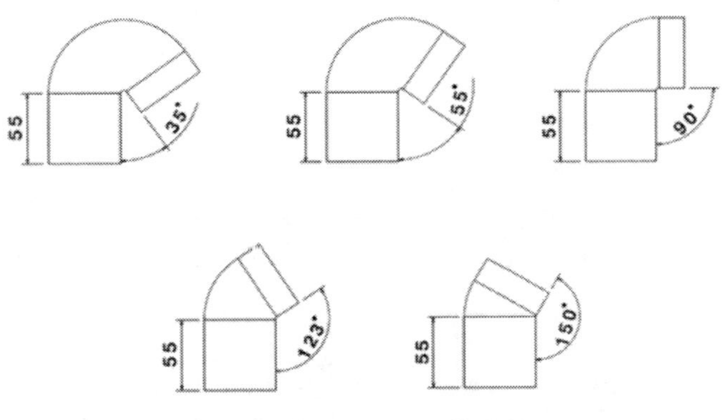

Figure . Intake Runner Bending Angle

Air flow behavior gives an important role as it to provide better air fuel mixture getting into combustion chamber. Five angles are selected for the analysis to find which angle is suitable to provide better engine performance. In this study, the intake runner are tested at five different angle, namely 35°,40°,45°,50° and 55° (figure 5). The bending angle are selected based on the allowable position of the intake manifold assemble on chassis following the rule stated in FSAE 2017 [1]. The analysis are all conducted on the fixed length and inlet pressure of 60 mm long pipe and 101325 Pa (ambient pressure at sea level) respectively. The results are plotted and presented in the following figures.

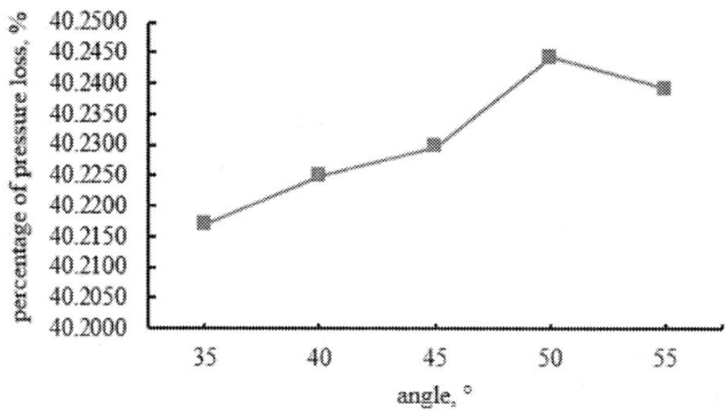

Figure (i) Graphical result for pressure loss at different bending angle

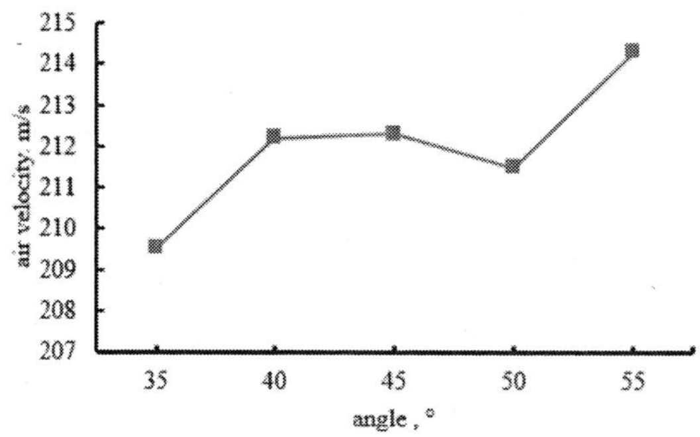

Figure (ii) Graphical result for air velocity at different bending angle

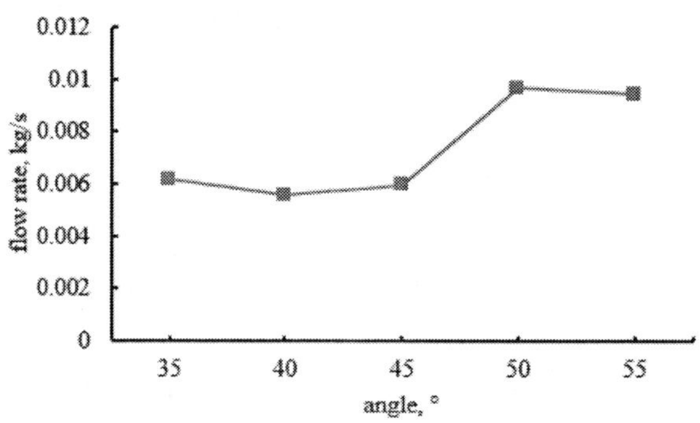

Figure (iii) Graphical result for mass flow rate at different bending angle

From figure 6(i), it can be seen that the pressure loss is the highest at 35° angle. The pattern of the pressure loss are inclined with the increment of bending angle until it reaches 40.2443 % of pressure loss at 50° angle, then it dropped to the 40.2391% of loss at 55°. The highest pressure loss occurs at 150° angle. This high value is because the air flow collides with the inner wall of the bending angle. This is due to narrow corner of the runner that causes the collision to happen. The lowest pressure loss occurs at 35° angle. The air travel shorter distance since the length is change with the change of the angle.

As for the effect of bending angel on the air velocity, figure 6(ii) depicted a different trend compared to the line of pressure loss percentage. Highest velocity occurs at 55° angle. The air flow travel inside the pipe with no resistance causes the velocity to shot up. There is less collision occurs between the air flow and the wall. The lowest velocity occurs at 35° angle. The narrow corner of the angle will provide the high collision between air flow and runner and reduced the air flow velocity [13]. Meanwhile, figure 6(iii) showed that the mass flow rate is the highest at 50° and the lowest at 35°. The mass flow rate is greatly related with the surface area. Theoretically, the highest angle tend to gives more surface area inside the runner pipe while decreasing in bending angle reduces the mass flow rate. Highest angle of runner inclination produce the highest turbulence kinetic energy [13].

3. Selected Runner Design on Current Intake Manifold Analytical Result

Based on the finding in analysis on the runner geometry, runner with the 60mm length, 55mm outlet diameter and 55° angle is selected to be applied on the current intake manifold. Table 2 shows the analytical result on the current intake manifold without installation of the runner. The current intake manifold is analyzed for comparison of the result. Meanwhile, Table 3 shows the analytical result of current intake manifold with application of selected data to study on the effect of the runner on the intake manifold. Both tables 2 and 3 show the comparison of analytical result between intake manifold without application of runner and intake manifold with installation of intake runner. The graphical results of the comparisons are depicted in figures 7(i), (ii) and (iii).

Table 2. Analytical result of current intake manifold without runner

	Outlet pressure (Pa)	Velocity (m/s)	Mass flow (kg/s)	Pressure Loss %
outlet 1	59068.3	41.5	0.0022	41.70
outlet 2	59047.2	52.9	0.0029	41.72
outlet 3	59090.7	57.1	0.0035	41.68
outlet 4	59062.1	40.2	0.0024	41.71

Table 3. Analytical result of current intake manifold with application of selected data to study on the effect of the runner on the intake manifold

	Outlet pressure (Pa)	Velocity (m/s)	Mass flow (kg/s)	Pressure Loss %
outlet 1	59030.4	33.1	0.00229	41.74
outlet 2	59036.0	39.6	0.00265	41.74
outlet 3	59038.6	38.2	0.00239	41.73
outlet 4	59033.0	40.7	0.00286	41.74

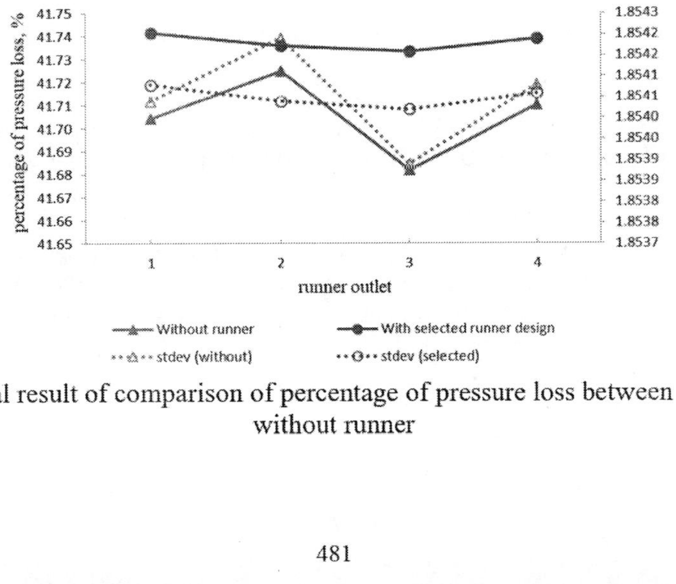

Figure (i) Graphical result of comparison of percentage of pressure loss between selected runner and without runner

From Figure 7(i), it can be seen that the percentage of pressure loss is high for the intake manifold with runner. But there's only a small difference between installed runner and without runner. This high in value may be cause by additional distance that the air needs to travel before exit the intake manifold. Although there is high pressure loss compared to without runner, is shows the uniform flow of air at the outlet of each runner. This uniform distribution of air is good which it can supply stable amount of air before entering the combustion chamber. The performance of the engine can be increase due to this effect. At outlet 1, percentage of pressure loss of intake manifold with runner is 0.04% higher than current condition of intake manifold. The highest difference shows at outlet 3 which is 0.05% more of pressure loss for intake manifolds with runner than without runner.

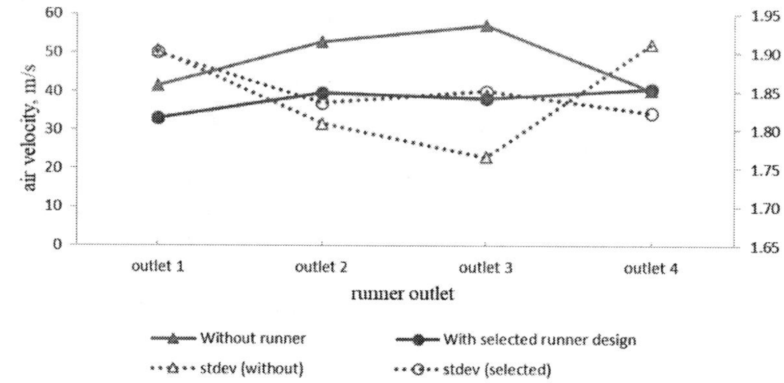

Figure (ii) Graphical result of comparison of air velocity between selected runner and without runner

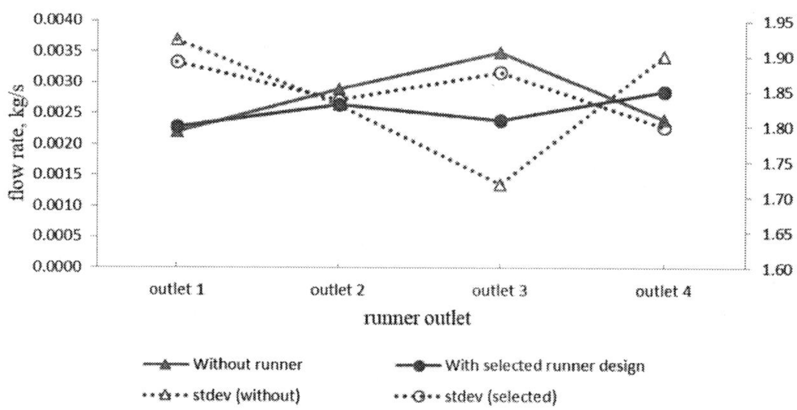

Figure (iii) Graphical result of comparison of air mass flow rate between selected runner and without runner

Figure 7(ii) shows the graphical result obtains from analysis of the air velocity on the current intake. The air velocity for application of runner is reduce compared to no runner attach on the manifold. As from previous study, the longer distance by the air flow will cause low in its speed. The bending angle also affects the reduction in speed of the air. The air need to be high in speed so more air can enter the combustion chamber. But there's only a small difference between the result and this difference will cause only in small percentage of the engine performance. Highest velocity difference for velocity occurs at outlet 3 with 33% more on current condition of intake manifold. Only 1.2% velocity difference between both intake manifold occurs at outlet 4.

IOP Conf. Series: Materials Science and Engineering **429** (2018) 012075 doi:10.1088/1757-899X/429/1/012075

The graphical result of the air mass flow rate shows in Figure 7(iii) explain that the application of runner on intake manifold cause the mass flow rate to be decrease compared to no runner attached. This occurs because of the sudden change in flow at the runner. To obtain high air mass flow rate, the flow need to move as smooth as possible before getting out from the manifold. Since the limitation on the rule of FSAE 2017, the intake runner must be applied for positioning the intake manifold on the chassis. From the result, highest difference of mass flow rate occurs at outlet 3 which intake manifold with runner gives about 38% less than current intake manifold.

. Conclusion

Based on result obtained, the application of intake runner on intake manifold give some effect in improving the distribution of air flow getting into the engine. Better air distribution in the intake manifold can help to improve the volumetric efficiency of an engine. The runner design need to meet requirement for increasing engine performance which is minimizing the pressure loss, increasing air velocity and improve amount of air mass flow rate flow into the engine. 60 mm runner length is selected as first consideration in runner design since it balance all requirement needed. Higher air velocity is better since the air can be forced into the engine before valve is closed for better compression of air fuel mixture but the diameter cannot be change due to limitation on fitting the runner with the engine. Bending angle of the runner helps for positioning the intake manifold on the chassis. The highest angle inclination is better for the air to flow into the engine. Lower angle can increase the resistance for the air flow in intake runner. Due to limitation of space, highest angle is gives by the 55° angle for the runner. Combining these three design consideration, new runner design is produced and analyzed along with the current intake manifold. The comparison of result between intake manifold with runner and without runner has proven that the runner provided uniform flow of the air distribution at the outlet of each runner. The improved design with even air distribution for better engine performance are ready to be tested on real running engine.

. References

[1] SAE International, 2017-18 Formula SAE ® Rules 2017.

[2] Hamid A F A, Rahman M T A, Khan S F, Adom A H, Rahim M A, Rahim N A, Ismail M H N and Norizan A 2017 Connected car: Engines diagnostic via Internet of Things (IoT) *Journal of Physics: Conference Series – IOPscience.* vol. 908 p. 12079

[3] Norizan A, Rahman M T A, Amin N A M, Basha M H, Ismail M H N and Hamid A F A 2017 Study of intake manifold for Universiti Malaysia Perlis automotive racing team formula student race car *Journal of Physics: Conference Series – IOPscience.* vol. 1, p. 12069

[4] Mohamad M L, Rahman M T A, Khan S F, Basha M H, Adom A H, and Hashim M S M 2017 Design and static structural analysis of a race car chassis for Formula Society of Automotive Engineers (FSAE) event *Journal of Physics: Conference Series – IOPscience.* vol. 908, p. 12042

[5] Azmeer M, Basha M H, Hamid M F, Rahman M T A, and Hashim M S M 2017 Design optimization of rear uprights for UniMAP Automotive Racing Team Formula SAE racing car *Journal of Physics: Conference Series – IOPscience.* vol. 908, p. 12051

[6] Zaidie M N A, Hashim M S M, Rahman M T A, Basha M H, Ibrahim I, Kamaruddin N S and Shahriman A B 2017 Analysis of a front suspension system for UniART FSAE car using FEA

Analysis of a front suspension system for UniART FSAE car using FEA, *Journal of Physics: Conference Series – IOPscience.* vol. 908, p. 12058

[7] Zaman Z I, Basaruddin K S, Basha M H, Rahman M T A, and Daud R 2017 Finite element prediction on the chassis design of UniART4 racing car *AIP Conference Proceeding.* vol. 1885, pp. 1–10

[8] Chaubey A and Tiwari A C 2017 Design and CFD Analysis of The Intake Manifold for the Suzuki G13bb Engine, *International Journal for Research in Applied Science & Engineering Technology (IJRASET)* 5(5) 1258–1276

[9] Hushim M F, Alimin A J, Razali M A, Mohammed A N, Sapit A and Carvajal J C M 2015 Air Flow Behaviour On Different Intake Manifold Angle for Small 4-Stroke PFI Retrokit Kit System. *ARPN Journal of Engineering and Applied Sciences* 1–7

[10] Moster D A 2012 Intake Manifold Design for An Air Restricted Engine Master Thesis, University of Cincinnati, USA, viewed 2017

[11] Ilardo R and Williams C B 2010 Design and Manufacture of A Formula SAE Intake System Using Fused Deposition Modeling And Fiber-Reinforced Composite Materials Plenum Plenum Elbow Cylinder Runner *Rapid Prototyping Journal* 16 (3), 174-179

[12] Heisler H 1995 *Advanced Engine Technology* (1st ed. Burlington: Elsevier.)

[13] Murali Krishna B, Bijucherian A, Mallikarjuna J M 2010 Effect of Intake Manifold Inclination on Intake Valve Flow Characteristics of a Single Cylinder Engine using Particle Image Velocimetry *International Journal of Engineering and Applied Sciences* 6(2)

Acknowledgment

The authors acknowledge technical support from the School of Mechatronic Engineering at the Universiti Malaysia Perlis (UniMAP). The authors are also grateful for financial support provided by the Ministry of Higher Education Malaysia via grant FRGS 9004-00063.

IOP Conf. Series: Materials Science and Engineering **429** (2018) 012076 doi:10.1088/1757-899X/429/1/012076

Fluid Flows for Heat Transfer Enhancement by Using ZnO Water Nanofluids with Different Concentrations

Nadhum H Safir[1,*], Z M Razlan[1], N A M Amin[1], Shahriman A B[1], W K Wan[1], I Zunaidi[1] and Nazih A Bin-Abdun[1]

[1]School of Mechatronic Engineering, Universiti Malaysia Perlis, Pauh Putra Campus, 02600 Arau, Perlis, Malaysia

*Email: naka82y@gmail.com

Abstract. An empirical study on forced convective heat transfer and flow thermal properties of a nanofluid consisting of three various concentrations of ZnO nanofluids (0.09, 0.2 and 0.44%) and water which flows in a cooling coil made of copper pipe (heat exchanger) under laminar flow conditions are investigated. The ZnO nanoparticles of 40 nm diameter are used in this study. Working fluid flow rate of 0.5 lpm to have the fully laminar regime (Re< 1325). The results prove that the heat transfer coefficient of ZnO/water nanofluid when concentration 0.44 % is higher than that of other concentrations and also the water at same conditions (temperature and velocity).

1. Introduction

A nanofluid is a fluid with a colloidal dispersion of man-sized particles of another substance in water or other base fluids. In a colloid two substances are distinguishable but can interact through weak surface molecular forces. The nanoparticles used in nanofluids commonly have an average size below 100 nm. Relatively, large surface area of nanoparticles increases the stability of nanofluid and reduces the problem of sedimentation of nanoparticles. Heat transfer increased with increase in surface area of nanoparticles compared to micro particles because of increased stability of nanofluids.

Nanofluids are gaining popularity among academic researchers and receiving more attention from industry as they continue to demonstrate improvements in heat transfer properties of liquid cooling processes.

Rana et al. (2013) performed experiments in subcooled flow boiling of ZnO/water nanofluids with different low particle concentrations (≤0.01 volume %) in horizontal annulus at heat fluxes from 100 to 450 kW/m2 and flow rates from 0.1 to 0.175 lps at 1 bar inlet pressure and constant sub cooling of 20°C to determine bubble behavior and heat transfer with flow rates of ZnO. They observed that increase in heat flux leads to increase in bubble diameter; the heat transfer coefficient increases with increase in heat flux and particle volume fraction of ZnO.

Based on the following key properties, ZnO nanomaterial is selected for experimentation:

(1) Corrosive resistant,
(2) Good thermal conductivity,
(3) Easy availability in purity ranges from 94% to 99.9%,
(4) Excellent size and shape capability,
(5) Low cost.

IOP Conf. Series: Materials Science and Engineering **429** (2018) 012076 doi:10.1088/1757-899X/429/1/012076

2. E perimental Setup

Preparation of nanofluid is the first key step towards using nanophase particles to enhance the heat transfer performance of conventional fluids. The nanofluid does not simply refer to a liquid-solid mixture of base fluid and nanoparticles. A nanofluid has some special requirements such as uniformity, stability, low agglomeration of particles, and no change in physical properties of the conventional fluid. In general, the following are the effective methods used for preparation of suspensions: (1) using surface activators and/or dispersants, and (2) using stirrer and ultrasonic vibrations. These methods can change the surface properties of the suspended particles and can be used to suppress the formation of particle clusters in order to obtain stable suspensions.
The experimental devices used includes fluid flow lines, a centrifugal pump, a flowmeter, a tank of working fluids, two-water bath (one to control the working fluid temperature and another one is for controlling the heat sources temperature, i.e., the water tank that simulate the battery), two cooling coils (heat exchanger), data logger and twelve-thermocouples (K-type) were used for temperature measurement. The working fluid was pumped to the heat exchanger in the test compartment from a tank that the temperature was controlled by the first temperature bath. A water that the temperature was being controlled from the second temperature bath was pumped to a tank (simulated battery) in the test compartment as the heat load of the battery. A heat exchange occurs in the test compartment and the temperature differences between inlet and outlet of the working fluid flows were measured. The details of tested conditions were shown in table 1. ZnO/water nanofluids had been prepared by two-step method so that dispersing ZnO nano-powder in water and stirring the mixture for approximately 2 hrs, then the mixture was placed in an ultrasonic bath for about 8 hrs to complete the mixing processes.

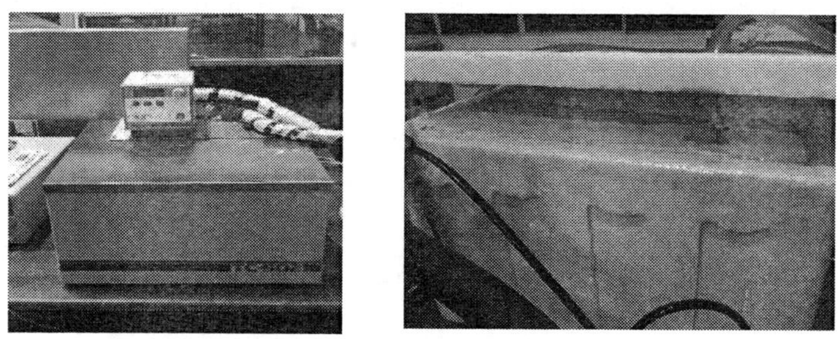

Figure 1. Cooling System

The purpose of the conducted experiment is to understand the heat exchange process and know heat removal rate from simulated battery model. Using water and ZnO/water as the working fluid in this study.

Table 1. Specifications and test conditions.

Elements	Specifications
Volume flow rate of the pump	0.5 lpm
The surface area of copper tube (one coil) ($\mathbf{A_{Si}}$)	0.0346 m^2
Cross-sectional area of copper tube (one coil) ($\mathbf{A_{ci}}$)	$1.9635 \times 10^{-5} m^2$
length of one coil	2.2 m
Number of coils	2
The ambient temperature in the compartment	40 °C
Working fluid temperature in tank	29 °C
Test compartment size	(43.5×27×7) cm^3

2.1 Shape and Size Nanoparticles

Should be investigate from the nanoparticles size and shape, as spherical shape particles give higher thermal conductivity improvement from cylindrical particles (Xie et al., 2003). Wu et al. (2010) proposed the Scanning Electron Microscope (SEM) to be used to study the shape and size of nanoparticles. Water was used as the base fluid. The nanoparticles (ZnO, 99.9% metals oxide) was purchased from US-Research Nanomaterial.

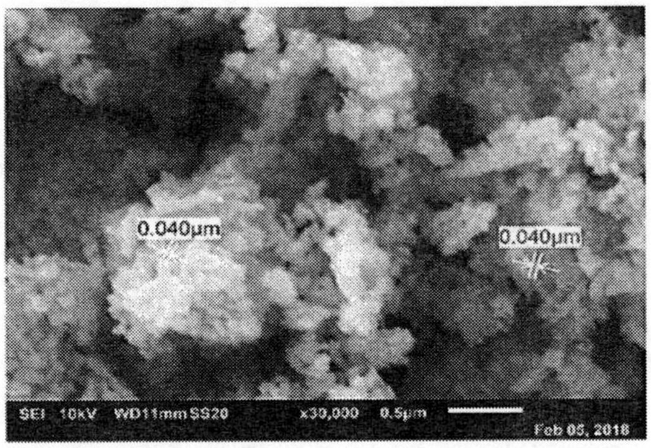

Figure 2. SEM image of dry ZnO nanopowders

Figure 2, the SEM image of dry ZnO nanopowders which shows the nanometre-sized (40nm), and most of the nanopowders are spherical in shape.

2.2 Thermophysical Properties of Working Fluids

The values for thermophysical properties of the nano-fluid (ZnO/water) at different concentrations are taken at the average temperature of working fluid in cooling coils.
The density of ZnO/water is the calculated values by using analytical models suggested by Pak and Cho (1998), while thermal conductivity, viscosity, and specific heat have been measured by using a KD2 Pro, viscometer (model LVDV- Pro, Brookfield Instruments) and differential scanning calorimetry (DSC) respectively.

2.3 Influence of particle concentrations and temperature on thermal conducti ity

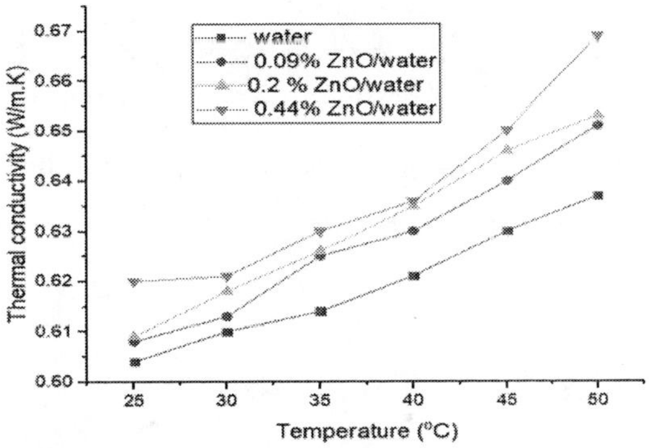

Figure 3. Effect of volume concentration and temperature on ZnO/water nanofluid thermal conductivity

In Figure 3, it is noted that the thermal conductivity increased with increasing volume concentration and temperature. The measured thermal conductivity of 0.09%, 0.2% and 0.44% nanofluid at 50°C is 6.1%, 6.2% and 6.22% respectively higher than the nanofluid at 25°C.

2.4 Influence of particle concentrations and temperature on dynamic iscosity

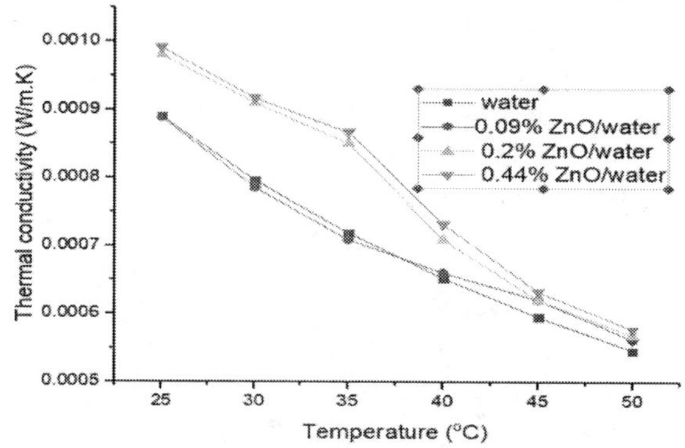

Figure . Effect of fraction concentration and temperature on ZnO/water nanofluid viscosity

Figure 4 shows that the nanofluid (ZnO/water) viscosity decreases with increasing temperature. The 0.09% nanofluid at 50°C is 36% lower than those at 25°C. The 0.2% and 0.44% nanofluid also show similar observation with 41.9% and 42% decrement respectively. And it is observed that the dynamic viscosity of nanofluid (ZnO/water) increases with increasing nanoparticle volume concentration in water.

2.5 Influence of particle concentrations and temperature on specific heat

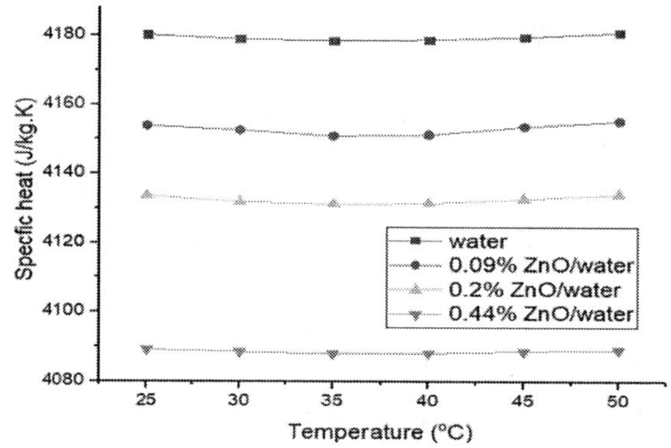

Figure . Effect of fraction concentration and temperature on ZnO/water nanofluid specific heat

In Figure 5, it is showed that the specific heat capacity of ZnO/water decreases moderately with increasing temperature and decreases highly with the particle volume concentration changes.

The density of nanofluid and base fluid are calculated at the average bulk temperature, which is proposed by Pak and Cho (1998) and Abbasian and Amani (2012) respectively.

$$\rho_{nf} = \phi \, \rho_{nf} + (1-\phi) \, \rho_{bf} \quad (1)$$

$$\rho_{bf} = -9.339158\times10^{-8} \, T_b^4 + 1.364893\times10^{-4} \, T_b^3 - 0.07714568 \, T_b^2 + 19.251515 \, T_b - 764.475639 \quad (2)$$

In Equations (1,2), ρ is density of working fluids, 'T_b' indicates temperature which is the average values of inlet and outlet temperature of the water moving through the heat exchanger, 'ϕ' indicates volume concentration, 'np' indicates (ZnO) nanoparticle, 'nf' indicates (ZnO/water) nanofluid and 'bf' indicates water (base fluid).

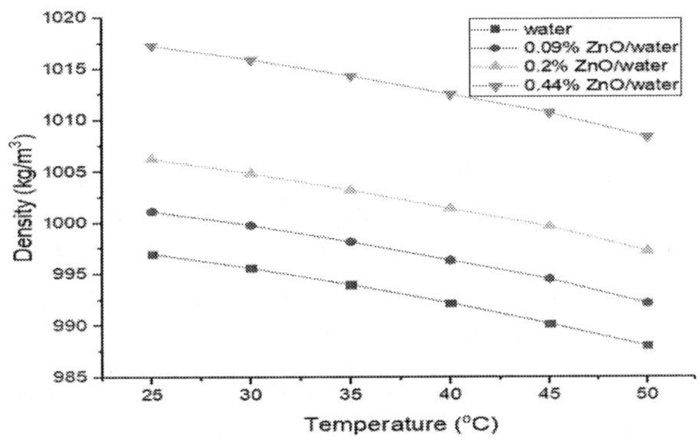

Figure 6. Effect of fraction concentration and temperature on ZnO/water nanofluid density

In Figure 6, it is noted that the density of ZnO/water increases substantially with increasing particle volume concentration and decreases with increasing temperature.

2.6 etermination of Heat Transfer Processes

To get coefficient of heat transfer (h) and Nusselt number (Nu), the following method has been performed. According to Newton's cooling law:

$$q = h(\Delta T_b - \Delta T_w) \tag{3}$$

Heat transfer rate can be defined as:

$$Q = m^{\bullet}\ Cp\,(T_{in} - T_{out}) \tag{4}$$

Regarding the equation of Q in the above equations:

$$h = \frac{Nu_{av}\ k}{d_i} \tag{5}$$

Nusselt number was determined from Kay's correlation which is referred than Hausen.

$$Nu_{av} = \frac{h\ d_i}{k} = 3.66 + \frac{0.0068\ Re\ Pr\ (d_i/L)}{1+0.04\left[(d_i/L)\ Re\ Pr\ \right]^{2/3}} \tag{6}$$

$$Pr = \frac{\mu \times C_P}{k} \tag{7}$$

Where,

Q… heat transfer rate, A… area of coil, ΔT… temperature difference of the cooling working fluids,

$m^{\bullet}$ … mass flow rate, T_{in} and T_{out} … inlet and outlet temperatures, T_w …tube wall temperature, T_b …bulk temperature, d_i… inside diameter of the tube, Cp…specific heat of working fluids, μ…dynamic viscosity of working fluid, k…thermal conductivity of working fluids, Nu_{av}…average Nusselt number for the heat exchanger.

3. Results and iscussions

Firstly, before conducting methodical experiments on the application of nanofluids as working fluid in the heat exchanger, some experiments were carried out on water for the verification of the heat exchanger and accuracy of the experimental setup. Noting that the Nusselt number increases with increasing Reynolds number.

IOP Conf. Series: Materials Science and Engineering **429** (2018) 012076 doi:10.1088/1757-899X/429/1/012076

Table 2 shows the calculation results of three important non-dimensional quantities for (ZnO/water) nanofluid and (water) base fluid. As expected, the Nusselt number is seen to rise with increasing in Reynolds number, while the Prandtl number is increasing with increase in volume fraction.

Table 2. Results of non-dimensional quantities for (ZnO/water) at bulk mean temperature with different concentrations.

Types	0 %	0.09%	0.2%	0.44%
Reynolds number	1324	1322	1318	1307
Prandtl number	5.401	5.399	5.400	6.16
Nusselt number	4.5243	4.5223	4.5199	4.5147

Table 3 shows the results of heat transfer coefficient and heat transfer rate for base fluid and nano-fluid.

Table 3. Summary of heat transfer coefficient, mass flow rate and heat transfer rate for nano-fluid (ZnO/ water) at bulk mean temperature with different concentrations.

Volume Concentrations (%)	The coefficient of heat transfer (h) $W/m^2.°C$	Heat transfer rate (Q·) W
0	558.32	30.9
0.09	559.4	34.8
0.2	560.7	46.5
0.44	563.6	66.3

4. Conclusions

In this paper, experimental facility was used to study heat transfer characteristics of ZnO/water nanofluid. The working fluid was cooled in tank. Flow heat transfer experiments in sub-heated region were carried out to study heat transfer characteristics of ZnO/water nanofluid. The results of the present investigation are summarized as follows.

- After preparation of the nanofluid without any surfactants, and then based on the photo capturing method to check stability, it was noted that ZnO/water nanofluid can be made partial stability (6 hrs) by using stirrer and ultra-sonication.
- Dispersion of the ZnO nanoparticles into the water increases the viscosity and thermal conductivity of the ZnO/water nanofluid, this augmentation increases with the increase in nanoparticle volume concentrations.
- Experimentally, dispersion of the ZnO nanoparticles in water significantly improves the heat transfer rate of the heat exchanger. The degree of the heat transfer improvement depends on the amount of ZnO nanoparticle added to water. When the volume concentration of 0.44 %, the heat transfer rate enhancement of 53.4 % compared to water was recorded.

References

[1] Rana K B, Rajvanshi A K and Agrawal G D 2013, Avisualization study of flow boiling heat transfer with nanofluids, *Journal of Visualization*, vol. 16, pp. 133–143.
[2] Xie H, Lee H, Youn W, and Choi M 2003, Nanofluids Containing Multiwalled Carbon Nanotubes and Their Enhanced Thermal Conductivities. *Journal of Applied Physics* 94: 4967–4971.
[3] Wu S, Zhu D, Zhang X, and Huang J 2010 Preparation and melting/freezing characteristics of Cu/paraffin nanofluid as phase-change material (PCM), *Energy and Fuels*, vol. 24, no. 3, pp. 1894–1898.
[4] Pak B C and Cho Y 1998 Hydrodynamic and Heat Transfer Study of Dispersed Fluids with Submicron Metallic Oxide Particles, *Exp. Heat Transfer*, 11, pp. 151–170.

[5] Abbasian Arani A and Amani J 2012 Experimental study on the effect of TiO_2–water nanofluid on heat transfer and pressure drop. *Experimental Thermal and Fluid Science*. 42107-115.
[6] Prasad P V D and Gupta A V S S K S 2016 Experimental investigation on enhancement of heat transfer using Al_2O_3/water nanofluid in a u-tube with twisted tape inserts. *International Communications in Heat and Mass Transfer*, 75, 154–161.

Acknowledgments
Thanks to all the authors and those who contributed and helped in completing this paper.

IOP Conf. Series: Materials Science and Engineering **429** (2018) 012077 doi:10.1088/1757-899X/429/1/012077

Development of an annular mist flow generating device for air-water flow

R Heng[1], Z M Razlan[1], I Zunaidi[1], W K Wan[1], Shahriman A B[1], M S M Hashim[1], M K Faizi[1], I Ibrahim[1], M S Muhamad Azmi[1] and N S Kamarrudin[1]

[1] School of Mechatronic Engineering, Universiti Malaysia Perlis, Pauh Putra Campus, 02600 Arau, Perlis, Malaysia.

Corresponding author: zuradzman@unimap.edu.my

Abstract. Annular mist flow generating device is being developed to generate annular mist flow of air and water to simulate two phase flow pattern in the evaporator. The annular mist flow generating device is use for experimental study of the air water flow behaviour that simulates the two phase flow of refrigerant. Annular mist flow at inlet of the dividing header will provide batter air water distribution to the branches. The annular mist flow generating device shall be designed using computer aided drawing (CAD) software. Then, fabricate the annular mist flow generating device. After that, the annular mist flow generating device will be test experimentally by visual observation of the air water flow pattern. The air superficial velocity and water superficial velocity use in this experiment is 1.0m/s - 5.0m/s and 0.015m/s - 0.045m/s respectively. The test channel had a horizontal header with a square cross section of 20mm x 20mm and length of 290 mm with 10 upward branches. In this experiment the air water flow patterns is observed and recorded using digital camera. It was found that the device generate annular mist flow at frontal of the dividing header. This experiment proved that the device is capable of generating annular mist flow within the design range of air and water superficial velocities.

1. Introduction

An annular mist flow generating device is a device use to generate an annular mist flow pattern. Annular mist flow is creating by high gas superficial velocity and low liquid superficial velocity [1].In annular mist flow, the gas phase is travel in the centre of the tube and surrounded by liquid phase as wavy film around the pipe wall [1].Annular mist flow is a very important flow regime in many industrial including chemical processing plants, air conditioning, refrigeration system and nuclear reactors [2].In this project, annular mist flow generating device is use to generate a simulated two phase flow pattern of refrigerant by using air water flow. Besides, annular mist flow generating device is fabricated to modelled the mist flow generated in compact evaporators used in automobile air conditioning system [3][4].Air water flow is spray from generator to the dividing header to generate a mist flow before the air water flow distribution to the branches (refrigerant tubes) [3][4].Gas-liquid distributions to the branches are greatly influenced by the flow pattern in the dividing header. The annular mist-flow inlets have batter uniformity of the water distribution compare with other flow pattern [5].

2. Literature review

2.1 Two phase flow pattern
The two phase flow pattern can be classifier base on the flow configuration, by their character, the distribution of the interfaces and the mechanisms dominating pressure drop and the flow heat and mass transfer [6].

Annular (A), High velocity of gas flow at the centre of the pipe is surrounded by the low velocity of liquid flow around the pipe wall with some liquid drop in the centre of the tube. Include annular mist and wavy annular [2].

2.2 Flow pattern detection
There are two main type of the method to detect the flow pattern. The first type is void fraction detection. Void fraction is using hot wire probe, photon attenuation technique and conductance probe to detect the flow pattern. Besides, void fraction also includes the visualization test, which is using a transparent pipe and high speed camera to capture the flow pattern and motion.

The second method is pressure fluctuation detection. This method is using spectral distribution of wall pressure fluctuations to determine the flow pattern. But this method can't clearly distinguish some of the flows.

The void detection method more accurate and batter then pressure detection method because void detection more closely related to fluid distribution.

2.3 Flow pattern prediction
To predict the flow pattern is map the experiment data and transition boundaries on two dimensional or dimensionless maps.

2.3.1 Dimensional coordinate's maps
The basic types of coordinates use is superficial velocities, mass superficial and momentum flux.

2.3.2 Dimensionless coordinate's maps
Dimensionless coordinate's maps use dimensional group use as coordinates. This is use to predict the physical variable that affect the flow pattern. The dimensional parameters of the isothermal incompressible condition as below: [7].

Gas and liquid superficial velocity (j_g,j_l)
Gas and liquid density (ρ_g, ρ_l)
Gas and liquid viscosity (μ_g, μ_l)
Pipe diameter (D)
Pipe roughness (ε)
Pipe inclination angle (β)
Acceleration of gravity (g)
Liquid surface tension (σ)
Solid liquid gas contact wetting angle (γ)

This all variable will affect the flow pattern but the entrance geometry does not affect the flow pattern.

*2.4 Gas superficial velocity*This gas superficial velocity (J_G) is referring to gas velocity in a pipe with no liquid phase present. If the liquid present in the pipeline will cause the velocity of the gas become higher due to the reduction area in the pipe [8].

$$J_G = \frac{\dot{Q}_G}{A} = \frac{G}{\rho_G}\chi = \varepsilon\, u_G \qquad (1)$$

2.5 Liquid superficial velocity

This liquid superficial velocity (J_L) is referring to liquid velocity in a pipe with no gas phase present. If the gas present in the pipeline will cause the velocity of the liquid become higher due to the reduction area in the pipe [8].

$$J_L = \frac{\dot{Q}_L}{A} = \frac{G}{\rho_L}\chi = \varepsilon\, u_L \qquad (2)$$

2.6 Spray nozzle

A spray nozzle is a device that uses to dispersion the liquid phase into a spray. Spray nozzle can classify as twin-fluid, swirl, hydraulic, rotary, ultrasonic, electrostatic, and many more. To create the annular mist flow can use atomization spray or two fluids nozzles [9].

2.6.1 Two fluid nozzles

The process of two fluid nozzles is when the high velocity gas is pressure and contact with liquid steam in order to atomize it [10]. After that the liquid jet breaks up due to momentum transfer from the gas to the liquid. This process is known as two fluid atomization and the nozzles of the process is known as two fluid nozzles [10]. There is much type of the two fluid nozzles such as air blast nozzle, prefilming air blast, plain jet air blast nozzle, air assist nozzles and effervescent nozzles. The gas and liquid phase can contact either inside or outside the atomizer which is classify as internal and external mixing nozzle respectively.

2.6.2 Nozzle material

Nozzle can made from metal and plastic such as brass, nylon, nickel alloys, stainless steel, ceramic, PTFE and PVC. To select the material to construct the nozzle is base on the properties of the liquid use and the temperature of the surrounding.

3. Research methodology

3.1 Design identification

External mix air blast atomization nozzle can create annular mist flow [11].The mechanic of air blast atomization is the liquid is injected into an annular co-flowing gas stream, which breaks up due to gas-liquid momentum transfer [12].The types of nozzles of interest in this research is air blast atomizers with the gas and liquid that external mixing.

From the equation of superficial of liquid and gas found that the volumetric gas and water flow rates ($\dot{Q}_G, \dot{Q}_L$) and area outlet of the atomization nozzle will influence the superficial of the gas and liquid. The diameter of gas and liquid of the atomization is design to have same cross sectional area for both gas and liquid outlet of the generator.

3.2 Material choose

Brass is choosing as material to fabricate the generator. This is because brass easy to machine, low cost of machining and good heat conductor.

3.3 Fabrication

The annular mist flow generator is fabricating by using CNC milling machine and lathe machine.

3.4 Flow inlet conditions

The air superficial velocity, J_A and water superficial velocity, J_W will affect the flow regime. The gas and liquid superficial velocities are determine based on the quality x and mass flow rate of the R-134a refrigerant in real evaporator and they are 1.0 -5.0 m/s and 0.015-0.045m/s [6].For the R-134 refrigerant flow in real evaporator, range of quality x =0.2-0.7 and mass flow rate M =40-160 kg/h. Besides, the temperature using in this research is room temperature which about 25 °c.

3.5 Test apparatus

The experiment apparatus to test the annular mist flow generator is show at Figure **(1)** Air and water flow rate is control by gas flow meter and liquid flow meter before supply to the test channel that show in Figure **(2)** The test channel is made of transparent acrylic plate. The test channel has square cross section of 20mm x 20mm and length 290mm which is set as horizontal with 10 branches upward.

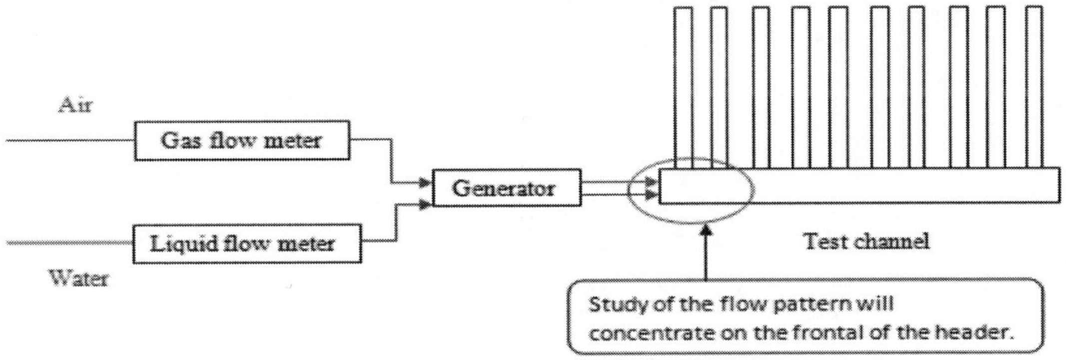

Figure 1. Experiment apparatus.

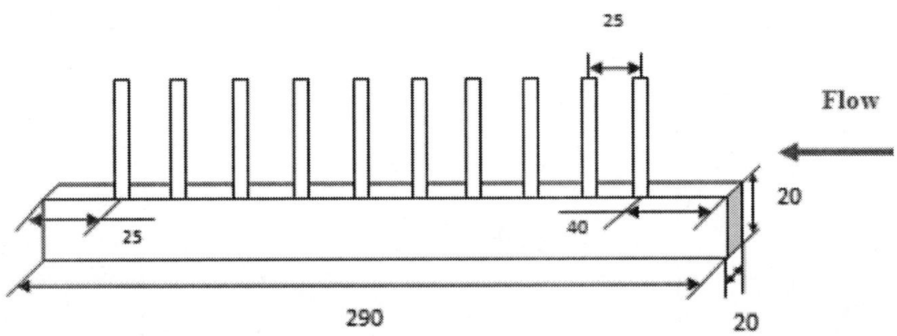

Figure 2. Test channel with 10 upward branches in (mm).

4. Result and discussion

4.1 Design and fabrication

Figure 3 is the outer 3D design of the annular mist flow generator by using CAD software. Figure 4 is the basic 2D drawing with dimension (mm) of the generator connecting with the air and water resources. Figure 5 is about the Detail 2D sketch with dimension (mm) of the generator which the precision is important on generating the the annular mist flow. Figure (6) is about the final product of annular mist flow generator after fabrication.

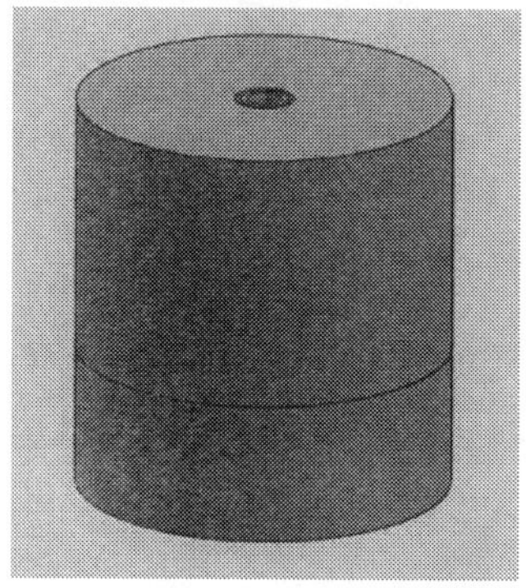

Figure 3. Annular mist flow generator in 3D sketch.

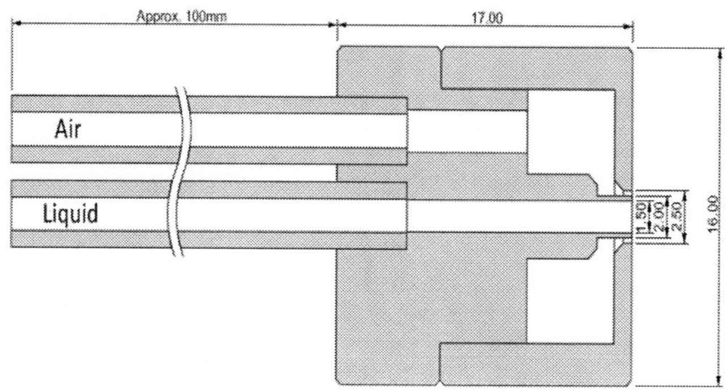

Figure 4. Basic 2D sketch with dimension (mm).

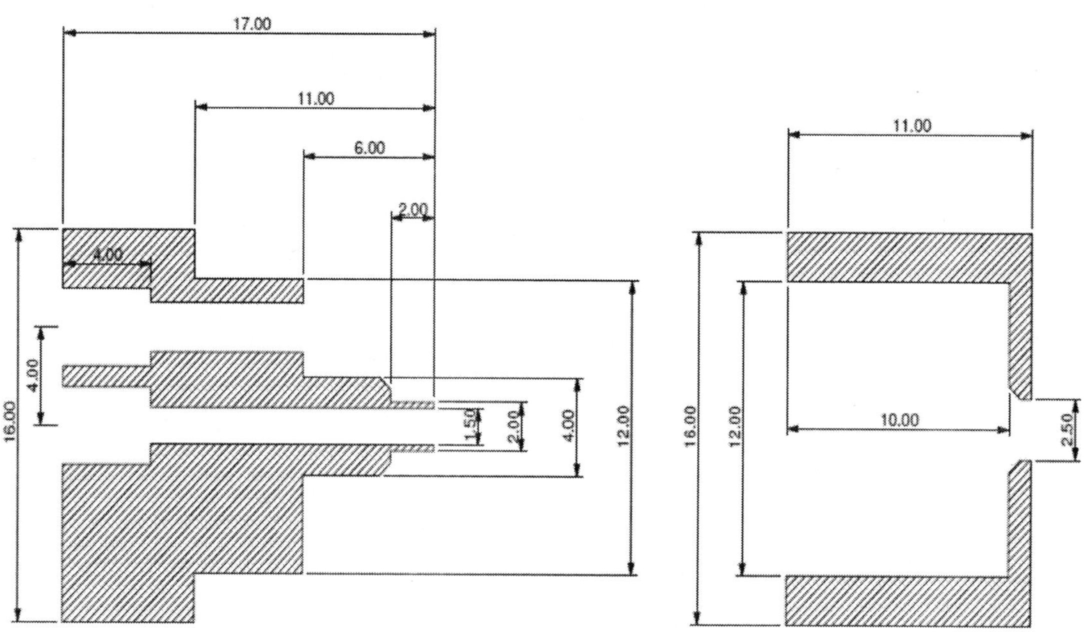

Figure 5. Detail 2D sketch with dimension (mm).

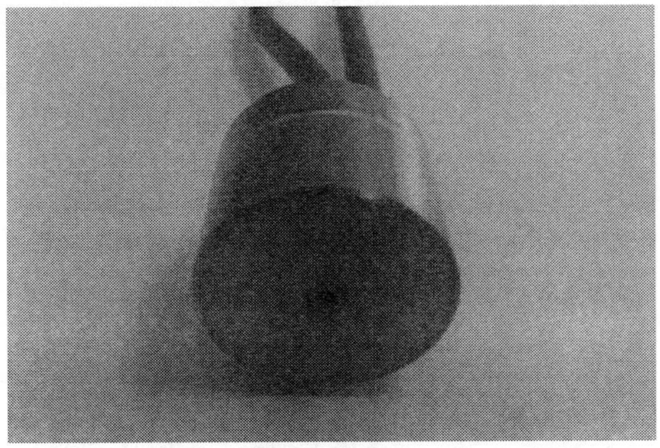

Figure 6. Annular mist flow generator after fabricate.

4.2 Experiment result

This experiment is carrying out to proof that the designed generator can generate the annular mist flow with different air and water superficial velocities. Figure 7 shows a typical visual data of annular mist flow pattern that air superficial velocity at 1.0m/s and water superficial velocity at 0.015m/s. The annular flow is form before the 1st branch of the header. Figure 8 shows the visual data of air superficial velocity at 3.0m/s and water superficial velocity at 0.015m/s, the annular flow is form at the 1st branch of the header. Figure 9 is when the air superficial velocity at 5.0m/s and water superficial velocity at 0.015m/s, the annular flow is form before the 2nd branch of the header. From the figures show that when increase the air superficial velocity the water film will become thin and more water will force to rear part of the header.

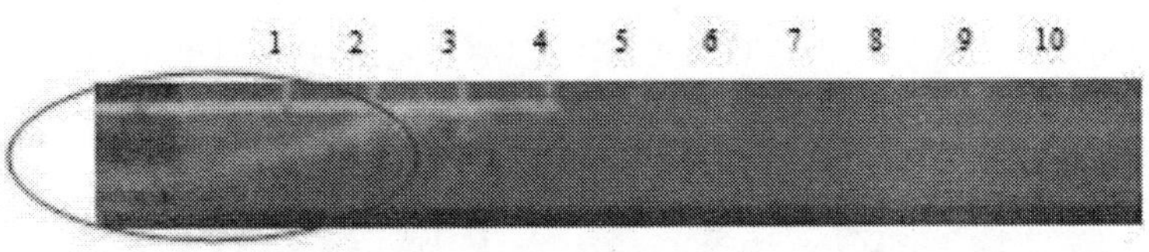

Figure 7. J_A=1.0m/s and J_W=0.015m/s.

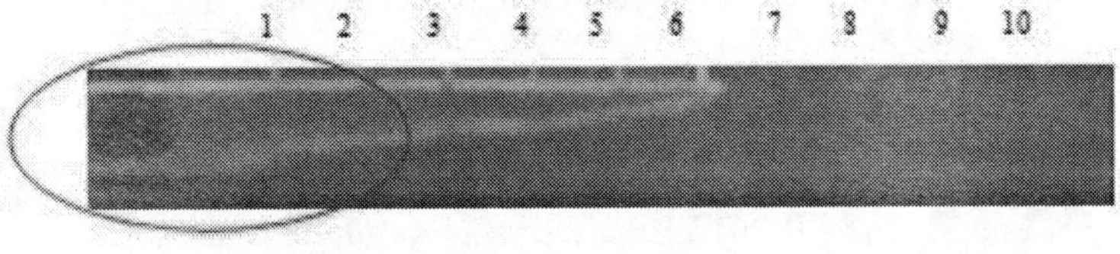

Figure 8. J_A=3.0m/s and J_W=0.015m/s.

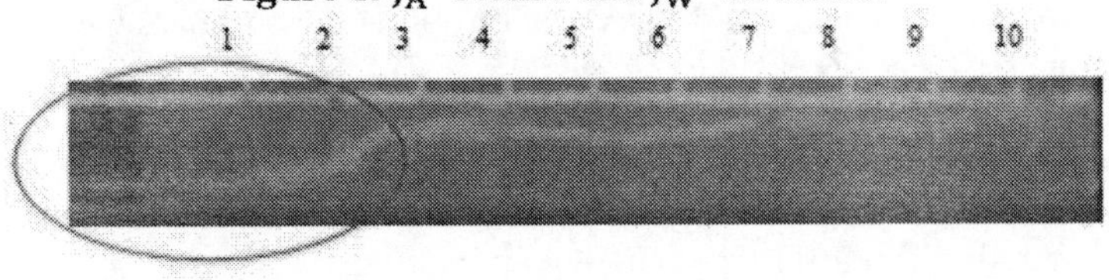

Figure 9. J_A=5.0m/s and J_W=0.015m/s.

4.3 Effect of the superficial velocity

Figure 10 the water superficial velocity is set as constant and the air superficial velocity is varies at 1.0m/s, 3.0m/s and 5.0m/s. To study effect of the air superficial velocity to the flow pattern when using the annular mist flow generator. When the air superficial velocity increase as the water superficial velocity is constant, air flow that reaches the no. of branch will also increase.

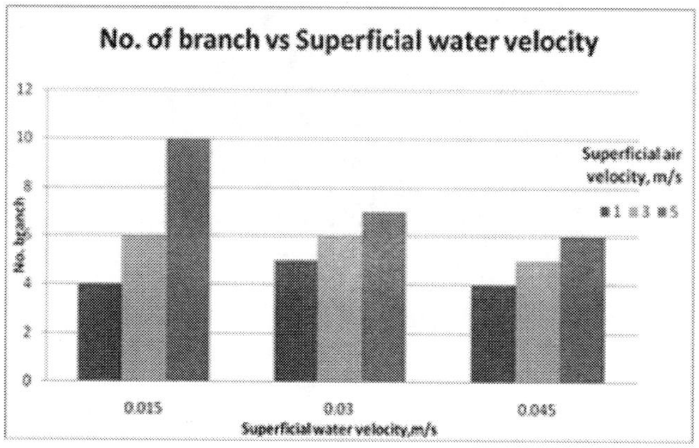

Figure 10. No. of branch versus superficial water velocity.

5. Conclusion

An annular mist flow generating device has been designed, fabricated and proved that can generate annular mist flow by experimental study. The air superficial velocity use in this research is 1.0m/s, 3.0m/s and 5.0m/s. The water superficial velocity use in this research is 0.015m/s, 0.03m/s and 0.045m/s.

When increase the air superficial velocity the water film will become thin and more water will force to rear part of the header

When the air superficial velocity increase as the water superficial velocity is constant, air flow that reaches the no. of branch will also increase.

Acknowledgments

Authors wish to acknowledge School of Mechatronic Engineering, Universiti Malaysia Perlis for the lab facilities. Special thanks to those who contributed to this project directly or indirectly.

References

[1] Al-yarubi Q S K 2012 *PHASE FLOW RATE MEASUREMENTS OF ANNULAR FLOWS*
[2] Cioncolini A and Thome J R 2013 Liquid film circumferential asymmetry prediction in horizontal annular two-phase flow *Int. J. Multiph. Flow*
[3] Razlan Z bin M 2011 Gas-Liquid Flow Distributions in Multipass Channels with Vertical Upward Branches *Open Transp. Phenom. J.* **3** 17–30
[4] Razlan Z M, Isobe R, Mizuno Y, Goshima H, Hirota M, Maruyama N and Nishimura A 2010 Gas-liquid distributions in upward multi-pass channels of compact evaporator *14th International Heat Transfer Conference* (Washington DC, United State) pp 913–21
[5] Omer K and Ashgriz N 2011 Chapter 24 - Spray nozzles *Handbook of Atomization and Sprays*
[6] Barnea D, Shoham O, Taitel Y and Dukler a. E 1980 Flow pattern transition for gas-liquid flow in horizontal and inclined pipes. Comparison of experimental data with theory *Int. J. Multiph. Flow*

[7] Marchitto A, Devia F, Fossa M, Guglielmini G and Schenone C 2008 Experiments on two-phase flow distribution inside parallel channels of compact heat exchangers *Int. J. Multiph. Flow*
[8] Steven R 2010 Wet Gas Measurement *CEESI*
[9] Hede P D, Bach P and Jensen A D 2008 Two-fluid spray atomisation and pneumatic nozzles for fluid bed coating/agglomeration purposes: A review *Chem. Eng. Sci.*
[10] Lyons O, Persoons T and Murray D 2010 Particle sizing and flow measurements in an atomizing mist jet nozzle: A shadowgraphy approach *23rd Annual Conference on Liquid Atomization and Spray Systems* pp 1–9
[11] Salque G, Gajan P, Strzelecki A, Couput J P and El-Hima L 2013 Atomisation rate and gas/liquid interactions in a pipe and a venturi: Influence of the physical properties of the liquid film *Int. J. Multiph. Flow*
[12] Lyons O F P, Quinn C, Persoons T and Murray D B 2012 Heat transfer and flow in an atomizing mist jet: A combined hot film and shadowgraph imaging approach *Journal of Physics: Conference Series* vol 395

Numerical investigation of the aerodynamic characteristics for the integrated vertical movable vanes with the airfoil straight blade vertical axis wind turbine

Kadhim H Suffer[1], Isam E Yousif[2], Zuradzman M Razlan[3]

[1] Mechanical Engineering Department, Al-Nahrain University, P.O. Box 64040, Baghdad, Iraq
[2]Fuel and energy Engineering Technologies Department, Northern Technical University, Kirkuk, Iraq
[3]School of Mechatronic Engineering, Universiti Malaysia Perlis, Pauh Putra Campus, 02600 Arau, Perlis, Malaysia

[1]kadhim_askar@yahoo.com, [2]iezzulddin77@gmail.com, [3]zuradzman@unimap.edu.my

Abstract. With the depletion of fossil fuels and the effects of fossil fuel emissions on climate change. The demands of the renewable energy is increasing. For these reasons, many researchers they are focused on harvesting renewable energy and reducing the use of fossil fuels. The vertical axis wind turbine (VAWT) is one type of the wind turbine that is widely used now a day. Low starting torque occurs when the airfoil is used in the straight blade vertical axis wind turbine as a blade. In order to increase the turbine starting torque, a flat plate vane type VAWT having movable vane is integrated with the airfoil straight blade VAWT in one rotor combination. In the present study, the numerical simulation is carried out to predict the aerodynamic characteristics of the combined three blades vertical axis wind turbine. A commercial available SolidWork2016 and computational fluid dynamic (fluent) software's with k-ω turbulence mode are used in this simulation. The predicted results from the simulation show that the turbine starting torque is increased when the vertical movable vanes are integrated with the straight blade in the vertical axis wind turbine. The flow of air passes freely through the movable vanes in case of the vanes are fully open at the negative side of the turbine, this case helps to increase turbine rotation and so, thus, the power of the turbine increases. Finally, the predicted results from the simulation give a good agreement with those results reported in the published literature for VAWT has different blade configurations.

Keywords. Renewable energy; Wind energy; VAWT; CFD; Straight blade rotor; k-ω turbulence model.

1. Introduction

Non-renewable resources of energy are limited in the world and are depleting at a faster rate due to rapidly growing demand. Wind energy is one of the renewable resources of energy. The wind turbine is the machine which converts the kinetic energy available in the wind into mechanical power, then to electrical energy by an electrical generator. A wind turbine is unable to capture all of the energy in the wind. Therefore, it is reasonable to assume an upper limit for the aerodynamic efficiency. The theoretical maximum power coefficient of wind turbine of any design not exceeding the 59 % it's mentioned as a Betz' limit [1]. The wind turbine has two type's Horizontal axis wind turbine (HAWT) and Vertical axis wind turbines (VAWT) depending on the shaft of the turbine rotor. HAWT more efficient than VAWT.

VAWT is further can be divided into lift driven VAWT (Darrieus) the earliest type and drag driven VAWT (Savonius type) [2]. The most common of the Darrieus wind turbine is the H-type VAWT.

Many researchers have been focusing their attention on improving the aerodynamic characteristics of VAWT, and these studies were summarized by laboratory measurements and theoretical of the flow in an around the blades of the wind turbine. Computational fluid dynamics (CFD) is the promising technique in simulation of wind turbine for many research to help many researchers for investigation the aerodynamic characteristics of the VAWT. Finite volume method is one of most of numerical methods which are used in CFD to solve the physical problems. The finite volume method is more efficient to solve the flow field around the vertical axis wind turbine rotor [3].

Suffer *et al.* [4] simulated numerically the aerodynamics and modal analysis for the three combined blades vertical axis wind turbine. Each blade has a vane type of horizontal flat plate (having three movable vanes) with the Darrieus helical airfoil (Gorlov). The aerodynamics results from the simulation show the power of the turbine increases. While the modal results show that until the 5th natural frequency the effect of damping can be neglected.

Rémi *et al.* [5], conducted a parametric study of vertical axis turbines by using the k-ω SST RANS model in its unsteady form. The best aerodynamic configuration is determined by studying the effect of solidity, number of blades, Reynolds number, blade pitch angle (fixed and variable) and blade thickness on the aerodynamic efficiency of the turbine.

Hussen *et al.* [6] Improved the performance of the vertical axis wind turbine by using CFD analyses to improve the performance of the designed rotor of the three blades have a section of the NACA 0018 airfoil straight and then twisted to 45° and 90°.

Ying *et al.* [7] studied the aerodynamics and motion performance of the floating of a 5 MW H-type VAWT. The floating foundation was a truss spar type is also studded. Theory of double-multiple-stream-tube is used also to calculate the aerodynamic loads which acting on the wind turbine considering the motions of the floating foundation. The results of that steady of a H-type turbine were give a better motion performance.

Suffer *et al.* [8] presented the numerical simulation of the two configurations of the vane type vertical axis wind turbine (VVAWT) rotors. The blades of the turbine have a cavity shape with three movable vanes, the first configuration has three blades with angles of 120° between each blade, and the second configuration has four blades with angles of 90° between each blade. The predicted results from the numerical simulation gives acceptance compared with this is reported for steadying VAWT having different blade designs.

Qasim *et al.* [9] designed vertical axis wind turbine movable horizontal vanes rotating instead of being fixed to reduce the negative torque on the form of non-work (against the direction of the wind). This is so that the frames vanes open and pass the wind through the frame freely without the resistance to the air. This idea has been put forward first by designing three vanes being horizontal in each frame and second by designing three vertical vanes in a single frame.

The effect of different design parameter for performance of the evaluation of SB-VAWTs were investigated by Khammas *et al.* [10]. During this investigation they are finding that the VAWT has better capability of extracting more power from low wind than HAWT. SB-VAWT having the effect of parameters like Blade aerodynamics, Solidity, Pitch angle, Number of blades and TSR on performance.

2. Current simulation

The objective of this study is to investigate numerically the aerodynamic characteristics of integrated vertical movable flat plate vanes with the H-type (Darrieus) airfoil in the one configuration of the blade for the three blades VAWT. The airfoil is a NACA 0012, and the characteristic geometry of the airfoil are Height (100mm), Chord (c = 40mm), Thickness (t = 4.8mm), while the flat plate vane has a height (96mm) and (18mm) width. The pre-processing steps of this study are carried out by using solidworks2016 software for modelling the geometry for three combined blades vertical axis wind turbine as shown in Figure1 (a). The dimensions and boundary conditions of the computational domain with a top section view are presented as shown in Figure 1 (b).

The second step of pre-processing is carried out by using Gambit software for machining the modelled geometries (single blade with closed vane, single blade with open vane, three blades at 90° angular position and three blades at 270° angular position. The number of mesh for each model are presented in table 1.

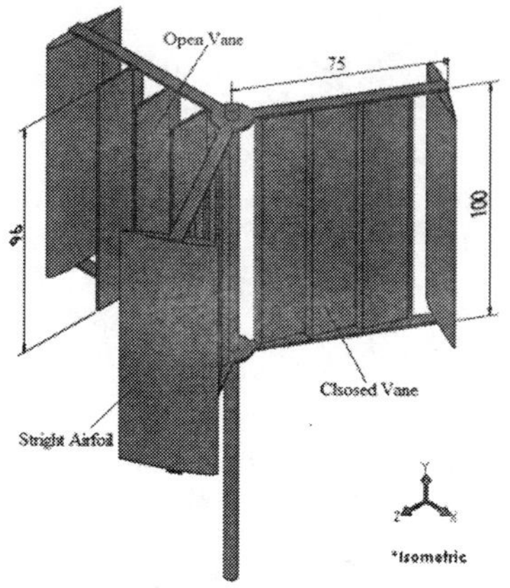

(a)

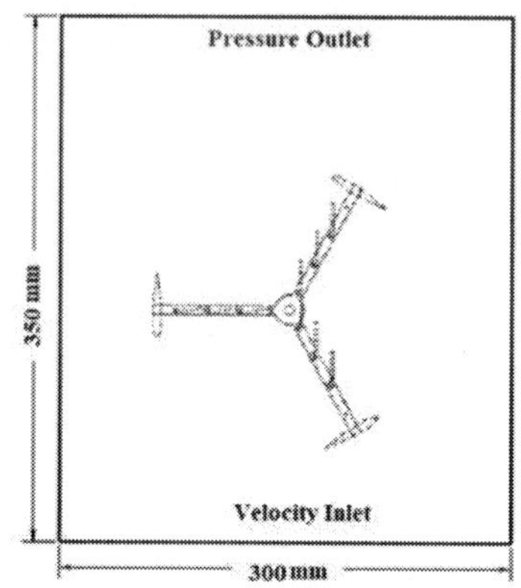

(b)

Figure 1. (a) Model of three combined blade VAWT, (b) Dimension and boundary of the computational domain.

Table1. Number of meshes for modelled geometries

Model Geometry	Number of Mesh Volume
Single blade with closed vane	2105801
Single blade with open vane	3348589
Three blades at 90° angular position	2407333
Three blades at 270° angular position	2446661

3. Result and iscussion

In the third step of pre-processing the available commercial ANSYS FLUINT 16.1 the Shear Stress Transport (SST) *k*-ω turbulence model is used to simulate the modelled geometries for different blade angular positions. Upstream air velocities 14 m /s is used in this simulation as a velocity inlet. The residual plots of the current study and accuracy is adopted at the accuracy 10^{-5} to achieve the best accuracy to prove the validity of the results. Therefore, according to the residual plots the accuracy is acceptable for the purpose of the results of this study.

The predicted drag coefficient (Cd) results and the projected areas from the current simulation for modelled geometries are mentioned in Table 2. From the Cd results, it can be observed that Cd is maximum at 90 ° blade angular position, while the Cd is minimum at 30° blade angular position. Figure 2 (a) shows the results of the static pressure distribution in and around of the combined single blade with a closed vane, while Figure 2 (b) shows these results of the combined single blade with a closed vane. From these results, it can be concluded that the static pressure is maximum in the upstream airflow direction in the front face of the blades, while the static pressure is minimum in the downstream airflow direction on the back face of the blades. The static pressure drop decreases when the blade rotates to the negative side and this helps to increase turbine power.

Figure 3 (a) and 3 (b) shows the contours of velocity distribution for the combined blade at 90° and 270° angular positions respectively. From these contours it can be seen that the flow of air passes through the blade freely because all vanes are opened and thus, there is no resistance to the flow in this side. This helps to increase turbine rotation and so the power of the turbine increases

Table 2. Values of drag coefficient and projected area for modelled geometries

Model Geometry	Projected Area (m^2)	Drag Coefficient
Single blade with closed vane	0.00697728	1.01
Single blade with open vane	0.00226818	0.6373
Three blades at 90° angular position	0.01126593	1.491
Three blades at 270° angular position	0.00766428	1.289

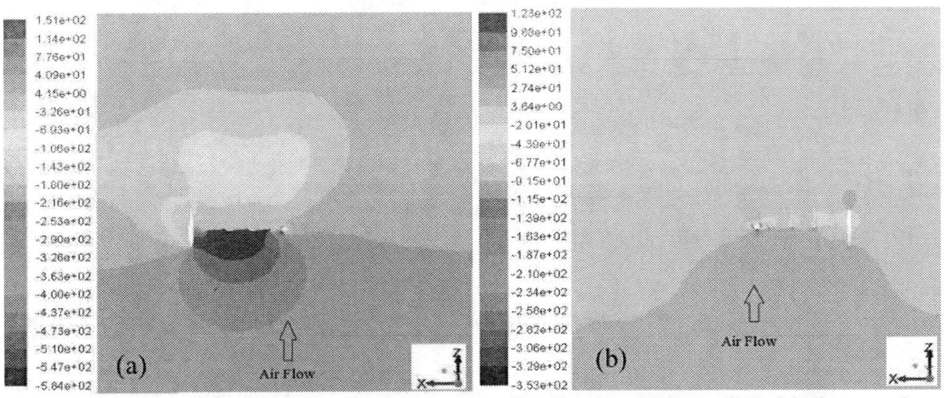

Figure 2. Contours of the static pressure distribution, (a) At 90° blade angular positions, (b) At 270° blade angular positions.

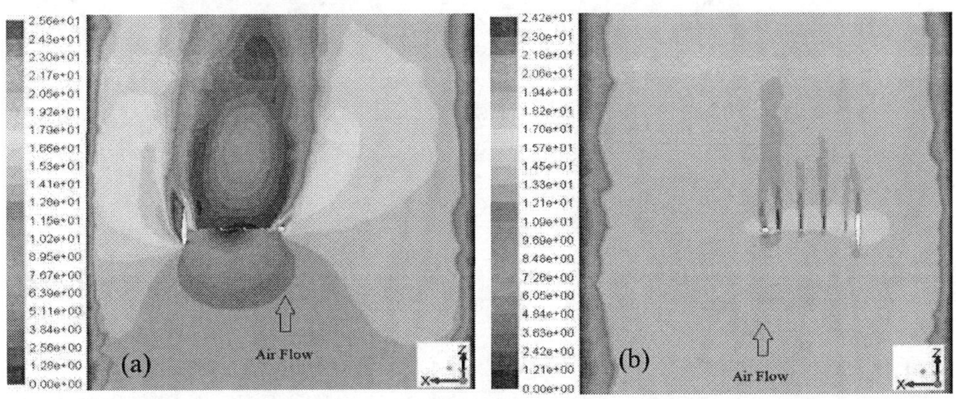

Figure 3. Contours of the velocity distribution; (a) at 270° blade angular positions, (b) at 90° blade angular positions.

Figure 4 shows the static pressure distribution inside and outside the rotor of the three combined vertical axis wind turbine. For the first turbine blade at 90° angular positions when the all the vanes are fully closed the maximum static pressure (1.86E+02 Pa) occurs at the upstream blade side, while minimum static pressure (-1.76E+02 Pa) occurs at the downstream blade side. While for the first turbine blade at 270° angular positions when the all the vanes are fully open the maximum static pressure (1.6E+02 Pa) occurs on the upstream of the third turbine blade.

Figure 5 (a) and 5(b) shows the contours of the distribution of the dynamic pressure between and around the combined blade in the flow domain at different case of angular positions. Figure 6 (a) and 6 (b) shows the contours of velocity distribution for the combined three blades VAWT at 90° positions in case of closed vanes, and at 270° blade angular positions in case of open vanes. From the results of Figures (5) and (6) it has been can be concluded clearly from the contour distribution the maximum vortex region occurs in positive side because of the vanes are closed. While the vortex region is a decrease in negative side, depending on the opening of the vanes with angular position and vortex region is disappearing when the vanes are fully open. From these results, it is concluded that the turbine negative torque is reduced and then the turbine output power is increased.

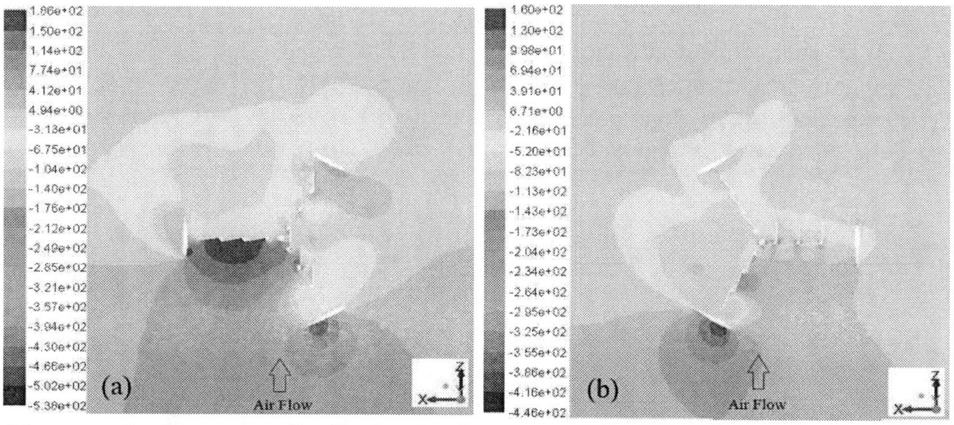

Figure . Static pressure distribution for three blades VAWT, (a) 90° angular position and, (b) 270° angular positions.

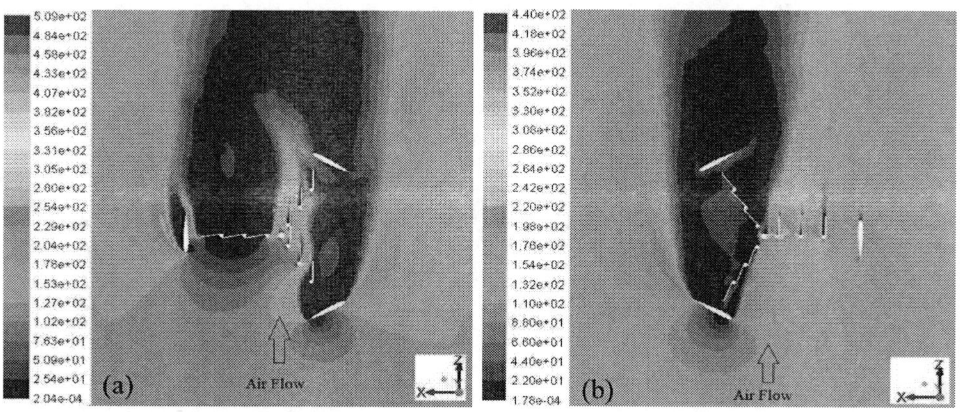

Figure . Dynamic pressure distribution of the three blades VAWT; (left), 90° angular position and, (right) 270° angular positions.

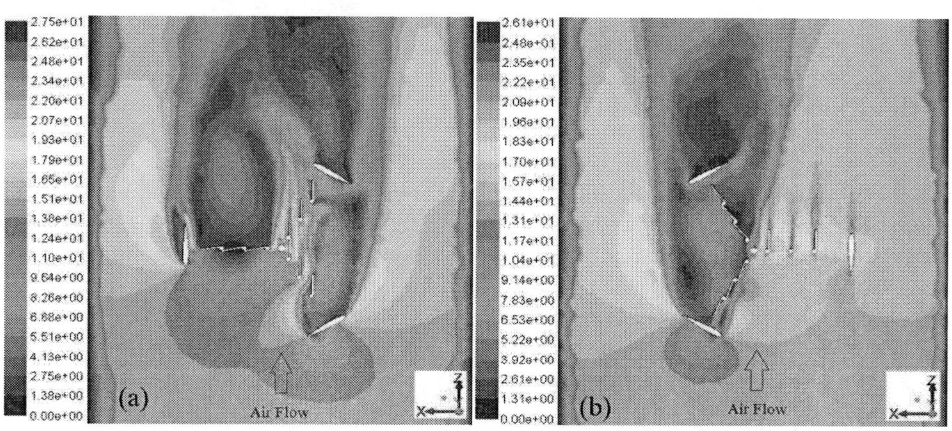

Figure : Velocity distribution of the three blades VAWT; (left), 90° angular position and, (right) 270° angular position.

. Conclusion

In this investigation the numerical simulation is carried out for the modeled geometry of the three dimensional combine blades (vertical flat plate vanes with the H-type airfoil) for the vertical axis wind to investigate the aerodynamics characteristics. ANSYS FLUENT16.1 software's is used for this investigation with k-ω turbulence model. The main conclusions from this simulation are

(1) The coefficient of drag C_d is high when the blade of the turbine in a positive side and will become maximized at the 90° blade angular position because of all vans are fully closed. While the minimum Cd at the negative side 270° blade angular.

(2) The static pressure drop is maximum occurring across the combined blade rotor sides with closed vane from the upstream to the downstream side at the 90° angular position. While minimum static pressure drop occurring across the combined blade rotor sides with open vane from the front side for the turbine rotor to back side of the turbine rotor at 270° blade angular position.

(3) The flow of air passes through the combined blade on the negative side when all the vanes are freely open. This indicated to the resistance of the blade to the flow is decreased. This case helps to increase turbine rotation and so the power of the turbine increased.

From above results, it can be concluded that the decreases in Cd and static pressure drop in the negative side helps to increase the rotation of the turbine and then the power of the turbine is increased. The predicted results from the numerical simulation gives acceptance compared with this is reported for steadying VAWT having different blade designs.

. References

[1] Sathyajith and Mathew S 2006 *Wind Energy-Fundaments, Resource Analysis and Economic* (Springer) Ed 1st

[2] Izli, N A V and F Kurtulmn 2007 A study on aerodynamic properties of some NACA profiles used on wind turbine blades *Applied Sci.* **7** 426-433.

[3] Mohamed M H, Janiga G, Pap E, Thevenin D 2011 Optimal blade shape of a modified Savonius turbine using an obstacle shielding the returning blade *Energy Conversion and Management* **52** 236 – 242.

[4] Suffer K H, Kahtan Y Y and Razlan Z M 2018 Aerodynamics and modal analysis for the combined Vane Type Vertical Axis Wind Turbine", *2nd Recent Trend in Science and Engineering Research International Conf.* (Penang: Malaysia)

[5] Rémi G Guy D and Matthieu B 2013 Parametric study of H-Darrieus vertical-axis turbines using RANS simulations. *21st Annual Conf. of the CFD Society of Canada* (Sherbrooke: Canada)

[6] Hussen M S, Rambabu K, Ramji M and Srinivas E 2015 Design and analysis of vertical axis wind turbine rotors *International Journal on Recent Technologies in Mechanical and Electrical Engineering (IJRMEE)* **2** 54-62

[7] Ying G, Liqin L, Xifeng G and Wanhai X 2018 Aerodynamics and motion performance of the H-Type floating vertical axis wind turban *Appl. Sci.* **8** 262

[8] Suffer K H, Quadir G A, Ismail K A and Usubamatov R 2016 CFD analysis of three and our blades movable vanes type vertical axis wind turbine having movable vanes *International Journal of Smart Grid and Clean Energy* **5** 259-263

[9] Qasim A Y, Usubamatov R and Zain Z M 2011 Design of vertical axis wind turbine with movable vanes *Australian Journal of Basic and Applied Sciences* **5** 896-902

[10] Khammas A F, Mustaffa M T, Usubamatov R, Askar K H, Qasim A Y and Guadir G A 2015 Investigate the effect of difference design parameter on performance evaluation of straight blade vertical axis wind turbine(SB-VAWT) *International Journal of Engineering Technologies and Management Research* **2** 9-18

IOP Conf. Series: Materials Science and Engineering **429** (2018) 012079 doi:10.1088/1757-899X/429/1/012079

Experimental Study on Relation of Small UAV Propeller Thrust Generation and The Mouth-Ring

M N K Othman[1], Z M Razlan[1], M S Muhamad Azmi[1], I Zunaidi[1], W K Wan[1], Shahriman A B[1], M S M Hashim[1], M K Faizi[1], I Ibrahim[1] and N S Kamarrudin[1]

[1] School of Mechatronic Engineering, Universiti Malaysia Perlis, Pauh Putra Campus, 02600 Arau, Perlis, Malaysia.

Corresponding author: zuradzman@unimap.edu.my

Abstract. In this study, the main focus is to study the thrust generation of various propeller blade width-spans with the mouth-ring application. By using the parameter, the thrust were calculated based on the value of velocity, voltage and RPM. The testing was performing in three different width-span that are; 10mm, 8mm and 5mm. There are also variable that be tested that cover the reducing the diameter of mouth-ring from 36 mm to 32mm. In addition, if compared the gaps and different position proved that centre position gives better thrust. Furthermore, the application of mouth-ring is evaluated to be good base on this study

1. Introduction

Quadcopters are typically class of Vertical Take-Off and Landing (VTOL) vehicles, that means it will take off and landing vertically by the ground. This idea was driven by Leonardo da Vinci attempts to design a vehicle that can moved up and forward. Quadcopters are categorizing by the six degree of freedom vehicle and the translational position (x, y, and z) is measured by internal coordinate of the system [1]. As the comprehension of inflexible body flight elements enhanced over time, researchers and architects could in the end outline a rotorcraft plane equipped for producing and keeping up lift over developed timeframes by mid twentieth century. These airships were kept an eye on and with the section of time alongside advances in innovation, the classification of kept an eye on air ship developed into unmanned air ship [2].

Quadcopters by and large utilize two sets of indistinguishable altered pitched propellers; two clockwise (CW) and two counter-clockwise (CCW). By changing the speed of each rotor it is possible to specifically generate a desired total thrust; to locate for the centre of thrust both laterally and longitudinally; and to create a desired total torque, or turning force. For best execution and most straightforward control calculations, the engines and propellers ought to be put equidistant.

An arrangement of propellers that being closed by shroud is called ducted fan or mouth-ring, this mouth-ring suitable to use for small type quadcopters. A typical mouth-ring that been used at the quadcopters propeller as the protection guard to protect the propeller when impact, the other function is to enhanced recovery the guard helps deflect the drone after a collision. It also gives better landing protection; the quadrotor will eventually land on its side. But with the guards, the props are protected [2].

2. Ducted fan (mouth-ring)

Ducted fan of unmanned ethereal vehicles (UAVs) give vehicles with the propeller encompassed by a duct and are fit for drifting flight for Vertical Take Off and Landing (VTOL). For a ducted-fan UAV, the duct expands the thrust by around 25% - 30%, and the surrounded propeller guarantees the security of the vehicle [3]. Ducted propellers consist of a combination of two principal components. The first is an annular wing which can be either symmetric with respect to the rotation axis or asymmetric to accommodate for the wake flow field variations [4].

Figure 1. Ducted propeller (mouth-ring) [5]

For the unique geometric structure, the duct produces an extra lift on the vehicle contrasted and a helicopter in drifting flight. In journey flight, the duct act as part of a ring air foil which delivers the dominant part aerodynamic lifts, but it additionally confronts a solid pitching moment that produce by an important aerodynamic drag on the vehicle [6].

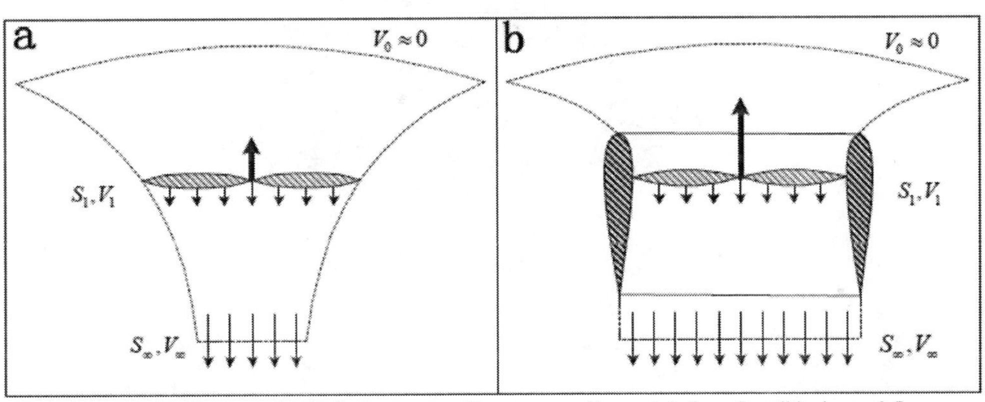

Figure 2. Contraction of air stream from propellers; (a) free fan (b) ducted fan

2.1. Measure the thrust generation

Thrust generation basically base on Newton's second law and third law, the generation of thrust that make the quadcopters to hovers and able to take off and landing. To measure the thrust generated by the quadcopters, first the will condition and the air flow shall be ignored. In mathematical terms;

$$T = v\,\frac{dm}{dt} \tag{1}$$

where T is thrust generated, v is the air velocity and $\frac{dm}{dt}$ is mass flow rate of the air, $\dot{m}$.

The mass flow rate of an air flow is denoted by;

$$\dot{m} = \rho A v \tag{2}$$

where ρ is air density and A is the flow area.

Substitute equation 2 into equation 1;

$$T = (\rho A v)v$$
$$T = \rho A v^2 \tag{3}$$

2.2. Assembly of experimental rigs

The type of propeller that will be used should be constant, thus the effect of thrust generation will be studied, and quadcopters that will be used also should be the same throughout this experiment to prevent any error or data that will affect the experiment later. The suitable length and diameter of mouth-ring will propose and the experiment will be conduct to study the relationship between thrust generation and the mouth-ring. Three propellers with different diameter were used in this research.

2.3. Method used

The position of apparatus must be as stated to prevent any disturbance during experiment. The propeller must be attach strongly to the motor to make sure that it will not break or will collide with the mouth-ring. For the positioning of mouth-ring, used the water balancer to make sure that mouth-ring is stable both ends. That makes the experiment more accurate. Mouth-ring also needed to be place first at the centre from its diameter and the propeller.

Figure 3. Apparatus setup for experiment

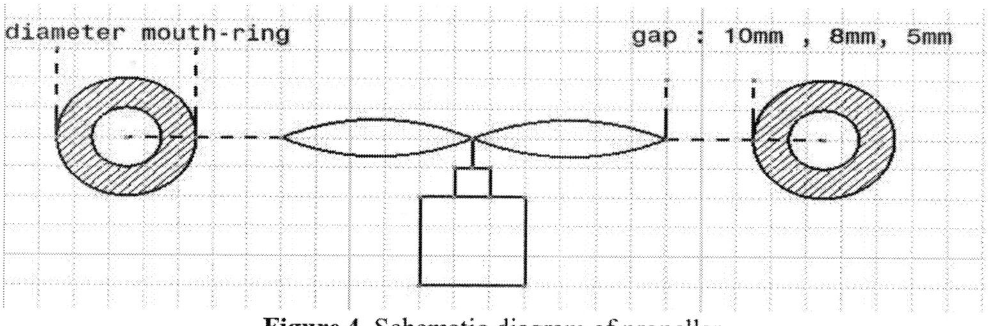

Figure 4. Schematic diagram of propeller

3. Results and Discussion

Figure 5 (a), (b) and (c) shows the comparison between 36mm diameter and 32mm diameter for position at centre, which is at 10, 8 and 5 mm gap from propeller, respectively. From Fig. 5(a), the 36mm diameter gives more thrust reading compare to the 32mm diameter. This is because the wider diameter conducts more air to the propeller thus gives higher thrust. In Fig. 5(c), thrust generated at the 36mm diameter gives higher reading compare to 32mm diameter but there is slightly increase value of thrust at 1.4635 N. compare to the 32mm diameter the thrust increase smoothly as the RPM increase. Figure 5(c) shows graph trends of 36mm diameter increase steadily as the RPM increase. Although for the 32mm diameter show that there are rapidly increase value of thrust at 1.0485 N, this rapidly increase value of thrust may be obtain by the ducted air is being disturbed or at certain RPM the Thrust value is high due to the air flow being conducted highly.

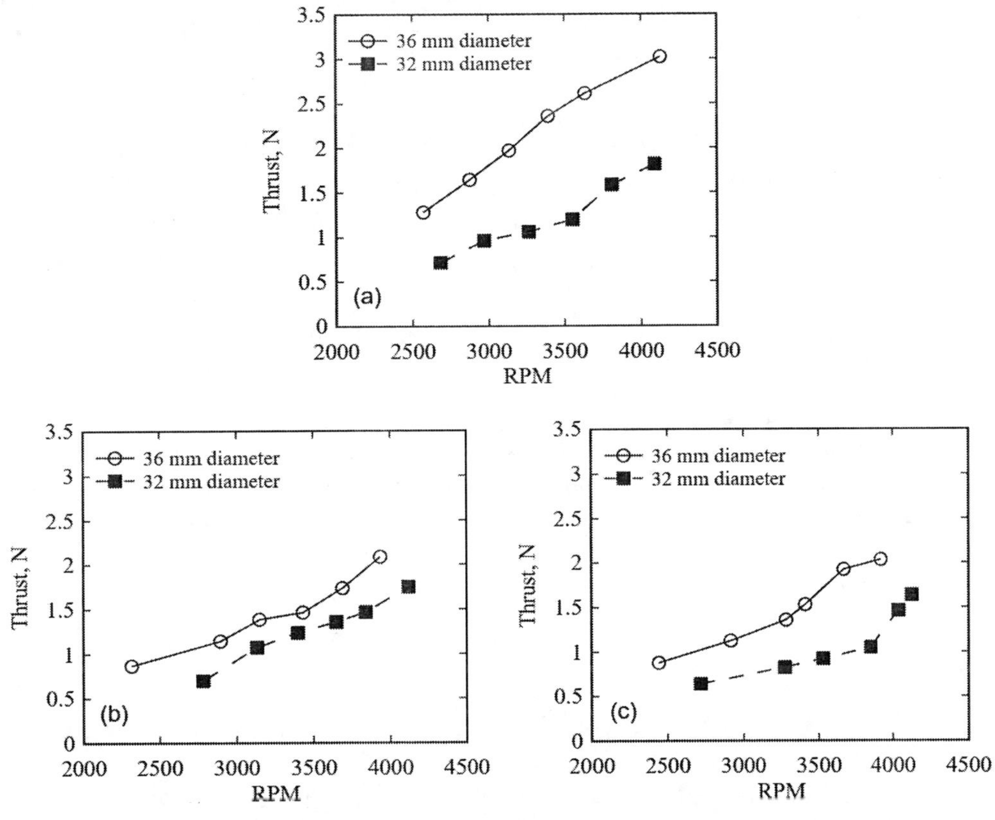

Figure 5. Graph of different diameters at centre for (a) 10 mm gap, (b) 8 mm gap and (c) 5 mm gap

Based on data above, thrust generated of 36 mm diameter mouth-ring are higher than the 32mm diameter mouth-ring for all gaps different. Furthermore, the longest gap (10 mm) between mouth-ring and propellers gives the highest reading that is 3.0206 N gives the highest reading of thrust 62.46% higher than 32 mm diameter mouth-ring.

From Figure 5, higher diameter of mouth-ring increase the thrust generated that's means the thrust can be optimize using much more wide mouth-ring. Although it increase the thrust, this experiment does not include the mass of mouth-ring its selves this results might be differs when the different mases is used.

The increase velocity in 10 mm gap is because of the broke down of tip vortex at the end of the blade tip. It happened that 10mm has the biggest gap that cross the tip vortex and make the air flow much more stable. Thus the air flow along the propeller stable and velocity increase.

The higher thrust by wider diameter mouth-ring is because the air flow from above the propeller can be utilized by much more wide diameter. This is because 36 mm diameter can assure that air flow from upper blade can optimize be across the centre much more efficient than 32 mm diameter. This wider diameter stabilizes air flow towards propeller's vortex that produces higher velocity.

From this study, this is the data form testing shows that in particular clearance, the air flow rate differ from each other. This is maybe because of the placing of mouth-ring have made turbulence at blade tips did not happened so that air flow would come straight downwards.

The wider diameter assures that air flow inside the propeller towards laminar flow, which makes that wider diameter mouth-ring gives higher reading. This also may because of wider diameter can ducted more air towards the blade tips compare to others. This ducted air flow much more stable and will produce thrust.

This research is typically based on experimental parts only; all data that get from experiment may be different from the theoretically. But, based on data that have been observe the experiment may be have some errors due to the position of the anemometer might be not at the best point to get the higher velocity. Furthermore the errors may be because of the air humidity might be different from time to time and also wind condition might be slightly effected the reading

4. Conclusion

Thrust generation of various propeller blade-span and additional condition of mouth-ring application was examined through this research. Both of tests with 36 mm mouth-ring and 32 mm of mouth-ring were tested also with different position and blade-span. The conclusions which can be getting from this research are;

a) The size of blade tip clearance also affect the thrust generation by reducing the turbulence air flow at the vortex. If the gap is too large, it rejects all the advantages of using the duct. If the gap is too close, it will reduce the flow of air because the air will be collides with the walls thus reducing the velocity.

b) The diameter of mouth-ring can be change by reducing it to study the effect towards thrust generation. By reducing the diameter of mouth-ring, it also reduce the thrust because of the air might be not enough be ducted into the blades. The wider diameter ensures that air is ducted through the blades thus produce higher thrust.

c) Application of mouth-ring might be importance with the suitable design and size. The uses of mouth-ring (ducted) helps in velocity increase with the good streamline of air flow through the duct.

5. References

[1] M. Hehn and R. D. Andrea, "Quadrocopter Trajectory Generation and Control," *IFAC World Congr.*, vol. **18**, no. 1, pp. 1485–1491, 2011.

[2] J. T. Khan, "Electrical and mechanical design of multipurpose ducted-fan type VTOL UAV," *Can. Conf. Electr. Comput. Eng.*, pp. 000015–000020, 2011.

[3] I. K. Peddle, T. Jones, and J. Treurnicht, "Practical near hover flight control of a ducted fan (SLADe)," *Control Eng. Pract.*, vol. **17**, no. 1, pp. 48–58, 2009.

[4] R. Bontempo, M. Cardone, M. Manna, and G. Vorraro, "Ducted propeller flow analysis by means of a generalized actuator disk model," *Energy Procedia*, vol. **45**, pp. 1107–1115, 2014.

[5] S. Yilmaz, D. Erdem, and M. S. Kavsaoglu, "Performance of a ducted propeller designed for UAV applications at zero angle of attack flight: An experimental study," *Aerosp. Sci. Technol.*, vol. **45**, pp. 376–386, 2015.

[6] Z. Wei, F. Ningjun, W. Zhengjie, and W. Yanxuan, "Modeling and Aerodynamic Analysis of a Ducted-Fan Micro Aerial Vehicle," *Conf. Model. Identif. Control (ICMIC), 2012*, pp. 768–773, 2012.

Numerical simulation of a new spoiler on upper surface of Clark Y14 wing

Mohammed Kheiraldeen Abbas Askar[1], Amer Q Hameed[2], Kadhim H Suffer[3], Zuradzman M Razlan[4]

[1, 3] Mechanical Engineering Department, Al-Nahrain University, P.O. Box 64040, Baghdad, Iraq
[2]Mechanical Engineering Department, Dyaala University, Dyalla, Iraq
School of Mechatronic Engineering Universiti Malaysia Perlis, Pauh Putra Campus, 02600 Arau, Perlis, Malaysia

[1]dr.mohammedaskar@yahoo.com, [2]amerkader93@gmail.com, [3]kadhim_askar@yahoo.com, [4]zuradzman@unimap.edu.my

Abstract. Spoilers are one type of devices that used to reduce the lift on the aircraft wing. A theoretical and numerical study involves simulation of the flow behaviour around of airfoil type (Clark y-14) for the wing section are carried out. ANSYS14 software is used to solve the governing equation in three dimensions steady incompressible turbulent regime with appropriate turbulent model ($k-\omega$, SST). In this study the lift and drag for different angles of attack, with and without spoiler effect on distribution of pressure and velocity distribution on the wing surface are steadied. A constant velocity 40 m/s is used on the wing with and without new spoiler have different cord (10% C) and (15% C) and angles of attack from -8° to +18°. The wing model is rectangular with (30 cm span and 9 cm cord). Wing has aspect ratio 30/9 and a maximum thickness of 14 % of the wing chord. A trial and error for five-gaps between spoiler and wing (0%, 2.77%, 3.33%, 3.88% and 6.66% of the wing cord)are carried out when is the wing cord is equal to 9 cm and for different gaps (0,2.5, 3, 3.5and 6 mm). Theoretical results indicated when is location spoiler 50%C, δ=0, new spoiler cord (HI) 10%C and gap between new spoiler and wing equal 3.33%C which increase maximum lift theoretically to (28.29%) compared with clean wing.

Keywords: Rectangular wing, new spoiler, ANSYS software, lift, drag, high drag device.

1. Introduction

Spoiler is the one type of high drag devices that used to reduce the lift on the aircraft wing [1]. It is like flap structure of rectangular cross-section whose leading edge is hinged on upper wings (at any angle) disturbing the streamlined flow of air there by changing the amount of lift. The net force on aircraft in the vertical direction is equal to weight minus lift. Thus, when the spoilers are employed lift gets reduced as a result the net downward force acting on the wing increases. There are two types of spoiler used in aircraft, *Flight Spoilers* and *Ground Spoilers* [2]. The flight spoilers are used to reduce lift and increase drag, without increasing speed to a very high rate this allows easier landing the aircraft, while ground spoilers also called lift dumpers (or called airbrake) deployed on landing of the aircraft. These types are slow down the aircraft by increasing the drag and also decrease the lift gradually. As a result, all the aircraft weight is moved from the wings to the undercarriage. This leads to ease of braking also reduces chances of skidding. To get a clear understanding of the work of the spoiler, and its relationship to lift and drag, and the development of mechanisms spoiler for the best result of the corruption of many of the research conducted in the past to investigate this matter. Some of studies are focused on the experimental work for the spoilers [3-6], while the other studies was carried out by use numerical simulations [7-13]. From previous studies, it can be seen that there are many types of investigations and studies for the effect of plain spoiler at different positions and different heights and some configurations of spoiler on the aerodynamic characteristics of different wing sections experimentally and numerically by use CFD software. There is no study include the gap between the spoiler and wing.

2. Current simulation

The aim of this study is to analysis the airflow around the wing with and without new spoiler to control the separation of the flow over the wing at high angle of attack. The new spoiler in this steady means to change the position (gap) and the inclination angle of spoiler with respect to wing for control the separation. Because of the complication of airfoil surfaces with spoiler configurations and the strong viscous fluid effects it is difficult to obtain an analytical solution of the Navier- stokes equation for such complex configurations. Thus the numerical techniques have been used to solve this equation. The analysis of the flow field around airfoil with and without new spoiler will be carried out by using a numerical techniques using a computational fluid dynamics (CFD).

AutoCAD software is used to modelling the 3-D wing by using the Clark y-14 airfoil data for 2-D and then extrude the profile of this data 30cm to get 3-D wing and then draw spoiler over this wing as shown in Figure 1. The computational domain of the test section is fixed as (30.48cm x 30.6 cm x 60.96cm) as shown in Figure 2. In the previous studies the wing model with and without spoiler were tested at different angels of attack ranged from (-8° to 18°) [14]. In this study the wing model used was a rectangular wing of Clark y-14 airfoil, and had a maximum thickness (14 % C) and an aspect ratio (δ) of (30/9), the wing model used has 30cm span and 9cm chord. The model is meshed as a fine mesh with min. size 0.0000002m and max. Size 0.000002m for the wing with and without new spoiler as shown in Figure 3 (a) and 3 (b) respectively.

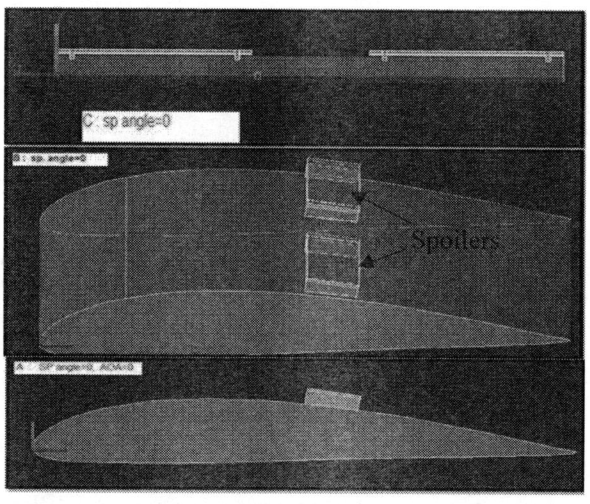

Figure 1. 3-D wing model by using the Clark y-14 data with spoilers.

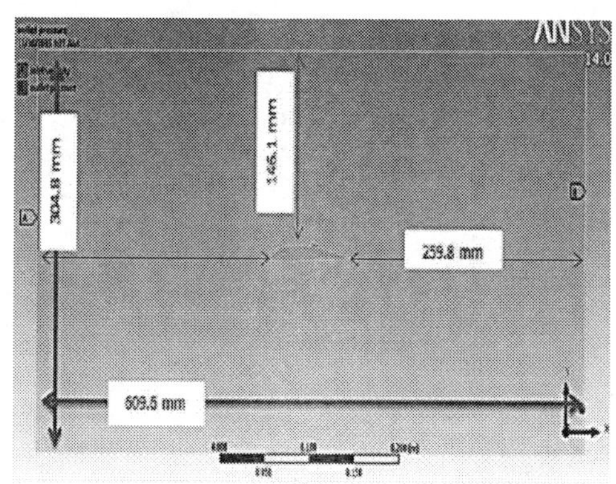

Figure 2. The domain of the test section.

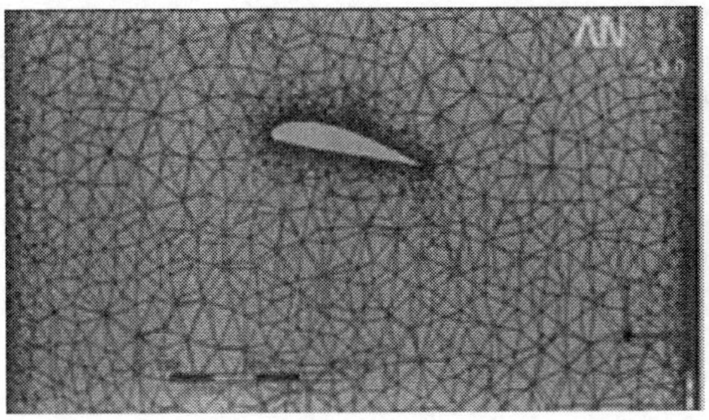

Figure 3. (a) Generated mesh for the clean wing.

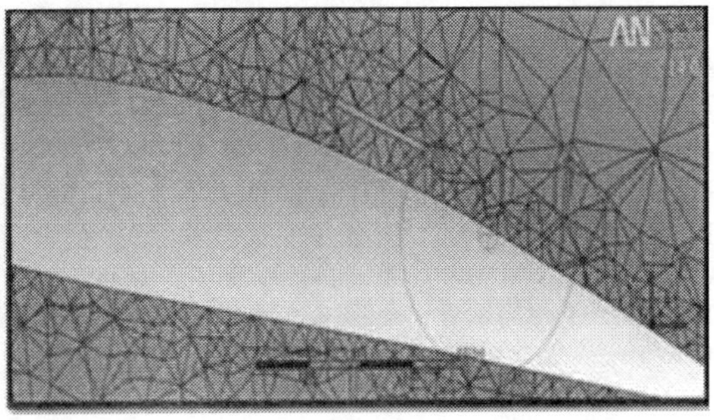

Figure 3. (b) Generated mesh for the wing with spoiler

3. Results and discussion

Fluid flow (fluent) software is used to solve the current study problem. In this analysis the governing partial differential equation in three dimensions, which based on conservation of momentum and conservation of mass equations will be solved. The (k-ω) turbulence model also will be used in this analysis to demonstrate the turbulence effect on the flow, because of its give a good behaviour in adverse pressure gradients and separating flow. The working fluid is air at temperature 17^0 C and the flow characteristics assumed to be as follows 3-D, incompressible flow, Newtonian and physical properties of the air are constant. The inlet is choice as inlet velocity and fixed to 40 m/s corresponding to Reynolds number of $8.24119 x 10^5$, and the outlet is a pressure outlet. No slip boundary condition is specified for the wing with and without winglet surfaces and is set as a wall. Figure 4 shows the scaled residuals of continuity, momentum, k and ω as the solution progresses to convergence history. In order to illustrate the effect of best gap in the improvement of aerodynamic performance, simulations were performed for wing without and with best gap for angles of attack (A.O.A.) from (0° to 10°) in this test to study for steadying all sides. The gaps between spoiler and wing increase the lift coefficient of Clark y-14 airfoil wing compared with clean wing (wing without spoiler). Figure 5 shows the results of lift against the angle of attack (A.O.A.) for clean wing and wing with gap 3mm, these results indicate the gaps worked like nozzle work in other word increase air.

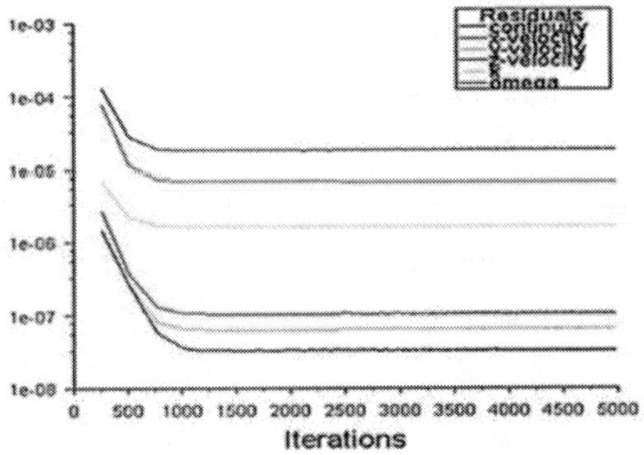

Figure 4. 3-D wing model by using the Clark y-14 data with spoilers.

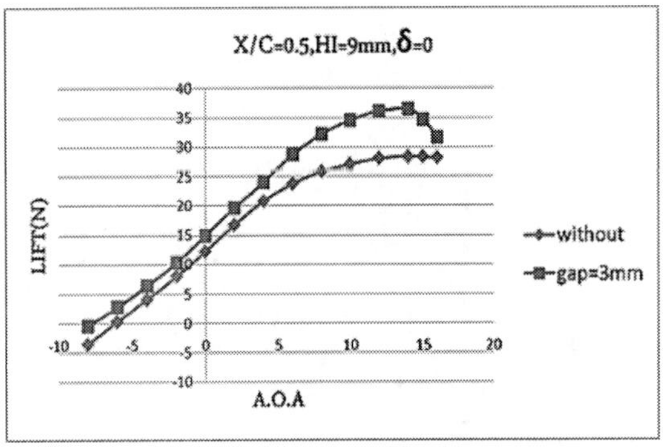

Figure 5. The domain of the test section.

Figures (6) and (7) show the contours of steam line for the flow of air over the clean wing and the wing with best spoiler gap (3mm), with aspect ratio (δ = 0) and A.O.A. = 0°, while Figures (8) and (9) for aspect ratio (δ = 0) and A.O.A. = 10°. The velocity magnitude has maximum value at the upper surface especially at the leading edge, and low value at the lower surface led to increase pressure, the streamline flow during the gap shows in Figure 8. In Figure 6 when clean wing used, it can see that the streamline driving away from the wing at trailing edge compare with the results of Figure 7, the flow streamline approach more the upper surface of the wing that effect allow flow rate remain laminar and prevent the separation. Frome these results it can be concluded that the effect of best gap and that effect more clear when the A.O.A. = 10° as shown in Figures (8) and (9). When spoiler angle = 90° we see the streamline before spoiler move up and down spoiler and after spoiler turbulent flow and random as seen in Figure 18. Velocity in the gap (compressed boundary layer) led to push any separation flow on upper surface of the wing to back the wing and worked to reduce separation of stream line on upper surface Figure (6) and (7). The addition of gap contributes to accelerate the flow on the upper surface of the wing all these reasons cause increases the maximum lift, compare with clean wing clear shown in Figure 5. Maximum lift can be get from spoiler position 50% Wing Chord, Spoiler height or spoiler cord 9% Wing Chord, and gap 3.33%.

Figures (10) and (11) show the contours of total pressure in (Pa) for the flow of air over the clean wing and the wing with best spoiler gap (3mm), with aspect ratio (δ = 0) and A.O.A. = 0°, while Figures (12) and (13) for aspect ratio (δ = 0) and A.O.A. = 10°. From the results of Figure 11 it can be seen that the total and static pressures for wing with and without spoiler decrease at upper surface of spoiler and increase pressure lower surface of the wing. When the spoiler at the angle of 90°, the increases of the pressure on upper surface of the spoiler led to increase the drag as shown in Figure 19.

Figures (14) and (15) show the contours of the effect of viscosity in (kg/m.s) for the flow of air over the clean wing and the wing with best spoiler gap (3mm), with aspect ratio (δ = 0) and A.O.A. = 0°, while Figures (16) and (17) for aspect ratio (δ = 0) and A.O.A. = 10°. In Figures (17) and (17) it's clearly can see the effect viscosity at trailing edge its small than when the best spoiler gap used, and very high after spoiler when spoiler angle 90° and very small at trailing edge as shows in Figure 20.

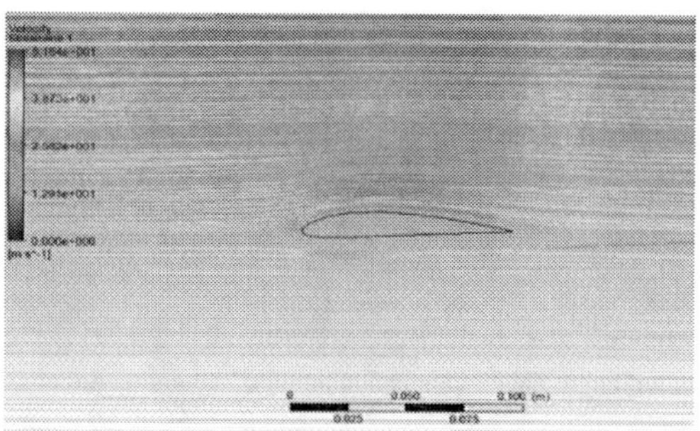

Figure . The contours of stream line on clean wing for δ=0 and A.O.A= 0

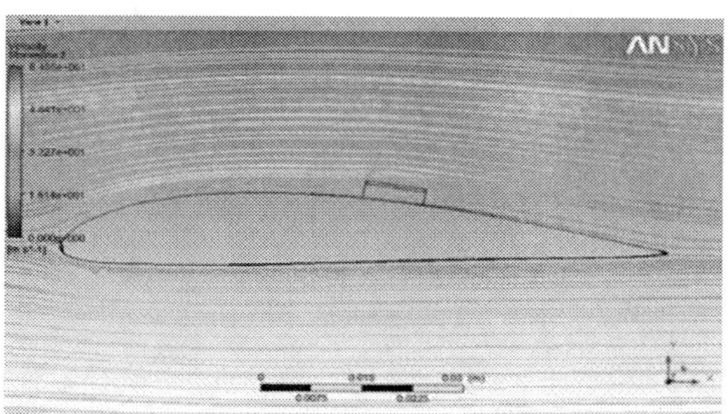

Figure . The contours of steam line on wing with new spoiler for δ=0 and A.O.A= 0

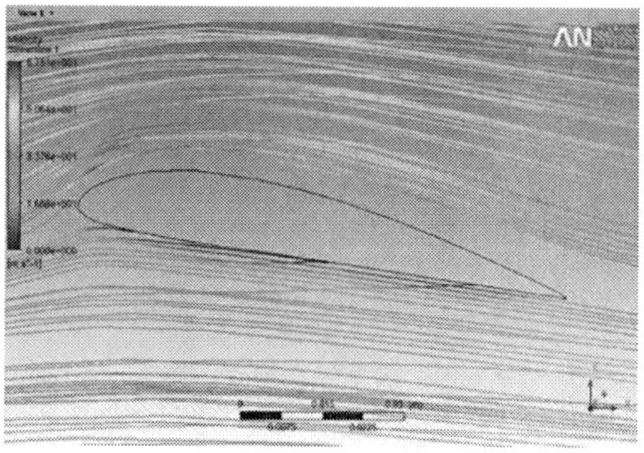

Figure . The contours of stream line on clean wing for δ=0 and A.O.A= 10

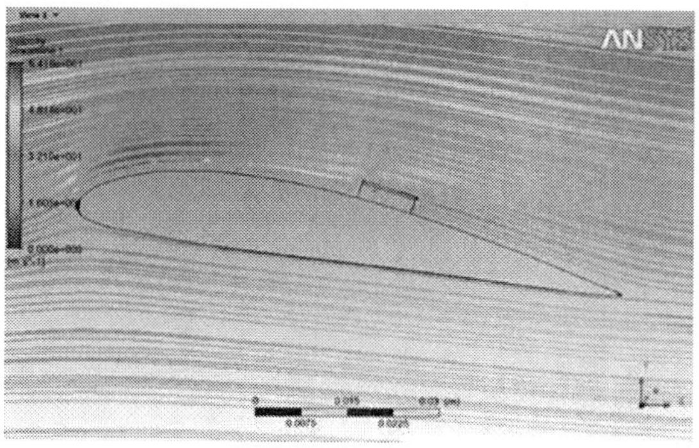

Figure . The contours of steam line on wing with new spoiler for δ=0 and A.O.A=10

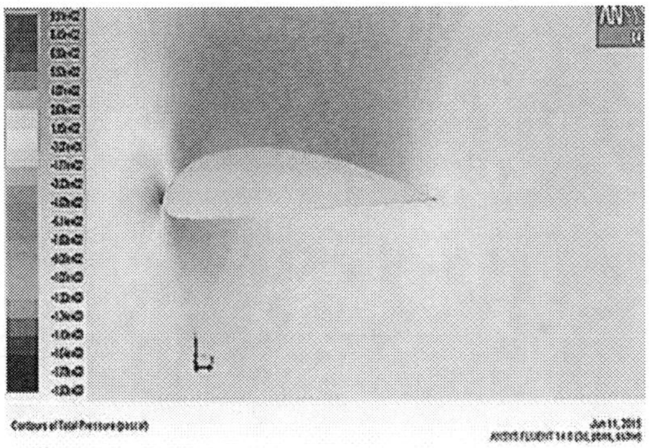

Figure 1 . The contours of the total pressure (Pa) on clean wing for δ=0 and A.O.A= 0

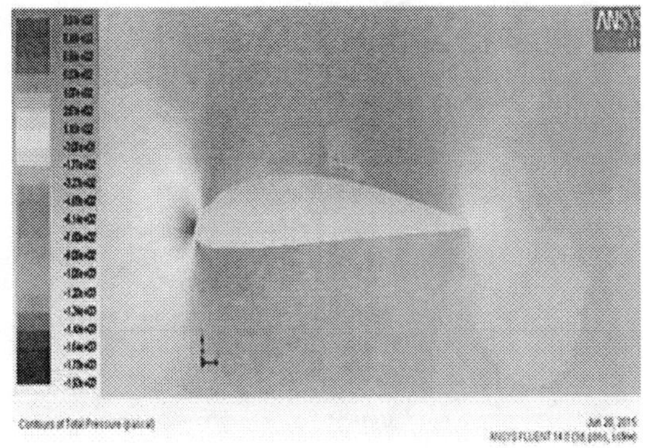

Figure 11. The contours of the total pressure (Pa) on wing with new spoiler for δ=0 and A.O.A= 0

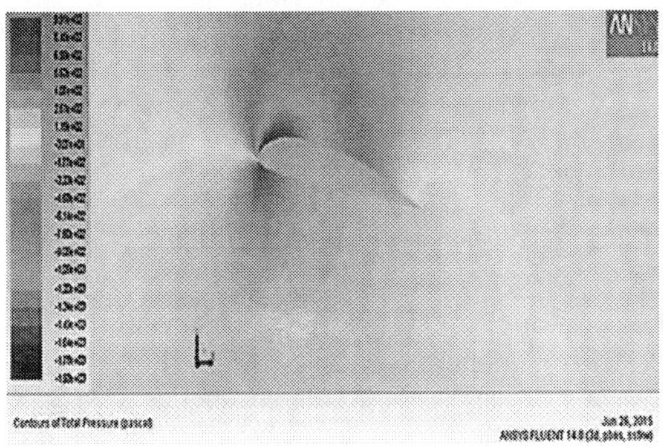

Figure 12. The contours of the total pressure (Pa) on clean wing for δ=0 and A.O.A= 10

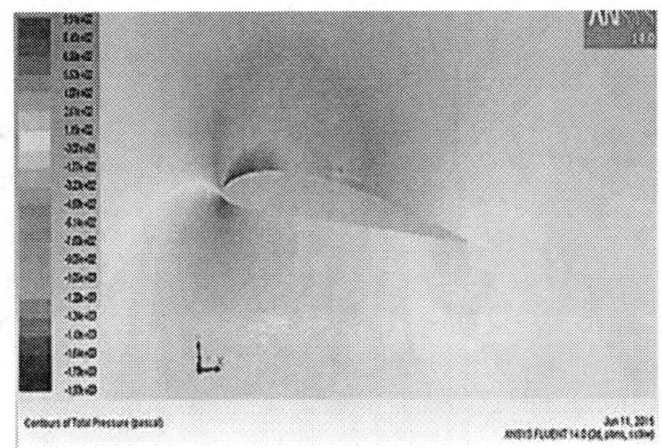

Figure 131 The contours of the total pressure (Pa) on wing with new spoiler for δ=0 and A.O.A= 10

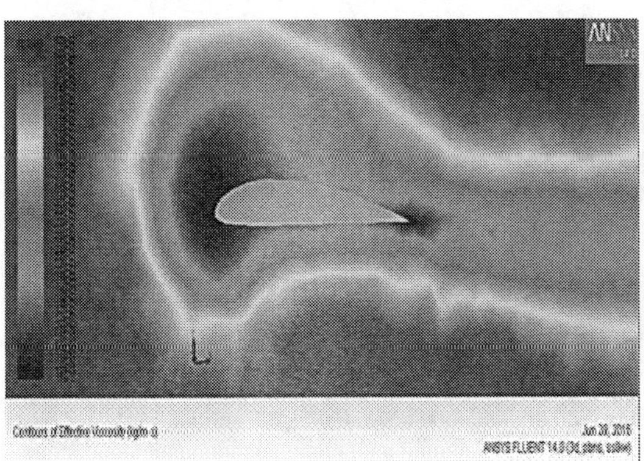

Figure 1 . The contours of the viscosity effect (kg/m.s) on clean wing for δ=0 and A.O.A= 0

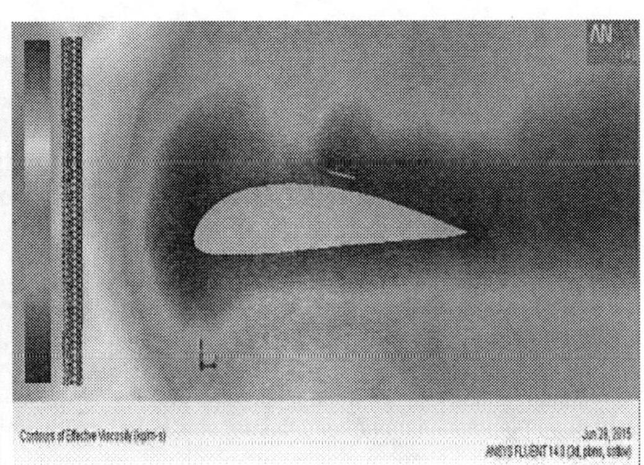

Figure 1 . The contours of the viscosity effect (kg/m.s) on wing with new spoiler for δ=0 and A.O.A= 0

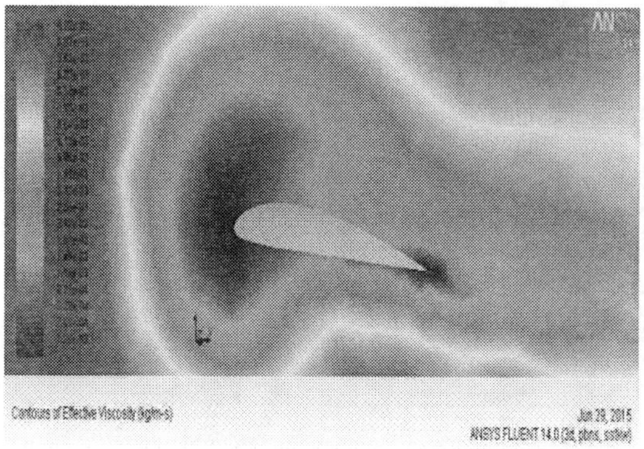

Figure 1 . The contours of the viscosity effect (kg/m.s) on clean wing for δ=0 and A.O.A= 10

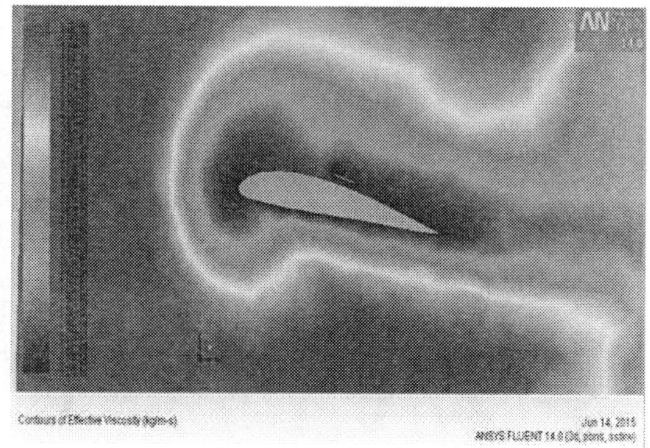

Figure 1 . The contours of the viscosity effect (kg/m.s) on wing with new spoiler for δ=0 and A.O.A= 10

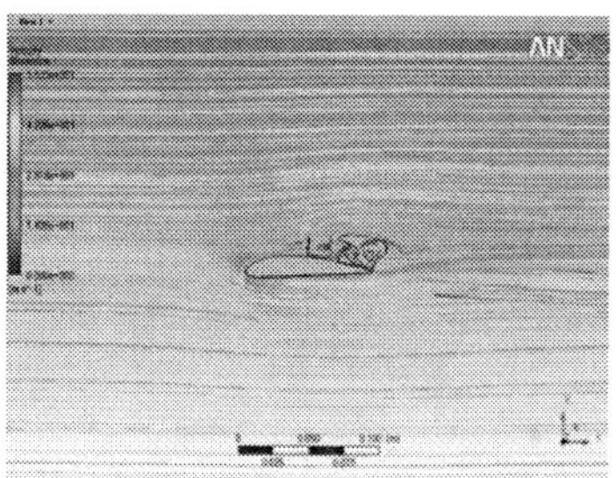

Figure 1 . The contours of stream line on clean wing for δ=90 and A.O.A= 0

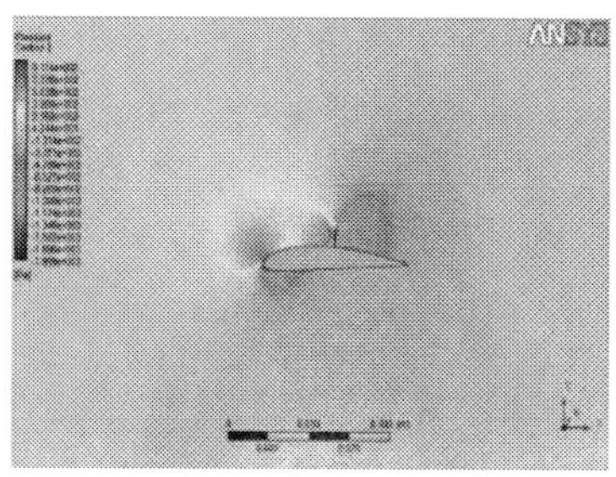

Figure 1 . The contours of total pressure (Pa) on clean wing for δ=90 and A.O.A= 0

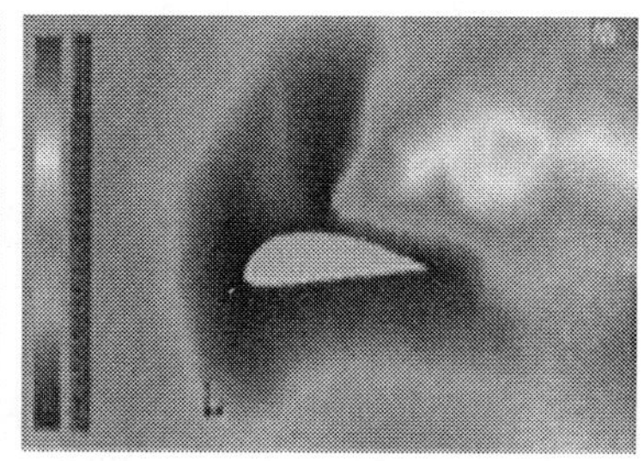

Figure 2 . The contours of the viscosity effect (kg/m.s) on clean wing for δ=90 and A.O.A= 0

4. Conclusion

In this investigation the numerical simulation is carried out for the clean wing and wing with the spoiler the three dimensional domain to investigate the aerodynamics flow. ANSYS FLUENT14. software's is used for this investigation with k-ω turbulence model. The main conclusions from this simulation are;

- The gaps between spoiler and wing increase the lift coefficient of Clark y-14 airfoil wing compared with clean wing (wing without spoiler). Maximum lift can be get from spoiler position 50% Wing Chord, Spoiler height or spoiler cord 9% Wing Chord, and gap 3.33%.
- Velocity in the gap (compressed boundary layer) led to push any separation flow on upper surface of the wing to back the wing and worked to reduce separation of stream line on upper surface and the addition of gap contributes to accelerate the flow on the upper surface of the wing all these reasons cause increases the maximum lift, compare with clean wing
- When the spoiler at the angle of 90°, the increases of the pressure on upper surface of the spoiler led to increase the drag.
- The effect viscosity at trailing edge its small than when the best spoiler gap used, and it's very high after spoiler when spoiler angle 90° and very small at trailing edge.
- Results reveal we can use Spoiler high lift device when spoiler angle 0° and high drag device when spoiler employed. From this study indicates to the first time in the history of aircrafts can be employed gap between spoiler and wing (used spoiler high lift device δ = 0 and high drag device when spoiler employed) to improve wing performance.

References

[1] Theodore A T 1975 *Introduction to the aerodynamics of flight* National Aeronautics and Space Administration (Washington).

[2] Nair S Nair and Nair Sy 2014 Aircraft Braking System *Indira Gandhi Delhi technical*

[3] Mashud M, Ferdous M and Omee S H 2012 Effect of spoiler position on aerodynamic characteristics of an airfoil *International Journal of Mechanical & Mechatronics Engineering* **12** 1-6

[4] Harley C D 2010 Aerodynamic performance of low from factor spoilers *School of Mechanical, Aerospace and Civil Engineering, University of Manchester, PhD Theses.*

[5] Doggett R V 1989 Some effects of aerodynamic spoilers on wing flutter *National Aeronautics and Space Administration* NASA, pp 193-209

[6] Blair G and Karamcheti K 1984 An experimental Study of airfoil-spoiler aerodynamics *Stanford University Department of Aeronautics and Astronautics Stanford, California 94305*

[7] Patel M P and Sowle Z H 2007 Autonomous sensing and control of wing stall using a smart plasma slat *Journal of Aircraft* 516-527

[8] Wentz H W 1975 Effectiveness of spoilers on the GA (W)-1 Airfoil with a high performance fowler flap *NASA Wichita state university Kansas 67208-1975.*

[9] C. S Barnes, Ph.D 1966 A developed theory of spoilers on airfoiles *Aeronautical research council curren papers London C.P.NO.887*

[10] Sherman A. Clevenson and John E 1957 Experimental investigation of the oscillating forces and moments on a two-dimensional wing equipped with an oscillating circular-arc spoiler *American society of civil engineers ISSN (print): 0893-1321-1957.*

[11] Kondwani Kanjere .et al. Aeroacoustic 2010 Investigation of Deployed Spoiler during Steep Approach Landing, University of Southampton, UK.- AIAA 2010-3992

[12] Guillaume F, Pape M C, and Montagnac M 2004 Numerical simulation around wing control Surfaces *24International congress of the aeronautical Sciences* 1-12

[13] Zhao M, Zhang M and Jianzhong X 2017 Numerical simulation of flow characteristics behind the aerodynamic performances on an airfoil with leading edge protuberances *Engineering Applications of Computational Fluid Mechanics* **11** pp 193-209

[14] Abbas M K and Hassan H A 2013 Numerical and Experimental study of the effect of winglet at subsonic flow *Al Anbar*

A comparative study on targeting CO_2 emissions reduction from small-scale utility system

Ahmed Mahmoud* **and Jaka Sunarso**

Research Centre for Sustainable Technologies and Faculty of Engineering, Computing and Science, Swinburne University of Technology, Jalan Simpang Tiga, 93350 Kuching, Sarawak, Malaysia

*Corresponding author: melhadi@swinburne.edu.my

Abstract. This study targets CO_2 emission reduction from a particular palm oil refinery scenario presented below where medium-pressure (MP) steam and low-pressure (LP) steam are supplied for process heating. Two different utilities configurations, i.e., steam boiler alone and steam boiler followed by steam turbine are evaluated and compared in terms of their CO_2 emission reduction target. Steam boiler alone achieved 20.9% CO_2 emission reduction compared to steam boiler followed by steam turbine at the same retrofit design with a minimum temperature driving force, ΔT_{min} of 17 °C. Such emission reduction is still considered attractive on global basis if the required electricity comes from the renewable energy resources.

1. Introduction

Emissions of greenhouse gases, such as carbon dioxide (CO_2), from the combustion of fossil fuels for process heating and electricity generation is currently the largest contributor to global warming. Improvement in energy system (such as retrofit of heating system), fuel switching to fuel with lower carbon to hydrogen ratio (such as natural gas), CO_2 capture and storage, and shift to renewable energy resources (such as solar, wind, and hydroelectric power) can allow significant reduction in global CO_2 emissions [1-4]. In this context, some countries like Chile has applied tax on CO_2 emissions to encourage the industries to reduce their emissions [5]. European Union, on the other hand, has proposed Emissions Trading System (ETS) to reduce CO_2 emissions from the industries through its economic incentives [6, 7]. The ETS provides right to under-emitters who produce carbon emissions less than the allowance allocation (i.e., allowable level of CO_2 emissions) to sell their surplus allowances for profit and over-emitters who produce more than their allowance allocation to buy additional allowances [8].

Considerable research attention has been devoted over the past decades to address issues related to CO_2 emissions from process industry. For example, Smith and Delaby [9] developed model that relates process minimum energy consumption to flue gas emissions from utilities systems such as furnace, boiler, and turbine. They also define the global counting of CO_2 emissions as the sum of emissions from the utilities on a site (i.e., factory) together with those from the outside of the factory fence. Additionally, they proposed a design methodology to minimise the emissions by fuel switching, utilities system design change, process modification, and applying end-of-pipe waste minimisation techniques [10]. Mahmoud et al. [1] developed graphical methodology to reduce CO_2 emission from chemical processes by utilising an alternative fuel with lower carbon to hydrogen ratio and reducing the energy requirement by retrofitting the existing heat exchanger network (HEN). In their work, for a given target of CO_2 emissions reduction, the fuel switching is initially performed. After the CO_2 emissions limit is reached, further CO_2 emission reduction is obtained by realising energy saving within the retrofitted HEN at a certain minimum temperature driving force (ΔT_{min}). Mahmoud and Sunarso [11] then extended the graphical method by aligning several different factors, i.e., fuel switching, energy saving, investment cost, carbon credit, and payback time to the CO_2 emissions reduction percentage in a single graphical framework that facilitates comprehensive accounting perspective to assist the decision makers in deciding their emissions reduction targets.

Little attention has been given to target CO_2 emissions from small-scale utility system that constitutes single boiler alone or single boiler followed by steam turbine. Majority of works in this topic have so far focus on targeting CO_2 emissions from the stand-alone furnace [1, 11, 12]. Thus, this work aims to investigate and compare CO_2 emissions from small-scale utility system that uses boiler alone or boiler followed by steam turbine.

2. Mathematical model for CO_2 emissions calculation

Carbon dioxide (CO_2) is produced mainly from the combustion of fossil fuels in furnace or boiler to provide heat and steam for process heating and electricity generation. Fossil fuels are combusted by oxygen (O_2) to produce CO_2 according to the following reaction:

$$C_nH_m + \left(n + \frac{m}{4}\right) O_2 \rightarrow nCO_2 + \frac{m}{2} H_2O \quad (1)$$

Where n and m represent the mole amount of carbon (C) and hydrogen (H) that are present in the fuel, respectively. Air is assumed to be provided in excess amount to ensure complete combustion so that carbon monoxide (CO) is absent.

CO_2 emissions rate (kg s^{-1}) is related to the heat duty of the combusted fuel using the following model [9].

$$CO_2 \; emissions = \frac{Q_{fuel}}{NHV} \times \frac{C\%}{100} \times \alpha \quad (2)$$

Where, Q_{fuel} is the heat duty from fuel (kW), NHV is the net heating value (kJ kg^{-1}), $C\%$ is the mass percentage of carbon in the fuel, and α is the ratio of the molar mass of the oxidised form (CO_2) to the non-oxidised form (C) of the pollutant.

2.1. CO_2 emissions from the steam boilers

Boilers produce steam from the combustion of fuel. A theoretical flame temperature and stack temperature of 1800 °C and 160 °C, respectively, are adopted as typical values. Steam is delivered to the process at the temperature required by the process. Alternatively, the steam can be delivered at a higher temperature and then subjected to the throttling process. In this case, the quality of the steam can be retained by adding boiler water feed following the expansion process; the process of which is defined as desuperheating (Figure 1).

IOP Conf. Series: Materials Science and Engineering **429** (2018) 012081 doi:10.1088/1757-899X/429/1/012081

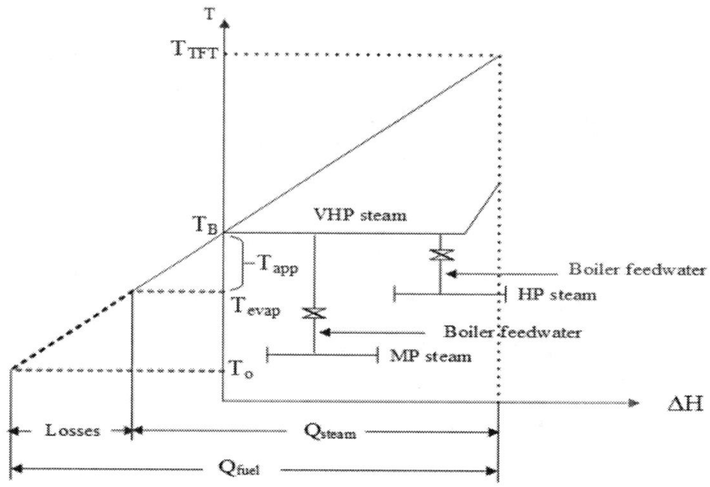

Figure 1. Temperature as a function of enthalpy for the steam generation in the boiler that also illustrates the effect of throttling process [9, 11].

The mass of the boiler feed water needed to desuperheat the steam can be calculated using heat balance by assuming that the boiler feed water is available at 100 °C (h = 419 kJ kg^{-1}) as expressed in Equation (3) [13].

$$x_{Desup} = \frac{h_{SUP} - h_{PROC}}{h_{SUP} - 419} \qquad (3)$$

Where x_{Desup} is the flow rate of desuperheated boiler feed water per kg of desuperheated steam, h_{SUP} is the enthalpy of superheated steam (kJ kg^{-1}), and h_{PROC} is the enthalpy of steam to process (kJ kg^{-1}).

If an approach temperature, T_{app} of 50 °C is assumed in the boiler, the fuel required in the boiler to fulfill the process duty can be calculated using Equation (4).

$$Q_{fuel} = \frac{Q_{PROC}}{L_{Proc}}\left(1 - x_{Desup}\right)\left(L_B + h_B\right)\left[\frac{T_{TFT_B} - T_0}{T_{TFT_B} - T_B - 50}\right] \qquad (4)$$

Where Q_{PROC} is the process heat duty (kW), L_{Proc} is the latent heat of steam delivered to the process (kJ kg^{-1}), L_B is the latent heat of steam under boiler condition (kJ kg^{-1}), h_B is the enthalpy of superheated steam under boiler condition (kJ kg^{-1}), T_{TFT_B} is the theoretical flame temperature in the boiler (°C), T_0 is the ambient temperature (°C), and T_B is the condensing temperature of steam under boiler condition (°C).

2.2. CO_2 emissions from the steam boilers followed by steam turbine

The combustion takes place in a high pressure boiler fired with any fuel to produce high pressure superheated steam. Then, the steam is passed through a steam turbine to generate lower steam level for process heating (Figure 2).

If an isentropic efficiency of the turbine is specified (e.g., 85%), the power generated and the heat left after the expansion can be calculated using Equations (5) and (6), respectively [9].

$$W_{ST} = 0.85\left[L_B \frac{(T_B - T_{PROC})}{T_B + 273} + h_{sup,B}\left[\frac{\frac{T_{SUP}-T_B}{\ln(T_{SUP}/T_B)} - T_{PROC}}{\frac{T_{SUP}-T_B}{\ln(T_{SUP}/T_B)} + 273}\right]\right] \quad (5)$$

Where W_{ST} is the actual power generated by the steam turbine (kW kg^{-1}), T_{PROC} is the temperature of the saturated process steam (°C), $h_{sup,B}$ is the enthalpy of superheat under boiler condition (kJ kg^{-1}), and T_{SUP} is the temperature of superheated steam from boiler (°C).

$$h_{out} = h_{sup,B} - \frac{W_{ST}}{0.85} \quad (6)$$

Where h_{out} is the enthalpy of the steam after expansion through a turbine (kJ kg^{-1}). If h_{out} is greater than h_{PROC}, it is necessary to desuperheat the steam. On the other hand, if h_{out} is less than h_{PROC}, then the wet fraction of the steam should be separated (Figure 2).

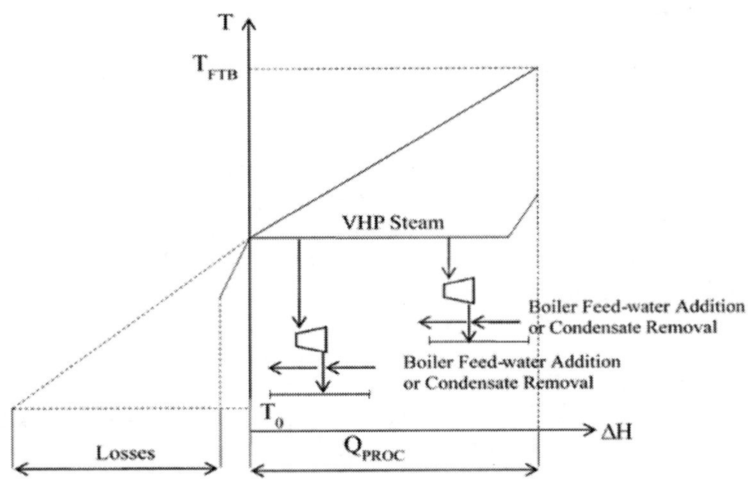

Figure 2. Temperature versus enthalpy profile of steam turbine and steam levels generation [9].

3. Case study

A heat exchanger network of palm oil refinery is considered in this study to demonstrate the calculation of CO_2 emissions from steam boiler alone and steam boiler followed by steam turbine. The work of Haslenda et al. can be consulted for the details of the process flowsheet for oil palm refinery [12]. The grid diagram of the existing HEN of palm oil refinery is shown in Figure 3, where the hot streams (H1 to H4) and the cold streams (C1 to C3) are grouped together in the upper and the lower parts of the grid, respectively. Medium and low pressure steams are used as the utilities to provide process heating.

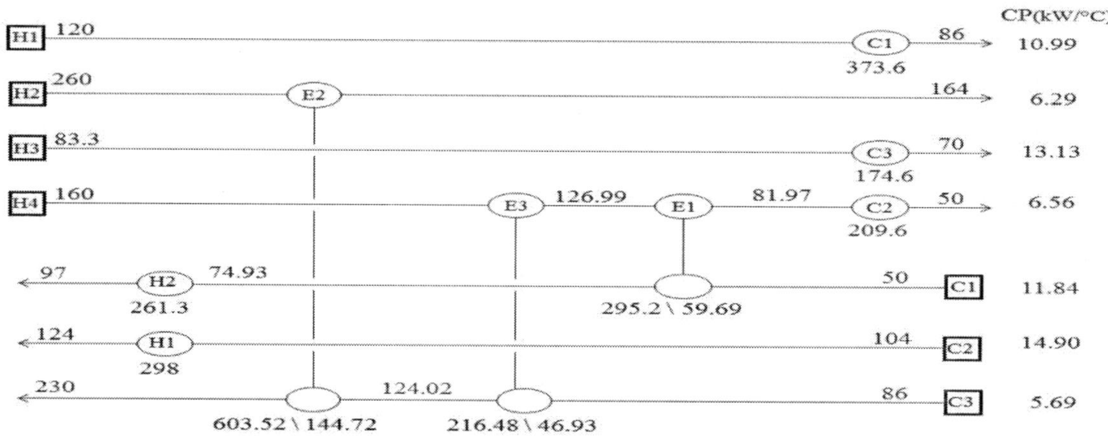

Figure 3. The grid diagram of the existing heat exchanger network for the palm oil refinery (Q (kW)\A (m^2)).

Table 1 lists the utility and the cost data for the case study. The total heat capacity (CP) of each stream is calculated by dividing its heat load with the temperature difference of the source and the target (Figure 3). A heat transfer coefficient of 0.2596 W m^{-2} K^{-1} was used for all streams [14]. The existing HEN configuration was designed at ΔT_{min} of 30 °C. This network experienced fouling after three months of operation due to the processing of edible oil. The fouling led to significant increase in the energy consumption. Figure 4 and Figure 5 show the balance grand composite curves of the HEN before and after the fouling, respectively. By comparing Figure 4 to Figure 5, it becomes apparent that the energy consumption after fouling (Figure 5) translates to the requirement for extra utilities as indicated by the dotted upper and lower lines in Figure 5. It is worth noting that fouling phenomenon is generally sensitive to the temperature variation, which in this edible oil processing case takes place significantly above 200 °C [12]. The existing HEN contains two streams that are exposed to temperature higher than 200 °C, i.e., the hot stream H2 and the cold stream C3 (Figure 3). The existing design of HEN at ΔT_{min} of 30 °C can always be retained by washing the fouled heat exchanger [11].

Table 1. The utility and cost data for 3 different utility streams.

Type	T_{supply} (°C)	T_{target} (°C)	H_f (kW m^{-2} K^{-1})	\$ kW^{-1} yr^{-1}
Cooling water	25	30	2.5	18.2
LP Steam	143	142	4.5	106.4
MP Steam	184	183	4.5	159.6

Note: heat exchanger capital cost = 1054 $A^{0.65}$ (A = heat exchange area).

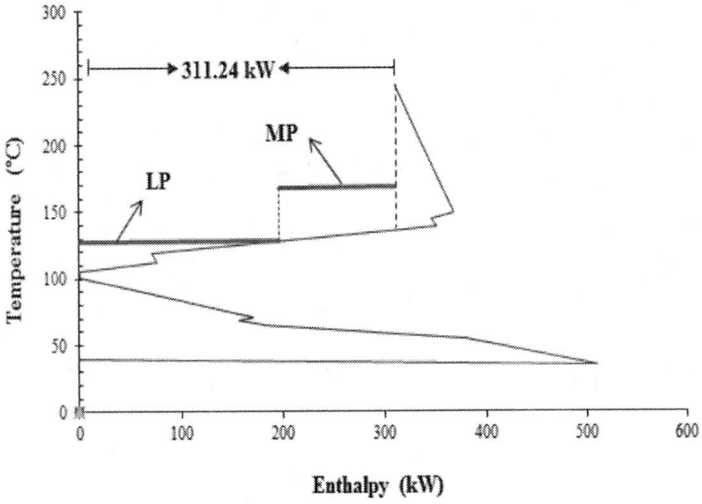

Figure 4. Balance grand composite curve at ΔT_{min} of 30 °C (before fouling).

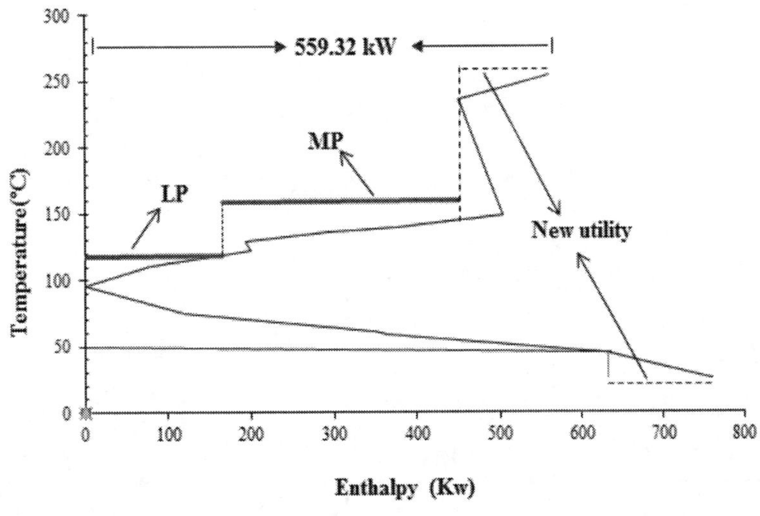

Figure 5. Balance grand composite curve at (ΔT_{min}) of 49.03 °C (after fouling).

Figure 6 displays the energy versus ΔT_{min} target curve of the existing HEN of palm oil refinery under study. The change in the energy requirement is quite sensitive to ΔT_{min} between 23.5 °C and 49 °C (Figure 6). The energy change with ΔT_{min} however becomes minor with an increase in ΔT_{min} between 16 °C and 23.5 °C (Figure 6). Thus, to maintain lower energy consumption, we selected ΔT_{min} of 17 °C, which is located in the less sensitive range as the target for the retrofit design of HEN. Table 2 presents the calculated CO_2 emissions from steam boiler alone and steam boiler followed by steam turbine at ΔT_{min} of 30 °C (existing HEN) and ΔT_{min} of 17 °C (target retrofit design of HEN). Both utility systems (i.e., steam boiler alone and steam boiler followed by steam turbine) provided emissions reduction of almost 50% from their existing emissions (at ΔT_{min} of 30 °C) due to the significant amount of energy saving from the retrofitting of the existing HEN at ΔT_{min} of 17 °C. In terms of CO_2 emission, Table 2 shows that using steam boiler alone is superior to using steam boiler followed by steam turbine. For

IOP Conf. Series: Materials Science and Engineering **429** (2018) 012081 doi:10.1088/1757-899X/429/1/012081

example, steam boiler alone offer 20.9% lower CO_2 emissions compared to the steam boiler followed by steam turbine. However, this reduction in CO_2 emissions may be altered if global CO_2 emission (i.e., total emission from the utilities on the palm oil refinery and from the external source to account for the imported electricity) is considered. In such a case, significant amount of power could be generated by using steam boiler followed by turbine, which may contribute to CO_2 emission reduction in the central power plant. However, using steam boiler alone can still be a competitive option if the imported electricity is generated from the renewable energy resources such as wind, solar, or hydroelectric power.

It is worth noting that for each option of the considered utility systems, further emission reduction beyond 50% can still be achieved if the existing fuel in the combustion devices is modified into fuel with lower carbon to hydrogen ratio and/or end-of-pipe waste minimisation technique such as capture and storage of CO_2 emissions is used.

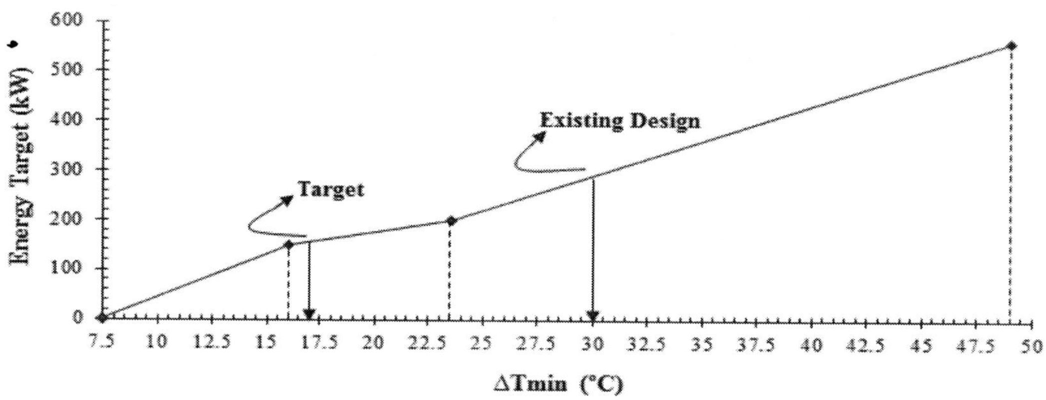

Figure 6. Energy versus ΔT_{min} target curve of the existing HEN of palm oil refinery.

Table 2. CO_2 emission from steam boiler alone and steam boiler followed by steam turbine.

ΔT_{min}	CO_2 emission, kg hr^{-1} (steam boiler)	CO_2 emission, kg hr^{-1} (steam boiler followed by turbine)
Existing design (30 °C)	84	103
Retrofit design (17 °C)	40.6	51.3

4. Conclusions

This work evaluated CO_2 emissions from small-scale utility systems (i.e., steam boiler alone and steam boiler followed by steam turbine) used to supply process heating (MP steam and LP steam) for palm oil refinery. The existing HEN of the palm oil refinery is retrofitted at ΔT_{min} of 17 °C, which is located at the less sensitive region on energy versus ΔT_{min} retrofit curve, to minimise the effect of fouled heat exchanger on the overall performance of palm oil refinery HEN. Such retrofitting of HEN led to almost 50% reduction in CO_2 emissions of both utility systems given the maximised heat recovery within HEN of palm oil refinery. However, the calculated CO_2 emission from steam boiler alone after the retrofit of HEN was lower by 20.9% compared to steam boiler followed by steam turbine on local basis. Steam boiler alone can still be superior to steam boiler followed by steam turbine on global basis if any required electricity by palm oil refinery is generated from the renewable resources. However, if the required electricity is obtained from non-renewable resources, then the trade-off between steam boiler alone case and steam boiler followed by steam turbine may occur due to the potential of cogeneration offered by the latter case.

References

[1] Mahmoud A *et al* 2009 A combined process integration and fuel switching strategy for emissions reduction in chemical process plants *Energy* 2009 **34** 190 – 195.
[2] Olajire A 2010 CO_2 capture and separation technologies for end-of-pipe applications – A Review *Energy* **35** 2610 – 2628.
[3] Varbanov P *et al* 2005 Synthesis of industrial utility systems: Cost-effective de-carbonization *Appl. Therm. Eng.* **25** 985 – 1001.
[4] Dominic C and Raymond R 2016 A review on process integration techniques for carbon emissions and environmental footprint problems *Process Saf. Environ.* **103** 291–307.
[5] Cristian M and Belén F 2018 Effectiveness of a CO_2 tax on industrial emissions *Energy Econ.* **71** 370 – 382.
[6] European commission, the EU Emissions Trading System (EU ETS), http://ec.europa.eu/clima/policies/ets_en (Accessed 26.05. 2018).
[7] Mona G *et al* 2013 Simultaneous optimization of CO_2 emissions reduction strategies for effective carbon control in the process industries *Chem. Eng. Res. Des.* **91** 1483 – 1498.
[8] Cohen M 2011 Is the UK preparing for "war"? Military metaphors, personal carbon allowances, and consumption rationing in historical perspective *Clim. Change* **104** 199 – 222.
[9] Smith R and Delaby O 1991 Targeting flue gas emissions *Trans IChemE* **69** 492 – 505.
[10] Delaby O and Smith R 1995 Minimization of flue gas emissions *Trans IChemE* **73** 21 – 32.
[11] Mahmoud A and Sunarso J 2018 A new graphical method to target carbon dioxide emission reductions by simultaneously aligning fuel switching, energy saving, investment cost, carbon credit, and payback time *Int. J. Energ. Res.* **42** 151 – 1562.
[12] Haslenda H *et al* 2012 *Operation system optimization. In: Zheng et al editors Handbook of CO_2 in power systems* (Heidelberg: Springer) chapter 17 pp 373 – 391.
[13] Townsend D and Linnhoff B 1983 Part II: Design procedure for equipment selection and process matching *AIChE J.* **29** 748 - 771.
[14] Miller K *et al* 1994 Viscosity and heat transfer coefficient for canola, corn, palm and soybean oil *J. Food Proc. Preservation* **18** 461 - 472.

IOP Conf. Series: Materials Science and Engineering **429** (2018) 012082 doi:10.1088/1757-899X/429/1/012082

Hydrogen Sulphide Emissions Reduction Using Hydrated Lime

Rajeb Salem Hwidi, Tengku Nuraiti Tengku Izhar, Farah Naemah Mohd Saad, Tijjani Adam

School of environmental Engineering, Univeristi Malaysia Perlis, Kompleks Pusat Pengajian Jejawi 3, Arau 02600, Perlis, Malaysia

Abstract. Hydrogen sulphide have become serious threat to the health management worldwide, the amount hydrogen sulphide emission keeps increasing due to several factors. The land field and industries are the major contributor of hydrogen sulphide emissions. Thus, this present the novel approaches to reduce hydrogen sulphide emission using hydrated lime. The hydrated lime was prepared from CaO with different of quick lime, several lime sample were prepared to determine the. The hydrated lime was prepared based on dosed with different amount of quick lime content. Significant reduction of hydrogen sulphide was observed over 15 days of treatment. The results further revealed that emission was not stopped by the low hydrated lime content indicating bacterial in the waste materials are resistant to low hydrated lime. Thus, 8% and above hydrated lime is needed to stop the gas emission.

1. Introduction

Gas such hydrogen sulphide emanating from land field is a major issue that threatens environmental quality in Malaysia. Gas resulted from the escalation of solid waste, which comprises unwanted matter produced by human activities which lead some heath issues, especially when the waste are very closed to residual areas [1-7]. In Malaysia, the increase in urban population reflects the escalation of solid waste generation [8-15]. Given its stable political conditions, Malaysia is one of the most economically successful developing countries in the world, hence the production of various products including food, automotive, crops, light and heavy industries are booming and expanding annually. The heavy development is boosted by the relative peace enjoyed and stability in both econmic and policies. There is rapid development due it regional advantages such good weather and etc [16-24]. These rapid economic transitions and increasing of urban population incur the escalation of solid waste generation[26-36]. The urban population Malaysia has increased by over 60% over the last few decades), and this increment has caused serious concerns regarding solid waste production {37-50]. Currently , there is no, gas treatment in various landfill in Malaysia, an approach such hydrated will novel approach to do away problem of the odor in Malaysia[50-55]. As of now, now any set up to manages odor in Malaysia. Equally, there no good landfill management in Malaysia. [56-60]. Therefore, the productions of waste are accelerating and normally the waste will end up in a landfill. As a result, the Malaysian government has imposed several rules, regulations and policies to underpin the management of waste in the country[61]. Urbanization and industrialization in Malaysia have changed the characteristics of solid waste. Thus, effective solid waste management practices must be updated to suit the current waste quantity is necessary [62]. Landfilling has been the primary method of waste disposal in Malaysia for many decades. The Solid Waste Management and Public Management Cleansing Corporation is one of Malaysian policy for solid waste management that was implemented in 2011 to manage all the waste produce in the country [63]. Although the policy is being implemented currently however we have yet to see stringent measures undertaken. Due to that,

Malaysia is currently facing serious landfill problems which include lack of capacity, overflowing landfill sites, serious odour without ventilation [64]. The nearby settlement It is necessary reduce the sulphate concentration in these waste to very low or eliminate completely using lime powdered can greatly reduce the emission of the gases [65]. The addition of this hydrated lime will suppress precipitates and agglomeration of waste element to generate sulphate [62, 66-74].

2. Method

5 research were developed for the experiment that varied in the hydrated lime % of 1%, 3%, 5% and 8%. All the sample were treated with similar particle size of quick lime. The columns were prepared with 40 % vegetables, 30% raw meat and 30% food remains of rice origin as main food for Malaysia. Prior to application of hydrated lime, the gas generated was monitored over 7 days. Where initial gas was observed for the three time of the day, morning, afternoon and evening. The hydrated lime $Ca(OH)_2$ was pure on the columns. The hydrated lime % of 1%, 3%, 5% and 8%of which were compared with standard column without any hydrated lime measurement were taken for 15 days using hand held gas analyser.

3. Results and Discussion

Initially the odor took 1 complete days before gas emission started this due to the decomposition process, the gas increase with time and reach it full peak in 3 days of decomposition, the increased as hydrated lime decreased the gas emission gradually with the lower hydrated lime. The section presents the results of meteorological measurements carried out in order to assess the odour reduction activities by the hydrated. From the subsequent figures that can be seen below, the results indicated that the potential capacity to reduce odour generation. Good amount of reduction can see in all results; however, a substantial generation can be seen in the evening. The decrease in H_2S concentration was observed, from the initial value of 160ppm recorded without the application of hydrated lime which name the standard column the which was observed for the next day and beyond. In the standard column, the gas emission was very rapid and compared decrease. The gas values 160ppm to 70ppm recorded on the last day that is 15 days of the treatment the lower hydrated lime content some closed pattern. The concentration decreased from 120ppm to 60ppm. Figure 1 show the

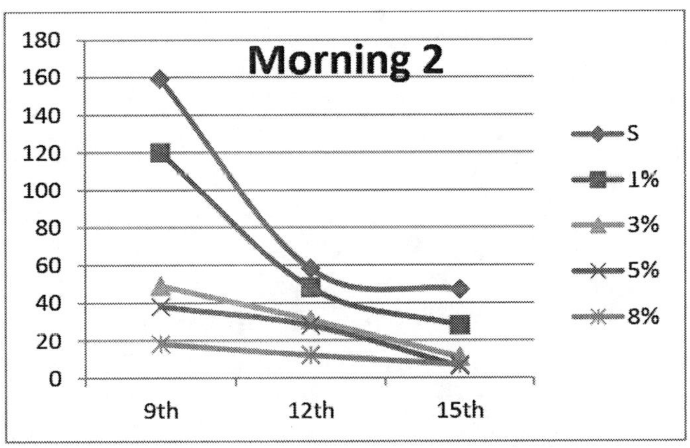

Figure 1: Hydrogen sulphide reduction with hydrated lime in the morning

Initially, the reduction was linearly decrease in a rapid form, all treatment sample were carried out under quite similar weather conditions for the whole periods of the treatment, the treatment temperature was ranged from 40°C to 55°C. The overall wind direction and wind velocity also same at the 15th days the measurement was done slightly with different temperature, was taken under slightly different weather conditions, the temperature increase to 43.1 to 45.5°C. From the figure 1 is obvious that, it is possible to observe the influence of the hydrated lime on the reduction of the concentrations of the gases. However, it can seen a very interesting trend of the quantitative reduction of the odorants with the application of the preparation was followed by almost fivefold decrease in the gas emissions.

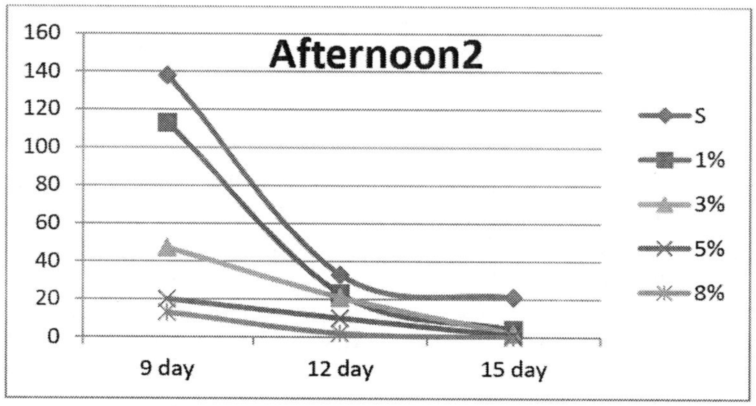

Figure 2: Hydrogen sulphide reduction with hydrated lime in the afternoon

The column responses to different waste composition and lime concentration are shown in figure 2 and the hydrogen sulphide reduction rate for different hydrated lime content can be seen. Research column show 8% hydrated lime, 5% hydrated lime, 3% hydrated lime and 1% hydrated lime. The 8% hydrated lime have the highest rate of reduction of the hydrogen sulphide, as can be seen here Initially, the reduction was linearly decrease in a rapid form, all treatment sample were carried out under quite similar weather conditions for the whole periods of the treatment, the treatment temperature was ranged from 40°C to 55°C. The overall wind direction and wind velocity also same at the 15th days the measurement was done slightly with different temperature, was taken under slightly different weather conditions, the temperature increase to 4.1 to 45.5°C. With low hydrated application, residual lime concentration did not show significant effect on the hydrated lime application. However, this trend changes with increase in hydrated lime content.

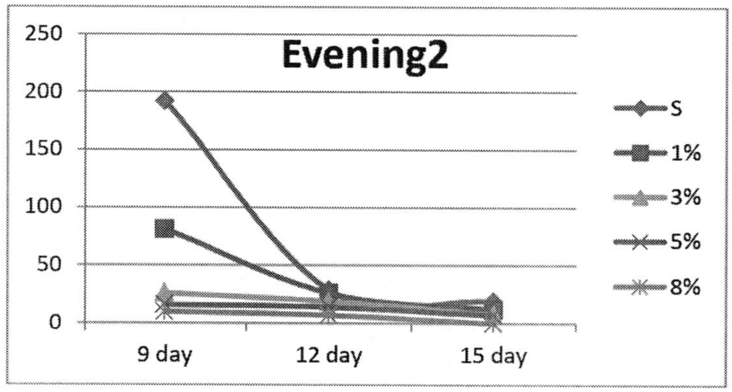

Figure 3: Hydrogen sulphide reduction with hydrated lime in the evening

From the figure3 can be seen that, increasing hydrated lime content can increase completely stop the gas emission this due to the change pH which lead to disassociation which lead to antibacterial activities within the sample columns. The column of 8% content with high lime trended to have greater pH values. In the treatment of hydrogen sulphide lime content should therefore be considered. Throughout, the reduction was linearly decrease in a rapid form, all treatment sample were carried out under quite similar weather conditions for the whole periods of the treatment, the treatment temperature was ranged from 40°C to 55°C. The overall wind direction and wind velocity also same at the 15th days the measurement was done slightly with different temperature, was taken under slightly different weather conditions, the temperature increase to 41.1 to 45.5°C. From the figure 1 is obvious that, it is not possible to determine unambiguously the influence of the hydrated lime on the reduction of the concentrations of the gases. However, a positive impact on the quantitative reduction of the odorants emitted by the lime reaction, especially at early stages where the application of the preparation was followed by almost five-fold decrease in the gas emissions.

4. Conclusion

The paper report possible to reduce hydrogen sulphide using hydrated lime. The hydrated lime was prepared and observed it anti-emission activities over 15 days. The observation was conducted during 3 time of the day morning, afternoon and evening. It was found that during the period, the decomposition processes are much less intense evening. This is might be due to the exothermic character of evening is higher which promote majority of processes which produce, among other compounds to generate hydrogen sulphide. It was equally found that anti emission activities of hydrated lime is high as the concentration increase

References

[1]. Anderson C., Malambo D. H.,M. Perez E. G. et al., 2015) Lactic acid fermentation, urea and lime addition: promising faecal sludge sanitizing methods for emergency sanitation," *International Journal of Environmental Research and PublicHealth*, vol. 12,no. 11, pp. 13871–13885

[2]. Anderson K., Spit J. Malambo, D., Gonzalez M. E., Nobela H. L. de Pooter (2014) Emergency Sanitation: Feacal SludgeManagement, WASTE, Gouda, The Netherlands

[3]. Adhikari, B., Dahal, K. R., & Khanal, S. N. (2014). A Review of Factors Affecting the Composition of Municipal Solid Waste Landfill Leachate, *3*(5), 273–281.

[4]. Adlan, M. N., Palaniandy, P., & Aziz, H. A. (2011). Optimization of coagulation and dissolved air flotation (DAF) treatment of semi-aerobic landfill leachate using response surface methodology (RSM). Desalination, 277(1), 74-82.

[5]. Ganesh, K. & Jambeck, J.R., (2013) Treatment of landfill leachate using microbial fuel cells: Alternative anodes and semi-continuous operation. *Bioresource Technology,* 139, 383–387.

[6]. Ghafari, S., Aziz, H. A., Isa, M. H., & Zinatizadeh, A. A. (2009). Application of response surface methodology (RSM) to optimize coagulation–flocculation treatment of leachate using poly-aluminum chloride (PAC) and alum. Journal of hazardous materials, 163(2), 650-656.

[7]. Guodong Zhang, Yan Jiao, Duu-Jong Lee (2015) A lab-scale anoxic/oxic-bioelectrochemical reactor for leachate Treatments, *Bioresource Technology* 186 (2015) 97–105

[8]. Handbook of Environment and Waste Management. (2012). Air and Water Pollution Control. Retrieved May 22, 2015,

[9]. Hariti Bouhezila, F., M., et al. (2011). "Treatment of the OUED SMAR town landfill leachate by an electrochemical reactor." Desalination. Vol 6(2) 23-30

[10]. Jiao Zhang, G.D., Zhao, Q.L., , Y., Zhang, J.N., Jiang, J.Q., Ren, N.Q., Kim, B.H., 2011. Improved performance of microbial fuel cell using combination biocathode of graphite fiber brush and graphite granules. J. Power Sources 196, 6036–6041.

[11]. Kargi, F., & Eker, S. (2007). Electricity generation with simultaneous wastewater treatment by a microbial fuel cell (MFC) with Cu and Cu–Au electrodes. Journal of Chemical Technology and Biotechnology, 82(7), 658-662.

[12]. Karmakar, S., Kundu, K., & Kundu, S. (2010). Design and development of Microbial Fuel Cell. Current Research. *Technology and Education Topics in Applied Microbiology and Microbial Biotechnology*, 1029-1034.
[13]. Katsoyiannis. A, Jun, W. & Zouboulis, (2003). Removal of humic acids by flotation, *colloids surfaces A: Physicochem. Eng. Aspects, 231,* pp. 181-193.
[14]. Khan, M. R., Bhattacharjee, R., & Amin, M. S. A. (2012). Performance of the salt bridge based microbial fuel cell. International Journal of Engineering & Technology, 1(2), 115-123.
[15]. Kirmizakis, P., Tsamoutsoglou, C., Kayan, B., & Kalderis, D. (2014). Subcritical water treatment of landfill leachate: Application of response surface methodology. Journal of environmental management, 146, 9-15.
[16]. Kolhe and Pawar (2011)Physico-chemical analysis of effluents from dairy industry *Recent Research in Science and Technology*, 3(5): 29-32
[17]. Krishnadas Ganesh, Jenna R. Jambeck (2013) Treatment of landfill leachate using microbial fuel cells: Alternative anodes and semi-continuous operation, *Bioresource Technology* 139, 383–387
[18]. Li, H., Zhou, S., Sun, Y., & Lv, J. (2010). Application of response surface methodology to the advanced treatment of biologically stabilized landfill leachate using Fenton's reagent. Waste Management, 30(11), 2122-2129.
[19]. Li, W. W., Sheng, G. P., Liu, X. W., & Yu, H. Q. (2011). Recent advances in the separators for microbial fuel cells. Bioresource technology, 102(1), 244-252.
[20]. Liu, H., & Logan, B.E (2004). Electricity generation using an air-cathode single chamber microbial fuel cell in the presence and absence of a proton exchange Membrane. *Environmental Science & Technology*, 38(14), 4040-4046.
[21]. Logan, B.E., Long and Bon (2008) Microbial Fuel Cells. John Wiley and Sons, Inc., New Jersey. Logan, B.E., Cheng, S., Watson, V., Estadt, G., 2007. Graphite fiber brush anodes for increased power production in air-cathode microbial fuel cells. Environ. Sci.Technol. 41, 3341–3346.
[22]. Maksudur R., & Ripon, M. S. A. Amin (1012)Performance of the Salt Bridge Based Microbial Fuel Cell International Journal of Engineering and Technology, 1 (2) 115-123
[23]. Maliana, J.F. F.G. Pohland. (1996). Design of anaerobic processes for the treatment of industrial and municipal wastes. *Water Qual Manage,* 7, 169-175.
[24]. Min, B., Cheng, S., & Logan, B. E. (2005). Electricity generation using membrane and salt bridge microbial fuel cells. Water research, 39(9), 1675-1686.
[25]. Mohajeri, S., Aziz, H. A., Isa, M. H., Zahed, M. A., & Adlan, M. N. (2010). Statistical optimization of process parameters for landfill leachate treatment using electro-Fenton technique. Journal of Hazardous Materials, 176(1), 749-758.
[26]. Muralidharan, A., Babu, O. A., Nirmalraman, K., & Ramya, M. (2011). Impact of salt concentration on electricity production in microbial hydrogen based salt bridge fuel cells. Indian Journal of Fundamental and Applied Life Science, 1(2), 178-184.
[27]. Nair, A. T., Makwana, A. R., & Ahammed, M. M. (2014). The use of response surface methodology for modelling and analysis of water and wastewater treatment processes: a review. Water Science and Technology, 69(3), 464-478.
[28]. Palaniandy Adlan, M. N., P., et al. (2011). "Optimization of coagulation and dissolved air flotation (DAF) treatment of semi-aerobic landfill leachate using response surface methodology (RSM)." *Desalination Vol 2(3) 33-42*

[29]. Palaniandy, P., Adlan, M.N., Aziz, H.A. and Murshed, M.F., 2010. Application of dissolved air flotation (DAF) in semi-aerobic leachate treatment. *Chemical Engineering Journal*, *157*(2), 316-322.

[30]. Petala, A.I. &, Zouboulis M. D. (2008). Performance of VSEP vibratory membrane filtration system during the treatment of landfill leachates. *Desalination, 222,* pp. 165–175.

[31]. Prasad Verma, S, B. & Mishra, I.M. (2010). Pretreatment of petrochemical wastewater by coagulation and flocculation and the sludge characteristics. *Journal of Hazardous Materials, 178,* pp. 1055-1064.

[32]. Puig, S. et al., (2011). Microbial fuel cell application in landfill leachate treatment. *Journal of Hazardous Materials*, 185(2-3), 763–767.

[33]. Puig, S., Serra, M., Coma, M., Cabré, M., Balaguer, M. D., & Colprim, J. (2011). Microbial fuel cell application in landfill leachate treatment. Journal of Hazardous Materials, 185(2), 763-767.

[34]. Qixing Wei Li, Zhou, and Tao Hua, (2010) Removal of Organic Matter from Landfill Leachate by Advanced Oxidation Processes: A Review, International Journal of Chemical Engineering, Vol. (2010), pp. 1-10

[35]. R.H. Kettunen, T.H. Hoililjoki & J.A. Rintala. (2009) Anaerobic and sequential anaerobic-aerobic treatments of industrial and municipal landfill leachate at low temperatures. *Bioresour. Technol,* 58, 40-41.

[36]. Raghab, S.M., Abd El Meguid, A.M. & Hegazi, H. a., (2013). Treatment of leachate from municipal solid waste landfill. *HBRC Journal*, 9(2), 187–192.

[37]. Rahimnejad, M., Adhami, A., Darvari, S., Zirepour, A., & Oh, S.-E. (2015). Microbial fuel cell as new technology for bioelectricity generation: A review. *Alexandria Engineering Journal*, *54*(3), 745–756.

[38]. Rahimnejad, M., Najafpour, G. D., Ghoreyshi, A. A., Shakeri, M., & Zare, H. (2011). Methylene blue as electron promoters in microbial fuel cell. international journal of hydrogen energy, 36(20), 13335-13341.

[39]. Renou, S., Givaudan, J. G., Poulain, S., Dirassouyan, F., & Moulin, P. (2008). Landfill leachate treatment: review and opportunity. Journal of hazardous materials, 150(3), 468-493.

[40]. Ringeisen BR, Henderson E, Wu PK, Pietron J, Ray R, Little B & Jones-Meehan JM. (2006). High power density from a miniature microbial fuel cell using Shewanella oneidensis DSP10. *Environ. Sci Technol*,40(80), 2629-2634.

[41]. Sani, a. et al., (2014) The influence of pH on the removal of ammonia from a scheduled waste landfill leachate. *Jurnal Teknologi (Sciences and Engineering)*, 68(5),. 25–28.

[42]. Schaetzle, O., Barrière, F., & Schröder, U. (2009). An improved microbial fuel cell with laccase as the oxygen reduction catalyst. Energy & Environmental Science, 2(1), 96-99.

[43]. Seow- Wee (2012) New Perspective of Integrated Solid Waste Management in Malaysia. In Proceeding 3rd International Conference on Human Habitat & Environment in the Malay World, 19-20 Jun 2012, UKM, Bangi

[44]. Serra Puig, S., M., Coma, M., Cabré, M., Balaguer, M.D., Colprim, J., (2011) Microbial fuel cell application in landfill leachate treatment. J. Hazard. Mater. 185, 763–767.

[45]. Sevda, S., & Sreekrishnan, T. R. (2012). Effect of salt concentration and mediators in salt bridge microbial fuel cell for electricity generation from synthetic wastewater. Journal of Environmental Science and Health, Part A, 47(6), 878-886.

[46]. Singhvi, P., & Chhabra, M. (2013). Simultaneous chromium removal and power generation using algal biomass in a dual chambered salt bridge microbial fuel cell. Journal of Bioremediation and Biodegradation, 4(5).
[47]. Solid Waste Management and Public Management Cleansing Corporation (2011) Implementation of Solid Waste and Public Cleansing Management Act (Act 672), Annual Report 2011 (Chapter 4). Kuala Lumpur: Solid Waste and Public Cleansing Management Corporation
[48]. Sztarr, Y. (2008). Measuring colour in trade waste. 2nd Annual WIOA NSW Water Industry engineers & Operators Conference. Jockey Club-Newcastle.
[49]. Teixeira, Cortez, S., P. et al. (2011). "Evaluation of Fenton and ozone-based advanced oxidation processes as mature landfill leachate pre-treatments." Journal of Environmental_Management. Vol 6(7) 333-339
[50]. Tizaoui Zahrim, A.Y, C. & Hilal, N. (2011). Coagulation with polymers for nanofiltration pre-treatment of highly concentrated dyes: A review. *Desalination, 266,* pp. 1-16.
[51]. Tong, Zhang, B., Liu, Y., S., Zheng, M., Zhao, Y., Tian, C., Feng, C. (2014). Enhancement of bacterial denitrification for nitrate removal in groundwater with electrical stimulation from microbial fuel cells. *Journal of Power Sources,* 268, 423-429
[52]. Trinh, T. K., & Kang, L. S. (2011). Response surface methodological approach to optimize the coagulation–flocculation process in drinking water treatment. Chemical engineering research and design, 89(7), 1126-1135.
[53]. Tugtas, A. E., Cavdar, P., & Calli, B. (2013). Bio-electrochemical post-treatment of anaerobically treated landfill leachate. Bioresource technology, 128, 266-272.
[54]. Umar, M., Aziz, H. A., & Yusoff, M. S. (2010). Trends in the use of Fenton, electro-Fenton and photo-Fenton for the treatment of landfill leachate. Waste Management, 30(11), 2113-2121.
[55]. Umar, M., Aziz, H. A., & Yusoff, M. S. (2011). Assessing the chlorine disinfection of landfill leachate and optimization by response surface methodology (RSM). Desalination, 274(1), 278-283.
[56]. Mohammad Razif Bin Mustafa, Tijjani Adam, U. Hashi, N. Azizah (2017) Specific and Selective Target Detection of Supra-genome 21 Mers Salmonella via Silicon Nanowires Biosensor, IConGDM2017-177, AIP Conference Proceedings ISSN: 1551-7616
[57]. B. Basri, Tijjani Adam and U.Hashim (2017) Synthesis of zinc oxide thin films prepared by sol-gel for specific bioactivity, IConGDM2017-178, AIP Conference Proceedings ISSN: 1551-7616
[58]. Mohammed Mohammed A.R Rozyanty Tijjani Adam, Bashir O. Betar (2017) Study of the Weathering Effect in a Natural Environment on the Hybrid Kenaf Bast /Glass Fibre-Filled Unsaturated Polyester Composite, ConGDM2017-181, AIP Conference Proceedings ISSN: 1551-7616
[59]. Neesyan A/L David Rajah, Tijjani Adam Uda Hashim (2017) Lattices Parameter Dependent Of Effective Atomic Interactive Coefficient For Unstressed And Stressed 1nm Nanostructure, ConGDM2017-198, AIP Conference Proceedings ISSN: 1551-7616
[60]. Mohamed Jamal Bin Amanullah, Tijjani Adam, Uda Hashim, (2017)Calculation of the Band Structure Parameter in Silicon Nanowires Using first Principle Analytical Method, AIP Conference Proceedings ISSN: 1551-7616
[61]. Mohammed Mohammed A.R Rozyanty, Tijjani Adam, Bashir O. Betar (2017)ZnO Characterization, UV and Photocatalysis Mechanism for Water Repellent Phenomena in

Polymer Poly Composite, IConGDM2017-211, AIP Conference Proceedings ISSN: 1551-7616

[62]. Shillan A/L Bathmanathan , Tijjani Adam, Uda Hashim (2017) Optimization and Validation of Highly Selective Microfluidic Integrated Silicon Nanowire Chemical Sensor, IConGDM2017-212, AIP Conference Proceedings ISSN: 1551-7616

[63]. Wali, F.K.M., 2015. Color Removal and COD Reduction of Dyeing Bath Wastewater by Fenton Reaction. *International Journal of Waste Resources*, *5*(171), 1-6.

[64]. Wang, Zhu, F., W., Zhang, X., Tao, G., 2011. Electricity generation in a membrane less microbial fuel cell with down-flow feeding onto the cathode. Bioresour. Technol. 102, 7324–7328.

[65]. Watanabe K., J. Biosci. (2006). Recent developments in microbial fuel cell technologies for sustainable bioenergy, *J. Biosci. Bioeng*, 106, 528-536.

[66]. Wein, Y.X, Li, Y.F. & Ye, Z.F. (2010). Enhancement of removal efficiency of ammonia nitrogen in sequencing batch reactor using natural zeolite. *Environ Earth Sci, 60,* pp. 1407-1413.

[67]. Woo, J.H. Im, H.J M.W. Choi, K.B. Han, C.W. Kim. (2001) Simultaneous organic and nitrogen removal from municipal landfill leachate using an anaerobic-aerobic system. *Water Res*, 35, 2403-2410.

[68]. Wu, Y., Zhou, S., Qin, F., Ye, X., & Zheng, K. (2010). Modeling physical and oxidative removal properties of Fenton process for treatment of landfill leachate using response surface methodology (RSM). Journal of hazardous materials, 180(1), 456-465.

[69]. WWidiastuti, N, Wu, H.W, Ang, M. & Zhang, D.K. (2008). The potential application of natural zeolite for greywater treatment. *Desalination, 218,* pp. 271–280.

[70]. You Zhang, J., Zhao, Q.L., S., Jiang, J., Ren, N.Q., 2008. Continuous electricity production from leachate in a novel upflow air-cathode membrane -free microbial fuel cell. Water Sci. Technol. 57, 1017–1021.

[71]. Young Li. Hon et al., (2010) Microbial fuel cells using natural pyrrhotite as the cathodic heterogeneous Fenton catalyst towards the degradation of biorefractory organics in landfill leachate. Electrochem. Commun. 12 (7), 944–947.

[72]. Zhu, X., Tian, J., Liu, R., & Chen, L. (2011). Optimization of Fenton and electro-Fenton oxidation of biologically treated coking wastewater using response surface methodology. Separation and Purification Technology, 81(3), 444-450.

[73]. Sohrabi, M.R., Khavaran, A., Shariati, S. and Shariati, S. (2014). Removal of Carmoisine edible dye by Fenton and photo Fenton processes using Taguchi orthogonal array design. *Arabian Journal of Chemistry*. 1-9.

[74]. Pawar, V. and Gawande, S. (2005). An overview of the Fenton Process for Industrial Wastewater. *IOSR Journal of Mechanical and Civil Engineering*, 127-136.

Development and investigation of thermoelectric cooling performance based on space scales

M Saifizi[1], T W Lee[1], S N N Anuar[1], I Zunaidi[2], N S Diana[1], W A Mustafa[1], W K Wan[2], Shahriman A B[2] and Z M Razlan[2]

[1]Faculty of Engineering Technology, University Malaysia Perlis, 02100 Padang Besar, Perlis.
[2]School of Mechatronic, University Malaysia Perlis, 02600 Arau, Perlis.

Abstract. The refrigerator is the key component to keep the medicine and biological sample in the hospital. However, the refrigerant that commonly used in the refrigerator will bring a negative effect to the environment. Other than that, the domestic refrigerator also has the problem of larger size and heavier weight since to the compact system like condenser, compressor, evaporator and expansion valve are assemble and use in the refrigerator. This study focused on the development of temperature control of the portable thermoelectric refrigerator for medical purpose. Thermoelectric cooling is using the thermoelectric refrigeration method through a thermoelectric module. This project is to analyze the space scales of the system which can perform a desired temperature control for the thermoelectric refrigeration. A commercial portable thermoelectric cooling is modified, which is able to achieve the temperature within the range of 8°C to 15°C by controlling the input current. Modification of this refrigerator is done by changing the original circuit of the refrigerator into a new circuit. The resistance temperature detector is used to replace the original thermostat that used to measure the inner temperature of the refrigerator. The input current is controlled by MOSFETs by using the pulse width modulation method. The temperature response of normal and small scale of thermoelectric cooling is 13.88°C and 10.91°C when input current 3A injected.

1. Introduction

In the hospital, the medicine entities such as blood sample, medicine and vaccine are the elements that very sensitive to the temperature. Most of the medical product will occur the changes in the chemical, physical and microbiological if the temperature keeps the product is exceed the desired temperature of that medical product [1]. For example, the blood bag shall maintain in +2°C to +10°C during transport and +2°C to +6°C for storage, vaccine maintain in -3°C to +8°C and the injection antibiotic should maintain in +8°C to +15°C [2–4]. The refrigerator used to store the medical product in the hospital usually is the domestic refrigerator which heavy and larger.

Hospitals often use domestic vapour refrigerators to store biomedical specimens. The vapor refrigerator uses Chlorofluorocarbon (CFCs) and Hydro Chlorofluoro (HCFCs) as a refrigerant that will cause the negative effect to the environment like derogation of ozone layer [5]. This kind of refrigerator is considering as not eco-friendly. The refrigerator only can operate by using AC power source and this also means that the refrigerator is not suitable to use in a rural area that may not have any AC power supply. Therefore, the thermoelectric refrigerator is the alternative choice for storing the sensitive material since it can be operated on either AC power source or DC power source. Since the refrigerator can be operated using a DC power source, it can be used to transport the medical product to other places out of the hospital. The thermoelectric module inside the refrigerator able to

IOP Conf. Series: Materials Science and Engineering **429** (2018) 012083 doi:10.1088/1757-899X/429/1/012083

cool down the refrigerator because the temperature differential(hot and cold) will occur between both sides of the thermoelectric modules when the DC current pass it and this is known as thermoelectric effect [6]. Since the refrigerator can be operated using a DC power source, it can be used to transport the medical product.

2. Methodology

This study is conducted on a portable thermoelectric refrigerator. The desired temperature range of 8°C to 15°C will able to achieve by the refrigerator for biomedical purposes like medical storage. There are few components used to develop this refrigerator.

2.1 Modification of Thermoelectric Refrigerator

The refrigerator is in a 12 liters portable refrigerator and the dimension is 41cm x 28cm x 29cm. The mass of this refrigerator is 5.3kg as shown in Figure 1. This refrigerator has a transformer and a full bridge rectifier to change the AC power source, 240VAC and 50Hz to a 24VDC power source to operate the thermoelectric devices. Other than that, this thermoelectric refrigerator can also operate by using the 24VDC power source such battery. Two thermoelectric devices are used in the refrigerator in order to improve the performance of the refrigerator. The desired temperature of the system is from 8°C to 15°C. Polystyrene is used to reduce the internal space of the refrigerator in order to reduce the time of cooling and increase the performance of the refrigerator.

Figure 1. Portable thermoelectric refrigerator

The thermoelectric refrigerator system is modified from the original refrigerator system. In order to achieve the desired temperature range which is +8°C to +15°C, the input current must be controlled appropriately. In order to control the input, MOSFETs have been used in the modified circuit to replace the relay as shown in Figure 2. The MOSFETs will be trigger by the Arduino Mega microcontroller using the pulse width modulation (PWM) method. In this section, the conventional circuit will be modified by using a resistance temperature detector (RTD) to replace the thermostat.

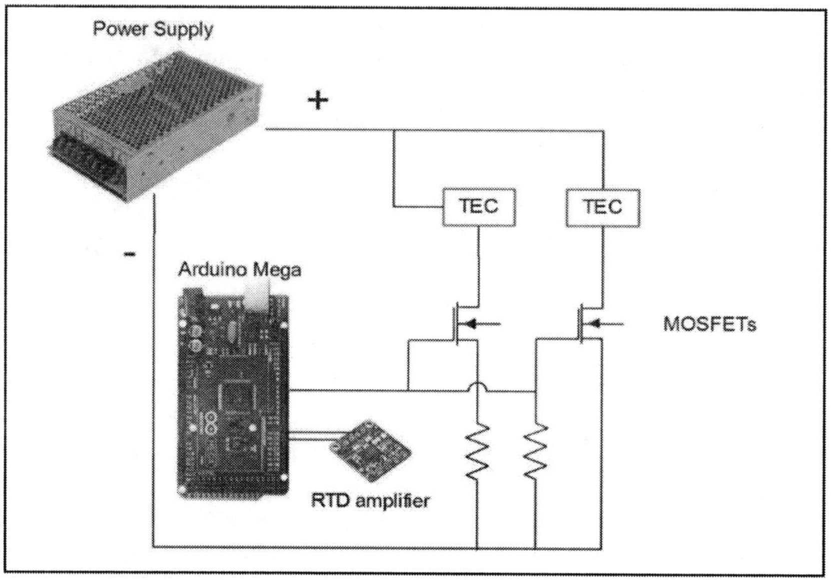

Figure 2. Schematic diagram of the modified electronic circuit of the thermoelectric refrigerator.

2.2 Time Response of Thermoelectric Refrigerator

The step response is an output of the system produced by a unit step input. It is a common analysis tool used to determine system performance. The time response experiment is conducted in two different scales of the thermoelectric chamber. The time response can be described by the following quantities which are steady state and settling time. The time response experiment is conducted based on open-loop. For this experiment, step rest introduced to the refrigerator system is the form of open-loop identification.

The program of the project is in the form of a block diagram. A duty cycle control of PWM and temperature display program during the testing of the performance modified refrigerator, the temperature of the RTD sensor measured will show in the display block diagram and the PWM of the Arduino board can be also controlled through adjusting the value of the slider gain in Figure 3. In order to determine which input current can optimize the performance of the modified refrigerator, there are 4 different input currents tested which are 2.5A, 2.6A, 2.8A and 3A. The steady state temperature and the settling time for each input current are measured and recorded and the comparison is made between the results. The ambient temperature during the experiment is within 28°C to 29°C and the inner space of the refrigerator is not reduce using the polystyrene. The duration of each experiment is 140 minutes and the data sample will be measured and recorded every 2 minutes. After the best input current to operate the refrigerator is decide, the internal space of the refrigerator is reduced. The internal space of the refrigerator is reduced by using the polystyrene to fill up space. The volume of the original refrigerator is 410 mm x 280 mm x 29 mm and the volume after fill in the polystyrene is 100 mm x 103mm x 1230 mm.

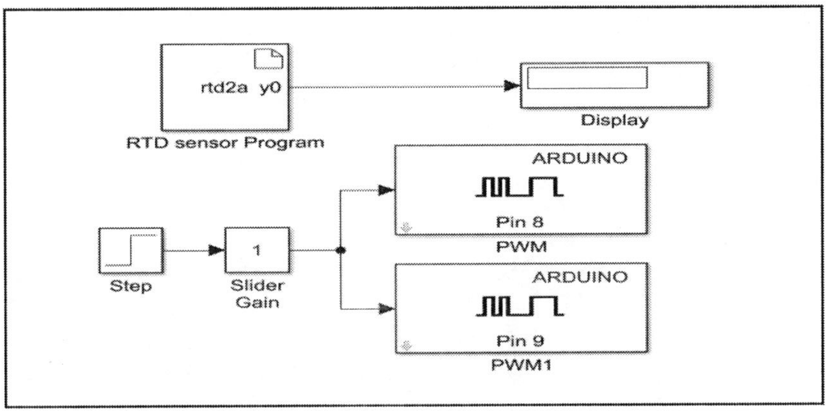

Figure 3. A simulink block diagram of the open-loop step response.

3 Result and Discussion

Experiment 1 is to decide the best input current for the modified refrigerator and the result is recorded in Table 1. From the Table 1, the input current able to achieve the desired temperature for the modified refrigerator is 5.2A-6.0A.

Table 1. Steady state temperature response of normal scale thermoelectric refrigerator due to the input current.

Input Current (A)	Temperature (°C)	Settling time (Min)
5.0	17.22	98
5.2	15.19	94
5.6	14.42	86
6.0	13.88	76

When the input current is 6.0A, the temperature measured and recorded is 13.88 and this is the lowest temperature obtained in experiment 1. The settling time when 6.0A input current used is 76 minutes. The refrigerator system has the worst response when 5.0A input current is injected as shown in Figure 4. The temperature of the refrigerator is higher compared to other value of input current and the settling time for the system is longer.

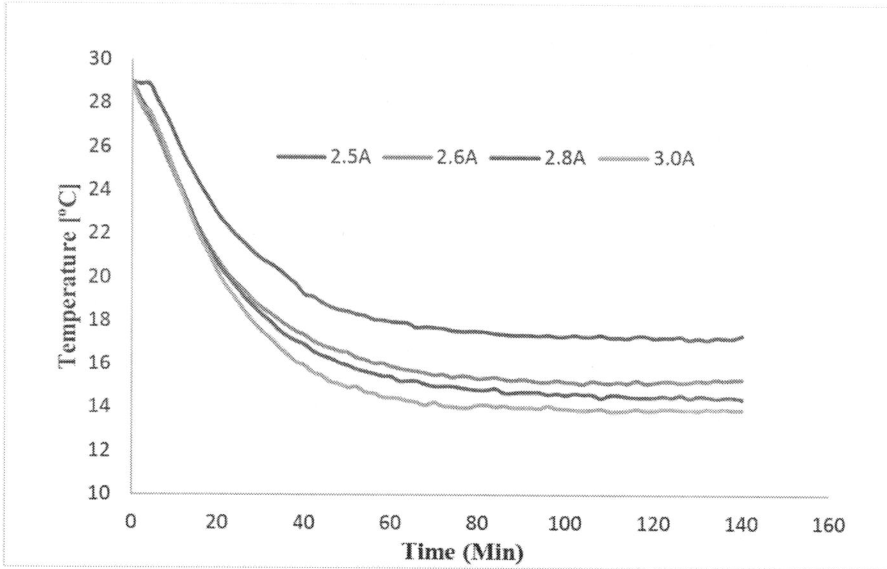

Figure 4. Temperature response of normal scale refrigerator due to the input current.

Next experiment is to decide whether the internal space of the refrigerator will affect the response of the system. Two different internal space refrigerators are tested using maximum input current 6A. Figure 5 shows the graphs of temperature vs. time of two different internal spaces.

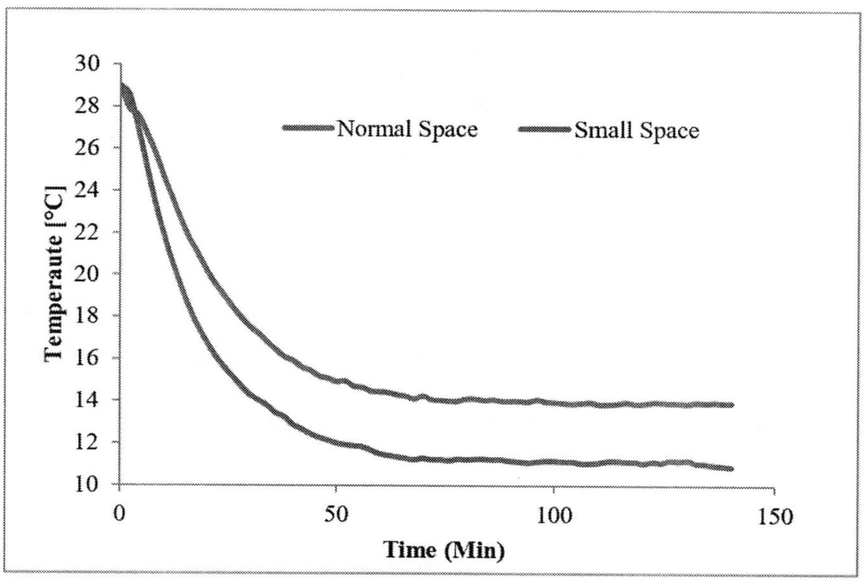

Figure 5. Temperature response of normal scale refrigerator due to input current.

The smaller internal space has faster response and the temperature is lower than the normal internal space. From the Table 2, the smaller internal space refrigerator has the faster settling time and the lowest temperature of this system is 10.91°C.

Table 2. Comparison steady state temperature response of thermoelectric refrigerator at 6 A.

Space scale	Temperature (°C)	Settling time (minutes)
Normal	13.88	76
Small	10.91	66

From the first experiment, the current input 5.6A to 6A able to operate for biomedical storage application. Other than that, the internal spaces of the refrigerator also affect the response of the refrigerator system. When the internal space is smaller, the refrigerator is able to obtain a lower temperature and the settling time of the refrigerator is much faster.

4 Conclusions

A thermoelectric refrigerator system is modified to design and develop a new control system to improve the performance of the refrigerator. The modified refrigerator system can vary the input current of the refrigerator in order achieve an inner temperature of the refrigerator in the desired range. This refrigerator is able to operate using the AC power source from a wall socket or the DC power source from the battery.

Acknowledgment
The authors are thankful to the Faculty of Engineering Technology, University Malaysia Perlis for the technical support provided for this research work.

5 References

[1] Martins R R, Farias A D, Oliveira Y, Diniz R D S and Oliveira A G 2017 Prevalence and risk factors of inadequate medicine home storage: a community-based study *Rev. Saude Publica* **51** 1–8

[2] WHO 2013 *National Standards for Blood Transfusion Service*

[3] Saritas I and Okay M 2014 Design of Portable Medical Cooler with Artificial Intelligent Control *International Conference on challenges in IT, Engineering and Technology (ICCIET'2014)* (Phuket, Thailand) pp 39–44

[4] Daphne Coleiro 2012 *Storage of Medicines and Medical Devices* (Msida)

[5] Rokde K, Patle M, Kalamdar T, Gulhane R and Hiware R 2017 Peltier Based Eco-Friendly Smart Refrigerator for Rural Areas *Int. J. Adv. Res. Comput. Sci. Softw. Eng.* **7** 718–21

[6] Patel J J, Patel M, Patel J J and Modi H 2016 Improvement in the COP of Thermoelectric Cooler *Int. J. Sci. Technol. Res.* **5** 73–6

Effect of Electrolyte Concentration during Solution Plasma on Copper Nanoparticle Size

M H S Al Anbouri, B W M Yii, P N Y Yek and M R M Julaihi

Swinburne University of Technology Sarawak, Jalan Simpang Tiga, 93350 Kuching, Sarawak, Malaysia

mjulaihi@swinburne.edu.my

Abstract. This research was conducted to investigate the effect of electrolyte concentration of solution plasma on copper nanoparticles size. Copper nanoparticles with the size between 1-100 nm have received a huge interest because of their optical, catalytic and electrical conducting properties. Electrolysis via solution plasma method has many advantages such as its cost effectiveness, simplicity, eco-friendliness, and less energy requirements compared to other methods. To synthesize copper nanoparticles via solution plasma, two wires of copper and platinum as cathode and anode respectively were immersed in a glass beaker of 300 ml of two electrolyte solution (K_2CO_3 and NaOH) with three different concentrations (0.1 M, 0.5M and 1M), and electrical current was applied similar to an electrolysis process. The images of the nanoparticles were obtained using scanning electron microscopy and their sizes were analyzed using ImageJ software. It was found that when the concentration of K_2CO_3 increased, the size of copper nanoparticle also increased. However, the size of copper nanoparticles decreased when the concentration of NaOH was increased. This research shows the potential of easily controlling size of copper nanoparticles by changing the electrolyte concentration used during solution plasma process.

1. Introduction

Nanoparticles are divided into two main groups, which are organic and inorganic. Organic nanoparticles mainly consist of carbon. Inorganic nanoparticles are split into three groups magnetic, metallic and semi-channel nanoparticles. Copper nanoparticles are classified under metallic group. Copper nanoparticles are recently becoming increasingly important due to their optical, catalytic and electrical conducting properties [1]. A variety of innovative application of nanoparticles has led to a variety of the copper nanoparticle's shapes and sizes. Due to that reason, the study on how to control the shape and size of copper nanoparticles have gained importance. Figure 1 shows that there are many methods to synthesize copper nanoparticles which can be categorized into three approaches which are chemical, physical and biological [2].

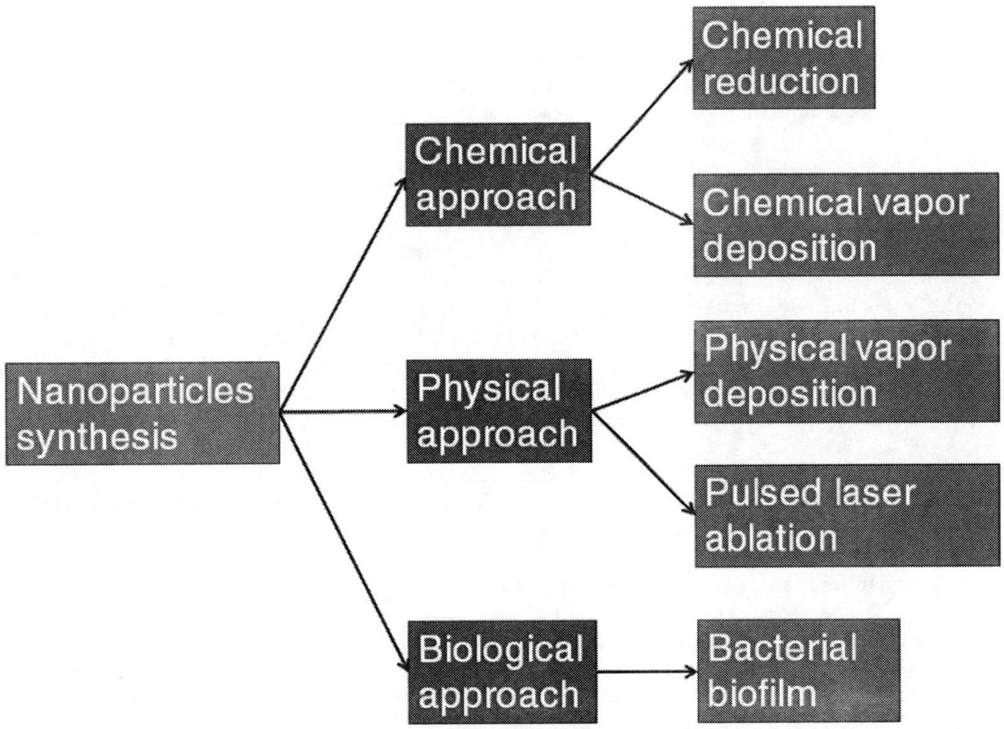

Figure 1. Typical methods of nanoparticles synthesis

During nanoparticles synthesis, the gases released and other emissions occurring tend to be polluting to the environment. As a result, green synthesis methods that do not involve toxic chemicals usage and are eco-friendly are needed to decrease the environmental pollution. Such green synthesis methods typically involve irradiation, polyoxometalates, toluene and biological methods which overcome the problem of using chemical agents. However, even these green synthesis methods present their own drawbacks. Such drawbacks are the selection of stabilizing and reducing nontoxic agents as well as the selection of solvent involved.

2. Solution plasma method

2.1 Advantages of selecting the synthesis method

There are many methods to synthesize copper nanoparticle, each of them requires different conditions and experimental setup. Among those methods, electrolysis via solution plasma method was chosen. Solution plasma is a type of hybrid liquid-gas phase plasma that is generated simply by conducting electrical current to the electrolyte.

Solution plasma process is a very simple and useful method for metal nanoparticles synthesis because it can provide extreme rapid reactions because of reactive radicals, chemical species and UV radiation generated in the atmospheric pressure plasma [3]. It can be applied to any metals or alloys. Also, it allows the control of nanoparticle size by changing the plasma initiation voltage or the electrolyte concentration. The nanoparticles produced from solution plasma process were usually in spherical shape [4]. This is a green approach of copper nanoparticles because it does not release any greenhouse gases and does not require any chemical addition. The process can be conducted under atmospheric pressure in an open system. Compared to chemical methods, it requires less processing time, low energy requirements as well as low cost. Because of all the advantages offered by solution plasma, electrolysis via solution plasma is the most appropriate method for copper nanoparticle synthesis.

2.2 Factors affecting nanoparticle sizes
There are various factors that influence the synthesis, applications and characterization of nanoparticles. Some researchers have reported that time, environment and usage of catalysts can affect the dynamic nature of synthesized nanoparticles. There are other important factors that affect the size of synthesized nanoparticles including discharge time, concentration, voltage, temperature and the method used for synthesis.

In liquid phase processes, the liquid concentration is considered as a major factor on controlling the size of the nanoparticles synthesized. The concentration of the synthesis medium, in this case electrolytes, can influence the size and morphology of the produced nanoparticles. Conventionally, in order to control the size of nanoparticles, additives were added such as reducing agents, surfactants and electrolytes. Firstly, the concentration of added reducing agent can decrease the desired size of copper nanoparticles as long as the precursor concentration is maintained [5]. Moreover, surfactants were also tested upon using the liquid plasma method for copper nanoparticles. It was found that the size decreased when the concentration of the surfactants increased [1].

2.3 Advantages of controlling nanoparticles size related to application
Copper nanoparticles are being increasingly used in different applications on fields of engineering and science. The properties of nanoparticles are massively different from bulk materials, such as the emission of purple light from gold nanoparticles. Controlling nanoparticles size will assists in enhancing the properties of the desired element.

Controlling copper nanoparticles size would have many advantages to its real world applications. One importance of controlling copper nanoparticles size is to reduce the emission of nitrogen oxide in automobiles. Using biodiesels in vehicles engines has lowered the performance and caused a high rate of nitrogen oxide emission. Copper nanoparticles can replace some of the fuel additives in biodiesel. However, this requires copper nanoparticles with size between 40-50nm, which can be difficult to obtain consistently using conventional methods [6].

In addition, selection of catalysts depends on the ratio of surface-to-volume ratio and copper nanoparticles are being used as a catalyst because of this unique property [7]. They are applied as vapour detoxification and liquid-vapour catalyst. Since catalyst performance depends on the ratio surface-to-volume, controlling the size is an important factor to meet this requirement. Compared with noble metals, copper nanoparticles are low in cost and more conductive. Therefore, they are more appropriate for micro-fabrication technologies such as ink printing services.

2.4 Electrolytes selection
Electrolytes can be categorized as strong electrolytes and weak electrolytes. Strong electrolytes have an ability to completely ionize in water which means a 100% of the dissolved compound will break into anions and cations and they are good conductors of electricity. However, weak electrolytes are partially ionised and dissolved and they are considered as poor conductors of electricity. In addition, the weak and strong electrolytes are classified into three groups which are acids, bases and salts. In this paper, K_2CO_3 and NaOH were selected as the working electrolytes. Both are strong electrolytes that dissolve completely in water, ensuring consistent results throughout the experiments.

3. Materials and methodology

3.1. Materials
The following chemicals and materials were used to conduct the solution plasma synthesis:

1) Two types of electrolytes (NaOH and K_2CO_3)
2) Platinum wire with length of 1000mm and a diameter of 0.5mm
3) Copper wire with 1.0 mm in diameter and length of 20mm

3.2. Experimental setup

The nanoparticles were synthesized using electrolysis via solution plasma process. Figure 2 shows a glass beaker of 300 ml capacity in which the cathode used was a copper wire 1.0 mm in diameter and had 20mm immersion depth. Meanwhile, the anode consisted of a platinum wire which had a length of 1000mm and a diameter of 0.5 mm. The surface area of platinum wire (anode) was 50 times larger than copper wire (cathode); this required the anode to be shaped into a coil. Table 1 shows the electrolyte types and the concentrations used. It was important to ensure enough distance between cathode and anode to avoid any unnecessary material failure, electrical shocks or explosions.

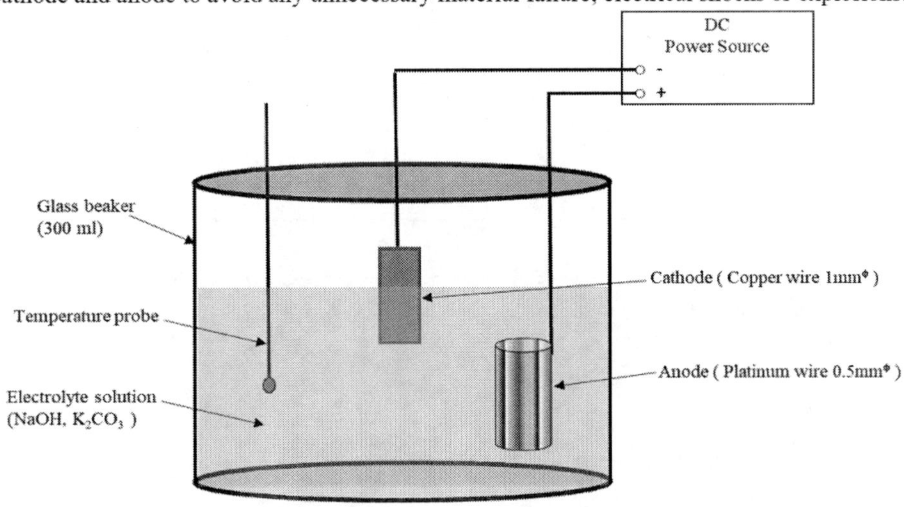

Figure 2. Schematic diagram of experimental setup of solution plasma for copper nanoparticle synthesis.

Table 1. Electrolyte types at different concentration

Electrolyte type	1st Concentration (M)	2nd Concentration (M)	3rd Concentration (M)
Potassium carbonate (K_2CO_3)	0.1	0.5	1
Sodium hydroxide (NaOH)	0.1	0.5	1

3.3. Experimental procedure

After setting up the experiment and all the wires are connected correctly, the DC power supply was turned on. Then the voltage increased in 5V increments every minute, warming the electrolyte, until the plasma in form of glow discharge plasma was reached. The glow discharge voltage was at least 110V. Then the voltage was maintained to sustain the plasma emission, allowing for the synthesis of nanoparticles. During this process, the current must be monitored closely to ensure it does not exceed 4A as it may result in excessive electrolyte bubbling. After 15 minutes, the electricity supply was turned off, and the glass beaker that contains the nanoparticles was left overnight to allow sedimentation of the suspended particles. This eased the process of collecting the nanoparticles synthesized and storing them into small centrifuged tubes to send them for further analysis.

3.4. Results reviewing

The images of the copper nanoparticles are obtained using a Scanning Electron Microscope (SEM). After the SEM images were obtained, the images were processed and the nanoparticles sizes were measured using ImageJ software. ImageJ is a publicly-available software that is recommended by the

IOP Conf. Series: Materials Science and Engineering **429** (2018) 012084 doi:10.1088/1757-899X/429/1/012084

National Institute of Health and has the ability to measure the size and angles of particles as well as evaluating the overall area of the sample.

4. Results

4.1. Using K_2CO_3 as electrolyte

Table 2. Experimental results using K_2CO_3 as electrolyte

Experiment No.	Concentration of electrolyte (M)	Voltage needed to reach glow discharge plasma (V)	Average size of copper nanoparticle (nm)
1	0.1	130	$6.99x10^3$
2	0.5	115	$23.26x10^{-3}$
3	1	95	$28.47x10^{-3}$

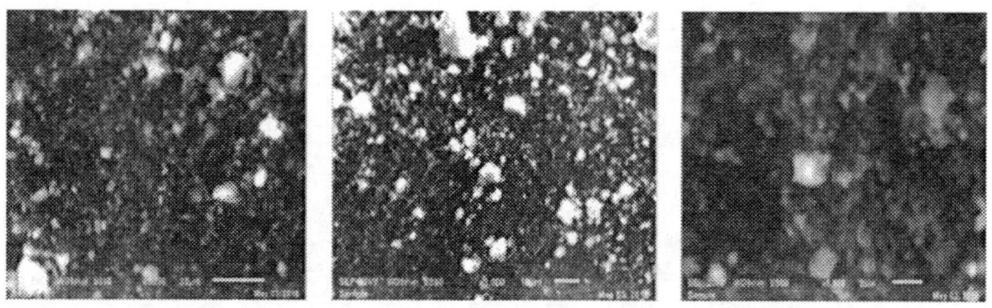

Figure 3. SEM images of copper nanoparticles using 0.1M K_2CO_3 electrolyte

Figure 4. SEM images of copper nanoparticles using 0.5M K_2CO_3 electrolyte

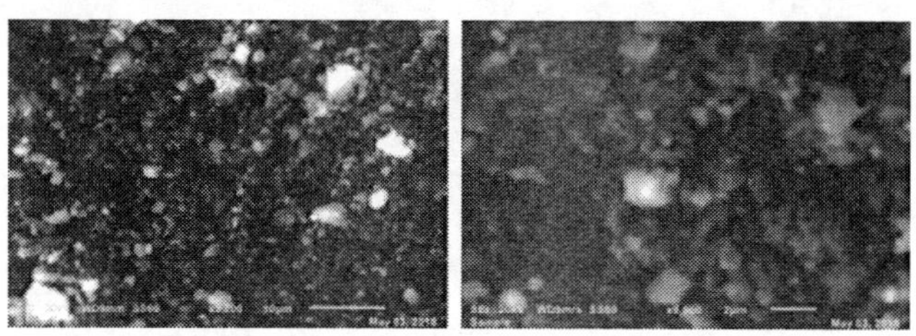

Figure 5. SEM images of copper nanoparticles using 1.0M K_2CO_3 electrolyte

4.2. Using NaOH as electrolyte

Table 3. Experimental results using NaOH as electrolyte

Experiment No.	Concentration of electrolyte	Voltage needed to reach glow	Average size of copper

	(M)	discharge plasma (V)	nanoparticle (nm)
4	0.1	150	$37.95x10^3$
5	0.5	120	$31.19x10^3$
6	1	115	$28.35x10^3$

Figure 6. SEM images of copper nanoparticles using 0.1M NaOH electrolyte

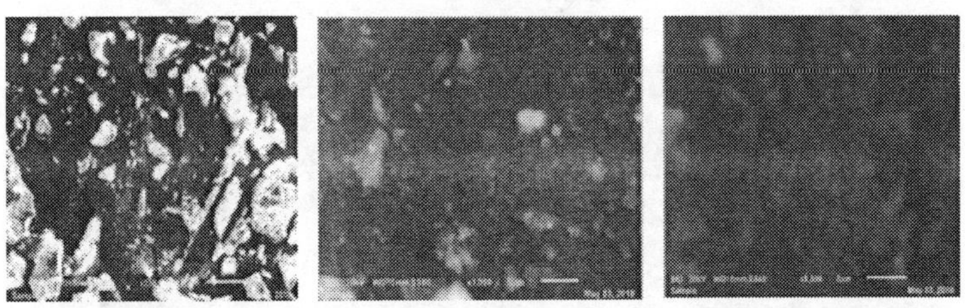

Figure 7. SEM images of copper nanoparticles using 0.5M NaOH electrolyte

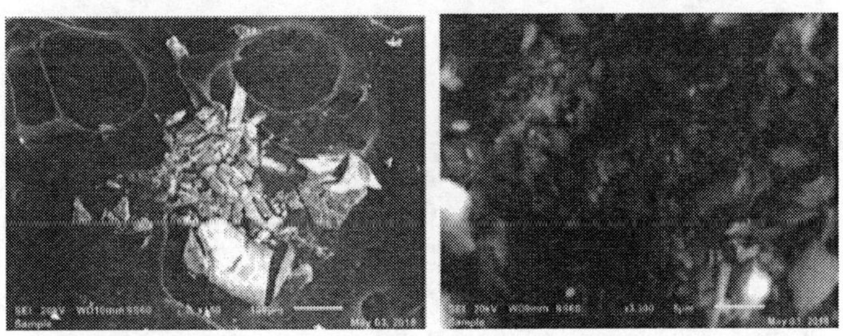

Figure 8. SEM images of copper nanoparticles using 1.0M NaOH electrolyte

5. Discussion

5.1. Influence of glow discharge plasma on amount of produced copper nanoparticles

In order to produce copper nanoparticles, the cathode wire must undergo glow discharge plasma and sustain it for at least 10 minutes. Figure 9 shows the glow discharge plasma is luminous which can be generated when the electrons gain enough energy to produce visible light by repeated collisions that form photons. This happens due to the interaction of both cathode and anode in the solution plasma.

Figure 9. Glow discharge plasma

When the copper wire emits glow discharge plasma, the amount of copper nanoparticles produced will be optimal. It can be seen from Figure 10 that specimens 1 to 4 were highly eroded, while specimens 5 and 6 were slightly eroded after plasma emission. This was because specimens 5 and 6 did not reach full glow discharge plasma, only partial plasma emission compared to specimens 1 to 4. As a result, a smaller amount of nanoparticles were produced when 0.5M and 1.0M NaOH electrolyte concentrations were used.

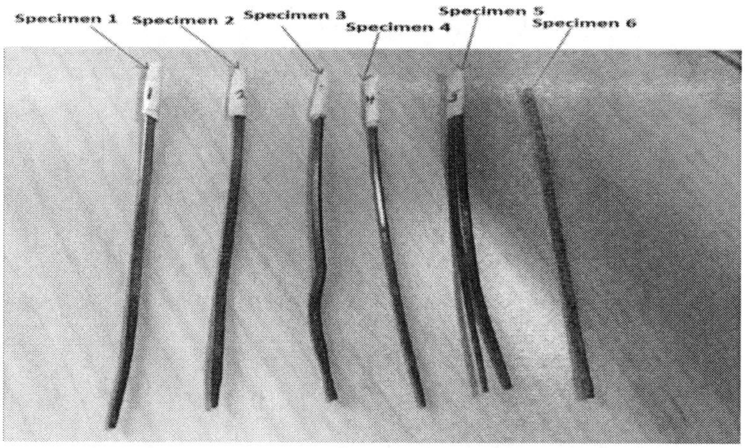

Figure 10. Copper specimens after the experiment

5.2. Relationship between electrolyte concentration on nanoparticles size

The relationship between the electrolyte concentration and copper nanoparticles size was observed in this research. The concentration of the electrolyte solution can influence the size and morphology of the produced nanoparticles. When using K_2CO_3 as an electrolyte, it was observed that the size of the produced nanoparticles increased as the electrolyte concentration increased. This gave a linear relationship between copper nanoparticles size and the concentration of K_2CO_3 as shown in Figure 11.

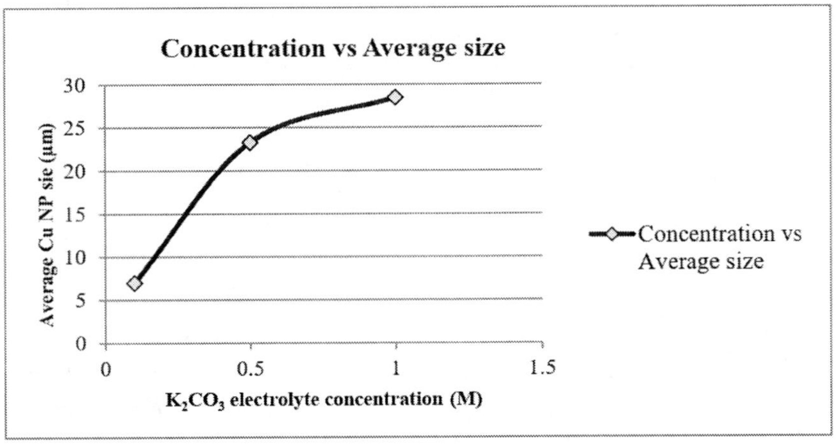

Figure 11. Relationship of concentration of K_2CO_3 and the average size of copper nanoparticles

However, in Figure 12, using NaOH showed an inverse relationship between the electrolyte concentration and the average nanoparticle size. Under this condition, in order to produce smaller copper nanoparticles, higher concentration of NaOH electrolyte was required. The reason behind this relationship was because of the lesser number of nanoparticles produced for 0.5 M and 1.0 M concentration of NaOH. Less nanoparticles produced would give a wider range of shape and sizes and thus accounting for the relationship.

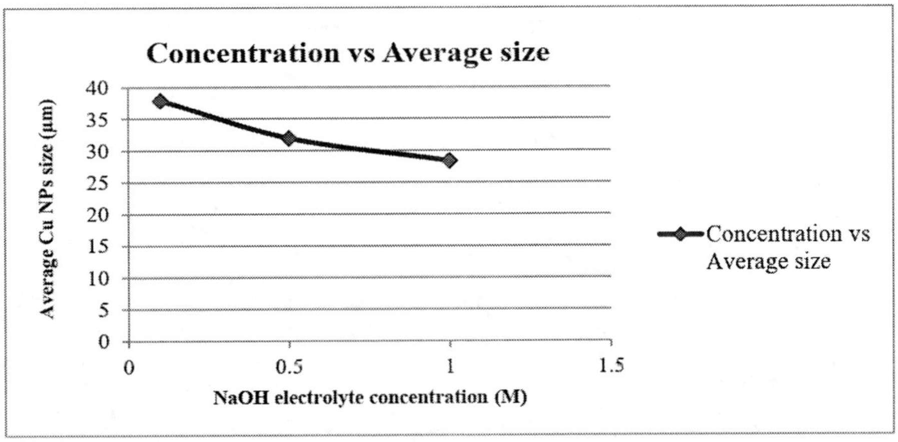

Figure 12. Relationship of concentration of NaOH and average size of copper nanoparticles

5.3. Influence of electrolyte concentration and conductivity on voltage needed to reach glow discharge plasma

It is known that electrolytes are solutions which have dissolved ions and cations which are capable of conducting electric current since they liberate charged particle inside the solution. To generate and sustain glow discharge plasma emission, the electrolyte has to be conductive. The conductivity increases when the quantity of dissolved particles increases. Also, it is known that concentration is the amount of dissolved particles or substances inside a solution. This means that higher concentration needs higher amount of dissolved substance per unit volume. Thus, higher concentrations of electrolytes lead to higher conductivities since the amount of substance in the solution increases.

It was observed that when the concentration of electrolyte added increases, the conductivity will increase which will lead to lower voltage to reach glow discharged plasma formation. Figure 13 shows an inverse relationship between the voltage and the concentration for both electrolytes used. This is because the increased conductivity will increase the interaction between the cathode and anode, which would in turn decreases the effort needed to reach glow discharge plasma. As a result, the voltage needed to obtain glow discharge plasma decreased when higher concentration electrolytes were used.

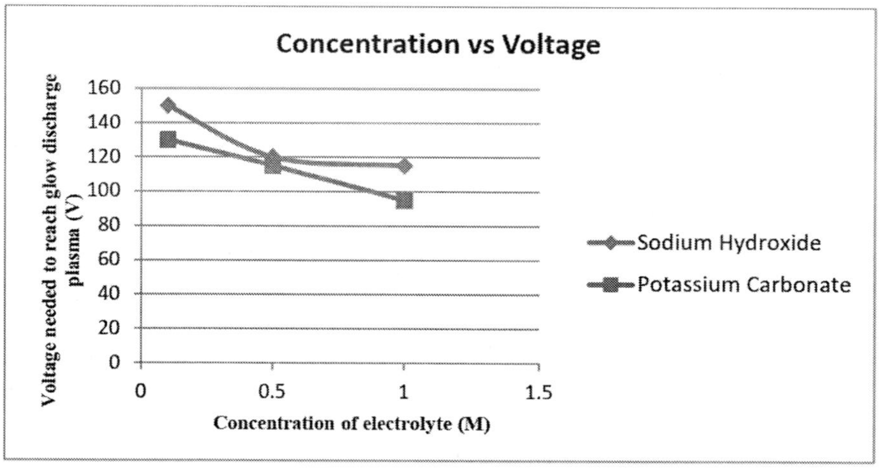

Figure 13. Relationship between concentration and voltage needed to reach glow discharge plasma

6. Conclusions

This research aims to control the size of the copper nanoparticles by changing the concentration of the electrolyte used. Two types of electrolytes were used and investigated in this case to confirm the relationship between electrolyte concentration and the average size of copper nanoparticles. It was observed that when the concentration of K_2CO_3 increases, the size of the produced nanoparticles also increases. In contrast, the size of the copper nanoparticles decreases when the concentration of NaOH increases. Moreover, the intensity of glow discharge affected the amount of synthesized copper nanoparticles in which and increases in the number will happen when the intensity increases. In addition, the conductivity showed a rise when both electrolyte are subjected to higher concentration which leads to lower glow discharge plasma initiation voltage.

The control of copper nanoparticle during solution plasma synthesis size allows them to be used in various applications.. Copper nanoparticles can be used as a fuel additive in biodiesels where the required size is around 50 nm. Other applications include being used as a catalyst due to its unique high-surface-to-volume-ratio. Since copper nanoparticles size is affected by many factors, the research topic can be further extended to study the effect of the supplied voltage on copper nanoparticle sizes using the same method of synthesis. Other types of electrolytes can be investigated such as using acids and bases. Meanwhile, other metal nanoparticles such as silver and gold are widely used in different application. Therefore, the effect of various electrolyte concentrations reported here have the potential to be applied in the field of nanotechnology and nanomaterials.

References

[1] Lee H, Park S H, Seo S G, Kim S J, Kim S C, Park Y K and Jung S C 2014 *Curr. Nanosci.* **3** 7
[2] Patra J K and Baek K H 2014 *J. Nanomater.* **2014** 1
[3] Bratescu M A, Cho S P, Takai O and Saito N 2011 *J. Phys. Chem.* **115** 24569
[4] Hu X L. Takai O and Saito N 2013 *J. Phys. Conf. Ser.* **417** 012030
[5] Ghorbani H R 2014 *Orient J. Chem.* **30** 1941
[6] Umer A, Naveed S, Ramzan N and Rafique M S 2012 *Nano* **7** 1230005
[7] Tamilvanan A, Balamurugan K, Ponappa K and Kumar B M 2014 *Int. J. Nanosci.* **13** 1430001

Acknowledgements

The authors would like to thank Mr. Peter Yek Nai Yuh from University College of Technology Sarawak for conducting the SEM observation for the nanoparticle samples.

IOP Conf. Series: Materials Science and Engineering **429** (2018) 012085 doi:10.1088/1757-899X/429/1/012085

Controlled nanocrystallites growth of plasma-treated Cu sheets

A M S M Al-Sawafi, T M Lim, P N Y Yek and M R M bin Julaihi

Swinburne University of Technology Sarawak, Jalan Simpang Tiga, 93350 Kuching, Sarawak, Malaysia

mjulaihi@swinburne.edu.my

Abstract. Controlling the growth size of nanocrystallites (NCs) has been a big challenge for previous researchers due to unpredictable growing rate and shape obtained during the synthesis. This paper demonstrates an efficient way to control the growth of copper (Cu) NCs using a combination of solution plasma, sunlight and water. The NCs are synthesized by using solution plasma method which has a lot of advantages in controlling NC growth sites. The material used in this paper is Cu sheet due to its antibacterial properties, high electrical and heat conductivity, and low cost. This method is followed by immersing the plasma treated Cu sheets in distilled water and exposing it to sunlight. This method is also known as submerged photosynthesis of crystallites (SPSC), which is a low-temperature and surfactant technique used for NCs fabrication. The photosynthesis-like method is used to allow for NCs growth. The NCs growth size and shape was observed using SEM and ImageJ software. The NCs shapes can be analyzed by observing the antibacterial properties of NCs which is related to the small size of the NCs. The beneficial outcomes of Cu NC shape with antibacterial properties are expected, which can be used in various applications without causing any environmental impact. In addition, using a combination of solution plasma and sunlight exposure, we demonstrated an easy and fast synthesis method to develop Cu NCs economically.

1. Nomenclature

NCs	Nanocrystallites
NPs	Nanoparticles
SEM	Scanning Electron Microscopy
SPSC	Submerged Photosynthesis of Crystallites

2. Introduction

In the past two decades, metallic nanocrystallites have been a huge subject for the in materials engineering field due to their special physicochemical and electrical properties. Research has been increasing for the nanomaterials that possess such special properties [1]. For instance, the physical properties of nanocrystallites can be improved in thermal stability and mechanical stiffness by using specialized synthesis methods [2].The pure metals and alloy NPs have been used for some applications such as magnetic, catalysis, optoelectronics and microelectronics. It is also used for conductive pastes, battery electrodes, and fuel cells. Among the different metals of NPs, copper NCs have gained a lot of attention because their applications in electrical devices such as multilayer ceramic and oriented circuit boards [3]. Cu is typically chosen because of its low cost and high conductivity [2].

The fabrication of the NPs is one of the most important topics of concern regarding nanotechnology since it is very important to have the ability to control the NPs shape, size and its distribution on the Cu sheet [4]. Also, there is a pressing need for a NC green synthesis methods which would result in their usage in different applications such as in the biomedical and antimicrobial usages [2].

Furthermore, Cu materials are commonly used methods since it shows high antimicrobial properties and cost-effectiveness compared with other metals with similar properties such as silver [2]. Typical methods to synthesize Cu NCs are pulsed sonoelectrochemical synthesis, microemulsion technique, polyol method, thermal decomposition method, laser ablation method and chemical reduction method [4]. In this paper, the method that was used is called solution plasma synthesis due to its simplicity to control size and morphology of the NCs [4].

3. Literature review

3.1 Plasma treatment and their applications

Firstly, plasma discharge is defined as the electrical unbinding of charged particles combined with free radicals generation. Since plasma discharge has multi-reactive constituents, its ability has the desirable effect such as materials surface modification [5]. The interaction of plasma-liquid produces energetic nucleation sites in materials surfaces. The usage of solution plasma rather than solid electrodes gives the ability to form the NPs on the surface of the plasma solution interaction region, rather than on the plasma layer itself which consists of ionized gases [6].

Usage of solution plasma in Cu NC synthesis is are due to its low cost, simple process and the ease in controlling experimental parameters on the liquid surface [7]. The other advantages are the ease of experimental setup, simple mass production due to high yield, the availability of the metallic sheet used as raw materials for NPs synthesis and the use of the gas supply is not required [7].

3.2 Types of synthesis method

There are many methods used to synthesize the NPs and NCs divided into three types. The first type is the physical methods that break down bulk material into nanosized materials by cutting, scratching or grinding techniques as shown in Figure 1 [2]. Physical method is applied by physical strategies such as using the laser excision or vacuum discharge to synthesize the nanomaterials.

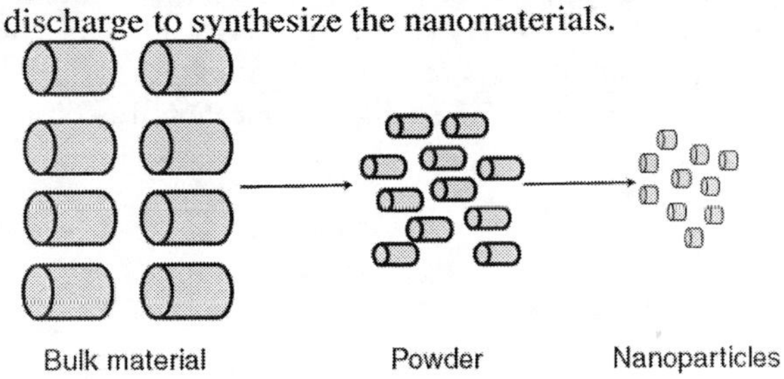

Figure 1. Physical synthesis methods of nanoparticles.

The other types of synthesis methods are the biological and chemical methods, which are used to develop the structure of NP atoms or molecules as shown in Figure 2 [2]. The chemical methods are

IOP Conf. Series: Materials Science and Engineering **429** (2018) 012085 doi:10.1088/1757-899X/429/1/012085

applied by using chemical reduction, electrochemical and microwave. Likewise, the other methods to synthesize NPs is biological synthesis, typically by extractingthe enzymes or bacterium from the plant sources [2].

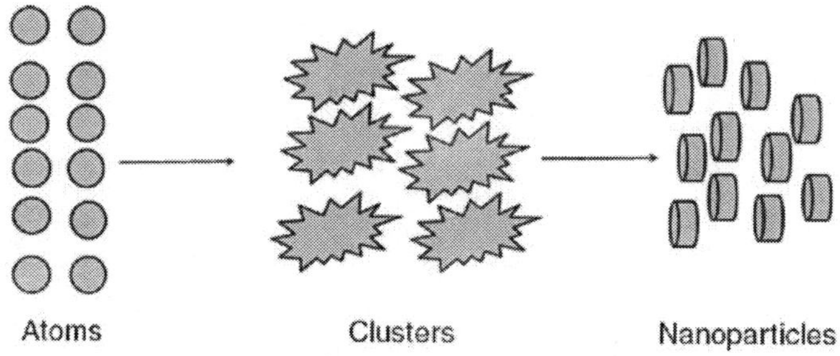

Figure 2. Chemical and biological synthesis methods.

However, some of these methods have high cost and complex setup. Also, some chemical and physical synthesis methods produce toxic by-products which are considered to be ecologically hazardous and it cannot be used in medical applications [2]. Therefore, the method used in this pper can be called a green synthesis method since its by-products are environmentally friendly.

Here, the proposed synthesis method is called solution plasma and the submerged photosynthesis of crystallites (SPSC) method, which means a surfactant-free, low temperature approach that is used for metal NCs fabrication [8]. SPSC is done by photosynthesis-like reaction with the irradiation treatment of UV light on the materials immersed in water to allow for NC growth. The NC growth was observed at room temperature without the need for additional chemicals besides water [8]. Apart from NC growth, another useful application of the SPSC method is hydrogen gas generation [8].

3.3 Effect of electrolyte concentration and plasma voltage on NC growth

A plasma field is maintained on the surface of the Cu sheets. The plasma layer is caused by heating of the solution plasma near the electrode, in this case Cu sheet, which is called "glow discharge" [7]. The glow discharge plasma causes nanosized surface protrusions to appear on the Cu sheet, increasing the surface roughnes. We identified four parameters that could show effects on the surface morphology on Cu sheet during the experiment, namely plasma voltage, light intensity, time of light exposure and solution temperature [8].

However, in solution plasma method, the electrolyte concentration could affect the NC size as well. The higher the electrolyte concentration, the lower the voltage for glow discharge plasma initiation while the lower the electrolyte concentration, the higher the voltage needed for plasma initiation. A low plasma initiation voltage results in the appearance of smaller surface protrusions on the Cu sheets, and thus smaller NC size after undergoing SPSC process.

Table 1. Experimental observation condition and outcomes morphologies at 200 V [5].

Electrolyte(K_2CO_3) concentration	Electrolysis Time (min)	Morphology Shape
0.5M	60	Flower-like
0.1M	60	Flower-like
0.01M	60	Flower-like
0.001M	60	Flower-like

The table above shows that the NC can be maintained at a specific time which is 60 minutes at low electrolyte concentration in a high voltage as discussed in the previous literature. The observed shape is a flower-like shape in 200 voltages. Therefore, the NC growth size can be controlled by low electrolyte concentration in a fixed time at a high voltage [5].

3.4 Evaluation of NC growth size

The NCs growth size can be evaluated before and after the experiment by using scanning electron microscopy (SEM). The NC shapes such as whisker, pentagonal and tetragonal were mostly observed in high concentration electrolyte and extended time plasma treatment. On the other hand, the spherical shape and dendrite shape of the NPs were commonly observed during short plasma discharge times. [10].

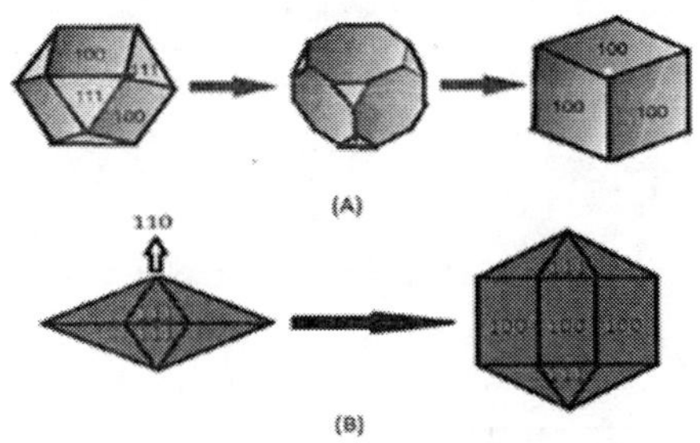

Figure 3. The growth mechanism for (A) cube shape and (B) for rod-shape particles [11].

The NC size growth is related to its antibacterial properties [12]. In order to improve the antibacterial application, the NC size in a Cu sheet should be small to provide a high surface area to interact with microbial layers [12]. Therefore the obtained size range of the NC is 50 nm which is the size needed to allow antibacterial properties to appear on the SPSC treated Cu sheet. The NPs work to reduce the growth of the bacteria by destroying their outer layer [13-14].

Also, our experiment aims to obtain a certain shape of NC in order to impart antibacterial properties on the Cu sheet, namely the hexagonal shape [15]. Therefore, the size and shape should be taken into consideration while observing the NC in order to observe the antibacterial properties related to the small size and a specific shape.

4. Methodology

4.1 Experimental apparatus

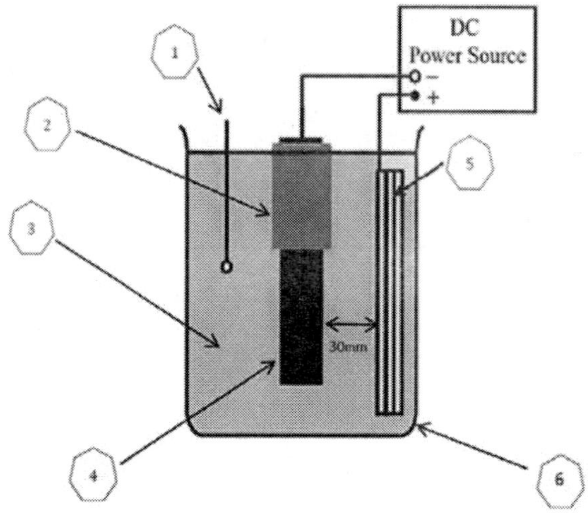

Figure 4. Diagram of experimental apparatus.

Table 2. Apparatus description.

No	Apparatus Description
1	Thermometer
2	Insulator
3	Electrolyte solution (K_2CO_3)
4	Cathode (Cu sheet with a size of 30×1×1 mm)
5	The anode of platinum wire with a diameter of 30 mm
6	Glass beaker with 250ml of capacity

4.2 *Experimental procedure*

4.2.1 Surface pretreatment. A Cu sheet was used as a cathode with a purity of 99.9% was cut into three pieces of 30×1×1 mm in size. The anode is a platinum wire 1 mm in diameter with purity of 99.9% curved into a coil shape. Both anode and cathode were washed with distilled water and then immersed in K_2CO_3 electrolyte. The contact area between Cu wire and Cu sheet is set with 20 mm of exposed area of Cu sheet. Firstly, the solution was preheated to 90 °C. After that, the applied discharge voltage was 140 V using DC voltage for 5 minutes to achieve plasma treatment on the sheet surface. Afterwards, the Cu sheet was collected and cleaned with distilled water and left to dry at ambient temperature. The surface pretreatment was repeated for subsequent Cu sheet specimens.

4.2.2 SPSC procedure. The Cu sheet placed inside glass test tubes filled with distilled water. Three Cu sheets were exposed to natural sunlight for 12, 24 and 36 hours respectively. After finishing exposure, the specimens were collected and let dry ambient temperature. The analysis of the surface morphology of NCs was observed using SEM.

5. Results and discussion

The results below show the SEM images of each of the plasma-treated Cu sheets after exposure to sunlight:

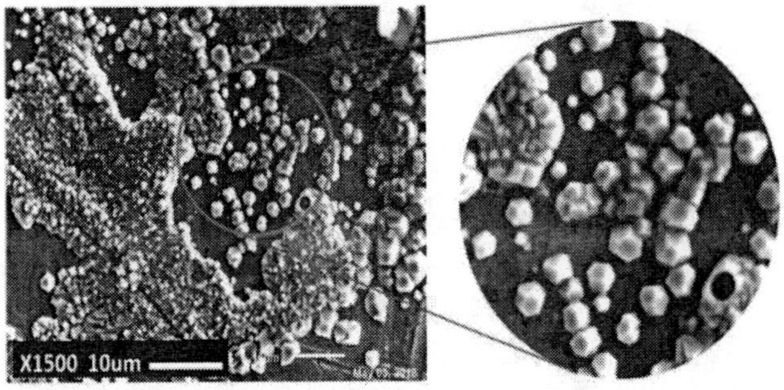

Figure 5. Surface of plasma-treated Cu sheet after 12 hours of SPSC.

According to the above figure, the growth of NCs occurred during the sunlight exposure. Using ImageJ software, the observed mean length of 20 NC shapes is around 293nm. Here, the NCs were shaped as tetrahexahedron-like after 12h of sunlight exposure.

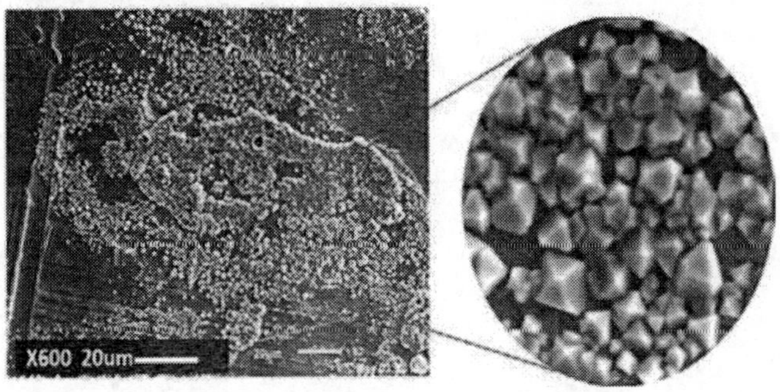

Figure 6. Surface of plasma-treated Cu sheet after 24 hours of SPSC.

After 24h of light exposure, the NCs length increased and the shape changed as well. The observed mean length of 20 NCs were around 3860 nm. The NCs shape changed to square pyramid-like after 24 hours of SPSC as shown in Figure 6.

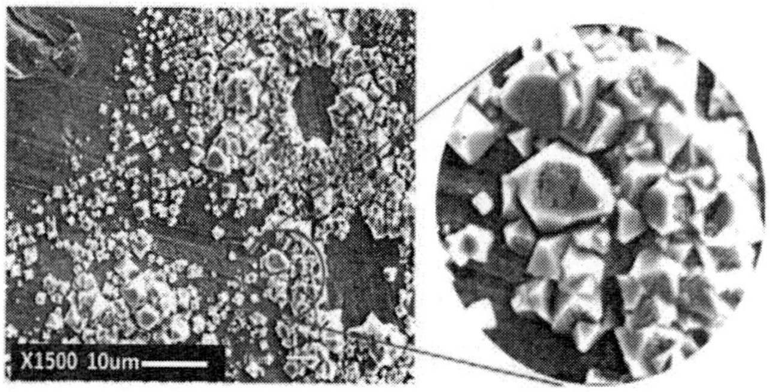

Figure 7. Surface of plasma-treated Cu sheet after 36 hours of SPSC.

In Figure 7, the NCs mean length of 20 NCs was around 4540nm after 36 hours of light exposure and the observed shape was decahedral-like shape.

Table 3. Summary of NC morphologies obtained.

Light exposure Times (hours)	NCs Mean Length (nm)	NCs Shape	Surface roughness
12	293	Tetrahexahedron-like	Fine
24	3860	Square pyramid-like	Fine
36	4540	Decahedral-like	Fine

The size and shape of NCs were highly dependent on the time of sunlight exposure to the plasma treated Cu sheets. There was a linear relationship between the exposure time and NCs size. Hence, the NCs growth can be controlled by adjusting the time of sunlight exposure. Also, the SEM images show that for each of the specimen the NCs have different shapes. The tetrahexahedron-like shape was observed in first 12 hours, square pyramid-like shape in 24 hours and decahedral-like shape in 36 hours of sunlight exposure. That means that as the NCs grow, the shapes also changed which allows for use in different applications depending on the shape and size requirements of the NCs. In addition, the plasma treatment of Cu sheet by using low electrolyte concentrations and high voltage resulted in fine structure and shapes of the NC nucleation sites, which in turn enhanced the growth of NCs.

A potential application of the Cu sheet after SPSC is that antibacterial properties can be obtained within the first 12 hours of SPSC, where showed the NCs size were around 293nm. The small size of NCs tends to have high surface volume, where they react to destroy bacteria by blocking the pores in bacteria layers. Thus the Cu NCs may possess antibacterial properties.

6. Conclusions

A new way of controlling NC growth size by using solution plasma and sunlight exposure in water has been successfully established. The size and shape of NCs depended on the time of sunlight exposure. The small size of NCs can be obtained in short time of sunlight exposure, while the large size NCs can be obtained in a long time of sunlight exposure. Likewise, high voltage, low concentration of K_2CO_3 solution during surface treatments of Cu sheet delivered a fine structure and new crystallites shape after SPSC method. The used of distilled water during SPSC resulted in a fine surface of NCs. The small size of NCs in short exposure time was shown, which in turn imparts the antibacterial properties of Cu sheet. The experimental parameters resulted in a simple control of size and shape of Cu NCs. Moreover, the easy and fast synthesis method of Cu NCs has been successfully obtained.

7. References

[1] Saito G, Hosokai S, Tsubota M and Akiyama T 2011 *J. Appl. Phys.* **110** 0
[2] Umer A, Naveed S, Ramzan N and Rafique M S 2012 *Nano* 7 1230005
[3] Rajesh K M, Ajitha B, Reddy Y A K, Suneetha Y and Reddy P S 2016 *Mater. Today Proc.* **3** 1985
[4] Lee H, Park S H, Seo S G, Kim S J, Kim S C, Park Y K and Jung S C 2014 *Curr. Nanosci.* 3 7
[5] Liu J, He B, Chen Q, Liu H, Li J, Xiong Q, Zhang X, Yang S, Yue G and Liu Q H 2016 *Electrochim. Acta* **222** 1677
[6] Graham W G and Stalder K R 2011 *J. Phys. D. Appl. Phys.* **17** 174037
[7] Saito G, Hosokai S, Tsubota M and Akiyama T 2012 *J. Appl. Phys.* **112** 1
[8] Jeem M, Julaihi M R M, Ishioka J, Yatsu S, Okamoto K, Shibayama T, Iwasaki T, Kato T and Watanabe S 2015 *Sci. Rep.* **5** 11429.
[9] Saito G, Nakasugi Y and Akiyama T 2014 *J. Appl. Phys* **116** 8
[10] Saito G, Nakasugi Y,. Yamashita Y and Akiyama T 2016 *Appl. Surf. Sci.* **290** 419
[11] Mott D, Galkowski J, Wang L, Luo J and Zhong C J 2007 *Langmuir* **23** 5740
[12] Ramyadevi J, Jeyasubramanian K, Marikani A, Rajakumar G and Rahuman A A 2012 *Mater. Lett.* **71** 114
[13] Kruk T, Szczepanowicz K, Stefańska J, Socha R P and Warszyński P 2015 *Coll. Surf. B Bioint.* **128** 17
[14] Dizaj S M, Lotfipour F, Barzegar-Jalali M, Zarrintan M H and Adibkia K 2014 *Mater. Sci. Eng. C* **44** 278
[15] Bogdanović U, Lazić V, Vodnik V, Budimir M, Marković Z and Dimitrijević S 2014 *Mater. Lett.* **128** 75.

Acknowledgements

The authors would like to thank Mr. Peter Yek Nai Yuh from University College of Technology Sarawak for conducting the SEM observation for the nanocrystallites samples.

IOP Conf. Series: Materials Science and Engineering **429** (2018) 012086 doi:10.1088/1757-899X/429/1/012086

Microkinetic deactivation studies on nickel supported catalyst for dry reforming of methane

P R K Fu[1], Twaiq F[1] and W L Chung[1]

[1] Faculty of Engineering, Computing and Science, Swinburne University of Technology Sarawak Campus, Jalan Simpang Tiga, 93350 Kuching, Sarawak, Malaysia

Abstract. In recent decades, greenhouse gases (GHG) such as CO_2 and CH_4 have been on the rise affecting climate change. Consequently, researchers have been studying on chemical reactions that are able to consume and convert GHG into useful resources. Dry reforming of methane is a heterogeneous catalytic reaction that consumes greenhouse gases such as CO_2 and CH_4 to produce syngas, CO and H_2, which later becomes an essential feedstock for Fischer-Tropsch process. Various researchers due to its affordability have studied nickel-based catalyst and it exhibits excellent performance in DRM reaction. However, coke formation is a major issue concerning catalyst deactivation during the dry reforming reaction. Therefore, thermodynamic study based on Gibbs energy minimization is conducted to identify the reaction routes that favour carbon deposition. In addition, the deactivation mechanism of coke formation together with microkinetic modelling of dry reforming of methane on nickel catalyst is being developed.

1. Introduction

Greenhouse gases (GHG) in the atmosphere has increased significantly since the beginning of Industrial Revolution. The main concerning issue associated with GHG is global warming which induces climate change. Dry reforming of methane (DRM) has drawn attention from researchers due to its benefit to be able to consume and convert GHG into useful resource, synthesis gas. DRM is a heterogeneous catalytic reaction that enables the production of valuable syngas, while consuming GHG such as methane and carbon dioxide as reactants for the catalytic process (1). The product from this catalytic process produces syngas or better known as synthesis gas is a mixture of hydrogen (H_2) and carbon monoxide (CO) that acts as a chemical intermediate for the production of Fischer-Tropsch synthesis, methanol and fine chemicals (2).

There are various types of catalysts used in the dry reforming reaction, such as noble metals and transition metals. Noble metals such as ruthenium and rhodium are comparatively more expensive to transition metal such as nickel. Therefore, nickel-based catalyst is more favourable for DRM process even at industrial scale. The type of supports may lead to different degree of dispersion of nickel in the support's pores support, which in turns affect catalytic behaviours and performance (3). The problem of coke formation on the catalyst's surface remains a major drawback in the dry reforming reaction. Accumulation of coke deposit on the surface of nickel catalyst causes inhibition of active sites that, eventually leads to catalyst deactivation. Hence, the formulation of kinetic model for dry reforming of methane is important to analyse the reaction mechanism of coke formation in accordance with the catalyst reaction activity.

Dry reforming of methane is a highly endothermic reaction, as shown in equation (1). This indicates that the large required energy input for methane and carbon dioxide to transform into syngas. Thus, the reforming reaction is carried out at high temperatures of above 1073K in order to produce H_2/CO ratio of unity.

$$CH_4 + CO_2 \rightarrow 2CO + 2H_2;\ \Delta H° = +247\frac{kJ}{mol} \quad (1)$$

$$2CO \rightarrow CO_2 + C;\ \Delta H° = -\frac{172kJ}{mol} \quad (2)$$

The Boudouard reaction in equation (2) occurs at temperatures below 800K that encourages carbon deposition on the catalyst surface. Ultimately, the amount of carbon deposits on Ni catalyst covering active sites of the nickel will deactivate the catalyst. This affects the overall DRM performance. The aim of this work is to carry out thermodynamic study to understand the reaction routes that favour carbon deposition, which later a microkinetic model is proposed.

2. Modelling Expression

The order of elementary reaction steps in a heterogeneous catalytic gas reaction comprises of adsorption and desorption of species, surface diffusion and chemical reaction of adsorbed species. There are different models that describe the reaction mechanism steps. The frequently used models in DRM are namely power law model, Langmuir-Hinshelwood-Hougen-Watson (LHHW) model and microkinetic (MK) model. This model expression together with the kinetic data gives a best fit in accordance to the experimental data and is used to describe the rate of reaction. This correlation allows further optimization for both the design of catalyst and reactor.

2.1. Power Law Model

The power law provides an evaluation of the overall rate of reaction with only consideration given to its reactants. This method of evaluation is applicable to initial guess estimates without consideration of the reaction mechanism in details (4). The power-law model as shown in equation (3), only evaluates the kinetics parameters of DRM reaction, where r represents the rate of reaction, m and n are the reaction orders with respect to the partial pressure, P of the reactant species, CH_4 and CO_2.

$$r = k[P_{CH4}]^m[P_{CO2}]^n \quad (3)$$

This analytical technique provides a rough estimation in evaluating reaction parameter, which does not represent the kinetics of reaction (5). Osazuwa et al (6) studied the reaction between CO_2 and CH_4 by fitting the reaction rates and partial pressure of reactants into the power law model. The rate equation for the kinetic model as shown in equation (4) enables the determination of the reaction order and pre-exponential factor. Figure 1 shows the estimated activation energy for methane and carbon dioxide obtained from Arrhenius plot.

$$r = Aexp(-\frac{Ea}{RT})[P_{CH4}]^{\alpha}[P_{CO2}]^{\beta} \quad (4)$$

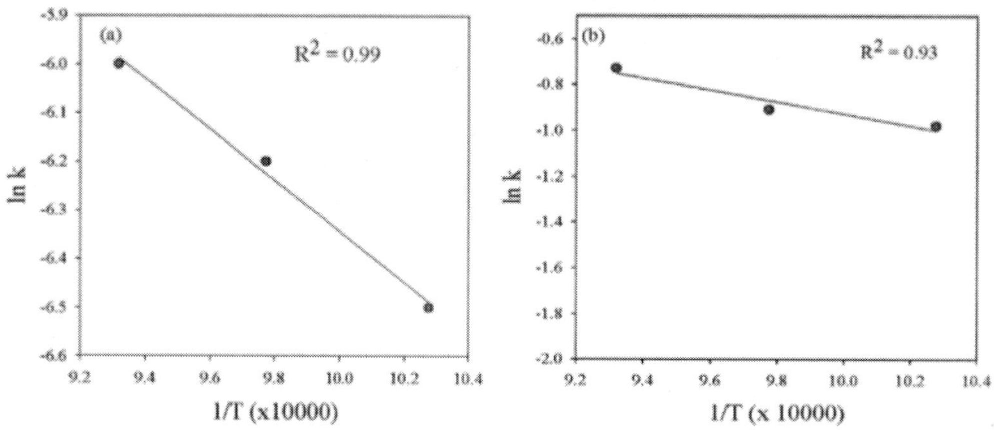

Figure 1. Estimation of activation energy, E_a for both methane (red line) and carbon dioxide (blue line) obtained via Arrhenius plot (6)

2.2. Langmuir Hinshelwood-Hougen-Watson Model

Langmuir-Hinshelwood kinetic describes the surface reaction of adsorbed species that consists of either a single-site or a dual-site mechanism (7). Single site mechanism only allows adsorbed reactants on the active site to take part in the surface reaction. A dual site mechanism happens when the absorbed reactants interact with its adjacent site; the adjacent site can be either vacant or occupied with different reactant species. All the potential reaction pathways that accounts for adsorption, surface reaction and desorption will be included in the model. The letter *S* in the elementary reaction below represents species adsorbed on the active site and the letter alone indicates vacant site.

$$A \cdot S \rightleftharpoons B \cdot S \quad (7)$$
$$A \cdot S + S \rightleftharpoons B \cdot S + S \quad (8)$$
$$A \cdot S + B \cdot S \rightleftharpoons C \cdot S + D \cdot S \quad (9)$$

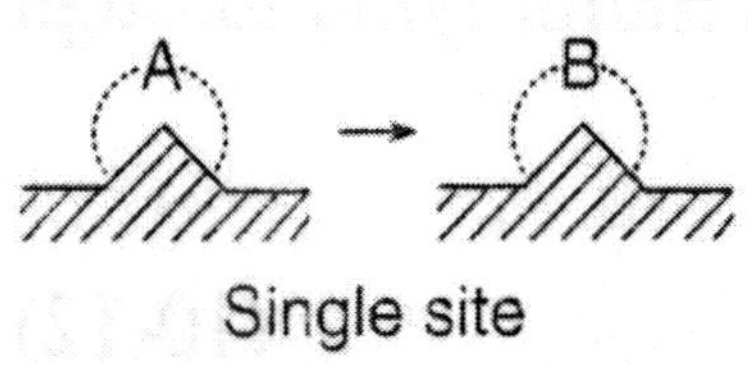

Figure 2. Single site mechanism

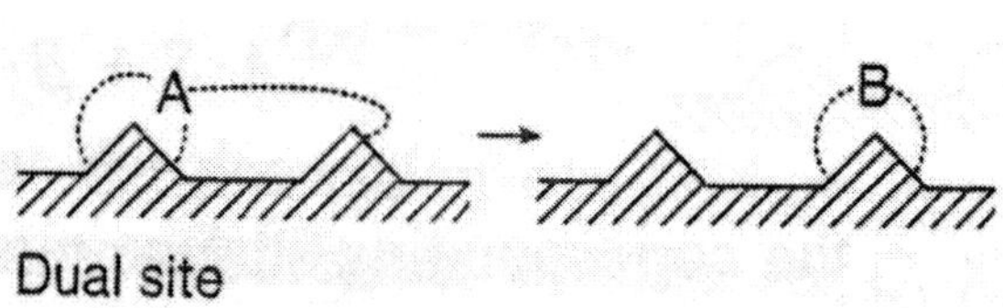

Figure 3. Dual site mechanism with vacant adjacent site

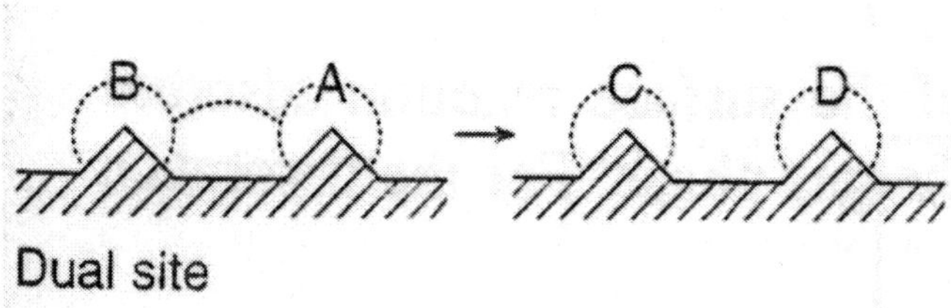

Figure 4. Dual site mechanism with occupied adjacent site

LHHW model requires a rate-determining step (RDS) in order to formulate the rate law. Sawatmongkhon et al (8) proposed a LHHW model for dry reforming of methane on Pt/Rh catalyst. The reaction mechanism consisted of five steps, which included adsorption of reactant species CO_2 and CH_4, surface reaction between adsorbed species and active sites at catalyst surface, and desorption of products, CO and H_2. Their reaction mechanism model is described as below.

$$CH_4 + * \rightleftharpoons CH_4 * \quad (10)$$
$$CO_2 + * \rightleftharpoons CO_2 * \quad (11)$$
$$CO_2 * + CH_4 * \rightleftharpoons 2CO * + 2H_2 * \quad (12)$$
$$2[CO * \rightleftharpoons CO + *] \quad (13)$$
$$2[H_2 * \rightleftharpoons H_2 + *] \quad (14)$$

2.3. Microkinetic Model

Microkinetic modelling is an approach to study and understand the catalytic processes at a microscale level (9). The model accounts for all possible elementary steps that sum up the catalytic cycle. Microkinetic model provides quantitative information of a catalytic reaction that includes absorption, surface reactions and desorption for a catalytic cycle (10). The prerequisite for this approach is to measure the reaction rate in the absence of heat and mass-transfer limitations. Dehimi et al (11) studied the microkinetic model of dry reforming of methane over Ni/γ-Al_2O_3. The model provides detail surface mechanism in DRM that includes the formation of carbon on the catalyst surface. The study is conducted under atmospheric pressure over a temperature range of 450°C to 650°C. The microkinetic model includes 34 irreversible elementary reactions, amounting to 12 surface species and 6 gas phase species.

3. Modelling Methodology

Thermodynamic equilibrium analysis on the proposed reactions is performed with ASPEN HYSYS based on direct minimization of Gibbs free energy method. All possible reaction pathways for DRM as listed, is studied at a temperature range of 600°C to 800°C, all at 1 atmospheric pressure. The evaluation of equilibrium constant, K for a reaction is done based on the Gibbs free energy of a reaction, as shown in equation (15). The enthalpy and entropy values are obtained from HYSYS simulation for the calculation of Gibbs free energy in Equation (16).

$$\text{Equilibrium constant, } K = \exp(\frac{-G}{RT}) \quad (15)$$
$$\text{Gibbs free energy, } \Delta G = \Delta H - T\Delta S \quad (16)$$

All possible reaction pathways in DRM reaction are described as below:

$$CH_4 + CO_2 \rightarrow 2CO + 2H_2 \quad (17)$$
$$H_2 + CO_2 \rightarrow CO + H_2O \quad (18)$$
$$CH_4 + H_2O \rightarrow CO + 3H_2 \quad (19)$$

$$CH_4 \rightarrow C + 2H_2 \quad (20)$$
$$2CO \rightarrow C + CO_2 \quad (21)$$

The rate equation for all the proposed reactions is done based on simple collision theory. The collision rate is to provide possibility of effective collision between reactants. The rate of collision for both unimolecular and bimolecular collision is calculated with equation (22) and equation (23), where d_{AB} is the arithmetic mean of two molecular diameter (m), $d_A$is the kinetic diameter (m), c'_A is the concentration of molecule A (m^{-3}), $c'_B$is the concentration of molecule B (m^{-3}), k_B is Boltzmann constant and T is the operating temperature.

Rate of collision for bimolecular, $Z_{AB} = d_{AB}^2 c'_A c'_B \left(8\pi k_B \frac{T}{\mu}\right)^{\frac{1}{2}}$ (22)

Rate of collision for unimolecular, $Z_{AA} = 2d_A^2 (c'_A)^2 \left(\pi k_B \frac{T}{m_A}\right)^{\frac{1}{2}}$ (23)

Reduced molecular mass, $\mu = \frac{m_A m_B}{m_A + m_B}$ (24)

4. Results and Discussion

4.1. Dry Reforming Reaction

Table 1 shows the equilibrium constant for dry reforming of methane. According to Le Chatelier's principle, as the K value decreases, the reaction is found to shift to the left. This indicates that more reactants are formed. Based on the result in Table 1, it is found that as the temperature increases, there is a slight decrease in equilibrium constant and rate of collision.

Reaction: $CH_4 + CO_2 \rightarrow 2CO + 2H_2$		
Temperature (°C)	**Equilibrium constant, K**	**Z_{AB} ($m^{-3}s^{-1}$)**
600	-2.85804	1.91E+31
650	-3.38482	1.81E+31
700	-3.90839	1.72E+31
750	-4.42375	1.63E+31
800	-4.92819	1.56E+31

Table 1. Equilibrium constant and rate of collision for the DRM reaction at various temperatures

4.1.1. CH_4 Conversion

Several feed ratio of CH_4:CO_2 is used to study the conversion of CH_4 in dry reforming of methane in a R-GIBBS reactor over a temperature range of 600°C to 1500°C at atmospheric pressure 1 atm. Figure 5 shows the CH_4 conversion against temperature. The highest CH_4 conversion is observed at CH_4:CO_2 feed ratio of 1 over the entire temperature range, whereas the lowest CH_4 conversion is observed at CH_4:CO_2 feed ratio of 2.

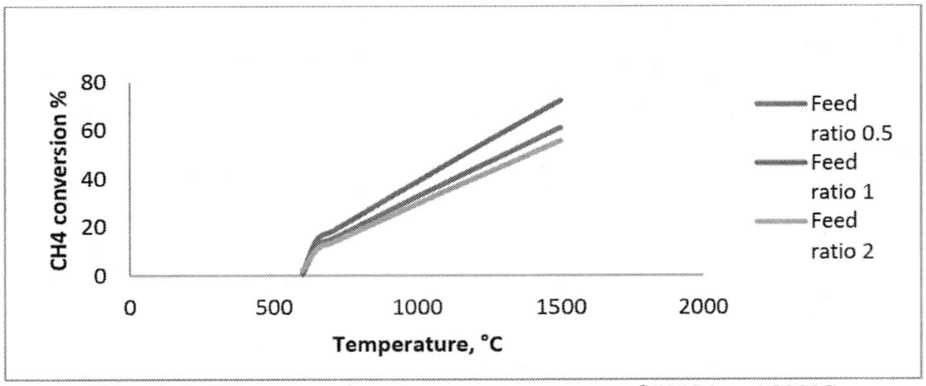

Figure 5. CH_4 conversion at temperature range of 600°C to 1500°C

4.1.2. CO_2 Conversion

Figure 6 shows that at CH_4:CO_2 feed ratio of 2, the CO_2 conversion reached 88%, whereas at CH_4:CO_2 feed ratio of 1 shows the lowest conversion. Based on the result obtained, the CO_2 conversion is found to be higher than the CH_4 conversion. As more CO_2 is consumed in the reverse water gas shift (RWGS) reaction, this will inevitably result in higher conversion of CO_2 as compared to CH_4.

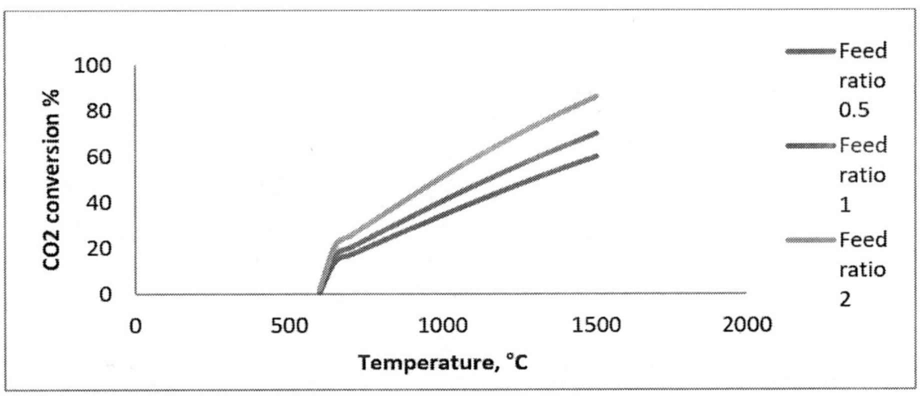

Figure 6. CO_2 conversion at temperature range of 600°C to 1500°C

4.2. Reverse Water Gas Shift (RWGS) Reaction

Table 2 shows the equilibrium constant and rate of collision for reverse water gas reaction (RWGS). Based on the result obtained, the K values decreased gradually and fewer products are expected to form.

Reaction: $H_2 + CO_2 \rightarrow CO + H_2O$		
Temperature (°C)	**Equilibrium constant, K**	**Z_{AB} (m^3s^{-1})**
600	2.730183	9.07E+31
650	2.148473	8.58E+31
700	1.633966	8.14E+31
750	1.195031	7.74E+31
800	0.846437	7.38E+31

Table 2. Equilibrium constant and rate of collision for the reaction, $H_2 + CO_2 \rightarrow CO + H_2O$ at various temperatures

IOP Conf. Series: Materials Science and Engineering **429** (2018) 012086 doi:10.1088/1757-899X/429/1/012086

4.3. Steam Reforming of Methane

Table 3 displays the equilibrium constant and collision rate for reaction between CH_4 and H_2O. As K value is found to decrease at elevated temperatures, very limited or no production of CO and H_2 can be obtained. The result in table 3 below shows that the overall collision rate is decreasing over this temperature range.

Reaction: $CH_4 + H_2O \rightarrow CO + 3H_2$		
Temperature (°C)	**Equilibrium constant, K**	Z_{AB} ($m^{-3}s^{-1}$)
600	-3.50283	1.93E+31
650	-3.9552	1.82E+31
700	-4.39927	1.73E+31
750	-4.83291	1.65E+31
800	-5.25372	1.57E+31

Table 3. Equilibrium constant and rate of collision for the reaction, $CH_4 + H_2O \rightarrow CO + 3H_2$ at various temperatures

4.4. Methane Decomposition Reaction

Table 4 shows the equilibrium constant and rate of collision for methane decomposition reaction. The equilibrium constant, K is relatively constant over the range of temperatures. The K value for this reaction is considered insignificant as compared to the K value of other reactions.

Reaction: $CH_4 \rightarrow C + 2H_2$		
Temperature (°C)	**Equilibrium constant, K**	**Rate of collision, Z_{AA} ($m^{-3}s^{-1}$)**
600	-26.6627	1.42E+37
650	-25.1647	1.34E+37
700	-25.1493	1.27E+37
750	-25.1544	1.21E+37
800	-25.1735	1.15E+37

Table 4. Equilibrium constant and rate of collision for methane decomposition, $CH_4 \rightarrow C + 2H_2$ at various temperatures

4.5. Boudouard Reaction

Table 5 shows the equilibrium constant and rate of collision for Bouduoard reaction. The highest K value is observed between 700°C and 750°C. Thus, carbon formation is expected between the temperatures of 700°C to 750°C. There is a slight decrement in rate of collision as the temperature increases.

Reaction: $2CO \rightarrow C + CO_2$		
Temperature (°C)	**Equilibrium constant, K**	**Rate of collision, Z_{AA} ($m^{-3}s^{-1}$)**
600	47.78895	7.93E+36
650	30.90787	7.50E+36
700	50.68433	7.11E+36
750	50.68945	6.77E+36
800	44.11203	6.45E+36

IOP Conf. Series: Materials Science and Engineering **429** (2018) 012086 doi:10.1088/1757-899X/429/1/012086

Table 5. Equilibrium constant and rate of collision for Boudouard reaction, $2CO \rightarrow C + CO_2$ at various temperatures

4.6 Mechanism for Coke Formation

The simulation is carried out for every possible reaction concerning both DRM's reactants, CO_2 and CH_4. The possible products from both reactants of DRM are; CO_2, CH_4, C, H_2O, CO and H_2. Methane decomposition is one of the major reaction pathways to coke formation. This endothermic reaction requires +74.87 kJ/mol of energy to initiate reaction (1). Figure 7 shows the result generated by HYSYS simulation. The carbon formation occurs rapidly as the temperature increases from 650°C onwards.

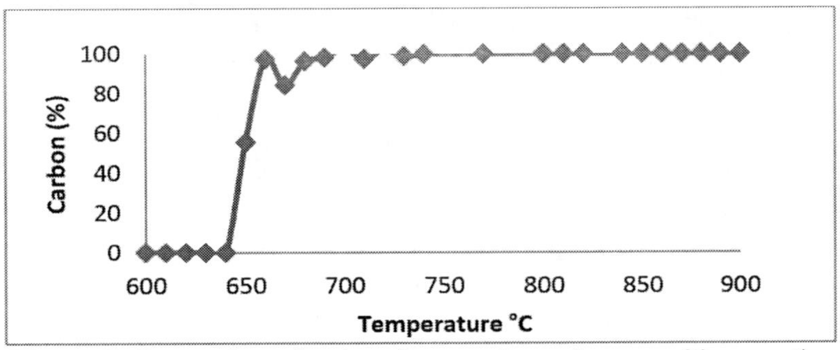

Figure 7. Conversion of carbon from methane decomposition reaction

Boudouard reaction is an exothermic reaction that requires -172 kJ/mol of energy to transform CO to CO_2 and C. Figure 8 shows the formation of carbon via Boudouard reaction. As the temperature rises from 600°C, carbon formation occurs drastically.

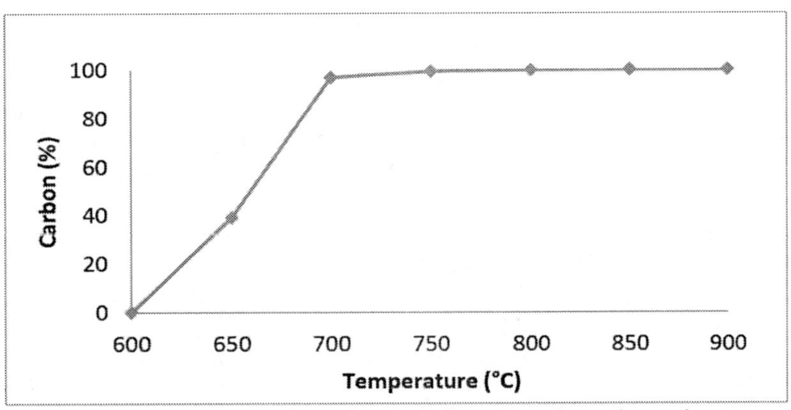

Figure 8. Conversion of carbon from Boudouard reaction

4.7 Proposed Reaction Mechanism

Considering the results from the thermodynamic equilibrium analysis, it is assumed that both CH_4 and CO_2 undergo dissociative adsorption and attachment onto the active site of nickel catalyst. Both CH_4 and CO_2 go through surface reaction, where other components such as C and H_2O are produced along with syngas. There are four reactions steps, which are required to complete the process of CH_4 dissociative adsorption.

$$CH_4 \cdot * + * \rightarrow CH_3 \cdot S + H \cdot * \quad (25)$$

$$CH_3\cdot * + * \rightarrow CH_2\cdot * + H\cdot * \quad (26)$$
$$CH_2\cdot * + * \rightarrow CH\cdot * + H\cdot * \quad (27)$$
$$C\cdot * + H\cdot * \rightarrow CH\cdot * + * \quad (28)$$

The reaction steps below describe the dissociative adsorption undergone by CO_2.

$$CO_2 + * \rightarrow CO_2\cdot * \quad (29)$$
$$CO_2\cdot * + * \rightarrow CO\cdot * + O\cdot * \quad (30)$$
$$CO\cdot * + * \rightarrow C\cdot * + O\cdot * \quad (31)$$

Both the dissociative adsorption steps for CH_4 and CO_2 has the possibilities to form C on the nickel surface. Another possible component that will form is H_2O. Dissociated O from CO_2 often reacts with H from CH_4 to form H_2O. The produced H_2O is then able to react with CO, which produces both CO_2 and H_2. The last step in the catalytic cycle of DRM is desorption, where products including H_2, H_2O, CO and CO_2 will be detached from the nickel surface.

$$OH + H \rightarrow H_2O \quad (32)$$
$$CO_2 + H_2 \rightleftarrows CO + H_2O \quad (33)$$

5. Conclusion

Based on direct minimization of Gibbs free energy method, the thermodynamic equilibrium analysis on the proposed reactions has been performed. All possible reaction pathways for DRM are studied at a temperature range of 600°C to 800°C and a microkinetic model has been proposed.

6. References

[1] Arora S and Prasad R 2016 An overview on dry reforming of methane: strategies to reduce carbonaceous deactivation of catalysts *RSC Advances* 6(110): 108668-108688

[2] Batiot-Dupeyrat C 2016 Dry Reforming of Methane. Perovskites and Related Mixed Oxides *Wiley-VCH Verlag GmbH & Co. KGaA*: 501-516

[3] Dehimi L, Benguerba Y, Virginie M, and Hijazi H 2017 Microkinetic modelling of methane dry reforming over Ni/Al2O3catalyst *International Journal of Hydrogen Energy* 42(30), 18930–18940

[4] Kathiraser Y, Oemar U, Saw E T, Li Z, and Kawi S 2015 Kinetic and mechanistic aspects for CO_2 reforming of methane over Ni based catalysts *Chemical Engineering Journal* 278, 62–78

[5] Iyer M V, Norcio L P, Kugler E L, and Dadyburjor D B 2003 Kinetic modeling for methane reforming with carbon dioxide over a mixed-metal carbide catalyst *Industrial & engineering chemistry research* 42(12): 2712-2721

[6] Osazuwa O S, Setiabudi H D, Abdullah S and Chin K C 2017 Syngas production from methane dry reforming over $SmCoO_3$ perovskite catalyst: Kinetics and mechanistic studies *International Journal of Hydrogen Energy* 42(15): 9707-9721

[7] Fogler H S 2006 *Catalysis and catalytic reactors. Elements of chemical reaction engineering.* Prentice Hall PTR: 645-756.

[8] Sawatmongkhon B, Theinnoi K, Wongchang T, Haoharn C and Tsolakisc A 2017 Combination of Langmuir-Hinshelwood-Hougen-Watson and microkinetic approaches for simulation of biogas dry reforming over a platinum-rhodium alumina catalyst *International Journal of Hydrogen Energy* 42(39): 24697-24712

[9] Maestri M 2012 Microkinetic Analysis of Complex Chemical Processes at Surfaces *Wiley-VCH Verlag GmbH & Co. KGaA*: 219-245.

[10] Polo-Garzon F, Scott J K, and Bruce D A 2016 Microkinetic model for the dry reforming of methane on Rh doped pyrochlore catalysts. *Journal of Catalysis*, 340, 196–204.

[11] Dehimi L, Benguerba Y, Virginie M, & Hijazi H 2017 Microkinetic modelling of methane dry reforming over Ni/Al_2O_3catalyst *International Journal of Hydrogen Energy* 42(30), 18930–18940

IOP Conf. Series: Materials Science and Engineering **429** (2018) 012087 doi:10.1088/1757-899X/429/1/012087

Analysis of Desired Air Flow Characteristic inside Dew Point Cooling system

N N A Noramzil[1], Z M Razlan[1,*], I Zunaidi[1,*], M Izham[1], Shahriman A B[1], A Harun[2], M S M Hashim[1], M K Faizi[1], I Ibrahim[1], N S Kamarrudin[1], W K Wan[1], M F H Rani[1] and M F A Zulkiflee[1]

[1]School of Mechatronic Engineering, Universiti Malaysia Perlis, Pauh Putra Campus, 02600, Arau, Perlis, Malaysia
[2]School of Microelectronic Engineering, Universiti Malaysia Perlis, Pauh Putra Campus, 02600, Arau, Perlis, Malaysia

*Corresponding author: zuradzman@unimap.edu.my

Abstract. Air cooling process is energy intensive processes in air conditioning which a lot of energy consumed due to usage of traditional systems based on the compressor cycle and the usage of refrigerants. Consequently, these system give harmful to the environment. These factors led to the active research and development of compressor free air conditioning. The former soviet scientist Valery Maisotsenko improved evaporative technology and called the resulting cycle as Maisotsenko cycle (M-cycle). This type of cycle is one type of dew point cooling system which is has very low power consumption and the ability to cool the air below the wet bulb temperature. The aim of this work is to analyze the desired air flow characteristic inside dew point cooling systems. Besides, the possibility to apply the dew point cooling system in Malaysia climate will be estimated.

1. Introduction

Remarkably, comfortable air conditions must be created for comfort living environment. In this regard, air condition systems such as vapour-compression cooling system, direct evaporative cooling system (DEC) and indirect evaporative cooling system (IEC) have become more prominent to supply comfortable environment. The energy used for air conditioning systems are able to reduce significantly when using evaporative air cooling system up to very high COP (up to 200), which is remarkably higher than typical air conditioning vapour-compression system [1].

The first type of evaporative cooling created was the direct evaporative coolers (DEC) were widely used in the dry climate regions, such as certain parts of India [2]. Direct evaporative air cooling system have direct contact of supply air with the water to utilize the latent heat of water evaporation. The results in evaporation of the water, will cools the air stream while in the same time increases air humidity. Humid supply air reduces the indoor air quality and makes room conditions more uncomfortable for humans. The most comfortable humidity level in the room varies from 40% to 60%. Humidity level below 30% and over 70% are uncomfortable [3]. Even the supply air is cool but the enthalpy of the air stream is constant.

While, an indirect evaporative air coolers (IEC) can supply sensible cooling of the supply air in the same time maintain the humidity. In the heat exchanger, the supply air stream passes over the dry channel while the working air passes over the opposite wet channel. The sensible heat flux from the dry channel is transferred to the wet channel, which results in cooling the supply air flow without adding moisture, because the latent heat of water evaporation is transferred to the working air stream [4]. In this

IOP Conf. Series: Materials Science and Engineering **429** (2018) 012087 doi:10.1088/1757-899X/429/1/012087

context, there is a kind of novel indirect evaporative cooling system which is the so-called "Maisotsenko cycle" or "M-Cycle". Besides, this type of cycle is one type of dew point cooling system.

2. Methodology

2.1. Data selection
Two weather data of any month in a year will be selected. April 2014 and October 2015 weather data will be obtained from the UniMAP greenhouse properties of Institute of Sustainable Agro Technology (INSAT). Those are weather data in October 2015 and February 2014. The purpose we use two types of data because here especially in Perlis between Januarys to May the weather is extremely hot and can almost achieved 40°C. Meanwhile, the October weather data act as normal data in Perlis.

2.2. Psychrometric analysis
Data obtain will be extracted to get only crucial information of desired air properties to use in psychrometric chart. From psychrometric chart, relative humidity and dew point temperature will emerged. After that, all those data will be sort out and plotted onto graph to see the pattern of Malaysia climate for M-cycle implementation.

3. Results
At this moment, average Malaysia temperature throughout a year have been recorded. During the sunny time the dry bulb temperature is 32°C to 34°C and the relative humidity is 72 to 78%. Meanwhile, in the night the ambient drops to 24°C and 26°C but the relative humidity pops up to 80 to 90%. From this information, all others surface condition, temperature and thermodynamic air properties are known from psychrometric chart, Figure 1 and recorded on Table 1.

Table 1 Temperature differences throughout the day

Observed information		Value, (°C)
Day time	Wet bulb temperature	29.1
	Dew point temperature	28.0
Night time	Wet bulb temperature	23.1
	Dew point temperature	22.3

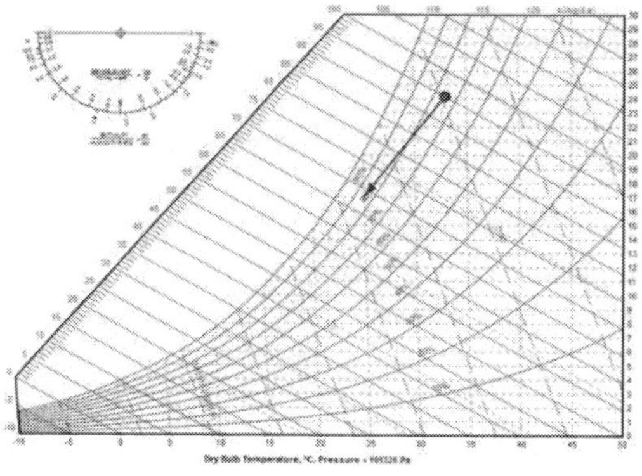

Figure 1 Results in Psychrometic Chart

All three air characteristic is related to each other. However, only two of the air characteristic were enough for us to plot on the psychrometric chart to get all the air characteristic information. In this case, all the air characteristic is obtained from the raw data taken by the automatic data sensor device in the greenhouse. By that, we can reduce the time to get the information from manual psychrometric chart. The relationship of air characteristic is shown in Figure 2 below.

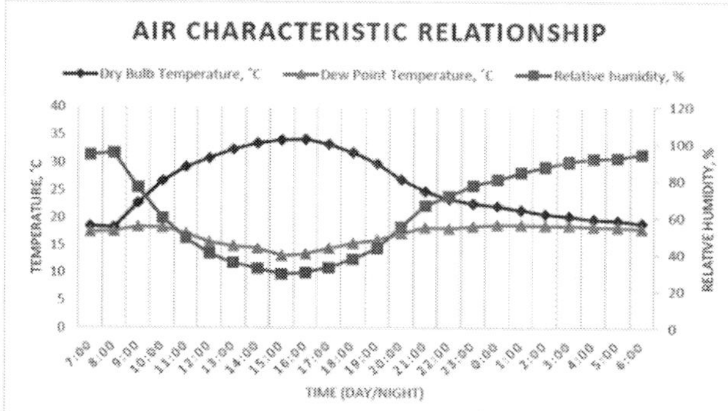

Figure 2 Air Characteristic Relationship

Meanwhile, for relative humidity, we can see that during peak of day time the percentage of relative humidity was about 30% only even though the dry bulb temperature was high that was 34.1 °C. Main reason why this situation happen because less water vapor contained in the air as normally Perlis in February was drought.

3.1. Result Comparison

Two types of air characteristic is taken for comparison purpose, (Figure 3) those were dry bulb temperature and relative humidity. Typically, dry bulb temperature for both month did not have any major changes. They were gradually increase and decrease at the same time line which were 28°C to 32°C during day time and 22°C to 25°C during night time.

Meanwhile, in term of relative humidity, there were huge differences between both months. In February the relative humidity was drastically dropped to minimum of 29.3% during day time cause of drought effect compare to October that was linearly throughout the day about 80% to 81%. Furthermore, because of this relative humidity problem, there were only slight temperature differences between dry bulb temperature and dew point temperature. So, it was meaningless as low dew point temperature is needed to cool the close space such as room unless we alter the relative humidity by drop down it value to human comfort range and the same time to get low value of dew point temperature.

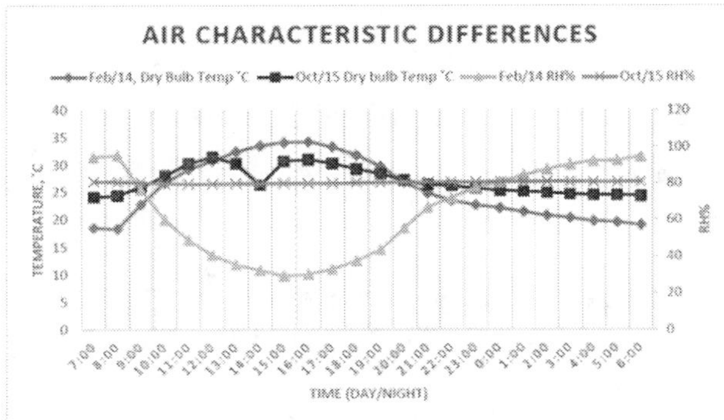

Figure 3 Air Characteristics Differences

4. Conclusion and recommendation

4.1. Summary

In this case, Malaysia does not give a favourable output product due to it climate. Modification is needed to use dew point cooling system based HMX on M-cycle cooling system efficiently. The cooling system need hybrid system to meet the comfort range for all time in the year due to different air characteristic at different time.

4.2. SWOT Analysis

SWOT analysis is the strategic planning method used to evaluate the strength, weakness, opportunity involve in the project. Figure 4 show the SWOT analysis for indirect evaporative cooler for dew point cooling system.

Strength	Weakness
• Low running cost • Environmental friendly • Low power consumption • Greater range of cooling	• Big size • Need air characteristic alteration
Opportunity	**Threat**
• Reduce waste of energy • No air pollution • Prototype not available in Malaysia yet • Everybody can use cooler	• Early stage of development • Product already have in other side of world

Figure SWOT Analysis

Acknowledgments

Thousands The authors acknowledge School of Mechatronic Engineering, Universiti Malaysia Perlis for the facilities. Special thanks to those who contributed to this project directly or indirectly.

References

[1] Tertipis D and E Rogdakis 2015 Maisotsenko cycle: technology overview and energy-saving potential in cooling systems *Energy Emiss. Control Technol.* **3** 15–22
[2] Wani C, S Ghodke, and C Shrivastava 2012 A review on potential of Maisotsenko cycle in energy saving applications using evaporative cooling *Int. J. Adv. Res. Sci. Eng. Technol.* **01** 15–20
[3] Gorshkov V 2012 SYSTEMS BASED ON MAISOTSENKO CYCLE Coolerado Coolers
[4] Anisimov S and D Pandelidis 2015 Theoretical study of the basic cycles for indirect evaporative air cooling *Int. J. Heat Mass Transf.* **84** 974–89

Development of High Efficiency Small Room Air Conditioner-Application of Dew Point Cooling System

N N A Noramzil[1], Z M Razlan[1,], I Zunaidi[1,], M Izham[1], Shahriman A B[1], A Harun[2], M S M Hashim[1], M K Faizi[1], I Ibrahim[1], N S Kamarrudin[1], W K Wan[1], M F H Rani[1] and A A A Alkaf[1]

[1]School of Mechatronic Engineering, Universiti Malaysia Perlis, Pauh Putra Campus, 02600, Arau, Perlis, Malaysia
[2]School of Microelectronic Engineering, Universiti Malaysia Perlis, Pauh Putra Campus, 02600, Arau, Perlis, Malaysia

*Corresponding author: zuradzman@unimap.edu.my

Abstract. The objective of this study is to design and simulate a vapour-compression refrigeration cycle and dew point cooling system. Besides, to identify air characteristic in Riyadh city which is has high temperature 44°C with relative humidity 25% the identification of climate is to create appropriate environment and high efficiency cooling. However, the cooling load for small room 20 m^2 reached to 3.9 kw because the temperature is very hot. The simulation for refrigeration and dew point cooling system shows the air distribution inside the room, from simulation the air distribution remains constant from 0s to 1 hour because it supplies air each cycle. The velocity for air conditioning is 2 m/s to provide comfort zone. The refrigerant R410a is taken as work fluid for refrigeration cycle.

1. Introduction

In recent years, the temperature getting hotter in summer, this increasing in temperature effect on the building which leads to increase the requirement for air conditioning. While vapour-compression refrigerant cycle is the most widely used system for air-conditioning, Refrigeration cycle is a close system process, a working fluid of the cycle is returned to initial state at the end of the cycle and recirculated. Refrigeration cycle consist of compressor and condenser, expansion and evaporator, the refrigerants carried in compressor cycle is very harmful to the environment and it consumes a large amount of energy [1]. These factors lead to developed of evaporative cooler. This evaporative cooling of the air consumes less energy compared with traditional cooling system and it is an environmental friendly system.

However, the state of the air is completely important for the efficiency of the evaporative air cooling system. Middle East region climate is hot and dry climate which is suitable conditions for evaporative air cooling system. This system gives sufficient cooling but the increasing of humidity gives the feeling of discomfort. However, evaporative cooling technology improved by Russian scientist Valery Maisotsenko which is Maisotsenko Cycle (M-Cycle), one of type of dew point cooling system. M-Cycle is a new cooling system with more efficiency and less energy consumption than the conventional cooling system, M-Cycle is friendly to environment and it does not use Freon gas or compressor. This cooling system use simple cross heat exchanger, the heat mass exchanger does not let humidity to be added into product air stream to give a comfortable feeling [2]. The objective of dew point cooling system to overcome the limitation of the indirect evaporative cooling by extracting a one channel of product air to become a working air and cooling the product air to its dew point [3].

2. Methodology

The heat load of the small room determined by using Heat Balance Method [4] where outdoor parameter using Riyadh city. The operational conditions of the compressor and the capillary tube should be designated. Designing each component one by one, and calculating the pressure drops in the condenser and the evaporator, a deviation from the ideal cycle line will occur. Determine heat transfer occur in heat mass exchanger. Therefore, a pre-determined state should be chosen and the design should be started from this state.

3. Results and Discussions

3.1. Cooling load results

External load result for small room area 20 m^2

Table 1. Results of external load

Heat Source	$Q\ (W)$
Wall	997.605
Roof	793.651
Floor	672.54
window	217.75
Door	38.91
Total	2720.375

While, internal load result for small room area 20 m^2

Table 2. Results of internal load .

Heat Source	$Q\ (W)$
People	343
Lights	516
Equipment	360
Total	1221

The total cooling load is sum of heat gained by external and internal load, therefore the heat gained should be removed from the space to maintain the temperature to provide acceptable range.

$$Q_{total} = Q_{external} + Q_{internal}$$

$$Q_{total} = 1221 + 2720$$

$$Q_{total} = 3941.357\ W$$

3.2. Simulation results by ANSYS CFX

3.2.1. Dew point cooling system air distribution & velocity

IOP Conf. Series: Materials Science and Engineering **429** (2018) 012088 doi:10.1088/1757-899X/429/1/012088

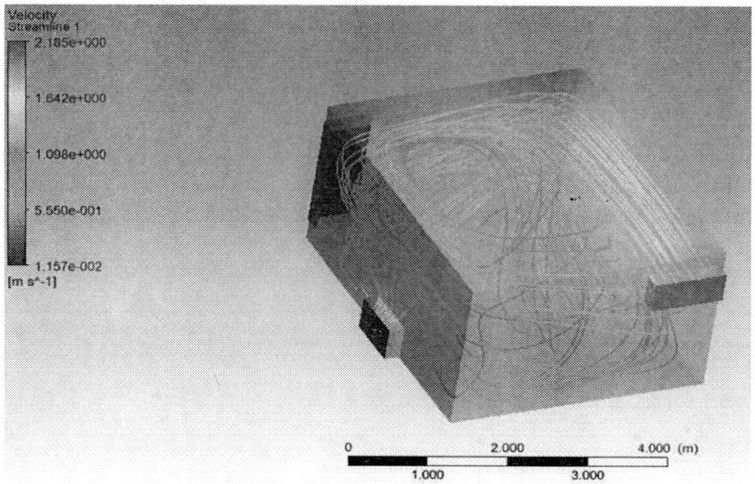

Figure 1. Air distribution and velocity for dew point cooling system

Figure 1 shows the dew point cooling system air velocity is 2 m/s and the air distribution inside the room and at the end the air will pass through the outlet.

3.2.2. Vapour-compression refrigeration cycle air distribution & velocity

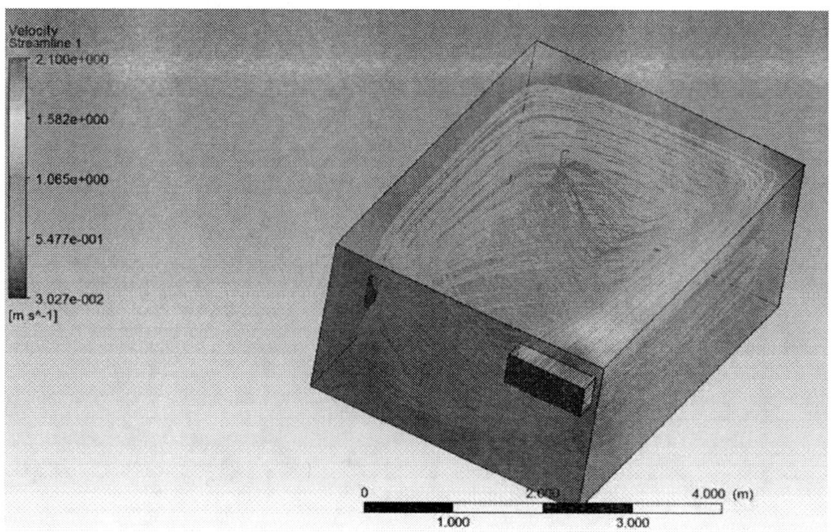

Figure 2. Air distribution and velocity for vapour-compression refrigeration cycle

3.2.3. Thermal distribution inside the room

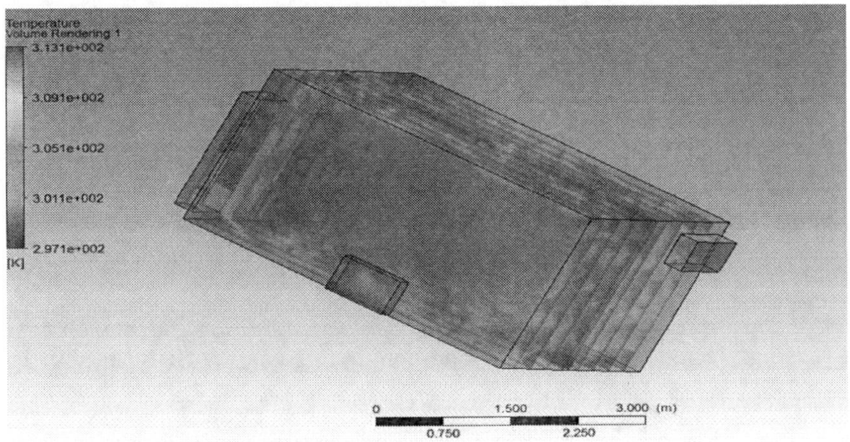

Figure 3. Thermal distribution inside the room

Figure 3 shows that the red colour in the figure is the heat transfer inside the room wall from outside temperature around 42℃ and inside temperature is 24℃ which is blue colour

3.3. Discussions

The cooling load has been done regarding to Riyadh climate for small room with 20 m^2, total cooling load is 3.9 KW. The outdoor design temperature is 44℃ where with this high temperature led to increase the total cooling load. The comfort zone temperature provided for small room is 24℃ while the relative humidity is 45%.

The working fluid to design the vapour-compression refrigeration cycle is R410a where the evaporator efficiency is 15% to ensure that the performance of evaporator remain constant and avoid the dust that may affect the evaporator. In addition, 6-15℃ is the best temperature to discharge air to space and 10℃ is the temperature of the evaporator. Temperature of compressor is 50℃ with 1Mpa as low pressure and 3 Mpa as high pressure. The compressor has 21.83 ml/ rev displacement volume with efficiency 85%.

The result shows the air distribution for dew point cooling system and refrigeration cycle where the air distribution for dew point cooling system remain constant after period time because for each cycle it provides new working fluid. While the result for refrigeration cycle, the air distribution change after period time because it circulates the working fluid. The result show that the heat transfer through the wall into the room with outside temperature 44℃.

4. Conclusion and recommendation

Through the calculation clarify that the maximum heat gained through, roof, window and wall by conduction heat transfer and this high heat gained due to the high temperature in the region. From detailed calculation it becomes clear the highest heat gained is through the wall. The room wall must contain more layer of material in order to prevent the heat enter through it. However, the total amount of heat gain to the room is 3.5 KW based on the maximum heat. The air-conditioning system type that will suitable to our case study is split air conditioner.

The efficiency of evaporator of refrigeration cycle can effect on cooling performance if the value of efficiency ignored, the size of evaporator related to the air velocity. High velocity gets less comfort with smaller size of evaporator. The air distribution clarify that the velocity has direct effect on the comfort zone.

The heat mass exchanger of dew point cooling system is only appropriate in dry climate because the system remains the moisture in the air constant. The feature of dew point cooling system is the dew point temperature of outdoor air is same with indoor air, also the specific humidity ratio same.

Acknowledgments
Thousands The authors acknowledge School of Mechatronic Engineering, Universiti Malaysia Perlis for the facilities. Special thanks to those who contributed to this project directly or indirectly.

References
[1] X. Yin, S. Li, and W. Cai, "Enhanced-efficiency operating variables selection for vapor compression refrigeration cycle system," *Comput. Chem. Eng.*, 2015.
[2] Z. Duan, "Investigation of a novel dew point indirect evaporative air conditioning system for buildings," 2011.
[3] S.-W. Ham and J.-W. Jeong, "DPHX (dew point evaporative heat exchanger): System design and performance analysis," *Energy*, vol. 101, pp. 132–145, Apr. 2016.
[4] ASHRAE, *2013 ASHRAE Handbook: Fundamentals*, SI Edition. Atlanta, GA: American Society of Heating, Refrigerating and Air-Conditioning Engineers (ASHRAE), 2013.

A Test of Possibility on Relative Humidity Function in Minor Operation Theatre

M N Rahman Y*, Z M Razlan, M Izham, M I Omar, N A A Zambri, Shahriman A B, I Zunaidi and W K Wan

School of Mechatronic Engineering, Universiti Malaysia Perlis, Pauh Putra Campus, 02600 Arau, Perlis, Malaysia.

*Corresponding author: nurrahman@unimap.edu.my

Abstract. High humidity air flow in Minor Operation Theatre (MOT) is undesired as it will cause discomfort to the occupant. Acceptable humidity air flow is important as it will affect the pureness of the indoor environment as well. The objective of this research work is to construct a realistic Computational Fluid Dynamics (CFD) model of MOT for investigation. The MOT which is occupied with one unit of Air Conditioner has been selected for this study. The RH of air is analysed. The experimental data is collected by applying a digital air flow anemometer with USB interface and the channels digital handheld temperature thermometer meter data loggers for validation purpose. CFD simulation is conducted by applying ANSYS Fluent software. The experimental result and simulation is compared and validated. Based on the simulation and experiment findings, it reveals that ANSYS Fluent can be used for RH in MOT.

Keywords. Computational Fluid Dynamics Simulation, Minor Operation Theater, Relative Humidity.

1. Introduction

A refinement research for better air quality in Minor Operation Theatre (MOT) currently is being conducted by many researchers. The ideal and comfortable environment in operation theater always being expected by the surgeons and patients as they spend most of their time in that room [1][2][3][4]. This research is an extension study from previous work which is to construct a realistic CFD model of MOT by investigating the Relative Humidity (RH) of air flow in the MOT of Universiti Malaysia Perlis (UniMAP)[2][5]. The RH of air flow in the MOT was numerically simulated and experimentally measured. It was also assumed that the lack of Interior Design (ID) of MOT will affect the RH of air flow and then influences the required range of RH which is between 20% and 60%. Therefore, an identification on achievement of RH in MOT needs a Computational Fluid Dynamics (CFD) method and then will be verified with experimental data. This has been supported by previous research works [6][7][8][9].

2. Methodology

2.1. CFD Model of MOT

Model of MOT has been constructed by CATIA (Figure 1) and numerically simulated by ANSYS Fluent. This commercial software has capability to predict the incompressible, compressible, laminar and turbulent fluid flow along with buoyancy and compressibility phenomena. Given Table 1 as the details of the room.

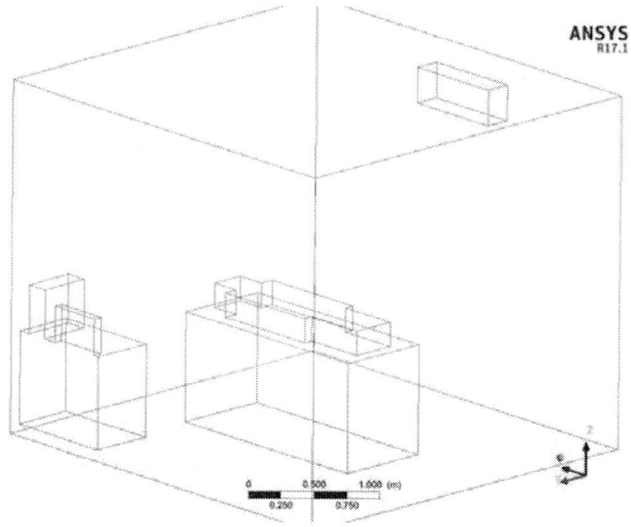

Figure 1. MOT Model (CATIA)

Table 1. The details of the room.

Items	Minor Operation Theatre
Dimension (m)	2.89 (H) x 3.21(W) x 3.4(L)
Air-conditioning Configuration	Single Unit

2.2. CFD Simulation Methodology

The commercial software named ANSYS Fluent 17.1 was utilized for simulation work in this study. In this study, the Minor Operation Theatre room (Figure 1) modelling must go through Pre-Processing, FLUENT-Solver and finally FLUENT Post-Processing for Minor Operation Theatre as shown in figure 2. The assumptions must be done while doing this simulation work due to commercial software limit and to reduce calculation time. The assumptions for this study are given as follows:

- The AC unit in MOT is steady condition.
- MOT space is completely sealed.
- The outdoor temperature on the surface of the room is constant
- Internal heat source emitted from the electronic devices and electric appliances will be neglected due to minimal effect on the temperature.
- The effects of physical condition of the room are neglected in this study.
- The furniture is included in this simulation.

The simulation details and meshed model are summarized in table 2 and shown in figure 3.

IOP Conf. Series: Materials Science and Engineering **429** (2018) 012089 doi:10.1088/1757-899X/429/1/012089

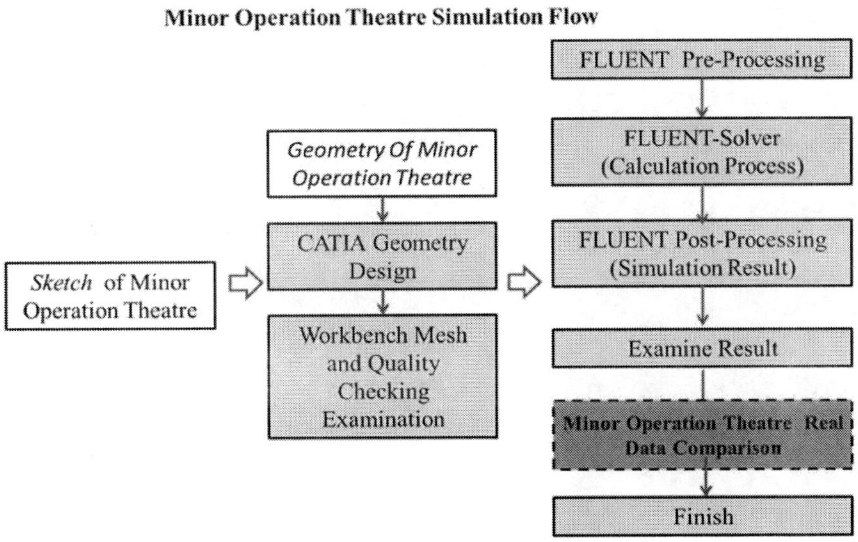

2

Figure 2. Flow chart of MOT Simulation Work.

Table 2. Summary of simulation work details.

Details	**Minor Operation Theatre**	
General	Gravity Force	On
MOT Models	Physical Model	3D
	Energy	On
Fluid Models	Viscous model	K-Epsilon (2 Equation)
	k-Epsilon Model	RNG
	Near-Wall Treatment	Standard Wall Functions
	AC Nozzle Velocity	2.8 m/s
	Air Temperature	6.9°C
Boundary Conditions	Mass Fraction of Air	99.2%
	Mass Fraction of H_2O	0.8%
	Relative Humidty	47%
	Type of Cells	Hexahedral
Mesh Size	Number of Nodes	167024
	Number of Cells	213814
	Gradient	Least Squares Cell Based
	Pressure	Second Order
Solution Method	Momentum	Second Order Upwind
	Turbulent Kinetic Energy	Second Order Upwind
	Turbulent Dissipation Rate	Second Order Upwind
	Energy	Second Order Upwind

IOP Conf. Series: Materials Science and Engineering **429** (2018) 012089 doi:10.1088/1757-899X/429/1/012089

Figure 3. Meshed Model.

2.3. Data Collection on Air Characters in MOT

Temperature distribution and RH data are experimentally measured in this study before furthering simulation of the MOT. The tools applied are summarized in table 3.

Table 3. Tools

No.	Material and Equipment
1	PVC Pipe
2	Pipe Stand
3	Thermocouple Wire (Type-T)
4	Data Logger
5	FLIR Ex-Series Infrared Cameras
6	Colemeter
7	Anemometer

For temperature distribution, the value is collected by using six panel of T-type thermocouple wire which four of the wire is label as T1, T2, T3, T4 and the other label as Inlet Temperature, T5 and Outlet Temperature, T6. Four labelled wire are assembled at the PVC pipe with stand and the inlet and outlet wire is placed at the inlet and outlet of the air conditioner. The value of temperature for each point is determined by using a data logger since all the wire is connected to it (figure 4). The pipe needs to be placed at each point in figure 5 for 10 minutes. At the same time, Colemeter which is the digital thermometer hygrometer humidity temperature sensor is used to measure the RH. This sensor is placed at the same point as in figure 6.

Figure 4. Channels Digital Handheld Temperature Thermometer Meter Data Logger

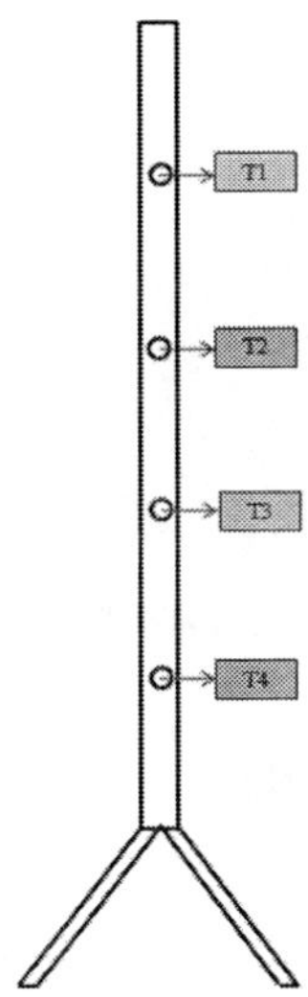

Figure 5. Four labelled wire in the PVC pipe

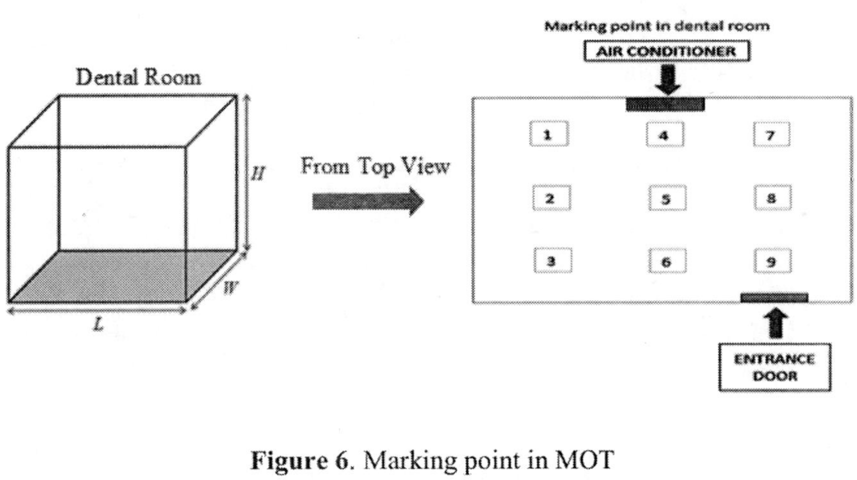

Figure 6. Marking point in MOT

IOP Conf. Series: Materials Science and Engineering **429** (2018) 012089 doi:10.1088/1757-899X/429/1/012089

For validation purpose on Temperature, RH and air flow velocity, the digital thermometer hygrometer humidity temperature sensor and the digital air flow anemometer with USB interface are applied (figure 7 and 8).

Figure 7. Digital thermometer hygrometer humidity temperature sensor

Figure 8. Digital air flow anemometer with USB interface.

3. Result and Discussions

Based on the experimental result as shown in table 4, the RH for MOT is 45.7% and the CFD Simulation for MOT is 47.1% respectively. The CFD approach was successfully performed to obtain the RH in MOT. The simulation results were related to the standard methods which are redefining the geometry of the MOT, optimization in mesh density, boundary conditions setting, solver setting, right solution methods and acceptable results were obtained from CFD post.

Figure 9 and 10 shows the CFD contour slices of the Mass Fraction of H_2O and Mass Fraction of Air inside Minor Operation Theatre. Based on the both figures (Figure 9 and 10). The ANSYS Fluent demonstrates 0.8% for Mass Fraction of H_2O and 99.2% for Mass Fraction of Air in MOT. Based on Total Mass-Weight Integral calculation, the RH is measured as 47.1%.

Table 4. Summary of results of Minor Operation Theatre

Details		Experiment Measurement o f MOT	CFD Simulation of M OT
	Relative Humidity %	45.7	47.1

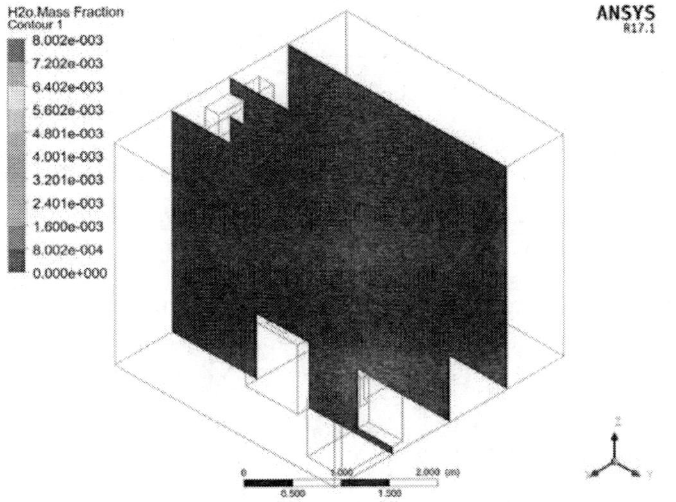

Figure 9. Slices of Mass Fraction of H_2O in Minor Operation Theatre based on location in figure 6.

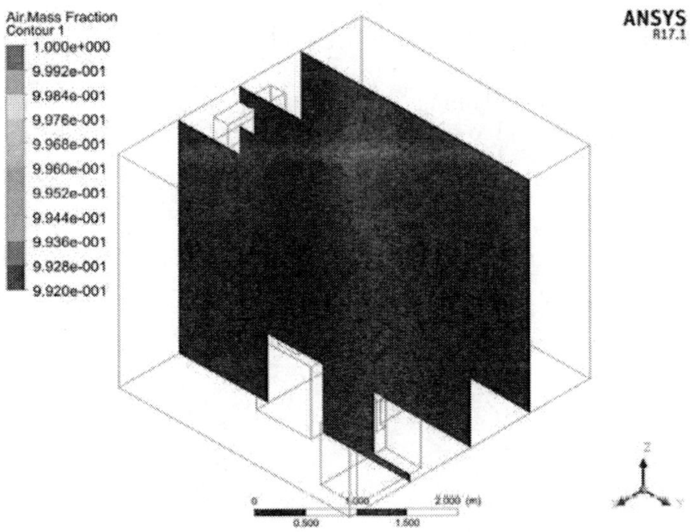

Figure 10. Slices of Mass Fraction of Air in Minor Operation Theatre based on location in figure 6.

4. Conclusion

The CFD simulation result for Minor Operation Theatre demonstrated differences in experimental result are about 3.06% for RH. The errors that occurred due to the uncontrolled mesh density and wall of Minor Operation Theatre in actual Minor Operation Theatre which hard to be quantified for CFD simulation. Based on the study observation, it reveals that ANSYS Fluent can be utilized for RH simulation in Minor Operation Theatre.

References

[1] R. Khatri, A. P. Singh, and V. R. Khare, "Identification of Ideal Air Temperature Distribution Using Different Location for Air Conditioner in a Room Integrated with EATHE - A CFD Based Approach," *Energy Procedia*, vol. 109, no. November 2016, pp. 11–17, 2017.

[2] A. M. Paul Danca, Florin Bode, Ilinca Nastase, "On the possibility of CFD modeling of the indoor environment in a vehicle," in *Sustainable Solutions for Energy and Environment, EENVIRO 2016, 26-28 October 2016, Bucharest, Romania*, 2017, pp. 656–663.

[3] S. Sadrizadeh, *Design of Hospital Operating Room Ventilation using Computational Fluid Dynamics (PhD thesis)*, no. February. 2016.

[4] S.A. Nada, H.M. El-Batsh, H.F. Elattar,N.M. Ali, "CFD investigation of airflow pattern, temperature distribution and thermal comfort of UFAD system for theater buildings applications," *J. Builiding Eng.*, vol. 6, pp. 274–300, 2016.

[5] T. C. Stevenson, "Experimental investigation of hospital operating room air distribution (MSc Thesis)," no. May, p. 97, 2008.

[6] S. Lin, B. T. Tee, and C. F. Tan, "Indoor airflow simulation inside lecture room: A CFD approach," *IOP Conf. Ser. Mater. Sci. Eng.*, vol. 88, no. 1, pp. 1–7, 2015.

[7] A. Abduladheem, K. S. M. Sahari, H. Hasini, W. Ahmed, and R. A. Mahdi, "Ventilation Air Distribution in Hospital Operating," vol. 2, no. 11, pp. 81–85, 2013.

[8] Yuanda Cheng, Jianlei Niu, Naiping Gao, "Stratified air distribution systems in a large lecture theater : A numerical method to optimize thermal comfort and maximize energy saving," *Energy Build.*, vol. 55, pp. 515–525, 2012.

[9] J. Lim, S. Chong, A. Husain, and T. B. Tuan, "Simulation of Airflow in Lecture Rooms."

IOP Conf. Series: Materials Science and Engineering **429** (2018) 012090 doi:10.1088/1757-899X/429/1/012090

Waste Heat Energy Recovery from Reciprocating Engine Using Thermoelectric Generator

L J Yong, N S Kamarrudin, M S M Hashim, K S Basaruddin and A A Rahman

School of Mechatronic Engineering, Universiti Malaysia Perlis, Pauh Putra Campus, 02600 Arau, Perlis, Malaysia

saifullah@unimap.edu.my

Abstract. This paper is covered waste heat energy recovery at SH1900 gasoline portable electric generator by using a thermoelectric generator (TEG). Thermoelectric is universally known as a device that can directly convert heat to the electricity. Voltage and current will be generated by the thermoelectric when temperature difference existing between the two surfaces of TEG. The exhaust system of the gasoline portable engine generator was used as the waste heat source. A single TEG1-199-1.4-0.5 Thermoelectric generator with the dimension of 40mm X 40mm was placed at the exhaust surface to harvest the waste heat from engine generator surface. Four tests were conducted to see the effect of the voltage and current generated on the temperature difference between cold source and hot source of TEG. The electric current generated through the temperature difference is then can be used to power an auxiliary device or to be stored.

1. Introduction

Increasing the concern of environmental issues of emission, limited availability of primary energy resources and rising worldwide demand toward preserving energy was encourage the researcher to search for new technologies in generating electrical power. Generally, this study covered about waste heat recovery using the thermoelectric generator (TEG) from a portable electric generator exhaust system. The amount of the energy used is transferred as heat to the exhaust gases over a standard driving cycle is 40% [1] which is expelled to the atmosphere and wasted as exhaust gases. Due to TEG potential to convert heat directly to electrical power with no pollution, it can be considered as an alternative green technology device. Several experiment conducted on TEG such as the placement of thermoelectric modules on water tube walls inside a boiler can effectively recover the waste heat at coal-fired power plants [2]. Also, the waste heat recovery by thermoelectric modules in automobile engine systems is a hot issue and has been widely investigated [3]. The TEG has the nature of heat engine governed by the laws of thermodynamics and the efficiency of energy conversion is limited by Carnot efficiency [4]. The TEG has many advantages over other conventional energy harvesting technologies, including simplicity, cleanliness, small size, long life span, high energy density, and quietness. TEG is currently widely used in applications ranging from power generators in space missions, common thermocouple sensors, small energy harvesters for self-powered sensors and to automobile exhaust energy harvesting.

IOP Conf. Series: Materials Science and Engineering **429** (2018) 012090 doi:10.1088/1757-899X/429/1/012090

2. Background

This paper presents the design and analysis if waste heat recovery energy using thermoelectric generator on exhausts surface of the portable generator. In the proposed solution, we make use of the thermoelectric compartment which is compact, small, able to withstand the vibration of the exhaust system. A single TEG is clamped on the exhaust system using compartment designed.

3. Research scope

The primary scope for this project is to design and analyses the TEG performance by utilizing waste heat recovery from the exhaust system in an internal combustion engine (ICE) which the engine that is used in this study was a portable generator. To achieve the scope, 4 tests carried out to determine the performance of TEG and the portable generator.

The temperature on aluminum heat sink and exhaust surface is determined using EXTECH VIR50 Infrared Thermometer. Voltage and current generated are determined by using a simple multimeter.

4. Working principle of TEG

Thermoelectric is universally known as a device that can directly convert heat to electricity. When there is temperature difference existing between the two surfaces of this device, voltage will be generated by the thermoelectric module. Conversely, the temperature difference is created when the voltage is transfer to it. The existing temperature different cause the charge carriers moving from the hot side to the cold side. This existing temperature gradient can be utilized to generate electricity. Due to the direction of the heating or cooling is regulated by the polarity of the voltage thermoelectric module can operate as a temperature controller device. Thermoelectric effect is referred to the three effects which are Seebeck effect, Peltier effect and Thomson effect. This classification was derived by Baltic German physicist Thomas Johann Seebeck and French physicist Jean Charles Athanase Peltier. Whenever an electric current passing through the resistive material Joule heating will generate, however, it is not same as the thermoelectric effect. The Thomson effects and Peltier–Seebeck are thermodynamically reversible, whereas Joule heating is not.

5. Prepare and assemble the Thermoelectric building block

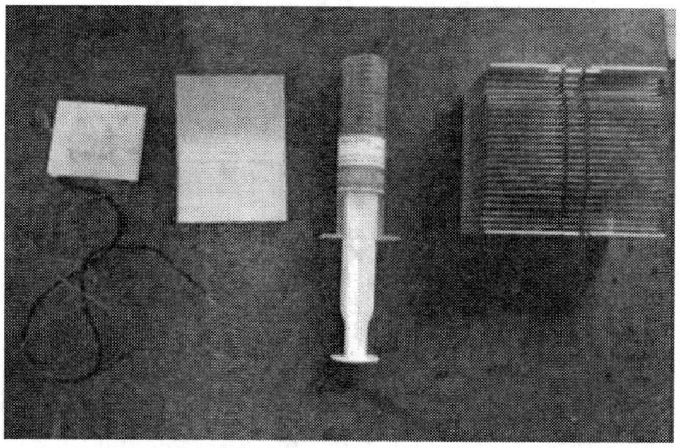

Figure 1 Sub part of the TEG module assembly

Figure 2 Thermoelectric module

Thermoelectric building block is build using 68mm × 76mm CPU heat sink, Heat Sink Compound ZP - 360 and TEGs 10 watt 5.6 volt with 44*40mm Thermoelectric Power Generation Peltier Module, as shown in Figure 1and Figure 2. The temperature of the exhaust surface is measured around 180oC and the maximum temperature that able to stand by these TEGs is not provided by the seller. Commercial TEGs module usually able to operate below 220 oC. Even though the exhaust surface temperature is low, but as a precaution aluminum plate is added to reduce contact temperature while preserving the TEGs from malfunction. The 68mm × 76mm aluminum plate is used as the heat conduction plate and clamped between TEGs and exhaust surface. The TEGs is clamped between the aluminum plate (hot surface) and heat sink (cold surface). 24-fins aluminum is used as a heat exchanger for this experiment and is mounted onto the compartment.

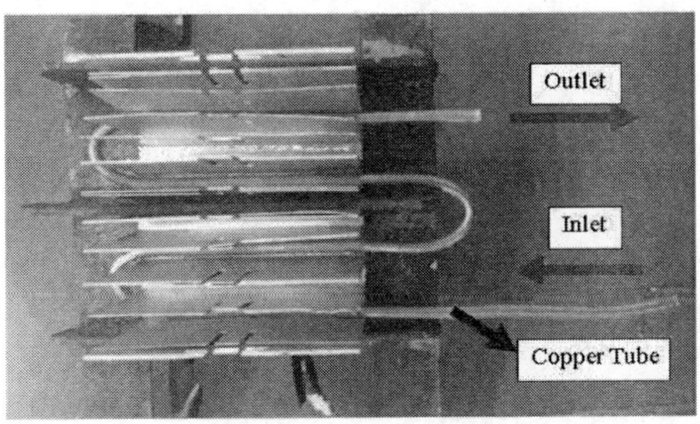

Figure 3 Final compartment with cooling system

Figure 3 shows the final compartment that used in experiments. Special compartment needs to be designed to hold the TEG module tightly at the surface of the exhaust silencer. The surface of the silencer is large than the TEG module and the main problem to place and hold the TEG is the exhaust surface vibration. The best design of jig can hold the TEG module without of any gap existing between the surface and the aluminum plate surface during the generator engine is running. In order to improve the heat conduction and heat removal from the aluminium heat sink a 3 mm inner diameter copper tube is installed onto the compartment The copper tube is bent and fixed on the aluminum sink to ensure the copper tube has the maximum coverage on the heat sink in order to remove and conduct heat away in a faster rate. The copper tube consists inlet which allows coolant to flow in and outlet which allow the coolant to flow out. A mini aquarium pump is connected to the inlet with aquarium tube to push the coolant to pass through the copper tube.

IOP Conf. Series: Materials Science and Engineering **429** (2018) 012090 doi:10.1088/1757-899X/429/1/012090

6. Complete experiment setup

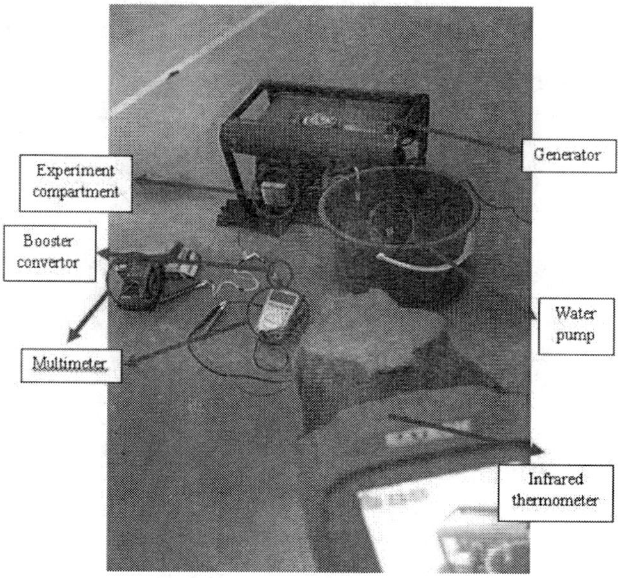

Figure 4 Complete experiment setup

Figure 4 shows the complete setup for experiments. The final compartment is mounted on the exhaust system. The complete setup is consist of a generator, experiment compartment, multimeter, water pump and infrared thermometer. Each of the experiment instrument will be explained in the following section.

6.1. Portable generator
SH1900 gasoline portable generator is used to carry out the experiments. Rated output of 1.0kW, rated voltage output of 230V and rated frequency of 50Hz.

6.2. Booster converter
The main function of a booster converter is to step up the voltage from input to output. The input voltage is 0.9V-5V DC , the output voltage is 5V and the output current is 450mA to 500mA.

6.3. Mini water pump
The main function of mini water pump is to pump coolant such as water and ice water into the copper tube that mounted on the aluminum heat sink. Type of mini water pump used is submersible, power rated of 8W and maximum flow rate of water is 600L/H.

7. Type of experiment

7.1. Control test
A control experiment is set up to test the hypothesis by looking for changes bought by alterations to a variable. The experiment is conducted with no coolant flow through the copper tube mounted on the aluminum heat sink and as a datum for wind test, water test and ice water test. The experiment is conducted to study the effect of natural convection on the temperature difference between the aluminum heat sink and exhaust surface. TEG and booster converter is connected to multimeter respectively to record the voltage and current generated. An infrared thermometer is used to measure the temperature on the aluminum heat sink and exhaust surface. Duration for the experiment is 30 minutes. Data is collected and recorded every minute. The data is then tabulated into graph for further analysis.

7.2. Wind test

Wind test is conducted by using a standing fan to generate air blow and air movement toward the aluminum heat sink. A standing fan is put at a distance of 1.5m away from the generator. This experiment is conducted with no coolant flow through the copper tube that mounted on the aluminum heat sink. The experiment is conducted to determine the effect of forced convection on the temperature difference between the aluminum sink and exhaust surface. TEG and booster converter is connected to multimeter respectively to record the voltage and current generated. The infrared thermometer is used to measure the temperature on the aluminum heat sink and exhaust surface. Duration for the experiment is 30 minutes. Data is collected and recorded every minute. The data is then tabulated into graph for further analysis.

7.3. Water test

The water test is conducted by using water as coolant the water coolant is pump into 4 mm outer diameter and 3 mm inner diameter copper tube that mounted on the aluminum heat sink using a 8 W mini water pump. 25 liter of 30oC water is pumped and circulate inside the copper tube throughout the experiment. The experiment is carried out to determine the effect of water cooling on the temperature difference between the aluminum heat sink and exhaust surface. TEG and booster converter is connected to multimeter respectively to record the voltage and current generated. An infrared thermometer is used to measure the temperature on the aluminum heat sink and exhaust surface. Duration for the experiment is 30 minutes. Data is collected and recorded every minute. The data is then tabulated into graph for further analysis.

7.4. Ice test

Ice test is conducted by using a mixture of ice water as a coolant. The ice water coolant is pump into 4 mm outer diameter and 3 mm inner diameter copper tube that mounted on the aluminum sink using a 8 W mini water pump. A mixture of 7oC ice water is pumped and circulate inside the copper tube throughout the experiment. The experiment is carried out to determine the effect of ice water cooling on the temperature difference between the aluminum heat sink and exhaust surface. TEG and booster converter is connected to multimeter respectively to record the voltage and current generated. An infrared thermometer is used to measure the temperature on the aluminum heat sink and exhaust surface. Duration for the experiment is 30 minutes. Data is collected and recorded every minute. The data is then tabulated into graph for further analysis.

8. Calculation for power generated

Electrical power production can be determined by using the equation below

$$P = IV$$

Where,
I – Current generated in Ampere
V- Voltage generated in Volts
P- Power generated in watt

9. Result and discussion

4 test was carried out which was control test, wind test, water test and ice test. The result and data collected are then tabulated into graph and bar chart for further analysis.

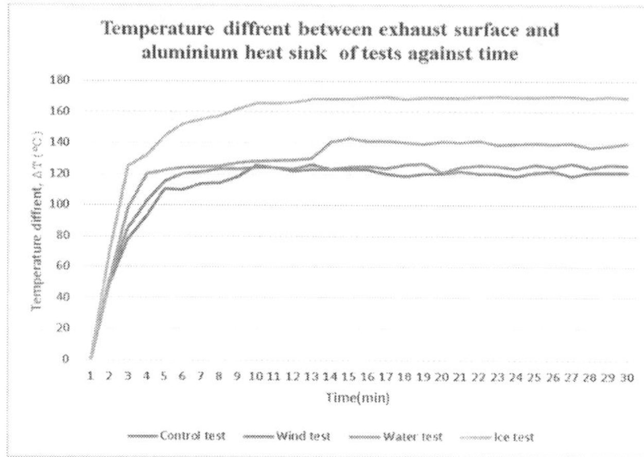

Figure 5 Temperature different between exhaust surface and aluminium heat sink of test against time

Figure 5 show graph of temperature difference between exhaust surface and aluminum heat sink of tests against time. The maximum temperature for exhaust surface achieved is around 180oC. Ice test has the highest ΔT compare to control test, wind test and water test due to 7oC of ice water mixture is run as coolant on the aluminum heat sink. The mixture of ice water is able to maintain the temperature of aluminum heat sink due to the heat absorbed is used to melt the ice. 25 liter of 30oC water is used in water test experiment as coolant on aluminum heat sink. The trend for ΔT value of water test is slightly decreased after 15 minutes due to the heat is absorbed and increase the temperature of the coolant. Therefore, the temperature on aluminum heat sink is increase and temperature different between aluminum heat sink and exhaust surface, ΔT is decreased. The overall trend for wind test is slightly higher than the control test. The fluctuation is caused by the air flow produced by standing fan blow toward aluminum heat sink and also exhaust surface. Control test has the lowest overall trend for ΔT value compare to wind test, water test and ice test. The fluctuation is caused by the irregular heat radiation from the aluminum heat sink which only guided by natural convection from the surrounding.

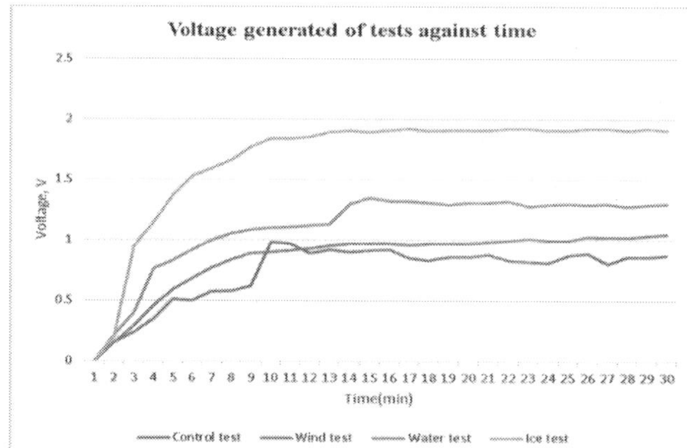

Figure 6 Graph of voltage generated of test against time

Figure 6 shown the graph of voltage generated for control test, wind test, water test and ice test against time. The yellow line indicated the voltage generated for ice test, the grey line indicated the voltage generated for water test, the orange line indicated the voltage generated for wind test and the blue line indicated the voltage generated for control test. The highest trend for voltage generated is ice test compare to water test, wind test and control test due to ice test have the greatest ΔT trend caused ice water mixture as a coolant. The maximum voltage generated for ice test is around 1.92V and remain stable according to ΔT. The second highest trend for voltage generated is water test due to it has the second highest ΔT trend. The maximum voltage generated for water test is 1.35V and decreasing slowly as the trend for ΔT decrease. The control test has the lowest voltage generated trend compare to wind test, water test and ice test. The maximum voltage generated is 0.98V and decreasing as the ΔT for control test decreasing.

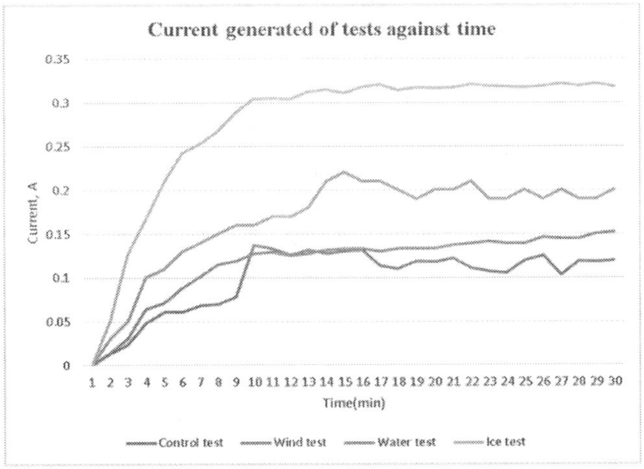

Figure 7 Graph of current generated of tests against time

Figure 7 shown the graph of current generated for control test, wind test, water test and ice test against time. The yellow line indicated the current generated for ice test, grey line indicated the current generated for water test, orange line indicated the current generated for wind test and blue line indicated the current generated for control test. The highest trend for current generated is ice test compare to water test, wind test and control test due to ice test has the greatest ΔT trend caused ice water mixture as coolant. The maximum current generated for ice test is around 0.32A and remain stable according to ΔT. The second highest trend for current generated is water test due to it has the second highest ΔT trend. The maximum current generated for water test is 0.22A and decreasing slowly as the trend for ΔT decrease. The control test has the lowest current generated trend compare to wind test, water test and ice test. The maximum current generated is 0.14A and decreasing as the ΔT for control test decreasing.

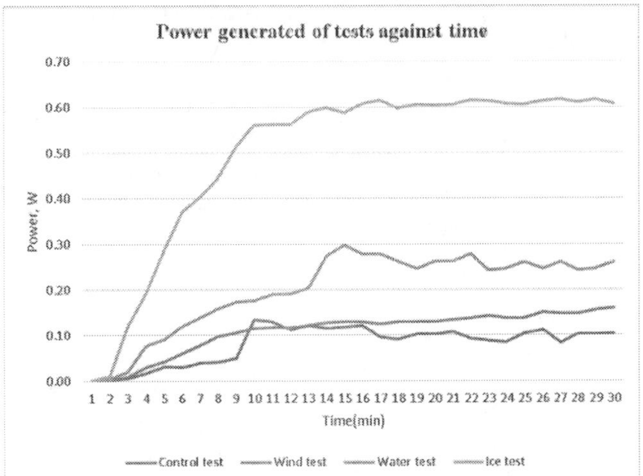

Figure 8 Graph of power generated of test against time

Figure 8 shown the graph of power generated for control test, wind test, water test and ice test against time. The yellow line indicated the power generated for ice test, grey line indicated the power generated for water test, orange line indicated the power generated for wind test and blue line indicated the power generated for control test. The highest trend for power generated is ice test compare to water test, wind test and control test due to ice test has the greatest ΔT trend caused ice water mixture can maintain low temperature on aluminium heat sink. The maximum power generated for ice test is around 0.62W and remain stable according to ΔT. The second highest trend for power generated is water test due to it has the second highest ΔT trend. The maximum power generated for water test is 0.3W and decreasing slowly as the trend for ΔT decrease. Wind test has maximum power generated of 0.16W. The control test has the lowest power generated trend compare to wind test, water test and ice test. The maximum voltage generated is 0.13W and decreasing as the ΔT for control test decreasing.

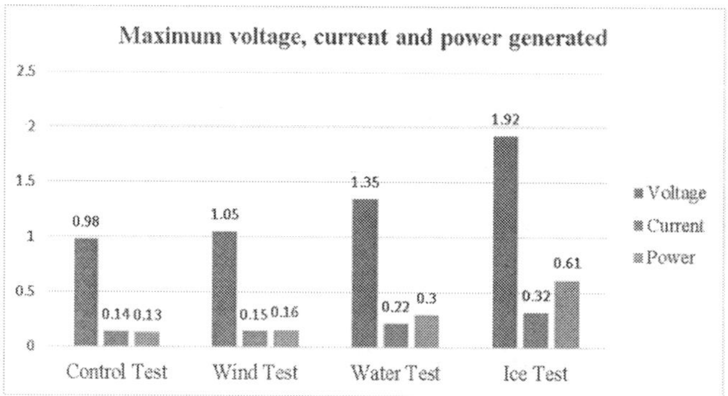

Figure 9 Bar chart for maximum voltage, current and power generated for 4 type of test

Figure 9 shown the bar chart for maximum voltage, current and power generated for control test, wind test, water test and ice test. The blue bar indicated voltage, orange bar indicated current and grey bar indicated power. For control test, the maximum voltage generated is 0.98V, maximum current generated is 0.14A and maximum power generated is 0.13W. The control test has the lowest value for maximum voltage, current and power generated amongst these 4 type of experiment due to it has the lowest temperature difference between exhaust surface and aluminum heat sink. For wind test, the maximum voltage generated is 1.05V, maximum current generated is 0.15A and maximum wind test is slightly higher than control test due to forced convection stimulate by standing fan in wind test conduct the heat from aluminium heat sink to surrounding and maintain a greater temperature different between exhaust surface and aluminium heat sink for wind test. For water test, the maximum voltage is 1.35V, maximum current generated is 0.22A and maximum power generated is 0.3W. The value for maximum voltage, current and power generated for water test is higher than control test and wind test due to coolant of water has higher and better heat conductivity. The ice test has the highest value for maximum voltage, current and power generated which is 1.92V, 0.32A and 0.61W. The highest maximum value is achieved by ice test due to ice water can maintain the temperature of aluminium heat sink for lowest (10oC) amongst 3 other test and achieved greatest temperature different between exhaust surface and aluminium heat sink.

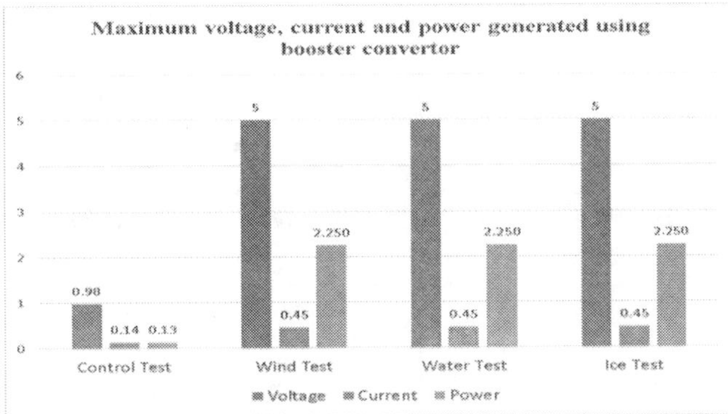

Figure 10 Bar chart for maximum voltage, current and power generated using DC booster

Figure 10 shown the bar chart for maximum voltage, current and power generated using DC booster. Due to the voltage, current and power generated are low for any application utilization, a 1V input and 5V output DC booster is used to step up and boost up voltage, current and power for each experiment. For control test, the voltage generated is below 1V (0.98V) thus the dc booster is unable to boost up the voltage, current and power. For wind test, water test and ice test voltage generated is greater than 1V thus the dc booster is able to boost up the voltage, current and power. The maximum voltage, current and power boost up by using dc booster is 5V, 0.45A and 2.25W.

10. Conclusion

In this project, we have been successfully developed waste heat recovery power generator by using the thermoelectric building block. Waste heat from the exhaust surface of the portable generator has been utilized to convert to electrical power by using the single thermoelectric generator module. The amounts of waste heat carried out from the exhaust surface of the generator toward the environment were analyzed through the experiment that was conducted. Four experiments that were conducted showed the positive effect of the temperature gradient toward the voltage harvesting from the waste heat. The ice test has the highest voltage, current and power generated without using booster converter is 1.92V, 0.32A and 0.61W. The maximum voltage, current and power generated by using booster converter is 5V, 0.45A and 2.25 for wind test, water test and ice test. Even though the amount of voltage harvested directly from the exhaust surface of temperature is too small, but this voltage has been stepped up using a commercial booster converter. Sufficient amount of voltage is achieved and able to directly powered any auxiliary device or it can be stored in the battery. The whole objective of the project to utilize waste heat using the thermoelectric is achieved.

REFERENCES

[1] Sze S M 1969 *Physics of Semiconductor Devices* (New York: Wiley–Interscience)

[2] Dorman L I 1975 *Variations of Galactic Cosmic Rays* (Moscow: Moscow State University Press) p 103

[3] Caplar R and Kulisic P 1973 *Proc. Int. Conf. on Nuclear Physics (Munich)* vol 1 (Amsterdam: North-Holland/American Elsevier) p 517

[4] Szytula A and Leciejewicz J 1989 *Handbook on the Physics and Chemistry of Rare Earths* vol 12, ed K A Gschneidner Jr and L Erwin (Amsterdam: Elsevier) p 133

Studies on water absorption of polypropylene/ recycled acrylonitrile butadiene rubber/ empty fruit bunch composites

N S Othman[1], R Santiagoo[1], Z Mustaffa[1], W A Mustafa[2], I Zunaidi[3], W K Wan[3], Z M Razlan[3] and Shahriman A B[3]

[1]School of Environmental Engineering, Universiti Malaysia Perlis, PusatJejawi 3, 02600 Arau, Perlis, Malaysia

[2]Faculty of Engineering Technology, Kampus Sg. Chuchuh, Universiti Malaysia Perlis, 02100 Padang Besar, Perlis, Malaysia

[3]School of Mechatronic, Universiti Malaysia Perlis, 02600, Arau Perlis.

Email: wanienurul91@yahoo.com

Abstract. Water absorption study is important in natural fibres polymer composites since they are hydrophilic in nature. In this study, water absorption properties of thermoplastic elastomer composites with the incorporation of empty fruit bunch (EFB) fibres were examined. The polypropylene (PP) matrix, recycled gloves (NBRr) and EFB fibre were mixed together using a heated two-roll mill machine at 180°C. A compatibilizer of polypropylene maleic anhydride (PPMAH) was added to improve the water resistant properties of the composites. Water absorption test was conducted by long-term immersion in distilled water at room temperature. From the analysis, the percentage of water absorbs for PP/NBRr/EFB composites was increase as the NBRr loading increases. However, it is interesting to observe that the percent of water absorbs from PPMAH compatibilized composites was decreases from 20 % to 15 % with the highest NBRr loading. It shows that PPMAH compatibilizer possibly acts as a protective barrier and enhances the interfacial adhesion between the fibre and matrices.

Keywords. Water absorption, thermoplastic elastomer, polypropylene maleic anhydride, recycled glove, empty fruit bunch

1. Introduction

The growing of environmental awareness has drive most of researchers to use the natural plant fibres in a polymer composite replacing the synthetic fibres. Natural fibres have their own applicability such as biodegradable, naturally less abrasive, vast availability and low-cost production [1–3]. Bamboo, sisal, jute, hemp, flax, rice husk, kenaf and oil palm fibre are the examples of plant fibres used as filler in polymer composite [4–6]. Malaysia as a tropical country generates about million tons per year of renewable biomass from oil palm industry such as oil palm trunks (OPT), oil palm fronds (OPF) and empty fruit bunch (EFB) [7]. EFB is preferable due to the excessive of biomass generated from the palm oil industries [8]. Several attempts have been made to utilized EFB in the fabrication of natural fibre polymer composites [9]. The most commonly used thermoplastic matrices to work with are polyethylene (PE), polystyrene (PS) and polypropylene (PP). Whereas thermoplastic elastomers (TPE) is the rapidly growing polymer types, that was fabricated from the combination of thermoplastic and the rubber elastomer [3,10]. Additionally, the incorporation of natural fibre as filler in TPE is not only reducing the disposal problem but also produce low-cost of the end-products.

Recycling of waste materials is an idealistic method to overcome the environmental issues. The abundance of rubber wastes is mainly obtained from tires and gloves industries. Rubber gloves are usually found at the industries of automobile, chemical and also medical purposes. Usually, this glove was discarded after a single used and after that, dumped at the landfill. Most of the rubber gloves is hard to degrade which then subsequently will increase the soil and water pollution [11]. Thus, the use of rubber glove waste in polymer composites seems important to reduce the environmental problems and also to produce the new novel sustainable products. In this work, nitrile gloves waste or known as recycled acrylonitrile butadiene rubber (NBRr) was used with the combination of polypropylene (PP) as a matrix and empty fruit bunch (EFB) fibre as a filler. However, the composites are usually weak at the interfacial bonding. This was due to the different polarity of hydrophobic polymer and hydrophilic natural fibre. Besides, the natural fibres are consists of highly moisture content which can affect the long-term performance of the composites.

Therefore, the objective of this research is to study the water absorption behaviour of develop thermoplastic elastomer, PP/NBRr/EFB composites with the addition of polypropylene maleic anhydride (PPMAH). Based on the literature, no research was reported on the study of water absorption of PP/NBRr/EFB composites with and without PPMAH which act as compatibilizer to improve the composites performance.

2. Material and Method

2.1. Material Preparation

The isotactic polypropylene (PP), grade 6331, from Titan PP Polymers (M) Sdn. Bhd, was used as the thermoplastic matrix. The melt flow rate of PP is 14g/10 min at 230 °C, and the density is 0.9 g/cm³. The rejected nitrile glove or known as recycled acrylonitrile butadiene rubber (NBRr) from Juara One Resources Sdn. Bhd. was used. Firstly, NBRr gloves were cut and masticulated into small particle size via two-roll mill machine. Then, NBRr particles were sieved into particular size about 150 μm to 300 μm. While for the filler, empty fruit bunch (EFB) fibre was collected from United Oil Palm Industries Sdn. Bhd. EFB was first cleaned to remove the impurities and dirt prior put in the oven with temperature 90°C for 24 hours to eliminate excessive moisture. Then, it was grinded into powder form and sieved into particular size of 150 μm to 300 μm. Whereas, for the PPMAH compatibilizer, it was gained from the Sigma Aldrich Chemical with the percentage of grafting 8-10% and melt flow index of 11.5 g/10 min.

2.2. Mixing compounding and sample preparation

The composition used in this study for PP, NBRr, EFB, PPMAH and the steps in preparing the compounds are shown in Table 1 and Table 2, respectively. A heated two-roll mill machine from Fang Yuan Instrument (DG) Co. was used to mix all the materials with temperature 180°C and rotor speed of 15 rpm. After mixing, the compounds were dried for 24 hours at 80°C under vacuum. The compounded samples were compression-moulded into 1 mm thin sheet at 180°C, using a hot press machine, model GT 7014 A with 6895 kPa. Seventh minutes of preheating followed by two minutes of compression were carried out. Then, samples were continued for two minutes of cooling compression. Based on the ASTM D638, a dumbbell shape samples was prepared using a Wallace die cutter model S6/1/6.A.

Table 1. Composition for PP/NBRr/EFB and PP/NBRr/EFB/PPMAH composite.

Materials	PP/NBRr/EFB (phr[a])	PP/NBRr/EFB/PPMAH (phr[a])
PP	100, 80, 70, 60, 50, 40	100, 80, 70, 60, 50, 40
NBRr	0, 20, 30, 40, 50, 60	0, 20, 30, 40, 50, 60
EFB	10	10
PPMAH[b]	-	5

[a]part per hundred resin
[b]The weight of PPMAH dependent on PP

Table 2. Mixing sequences.

PP/NBRr/EFB		PP/NBRr/EFB/PPMAH	
Time (min)	Materials	Time (min)	Materials
0	PP	0	PP
4	NBRr		PPMAH
6	EFB	4	NBRr
9	Discharging	6	EFB
		9	Discharging

2.3. Water Absorption Test

For the water absorption test, 5 dumbell specimens of each formulation were weighed (W_0, initial weight) prior to immersed in distilled water at room temperature. The specimens were taken out from water at regular intervals and wiped with filter paper. The weight gains after exposure (W_m) were recorded using a weighing balance with a precision of ± 0.001 (1mg). The samples continue re-immersed until the saturation limit was reached. The moisture content at any time t, (M_t) as a result of moisture absorption, was calculated by using Eq. 1 [12],

$$M_t = \frac{W_m - W_o}{W_o} \times 100 \quad (1)$$

3. Result and discussion

Figure 1 shows the water absorption percentage as a function of the square root of time for PP/NBRr/EFB composites. At the initial stage, water is rapidly absorbs into the composites. Then, water slowly absorbs with the prolong immersion time until it attained to a saturation level. It was observed that the percentage of water absorption of PP/NBRr/EFB composites is varied from 5.3 to 20 %. When PP/NBRr/EFB composite is exposed to the wet condition, water are easily reacted and penetrated onto the hydrophilic structure of fibre. Hemicellulose is the main fibre component that are responsible for the water absorption in natural fibres [13]. The process of fibre swelling due to water penetrate will develop stress at the interface between fibre and matrix which then, leads to microcracking mechanisms. This will increase the capillary of water transport into the composites. From the result obtained, percentage of water absorption increases as more NBRr loading increase. This trend is due to the less embedded of NBRr particles to the PP matrix. Thus, the composites will presented more voids and gaps as more NBRr added. A sufficient amount of EFB fibre and NBRr matrix is required to avoid the poor wettability and non-homogeneous of composites. Besides, the voids formation are also due to the water vapour generates during the mixing process [14]. Figure 2 shows the illustration of water retains in PP/NBRr/EFB composite. Numerous studies have attempted to explain the water absorption effects on natural fibre reinforced polymer composites [9,15,16].

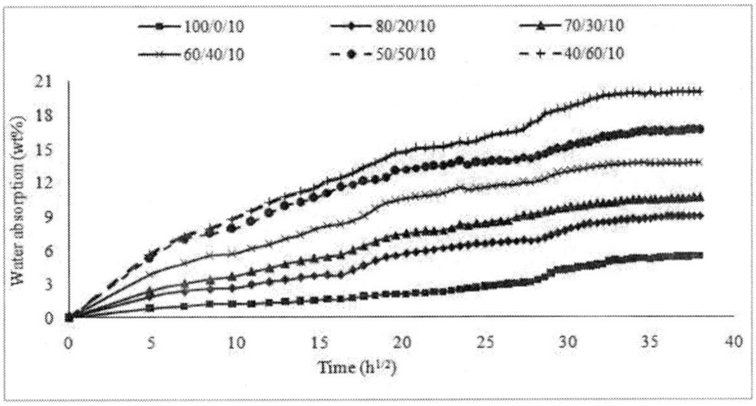

Figure 1. Percent increase in water absorption of PP/NBRr/EFB composites.

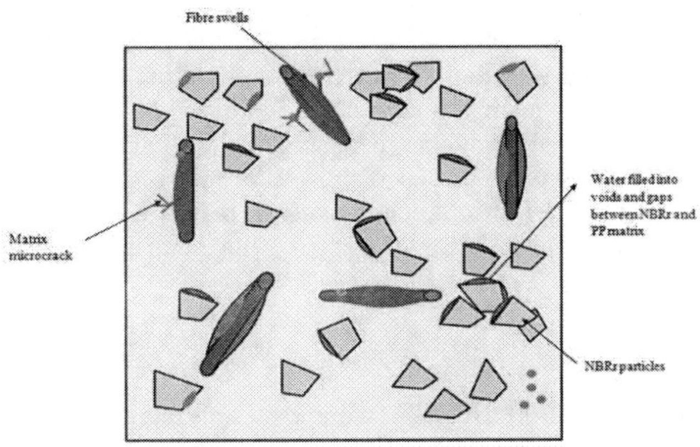

Figure 2. Illustration of water retains in the PP/NBRr/EFB composites.

However, further reduction of water absorption percentage of PP/NBRr/EFB/PPMAH composites is observed (as shown in Figure 3). The water absorption varies between 2.8 to 15 %. PPMAH compatibilizer has a tendency to reduce the water absorption rate. The anhydride group from PPMAH may react with the freely hydroxyl groups of the EFB fibres via C-O-C esterification reaction [17] by blocking some of the EFB hydrophilic ends. The PPMAH compatibilizer also functionalizes as a protective barrier by encapsulating the EFB fibres with the PP matrix. Besides, data obtained shows that, the saturation time of PPMAH compatibilized composites shows slightly increases. This prove the PPMAH compatibilizer gains the ability to resist water uptake and also lowering the velocity of moisture diffuse into the composites [18].

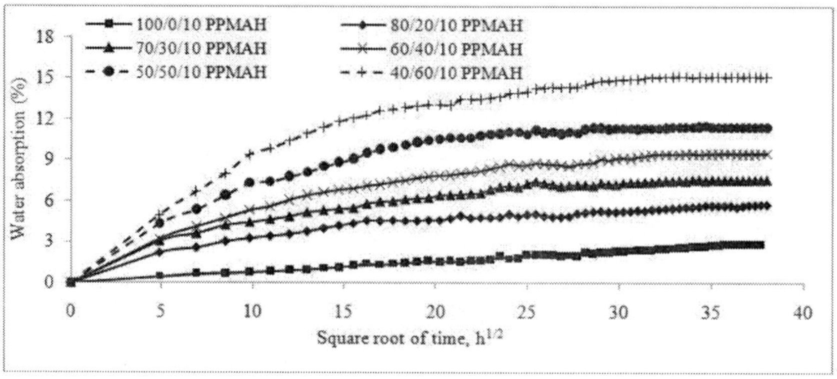

Figure 3. Percent increase in water absorption of PP/NBRr/EFB/PPMAH composites.

In short, there are two major mechanisms relates to water uptake of PP/NBRr/EFB composites. The first mechanism was due to the existence of natural fibre, which increases the hydrophilicity of the composite. When PP/NBRr/EFB composite is exposed to the moisture, water easily penetrates into the hydrophilic groups of natural fibre, which can reduce the interfacial adhesion between the fibre and matrix. Swelling of cellulose fibres leading to micro-cracking in the matrix which promotes capillarity transports [19]. The second mechanism was due to the additional of NBRr rubber particles which then disturbing the structural homogeneity of the compound. A sufficient amount of NBRr is needed to avoid the agglomeration. More voids formed if the NBRr matrix particles is too much, which will increase the ability of water uptake into the composites. Therefore, the addition of compatibilizer or other chemical reagent is required in order to improve fibre-matrix adhesion and reduce the water absorption by the composites.

4. Conclusion

Water absorption percentage of PP/NBRr/EFB composites is increasing as NBRr content increase. This was due to the voids formation during the compounding process. However, for the PPMAH compatibilized composites, the water absorption percentage is reduced. The presence of MAH moieties of PPMAH gave better adhesion between the fibre and matrices. PPMAH is one of the compatibilizer that can be used to resist water absorbs from the surrounding and improve the composites performance.

5. References

[1] Summerscales J, Virk A and Hall W 2013 A review of bast fibres and their composites: Part 3 - Modelling *Compos. Part A Appl. Sci. Manuf.* **44** 132–9

[2] Zainal M, Aihsan M Z, Mustafa W A and Santiagoo R 2018 Experimental Study on Thermal and Tensile Properties on Polypropylene Maleic Anhydride as a Compatibilizer in Polypropylene / Sugarcane Bagasse Composite *J. Adv. Res. Fluid Mech. Therm. Sci.* **43** 141–8

[3] Zainal M, Santiagoo R, Ayob A and Mustaffa W A 2017 Mechanical properties and chemical reaction of 3-aminopropyltriethoxysilane of polypropylene, recycle acrylonitrile butadiene rubber and sugarcane bagasse composites *Int. J. Microstruct. Mater. Prop.* **12** 55

[4] Shinoj S, Visvanathan R, Panigrahi S and Kochubabu M 2011 Oil palm fiber (OPF) and its composites: A review *Ind. Crops Prod.* **33** 7–22

[5] Mustafa W A, Saidi S A, Zainal M and Santiagoo R 2018 A Proposed Compatibilizer Materials on Banana Skin Powder (BSP) Composites Using Different Temperature *J. Adv. Res. Fluid Mech. Therm. Sci.* **43** 121–7

[6] Mustafa W A, Saidi S A, Zainal M and Santiagoo R 2018 Experimental Study of Composites

Material Based on Thermal Analysis *J. Adv. Res. Fluid Mech. Therm. Sci.* **43** 37–44

[7] Rozman, H.D., Z.A. Mohd Ishak and U S I 2005 *Natural Fibers: Biopolymers and Biocomposites*

[8] Abdul Khalil H P S, Issam A M, Ahmad Shakri M T, Suriani R and Awang A Y 2007 Conventional agro-composites from chemically modified fibres *Ind. Crops Prod.* **26** 315–23

[9] Razak N W a. and Kalam a. 2012 Effect of OPEFB Size on the Mechanical Properties and Water Absorption Behaviour of OPEFB/PPnanoclay/PP Hybrid Composites *Procedia Eng.* **41** 1593–9

[10] Amin S and Amin M 2011 Thermoplastic elastomeric (TPE) materials and their use in outdoor electrical insulation *Rev. Adv. Mater. Sci.* **29** 15–30

[11] Yehia A A 2004 Recycling of Rubber Waste *Polym. Plast. Technol. Eng.* **43** 1735–54

[12] Kim J K, Hu C, Woo R S C and Sham M L 2005 Moisture barrier characteristics of organoclay-epoxy nanocomposites *Compos. Sci. Technol.* **65** 805–13

[13] Dhakal H N, Zhang Z Y and Richardson M O W 2007 Effect of water absorption on the mechanical properties of hemp fibre reinforced unsaturated polyester composites *Compos. Sci. Technol.* **67** 1674–83

[14] Dastoorian F and Tajvidi M 2008 Influence of strain rate on the flexural properties of a wood flour/HDPE composite *J. Reinf. Plast. Compos.* **27** 1701–8

[15] Saputra O Z I A, Rinawati L, Rini K S, Saputra D A and Pramono E D I 2016 Effect Of Fiber Size On Mechanical and Water Absorption Properties Of Recycled Polypropylene / Empty Fruit Bunches (rPP / EFB) Bio- composites *Appl. Mech. Mater.* **842** 7–13

[16] Fatra W, Rouhillahi H, Helwani Z, Zulfansyah and Asmura J 2016 Effect of alkaline treatment on the properties of oil palm empty fruit bunch fiber-reinforced polypropylene composite *Int. J. Technol.* **7** 1026–34

[17] Ragunathan, S., Othman N.S., Mohamed A.K. and I H 2016 Effect of maleic anhydride-grafted polypropylene on polyropylene/ recycled acrylonitrile rubber/ empty fruit bunch composite

[18] Kittikorn T, Strömberg E, Ek M and Karlsson S 2013 Comparison of water uptake as function of surface modification of empty fruit bunch oil palm fibres in PP biocomposites *BioResources* **8** 2998–3016

[19] Joseph P V., Rabello M S, Mattoso L H C, Joseph K and Thomas S 2002 Environmental effects on the degradation behaviour of sisal fibre reinforced polypropylene composites *Compos. Sci. Technol.* **62** 1357–72

IOP Conf. Series: Materials Science and Engineering **429** (2018) 012092 doi:10.1088/1757-899X/429/1/012092

Overview of Catalytic Methanation in Hot Mixing Asphalt Plants Towards Greenhouse Gas Emissions

T F Nashruddin[1], M M A Aziz[1*], W A W A Bakar[2*], S A F S Muhammad[3], You Kok Yeow[4], M Mailah[5]

[1]Department of Geotechnics and Transportation, Faculty of Civil Engineering, Universiti Teknologi Malaysia, 81310 UTM Skudai, Johor Bahru, MALAYSIA.
[2]Department of Chemistry, Faculty of Science, Universiti Teknologi Malaysia, 81310 UTM Skudai, Johor Bahru, MALAYSIA.
[3]Department of Bioprocess and Polymer Engineering, Faculty of Chemical and Energy Engineering, Universiti Teknologi Malaysia, 81310 UTM Skudai, Johor Bahru, MALAYSIA.
[4]Department of Communication Engineering, Faculty of Electrical Engineering, Universiti Teknologi Malaysia, 81310 UTM Skudai, Johor Bahru, MALAYSIA.
[5] Department of Applied Mechanic and Design, Faculty of Mechanical, Universiti Teknologi Malaysia, 81310 UTM Skudai, Johor Bahru, MALAYSIA.

*Email: t_filza88@unimap.edu.my

Abstract. Road works mainly involve the use of asphalt for pavement construction. Among Hot-Mix asphalt (HMA) production processes, HMA plants are the most commonly used for asphalt production in world. HMA is shown as an industrial activity in which energy consumption can reach 60% of the total energy consumed by the construction and the maintenance of a road during 30 years of service life. Rate of production, bitumen temperature, burner flow rate and type of fuel consumption are the parameters which directly link to energy consumption and greenhouse gas (GHG) emission during HMA plant operation. Carbon dioxide (CO_2) in GHG has increased the temperature of the earth. Introduction of catalytic methanation as a new method in the HMA production was introduced as an effective way to convert CO_2 released from HMA plants to methane (CH_4). The optimum parameters were found through the experimental result which showed that Ru/Cr/Mn (5:10:85)/Al_2O_3 calcined at 700°C with 10 g of catalyst dosage gave 100% of CO_2 conversion, determined using FTIR, and yielded about 80% of CH_4 at reaction temperature of 300°C. Field Emission Scanning Electron Microscope (FESEM) show the presence of dispersed particles with undefined shape covering the catalyst surface. EDX analysis revealed that the mass ratio of Mn was high compared to the other elements. Thus, this study has proven that the catalytic methanation technology as the new effective method in HMA production plants can convert CO_2 in GHG emission to methane (CH_4). In conclusion, it is highly advisable to adapt these alternatives as the green roads and highways industry.

1. Introduction

Nowadays, injection moulding technology has been widely used in modern industry activity to In HMA plants, two main sources that produce organic emissions at the chimney are coming from combustion and Hot-Mix asphalt [1]. From the environmental aspect, the HMA plants (i.e. drum mixer process) can be classified by five parts which are (Figure 1): (1) energy flame, (2) material resources (i.e. aggregate and bitumen), (3) coating process, (4) HMA manufactured, and (5) emissions to the atmosphere.

The warming and drying of aggregates move into the upper part of the drum using the main conveyor belt. The bitumen is put into the remaining half of the drum, which is the flame end, to avoid excessive oxidation [2]. The final HMA mixture is produced after the aggregates are mixed with hot asphalt cement, then the HMA is loaded into trucks for hauling to the job site. The primary emission sources associated with HMA production are the dryers, hot bins, and mixers, which emit a variety of gaseous pollutants. HMA plant's emissions are not constant over time. There are fluctuations due to the process used, under or out of control [1, 3, 4]

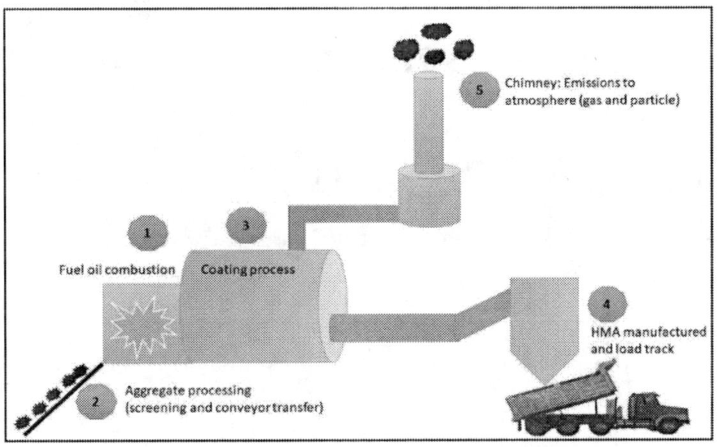

Figure 1. HMA plant operation[1].

Various methodologies and technologies have been developed by the road pavement industry to reduce the environmental impact: the use of warm mix asphalt (WMA) are considered as a cleaner practice than traditional HMA[5] due to the lower production and application temperatures (100-140°C) incorporating between asphalt plants various percentage of reclaimed asphalt pavement (RAP) blended together with virgin aggregate to conserve the natural aggregate resources [6] and use of by product material (i.e. slag, rubber) to prolong the service life of a pavement [7, 8]. Those techniques are focused more on sustainable environment during road construction. However, GHG emissions from asphalt production cannot be stopped.

In this paper, a chemical conversion technique known as catalytic methanation has been studied to convert CO_2 emissions to the higher value-added product, which is natural gas of methane (CH_4) [9]. This process does not only create a green and sustainable environment in HMA production but produced valuable methane gas that can also be used as a fuel to run turbine for electricity [10]. Thus, this study has two (2) main objectives, which are to determine the gas composition produced in HMA production plant and to develop the catalyst which can convert CO_2 in GHG emission from HMA plants to the natural gas of methane.

2. GAS COMPOSITION FROM HMA PLANT

GHG emissions have been monitored and controlled under binding environmental regulations and legislation. Operating of HMA plants require the agreement of local authorities which is given according to pollutants thresholds: CO, SO_2, NO_x, and total organic compounds (TOC)[1]. Gas/fumes from asphalt plants is a mixture containing different chemicals. The chemicals in asphalt vary depending on the source of the crude oil, the type of asphalt being made, and the process used. Figure

2 shows GHGs emissions that had been collected from HMA plants by using gas analyser (GA). Based on GA reading, hydrocarbon, carbon dioxide (CO_2), nitrogen oxides (NO_x), carbon monoxide (CO) and Air fuel oil have been detected from HMA plants. The amount of CO_2 released was about 6% because the type fuel was already filtered by following the Environment Act (Figure 3). Generally, CO_2 was produced from the type of fuel used and at the same time, the energy consumption needed for the heating process of the mineral aggregates and bitumen tanks. Other CO gas released was due to the combustion process in the dryer and was strongly influenced by the fine content of the mineral, and the water vapour content in the drum. Usually, NO_x emission depends on the nitrogen content of the fuel and was originated from the burner in the drying drum[11].

Figure 2: HMA plant and Gas analyser (GA).

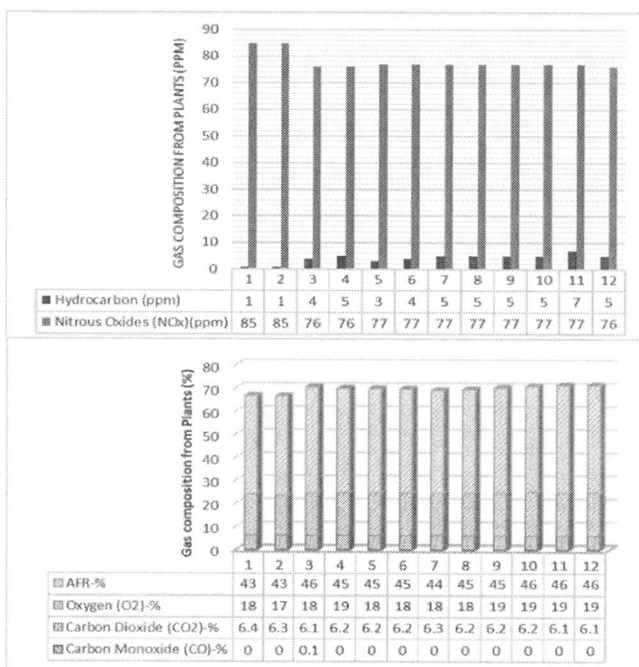

	1	2	3	4	5	6	7	8	9	10	11	12
Hydrocarbon (ppm)	1	1	4	5	3	4	5	5	5	5	7	5
Nitrous Oxides (NOx)(ppm)	85	85	76	76	77	77	77	77	77	77	77	76

	1	2	3	4	5	6	7	8	9	10	11	12
AFR-%	43	43	46	45	45	45	44	45	45	46	46	46
Oxygen (O2)-%	18	17	18	19	18	18	18	18	19	19	19	19
Carbon Dioxide (CO2)-%	6.4	6.3	6.1	6.2	6.2	6.2	6.3	6.2	6.2	6.2	6.1	6.1
Carbon Monoxide (CO)-%	0	0	0.1	0	0	0	0	0	0	0	0	0

Figure 3: GHGs gas composition of HMA plants from Gas Analyser (GA).

IOP Conf. Series: Materials Science and Engineering **429** (2018) 012092 doi:10.1088/1757-899X/429/1/012092

3. Experimental

In this study, catalytic methanation system was used to utilize the GHG emissions from HMA plants. The methanation reaction is the catalytic reaction where carbon dioxide reacts with hydrogen to produce methane and water as described in Equations 1 and 1 .

$$CO_2(g) + 4H_2(g) \rightarrow CH_4(g) + 2H_2O(l) \quad (1)$$
$$CO(g) + 3H_2(g) \rightarrow CH_4(g) + H_2O\ (l) \quad (2)$$

The catalytic methanation conversion system uses rare earth oxides. Rare earth oxides are expected to adsorb significant amount of the acidic CO_2 gases on the catalyst surface. Catalyst act as the conversion towards CO_2 that is released from asphalt production to form methane (CH_4) which is one of the components of natural gas. Methane gas production is an environmentally friendly approach for purification of natural gas [12]. Natural gas is an extremely important source of energy for reducing pollution and maintaining a clean and healthy environment [13].

3.1. Preparation of Chemical and Catalysts

Manganese (Mn) was used as the base in this study. Meanwhile Chromium (Cr), Nickel (Ni) and Cobalt (Co) were used as dopants and Ruthenium (Ru) as co-dopant. In addition, aluminium oxide beads (Al_2O_3) were used as the support materials for the preparation of catalysts. All the catalysts were prepared by aqueous incipient wetness impregnation method. 85 wt% 5 g of Manganese (Mn) salts was dissolved in small amount of distilled water, then mixed with a dopant and co-dopant solution in a beaker, according to the desired ratio. A homogeneous mixture was obtained by electromagnetic stirring at room temperature for 30 minutes. Alumina beads with diameter of 3 mm were used as support material in this study. The support was immersed into the catalysts solution until the solution was evenly absorbed on the surface of the support. It was then aged inside an oven at 90°C for 24 h followed by calcinations using a furnace in the air atmosphere at 400°C, 700°C and 1000°C for 5 h using a ramp rate of 10°C/min to remove all the metal counter ions and water present in the catalyst.

3.2. Catalytic Activity Measurement

The catalytic CO_2 methanation reaction was performed under atmospheric pressure in a fixed microreactor and analysed via online FTIR Nicolet Avatar 670 DGTS. 10 g of supported catalyst was placed in the middle of a Pyrex glass tube of diameter 10 mm and length 520 mm. It was then secured with glass wool at both ends. The reaction gas mixture (inlet) consisting of CO_2, Hydrogen (H_2), Nitrogen (N_2), (10/40/39 mL/min) was passed continuously through the catalyst and was heated in an isothermal tube furnace. The feed gas flow rate was adjusted with a mass flow controller and the reaction temperature was performed from 50°C up to 400°C with the increment temperature rate of 5°C/min (Figure 4).

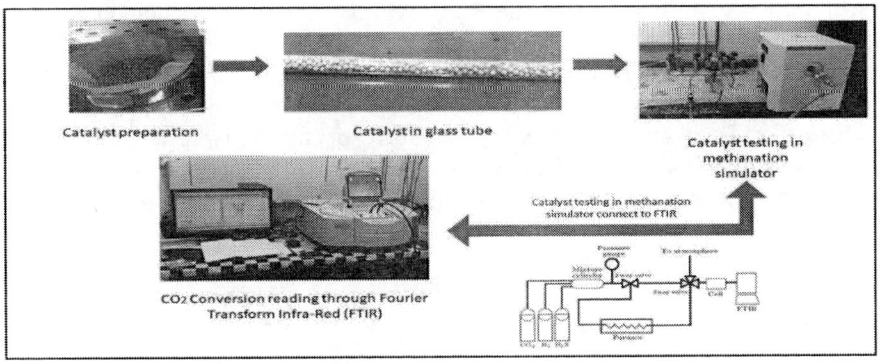

Figure 4. Catalytic Methanation System

4. Results

4.1. Catalytic Performance Of Mn Based Catalyst Over Various Dopant

Table 1 illustrates the catalytic performance of Mn based catalysts, which were calcined at 700 °C for 5 h. According to the results, a Mn (100%)/Al_2O_3 catalyst gives only 22.36% conversion of CO_2 at a reaction temperature of 400°C. Three types of chemical, which are Chromium (Cr), Nickel (Ni) and Cobalt (Co), were used as a dopant catalyst.

Incorporating Cr/Mn (15:85)/Al_2O_3, Ni/Mn (15:85)/Al_2O_3 and Co/Mn (15:85)/Al_2O_3 as dopant catalyst have increase the catalytic activity of Mn base to 26.43 %, 4.79% and 33.31 %, respectively, at 400°C reaction temperature. Combination of transition metal (i.e Mn, Cr, Nickel and Co) was found to accelerate the Methanation process [14].

The addition of Ru into Cr/Mn (15:85)/Al_2O_3, Ni/Mn (15:85)/Al_2O_3 and Co/Mn (15:85)/Al_2O_3 as co-dopant catalyst to form Ru/Cr/Mn (5:10:85)/ Al_2O_3, Ru/Ni /Mn (5:10:85)/ Al_2O_3 and Ru/Co /Mn (5:10:85)/ Al_2O_3 catalyst coincidentally enhanced catalytic activity for the conversion of CO_2 of the prepared catalysts. The small amounts of Ru addition can play an important role in the improvement of catalyst activity. The presence of Ru as a co-dopant catalyst to Mn based catalyst also has enhanced and eased the chemisorptions of the reactants (CO_2 and H_2) on their surfaces [14]. Combination of Ru/Cr/Mn (5:10:85)/ Al_2O_3, presented the highest combination of conversion CO_2 between other catalyst, which are fully 100% of CO_2 conversion at reaction temperatures of 300 °C.

Table 1. Percentage conversion of CO_2 from methanation reaction over Manganese based catalysts with various dopant.

Alumina supported catalyst	**Reaction Temperature (°C)**							
	50	**100**	**150**	**200**	**250**	**300**	**350**	**400**
Mn (100%)/Al_2O_3	6.22	6.19	6.51	4.94	7.87	6.22	9.21	22.36
Cr/Mn (15:85)/Al_2O_3	6.40	10.11	12.72	16.31	22.36	22.64	23.89	26.43
Ni/Mn (15:85)/Al_2O_3	5.98	12.03	14.49	25.78	29.05	31.45	30.98	41.79
Co/Mn (15:85)/Al_2O_3	10.64	6.60	6.68	10.26	11.32	12.44	21.01	33.31
Ru/Cr/Mn (5:10:85)/ Al_2O_3	**0.40**	**2.41**	**15.70**	**30.50**	**100.00**	**100.00**	**100.00**	**100.00**
Ru/Ni /Mn (5:10:85)/ Al_2O_3	8.61	14.83	23.14	46.51	78.52	80.97	80.57	80.39
Ru/Co /Mn (5:10:85)/ Al_2O_3	4.15	9.32	2.39	19.46	82.38	83.31	82.98	83.15

4.2. Formation of methane (CH_4)

Figure 5 illustrated the detection of the formation of CH_4 after CO_2, hydrogen (H) and Nitrogen (N) (simulate: CO_2, H and N was GHG gas in laboratory) was flowed through to Ru/Cr/Mn(5:10:85)/Al_2O_3 catalysts. The Ru/Cr/Mn(5:35:60)/Al_2O_3 calcined at 700°C was considered as the potential catalyst for CO_2 methanation reaction since it gave 100% of CO_2 conversion. Based on the results, CH_4 was not produced at initial temperatures of 50°C, 100°C and 150°C. As the reaction temperatures were elevated to 200°C, 250°C, 300°C, 350°C and 400°C, about 19.59%, 58.67%, 80%, 80%, 80% of methane formation were detected, respectively, by using GA. The high formation of methane at high reaction temperature was possibly due to its highly exothermic reaction where the methanation reaction was favored at high temperature [14].

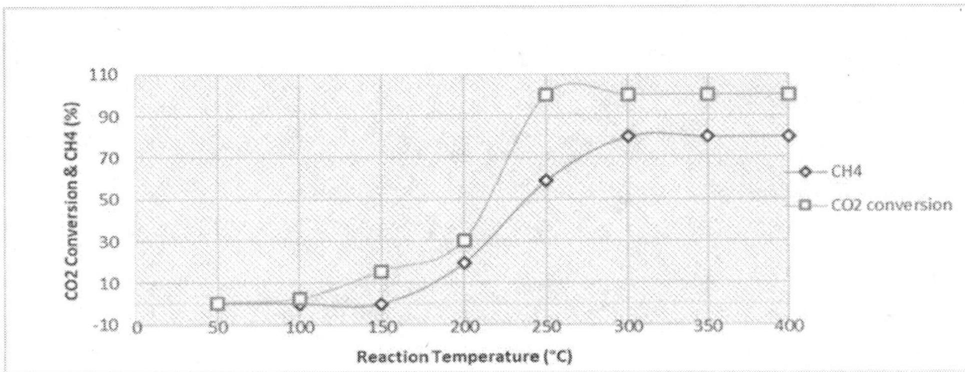

Figure 5: CO_2 Conversion and Formation of CH_4

4.3. Characterization: FESEM-EDX

Characterization: Figure 6 (i,ii,&iii) shows the FESEM micrographs of fresh and spent Ru/Cr/Mn(5:10:85)/Al_2O_3 catalysts, calcined at 700°C for 5 h with magnification of 20000× and 50,000×. The fresh Ru/Cr/Mn(5:10:85)/Al_2O_3 catalyst showed rough surface morphology with a homogeneous spherical shape, and comes with small particle sizes. In this study, it was found that the particle size of fresh Ru/Cr/Mn(5:10:85)/Al_2O_3 catalyst is categorized at the nano (<100 nm) level, which varies from 67 to 147 nm. The smaller particle size plays an important role in exhibiting higher catalytic activity. The morphology of the fresh catalyst changed significantly after the hydrogenation of the methanation reaction, which showed the formation of aggregated and agglomerated undefined shapes on the surface of the spent catalyst (Figure 6(ii)) due to the heat generated during the catalytic reaction and which caused the catalyst to agglomerate [15].

Furthermore, Table 2 shows the EDX analysis for the fresh and spent Ru/Cr/Mn(5:10:85)/Al_2O_3 catalysts. The elemental analysis performed by EDX confirmed the presence of Ru, Cr, Mn, Al and O in the potential catalyst. From EDX analysis, it can be detected that the composition of Mn was higher than the other composition of Ru and Cr on the fresh and spent Ru/Cr/Mn(5:10:85)/Al_2O_3 catalysts, which are 18.1 % and 10.7%. Moreover, the spent Ru/Cr/Mn(5:10:85)/Al_2O_3 catalyst confirms the existence of 3.40% carbon on the surface of the catalyst. The difference of fresh and spent catalyst surfaces was due to the presence of carbon. From the EDX analysis, the weight ratio of Aluminium (Al) was the highest, indicating that the catalyst surface was dominated by the Al from the support. This may be due to the noble metal and Mn being absorbed into the high porosity and high surface area of alumina beads[14].

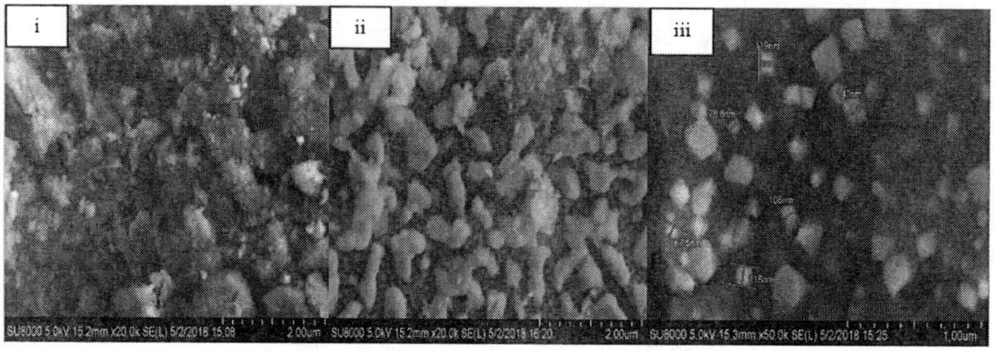

Figure 6: i) Fresh Ru/Cr/Mn(5:10:85)/Al_2O_3 catalyst with magnification 20000x, ii) spent Ru/Cr/Mn(5:10:85)/Al_2O_3 catalysts with magnification 20000x, iii) particle size of fresh Ru/Cr/Mn(5:10:85)/Al_2O_3 catalyst with magnification 50000x

Table 2.EDX analysis

Weight ratio (%)	**Ru/Cr/Mn (5:10:85)/Al_2O_3 catalysts, calcined at 700 °C**	
	Fresh	**Spent**
O	35.0	42.3
Al	32.3	29.6
Mn	18.1	10.7
Cr	8.9	4.0
Ru	3.1	1.0
C	-	3.4

5. Conclusion

Gas composition from HMA plants has successfully been detected by using GA. In general, gas that has been released to atmosphere are categorized as GHGs emissions. Therefore, this study tries to convert CO_2 emissions to natural gas by applying chemical conversion system. This method can also be called catalytic methanation system. Overall, the catalytic activity studies can achieve almost 100% of CO_2 conversion at low reaction temperature and produced 80% yield of methane (CH_4) at 300°C of reaction temperature. Catalyst of Ru/Cr/Mn (5:10:85)/Al_2O_3 calcined at 700°C was assigned as the potential catalyst that had successful been developed due to the contribution of 100% of CO_2 conversion at the reaction temperature of 250°C. Field Emission Scanning Electron Microscopy (FESEM) analysis illustrated the surface of the catalysts were covered with small and dispersed spherical particles. Energy Dispersion X-ray (EDX) analysis revealed that the weight ratio composition of Mn was high in the fresh Ru/Cr/Mn (5:35:60)/Al_2O_3 but reduced after Ru/Cr/Mn (5:35:60)/Al_2O_3 catalysts was used.

Acknowledgements

Without the support of Ministry of Education and Universiti Teknologi Malaysia (UTM), in the form of a research grant (Vote No. 15H32 and 19H98), this study would not have been possible. Author shall remain indebted to them for their generosity.

References

[1] Paranhos, R.S. and C.O. Petter, *Multivariate data analysis applied in Hot-Mix asphalt plants.* Resources, Conservation and Recycling, 2013. **73**: p. 1-10.

[2] Paranhos, R., A. Ventura, P. Monéron, and A. Jullien. *Smoke-stack airborne emissions of a hot mix asphalt plant for life cycle inventory available data.* in *2nd International Conference on Green and Sustainable Chemistry of American Chemical Society, Washington DC (USA), June 20.* 2005.

[3] Jullien, A., V. Gaudefroy, A. Ventura, C. De La Roche, R. Paranhos, and P. Monéron, *Airborne emissions assessment of hot asphalt mixing: methods and limitations.* Road Materials and Pavement Design, 2010. **11**(1): p. 149-169.

[4] Gaudefroy, V., V. Viranaiken, R. Paranhos, A. Jullien, and C. de La Roche, *Laboratory assessment of fumes generated by bituminous mixtures and bitumen.* Road Materials and Pavement Design, 2010. **11**(1): p. 83-100.

[5] Rubio, M.C., G. Martínez, L. Baena, and F. Moreno, *Warm mix asphalt: an overview.* Journal of Cleaner Production, 2012. **24**: p. 76-84.

[6] Alavi, Z., Y. He, J. Harvey, and D. Jones, *Evaluation of the combined effects of Reclaimed Asphalt Pavement (RAP), Reclaimed Asphalt Shingles (RAS), and different virgin binder sources on the performance of blended binders for mixes with higher percentages of RAP and RAS.* 2015.

[7] Lin, D.-F., L.-H. Chou, Y.-K. Wang, and H.-L. Luo, *Performance evaluation of asphalt concrete test road partially paved with industrial waste–basic oxygen furnace slag.* Construction and Building Materials, 2015. **78**: p. 315-323.

[8] Farina, A., M.C. Zanetti, E. Santagata, and G.A. Blengini, *Life cycle assessment applied to bituminous mixtures containing recycled materials: Crumb rubber and reclaimed asphalt pavement.* Resources, Conservation and Recycling, 2017. **117**: p. 204-212.

[9] Toemen, S., W.A.W.A. Bakar, and R. Ali, *Copper/nickel/manganese doped cerium oxides based catalysts for hydrogenation of CO2.* Bull. Korean Chem. Soc, 2014. **35**(8): p. 2349.

[10] Bakar, W.A.W.A., S. Toemen, and R. Ali, *Investigation of Carbon Dioxide Methanation over Ceria Based Catalysts.* 2014.

[11] EPA, *Hot Mix Asphalt Plants Emissions Assessment Report.* 2000, Air Quality Planning and Standards United States Environmental Protection Agency Research Triangle Park, NC.

[12] Zamani, A., R. Ali, and W. Bakar, *The investigation of Ru/Mn/Cu–Al 2 O 3 oxide catalysts for CO 2/H 2 methanation in natural gas.* Journal of the Taiwan Institute of Chemical Engineers, 2014. **45**(1): p. 143-152.

[13] Alliance., A.P., *Carbon Footprint: How does asphalt stack up? (IM-46).* 2010.

[14] Rosid, M., S. Jamal, W.A. Bakar, W. Azelee, and R. Ali. *Catalytic CO2/H2 Methanation Reaction over Alumina Supported Manganese/Cerium Oxide Based Catalysts.* in *Advanced Materials Research.* 2015. Trans Tech Publ.

[15] Toemen, S., W.A.W.A. Bakar, and R. Ali, *Investigation of Ru/Mn/Ce/Al2O3 catalyst for carbon dioxide methanation: Catalytic optimization, physicochemical studies and RSM.* Journal of the Taiwan Institute of Chemical Engineers, 2014. **45**(5): p. 2370-2378.

Experimental and numerical analysis on the adhesive T joint using high temperature adhesive

N F Bahri , M Afendi, Z M Razlan, I Zunaidi, W K Wan and Shahriman A B

School of Mechatronic Engineering, Universiti Malaysia Perlis, Pauh Putra Campus, 02600 Arau, Perlis, Malaysia.

Abstract. The adhesive T-joint is most widely used in automotive, aircraft and fluidization bed system. In order to determine the effect of the temperature and adhesive thickness on the joint strength, a series of a tensile test and numerical analysis were conducted in this study. The T-joint specimens were fabricated using 304 stainless steel plate (SS 304), 304 stainless steel perforated plate and high-temperature epoxy adhesive. The series of adhesive T-joints with different adhesive thicknesses (0.5, 1.0, 1.5 and 2.0 mm) were tested in tension loading at room temperature (RT) and elevated temperature (55 °C, 75 °C, 100 °C and 125 °C). Additionally, the T-joint structure was modelled, simulated and analysed for stress assessment to predict the strength of adhesive T-joint at the various temperature and adhesive thicknesses. The average failure stress decreases with the increasing of the temperature. Nevertheless, the failure stress of the T-joint decreased with the increasing of the adhesive thickness. The finite element prediction for T-joint strength is in a good agreement with the experimental failure stress results.

Keywords. T-joint, temperature, adhesive thickness, finite element prediction.

1. Introduction

Historically, mechanical joining techniques were involved nails, screws, bolts and rivets. The adhesive bonding is inadequate compared to normal adhesive for binding the book, paper, woodworking, and packaging process [1]. The adhesive was based on the animals, rubber from the wood and tree resins. In 19th-century, due to the technical and systematic interest, the researchers improved the adhesion properties. Sometimes, materials or parts were operated in high-temperatures, high-pressures, high speeds or highly- erosive environment. Thus, the adhesive behaviour towards the applied situation needs to be characterized in order to get a properly designed joint. Another important parameter that needs to be emphasized is the effect of the adhesive thickness at the different temperatures on the joint strength [2].

Adhesive property is one of the basic processes in adhesion where it heavily influencing the strength of the joint [1]. Joint performance (operating environment, load), adhesive type, substrates type, cost, aesthetics, health necessities, application, manufacturing process and pre-treatments are several important aspects in selecting the adhesives in the fabrication process. Other than that, testing and validation are required in order to verify process quality [3].

Numerous studies on the effect of temperature on the adhesive behaviour have been conducted. The tensile test experiment on a urethane structural adhesive at different temperature i.e (50 °C, 60 °C and 70 °C) was investigated by [4]. The experimental results have demonstrated that stress-strain behaviour is dependent on the temperature. Apart from that, [5] investigated the effect of temperature on the cohesive properties of an epoxy adhesive. It was reported that the fracture toughness is fairly unaffected by the temperature from -40 °C to 80 °C. The glass transition of the epoxy adhesive investigated is 90 °C meanwhile the peak stress in peeling loading decreased monotonically with increasing temperature within this temperature range. At glass transition temperature, Tg there is a rapid reduction in both modulus and strength when the temperature increases, and leads the adhesive to no longer carry a substantial load.

2. E perimental detail

2.1. Materials

The adhesives used in this study was E 214 HP, which is one of the component structural Hysol line by Henkel. The adhesives recommended to be used with metals for high temperatures application. The glass transition temperature, Tg was provided for E 214 HP: 120 °C. The curing schedule for two hour at 120 °C was follow for adhesive. The properties for material used to fabricate the adhesive T-joint presented in Table 1.

Table 1: Material properties of adhesive and adherent

Material properties	Elastic modulus (MPa)	Poisson's ratio
304 Stainless steel	193000	0.31
Hysol Adhesive	1887	0.39

2.2. Specimen Fabrication

2.2.1. Adhesive T-joint Fabrication

In this research, a jig was prepared for the T-joint fabrication process. The jig was used to assist the alignment of the base plate and perforated plate with respect to each other during curing process and functions as a fixture to control the adhesive thickness such as shown in Figure 1. The thicknesses of T-joint adhesive are 0.5, 1.0, 1.5 and 2.0 mm respectively. All fabricated adhesive T-joint specimens were cured at 120 °C using a furnace for two hours as illustrated in Figure 2.

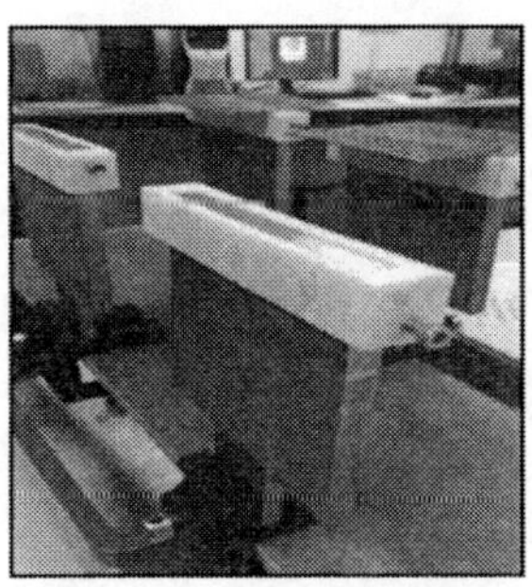

Figure 1. T-joint preparation

Figure 2. Curing of adhesive T-joint in furnace

2.3. Test Procedure

Tensile test for fabricated adhesive T-joint specimens was conducted by using Shimadzu AG-IS MS Universal Tensile Testing Machine (UTM) with a load capacity of 250 kN at temperature of RT, 55 °C,

75 °C, 100 °C and 125 °C at a constant crosshead speed of 1 mm/min. The thermostatic chamber model of TC- N300 SHIMADZU was used to retain the surrounding temperature for applied temperature test. The test temperature was set at the chamber, and a thermocouple was applied to the specimen to obtain an equilibrium of the sample. The specimen was then kept in the chamber at a chosen constant temperature for 30 minutes before conducting the tensile test. Five specimen of the adhesive T-joint with difference adhesive thicknesses; 0.5, 1.0, 1.5, 2.0 mm were undergo the tensile test at this elevated temperature.

3. Finite element modelling of Adhesive T-joint

Similar T-joint geometry model with various adhesive thicknesses used in the experimental testing were modelled using a SolidWork simulation software. The drawings of the T-joint models with selected adhesive thicknesses imported to ANSYS 14.5 software. The linear elastic isotropic and Multi-linear Isotropic Hardening (MISO) was set for adherent and adhesive respectively. In finite element (FE) analysis, the part of the modelling was divided into adherent (304 stainless steel) and adhesive (E 214HP Hysol). The mechanical properties of the adherent is referring to the Table 1. For the meshing size, the maximum sizes of adhesive and perforated plate are 0.5 mm and 1.0 mm for base plate (as shown in Figure 3) respectively. All the boundary condition of the modelling was set same as the experimental condition before run the finite element simulation (as shown in the Figure 4). In order to predict the failure of the adhesive T-joint, the threshold of the stress concentration was identified by using FE model analysis. The maximum Von Misses Stress of the adhesive T-joint was obtain by referring to the technical data from the experiment. The distribution of Von Mises Stress and deformation were calculated.

SS 304 plate

Adhesive layer

SS 304 perforated plate

Figure 3. Meshing

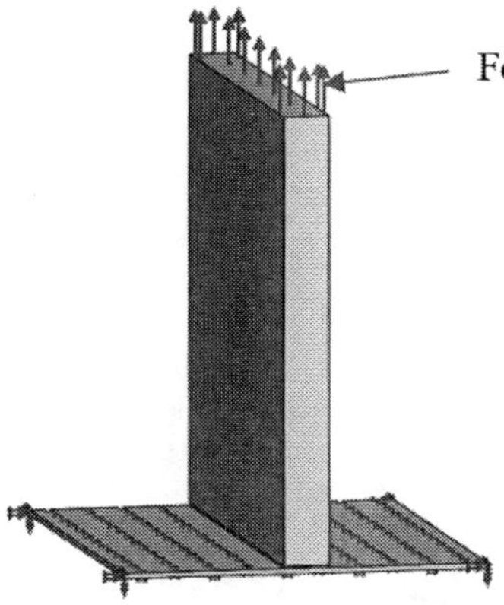

Figure . The loading condition force at T-joint

IOP Conf. Series: Materials Science and Engineering **429** (2018) 012093 doi:10.1088/1757-899X/429/1/012093

4. Result & Discussion

4.1. Results of the numerical analysis and the comparison between experimental results

There was a general trend of decrease in T-joint strength with increasing adhesive thickness. Meanwhile, the strength of the T-joint also significantly decreased with the increasing temperature. The optimum thickness is 0.5 mm while the T-joint being the least strong at 2.0 mm adhesive thickness. Hence, 0.5 and 1.0 mm adhesive thickness only have a small difference of failure load. Despites, larger flaws probability as thicker adhesive means potentially more defect in joint such as entrapped air, a gap at edges due to difficulty to apply adhesive at that area, not consistent mixing and many more flaws can possibly reduce overall strength of the joint seating [6]. In Figure 5, the results for the average experimental for the T-joint specimens at RT, 55 °C, 75 °C, 100 °C and 125 °C are depicted. The parameter attained it maximum failure stress at RT and the lowest is at 125 °C. It should be mentioned that at 125 °C, is exceeded the Tg values of the adhesive, which is the factor to explains the clear drastic drop in tensile strength of the T-joint. The results from the numerical analysis showed almost similar pattern with the experimental results.

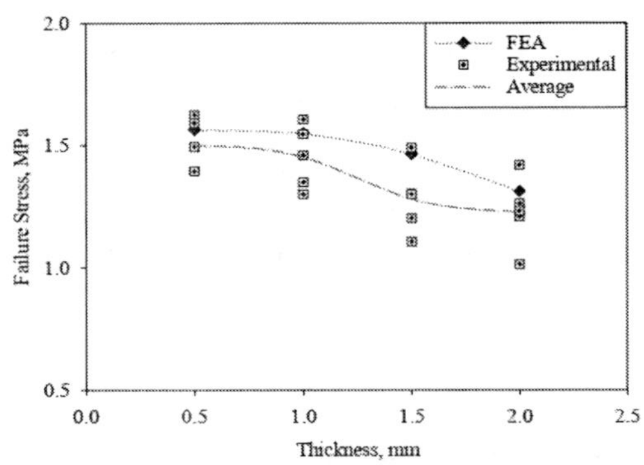

(a) RT

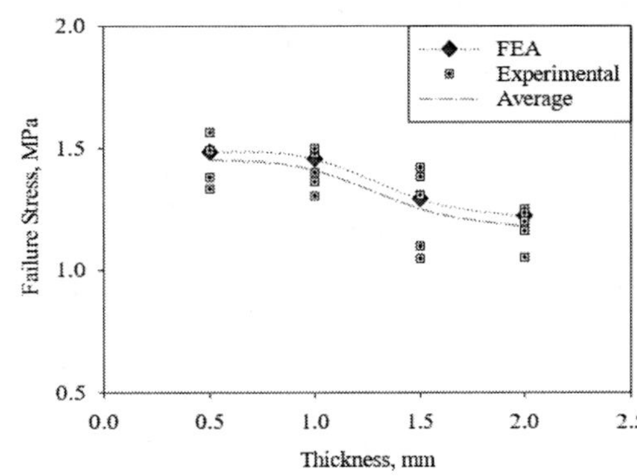

(b) 55 °C

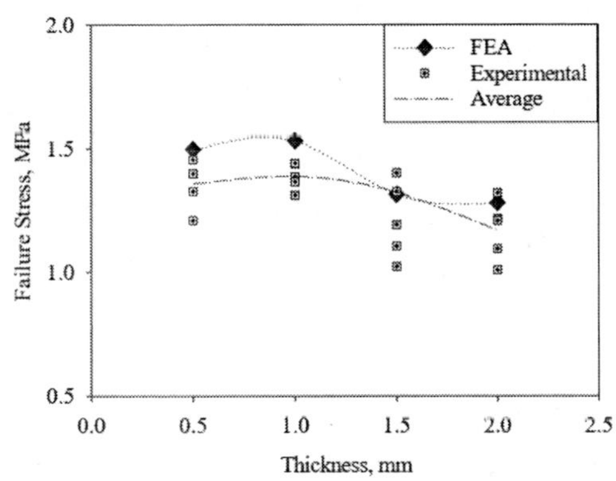

(c) 75 °C

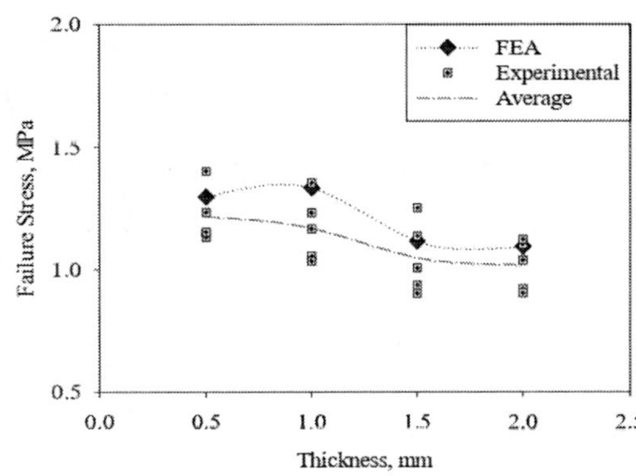

(d) 100 °C

IOP Conf. Series: Materials Science and Engineering **429** (2018) 012093 doi:10.1088/1757-899X/429/1/012093

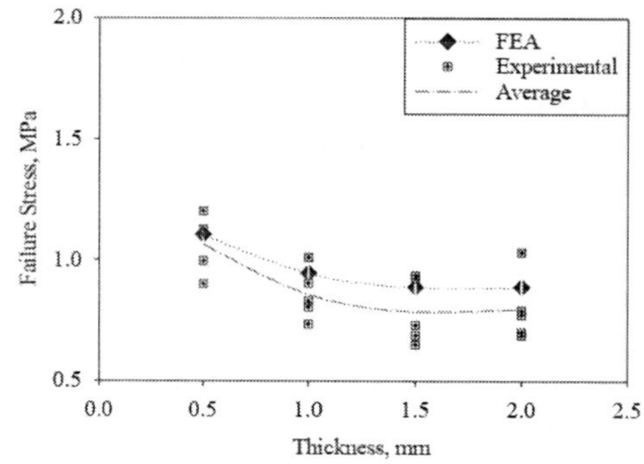

(e) 125 °C

Figure 5: Comparison of failure stress against adhesive thicknesses at a) RT, b) 55 °C, c) 75 °C, d) 100 °C and e) 125 °C

5. Conclusions

As a conclusion, the failure stress was found to be greatly related to the temperature and adhesive thickness. There are only small reductions of T-joint failure stress at RT until 100 °C. However, the failure stress of adhesive T-joint decreases significantly at temperature 100 °C due to the testing temperature already overpasses the Tg of the adhesive. Besides, the adhesive T-joint with adhesive thickness of 0.5 mm to 1.0 mm revealed the optimum value of adhesive T-joint strength for entire bond-line thickness. The FE analysis results for adhesive T-joint shows good agreement by having the same pattern in terms of failure stress in experimental results.

References

[1] A. Rudawska, "Adhesive Properties," 2012.

[2] Harris, J. A., & Fay, P. A. (1992). Fatigue life evaluation of structural adhesives for automative applications. *International Journal of Adhesion and Adhesives*, *12*(1), 9–18.

[3] R. D. Adams, F. M. Lucas, and S. Andreas, *Introduction to Adhesive Bonding Technology*. 2011.

[4] S. O. O. J. A. E. Park and K. M. Liechti, "Rate-Dependent Large Strain Behavior of a Structural Adhesive," pp. 143–164, 2003.

[5] T. Carlberger, A. Biel, and U. Stigh, "Influence of temperature and strain rate on cohesive properties of a structural epoxy adhesive," *Int. J. Fract.*, vol. 155, no. 2, pp. 155–166, 2009.

[6] G. Ji, Z. Ouyang, G. Li, S. Ibekwe, and S.-S. Pang, "Effects of adhesive thickness on global and local Mode-I interfacial fracture of bonded joints," *Int. J. Solids Struct.*, vol. 47, no. 18–19, pp. 2445–2458, 2010.

IOP Conf. Series: Materials Science and Engineering **429** (2018) 012094 doi:10.1088/1757-899X/429/1/012094

Validation of auto-generated material removal volume for regular form milling part

Pramod S Kataraki[a] and Mohd Salman Abu Mansor[a,*]

[a]School of Mechanical Engineering, Engineering Campus, Universiti Sains Malaysia, 14300 Nibong Tebal, Pulau Pinang, Malaysia.

*Corresponding author: mesalman@usm.my

Abstract. Feature recognition is the primary objective to be achieved for the attainment of generative CAPP system. The feature recognition works performed so far have been successful in recognizing and generating material removal volume for regular form features of regular form parts. These material removal volumes provide detailed information about the amount of material to be machined in each process thereby eliminating the detailed designing stage employed in conventional machining process. But the generated material removal volume (MRV) were not validated by real time mill machining, so an effort is made to validate the feature recognition works by real time mill machining using the material removal volume. From the real time mill machining it can be said that there is zero wastage of material and part obtained is of accurate dimensions.

1. Introduction

The integration of computer-aided design (CAD) and computer-aided manufacturing (CAM) has led the production industries into a new era of manufacturing, but still there are real time mill machining problems faced in manufacturing. The further integration of computer-aided process planning (CAPP) into CAD/CAM will be vital to perform process planning, process selection, tool selection [1,2] and achieve an efficient manufacturing process. The idea generation, design of part model, CAPP, CAM and CNC all together forms a CAPP system. The CAPP system can be of two types (i) variant type; (ii) generative type. The variant CAPP needs human intervention to perform part input and necessary modifications in retrieved process plans, and in generative CAPP very little human intervention is required to generate process plans [3]. Machining feature recognition, machining operation type selection, and process optimization are the functions of generative CAPP system [4]. The feature recognition has been the primary method being applied to recognize features [5] and feature based CAPP works have been widely performed to produce material removal volume to be input to CAM software for numerical code generation. The geometrical shape of B-rep parts were studied and cone faces having a hole feature in them were identified [6]. The chamfer, fillet, hole and pocket features present on symmetrical and non-symmetrical parts were also identified [7,8]. The features of regular form B-rep parts were recognized using feature recognition method and material removal volume data were generated in .SAT file format [9–11]. In the feature recognition of regular form B-rep parts conducted so far, there is a need for real time mill machining in order to validate the stock model volume and material removal volume generated by recognition work. In this paper an effort is made to perform real time mill machining so as to validate the feature recognition works performed so far.

2. Methodology

The following flow chart (Figure 1) has been adopted to perform real time mill machining of regular form milling part.

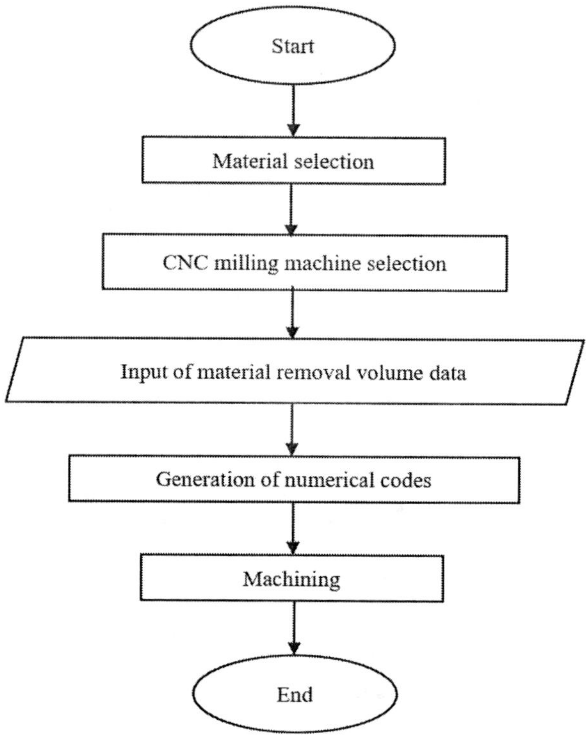

Figure 1. Flow chart.

2.1. Material and CNC machine selection

The material used for machining can be of aluminium or any other material based on the user requirements and the CNC milling machine can be 3 axis or 4 axis or 5 axis based on the complex shape of the part to be produced. In this work, aluminium material and 3-axis CNC milling machine are considered for machining of regular form part.

2.2. Input of MRV to generate numerical codes

In this section, an opener CAD part model (Figure 2) is considered for validation. The opener CAD part model is input to the algorithm developed by Kataraki and Abu Mansor [9] and material removal volume for various machining processes such as roughing and finishing were obtained (Figure 4). Figure 2 illustrates the CAD part model of opener. Figure 4 illustrates the material removal volume for roughing process and finishing process. These material removal volume data are input to AUTODESK FUSION 360 (free version for education purpose) software to generate numerical codes.

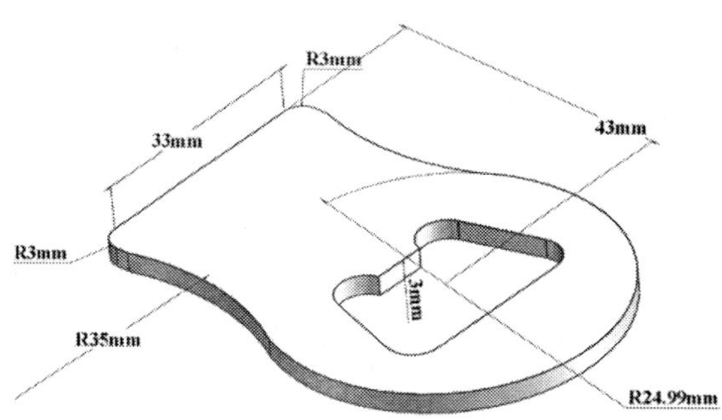

Figure 2. CAD part model of an opener.

IOP Conf. Series: Materials Science and Engineering **429** (2018) 012094 doi:10.1088/1757-899X/429/1/012094

The flat end mill tool is selected to mill the material removal volume for roughing and finishing process as shown in generated N codes-1 .TXT file format of Figure 3. The flat end mill tool is also selected to mill material removal volume for pocket (shown in N codes – 2 .TXT file of figure 3).

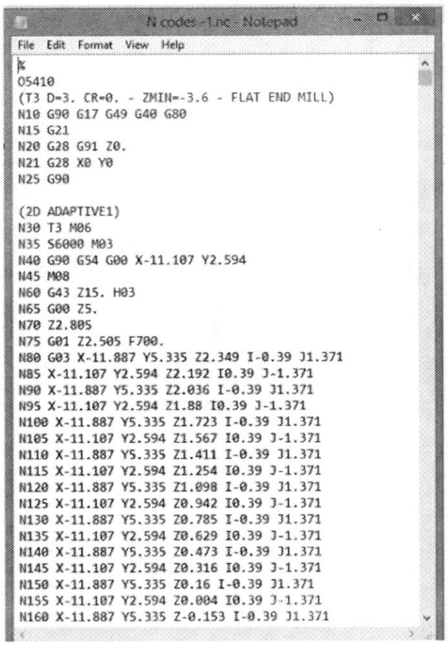

N codes -1.nc - Notepad

File Edit Format View Help

```
%
O5410
(T3 D=3. CR=0. - ZMIN=-3.6 - FLAT END MILL)
N10 G90 G17 G49 G40 G80
N15 G21
N20 G28 G91 Z0.
N21 G28 X0 Y0
N25 G90

(2D ADAPTIVE1)
N30 T3 M06
N35 S6000 M03
N40 G90 G54 G00 X-11.107 Y2.594
N45 M08
N60 G43 Z15. H03
N65 G00 Z5.
N70 Z2.805
N75 G01 Z2.505 F700.
N80 G03 X-11.887 Y5.335 Z2.349 I-0.39 J1.371
N85 X-11.107 Y2.594 Z2.192 I0.39 J-1.371
N90 X-11.887 Y5.335 Z2.036 I-0.39 J1.371
N95 X-11.107 Y2.594 Z1.88 I0.39 J-1.371
N100 X-11.887 Y5.335 Z1.723 I-0.39 J1.371
N105 X-11.107 Y2.594 Z1.567 I0.39 J-1.371
N110 X-11.887 Y5.335 Z1.411 I-0.39 J1.371
N115 X-11.107 Y2.594 Z1.254 I0.39 J-1.371
N120 X-11.887 Y5.335 Z1.098 I-0.39 J1.371
N125 X-11.107 Y2.594 Z0.942 I0.39 J-1.371
N130 X-11.887 Y5.335 Z0.785 I-0.39 J1.371
N135 X-11.107 Y2.594 Z0.629 I0.39 J-1.371
N140 X-11.887 Y5.335 Z0.473 I-0.39 J1.371
N145 X-11.107 Y2.594 Z0.316 I0.39 J-1.371
N150 X-11.887 Y5.335 Z0.16 I-0.39 J1.371
N155 X-11.107 Y2.594 Z0.004 I0.39 J-1.371
N160 X-11.887 Y5.335 Z-0.153 I-0.39 J1.371
```

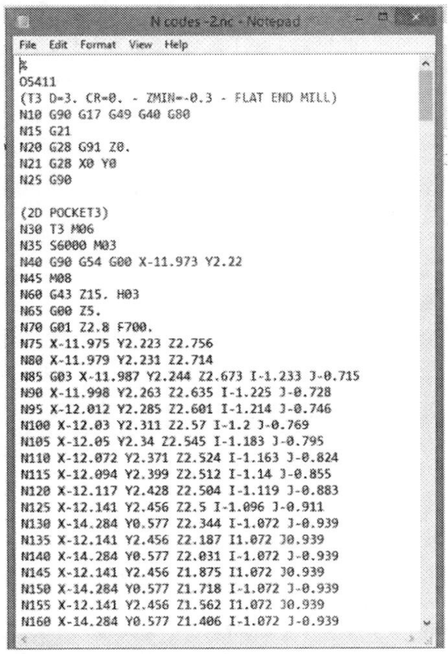

N codes -2.nc - Notepad

File Edit Format View Help

```
%
O5411
(T3 D=3. CR=0. - ZMIN=-0.3 - FLAT END MILL)
N10 G90 G17 G49 G40 G80
N15 G21
N20 G28 G91 Z0.
N21 G28 X0 Y0
N25 G90

(2D POCKET3)
N30 T3 M06
N35 S6000 M03
N40 G90 G54 G00 X-11.973 Y2.22
N45 M08
N60 G43 Z15. H03
N65 G00 Z5.
N70 G01 Z2.8 F700.
N75 X-11.975 Y2.223 Z2.756
N80 X-11.979 Y2.231 Z2.714
N85 G03 X-11.987 Y2.244 Z2.673 I-1.233 J-0.715
N90 X-11.998 Y2.263 Z2.635 I-1.225 J-0.728
N95 X-12.012 Y2.285 Z2.601 I-1.214 J-0.746
N100 X-12.03 Y2.311 Z2.57 I-1.2 J-0.769
N105 X-12.05 Y2.34 Z2.545 I-1.183 J-0.795
N110 X-12.072 Y2.371 Z2.524 I-1.163 J-0.824
N115 X-12.094 Y2.399 Z2.512 I-1.14 J-0.855
N120 X-12.117 Y2.428 Z2.504 I-1.119 J-0.883
N125 X-12.141 Y2.456 Z2.5 I-1.096 J-0.911
N130 X-14.284 Y0.577 Z2.344 I-1.072 J-0.939
N135 X-12.141 Y2.456 Z2.187 I1.072 J0.939
N140 X-14.284 Y0.577 Z2.031 I-1.072 J-0.939
N145 X-12.141 Y2.456 Z1.875 I1.072 J0.939
N150 X-14.284 Y0.577 Z1.718 I-1.072 J-0.939
N155 X-12.141 Y2.456 Z1.562 I1.072 J0.939
N160 X-14.284 Y0.577 Z1.406 I-1.072 J-0.939
```

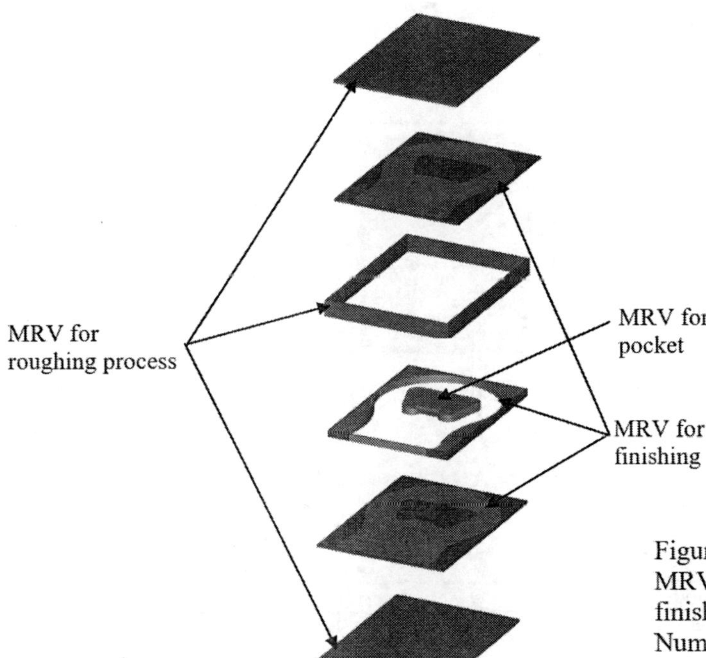

Figure 3. Illustrates the MRV data for roughing and finishing process, and their Numerical codes.

2.3. Machining

As per the stock model details generated by the algorithm [9] the stock model volume required for opener CAD part model is 18190.3 mm^3 (Figure 5). The .SAT files of machining processes are converted to .IGES format to be input to AUTODESK FUSION 360 (free version for education purpose) software so as to generate NC codes. The work piece of aluminium material is machined in 3-axis CNC milling machine. The end milling process is done to remove bottom and top layers of MRV for roughing and finishing. Then the pocketing is performed (Figure 4) to obtain pocket feature and produce opener.

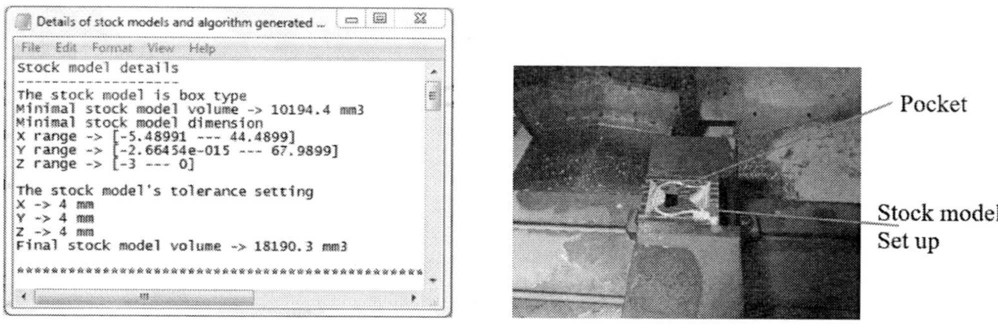

Figure 4. Details of stock model (.TXT file format) and Stock model set up.

3. Results

Based on the suggested stock model volume, the aluminium material volume is taken and milled in 3-axis CNC milling machine using the numerical codes generated from material removal volume. The produced opener is shown in Figure 5. The dimensions of the opener were measured to be 68mm length, 50mm width and 3mm thickness. The dimensions of CAD part model of opener input to algorithm to generate material removal volume are shown in Table 1. The dimensions of stock model generated by algorithm are also shown in Table 1.

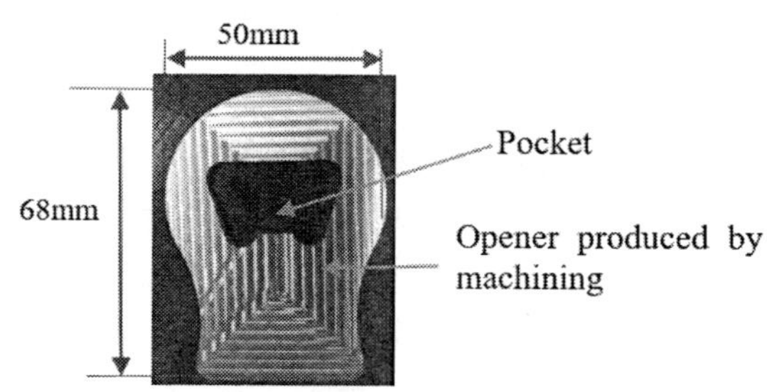

Figure 5. Final product of opener.

Table 1. Illustrates the dimensions of opener CAD part model, stock model, and final product opener.

Models	Dimensions (mm)
Opener CAD part model	68mm length × 50mm width × 3mm thick
Stock model	71.35mm length × 51mm width × 5mmthick
Final product opener	68mm length × 50mm width × 3mm thick

4. Conclusion

The time taken to generate MRV data is approximately 12 s and the machining time to produce opener is approximately 10 min, so the overall time taken to produce an opener is less than 30 min. The results show dimensions (68mm length × 50mm width × 3mm thick) of opener produced by milling are same as its CAD model input to the algorithm to obtain stock model and material removal volume. Hence, it can be concluded that the regular form part can be produced as per the required volumetric dimensions by using auto-generated material removal volume. Thus eliminating the detailed designing stage which is adopted in conventional machining.

Acknowledgment

This research is supported by the Ministry of Higher Education Malaysia under the Fundamental Research Grant Scheme (FRGS) (Reference No. 6071369). The first author also would like to thank the support of National Overseas Scholarship by Ministry of Tribal Affairs, New-Delhi, India.

References

[1] Chougule P D, Kumar S and Raval H K 2014 Relating Product Manufacturing Decisions to Environmental and Cost Performance Using CAPP *Procedia Mater. Sci.* **6** 476–81

[2] Peng Q, Hall F R and Lister P M 2000 Application and evaluation of VR-based CAPP system *J. Mater. Process. Technol.* **107** 153–9

[3] Xu X, Wang L and Newman S T 2011 Computer-aided process planning – A critical review of recent developments and future trends *Int. J. Comput. Integr. Manuf.* **24** 1–31

[4] Su Y, Chu X, Zhang Z and Chen D 2015 Process planning optimization on turning machine tool using a hybrid genetic algorithm with local search approach *Adv. Mech. Eng.* **7** 1–14

[5] Zhang X, Nassehi A and Newman S T 2014 Feature recognition from CNC part programs for milling operations *Int. J. Adv. Manuf. Technol.* **70** 397–412

[6] Zubair A F and Abu Mansor M S 2018 Hole Feature on Conical Face Recognition for Turning Part Model *IOP Conf. Ser. Mater. Sci. Eng.* **328** 12026

[7] Zubair A F 2018 Automatic Feature Recognition of Chamfer and Fillet for Turning Part Model by Volume Decomposition Method *J. Mech. Eng.* **5** 204–16

[8] Zubair A F and Abu Mansor M S 2018 Automatic feature recognition of regular features for symmetrical and non-symmetrical cylinder part using volume decomposition method *Eng. Comput.* **0** 1–21

[9] Kataraki P S and Abu Mansor M S 2017 Auto-recognition and generation of material removal volume for regular form surface and its volumetric features using volume decomposition method *Int. J. Adv. Manuf. Technol.* **90** 1479–1506

[10] Kataraki P S and Abu Mansor M S 2018 Auto-recognition of surfaces and auto-generation of material removal volume for finishing process *IOP Conf. Ser. Mater. Sci. Eng.* **328** 12027

[11] Kataraki P S and Abu Mansor M S 2018 Identification of Freeform Depression Feature in a Part Using Vertex Attributes From Feature Volume *J. Mech. Eng.* **5** 217–31

IOP Conf. Series: Materials Science and Engineering **429** (2018) 012095 doi:10.1088/1757-899X/429/1/012095

Characteristics of Iron powder when Pressed using Explosive Pressing method

Nor Azmaliana Azmi[1, a], Ahmad Humaizi Hilmi [1, b], M. Alias Yusof [2] and Ariffin Ismail[2]

[1]School of Manufacturing Engineering, Universiti Malaysia Perlis
[2]Faculty of Defence Studies & Management, Universiti Pertahanan Nasional Malaysia

[a] norazmaliana@studentmail.unimap.edu.my

[b] humaizi@unimap.edu.my

Abstract. Fine iron powder was pressed by Explosive Pressing using type of explosive; emulsion explosive (Emulex grade 150). The green density, hardness, strength and microstructure of iron compact was studied. The explosive is molded into shape charge to give more pressing energy. The underground explosion was conducted since confined space can give more blast energy. Green compact was sintered at 1000°C to strengthen the bonding between particles. Analysis of compact powder microstructure was performed using scanning electron microscope (SEM). The hardness of compact powder was tested using Rockwell Digital. Universal testing machine was used to conduct the tensile test. The result shows the hardness of compact powder for explosive pressing was 45.8 HRA to 50.2 HRA. Microstructure results show lesser porosity when the density was increased from 6.046 g/cm^3 to 7.298 g/cm^3. There was a presence of cold welding and mechanical interlocking phenomena at surface of specimen.

1. Introduction

Powder metallurgy is a rapidly growth technology. Development of Powder Metallurgy was growing rapidly along with the growth of automotive, machinery, and the other high end equipment industries. The fabrication technology of Powder Metallurgy iron base parts with high density, high strength and high precision has make a development trend in Powder Metallurgy industry [1][2][3]. Recent development of Powder Metallurgy technology such as, warm compaction (WC), surface densification (SD), hot isostatic pressing (HIP), High Velocity Compaction (HVC) and explosive compaction contributes to produce Powder Metallurgy components with improved mechanical properties and microstructure. Some researches, J.Wang et.al and Khan et al., stated that, the explosive compaction method is restricted due to problems to get the explosive material and low productivity [4][5]. The explosive material can be obtained from Tenaga Kimia Sdn Bhd in Rawang which is the sole company in Malaysia that produces emulsion explosive (Emulex grade 150). As stated, people are not easy to get these explosive materials if does not have a valid license. Therefore, the Institute of Quarrying Malaysia has created a Short Firer course that enables people to obtain a short firer license to get explosive materials. The acquired license may only be used to purchase explosive materials for certain purposes (research study) only and there will be certain permitted conditions.

2. Experimental Procedure

In this research work, pure iron powder without any addictive or lubricant was used to produce iron compact using explosive pressing method. Iron powder was supplied by Sigma-Aldrich, Sweden. The characteristics of this powder was given in Table 1. Then, mass of iron filled in die during pressing was 59.71g for each sample. The iron powder was packed in the die with height 10mm and diameter 50mm of each sample.

Figure 1 shows the graphic illustration setup of explosive pressing method. The explosive pressing setup consists of three main parts; shape charge (explosive), plunger and die. All part of explosive pressing except shape charge was made of P20 steel. The explosive material used in this research was emulsion explosive (Emulex grade 150) with a detonation velocity of 4000 to 5500 m/s. The explosive material was moulded into shape charge since more of the explosive acts in the desired direction, increasing effectiveness, reducing materials and decreasing protection requirement [6]. By ignition the detonator, pressing process was done successfully. Several samples were produced by this method. In order to bring out the sample from the die, the Universal Testing Machine was used to eject the sample from the die. After obtaining the sample, some of their properties are measured.

The green density of the sample was taken using Archimedes principle according to ASTM standard B962 [7]. Then, green samples were sintered at 1000°C respectively. The morphology of the sintered and polished sample was analysed using Scanning Electron Microscope (SEM) model (*Hitachi TM-3000*) Tabletop Microscope. The hardness of the samples was tested using Rockwell Digital Hardness machine. The force that applied to the specimen with a load of 60kg at four parts of the surface area of the specimen using diamond cone indenter. The indent time is 10s.

Table 1. Characteristic of Iron powder

Density (g/cm^3)	Particle size (μm)	Melting point (°C)	Boiling point (°C)
7.86	<10	1535	2861

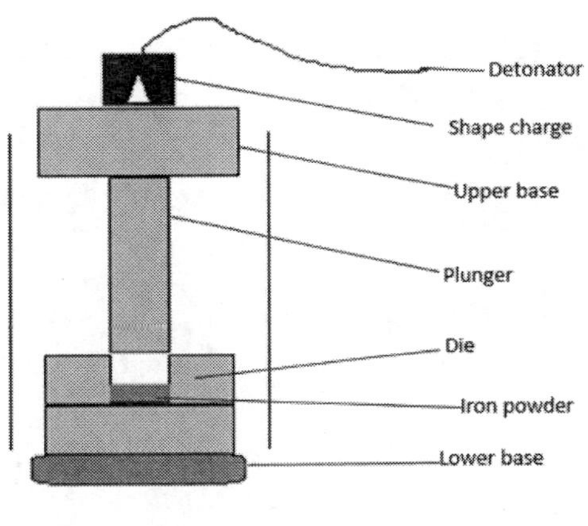

Figure 1. Schematic illustration of explosive pressing apparatus

3. Results and Discussions

3.1. Densification behavior

There are three specimens that had been pressed using Explosive Pressing. The first specimen is pressed with 500g of emulsion explosive (Emulex), while the second specimen is pressed using 750g of emulsion explosive (Emulex) and the third specimen is pressed using 1000g of emulsion explosive (Emulex). Figure 2,3 and 4 shows the green compact of Iron powder after blast test. It can be observed that; the Iron powder had changed its shape from powder to solid state for all specimens.

Specimen in Figure 2 shows that the green compact is fully compact and does not had crack, while the specimen in Figure 3 shows the surface is smooth and fully compact and specimen in Figure 4 indicates that the green compact also is fully compact during the explosion. The result proved that, the force generated during the experiment is very high until it can press the specimen into a solid metal.
Figure 5 displays the relation of pressing force with green density of specimen. The green density of iron compact increase with the increasing of pressing force. From the graph, at 500g of explosive, the green density of specimen was 6.046 $g/cm^3$with pressing force 13292.252 tonne. The green densities of specimen rise by 36.74 % at 750g of explosive with pressing force 33377.738 tonne. It also can be seen that, the maximum green density that can be obtained from this experiment was 7.298 g/cm^3 with amount of explosive 1000g and pressing force 33991.911 tonne.
As a conclusion, the greater the amount of explosive used, the greater the green densities of Iron compact due to the higher of force generated. Furthermore, the rise in pressing force leads to rise of densification of the Iron compact (Khan et al., 2014) (J Z Wang, Qu, et al., 2009). The densification of Iron compact from this experiment was achieved through a pressing force that created by the explosion energy that transferred to plunger to press the Iron powder. Obviously, the explosive pressing method was better than High-Velocity compaction method since the force generated from explosive pressing is higher than force generated by High-Velocity Compaction which was 1615 N/mm^{-2} that equal to 80.81 ton(Khan et al., 2014).

Figure 2. Green compact of Iron (Specimen 1)

Figure 3. Green compact of Iron (Specimen 2)

Figure 4. Green compact of Iron (Specimen 3)

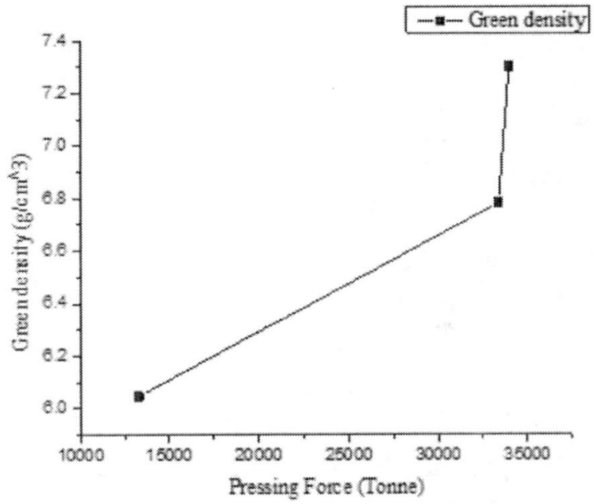

Figure 5. Graph of Green density vs Pressing force

3.2. SEM microstructure

The morphology of the SEM micrograph of Iron was presented in Figure 6,7 and 8. It can be seen that the particles are linked together and show a few pores. The SEM micrograph also shows low porosity since the particles are strongly bond after the sintering process. Besides, the phenomena of cold welding also appear among powder particles due to the less visible of grain boundaries after sintering process. Cold welding in the particle contact contributes to the development of strength in the compact [8].

From Figure 6 to 8, It can be clearly seen that, as the amount of explosive was increased, the porosity decreases, because high pressing force can produce the effective binding of particles. At 500g of explosive, there are many big pores and small pores appears on the SEM image. When the amount of explosive was increased to 750g, the pores are reduced. Then, at the highest amount of explosive which was 1000g, there was only small pores exist on the SEM image.

The strength of compact powder was depending on two phenomena which were cold welding and mechanical interlocking. The phenomena of cold welding occur at high pressing pressure. From figure, it can be obtained that, at 500g of explosive, the cold welding phenomena not appears since the grain boundary is visible. Then, at 750g and 1000g of explosive shows less visible of grain boundaries, indicating the present of cold welding phenomena.

As a conclusion, as the amount of explosive increased, the pressing force increased which resulted to high hardness of specimen. The density of specimen increased with reduction of pores and increment of cold welding effects. Thus, specimen gain strength with better resistance to deformation after the pressing forces was applied.

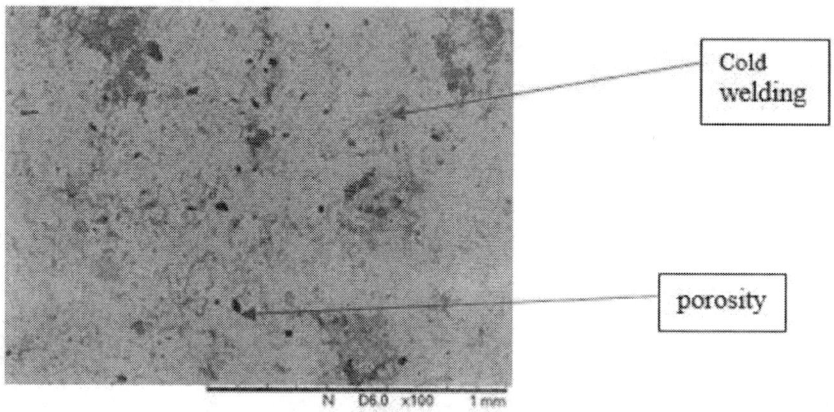

Figure 6. SEM microstructure of 500g of explosive

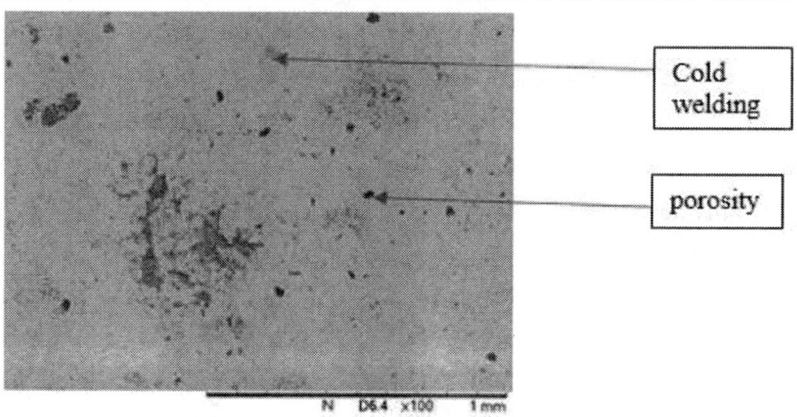

Figure 7. SEM microstructure of 750g of explosive

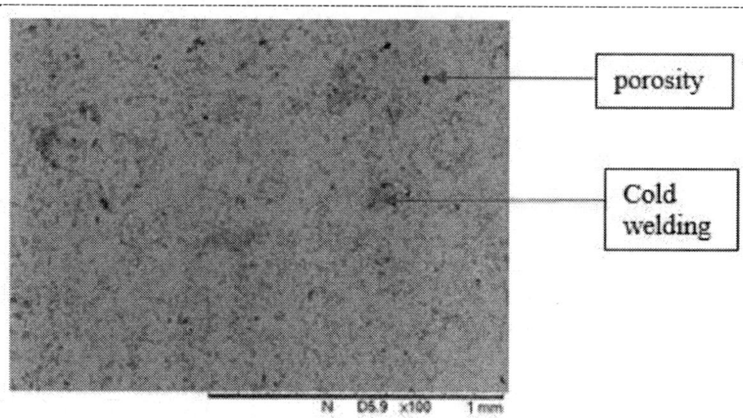

Figure 8. SEM microstructure of 1000g of explosive

3.3. Effect of explosive on Rockwell Hardness

Figure 9 shows the relationship between pressing force and the hardness of sintered specimen. From the graph, the hardness of sintered specimen increased proportianally with an increased in pressing force. The increasing of hardness shows a good result of compressibility of the Iron powder particles during pressing process (H. Li et al., 2014).

At 500grams of explosive, the hardness of sintered specimen was 45.8 HRA with pressing force 13292.252 tonne. When the explosive was increased to 750grams, the hardness of sintered specimen was increased to 46.7 HRA with pressing force 33377.738 tonne. The maximum hardness of sintered specimen was obtained for this experiment was 50.2 HRA at 1000grams of explosive and pressing force

33991.911 tonne. Figure 10 shows the graph of Hardness of sintered specimen vs Green densities of specimen. The increasing of hardness accompanied with an increase of pressing force is due to the increasing of density of specimen (Khan et al., 2013).

It can be concluded that, the higher the amount of explosive, the higher the hardness of Iron that formed due to its good compressibility of the powder particles during pressing. The increased in hardness accompanied with an increased in pressing force is to be attributed with the decreasing of porosity and increases in density (Khan et al., 2013). Besides, there is also some deformation in the compaction or pressing process that cause the hardness properties of the specimens increased.

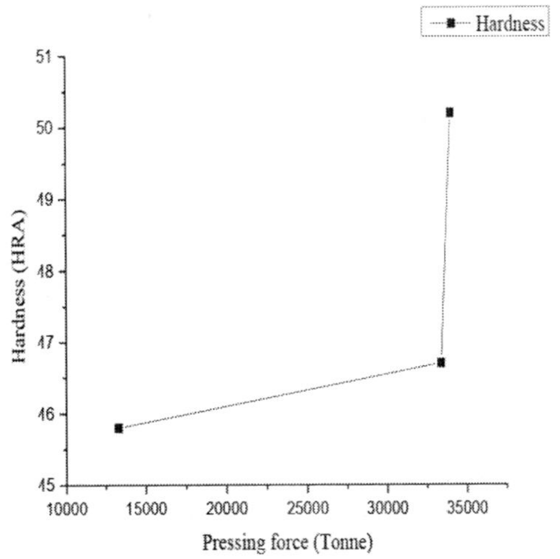

Figure 9. Graph of Hardness vs Pressing force

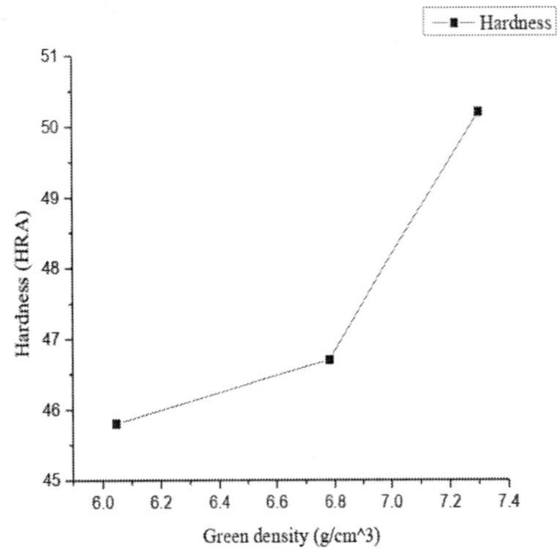

Figure 10. Graph of Hardness vs Green density

4. Conclusion

This research shows that it is possible to produce high density and high hardness of compact powder using Underground explosion. Since the results indicated the high density of compact powder which is $7.298g/cm^3$, high hardness of compact powder which is $50.2HRA$ at pressing force 33991.911 tonne and the SEM microstructure of compact powder shows the high quality bonding between the powder particles. Therefore, it can be concluded that, Underground explosion pressing is better to produce high quality of compact powder than conventional pressing.

5. References

[1] H. Zhang, L. Zhang, G. Dong, Z. Liu, M. Qin, X. Qu, and Y. Lü, "Effects of annealing on high velocity compaction behavior and mechanical properties of iron-base PM alloy," *Powder Technol.*, vol. 288, pp. 435–440, 2016.

[2] G. Sethi, E. Hauck, and R. M. German, "High velocity compaction compared with conventional compaction," *Mater. Sci. Technol.*, vol. 22, no. 8, pp. 955–959, 2006.

[3] B. Bos, C. Fors, and T. Larsson, "Industrial implementation of high velocity compaction for improved properties," *Powder Metall.*, vol. 49, no. 2, pp. 107–109, 2006.

[4] J. Z. Wang, X. H. Qu, H. Q. Yin, M. J. Yi, and X. J. Yuan, "High velocity compaction of ferrous powder," *Powder Technol.*, vol. 192, no. 1, pp. 131–136, 2009.

[5] D. F. Khan, H. Yin, Z. Usman, M. Khan, X. Yuan, W. Wang, and X. Qu, "Improvement of a high velocity compaction technique for iron powder," *Acta Metall. Sin. (English Lett.*, vol. 26, no. 4, pp. 399–403, 2013.

[6] B. Burch, "Determining and mitigating the effects of firing a linear shaped charge under

water," 2014.

[7] M. M. M. M. A. FATHY, Omyma EL-KADY, "Effect of iron addition on microstructure, mechanical and magnetic properties of Al-matrix composite produced by powder metallurgy route," *Trans. Nonferrous Met. Soc. China*, vol. 25, no. 1, pp. 46–53, 2015.

[8] P. Jonsén, "Fracture and Stress in Powder Compacts," 2006.

Acknowledgement

This work is funded by FRGS grant name Characteristic of powder surpassing 160,000-ton press using explosive pressing (FRGS 9003-00542). This project was done with the help of Lt Col Associate Professor Ariffin bin Ismail and blast research team from Universiti Pertahanan Nasional Malaysia.

Characteristic of Solid Metal using Underground Explosion Pressing

Nor Azmaliana Azmi[1, a)], Ahmad Humaizi Hilmi [1, b)], M. Alias Yusof [2,] and Ariffin Ismail[2]

[1]School of Manufacturing Engineering, Universiti Malaysia Perlis
[2]Faculty of Defence Studies & Management, Universiti Pertahanan Nasional Malaysia

[a)]norazmaliana@studentmail.unimap.edu.my

[b)]humaizi@unimap.edu.my

Abstract. This paper presents the results of an experimental study aim at enhancing the understanding of some characteristics of an interior explosion within a confined underground area on specimen used, as well as the effect of the explosive charge size on the pressure distribution along the compacted specimen which is placed in the apparatus. The study involved blast test which provides a measured pressure based on the amount of explosive used. The type of explosive used for this experiment is Emulex with detonation velocity range from $4500m/s$ to $5400m/s$. The microstructure of specimen A and B is analyzed using Scanning Electron Microscope (SEM). Then, the hardness of both specimens is tested using Rockwell Digital Hardness and the result for specimen A is $33.8HRA$, while for specimen B is $43.7HRA$. Besides that, the bending strength of specimen A and B had been tested using Calcium Carbide Explosive Shape charge and the force that applied to the specimen B during the bending strength is $243{,}160.68kN$, while for specimen A, there is no effect of bending strength test.

1. Introduction

Heavy duty press machine commonly used to compress heavy components for aircraft, spacecraft, and power-generation facilities. China company has built a 160,000tons heavy duty press machine which absolutely leads to high cost. Other modern countries that have heavy duty press exceeding 50,000 tons are Germany, Korea, Russia and Japan This research aim is to create a new method for pressing that used commercial explosive (Emulex). The energy that produces from the explosive can replace the force from the heavy duty press machine. Ideally, the more energy produced from the explosive, the stronger the material will be.

This research used Underground explosion method to press the solid metal (mild steel). Underground explosion method is selected since confined space can give more blast energy. An underground explosion of compact materials has been studied by the previous researcher[1,2]. Several things that have been investigated including the dynamic behaviors of concrete at high pressures and craters of soil that formed. But, the underground explosion comes with the problem of shock loading and high cost. Underground explosion method has been used in the study on dynamic behavior of concrete at high strain and pressure.

Confined explosions have an extreme effect which can lead to serious damage compare to the open air explosion. The geometrical factors of the space where the explosion occurs such as the opening size of

confined space, geometrical dimensions, and the explosive location influence the effect of the confined explosion [3,4]. The analysis of internal explosion in confined space has been studied by the previous researcher that focus on the pressures at the inner space boundary and the response analysis of the boundary structural elements. This analysis can assess the induced damage and predict the elements' response in confined space. There is one problem in confined space that is the shock wave interaction with the boundaries of the confined space. The force from an explosive charge detonated within a confined space consists of two stages. The first stage consists of the initial high pressure, a short period of reflected wave and several following reflected shocks while the second stage consists of a slowly decaying pressure. Previous research has presented the importance of the role of confined space with various geometries in an explosion [5,6,7]. The explosion of shaped charges inside the confined space creates conditions for combustion of aluminum particles. After the explosion of charges in the confined chamber filled with air and argon, different forms of the spectra were obtained. The explosion effect is strongly depending on the confinement volume and the charge type. Other researcher studied the suppression consequence of water mist for explosion fire in a confined space through experiment and the results shows that the accelerating flame overpressure can be reduced by water mist droplets.
Besides that, a shaped charge explosive method also takes into account as a method to generate more pressure impact during the blast test. Shaped charge is explosive with high penetration ability and is used mainly for military purpose. In military applications, shaped charges used against various types of armor, particularly as anti-tank devices. Shaped charges come in a wide variety of geometry and the geometric most common charge is conical shaped and hemispherical shape, but the best use is a conical shape, as it will provide greater penetration ability[8,9].

2. Experimental Procedure

2.1. Material selection

Explosive blast generates a heavy duty pressing compaction. For that reason, it must take into relevant aspects of the parameter apparatus to ensure that it can withstand the blast load from the explosion. Moreover, an important role is the thickness, size, and amount of the materials to be used also acts as the appropriate of a parameter to accommodate heavy duty pressing. The material use for this explosion compaction are mild steel for the cylinder and the upper and lower part is P20. Steel are containing a small percentage of carbon, strong and tough.
In the underground explosion, the apparatus used to consist of several parts which are the lower base, bolt and nut, mold, piston and upper base. The design of this apparatus was created by the previous designer. Figure 1 shows the main assembly of the apparatus for underground explosion while Figure 2 display the explode diagram of an assembly for the underground explosion apparatus.

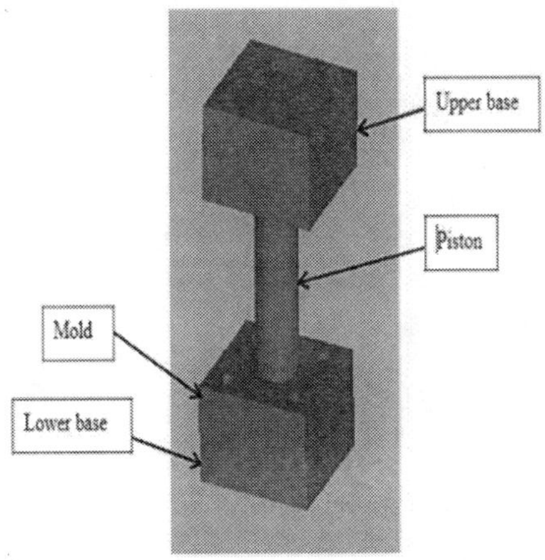

Figure 1. The main assembly of apparatus

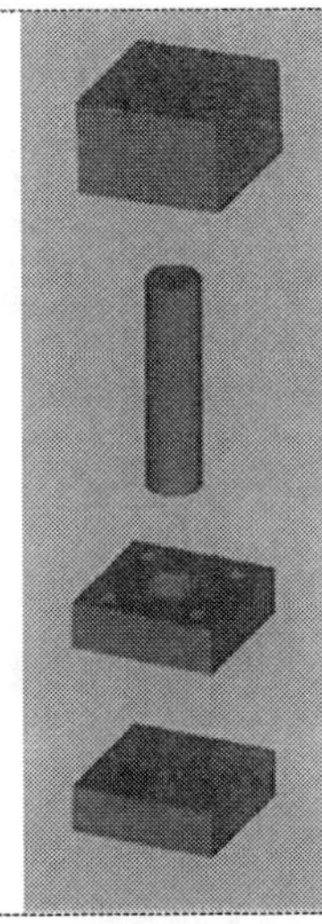

Figure 2. Explode diagram of assembly apparatus

2.2. Preparing shape charge

Type of explosive that are used for this experiment is commercial explosive (Emulex) with weight $500g$ and $750g$. The first step to prepare the shape charge is by weight the Emulex using kitchen scale. Then, Emulex explosive has been placed and compacted in the pipe tube with same diameter $21.9mm$ but different height which is $65mm$ for $500g$ and 150 mm for $750g$. Figure 3 and 4 shows the preparation of Emulex shape charge. The type of shape charge used is conical shape charge, since conical shape provide greater penetration ability.

Figure 3. Preparation of shape charge (a)

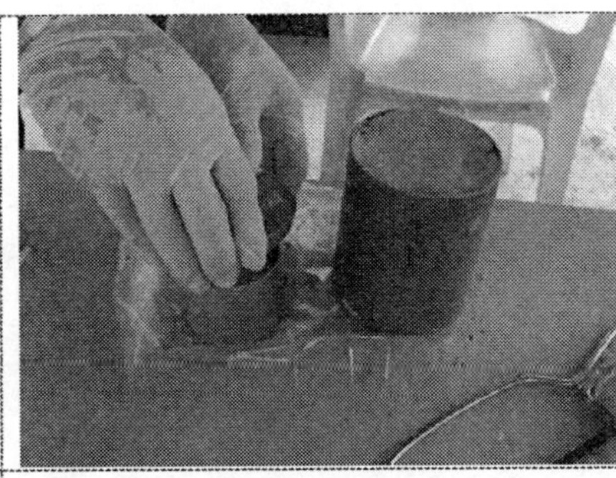

Figure 4. Preparation of shape charge (b)

2.3. Method of Underground Explosion

Preparation of underground explosion is start with the specimen that is used to press using underground explosion. There are two specimens (A and B) with both height $10mm$and diameter $25mm$. The material for the specimens is mild steel. Figure 5 shows the setup of the underground explosion. During the blast test, a hole with depth $0.5m$ is dig using excavator. Then, placed the underground explosion apparatus in the hole and make sure the apparatus is set up in vertical position, and at the same time, the specimen is put in the mold. After that, the shell cover is placed to prevent the apparatus get away from the original position. The conical shape charge is located on the top of underground explosion apparatus and at same the time, the detonator wire is connecting to shape charge. Next, soil is land fill until it fully covers the apparatus. Lastly, three sand bang with weight $50kg$ each is place on the soil to give more impact to the explosion. Then, high speed camera is set to capture the moment of blast test.

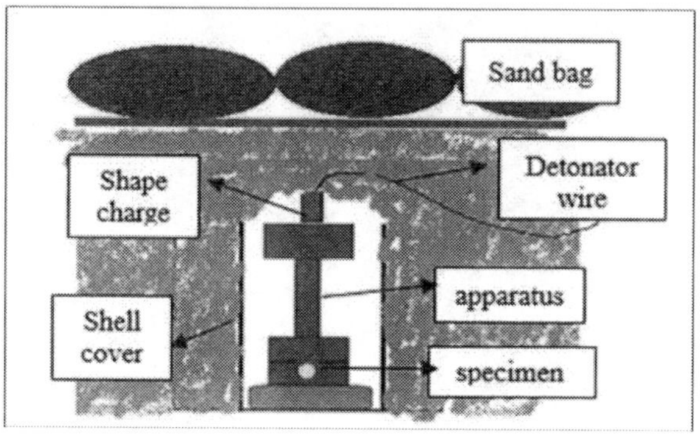

Figure 5. Schematic diagram of Underground Explosion pressing

2.4. Test method

Universal Testing Machine is used to eject the specimen that is overly attached to the mold. After that, the specimen had been cleaned and polish using sand paper. The diameters and thickness of the specimen were measured using Vernier calipers. The hardness of the specimen was tested using Rockwell Digital Hardness machine. Then, the microstructure of the specimen surface was analyzed using Scanning Electron Microscope (SEM). The last method to test the specimen is bending strength. Bending strength is done by using Calcium Carbide Explosive Shape Charge. Figure 6 shows the Calcium Carbide Explosive Shape Charge. The first step to carry out this experiment is the preparation of shaped charge explosive which is made of carbide fill up inside plastic with a cone shape and cover with modelling clay as shown in Figure 6(a). The amount of calcium carbide used to fully fill the shape charge mold is 16g for specimen A and B. Figure 6(b) shows specimen is placed in simply supported condition so that the metal plate get space to bend between the gap after experienced energy from the explosion. Figure 6(c) shows the shaped charge explosive, is placed on top of the specimen to get the result of the explosion.

Bending strength for both specimens can be calculated using Eq. (1):

$$\Delta \max = \frac{-PL^3}{48EI} \tag{1}$$

Where Δmax is deflection, P is the load, N, L is the length of the specimen, m, E is the Young Modulus of mild steel $(210 \times 10^9 Pa)$, I is the moment of Inertia of specimen, m^4

Moment of Inertia of the specimen can be calculated using Eq. (2):

$$I = \frac{1}{4}\pi r^4 \tag{2}$$

Where I is the moment of Inertia of specimen, m^4, r is the radius of the specimen, m.

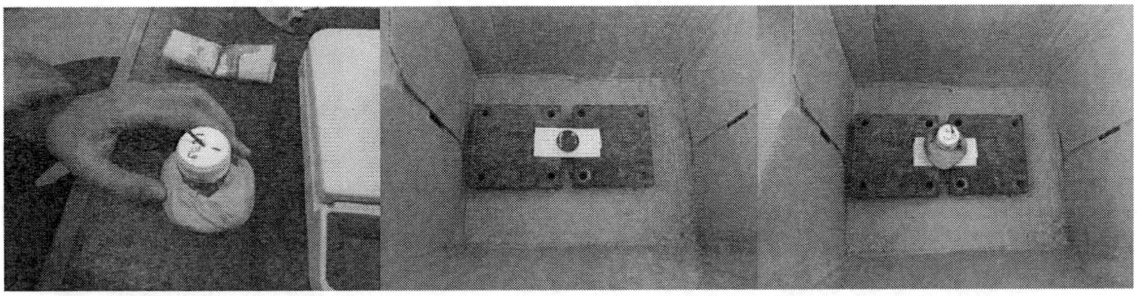

(a) (b) (c)

Figure 6. (a)Calcium carbide explosive shape charge (b)Specimen placed in simply supported condition (c)Shape charge is placed on top of specimen

3. Results and Discussions

3.1. Hardness of specimen

The hardness result for specimen A is $33.8HRA$, while the hardness for specimen B is $43.7HRA$. Specimen B shows the highest hardness since specimen B get more impaction during the explosion. Table 1 shows the result of both specimens before and after explosion. Specimen A with weight of explosive used, $500g$ shows the thickness is decrement from $10mm$ to $5.7mm$, while the diameter is increase from $25.5mm$ to $33.5mm$. For specimen B, with weight of explosive $750g$, shows the thickness of specimen is decrease from $10mm$ to $3.9mm$, while the diameter is increase from $25mm$ to $41.6mm$. So, as a conclusion, when more explosive is used, the thickness of specimen will be decrease and the diameter of the specimen will be rise.

Table 1. Result of both specimens before and after explosion

Specimen	Explosive used	Before explosion	After explosion
A	500grams	 Thickness: $10mm$ Diameter : $25mm$	 Thickness: $5.7mm$ Diameter: $33.5mm$
B	750grams	 Thickness: $10mm$ Diameter : $25mm$	 Thickness: $3.9mm$ Diameter: $41.6mm$

3.2. SEM microstructure

Figure 7 shows the SEM microstructure of mild steel, specimen A, and B after the explosion. From Figure 7(a) and Figure 7(b) there are slightly difference in the microstructure of the steel surface after pressed using $500g$ and $750g$ of Emulex. From Figure 7(a), the porosity is more on the surface of the specimen compared to Figure 7(b). The surface of the metal is dusty with the presence of impurities such as the rusting effect and scratches effect after polishing process. The white area represents the indention point of the pressing when performing the blast test. The white area on specimen B in Figure 7(b) is larger than the white area on specimen A in Figure 7(a) because the impact force is larger on specimen B in Figure 6(b).

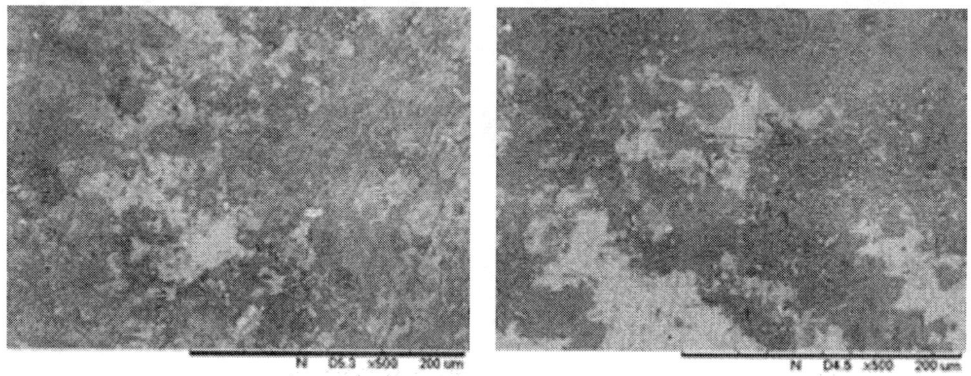

(a) (b)

Figure 7. (a)SEM microstructure of specimen A after explosion (b) SEM microstructure of specimen B after explosion

3.3. Bending strength

Figure 8(a) is the specimen A after bending strength test, it shows no changed after the bending strength test. Then, Figure 8(b) is the specimen B after bending strength test, as shown, the specimen is bent after the test. From the calculation using Eq. (1) and Eq. (2), the force that applied to the specimen B during the bending strength is 243,160.68kN. So as a conclusion, the specimen A is not bend after the test method because, the diameter and thickness of the specimen A is higher than specimen B.

(a) (b)

Figure 8. (a)Specimen A after bending strength (b)Specimen B after bending strength

IOP Conf. Series: Materials Science and Engineering **429** (2018) 012096 doi:10.1088/1757-899X/429/1/012096

4. Conclusion

As a conclusion, from this research, the method for Underground Explosion Pressing is suitable for pressing the solid metal (mild steel), since the result for specimen A is decrement from $10mm$ to $5.7mm$, while specimen B is decrease from $10mm$ to $3.9mm$. The hardness result for specimen A is indicated clearly to be $33.8HRA$, while the hardness for specimen B is $43.7HRA$. Then, the SEM microstructure for specimen A shows the porosity is more than specimen B. Last but not least, the bending strength result shows that, the specimen A, has no change since the specimen is not bend after the test, while the specimen B is bend after the test. The force that applied to the specimen B during the bending strength is $243{,}160.68kN$.

5. References

[1] B. Mobaraki and M. Vaghefi, "Numerical study of the depth and cross-sectional shape of a tunnel under surface explosion," *Tunn. Undergr. Sp. Technol.*, vol. 47, pp. 114–122, 2015.

[2] D. Hu, T. Long, C. Liu, G. Yang, and X. Han, "Swelling movement induced by underground explosion of aluminized explosive in multilayered compact material," *Int. J. Rock Mech. Min. Sci.*, vol. 71, pp. 330–339, 2015.

[3] V. R. Feldgun, Y. S. Karinski, I. Edri, and D. Z. Yankelevsky, "Prediction of the quasi-static pressure in confined and partially confined explosions and its application to blast response simulation of flexible structures," *Int. J. Impact Eng.*, vol. 90, pp. 46–60, 2016.

[4] M. Chul, Y. Seok, and D. Ho, "Analysis of explosion characteristics of combustible wood dust in confined system using the thermal decomposition rate and mass loss rate," *Appl. Therm. Eng.*, vol. 109, pp. 432–439, 2016.

[5] B. He, X. Jiang, G. Yang, and J. Xu, "A numerical simulation study on the formation and dispersion of flammable vapor cloud in underground confined space," *Process Saf. Environ. Prot.*, 2016.

[6] W. A. Trzci and J. Paszula, "Optical spectroscopy to study con fi ned and semi-closed explosions of homogeneous and composite charges," vol. 88, pp. 111–119, 2017.

[7] Y. Xu, L. Wang, M. Yu, and S. Wan, "Journal of Loss Prevention in the Process Industries Study on the characteristics of gas explosion affected by induction charged water mist in con fi ned space," *J. Loss Prev. Process Ind.*, vol. 40, pp. 227–233, 2016.

[8] B. Burch, "Determining and mitigating the effects of firing a linear shaped charge under water," 2014.

[9] W. Walters, "A Brief History of Shaped Charges," *24th Int. Symp. Ballist.*, vol. 1, no. December, pp. 3–10, 2008.

Acknowledgments

This research is fully funded by FRGS grant name Characteristic of Powder Surpassing 160,000 tons Press using Explosive Pressing (FRGS 9003-00542). Besides, the blast test of this research was done with the help of Lt Col Associate Professor Ariffin Bin Ismail and blast research team from Universiti Pertahanan Nasional Malaysia

IOP Conf. Series: Materials Science and Engineering **429** (2018) 012097 doi:10.1088/1757-899X/429/1/012097

Effect of annealing temperature on construction of CuO layer on electrodeposited Cu_2O layer by annealing

S M Zainal[1], M.Zamzuri[1,2], M.Hasnulhadi[1], Z.Nooraizedfiza[1], M.Marina[1], Fariza Mohamad[3], N. Hisyamudin[3] and M.Izaki

[1] School of Manufacturing, Universiti Malaysia Perlis, Pauh Putra Main Campus, 02600, Arau, Perlis, Malaysia.

[2] Institute of Nanoelectronic Engineering,Universiti Malaysia Perlis, Block A, Taman Pertiwi Indah, Seriab, 01000, Kangar, Perlis, Malaysia

[3] Faculty of Electrical & Electronic Eng. Universiti Tun Hussein Onn Malaysia, 86400, Parit Raja, Batu Pahat, Johor, Malaysia

[4] Department of Mechanical Eng., Toyohashi University of Technology, 1-1 Hibari Gaoka, Tempaku, Toyohashi, Aichi 441-8580, Japan

E-mail: fuzah2525@gmail.com

Abstract. We report the annealing effects on the morphological and structural properties of Cu_2O films electrodeposited on Cu substrates from an aqueous solution containing copper acetate and lactic acid. The Cu_2O/CuO layers has been constructed by electrodeposition of p-Cu_2O layer on the modified surface of Cu substrate followed by formation of the CuO layer by annealing from 100°C to 300°C. The morphology and structural analysis was performed using optical microscope, AFM and XRD respectively. The result showed that increasing the annealing temperature resulted in the increment of the Cu_2O grain size from about 96 to 201 nm. AFM reveals the precipitation of CuO grains on the Cu_2O layer by annealing in air at 300°C, and completely covered the surface of Cu_2O layer.

1. Introduction

Cuprous oxide (Cu_2O) is naturally p-type semiconductor with a direct band gap of 2.1eV [1], that acts as a light absorber layer in the solar cell. Over the year, Cu_2O has gained a broad attention due to its excellent optical and electronic properties [2,3]. The advantage of Cu_2O in optoelectronic applications includes its direct-transition band structure that suitable for solar radiation absorption with the theoretical photovoltaic conversion efficiency approximately 20% [4]. Cu_2O thin film can be prepared by various deposition techniques such as atomic layer deposition (ALD) [5], thermal oxidation of Cu sheets [6], molecular beam epitaxy [7], sputtering [8] and electrodeposition technique [9,10]. Electrodeposition is a common method used for deposition of Cu_2O thin film on a substrate. It is a low cost and simple electrochemical technique that give the opportunity for the broad area for thin film manufacturing [11].

On the other hand, cupric oxide (CuO), also a p-type semiconductor is used as another metal oxide candidate for solar applications. CuO has a direct band energy of 1.4 eV to 1.7 eV [12], with a large absorption coefficient that makes them active in the wide range of visible light. CuO has different crystal structures, colors, and physical properties with the Cu_2O [13]. Both the Cu_2O and CuO are used as a good candidate for photovoltaic applications because of their low cost, nontoxic, abundantly available on earth, excellent electrical and high optical properties [14]. The high-quality of CuO as well as Cu_2O can be prepared by thermal oxidation because of it is simplicity and low-cost [15].The thermal oxidation process is usually performed by annealing the Cu sheet at the temperature range from 300°C to 1000°C in air or oxygen gas (O_2). The two major products are CuO and Cu_2O, whose their formation depends on conditions under which the oxidation takes place. Fan et al. successfully prepared the CuO/Cu_2O on fluorine-doped tin oxide (FTO) substrate by simply annealing the electrodeposited Cu_2O layer [13]. They reported that the photocurrent density and photoelectrochemical (PEC) stability of the p-type heterostructure CuO/Cu_2O was improved greatly compared with the pure Cu_2O, which was greatly affected by annealing time and temperature.

In solar cell system, particularly the photovoltaic device, carrier mobility refers to both electron and hole mobility where this mobility depends on several factors such as the temperature, electron and hole concentrations. The low mobility relating to the diffusion length of minority carrier can affect the performance of the solar cell by reducing the conversion efficiency [16]. Also, the smaller grain size and inferior crystalline structure of the photovoltaic cell can affect the movement of an electron which directly reduced the efficiency of the device.

Here, we demonstrate the preparation of the CuO layer on electrodeposited Cu_2O layer. We applied the electrodeposition method for fabrication of Cu_2O layer followed by annealing method for construction of CuO layer, and report the effects of the annealing temperature towards the properties of the Cu_2O/CuO layer on modified Cu surface. The structural and morphology characterizations for Cu_2O/CuO layer were carried out by X-ray diffraction (XRD), optical microscope and atomic force microscopy (AFM).

2. Experimental Section

The Cu sheet (Emsys Sdn. Bhd. Malaysia) with a dimension of 2 cm (length) x 1 cm (width) was used as a substrate, and prior to the electrodeposition, the Cu sheet surface was polished with 2000-grit silicon carbide sandpaper using (MoPao TM 260E) Grinder Polisher machine to modify the surface before immersed with acetone for 3 minutes and then rinsed with distilled water. Next, the Cu_2O layer was deposited potentiostatically at 0.5V referenced to Ag/AgCl electrode on the Cu substrate at a constant electric charge of 1.7 C/cm^2 with a potentiostat (HOKUTO DENKO HABF-501A) in an alkaline aqueous solution containing a 0.4M copper(II) acetate monohydrate, $Cu(CH_3COO)_2$ and 3M of lactic acid, $C_3H_60_3$ (Kanto Chemical, Co., Inc.) at 328K. The solution was prepared with 500ml of deionized (DI) water with a resistivity of 18 MΩ.cm and potassium hydroxide,KOH was added for the pH adjustment to 12.5. A platinum, Pt plate was used as the counter electrode. Annealing of the electrodeposited Cu_2O layer was performed at 100°C, 200°C and 300°C for 2 hours in air condition using a heating furnace (NABERTHERM N 11/H Furnace). During annealing, the Cu_2O samples were placed on an alumina plate.

The surface roughness of the Cu sheet was determined with MITUTOYO FORMTRACER CS-3100 Surface Roughness Tester with a scan speed of 0.02mm/s and sample length of 4mm. The appearance of the Cu substrate before and after surface modification was observed by optical microscopy (MOTICAM 1000 1.3M Pixel Optical Microscope) with a 20x magnification. Structural characterization was carried out by X-ray Diffractometer (XRD BRUKER D2 PHASER), operated at 30kV and 10mA using Cu Kα radiation (wavelength of $K\alpha_1$=1.5406Å) that linked with DIFFRAC.SUITE EVA Software. The crystallite size (D) can be calculated using Debye-Scherrer equation ($D = 0.9\lambda/(\beta \cos \theta)$, where λ is the wavelength of X-radiation,β is the full width at half maximum (FWHM) of the highest intense peak and θ is the diffraction angle in radian.

3.Results and discussion

3.1 Characteristic of Cu_2O/CuO layer prepared on the Cu substrate

Fig.1(a) shows the schematic illustration of the cell configuration of Cu_2O/CuO on Cu substrate. The Cu_2O layer was deposited on Cu substrate followed by construction of CuO layer by annealing in air. The surface of the as-received Cu substrate before and after surface modification are shown in **Fig.1(b,c)**. The surface roughness of the Cu substrate was modified from 0.369 μm to 0.077 μm after polishing. **Fig.2** shows the transition of the physical appearance of the Cu_2O layer on Cu substrate before and after annealing at different temperature. The red brown color of Cu substrate changed to maroon color after deposition of Cu_2O layer. Furthermore, the color of Cu_2O layer became darker as the temperature increases. At 300°C, the Cu_2O layer changed to black color film indicating the formation of CuO layer [3]. The surface morphology of the Cu_2O/CuO layers was further examined by AFM.

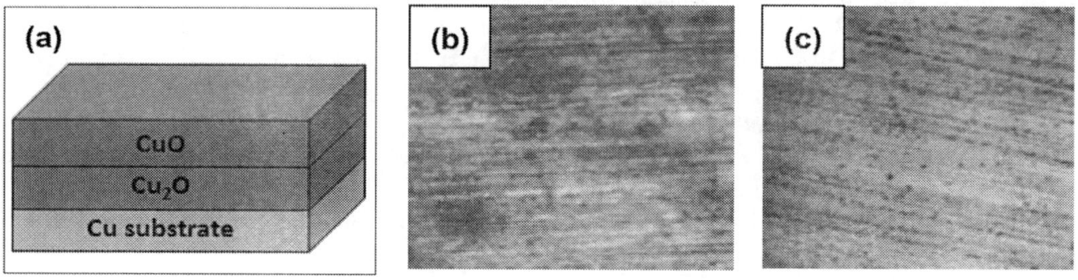

Figure 1: (a) Schematic illustration of cell configuration of Cu_2O/CuO on Cu substrate and (b) optical micrograph of the Cu sheet before and (c) after polishing.

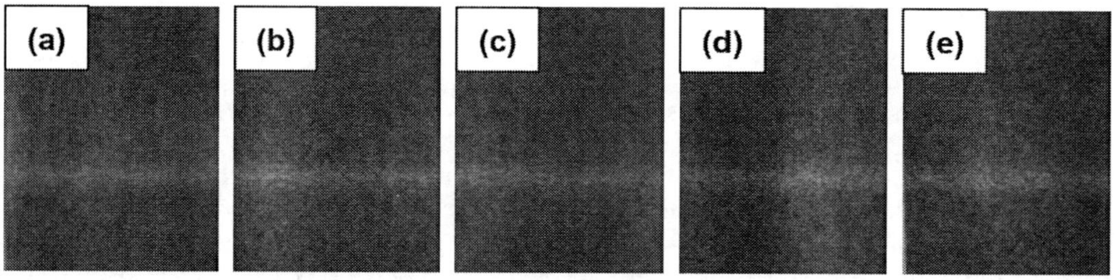

Figure 2 : The physical appearance of (a) as-received Cu substrate, (b) as-deposited Cu_2O layer on Cu substrate, and Cu_2O layer on Cu annealed at (c) 100°C, (d) 200°C and (e) 300°C respectively.

Fig.3 (a,b) displays the three-dimensional and plan view topography of electrodeposited Cu_2O on Cu substrate, respectively. The Cu_2O layers deposited on the Cu substrate were composed of pyramid-like Cu_2O particles grown in a direction perpendicular to the Cu substrate surface [17]. The triangular facet of the Cu_2O crystal was obviously observed in the plan view **(Fig.3b)**. Meanwhile, the annealed Cu_2O layers at different temperature are shown in **Fig. (a f)**. From **Fig.** , it shows that Cu_2O grains had changed significantly with the annealing temperature. The correlations between Cu_2O grain size and surface roughness of the as-deposited Cu_2O layer before and after annealing are presented in **Fig.** . As the annealing temperature increase up to 200 °C, the Cu_2O grain size has increased gradually from 96.151 nm to 201.206 nm and the surface roughness also increased from 0.023 µm to 0.034 µm. However, the Cu_2O grains size as well as surface roughness decreased to 173.440 nm and 0.017 µm respectively as the annealing temperature reached 300°C. Therefore, the thickness of CuO layer was estimated to be around 20 nm. At 300°C, the surface of the pyramid-like structure was covered completely to a compact irregular structure indicating the formation of CuO grains as shown in **Fig. (c,f)** [18].

IOP Conf. Series: Materials Science and Engineering **429** (2018) 012097 doi:10.1088/1757-899X/429/1/012097

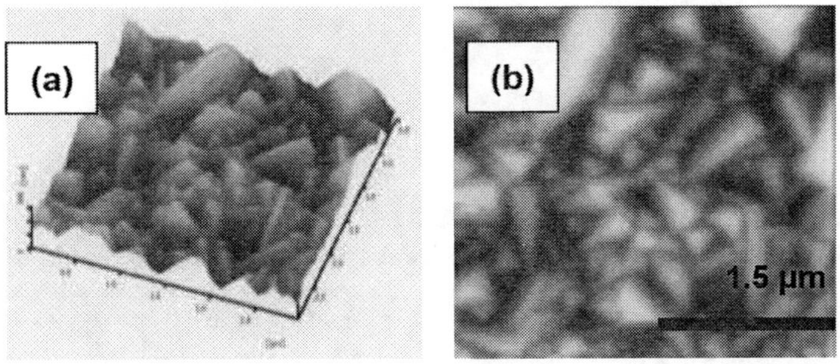

Figure 3: The surface topography of the deposited Cu_2O on Cu substrate without annealing (a) 3D View (b) plan view

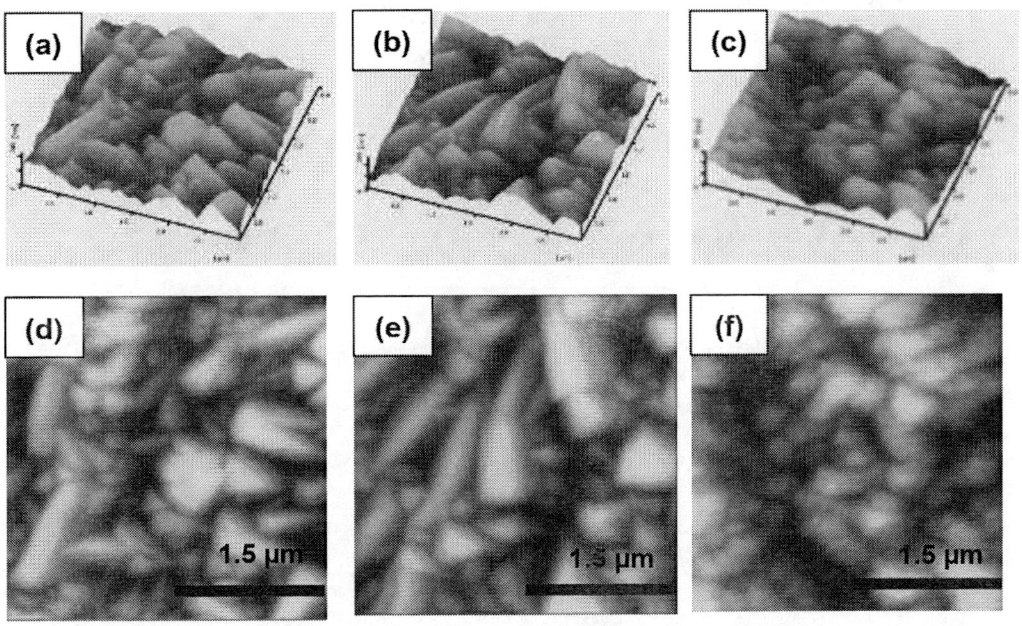

Figure : The 3D view and plan view of AFM images of the deposited Cu_2O on Cu substrate annealed at (a,d) 100°C, (b,e) 200°C and (c,f) 300°C.

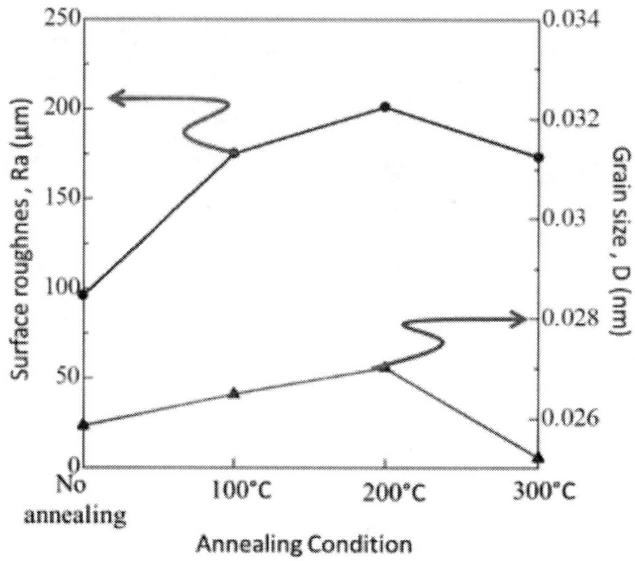

Figure : The surface roughness and grain size of the deposited Cu_2O on Cu substrate.

3.2. Structural analysis of the Cu_2O/CuO layer prepared on the Cu substrate

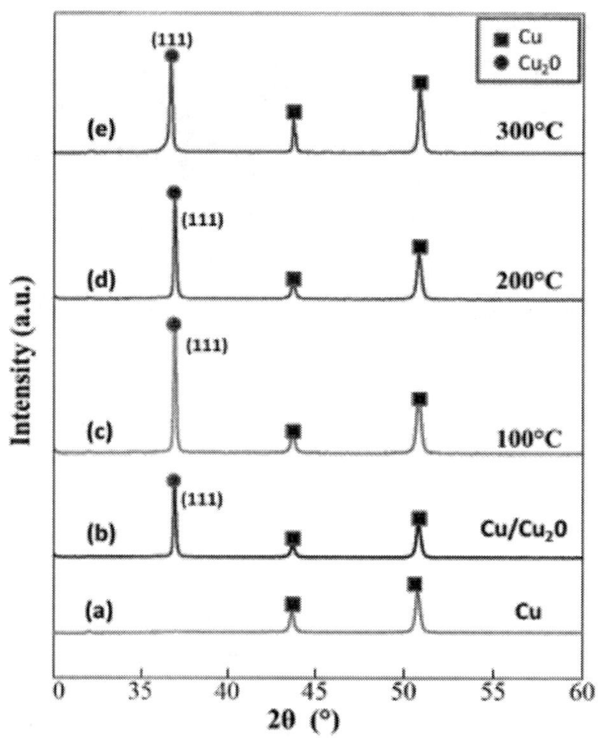

Figure : XRD patterns of the (a) Cu and (b) deposited Cu_2O layer prepared on Cu substrate without annealing, Cu_2O annealed at (c) 100°C, (d) 200°C and (e) 300°C.

X-ray diffraction (XRD) analysis was carried out to examine the orientation and crystal structure of the Cu_2O/CuO layers. **Fig.** shows the XRD patterns for as-deposited Cu_2O layer and annealed Cu_2O layer at different temperature from 100°C until 300°C. The as-deposited Cu_2O layer revealed single orientation which was indexed to (111) at diffraction peaks of 36.4°. The Cu_2O orientation was corresponding to cubic Cu_2O pattern (PDF 01-071-3645) and the intensity increase as the annealing temperature increased up to 300°C. No obvious peak of CuO layer could be seen as the Cu_2O layer was annealed in air at 200°C and 300°C irrespective with the changed in the microstructure as shown in the **Fig. (c,f)**.

However, the peaks of CuO layer become visible in the diffraction pattern when the annealing temperature was slightly increased to 350°C [13]. Based on the changes in the appearance (**Fig. 2e**) and microstructure of annealed Cu_2O layer observed by AFM (**Fig. (c,f)**), it is speculated that the CuO layer could be constructed by annealing of Cu_2O layer in air at temperature as low as 300°C.

. Conclusions

The Cu_2O/CuO layer has been constructed by electrodeposition of the p-Cu_2O layer on the Cu substrate followed by formation of the CuO layer by annealing at different temperature. The Cu_2O layer was prepared by electrodeposition on Cu substrate in an alkaline aqueous solution containing copper (II) acetate and lactic acid. The thickness of Cu_2O layer was maintained by controlling the electric charge during the electrodeposition process. The results obtained from AFM showed that increasing the annealing temperature resulted in the increment of grain size of Cu_2O layer and led to the formation of CuO layer on its surface. The thickness of CuO layer was found to be around 20 nm. Although further improvement is needed, these results show the importance of the crystalline structure of the Cu_2O/CuO layer to the improvement of the photovoltaic device performance.

Acknowledgements
This work was supported by the Fundamental Research Grant Scheme (FRGS) from Ministry of Higher Education (MOE) Malaysia under Grant No. FRGS/1/2017/TK03/UNIMAP/02/6 and Short Term Grant Universiti Malaysia Perlis under Grant No. 9009-00042.

References

[1] Nishi Y, Miyata T and Minami T 2016 *Sol. Energy Mater. Sol. Cells* **1** pp 405–410
[2] Lahmar H *et al* 2017 *J. Alloys Compd.* **1** pp 36–45
[3] Hussain S *et al* 2017 *Opt. - Int. J. Light Electron Opt.* **13** pp 372–377
[4] Shinagawa T *et al* 2013 *J. Mater. Chem. A* **32** 9182
[5] Sinha S, Nandi D, Kim S, and Heo J 2018 *Sol. Energy Mater. Sol. Cells* **1** pp 49–68
[6] Guo D and Ju Y 2016 *Mater. Today Proc.* **3** pp 350–353
[7] I. Sakellis *et al* 2014 *Energy Procedia*, pp 37–42
[8] Cheng K *et al* 2013 *Sol. Energy Mater. Sol. Cells* **11** pp 120–125
[9] Zamzuri M *et al* 2015 *Thin Solid Films* pp 136–141
[10] Lee D and Ebong A 2017 *Renew. Sustain. Energy Rev.* pp 1286–1297
[11] Kanda H *et al* 2017 *Mater. Today Energy* pp 190–198
[12] Shaislamov U *et al* 2016 *Int. J. Hydrogen Energy* **1** pp 2253–2262
[13] Fan D, Chen Q and Wang Y 2017 *J. Phys. Chem. Solids* **1** pp 139–144
[14] Ali N *et al* 2016 *Renew. Sustain. Energy Rev.* pp 726–737
[15] Chen L 2013 *Mater. Sci. Semicond. Process.* **1** pp 1172–1185
[16] Izaki M *et al* 2007 *J. Phys. D. Appl. Phys.* pp 3326–3329
[17] Hossain M *et al 2017 Mater. Sci. Semicond. Process* **3** pp 203–211
[18] Khoo P *et al* 2018 *Thin Solid Films* ***653*** *pp 158-164*

IOP Conf. Series: Materials Science and Engineering **429** (2018) 012098 doi:10.1088/1757-899X/429/1/012098

Investigation of Adhesive Joint Strength for Dissimilar Metal Adherends by Using Taguchi Method

N Atikah[1], M Afendi[1], Shahriman A B[1], I Zunaidi[1], Z M Razlan[1], W K Wan[1], M N Mazlee[2] and M S A Majid[1]

[1] School of Mechatronic Engineering, Universiti Malaysia Perlis, Kampus Pauh Putra, 02600 Arau, Perlis, Malaysia.

[2] Centre for Frontier Materials Research, Universiti Malaysia Perlis, 02600 Arau, Perlis, Malaysia

Abstract. The dissimilar metal joints of stainless and medium carbon steel have been emerged as a structural joint for various industrial applications which provides good combination of mechanical properties like strength, corrosion resistance with lower cost. This paper concerned with the growth of adhesive joint technique since adhesive joint has the applicability to joint wide variety of engineering materials. This study is to investigate the significance of factors in dissimilar AISI 304/1045 butt-joint specimen tensile strength by using Taguchi method and Analysis of Variance (ANOVA). The joining strength was evaluated with manufacturing factors of abrasive paper, heat treatment and adhesive thickness. It was obtained that abrasive paper has significant effect on adhesive joint strength while heat treatment effect is insignificant. Nevertheless, the joint strength decreased with increment of adhesive thickness.

1. Introduction

Dissimilar metal joint can be complicated and extents a wide-ranging of methodologies, materials and procedures [1]–[3]. It is often more difficult than joining the same material. Adhesive joint has a good spot in present industry. There are also a lot of comprehensive study on adhesive joint available yet not much study was done on the behaviours of adhesive bonding strength on heat treated metals specifically stainless steel and carbon steel. Statically design of experiment (DOE) is regularly seen as crucial part of experiment. Taguchi method is a simple, systematic, and efficient approach for optimizing the process parameters [4], [5].

2. Design of experiment (Taguchi method)

There are two groups of specimen studied which are as received (AR) and heat treated sequence 2 (SQ2). The experiment matrix in Taguchi method were formed after deciding on the control factors and orthogonal array. Table 1 is the studied control factors and its level. L9 orthogonal array was chosen in this study. The experiment matrix are shown in table 2. Three specimens were replicates for every experiment condition.

Table 1. Control factors and levels

Control factors	Level 1	Level 2	Level 3
Abrasive paper (grit)	180	500	1000
Heat treatment	Annealing	Normalizing	Quenching
Adhesive thickness (mm)	0.5	1.0	1.5

Table 2. Experiment matrix

Experiment conditions	Control factors (AR)		Control factors (SQ2)		
	Abrasive paper (grit)	Adhesive thickness (mm)	Abrasive paper (grit)	Heat treatment	Adhesive thickness (mm)
1	180	0.5	180	Annealing	0.5
2	180	1.0	180	Normalizing	1.0
3	180	1.5	180	Quenching	1.5
4	500	0.5	500	Annealing	1.0
5	500	1.0	500	Normalizing	1.5
6	500	1.5	500	Quenching	0.5
7	1000	0.5	1000	Annealing	1.5
8	1000	1.0	1000	Normalizing	0.5
9	1000	1.5	1000	Quenching	1.0

3. Specimen and testing

3.1. Material

The adhesive joint specimens were prepared with accordance to ASTM D2094-00. The specimen are made of circular rod medium carbon steel (AISI 1045) and stainless steel (AISI 304) cut into 12.7 mm diameter and 38.1 mm length as adherends and a slow-setting epoxy adhesive, Araldite Standard 90 Minutes as adhesive. The adhesive consisted of two components which are resin and hardener. The mechanical properties of adherends and adhesive are shown in table 3.

Table 3. Mechanical properties of adherends and adhesive

Material	Young's modulus	Poisson ratio
AISI 1045 (adherend 1)	206 GPa	0.290
AISI 304 (adherend 2)	210 GPa	0.290
Araldite Standard 90 Minutes (adhesive)	1850 MPa	0.396

3.2. Specimen fabrication

The surface treatment were applied by grinding the bonding area with Silicon Carbide (SiC) abrasive paper by using Metco grinding and polishing machine. Heat treatment were done on both adherends. AISI 1045 were heat treated at 925 °C for 90 minutes and subsequently cooled by three different cooling rates namely as furnace condition cooling (annealing), air condition cooling (normalizing) and room condition water quenching (quenching). AISI 304 were heat treated at 1050 °C for 90 minutes and subsequently cooled in room condition water quenching (annealing). The adherend's surface were acetone wiped prior to bonding. The adhesive bonding process was prepared by mixing resin and hardener with weight ratio of 1:0.8 respectively. Adhesive were stirred for 3 minutes ahead of applying on adherend's surface. The adherends were then fitted into the alignment jig and the adhesive thickness was controlled using the micrometer head as shown in figure 1. Lastly the specimen was cured fully in 12h at room temperature with handling strength in about 6h in the alignment jig.

Figure 1. Specimen placed in an alignment jig

3.3. Tensile test and surface roughness examination

The tensile test was conducted using Universal Testing Machine (UTM). Figure 2 shows the fixture set-up for tensile test of dissimilar AISI 304/1045 butt-joint specimen. The specimen was hold by a pair of holding jig. The test were conducted at a crosshead speed of 1.0 mm min^{-1}. Surface roughness of bonding area were measured by using a surface profilometer, F-Mitutoyo. The profiler moves along a single direction with a scanning length of 8.0 mm.

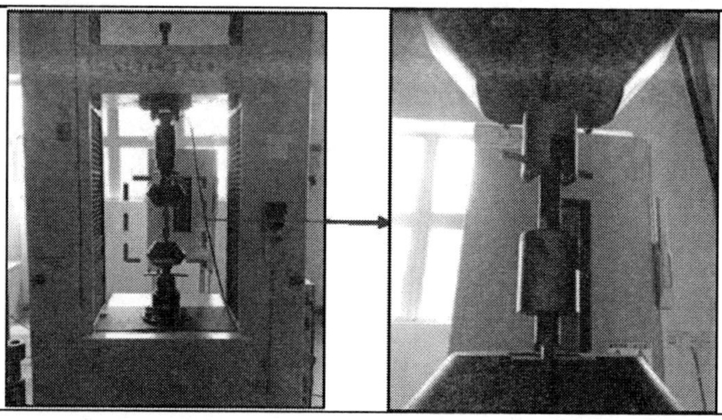

Figure 2. Tensile test experiment set-up

4. Results and discussion

The average tensile strength obtained were computed and presented in table 4. Based on the joint strength, statistical analysis by ANOVA was performed.

4.1. Statistical analysis by ANOVA

Table 5 presents the analysis results for the two groups. For AR, the contribution of adhesive thickness is 59.38%, higher than abrasive paper, 28.76%. In addition to this, the change of level within the range investigated in abrasive paper and adhesive thickness does have significant effect on the joint strength since the F ratio for both factors are more than 4. There are external factors that affect the joint strength by 11.86%. Meanwhile for SQ2, the main factor influencing the joint strength in this group is adhesive thickness, 67.51% followed by the abrasive paper with 12.80% and heat treatment with 11.00%. This result shows that the adhesive thickness is more significant than abrasive paper. Statistically, heat treatment have very little influence to joint strength in SQ2. The change of level within the range of adhesive thickness investigated seems to have significant effect on the joint strength where the *F* ratio is bigger than 4.

Table 4. Joint strength of dissimilar AISI 304/1045 butt-joint specimen

Experiment conditions	Joint strength (N/mm^2)	
	AR	SQ2
1	19.41	6.42
2	8.69	2.95
3	7.74	2.53
4	11.58	4.87
5	8.04	5.87
6	5.20	7.21
7	9.28	2.19
8	6.07	8.03
9	3.72	8.38

Table 5. ANOVA results for joint strength of AR and SQ2

Group	Factors	*DOF*	SSA_i	*MS*	Contribution (%)	F
AR	Abrasive paper (A)	2	48.42	24.21	28.76	4.85
	Adhesive thickness (B)	2	99.96	49.98	59.38	10.01
	Noise	4	19.97	4.99	11.86	
	Total	8	168.36			
SQ2	Abrasive paper (A)	2	6.24	3.12	12.80	1.47
	Heat treatment (B)	2	5.36	2.68	11.00	1.27
	Adhesive thickness (C)	2	32.89	16.44	67.51	7.77
	Noise	2	4.23	2.12	8.69	
	Total	8	48.72			

4.2. Effect of abrasive paper, heat treatment and adhesive thickness.

Surface roughness of adherends decrease as abrasive paper grit increase for both AR and SQ2 group. Joint strength of AISI 304/1045 against surface roughness for AR and SQ2 group at 1 mm adhesive thickness were plotted in figure 3. It was observed that for AR group, the increased roughness of both AISI 304 and AISI 1045 improves the joint strength of adhesive AISI 1045/304 butt joint specimen. Some studies have verified that roughness effect for stainless steel and carbon steel differ to one another. For stainless steel, as roughness increases, adhesive bonding strength increases. But for carbon steel, as the roughness increases, bonding strength decreases. Thus, the observation for AR group indicates that roughness of AISI 304 plays the role in strengthen the joint strength.

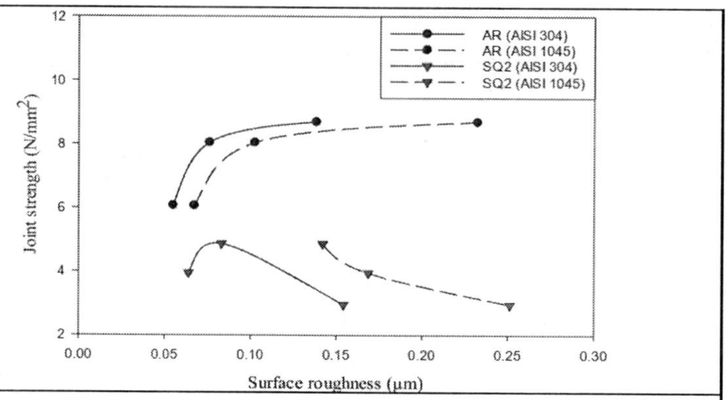

Figure 3. Surface roughness against joint strength plot

The opposite results occurred in SQ2 group where increased roughness of AISI 1045 lower the joint strength. For AISI 304, the joint strength initially increases and then decreases with roughness. In addition for SQ2, the surface roughness of both adherends increased slightly after heat treatment Noted that in this group, the microstructure and mechanical properties of each steels have changed because of the heat treatments processes. The mechanical properties for AISI 304 at three different roughness is same but the properties for AISI 1045 is different. The optimum surface roughness exists for a maximum joint strength and the roughness range depends on the adherend which explains why the behaviour of SQ2 specimen is not in good agreement with AR group [6]. The effect of adhesive thickness is similar for AR and SQ2 where both groups exhibit the same pattern. The joint strength decreases with an increase in adhesive thickness. These adhesive thickness and tensile bonding strength trends are similar to those obtained in a study by Naito et al. [7] . This might be due to the less flaws such as voids contained by the thin adhesive layers [8]–[10].

4.3. Main effects plot

The value for average joint strength for each factor at each level was obtained and the result were plotted in figure 6 for AR and figure 7 for SQ2. It can be seen that the optimum combination of factors and their levels A_1B_1 yield the highest joint strength for AR. While for SQ2, the optimum factors and their levels combination is $A_2B_2C_1$.

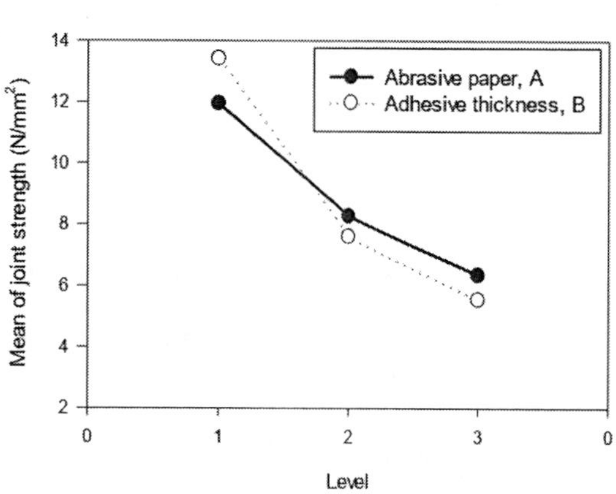

Figure 4. Main effect plot for AR group

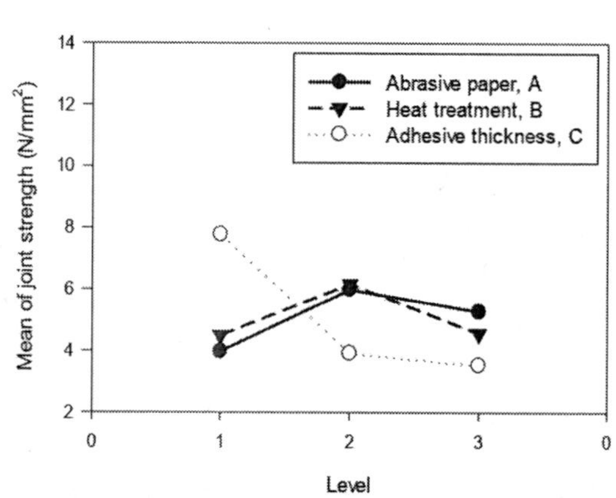

Figure 5. Main effect plot for SQ2 group

4.4. Confirmation test

The last step was to validate the estimated result against experimental value. Based on the main effect plot, the confirmation test was required because the obtained optimum combination of factors and their levels for SQ2 did not correspond to any of experiment condition in the experiment matrix except for AR group. The estimated value for optimum combination of factors and their levels, Y_{opt} is calculated by using equation (1) and (2).

$$Y_{opt} = m + (m_A - m) + (m_B - m) + (m_C - m) \tag{1}$$

$$m = \frac{T}{n} \tag{2}$$

where, m is the average joint strength, m_A is the average joint strength for factor A, m_B is the average joint strength for factor B and m_C is the average joint strength for factor C, T is sum of all responses, n is the number of response.

The estimated joint strength is 9.763 N/mm^2 while the experimental value is 9.415 N/mm^2. The experimental value of adhesive butt joint of AISI 304/1045 is close to the estimated value and verifies that the experimental result for SQ2 group is strongly correlated to estimated result.

5. Conclusion

Abrasive paper and heat treatment increased the adherend surface roughness. The optimum surface roughness exists for a maximum joint strength and the roughness range depends on the adherend mechanical properties Heat treatment changed the microstructure but lowered the joint strength of AISI 304/1045. ANOVA proved that heat treatment effect is insignificant. Joint strength decreased with increment of adhesive bond thickness. In this study, 0.5 mm adhesive thickness resulting the highest joint strength for both as received and heat treated group. Lastly, it is evident that Taguchi method provides simple and systematic methodology for optimizing the process parameters.

6. References

[1] Ciupack Y, Pasternak H, Mette C, Stammen E and Dilger K 2017 Adhesive bonding in steel construction-challenge and innovation *Proc. Eng. on Mod. Build. Mater., Struct. and Tech. (Germany)* vol 172 p 186–193

[2] Mishra R R, Tiwari V K, and Rajesha S 2015 A study of tensile strength of mig and tig welded dissimilar joint of mild steel and stainless steel *Int. J. Adv. Mater. Sci. Eng.* **3** 23–32

[3] Farren J D, Dupont J N and Noecker F F 2007 Fabrication of a carbon steel-to-stainless steel transition joint using direct laser deposition — a feasibility study *Suppl. to Weld. J.* 55–61

[4] Athreya S and Venkatesh Y D 2012 Application of taguchi method for optimization of process parameters in improving the surface roughness of lathe facing operation *Int. Ref. J. Eng. Sci* **1** 13–19

[5] Kamaruddin S, Khan Z A, and Foong S H 2010 Application of taguchi method in the optimization of injection moulding parameters for manufacturing products from plastic blend *Int. J. Plast. Technol.* **4** 152–166

[6] Ghumatkar A, Budhe S, Sekhar R, Banea M D, and De Barros S 2016 Influence of adherend surface roughness on the adhesive bond strength *Lat. Am. J. Solids Struct.* **13** 2356–70

[7] Naito K, Onta M and Kogo Y 2012 The effect of adhesive thickness on tensile and shear strength of polyimide adhesive *Int. J. Adhes. Adhes.* **36** 77–85

[8] Ikegami K, Fujii T, Kawagoe H, Kyogoku H, Motoie K, Nohno K, Sugibayashi T, and Yoshida F 1996 Benchmark tests on adhesive strengths in butt, single and double lap joints and double-cantilever beams *Int. J. Adhes. Adhes.* **16** 219–226

[9] Arenas J M, Narbo´n J J and Alı´a C 2010 Optimum adhesive thickness in structural adhesives joints using statistical techniques based on Weibull distribution *Int. J. Adhes. Adhes.* **30** 160–165

[10] Banea M D, Silva L F M and Campilho R D S G 2015 The effect of adhesive thickness on the mechanical behavior of a structural polyurethane adhesive the effect of adhesive thickness on the mechanical behavior of a structural polyurethane adhesive *J. Adhes* **91** 331–346

IOP Conf. Series: Materials Science and Engineering **429** (2018) 012099 doi:10.1088/1757-899X/429/1/012099

Effect of processing parameters on the shrinkage behaviour of injection moulded polypropylene gear

A A Jamaluddin[1], N M Mehat[2] and S Kamaruddin[1]
[1]Mechanical Engineering Department, Universiti Teknologi PETRONAS, 32610 Seri Iskandar, Perak Darul Ridzuan, Malaysia
[2]Department of Mechanical Engineering Technology, Faculty of Engineering Technology, Uniciti Alam Campus, Universiti Malaysia Perlis, Sungai Chuchuh, Padang Besar, 02100, Perlis, Malaysia.

Abstract. The polymer gears are widely used as basic motion transfer elements in many industries. Continuous improvement on the quality performance of polymer gears is very important for their ability to meet present demand. So that, controlling the parameters setting in polymers gears are very importance in order to get the best gears quality. In this study, the influences of injection moulding processing parameters on shrinkage of polypropylene gears was investigated by using Taguchi approach. It was found that the optimal processing parameters for addendum circle shrinkage, dedendum circle shrinkage and tooth thickness shrinkage are A1B1C3D3E1F2G3, A3B1C3D3E3F1G3, and A1B3C3D2E3F2G1 respectively. This results showed that pattern of fluctuation showed either an improvement or reduction in quality characteristics upon the change of process parameters setting.

1. Introduction

An increasing demand of polymer gears used as a basic and motion transfer components make it is widely used in many industries starting from lightweight industries such as copy machine, printers and facsimiles until to heavyweight industries such as automotive, robotics and starter motors manufacturing [1]. The light weight, noise reduction, anti-corrosion, oil free condition as well as low materials and manufacturing cost advantages that owned by polymers gear have contribute to their increasing of application compare to metal gears [2]. Nowadays, an injection moulding process offer the shorts product cycle, excellent aesthetics appearance and high flexibility of part geometry for polymer gears production with low cost and high productivity rate. Although injection moulding is the most widely used process in polymers gear manufacturing, there are still needs tremendous efforts to solve the problems especially regarding to shrinkage in polymer gears. Over the past few decades, a lot of studies are conducted with purpose to enhance the performance and durability of polymer gears by means of internal structure enhancement [3], [4] and design modification [5], [6]. Apart of material enhancement, another consideration for the possibility of optimizing the injection moulding process from the manufacturing point of view should not grossly underestimated. It could be an appealing approach to enhance the polymer gear's quality especially when it relates to shrinkage problem, considering no extra processing cost is required. Generally, there are numerous processing parameters that need to be controlled during injection moulding which can be classified into four namely, temperature, pressure, time, and distance [7]. With plastic materials exhibiting extremely convoluted properties, and the intricacy of the moulding process, attaining the desired final moulded gear quality particularly when it comes into shrinkage, is a very challenging task. Thus, the aim of this study is to investigate the impact of variation in processing parameters on shrinkage behaviour in moulded polymer gears as well as to determine the optimal processing parameters that can minimize the shrinkage problem in polymer gear. Seven processing parameters including melting temperature, mould temperature, injection pressure, injection time, packing pressure, packing time and cooling time with three levels of each was systematically analysed and investigated in this study via the Taguchi method.

2. Methodology

Polypropylene (PP) was used as gear material in this study, denoted as Titanpro Polypropylene 6331 by Titan Chemicals. Prior to injection moulding processs, the PP was dried at 80 °C for 2 hours to remove moisture for better moulding quality. The spur gear with module 1.5, pressure angle 20°, 20 teeth and 10mm thickness compliance with American Gears Manufacturers Association (AGMA) was the part studied (Figure 1).

Figure 1. Spur Gear

2.1. Design of experiment

The experiment was conducted by using Taguchi method approach in order to study the influences of processing parameters on polymer gears shrinkage. Seven parameters of melting temperature, mould temperature, injection pressure, injection time, packing pressure, packing time and cooling time with three levels of each factor was selected, as shown in Table 1. Shrinkage in addendum circle, dedendum circle and tooth thickness of the gear were studied. The orthogonal array (OA) L_{18}, as tabulated in Table 2 are used in designing the experiments. All the trials in L_{18} OA was conducted by using Battenfeld TM750/210 injection moulding machine with five repetitions of each trial.

Table 1. Factors and levels selected in experiments

Column	Parameters	Level 1	Level 2	Level 3
A	Melting temperature (°C)	220	230	240
B	Mould temperature (°C)	40	50	60
C	Injection pressure (bar)	80	90	100
D	Injection time (s)	1	2	3
E	Packing pressure (%)	70	80	90
F	Packing time (s)	5	10	15
G	Cooling time (s)	50	60	70

2.2. Shrinkage Measurement.

Rax Vision DC 3000 Mitutoyo profile projector was used to measure the shrinkage rate of moulded gear addendum and dedendum circle as well as tooth thickness by measuring their two dimensional contours. To reduce the variance in measurement, five samples of each injected gear are measured and their relative shrinkage rate are calculated based on following equation:

S = (D - Dm)/Dm x 100% (1)

Where S = shrinkage, D = injected gear dimension and Dm = mould dimension

Table 2. L_{18} OA of experimental study

Trials	A	B	C	D	E	F	G
1	1	1	1	1	1	1	1
2	1	2	2	2	2	2	2
3	1	3	3	3	3	3	3
4	2	1	1	2	2	3	3
5	2	2	2	3	3	1	1
6	2	3	3	1	1	2	2
7	3	1	2	1	3	2	3
8	3	2	3	2	1	3	1
9	3	3	1	3	2	1	2
10	1	1	3	3	2	2	1
11	1	2	1	1	3	3	2
12	1	3	2	2	1	1	3
13	2	1	2	3	1	3	2
14	2	2	3	1	2	1	3
15	2	3	1	2	3	2	1
16	3	1	3	2	3	1	2
17	3	2	1	3	1	2	3
18	3	3	2	1	2	3	1

3. Results and Discussion

3.1 Signal-to-Noise Ratio (S/N)

Taguchi method introduced the S/N ratio approach in order to measure the deviation of quality characteristics from their target value. The higher value of S/N ratio indicate the minimization of noise sensitivity of any process. The S/N ratio can be divided into three stages: the-lower-the-better, the-nominal-the-better and the-bigger-the-better. The objective of this study is to minimize the value of shrinkage in addendum and dedendum circle, as well as in tooth thickness of the PP injected moulded gears. The S/N ratio for all 18 trials are reported in Table 3.

Table 3. The S/N ratio of quality characteristics

Trials	Addendum circle (dB)	Dedendum circle (dB)	Tooth thickness (dB)
1	33.800	32.821	25.347
2	34.258	33.043	25.783
3	34.659	33.900	26.577
4	33.741	32.753	22.275
5	34.511	33.176	25.398
6	34.740	33.559	25.955
7	34.398	33.287	22.150
8	34.882	33.506	22.286
9	34.227	33.038	22.298
10	34.327	33.659	22.550
11	33.825	32.649	25.906
12	34.166	33.059	26.595
13	34.636	33.463	22.529
14	35.217	33.983	22.258
15	34.020	32.831	25.636
16	31.803	34.140	22.639
17	34.626	33.491	22.142
18	33.627	32.845	21.837

3.2 The effects of processing parameter on PP gears shrinkage

For better interpretation of the S/N results tabulated in Table 3 in determining the optimal levels of each processing parameters studied in this research, the results in Table 3 can be presented graphically. Figure 3 demonstrates the impact of variation in melting temperature (A), mould temperature (B), injection pressure (C), injection time (D), packing pressure (E), packing time (F) and cooling time (G) on the shrinkage behaviour in addendum and dedendum circle as well as in tooth thickness of the PP gears.

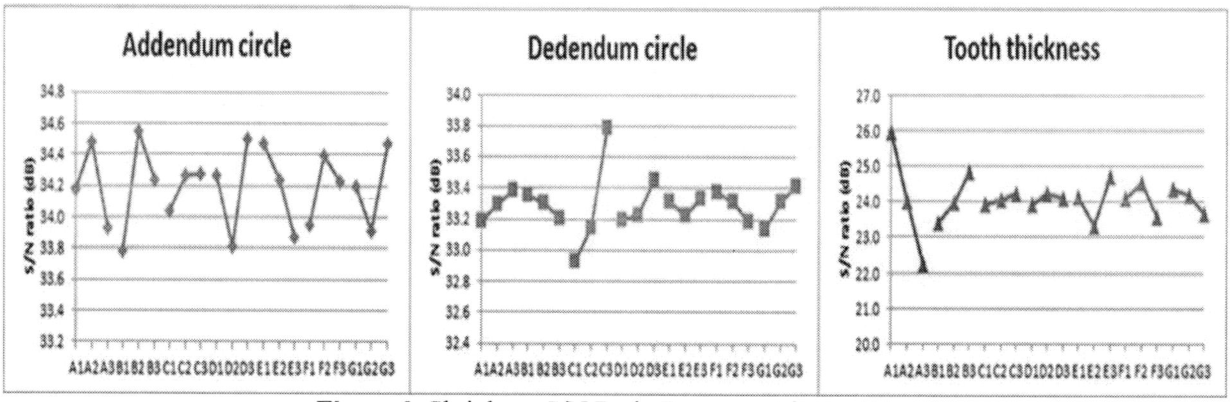

Figure 2. Shrinkage S/N Ratio Response Plot

From Figure 2, it is clearly shown that shrinkage behaviour in addendum and dedendum circle as well as in tooth thickness of the injected PP gears, which in this case, are presented by the S/N ratio, are affected significantly by variations in processing parameters. It can be observed that for the shrinkage in addendum circle of the moulded PP gear, the increment of melting temperature, mould temperature, injection pressure and packing time levels, the shrinkage rate has been minimized. However, the shrinkage rate seen to be increased upon the increment of injection time, packing pressure and cooling time levels. For shrinkage in dedendum circle of the moulded PP gear, the increasing of melting temperature, injection pressure, injection time and cooling time is observed in minimizing the shrinkage rate. Otherwise, the increasing of mould temperature, packing pressure and packing time levels have increased the shrinkage rate. Meanwhile, for the case of tooth thickness, the increment of mould temperature, injection pressure, injection time and packing time levels has reduced the shrinkage rate of tooth thickness. In contrary, the increment of melting temperature, packing pressure and cooling time is observed to increase the tooth thickness shrinkage.

The best combination of processing parameters and levels can be easily obtained from Figure 2 by selecting the level of each parameters with the highest value of S/N response. For addendum circle shrinkage minimization, A1B1C3D3E1F2G3 is identified as an optimal parameters combination, represent 220°C melting temperature, 40°C mould temperature, 100bar injection pressure, 3s injection time, 70% packing pressure, 10s packing time and 70s cooling. Meanwhile, A3B1C3D3E3F1G3 is identified as an optimal parameters combination in minimizing the shrinkage in dedendum circle represent 240 °C melting temperature, 40 °C mould temperature, 100 bar injection pressure, 3 s injection time, 90 % packing pressure, 5 s packing time and 70 s cooling time. On the other hand, A1B3C3D2E3F2G1 is identified as optimal processing parameters in minimizing the tooth thickness shrinkage of the moulded PP gear. The parameters setting are melting temperature of 220 °C, mould temperature of 60 °C, injection pressure of 100 bar, injection time of 2s, packing pressure of 90 %, packing time of 10s and cooling time of 50s.

3.3 Analysis of Variance (ANOVA)

To examine the extent to which processing parameters affects the shrinkage in addendum and dedendum circle as well as in tooth thickness of the moulded PP gear, ANOVA of the Taguchi method for the S/N ratios (Table 3) is performed. ANOVA enables engineers to quantitatively estimate the relative contribution of each processing parameter to the overall measured quality characteristics. Table 4 presents a summary of percentage contribution, denoted as "P" of all seven parameters on shrinkage in addendum and dedendum circle as well as in tooth thickness of the moulded PP gear. Roy [8] suggested of using the 10% rule, where, a parameter is considered insignificant when its influence is less than 10% of the highest parameter influence.

Table 4. ANOVA of percentage contribution

	Addendum circle	Dedendum circle	Tooth thickness
Parameters	P	P	P
Melting temperature °C)	9.960	3.440	69.549
Mould temperature (°C)	19.612	1.985	9.941
Injection pressure (bar)	2.293	70.533	0.385
Injection time (s)	16.002	7.238	0.434
Packing pressure (%)	12.202	1.230	9.639
Packing time (s)	6.503	3.050	4.675
Cooling time (s)	10.029	6.628	2.495

From the results of percentage contribution (P) in Table 4, the mould temperature appears to be the most decisive processing parameter in reducing the shrinkage in addendum circle of the moulded PP gear. The percentage contribution of the mould temperature is 19.612%, outweighing the other processing parameters followed by injection time, packing pressure, cooling time, melting temperature, packing time and injection pressure for the case of addendum circle. For dedendum circle, only two out of seven processing parameters, namely injection pressure and injection time, are considered to be significant, following the 10% rule. Meanwhile for tooth thickness, the most decisive processing parameter in reducing the shrinkage is found to be melting temperature with the percentage of contribution (P) of 69.549 %. By using 10% rule, the second significant parameter is mould temperature followed by packing pressure.

4. Conclusion

In this study, the influence of processing parameters on shrinkage in addendum and dedendum circle as well as in tooth thickness of PP injection moulded gears was investigated by using Taguchi approach. The findings from the analysis of the results, obviously showed that the variation in parameters setting have significant impact in minimizing the shrinkage in addendum and dedendum circle as well as in tooth thickness of the moulded PP gear. The set of optimal processing parameters that predicted in minimizing the shrinkage in addendum circle is identified as melting temperature of 220 °C, mould temperature of 40 °C, injection pressure 100 bar, injection time 3 s, packing pressure of 70%, packing time of 10s and cooling time of 70s. For the shrinkage in dedendum circle, the set of optimal processing parameters is predicted to be melting temperature of 240 °C, mould temperature of 40 °C, injection pressure 100 bar, injection time 3 s, packing pressure of 90%, packing time of 5s and cooling time of 70s. As for tooth thickness, the set of optimal processing parameters that predicted in minimizing the shrinkage is identified as melting temperature of 220 °C, mould temperature of 60 °C, injection pressure 100 bar, injection time 2 s, packing pressure of 90%, packing time of 10s and cooling time of 50s.

References

[1] Singh PK, Akant Kumar Singh S 2018 *Trib. Int.* **118** 264.
[2] Senthilvelan S, Gnanamoorthy R 2007 *Mat. and Desig.* **28** 765–772
[3] Hirogaki T, Aoyama E, Katayama T, Iwasaki S, Yagura Y, Sugimura K 2004 *Comp. Struc.* **66** 47.
[4] Kurokawa M, Uchiyama Y, Iwai, T, Nagai S 2003 *Wear* **254** 468.
[5] Düzcükoğlu H 2009 *Mat. and Desig* **30** 1060.
[6] İmrek H 2009 Trib. Int. **42** 503.
[7] Fei NG, Mehat NM, Kamaruddin S 2013 *ISRN Ind. Eng.* 2013 462174
[8] Roy RK 2001 Design of experiment using the Taguchi approach: 16 steps to product and process improvement, John Wiley and Sons, New York.

Dimensional accuracy and surface finish of investment casting parts by indirect additive manufacturing from fused filament fabrication

P Y Cheng[1], S F Khan[1,*]

[1] Additive Manufacturing and Design Lab, School of Mechatronic Engineering, Universiti Malaysia Perlis, Kampus Tetap Pauh Putra, 02600 Arau, Perlis, Malaysia.
[*] coresponding author, sfkhan@unimap.edu.my

Abstract. This article discussed the utilization of open-source 3D printer to manufacture master pattern for investment casting in term of geometry accuracy and surface finish. Two sets of 3DP parts comprising of one set with post-processing and another set without post-processing were used to make RTV silicone moulds. The wax patterns produced from RTV silicone moulds are used as patterns for making investment casting in Plaster of Paris mould. The results show that the post processing does not affect or improve the casted part as the final casted parts will undergo post-processing. This indirect approach investment casting using open-source 3D printer can be seen as new cost effective solution for small batch production of part which does not required allowable tolerance below ±1.14mm.

1. Introduction

Additive manufacturing (AM) or 3D printing is a process where virtual model created from computer aided design (CAD) software can be built into physical 3D part by adding materials in a layer-by-layer manner. This process can make model making or prototyping processes to become rapid. Fused filament fabrication (FFF) or also known as fused deposition modelling (FDM) is a material extrusion method in one of the seven methods in AM. FFF deposited materials onto a build platform layer by layer until the physical is built up and solidified. FFF can improved investment casting process by printing the 3D part as master-pattern to replace the wax pattern which is produced by injection moulding in die casting for investment casting. Study on lead time and cost comparison of fabricated sacrificial pattern used in investment casting showed that the AM technology could significantly reduce the lead time and cost as compared to conventional method for pattern production[1-2].

However, the study [1-2] used commercial 3D printer to build up the 3D printed patterns for investment casting. Thus the cost is relatively high for cottage industries implementation. Hence, low cost open-source 3D printer such as RepRap and Fab@Home is seen as a viable 3D printer to be used in investment casting for cottage industry or as a hobbyist tool in a person's home [3-4].

In indirect method, the master pattern of the desired casting is used to make the soft tooling such as room temperature vulcanized (RTV) silicone rubber moulding, epoxy mass casting and polyurethane face casting. For indirect AM method, the mould can used to generate multiple wax patterns to produce a low number of casting. However, the indirect method is not suitable when the geometry of the casting is complex and having internal cavities, because it would not be possible to remove the pattern without damaging the mould [5].

This paper discussed the geometrical accuracy and surface finish of investment casting parts produced by indirect additive manufacturing approach using post-process and without post-process 3D printed part from RepRap 3D printer.

2. Methodology

A CAD model is designed using CATIA software. The design model is shown in Figure 1 and the benchmarking parameter is shown in Figure 2 and Table 1. This work is focused on the deviation and shrinkage rate between the CAD model and the cast part. The CAD model file was converted into STL file and uploaded into Slic3r software to generate G-code. The calibration in Slic3r improves the quality of 3D printed part [6] and the setting is shown in Table 2. Proterface software acts as the 3D printer interface controller.

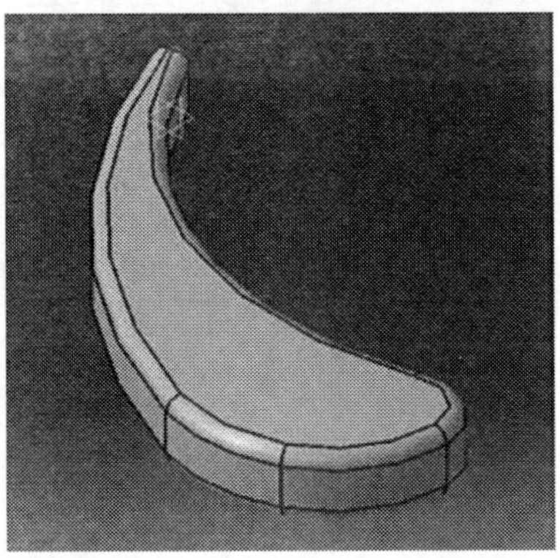

Figure 1. CAD Design Model

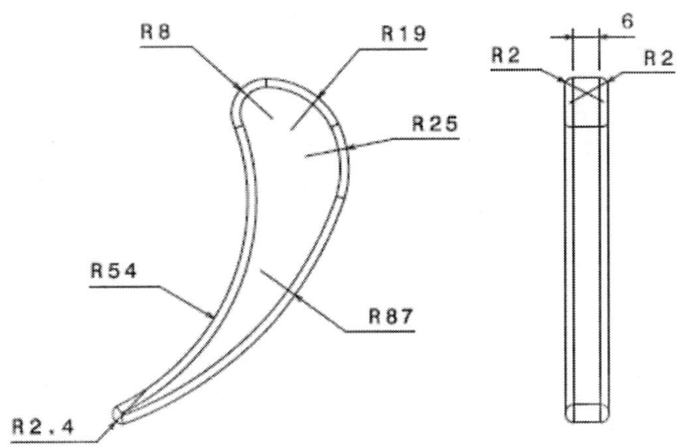

Figure 2. Design parameter

Table 1. Parameter of Design

Features Parameter	Line 1 (T1) Height	Curve 1 (C1) Radius	Curve 2 (C2) Radius	Curve 3 (C3) Radius	Curve 4 (C4) Radius	Curve 5 (C5) Radius	Curve 6 (C6) Radius	Fillet 1 (F1) Radius	Fillet 2 (F2) Radius
Value (mm)	10	19	25	87	2.4	54	8	2	2

Table 2. Slic3r Print Setting

Setting parameter	Layer height	First layer height	Horizontal shells solid layers (top)	Horizontal shells solid layers (bottom)	Fill density	Fill pattern	Speed for print moves	Speed for non-print moves	Extruder temperature	Bed temperature
Value (mm)	0.4 mm	0.35 mm	5 layers	5 layers	40%	Recti-linear	30 mm/s	130 mm/s	190 °C	60 °C

The printing filament material is Poly-lactic Acid (PLA). The recommended temperature range for printing using PLA filament is 185°C to 220°C. Two set of master-pattern was printed using PLA filament. One set of master-pattern is without post processing condition and the other set is with post-processing. Post-processing method used vapour smoothing process which is applied to improve the surface finish. In this process, acetone solution (C3H60) is used and is poured into a beaker and heated with 80°C. The beaker is sealed completed and wait the acetone vapourized completely. The part printed is put into the beaker. The process is carried out for 3 to 5 minutes so that the heated acid can be fully vapourized on the part.

2.1. Specimens preparation Soft tooling and investment casting

Room Temperature Vulcanizing (RTV) silicone is chosen to produce wax pattern for investment casting. Two piece mould method is selected so that it can be assembled and disassembled. Parting line was drawn onto the master-pattern such that modelling clay be placed to create the first part of the mould. The second part of the mould was created the same way as the first part but with sprue and air vents added in to make a gating system.

RTV silicone mould was created using a mixing ratio of curing agent and silicone rubber at 1.5:98.5 by weight. The mixture was stirred evenly and de-gas before pouring into the mould box to avoid air bubbles. The mixture was allowed to be cured overnight at room temperature.

Paraffin wax was used for investment casting and was heated in an oven at 90°C to melt it. The melted wax was poured into the RTV silicone mould. The wax solidified after 30 minutes and was removed from the mould.

2.2. Plaster mould and casting

Plaster mould was fabricated by mixing plaster of Paris powder and water with a ratio of 1:1 by weight. The mixture was poured into the mould box containing the wax pattern and left for one day for drying before de-waxing. The plaster mould is heated in an oven at 75°C to melt out the wax before casting of a low melting point aluminium alloy, Zamak. Zamak is melted at 400°C using gas metal melting furnace (TEK - 160 Gas Metal Melting Furnace 110V). The molten Zamak is poured into the plaster mould. After the cast is solidified, the casted prototype is removed from the mould using a hammer. The sprue and vents are cut off and the surface of casted prototype is finishing using angle grinder.

2.3. Parameter measurement

The dimensional accuracy measurements on cast parts were measured using Mitutoyo Beyond Plus GR2008-09571 Coordinate-measuring machine (CMM) with accuracy of 1μm (Figure 3). Surface Roughness Tester Mitutoyo CS-3000 525-780E-1 was used in surface roughness measurement of cast part (Figure 4). The results are given in term of Ra with μm unit. Table 3 shows the setting for the surface roughness tester.

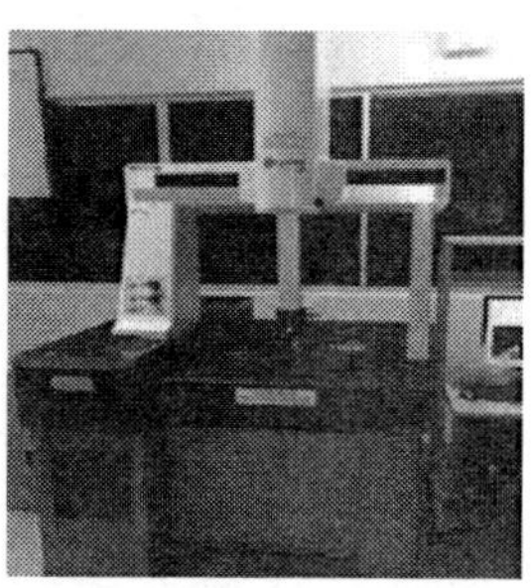

Figure 3. Coordinate Measuring Machine (CMM)

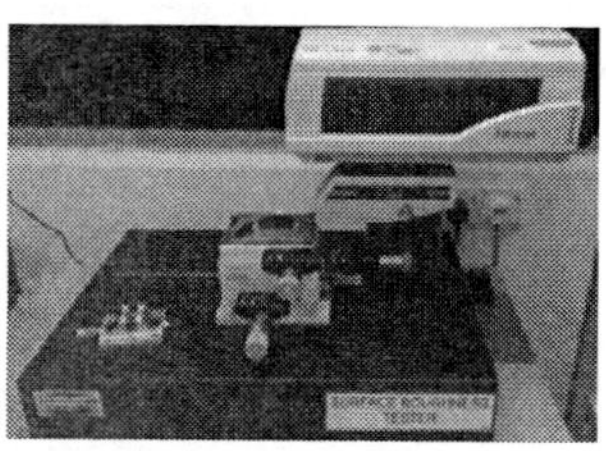

Figure 4. Surface Roughness Tester

Table 3. Surface Roughness Tester Setting

Feature	Measuring Speed (mm/s)	Measuring Length (mm)	Sampling Length (mm)
Top / Bottom	0.05	7.5	5.0
Side	0.05	1.8	1.0

3. Results and discussion

The analyses include comparison of parameter measurement between the master pattern and the cast part. The difference in parameters is calculated in length deviation and percentages.

$$Parameter\ difference = \frac{New\ parameter - Original\ parameter}{Original\ parameter} \times 100\%$$

The 3D printed part and the wax pattern is shown in Figure 5 and Figure 6. The Fillet 2 (F2) radius are not measured and recorded because the break-away support structure is built on that fillet. The fillets are not in proper shape after the supporting structure is removed. However the problem can be solved if dissolvable support material is used.

In Figure 7, from the visual inspection; the casted part have smooth surface area, however, there are minor defects on the bottom surface. This could be due to incomplete wax burn out or trapped air during casting.

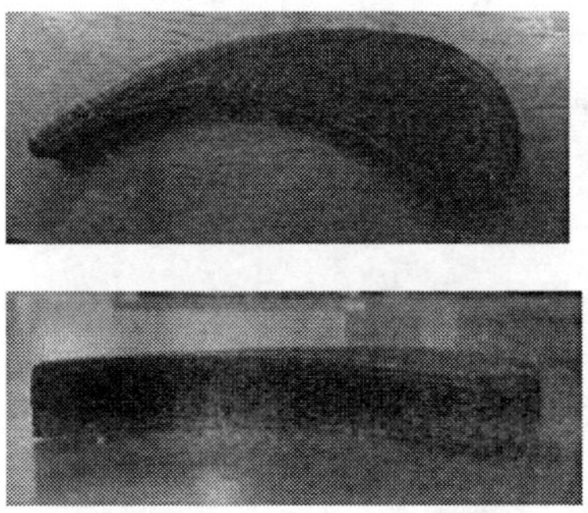

Figure 5. 3D Printed Part (Top View) (Side View)

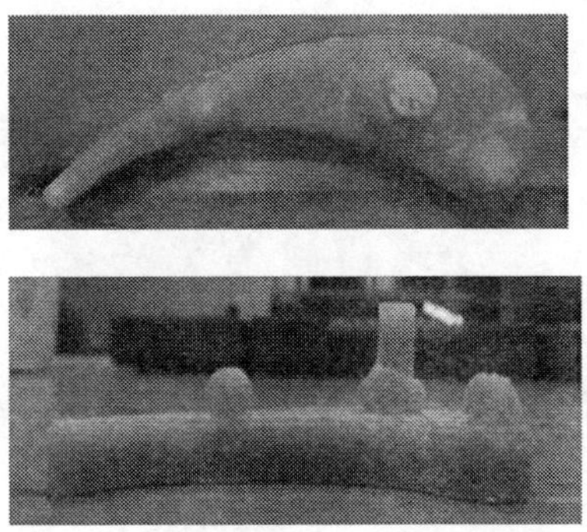

Figure 6. Wax Pattern (Top View) Wax Pattern (Side View)

Figure 7. Casted Part (Top View) Casted Part (Side View) Casted Part (Bottom View)

3.1. Geometrical accuracy and surface rough analysis

The comparison of deviation between CAD model and cast part is shown in Figure 8. From the result, it show that the deviation between casted prototype which the FFF prototype with and without post processing are similar. This is because the difference between with and without post processing FFF prototype is negligible. The maximum deviation between CAD model and casted prototype is +1.14mm which is the Fillet 1 feature.

The surface roughness for casted part which the FFF Model with and without post processing is shown in Figure 9. It shows that the post processing process does not affect the surface finish of casted part because the surface roughnesses for both casted parts are similar. Besides that, the casted parts have better surface finish compared to FFF model after they undergoes the post-processing such as part removal and finishing.

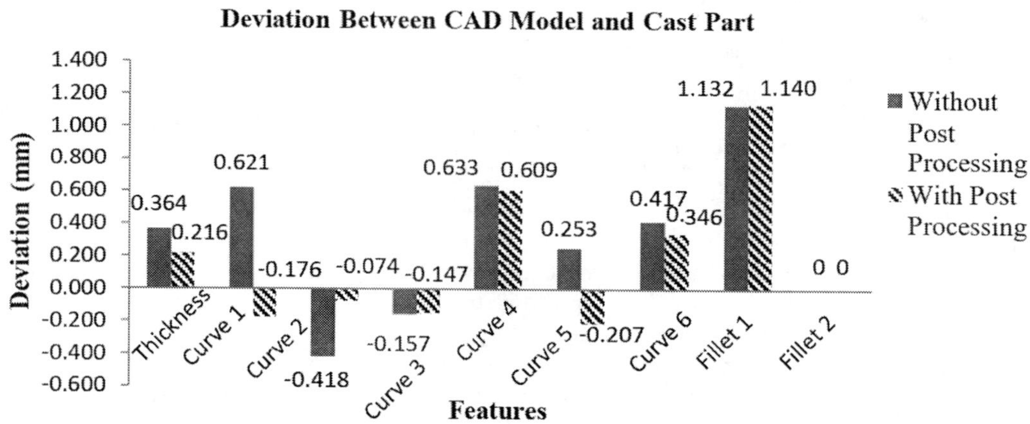

Figure 8. Comparison of features deviation between CAD Model and Cast Part

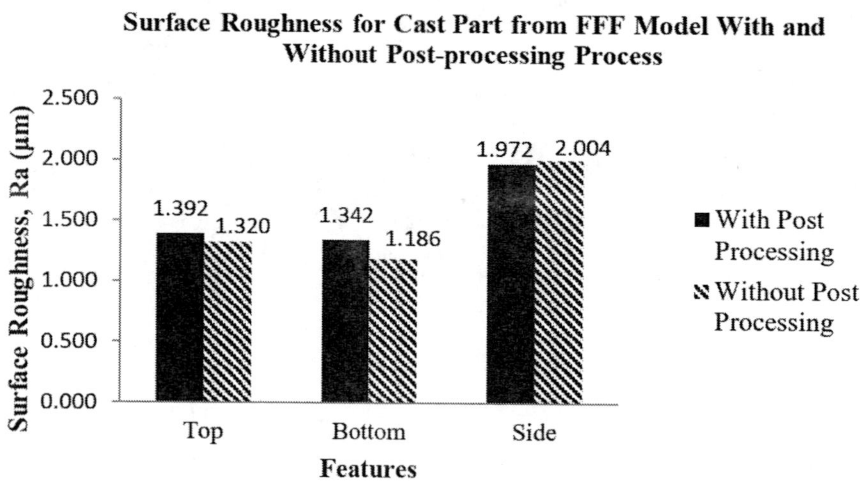

Figure 9. Surface Roughness for Cast Part from 3D Printed Part with and without Post-processing.

4. Conclusion

The experiment show that the post processing process on the FDM model does not affect or improve the dimension accuracy and surface finish of casted part in investment casting. The maximum deviation of casted metal built by 3DP model which does not undergoes the post processing process is +1.132mm while the maximum deviation of casted metal built by 3DP model which undergoes the post processing process is +1.140mm.

On the other hand, the highest surface roughness for casted metal built by 3DP model which does not undergoes the post processing process is 2.004 µm while the highest surface roughness for casted metal built by 3DP model which does not undergoes the post processing process is 1.972 µm. Therefore, post processing process on the FDM model is unnecessary in investment casting.

The result shows that the dimensional accuracy in indirect investment casting using RepRap 3D printer is low. The maximum deviation of casted metal is +1.14mm. However, the surface finish of casted part is good. The surface roughness is ranged from 1.186 to 2.004µm. The shape of 3DP printed part is also limited as the printed part will damaged during the support material removing process.

IOP Conf. Series: Materials Science and Engineering **429** (2018) 012100 doi:10.1088/1757-899X/429/1/012100

However, this problem can be solved by using other open source 3D printer which the supporting material is different material with building material.
This indirect approach investment casting using RepRap 3D printer can be used to manufacture the simple part which the allowance tolerance is ±1.14mm. This method is suitable for high batch production of metal artificial stature or souvenir in cottage industry because the manufacturing cost is low.

5. Acknowledgement

The authors would like to express profound gratitude and deep appreciation to the School of Mechatronics Engineering and Universiti Malaysia Perlis in providing research platform well as providing the facilities and equipment.

6. References

[1] Chua C K, Feng C, Lee C W and Ang G Q 2005 *The International Journal of Advanced Manufacturing Technology*, Vol. **25**, No. 1-2, pp. 26-32.
[2] Khan S F, Dalgarno K W and Siregar R A 2015 *Applied Mechanics and Materials*, Vols. **786**, pp. 354-360,
[3] Pearce J M 2010 *Journal of Sustainable Development*, vol **3**, no. 4, pp 17-29
[4] Pramod V, Baiju A, Babuji A, Plathanam R, and Abraham A K 2015 *International Journal of Emerging Technology and Advanced Engineering*, Volume **5**, 199-202,.
[5] Chua, C.K., Chew, T.H. and Eu, K.H 1998*The International Journal of Advanced Manufacturing Technology*, Vol. **14**, No. 9, pp.617-623,
[6] Lanzotti A, Martorelli M, and Staiano G 2015 *Journal of Manufacturing Science and Engineering*, Vol. **137**.

IOP Conf. Series: Materials Science and Engineering **429** (2018) 012101 doi:10.1088/1757-899X/429/1/012101

Effect of infill on tensile and flexural strength of 3D printed PLA parts

S.F.Khan[1,*], H. Zakaria [1] ,Y.L.Chong[1], M. A. M. Saad[1], and K. Basaruddin[1]
[1] School of Mechatronic Engineering, Universiti Malaysia Perlis, Kampus Tetap Pauh Putra, 02600 Arau, Perlis, Malaysia.
[*]Corresponding author: sfkhan@unimap.edu.my

Abstract. Development in open source three-dimensional (3D) printing has generated interest for creating low-cost customised objects; in both personal and commercial. In the 3D printing process, the quality and cost of the objects produced are normally influenced by the processing parameter. One of the parameter is infill patterns. Infill patterns can be used to reduce material usage and printing time. Various infill patterns are available in the 3D printing software. Four different infill patterns were used to evaluate the mechanical properties of fabricated parts. Open source 3D printer was used to print five PLA specimens for each different infill pattern. Tensile and 3-point bending tests were performed on the printed parts in order to determine the tensile and flexural strength. The mechanical strength of printed parts is influence by different infill patterns. The results obtained indicated that Rectilinear pattern has the highest tensile and flexural strength. The average tensile strength of Rectilinear is 19.1 MPa with average modulus elasticity of 10.51 GPa, whereas the average flexural strength of Rectilinear is 24.4 MPa with average tangent modulus elasticity of 0.359 GPa.

1. Introduction

Additive Manufacturing (AM) technology is utilized in making physical objects from three-dimensional (3D) models generated by computer-aided design (CAD) software. In the late 1980's the method or technology of fused deposition modeling (FDM) was invented and expanded by S.Scott Crump and was commercialized in 1990. After the patent expiration of FDM innovation, a vast open-source advancement group was formed and both businesses and DIY variations using FDM for 3D printing showed up [1]. RepRap, a short form for Replicating Rapid Prototyping, is a kind of 3D printer that was developed after the patent of FDM expired and released with concept of open free software license. RepRap 3D printer basically able to print out different type of pattern fills of the internal structure of the printed part and it will generate different of mechanical properties.

However, the printed pattern fills characteristics of the internal structure of the printed parts is not fully defined and studied yet. It is desirable to include others parameters to support its printing process such as control of temperature and internal supporting structure.

2. Methodology

In order to test the specimen's mechanical properties, a series of samples is required to be built with different infill patterns. The test specimens were design using CATIA V5R20 computer-aided design (CAD) software to ASTM D638-14 and ASTM D790-10 for tensile and bending test. The geometrical dog-bone and rectangular bar test specimen model were representation as solid model in CATIA. The solid CAD models were saved in Standard Triangular Language (STL) file format which is the file format that can be read by 3D printer. STL file format interpret the external closed surfaces from solid CAD model and divide the model into slices for printing purpose. However, some general adjustments are required of the STL file such that it is correct in size, position, and orientation for test specimen printing[2]. The adjustments were made using Slic3r software. The G-code file was decoded by the Pronterface software and printing time was estimated, the designed specimen were shown in the software. After confirming the proper printing setup, the printer can start printing. Polylactic Acid (PLA) filament was used as obtained from supplier to print the specimens.The selected PLA was extruded from the heated nozzle and build the specimen layer by layer from bottom until the top according to the coordinate as generated in the G code until the specimen is completely formed as drawn in the CAD file. Six specimens are printed for each infill pattern from dog-bone shape and rectangular bar shape respectively.

2.1. Printing Setup

Slic3r was also used to setup the printing parameter in term of material usage, temperature, nozzle movement speed, nozzle flow rate and the machine stability besides infill patterns. The dimensional accuracy and surface roughness of the specimens can be influence by these parameters. Set of testing specimens were produced selected infill patterns of rectilinear, concentric, honeycomb and hilbert curve. The Slic3r software then generates G-code from the STL files with the setup parameters. Open-source 3D printing machine based on REPRAP 3D Printer used this G-code to print 3D specimens

2.2. Tensile and Flexural Test

The mechanical behaviour for a particular part is the reaction of the material to a mechanical stress. The deformation causes from the applied force of a component relying on the direction of the applied force and the size of the component design or mechanical properties [2]. The mechanical properties of printed part from PLA such as tensile strength and flexural strength were determined according to American Society for Testing and Materials (ASTM) standard.

Tensile testing of the specimens was performed on a Universal Testing Machine (UTM) by using the crosshead testing rate of 5mm/min. In order to obtained the tensile properties of the testing specimens [4] accurately, the printed specimens was prepared based on the designed dimensional according to the ASTM D638-14. There were at least five dumbbell or dog-bone shape specimens have to be tested for each of the type of infill pattern so that results will be more precise [5].

Flexural testing of the specimens was performed on a Shimadzu Universal Testing Machine using a particular crosshead testing rate of 1.28 mm/min. In order to identify the flexural properties of the testing specimens, the printed specimens was prepared based on the designed dimensional according to the ASTM D790-10. There were at least five rectangular bar shape specimens have to be tested for each of the type of infill pattern.

The drawing file was input and the details parameters such as drawing dimension and geometry design. In this tensile properties testing would focus on determining the tensile strength, elongation and modulus of elasticity or Young's Modulus as well as respectively standard deviation of the printed part with four different types infill patterns. The tensile testing was conducted on those specimens with setup the crosshead speed of 5mm/min and applied force of 100kN. All the specimens were designed according to the ASTM D638-14 in category of Type I specimen.

3. Results and discussion

3.1. Comparison of Stress-strain Among Infill Patterns

Figure 1 show the details scattered plot for different infill patterns which showed that the Rectilinear infill pattern contains highest tensile strength compared to Concentric, Honeycomb and Hilbertcurve. While for Hilbertcurve infill pattern was lowest tensile strength or can be classified as weakest in tensile strength among these four types of infill patterns. The difference of tensile strength between Rectilinear and Hilbertcurve was 14.93 MPa. This implies that Rectilinear can withstand extra 78.17% of load compare to Hilbertcurve before going to rupture.

From the results of maximum strain showed that Rectilinear and HoneyComb are highest and closed to each other which contribute 3.45% and 3.66% respectively. The difference of maximum strain between both of these infill patterns was 0.21%. Hence, this indicate that the force distribution among the intra bonding of HoneyComb available to flow much more fluent and less concentrate in the linkage chain as a result it yields a slightly higher strain of extra 6.09% than Rectilinear.

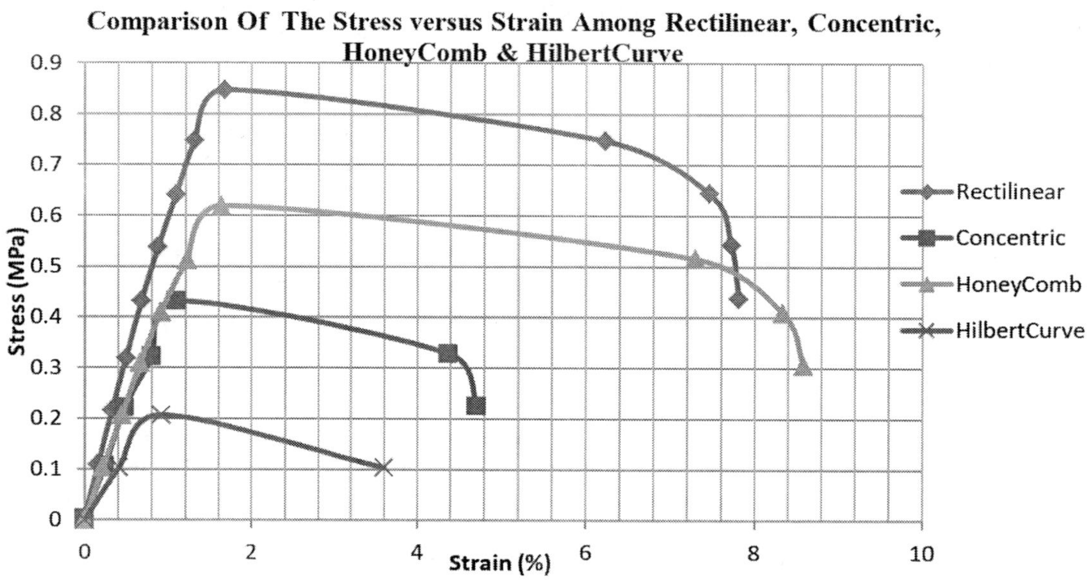

Figure 1. Scattered plot of infill patterns

Modulus of Elasticity among these four types of infill pattern was obtained by plotted a slope in the elastic region of the graph and identified that Rectilinear infill pattern was highest which achieved 10.51 GPa, where Concentric was 7.34 GPa, HoneyComb was 6.92 GPa and HilbertCurve was 3.27 GPa which is the weakest among these four infill patterns. The difference of modulus of elasticity among the strongest and weakest was 7.24 GPa and this possible to indicate that the different infill patterns have a considerable influence on strength and modulus tensile properties of the material.

It is possible to conclude that the Rectilinear infill pattern overall contains the strongest tensile mechanical properties among the four different type infill pattern. Although, the HoneyComb infill pattern attain a slightly higher 0.21% of maximum strain, this is possible due to fabrication of printed part where the excess material extruded from the nozzle and causes the orientation of the molecule chains which decreases the elongation characteristics of the infill pattern [5]. In Table 1 is the summary of tensile properties for those infill patterns have been tabulated.

Table 1. Summarize Of Tensile Properties against the Infill Patterns

	Infill Pattern			
Tensile Properties	Rectilinear	Concentric	HoneyComb	HilbertCurve
Young's Modulus, E (GPa)	10.51	7.34	6.92	3.27
Tensile Strength, σ (MPa)	19.1	10.1	13.2	4.17
Maximum Force, F (kN)	0.799	0.463	0.611	0.207
Maximum Strain, ε (%)	3.45	2.48	3.66	1.74

3.2. Comparison of Infill Patterns

The data showed the details scattered plot, Figure 2 showed that the Rectilinear infill pattern contains the highest flexural properties compared to concentric, honeycomb and hilbertcurve. While for hilbertcurve infill pattern was lowest flexural properties or can be classified as weakest in flexural properties among these four types of infill patterns. The difference of flexural strength between Rectilinear and Hilbertcurve was 14.7 MPa. This implies that Rectilinear can withstand extra 14.7 MPa of maximum flexural stress compare to Hilbertcurve before going to rupture. Tangent modulus of elasticity among these four types of infill pattern was obtained by plotted a slope in the elastic region of the load-deflection curve with specific formula calculation and identified that Rectilinear infill pattern was highest which achieved 0.359 GPa, where Concentric was 0.263 GPa, HoneyComb was 0.287 GPa and HilbertCurve was 0.236 GPa which is the weakest among these four infill patterns. The difference of modulus of elasticity among the strongest and weakest was 0.123 GPa.

Although, the flexural strain showed that HilbertCurve was 0.068% higher than the other three infill patterns, but overall Rectilinear was indicated that better results than Concentric, HoneyComb and HilbertCurve regardless whether it is for tangent modulus of elasticity, flexural stress, flexural strength and maximum force as shown in Table 2.

It is possible to conclude that the Rectilinear infill pattern overall contains the strongest flexural mechanical properties among the four different type infill patterns. However, HilbertCurve considered its flexural properties are advance in term of flexural strain. Hence, it is possible to indicate that the different infill patterns have a considerable influence on strength and modulus of flexural properties of the material. In Table 2 is the summary of flexural properties for the infill patterns.

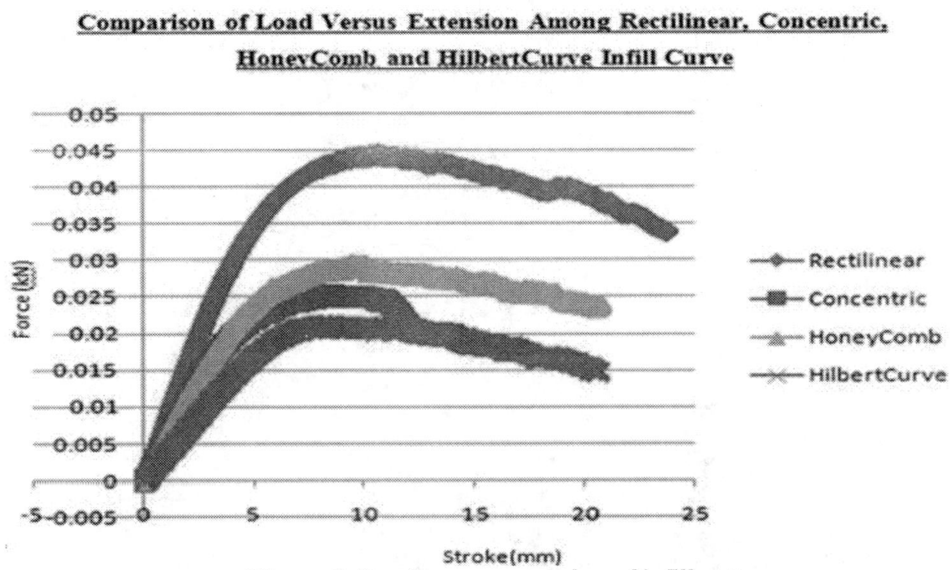

Figure 2. Load versus extension of infill patterns

Table 2. Summarize of Flexural Properties against the Infill Patterns

	Infill Pattern			
Properties	Rectilinear	Concentric	HoneyComb	HilbertCurve
Tangent Modulus of Elasticity, E_B (GPa)	0.359	0.263	0.287	0.236
Flexural Stress, σ_{fB} (MPa)	24.4	12.8	14.5	10.6
Flexural Strength,	24.4	12.5	14.0	9.7

IOP Conf. Series: Materials Science and Engineering **429** (2018) 012101 doi:10.1088/1757-899X/429/1/012101

σ_{fM} (MPa)				
Maximum Force, P (kN)	0.042	0.025	0.030	0.020
Flexural Strain, ε_f (%)	0.056	0.054	0.061	0.068

3.3. Comparison of Mechanical Properties of Infill Patterns

The best modulus of elasticity was plotted in graph as shown in Figure 3 below. It can be observed that Rectilinear infill pattern was highest whereas HilbertCurve infill pattern was lowest in term of tensile and flexural test as well. Meanwhile, the best strength graph was plotted as shown in Figure 4 below. Obviously we able to define that Rectilinear infill pattern consist of highest strength while HilbertCurve was lowest in both tensile and flexural test as well.

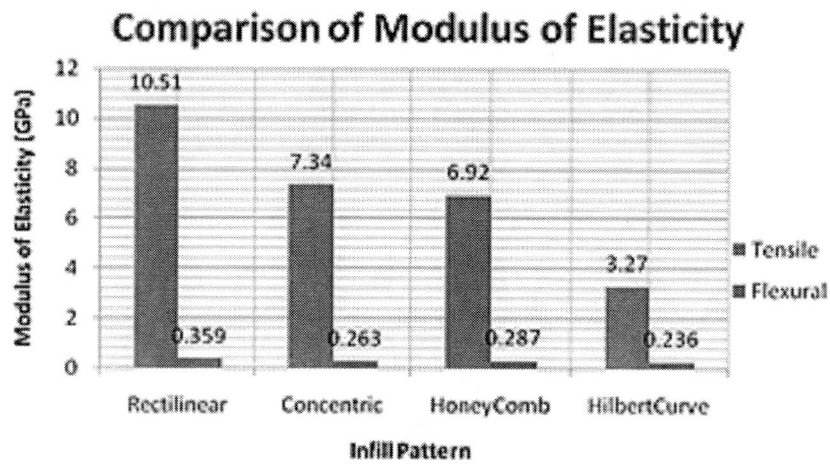

Figure 3. Modulus of elasticity in different infill

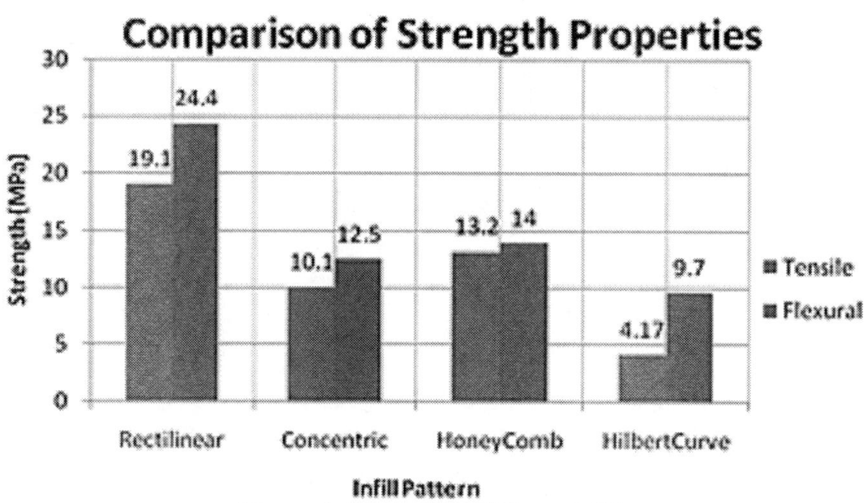

Figure 4. Strength in different infill

4. Conclusion

Experimental testing was conducted in order to identify the mechanical properties of 3D printed part that printing by RepRap 3D printer meanwhile to select the desired infill pattern. There are some conclusions found out from this research. Firstly, the setting of the parameters of the RepRap 3D printer such as density, nozzle temperature and movement speed and stability of the machine as well was important, this can be affect the printing quality of the specimen and causes the deviation towards the testing results.

Secondly, from the experimental testing results showed that the specimens generated with different infill pattern was given different prominence results, therefore can be concluded that the mechanical strength properties was depend on the given infill pattern that printed by RepRap 3D printer.

Thirdly, throughout the experimental testing results, Rectilinear infill pattern was achieved the best results among the four different type of infill patterns such as Rectilinear, Concentric, HoneyComb and HilbertCurve.

Fourthly, in comparison with computer simulation and experimental testing results showed that the tensile analysis was as expected achieved the same outcome with experimental testing results, where flexural analysis was beyond expectation and showed that HoneyComb as strongest strength, therefore computer simulation was only as a reference since we unable exactly duplicate whichever parameters, conditions, or influences into the virtual simulation software, consequence produce the divergent results.

Fifthly, from the results of experimental testing results compare with manufacture's bulk material datasheet showed that the strength of printed part was weaker compare to original bulk material, this is possible to conclude that the bulk material after undergone heated and melted was definitely changed its properties.

5. Acknowledgment

I would like to thank the School of Mechatronic, University Malaysia Perlis for the support and facilities provided in completing this research.

6. References

[1] Fused Deposition Modeling. (n.d.). Retrieved November 2014, from Wikipedia: http://en.wikipedia.org/wiki/Fused_deposition_modeling

[2] Villalpando L., Eiliat H., Urbanic R.J. 2014 *Procedia CIRP*, Vol. **17** pp. 880-805.

[3] Davis J. R. 2004 *ASTM International*, pp. 1. ISBN 978-0-87170-806-9

[4] Bagsik A., Schöppner V. 8–10 May 2011, *Proceedings of the Society of Plastics Engineers* ANTEC Conference, (Anaheim); Boston, MA, USA.; pp. 1–5..

[5] Bellini A. and Guceri S. 2003 *Rapid Prototyping Journal*, Vol. **9** Issue: 4, pp.252-264.

IOP Conf. Series: Materials Science and Engineering **429** (2018) 012102 doi:10.1088/1757-899X/429/1/012102

The Challenges of Space Management in Higher Education Institution

M N F Saaid[1*], M A Ayob[1], A I A Yunus[1], M A Ayob[1], A W Razali[2], K A Maarof[1]

[1]Faculty of Architecture, Planning and Surveying, Universiti Teknologi Mara, Kampus Kota Samarahan, Sarawak, Malaysia.
[2]Faculty of Engineering, Universiti Malaysia Sarawak, Kota Smarahan, Sarawak, Malaysia

*Email: mohammad928@sarawak.uitm.edu.my

Abstract. Space management is a process to determine efficiency, allocation of space, problems and issues including monitoring the current usage of space. Space Management in Higher Education is an important subject as providing a space is expensive and university itself should be able to provide the best service to support the core business of the universities. However, university commonly faced problem of limited space to cater universities programs and some spaces are not fully utilized. The challenges during implementation of space management in higher education institution are availability of information, personnel experience, availability of space, different interest and objective, lack of research and ineffective systems implemented. Thus aim for this paper to highlight the barriers and challenges faced on the space management implementation. A thorough literature review conducted to gauge potential of space management in Higher Education. The questionnaire distributed among academic staff of Higher Education in Sarawak to determine the challenges towards the space management. The finding revealed that space limitation in the higher education institution is the main challenges towards efficient implementation of space management. The outcome of this study will help to provide some insight for Higher Education in Sarawak to plan a better space management.

1. Introduction

Increase in Malaysian population creates demands for facilities especially the space for education institution and it leads to growth of number of universities each and every year in order to cater the enrolment of new students. Universities expected to be able to provide the best facilities in term of space to support universities' core business. For universities, it is crucial to provide space which enlisted as among of the important element in related for the activities to take place on the specific work to be held [14]. Teaching and learning including research activities take place and it is proven that space is one of the important elements to accommodate the needs of education system [13]. In addition, scarcity in term providing excellent space in higher education institution all over the world become one of the top ten issues faced by system [12]. University should be able to provide an efficient and excellent of the space resources [10]. According to Space Management Project [1], space utilization being defined as how rooms and spaces are being used in term of how they are in use and how the in use. In addition, space within its organization is the most expensive assets own [4].

1.1. Space management

Space management is one of the multi-disciplinary elements in facilities management and the concept of space management is to deliver service and how to manage the sources effectively. Space management includes alternative officing, free addressing, group addressing, group addressing, virtual office, space sharing and on and of premises management [5]. The effectiveness of space management must be emphasized by the top management itself and according to NAO [5], there are three essential elements that must be well addressed which area to develop management committee members, model or technique to manage space and knowledge of people responsible for the space management. Space considers as valuable and expensive as it cost a lot of money to create proper and suitable spaces and it also possible to generate income for universities. [3, 4, 6]. Thus, spaces and rooms must be fully utilized to prevent problems in space management and if not been managed properly, it will cause waste of space. Expenditure cost and level of productivity affected by efficiency and effectiveness of space usage [5]. When space is not fully utilized, it will induce cost of operation, energy, cleaning and maintenance [2]. Despite or providing new spaces for teaching and learning, existing space resources must be fully utilized as to address new ways of teaching and learning adopted in higher education institutions. Government itself invested billions of ringgit in order to provide facilities for education as an efficient and effective space usage will be control and expenditure cost and level of productivity [9].

1.2. Challenges of Space management

There are several constraints in effective space management such as the lack of information on the predicted and observed is not being identify at early stage during strategic planning regarding types spaces and hours of activity going to be conducted [1, 7]. If the utilisation of space is not optimize, it is possible to reconfigure by means of redesign spaces, rooms sizes but restricted by cost. The utilisation space can be measured on frequency and occupancy rates, but the frequency rate is easier to address compared to occupancy rate. Frequency can be measured on how many times a space or room being use but occupancy more to the numbers of occupant vs capacity of the space itself. The second challenges are partial information on the spaces going to be used for how many hours and size of group [1, 3, 7]. The person in charge for the timetabling does not have full information on the room capacities, exact number of occupants (size of group), types of rooms, available equipment, hours of spaces going to be used and others. The third challenges are different interest and objectives for the institution, academic staff, administrator and students will become one of constraints, [1]. Top management interest more to financial health for its institution as space can generate income for university. Although space considered as assets that can generate income, the interest for the academic staff, and students itself might be differ. For example, academic staff promote flexible course as to attract enrolment of students but is care less on the utilisation of the space itself. The fourth challenges are lack of research on space management. Previous researchers found that researching in space management area was quite difficulties [6]. Lack of research on space management or space utilization become one of the constraints in which the data collected can be used as a guideline [7]. This statement supported by [4] although lacking in term of research, it is balanced by numerous technical reports throughout the world. The fifth challenges are person in charge is not an expert in space management and optimization of space was restricted due to too many parties involve but there is no centre specializes in managing space [2]. These statement supports by [7] that the person in charges in space is have no background of this field. The sixth challengers are limited space where as the limitation of space restricted the space management system to be implemented efficiently [8] as the space itself is not being well equipped in accordance with the teaching and learning requirements [7]. HEIs facing common space management problems such as low utilization rate for teaching space and usage of space mismatch with its design [4]. The limited space mention by previous researcher is not mainly on limitation on the space but it includes restriction of spaces to be used by occupants. The seventh challenges are top management support. According to [10], top management support gives significant impact towards effectiveness of space management implementation. As a result, lack of top management support contributes to low utilization of space even though best planned and technique been applied on managing space. Top management may establish space management committee [5]. The eight challenges are data from the survey that been collected does not apply to the space management policy. This scenario happened due to the difficulties to interrogate and analyse data to apply it into information management system. Furthermore, it is proven to be costly and time consuming [1].

2. Research Methodology

This paper is to identify the challenges of space management implementation at UiTM Kota Samarahan and two blocks has been selected as the case study (Block M and Blok G). The methods used for this paper is based on both primary and secondary data collection. To achieve the objective, researchers provided sets of questionnaires among academic staff in related to space management. Sets of questionnaires distributed among the academic staff that occupy these 2 blocks. This paper was using Likert scale as 1 indicated as "Strongly Disagree" until 5 which indicated as "Strongly Agree".

3. Results and Discussion

This paper is to identify the challenges of space management implementation at UiTM Kota Samarahan and two blocks has been selected as the case study (Block M and Blok G). The methods used for this paper is based on both primary and secondary data collection. To achieve the objective, researchers provided sets of questionnaires among academic staff in related to space management. Sets of questionnaires distributed among the academic staff that occupy these 2 blocks. This paper was using Likert scale as 1 indicated as "Strongly Disagree" until 5 which indicated as "Strongly Agree".

TABLE 1. Summarization of Challenges of Space Management

No	Challenges	Mean
1	Limited Space	4.35
2	Person in charge have no background in space management	4.20
3	Lack of research	3.95
4	Lack top management support	3.80
5	Different interest and objective	3.75
6	Data collected not utilized	3.75
7	Partial information on spaces	3.70
8	Predicted and observed	3.70

From the Table 1 as mentioned above, it shows that the highest rate of challenges in space management is limited space. limitation of space restricted the space management to be implemented efficiently. 40% of respondents answered strongly agree and 55% answered agree that limited space is one of the challenges in space management due to every faculty has been allocated with designated spaces or rooms and its quite difficult to used undesignated space. For example, Faculty of Architecture, Planning and Surveying has been allocated with 2 blocks academic which are block M and Block G, but in these two (2) blocks, there are certain room for the use of faculty. Limited space is not only on the number of spaces can be used but it includes the suitability of the spaces for teaching and learning, for example big theatre hall must be equip with speaker, projector, very big white board and ventilation system. Low utilisation rate of space be affected by limitation of space. According to [7, 8] limited space is not only on the spaces or rooms, but it includes restriction to use the spaces and rooms be used. Thus, limited space means restriction to use another space even though it is unused. The lowest challenges are share by 2 which are predicted and observed and partial information. 50% of respondents answered Agree there is gap between predicted and observed as one of the challenges in space management. During strategic planning, it must be set up how many rooms, students inducing suitable equipment for the room to be used during teaching and learning but normally it is inevitable to predict exact number of rooms, capacity of the room including size of groups going to use the rooms. According to [1, 7], it is not being identified at the early stages of strategic planning in term of types of spaces, hours of activity including size of group. While for the partial information, 60% answered Agree and supported by [1, 3, 7] that the person in charge have incomplete information regarding room capacity, functionality, exact number of occupants, available equipment, intended hour of usage become one of the setback to do timetabling and schedule.

4. Conclusion

In conclusion, space management is one of the significant element in facilities management. Space are considered as valuable assets that can give direct impact to financial health of higher education institution. Abundance of unused space lead toward waste of money as space need to be operate, maintaining, energy and cleaning. If it is not being manage effectively and efficiently, it will create extra expenses at the end. Therefore, the need of finding solution to counter the setbacks are required and it involved a lot of parties to play their roles as to create sustainability in space management. This paper is helpful to highlight challenges in enhancing the quality of space management in higher education institution. Based on the results, there are eight challenges obtain from the questionnaire which are limited space ranked highest and predicted and observed and partial information ranked lowest.

References

[1] Space Management Project 2006 Space Utilisation Practice Performance and Guidelines
[2] Ibrahim I, Yusuff W Z W, Bilal K 2012 Space Management A study on a space usage level in higher education institution *Social and Behavioral Science*
[3] Aziz A A, Hashim A E, Baharum Z A 2013 Space Inventory Management in the Malaysian Public Universities *Social and Behavioral Science*
[4] Abdullah S, Ali H M, Sipan I, Awang M, Rahman M S A, Jibril J D 2012 Classroom Management Measuring Space Usage *Social and Behavioral Sciences*
[5] NAO Space Management in Higher Education 1996 A Good Practice Guide National Audit Office London
[6] Wahab A A 2005 Pengurusan Sumber Fizikal IPT Pengurusan Ruang *Universiti Teknologi Malaysia*
[7] Rahman A M S, Abdullah S, Ali H M 2012 Space Utilisation Survey (SUS) in Malaysia HEI's Towards Sustainable of Existing Building Assets *Social and Behavioral Science*
[8] Zainuddin K, Mokhtar E S, Wan Yusoff K 2011 Developing a UiTM (Perlis) Web-Based of Building Space Management System A Preliminary Study in Locating a Specified Space/Room Area Using Open Source GIS Tool *Engineering 20*
[9] Ibrahim I, Yusoff W Z W, Sidi N S S 2011 A Comparative Study on Elements of Space Management in Facilities Management at Higher Education Institutions *Social and Economic Development*
[10] Amaratunga D, Baldry D 2000 Assessment of Facilities Management Performance in Higher Education Properties *Facilities Management 18*

IOP Conf. Series: Materials Science and Engineering **429** (2018) 012102 doi:10.1088/1757-899X/429/1/012102

[11] New Castle University 2007 Report of Space Management in Higher Education Newcastle University
[12] Marmolejo F 2007 Higher Education Facilities, Issues and Trends PEB Exchange
[13] Steven S 2009 No more room abroad the ark A UK higher education perspective on space management Interlending & Document Supply
[14] Burke E, Varley D 1996 Space Allocation An analysis of Higher Education Requirements Nottingham UK

The Insight on Quality Assessment System in Construction (QLASSIC) Implementation in Sarawak

K M A K Azir[1,a], **W M N Muhammad**[1,b], **M N F Othman**[1,c] and **D H Daeng**[1,d]

[1]Faculty of Architecture, Planning and Surveying, Universiti Teknologi MARA, Cawangan Sarawak, Kota Samarahan, Malaysia

*Email: asyraf765@sarawak.uitm.edu.my

Abstract. Construction industry is one of the industries that contributes significantly to negative impact in terms of material, workmanship and quality of building. To overcome this issue, the Malaysian Construction Industry Development Board (CIDB) suggested a quality assessment system using a point monitoring checklist called Quality Assessment System in Construction (QLASSIC). QLASSIC is an effective tool to provide a criterion quality workmanship for contractors, hence allowing the building occupier getting the most quality building product from the developers. This initiative will be aligned with the national quality and safety aspiration as stated under the Construction Industry Transformation Programme (CITP) 2016-2020. Furthermore, in Sarawak with rapid construction industry growth of 13 percent over the year, the numbers of construction project assessed by QLASSIC is still considered at a low pace. Thus, this study aims to deliberate the insight of contractors and developers on the assessment system. Ample literature review was conducted on building quality assessment's benefits and its challenges, and later, a pilot survey were conducted among the contractors and developers in Sarawak for validation. In conclusion, this study revealed that Sarawak construction industry is slightly increasing towards the usage of quality assessment. However, lack of knowledge and getting a lower score of assessment may discourage the continuity of the implementation. The research findings may be a basis of references and necessary measures to promote QLASSIC implementation in Sarawak.

1. Introduction

Nowadays, buildings and infrastructure developments are being rapidly constructed in Malaysia especially in Sarawak, which plans to achieve the developed state status in 2030. In the last six years, the Sarawak construction industry continued to boom at over 13 per cent over the year and it is estimated about RM168 billions worth of mega projects are to be on the cards in the near term [1]. Therefore, to keep ahead in today's competitive project bidding scenario, the construction companies have to produce good quality of work so that it will increase the level of satisfaction of the end users and build the clients' confidence towards the quality of the workmanship [2]. Quality assessment building system on the other hand plays a very important role in construction industry because it is used to measure the quality standard and to achieve the marketability of the building constructed [3]. Back in November 2006, the Technical Committee on Quality Assessment in Construction with the assistance of Construction Industry Development Board (CIDB) Malaysia has developed a system called Quality Assessment System in Construction (QLASSIC) as a quality assessment system for building construction work standard through a scoring system based on the Construction Industry

Standard (CIS 7:2006) [4]. There are several categories of buildings that can be accessed using QLASSIC such as residential, public, commercial and industrial buildings. The total score of building quality is divided between the cost proportions of four components which are the structural work, architectural works, mechanical and electrical works, and external works [4]. This system is designed as a benchmark of the quality level for various construction elements, to evaluate the performance of the contractors according to the quality of workmanship and used as a specification of quality criteria for a contractor's performance scorecard. The challenge now is to achieve a widespread adoption of this quality assessment system, because there are a few numbers of developers and contractors practicing QLASSIC in their projects. According to the statistics by CIDB, in 2015 only 270 out of 4,512 projects adopted QLASSIC to assess the quality of the projects [5]. This paper aims to provide an overview of contractors and developers on the QLASSIC assessment system by identifying its possible benefits and challenges of its implementation. The scope of this paper provides a guideline in obtaining the data required for the research. The respondents are contractors and developers, who may have or may not have the experience in implementing QLASSIC and the area is in Sarawak. Therefore, the significance of this research is to assist the contractors and developers to understand the impact of QLASSIC and to assist the authority to overcome the implementation challenges.

2. Quality Assessment System in Construction Industry (QLASSIC)

In the last few decades, the Malaysian construction industry has been criticized for its performance and productivity and it was proved that the Malaysian construction industry carries poor quality on its projects. In November 2008, the Malaysian Housing and Local Government Minister stated that there are approximately 250 complaints from the buyers regarding building defects every year [6]. Therefore, the quality of the buildings becomes an issue because there were many accidents occurred related to building failure and most of the critics received from the public were related to common problems such as leaky roof, leaky walls, dampness, water penetration from upper toilet, and crack at floor slab [7]. It is common for construction projects to have some failure and defects, but it may bring negative impact toward cost, duration and construction parties' status [8]. Therefore, it is necessary to investigate the importance of quality for construction project success. The implementation of QLASSIC started when the previous system from the International Organization for Standardization (ISO) 9000 was not applicable to the construction industry because it was very difficult to be conducted and only suitable to be used for manufacturing industry. In order to cope with that issue, CIDB introduced QLASSIC through Construction Industry Standard (CIS) 7:2006 which was adapted from Construction Quality Assessment System (CONQUAS) Singapore to meet the requirement with the Malaysian construction industry [9]. It had also been modified with several adjustments in CONQUAS and changed into QLASSIC in 2007 [10]. The system was created by the Technical Committee on Quality Assessment in Construction together with CIDB and the relevant industry stakeholders in 2006 [11,12]. The main roles of CIDB in the development of QLASSIC are to assist and act as a moderator and facilitator for the technical committee throughout the development process of this standard so that, this system will be conducted in an organized way [4].

Referring to MBAM [13], there are several benefits of QLASSIC system. These benefits are to evaluate the level of quality in the construction industry, to have a standard quality assessment system as a point of reference for quality in the construction works, to assist contractors to accomplish defect-free while carrying out the construction work, to be used as criteria to evaluate the performance of contractors based on the quality of workmanship and to be used for data compilation for statistical analysis in estimating the level of quality and productivity of the construction industry. Nevertheless, there are still some challenges in implementing QLASSIC. According to the research made by [14], QLASSIC causes delay in construction. The statement was also being supported by [15], whereby the process of assessment may cause delay in construction activities. An additional cost due to requirement fees for the quality assessor was also becoming one of the challenges needed to be faced by the contractors and developers [16]. In fact, there is still a large number of developers and contractors in Malaysia that are not aware with the quality assessment system even the big construction companies [9]. This issue was justified through a report prepared by [17], stating that 60% of G7 contractors are not aware on the QLASSIC implementation. Besides, in 2014, only 264 projects were assessed with this system compared to the huge number of developers [18]. The scenario was similar to that in Sarawak where the implementation of QLASSIC percentage was still considered low compared to other regions as in Figure 1 [16]. Generally, QLASSIC was not widely implemented and often taken lightly by the developers and contractors [10]. From 2015 until 2017, only 13 out of 120 projects were evaluated using QLASSIC where 10 projects were located in Kuching and three others in Miri [19]. According to S. Katessan [20], it is compulsory for all government projects exceeding RM50 million in amount to implement the QLASSIC assessment system starting from 2018. A decade after QLASSIC was first introduced to the construction industry, only 6% of the total projects in Malaysia are using it [5].

FIGURE 1. QLASSIC Assessment Distribution by State in 2014 [16].

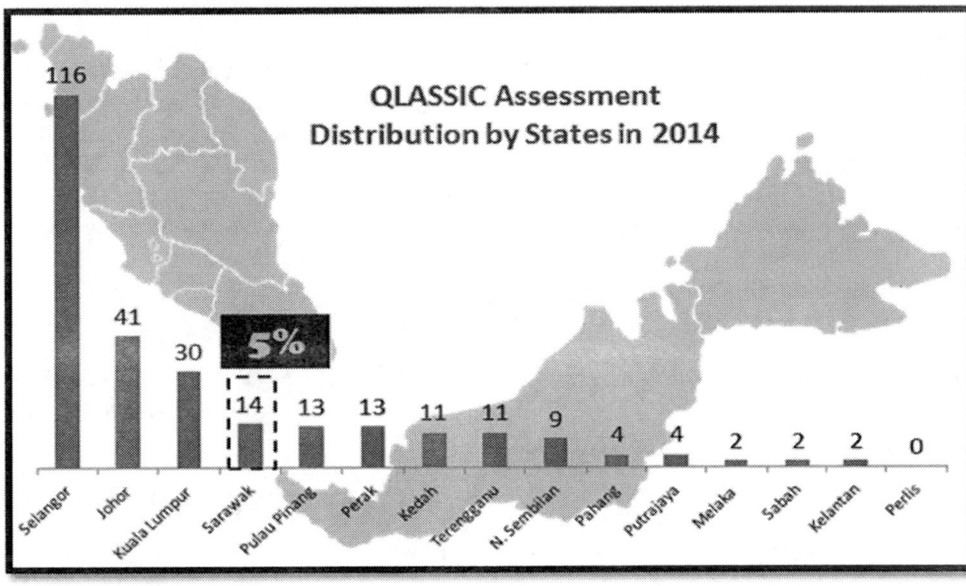

3. Methodology

An intensive literature review was conducted to explore the information related to QLASSIC implementation. The aim of this research is to highlight the insight of QLASSIC implementation by identifying its challenges. The paper began with a review of QLASSIC and its benefits in the construction industry. Subsequently, a structured questionnaire was used to obtain the data. The research focused on the construction industry in the state of Sarawak. Sarawak is the biggest state in Malaysia, where major construction projects take place in Kuching and Miri. The questionnaire was performed primarily to collect the representative data from two main parties, i.e., the contractors and the developers. 40 sets of questionnaires were distributed. The questionnaire was divided into three sections where each section focused on one objective. Section A briefly exhibit the demographic information such as the company name, the nature of the company's business, the designation of the respondent and the type of project being assessed by using QLASSIC. Section B comprised of the benefits of QLASSIC and lastly, Section C was designed to ascertain the challenges in QLASSIC implementation. These sections were formatted in an attitude scale as Likert scale type. The attitudinal scale is used to measure the respondents' attitude towards the various aspects and to provide the techniques to combine the answer into one overall indicator. The data received were analyzed using the descriptive analysis which showed the unvaried summary statistics such as mean for several

IOP Conf. Series: Materials Science and Engineering **429** (2018) 012103 doi:10.1088/1757-899X/429/1/012103

variables in a single table and calculated the standardized values. This method was achieved using the Statistical Package for Social Science (SPSS) version 14.0 software. The results were used to get an insight on the current practice of QLASSIC through acquiring their perception on the benefits and challenges related to the implementation of QLASSIC.

4. Results and Analysis

An In this study, 40 sets of questionnaires were distributed across Kuching and Miri. These sets were distributed face-to-face and via e-mail to contractors and developers. However, only 37 sets of questionnaires were returned within the stipulated period. The percentage of the response received over the number of questionnaires distributed is 92.5%. Furthermore, the feedback from the 37 respondents in this study were sufficient to make a statistical analysis for the research. Section A underwent the demography of the respondents' business. The results show that most of the respondents are from the contractor companies with the percentage of 75.6%, equal to 28 number of respondents. Meanwhile, 24.4% of the respondents were the developers with a number of 9 respondents only. This shows that the sampling for this research are mostly responded by the contractors. Furthermore, the analysis of Section A found 59% of the respondents have implemented and have the knowledge regarding QLASSIC. The rest of the result which is 41% has never used or are unfamiliar of QLASSIC (Figure 2). According to the short interview with the respondents, most of them are not familiar in using QLASSIC as they are using the other assessment methods such as ISO, QAQC and CONQUAS. On the other hand, those companies implementing QLASSIC is due to the requirement by the authority or the client. From 24 contractor-developers, 39% of QLASSIC implementation is on residential project. Then, followed by the commercial building with a percentage of 29% and industrial building 18%. The least type of project implementing QLASSIC is public building which contributes only 14% as per Figure 3.

FIGURE 2. Demographic distribution of QLASSIC Implementation.

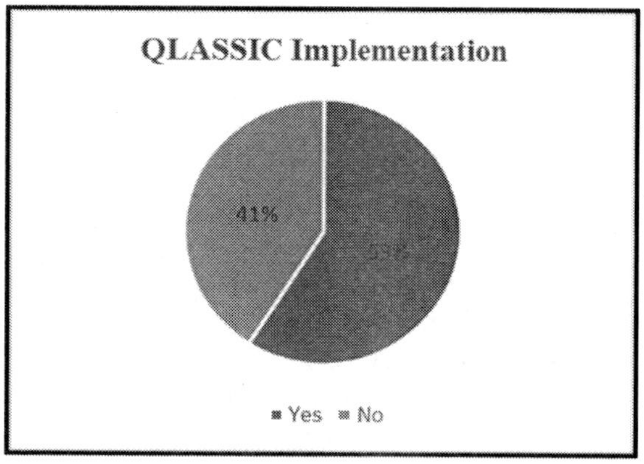

FIGURE 3. Demographic distribution of Project Assessed by QLASSIC.

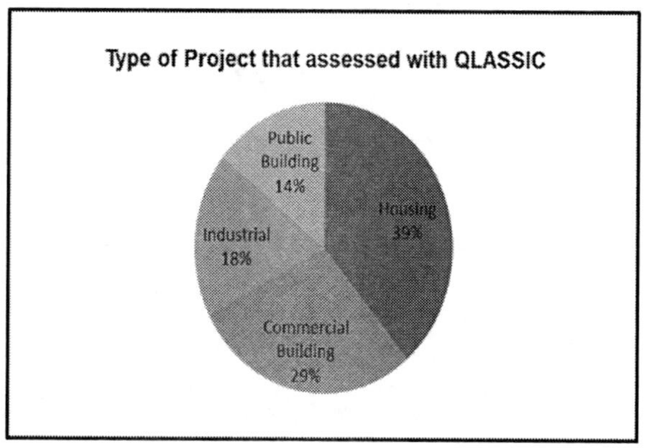

Under Section B, the data is analyzed by using the descriptive analysis. The first data analyzed is the perception on the benefits of QLASSIC. According to Table 1, the highest mean is 4.43 and its standard deviation is 0.558, pointing to the benefit of QLASSIC to measure the quality of workmanship in the construction work. This is followed by the benefit to build contractor-developer reputation and to improve end-user satisfaction, both with the mean of 4.37 with standard deviation 0.490 and 0.598 respectively. The third highest benefit is to use as a requirement to select the successful contractor-developer with the mean score at 4.31 and standard deviation of 0.676. The fourth highest benefit is to ensure the defect issues are identified for future improvement with the mean score of 4.29 and standard deviation 0.572. Two other benefits are to improve the company productivity and to use QLASSIC as marketing tools, ranked the two lowest with the mean distribution of 4.09 and 3.94 respectively. The scores indicated that most respondents agree that the major benefit of QLASSIC implementation is to measure the quality of workmanship.

TABLE 1. Distribution of mean and standard deviation for the Perception on Benefit of QLASSIC.

Rank	Description	Mean	Standard Deviation
1	To measure the quality of workmanship in the construction work.	4.43	.558
2	To build the reputation of the contractor/developer firm.	4.37	.490
3	To improve end-user satisfaction.	4.37	.598
4	To use as a level of standard to be awarded to the successful developers and contractor.	4.31	.676
5	To ensure defect issues are identified for future improvement.	4.29	.572
6	To improve the company productivity.	4.09	.612
7	To use QLASSIC as marketing tools.	3.94	.765

Moreover, Section C proceeded with evaluating the perception of contractor-developers on the challenges in QLASSIC implementation. The descriptive statistics are summarized in Table 2.

TABLE 2. Distribution of mean and standard deviation for the Perception on Challenges in QLASSIC Implementation.

Rank	Description	Mean	Standard Deviation
1	Contractor/developer not aware on QLASSIC.	4.03	.747
2	Level of knowledge on QLASSIC is still low.	4.00	.840
3	Process in applying and implement QLASSIC is complicated.	3.94	.765
4	Previously gained lower QLASIC score may discourage future implementation.	3.83	.747
5	QLASSIC has effect on construction cost.	3.63	.843
6	QLASSIC may cause delay in construction activities.	3.17	.707

It was found that being unaware of QLASSIC assessment among respondents contribute to the highest challenges with the mean of 4.03 and 0.747 in standard deviation, followed by lack of knowledge on QLASSIC, with the mean of 4.00 and 0.840 standard deviation. Then, the complicated process in applying and implement QLASSIC becomes the third highest issue scoring at 3.94 mean and 0.765 standard deviation. The fourth challenge is where the previous low QLASSIC score may discourage future QLASSIC implementation with 3.83 mean. Nevertheless, the challenges of that the QLASSIC has the effect on construction cost and may cause delay in construction activities are the two lowest perceived by the respondents with the mean distribution of 3.63 and 3.17 respectively.

5. Conclusions

IOP Conf. Series: Materials Science and Engineering **429** (2018) 012103 doi:10.1088/1757-899X/429/1/012103

Based on the research findings, it can be concluded that the QLASSIC implementation in Sarawak and the knowledge level of professional players toward QLASSIC are slightly increasing. This was supported by more than 50% respondents who implemented and are aware of the purpose of QLASSIC. Majority these respondents agreed that the implementation of QLASSIC can measure the quality of workmanship. It was also found that most of the respondents agreed that through high QLASSIC recorded scores, business reputation can be built in construction industry. However, throughout this study it was also found that nearly half of the respondents are neither having any experience in implementing QLASSIC nor aware of the quality assessment. The main challenge faced by the contractors and developers is not being aware of the existence of QLASSIC, followed by other challenges such as low level of knowledge in QLASSIC, complicated process in applying and implementing QLASSIC, and the low previous scores in QLASSIC marks, hence discouraging further implementation. Thus, a few strategies are suggested in this research to assure the implementation of the quality assessment system. One of the suggestions received from the respondents is to appoint a third party monitored by CIDB to conduct QLASSIC. It is to make sure that the QLASSIC implementation is expedited. Next, the process flow in QLASSIC application can be revised to become more customer friendly. Furthermore, the QLASSIC scoring system needs to be improved and persistently checked for the quality of the project. In addition, promoting QLASSIC through organizing more roadshows, seminars, trainings, and reducing the fees for its registration may also attract more construction players, hence helping to achieve the objectives of Construction Industry Transformation Program 2016-2020.

References

[1] R Lau 2018 Construction Sector Boom to Continues into 2018 (The Borneo Post)
[2] Janipha N A I and Ismail F 2013 Conceptualisation of Quality Issues in Malaysian Construction Environment *Procedia - Social and Behavioral Sciences* 53–61
[3] Ahmad N, Sabli M N and Othman A A 2014 Assessing the implementation of conquas standard amongst contractors towards improving quality of workmanship The Institution of Engineers Malaysia
[4] Iva P 2013 Cis 7:2006 Quality Assessment System for Building Construction Work Fluids *11th Offshore Mediterranean Conference and Exhibition* Ravenna Italy
[5] V. Nair 2016 QLASSIC Case of Setting High Standards *The Star*
[6] R K Bhatt 2016 QLASSIC: Doing it Right the First Time *iProperty Magazine*
[7] Wahab S N A and Hamid M Y 2011 A review factors affecting building defects of structural steel construction. Case study: Student accommodation in UiTM Perak *Procedia Engineering* 174–179
[8] Ahzahar N et al. 2011 A Study of Contribution Factors to Building Failures and Defects in Stabilized Construction Industry *Procedia Engineering the 2nd International Building Control Conference* 249–255
[9] Kam K J Hilmy A and Hamid A 2012 Elixir International Journal **46** 8525–8529 Fluids *11th Offshore Mediterranean Conference and Exhibition* Ravenna, Italy
[10] M A Malek and A Norizam 2013 *Malaysian Construction Research Journal* **13** 69-77
[11] Information on https://www.scribd.com/document/343958042/September-2015-QLASSIC-the-Way-Forward-in-the-Construction-Industry
[12] CIDB 2017 Quality Guidebook for Homeowners; First Edition Construction Industry Development Board Malaysia
[13] MB Association 2017 QLASSIC Awareness and Assessor Training MBAM Training Centre
[14] N Manap, Y K Goh and N Syahrom 2017 The International Conferences on Eco Engineering Development 1-12
[15] Bernama., Government Construction Projects Must Apply QLASSIC The Borneo Post, 2017
[16] Chuang, QLASSIC Implementation in Kuching [Interview] February 2018

[17] I Loyalty 2016 Understanding Customer Satisfaction on CIDB & Construction Products and Services Ipsos
[18] G. Tan, "CIDB to Make QLASSIC Requirement The Borneo Post.
[19] CIDB, "QLASSIC Projects in Sarawak 2006-2017 Construction Industry Transformation Programme, 2018
[20] S. Katessan 2017 QLASSIC Awards 43 Quality Building Projects Starproperty *Meeting* San Francisco

IOP Conf. Series: Materials Science and Engineering **429** (2018) 012104 doi:10.1088/1757-899X/429/1/012104

Low-cost impedance approach using AD5933 for sensing and monitoring applications

Rudy Tawie[1] and **Embam Da'ud**[1]

[1]Faculty of Architecture, Planning and Surveying, Universiti Teknologi MARA, Cawangan Sarawak, Kota Samarahan, Malaysia

*Email: rudy@sarawak.uitm.edu.my

Abstract. The use of electro-materials such as piezoelectric, thermoelectric and pyroelectric has gained popularity due to their unique properties and can be deployed in a system for sensing and monitoring applications. Due to advancement in technology, Analog Devices has introduced in the market the first integrated impedance device, AD5933. The AD5933 is relatively inexpensive compared to an impedance analyzer such as HP 4192A, portable and has a 1MSPS sampling rate, fast enough for frequency sweeping. A sensing technique known as impedance approach has been commonly used to utilize the coupling property of the electro-materials by measuring their electrical impedance directly. The changes in the impedance response signals at certain frequencies can be used to detect changes in the integrity of a structure. However, there is still need for a feasibility study to investigate the use of low-cost impedance approach and it is necessary to carry out an experimental validation of a proposed system for sensing and monitoring applications.

1. Background and Recent Studies

The technology breakthroughs in electro-materials have a direct impact on some of today's most challenging global problems. The use of electro-materials such as piezoelectric, thermoelectric and pyroelectric has gained popularity due to their unique properties. Many researchers e.g. [1-3] have taken advantage of the coupling property of these electro-materials by measuring their electrical impedance directly via a sensing technique known as impedance approach. At high frequencies (typically > 30 kHz), changes in integrity of a structure can be detected via the changes in impedance response signals of the electro-materials. To date, development of impedance measuring devices has allowed a sensing system that can be adapted in various fields. Furthermore, they are low cost, portable and easily used.

Na and Baek [4] applied multiple monitoring approach using piezoelectric transducers via AD5933 evaluation board [5]. The researchers successfully showed the advantage of the low-cost impedance technique to monitor adhesive defects of a glass-fiber epoxy plate.

Al-Ali et al. [6] designed and implemented a low-cost, portable impedance analyzer which can be used to measure impedance up to 100 kHz and from 10 Ω to 100 kΩ. The proposed system was used to monitor growth of fruits at different ages by using bio-impedance data.

Tambara and Bulhões [7] introduced a low-cost impedance analysis system that can be used to estimate the electrode-electrolyte parameters. The proposed system, which can be carried out using frequency sweep, allowed the resistance and capacitance of the equivalent model to be measured using only voltage outputs. The system is also flexible in application that it can be easily modified to suit user's requirements. Their results showed good accuracy for a system that is lower in price and smaller in size.

Meanwhile, Potdar et al. [8] adapted the AD5933 for determining respiration rate by transmitting current via two electrodes and then measuring the output impedance signals through two other electrodes. The impedance signals can be calculated using the magnitude and phase shift at different frequencies from 0.1Hz to 100KHz. Using signal processing technique, the impedance can also be converted into voltage and can be separated into real and imaginary parts using the Discrete Fourier Transform (DFT) algorithm.

2. Impedance Measurement and Proposed Setup

The There are a number of methods that can be chosen to measure impedance. Each of the method has its own advantages and disadvantages and in order to choose the most suitable method depends on the measurement requirements and factors such as frequency range, bandwidth, accuracy and ease of use [9] i.e. it is not possible to consider all these with only one method. Conventionally, impedance can be measured with the use of an impedance analyzer. However, the conventional impedance analyzers in the market are very expensive and not suitable for field applications since they are bulky [10]. Due to advancement in technology, Analog Devices has introduced in the market the first integrated impedance device, AD5933. The AD5933 is relatively inexpensive compared to an impedance analyzer such as HP 4192A. It is also portable and fast enough for frequency sweeping at a 1MSPS sampling rate. It is easy to use with any computer using a USB cable.

The low-cost impedance method with a AD5933 evaluation board is shown in Figure 1. The electrical impedance can be measured by the frequency domain ratio of the voltage to the current. Impedance is a complex number consist of real part or resistance, R and imaginary part or reactance, X. The R and X parameters can be converted into desired parameters such as impedance magnitude, |Z|; phase, θ; admittance magnitude, |Y|; conductance, G; and susceptance, B.

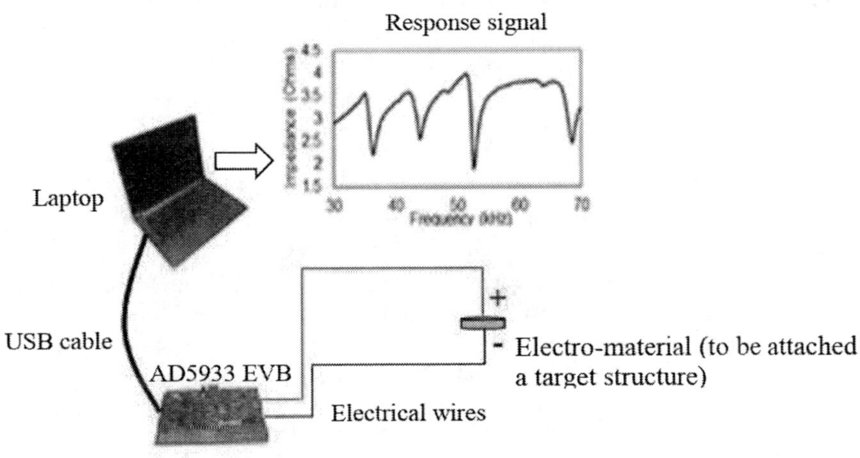

Figure 1. Proposed low-cost impedance method

3. Data Analysis

There are a number of methods that can be chosen to measure impedance. Each of the method has its own. There are three commonly used statistical metrics for quantifying the changes in response signals, namely, root mean square deviation (RMSD), mean absolute percentage deviation (MAPD) and correlation coefficient deviation (CCD):

$$\text{RMSD} = \sqrt{\Sigma_N[Re(Z_i) - Re(Z_i^o)]^2 \,/\, \Sigma_N[Re(Z_i^o)]^2} \quad (1)$$

$$\text{MAPD} = \Sigma_N|[Re(Z_i) - Re(Z_i^o)]/Re(Z_i^o)| \quad (2)$$

$$\text{CCD} = 1 - \text{CC, where CC} = \frac{1}{\sigma_Z\sigma_{Z^o}}\Sigma_N[Re(Z_i) - Re(\bar{Z})].[Re(Z_i^o) - Re\bar{Z}^o)] \quad (3)$$

In the equations, $Re(Z_i)$ and $Re(Z_i^o)$ represents real part of the reference or baseline and the corresponding impedance signals, respectively, N is the total number of signals and symbols $\bar{Z}$ and σ_Z signifies mean values and standard deviations, respectively. The statistical metrics yield a numerical value that represents changes in frequencies and changes in amplitudes of the reference signal as a sign that damage has occurred to a structure.

4. Conclusions

The idea is to propose an impedance system that is low cost, portable and easy to be used for field applications, hence to replace the bulky and expensive conventional impedance analyzers. In recent years, several researchers have used AD5933 applied to electro-materials and gained recognition by the research communities for their contribution in advancing the low-cost impedance approach. However, there is still need for a feasibility study to investigate the use of low-cost impedance approach due to its limited frequency range (up to 100 kHz), and it is necessary to carry out an experimental validation of a proposed system for sensing and monitoring applications.

References

[1] Mazlina M H, Sarpinah B, Tawie R, Daho C D and Annuar I 2016 Monitoring of bone healing by piezoelectric-EMI method. *AIP Conference Proceedings* DOI: 10.1063/1.4940256

[2] Tawie R and Lee H K 2011 Characterization of cement-based materials using a reusable piezoelectric impedance-based sensor. Smart Materials and Structures **20** 8 9 doi.org/10.1088/0964-1726/20/8/085023

[3] Park S, Ahmad S, Yun C B and Roh, Y 2006 Multiple crack detection of concrete structures using impedance-based structural health monitoring techniques *Exp. Mech.* **46** 609–618

[4] Na W S and Baek J. 2017 Adhesive Defect Monitoring of Glass Fiber Epoxy Plate Using an Impedance-Based Non-Destructive Testing Method for Multiple Structures *Sensors* **17** 1439

[5] Analog Devices. http://www.analog.com/en/index.html

[6] Al-Ali A, Elwakil A, Ahmad A and Maundy B 2017 Design of a Portable Low-Cost Impedance Analyzer *In Proceedings of the 10th International Joint Conference on Biomedical Engineering Systems and Technologies (BIOSTEC 2017)* 104-109 DOI: 10.5220/0006121901040109

[7] Tambara R V, Bulhões L O S 2016 Impedance Analysis for Medical and Electrochemical Applications Using a Low Cost Instrumentation. In: Simini F. Bertemes-Filho P. (eds) II Latin *American Conference on Bioimpedance. IFMBE Proceedings* **54** Springer, Singapore

[8] Potdar S V, Panditrao A, Khambete N D 2016 Monitoring breathing rate using bio-impedance technique *International Conference on Computing Communication Control and automation (ICCUBEA)* doi.org/10.1109/ICCUBEA.2016.7860143

[9] Holder D S editor 2005 Electrical Impedance Tomography: Methods, History and Applications Institute of Physics Publishing, London

[10] Park S, Grisso B L, Inman D J and Yun C B 2007 MFC-Based Structural Health Monitoring Using a Miniaturized Impedance Measuring Chip for Corrosion Detection. Research in Nondestructive Evaluation **18** 2 139-150

IOP Conf. Series: Materials Science and Engineering **429** (2018) 012105 doi:10.1088/1757-899X/429/1/012105

The Perception on the Importance of Construction Industry Payment and Adjudication Act (CIPAA) 2012 Towards Remedying Payment Disputes: A Research on Subcontractors in Kuching, Sarawak

N A Hadi[1,a)], M K F Othman[1,b)], A M Dadi[1,c)]

[1]Faculty of Architecture, Planning and Surveying, Universiti Teknologi MARA Cawangan Sarawak, 94300 Kota Samarahan, Sarawak, Malaysia.

[a)]Corresponding author: nurulhudaya@sarawak.uitm.edu.my
[b)]mohd_khairul135@ sarawak.uitm.edu.my
[c)]annishadadi@gmail.com

Abstract. The Construction Industry Payment and Adjudication Act (CIPAA) 2012 seek to provide the construction industry players with an alternative avenue to resolve the disputes quickly and at affordable costs. This Act can be used for any written construction contract that relates to any construction work in which includes local and international contracts. The current modes of construction dispute resolution by way of arbitration or court litigation take months and quite typically years to complete. Nevertheless, the implementation of CIPAA is relatively new to the construction industry, thus the possibility of construction players to be lack of knowledge and awareness about CIPAA is greater. Furthermore, payment problem normally occurs between main contractor and subcontractor. Therefore, this paper aims to measure the perception of subcontractors towards CIPAA in solving payment dispute. The pertinent features of CIPAA as well as its importance have been explained through an extensive literature review. In addition, a questionnaire survey was adopted to elicit the perception of subcontractors towards CIPAA. It is found that lack of awareness among respondents towards the implementation of CIPAA in construction industry is relatively high. Most of the respondents perceived CIPAA positively and expecting its can help to ease the cash flow. This research is expected to increase the subcontractor's understanding on payment right and their legal position if payment is in default. It is also expected to grab the attention of the employers in making their payment promptly and timeliness. In conclusion, it is aimed that the construction industry can be benefited from the enactment of CIPAA.

1. Introduction

In Malaysia, construction industry contributes a significant amount to the Gross Domestic Product at 6% [1]. It has grown day by day since the new development process and technology in construction has been introduced to the industry. Since Malaysia is getting well developed nowadays, there are a lot of disputes or issues occurring in the projects [2]. Disputes and issues in construction escalate due to many reasons as reported in the cases. A survey by Abidin [3], highlighted eight areas or nature of construction disputes, which are payment (51%), delay (19%), termination (18%), variation (13%), damages (11%), performance bond (8%), default (8%), and defect (1%). From the analysis, payment problems are the highest percentage in contributing to construction disputes. The construction industry has been facing payment dispute for years and is well aware in precedent cases as reported in court proceedings, various law journals and law reports [4].

Payment has been believed to be the lifeblood of the construction industry; there are fundamental to the process of construction, they are the root of many of its problems [5]. Construction projects involve long durations and multiple phases of works, not like many other industries. Late payment may affect the cash flow severely, as well as the timely schedule and quality of the project [6]. Since construction projects involve many parties, any payment problem encountered at the higher end of the hierarchy may give a huge impact on the cash flow through the chain of contracts.

In order to accomplish a construction project objective, the relationship between the main contractor and sub-contractor is significance. Hence, both parties need to acquire good relationship to achieve a smooth construction process [7]. However, the payment issues are always arising between these two parties such as non-payment, delayed payment as well as underpayment [6]. May & Siddiqi [8] also stated that the usually the main contractors will find some ways to transfer the risk to the subcontractor in case of non-payment from the client. Basically, this is done by inserting the clause of pay-when-paid or pay-if-paid in the contracts [7,9].

When such case arises, limited remedies and security are available pending dispute resolution. The standard forms of contract in the main contract are not clear regarding payment provisions on domestic subcontractors and suppliers. No protection for them against this unnecessary financial risks burden [5]. The Government has been in the know of the problems faced by the construction parties in rightfully receiving a regular and timely payment, which impacts the cash flow during project implementation. To ensure a smooth construction operation, this problem must be tackled [1]. Hence, the government has gazetted one payment act that could help to fix the issues, namely Construction Industry Payment & Adjudication Act (CIPAA) 2012.

2. Construction Industry Payment and Adjudication Act (CIPAA) 2012

Construction Industry Payment & Adjudication Act 2012 (CIPAA) was first published in Malaysia to overcome disputes regarding payment and cash flow in construction projects. It was first introduced and gazetted by the Parliament on 22 June 2012 and has been legislated and operated on April 15, 2014 [10]. This Act can be used for any written construction contract that relates to any construction work in which includes local and international contracts. This Act is made under strategic planning to enhance the construction industry image [9]. As for the authorisation, The Kuala Lumpur Regional Centre for Arbitration (KLRCA), currently known as Asia International Arbitration Centre (AIAC) has been chosen to be as the adjudication authority to appoint adjudicator, manage adjudicator's appointment procedure and giving practical training courses for statutory adjudication. In addition, CIPAA helps to achieve smooth disputes resolution to the construction projects [2].

Among the pertinent features of the act are outlawing the practices of "Pay-When-Paid" and "Conditional Payment" from the construction contract [11]. With a 'pay-when-paid' provision in a construction contract, the sub-contractor would only receive his claim after the main contractor has his payment from the client. The used of the terms or culture in the industry has affected the duration of the projects as it slows down the payment process in which the cash flow will be in a bad term [12]. Now is the right time for the construction industry to cut this unfavourable and unfair practice. Every work and service must be paid at the time it is being completed. If any default at any part of the contract chain, the parties at the chain shall find the solution themselves and not and shall not pass the task to any third parties [1].

Next, CIPAA also offer streamlining payment procedures by hidering payment uncertainties among all parties in construction contracts. CIPAA recommends providing statutorily implied terms of progress payment if there are no express terms regarding the matter [6,13,14]. It means the parties may hold a wide range of payment terms, but if there is no such provision in the contract, then only default mechanism will take place. The default mechanism by CIPAA establishes a payment process and timeframes for contracts without suitable payment terms. In addition, CIPAA also provides procedures

IOP Conf. Series: Materials Science and Engineering **429** (2018) 012105 doi:10.1088/1757-899X/429/1/012105

on responses according to payment claims when payment is in default with or without express contractual terms on payment.

Another pertinent feature of CIPAA is establishing a cheaper and speedier dispute resolution mechanism. CIPAA have introduced a new mechanism for settling construction disputes through an interim but binding dispute resolution process called adjudication [2]. The emphasis of adjudication is on speed but not finality. Under Clause 12 (2), Part II of CIPAA, the adjudicator must make decision on the disputes within 45 days; therefore, ensures the speedy resolution [14]. Adjudication procedure can be started at any time during construction once dispute relating payment arises. The affected parties shall not carry the burden themselves up until completion, and after that, file the dispute in arbitration or litigation. Dispute can be resolved efficiently and effectively through adjudication while a project still running.

Finally, provides security and remedies for payment recovery according to the decision made by the adjudicator. The successful aggrieved party will be able to recover its past debts and damages and at the same time prevent from further exposure [6]. Under the security provisions, a payment bond mechanism is recommended at the apex level within the hierarchy of the construction project implemented. The employers will have to issue the payment bond to the main contractors as the payment guarantee [1]. As for remedial provisions under the Act, remedies available include interest on late payment, suspension of works, direct payment from principal, judgment debt recovery, and other rights or remedies available in Construction Contract or any other written law such as deregistration of business licenses. From all the pertinent features of CIPAA that has been highlighted, it is definitely can give benefit to the subcontractors particularly in the situation where payment problem occurs.

3. The Importance of CIPAA Implementation

The whole construction industry and the nation will benefit by successful and efficient delivery of projects in terms of financial or cash flow, time and schedule, and dispute resolution mechanism. Establishing CIPAA is important to enable a speedy, time-bound, contemporaneous, cheap and binding dispute resolution mechanism by an independent and accredited adjudicator [6]. This is parallel with Clause 12 (2), Part II of CIPAA, the adjudicator must make a decision within forty five (45) working days ensuring a speedy adjudication process [14]. Besides that, CIPAA can provide a legal remedy to non-payment as by way of suspending or reduce the progress of performance or direct payment from the principal [2]. CIPAA permits the contractors to slow down or suspend the work progress if the adjudicated amount is not being paid, either in part or in whole. In addition, CIPAA permits the successful claimant to directly request for payment from the principal (client) if the adjudicated amount is not paid.

The next importance of CIPAA implementation is the improvement of the delivery system by providing quicker dispute resolution. This will reduce the unnecessary wastage through inefficient dispute resolution methods, which are only superficial to the core business of construction [1]. With the use of adjudication as a dispute resolution method, CIPAA is expected to have lower cost due to short time frame involved compared to arbitration or litigation. It offers a relatively cheaper process by express provisions prescribed by the proposed Act itself [14].

The objective of CIPAA is to ease the cash flow of the contractors. Due to that, CIPAA is important to help the industry in sustaining the competitiveness of all parties to implement construction projects [2]. Timely payment leads to ease of cash flow, and has a positive impact on the quality of construction works because the parties can concentrate on their works and services effectively and efficiently [5,6]. Other than that, the importance of CIPAA is to have a dispute remains confidential as adjudication is not conducted in an open manner like litigation. The adjudication process from beginning to the end is held privately to ensure its confidentiality. The confidentiality as stipulated in section 20 may add attractiveness to this dispute resolution process, though, parties must be aware that given the possibility of litigation regarding the adjudication process and decision, confidentiality in adjudication may not be guaranteed [15].

Last but not least, CIPAA can improve the image of the construction industry image by promoting professionalism and integrity practices among the contractors, construction professionals, and clients. From there, the value of human capital in the construction industry may as well be enhanced [1,10].

4. Research Methodology

The focus of this paper is to investigate the perception of subcontractors toward Construction Industry Payment and Adjudication Act (CIPAA) 2012 in Kuching, Sarawak. First, related theories and studies were reviewed to derive the possible variables related to the importance of CIPAA to the construction industry. The quantitative research method was adopted to confirm the respondents' perception based on seven (7) variables obtained in the literature review.

The overall population of subcontractors in Sarawak has been identified by using the statistics of List of Contractors from the Centralised Information Management System (CIMS) from the Construction Industry Development Board (CIDB) website. All companies have been classified into CIDB grade, type of works and location of the company. The research focus in Kuching, Sarawak is due to majority of the contractor in Sarawak located in Kuching division. It is also found in CIMS that the majority of construction company undertaking subcontractor's works are Small and Medium Enterprise (SME). As for the requirement for SME, Contractor with CIDB Grade G6, G5 and G4 seem to have fulfilled the criteria state in the SME Corporation Malaysia policies. The construction sector is under the services and other sectors groups of companies in SME policy. Therefore, the potential respondents are narrowed down to company located in Sarawak with registered for subcontracting work.

From CIMS, it identified 176 potential respondents for the questionnaire survey. Therefore, 176 sets of questionnaires have been distributed to respondents of subcontractors through e-mail and by hand. Subsequently, a total of 54 returned questionnaires from the subcontractors, giving the approximate response rate of 31%.

These data obtained were analysed by using SPSS 23.0 Analytical Software through descriptive analysis and correlation test.

5. Result and Discussion

As shown in Figure 1, the result shows that the majority of the respondents are lacking in awareness of CIPAA with 37% from total 54 respondents. The result is supported by the previous study conducted in Malaysia among G7 contractors which found that only 43% out of 30 respondents aware on the implementation of CIPAA [16]. In comparing with the previous study, the lower number of awareness in this research shows lack of interest of SME subcontractors in obtaining recent information related to construction issues, hence, is shown in the result. Further research needs to be done to investigate the reasons behind this.

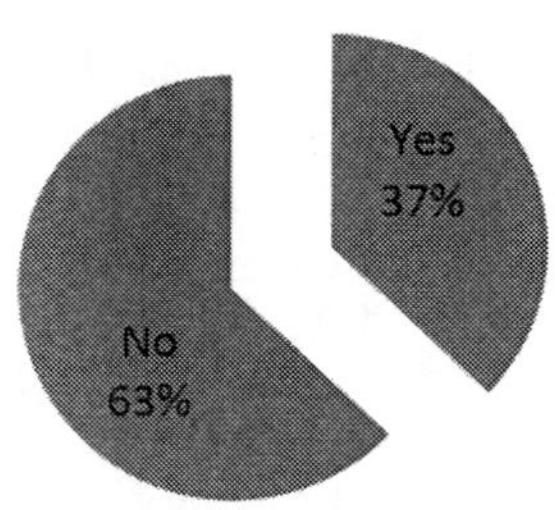

FIGURE 1: The Awareness of CIPA implementation (N = 54)

Table 1 shows a ranking of the perceived importance of the implementation of CIPAA. The result indicates that CIPAA can help to ease the cash flow of the contractors was considered the most important towards CIPAA implementation among subcontractors. In contrast, CIPAA will improve delivery system provides quicker and cheaper disputes resolution perceived the lowest among all the perceived significance of CIPAA. The result is expected because CIPAA is intended to inject much needed cash flow into the contractual arrangements between parties that saw progressive claims as the recognised and accepted way of doing business in construction contracts [17]. A prohibition 'pay-when-paid' clause in construction contracts leads for better cash flow for all parties in construction project including the subcontractor. Furthermore, it is also suggested that CIPAA could be the solution to most of the subcontractors payment problems [9].

Parties in dispute are able to remain confidential compared to litigation ranked second in the analysis which considered an importance CIPAA implementation. CIPAA offers important feature of resolving dispute which are confidentiality. Parties wanting a quiet settlement will find adjudication attractive. The confidentiality requirement applies to the adjudicator and all parties in dispute [15].The high preference for confidentiality indicates that good business reputation is important to the subcontractor. The litigation process held in the open court resulted in sour business relationship and bad for future trades, thus, depriving business opportunity [18].

CIPAA encourages professionalism and promotes integrity amongst construction parties ranked third in the analysis which considered an importance CIPAA implementation. Previous research found that the most significant factors of late and non-payment in the Malaysian construction industry are the attitude and culture of the main contractors and delay in certification on the consultants' part [12]. Moreover, it is also being identified that among the top factors for late payments are that the clients deliberately delay the payment for their own financial benefit, purposely delay in releasing the retention monies to the contractors and are likely to hold the payment for personal reasons [19]. By implementation of CIPAA, it provides a mechanism for the recovery of payment upon the conclusion of the adjudication process in addition to a host of other remedies such as a right to reduce the rate of work progress or to suspend work (Clause 29, Part IV) or even to secure direct payment from the principal (Clause 30, Part IV) [20]. Therefore, the mechanism introduced in CIPAA encourages construction parties to perform their responsibilities accordingly to protect their rights of payment in order to reduce disputes.

Interestingly, CIPAA offers lower fees by having adjudication as a method of dispute resolution and CIPAA improves delivery system provides quicker disputes resolution were found to be least important in CIPAA implementation. This result contradicted pertinent features of CIPAA which is cheaper and speedier dispute resolution mechanism called adjudication [12]. The contradictory result indicates that the possibility of the subcontractors' lack of experience in adjudication process introduced in CIPAA. The subcontractors may not aware on the lower cost and short time frame comparing with litigation process as dispute resolution method.

TABLE 1: The importance of CIPAA implementation

The importance of CIPAA Implementation	Mean	Std. Deviation	Rank
CIPAA helps to ease the cash flow of the contractors.	3.7778	.60397	1
Parties in dispute are able to remain confidential	3.6111	.56357	2
CIPAA encourages professionalism and promotes integrity amongst construction parties	3.5741	.49913	3
CIPAA provides Legal Remedy to Non-Payment	3.5370	.53950	4
CIPAA enhances the value of human capital in the construction industry	3.5185	.54047	5
CIPAA offers lower fees by having adjudication as a method of dispute resolution	3.4630	.57340	6
CIPAA will improve delivery system provides quicker disputes resolution.	3.4259	.49913	7

TABLE 2: The Correlation Matrix between the awareness of CIPAA and the importance of CIPAA implementation

The importance of CIPAA implementation	Correlation
CIPAA helps to ease the cash flow of the contractors.	**0.413****
Parties in dispute are able to remain confidential	0.122

CIPAA encourages professionalism and promotes integrity amongst construction parties	-0.115
CIPAA provides Legal Remedy to Non-Payment	0.019
CIPAA enhances the value of human capital in the construction industry	-0.027
CIPAA offers lower fees by having adjudication as a method of dispute resolution	0.050
CIPAA will improve delivery system provides quicker disputes resolution.	0.037

*** Correlation at 5% significance level**
**** Correlation at 1% significance level**

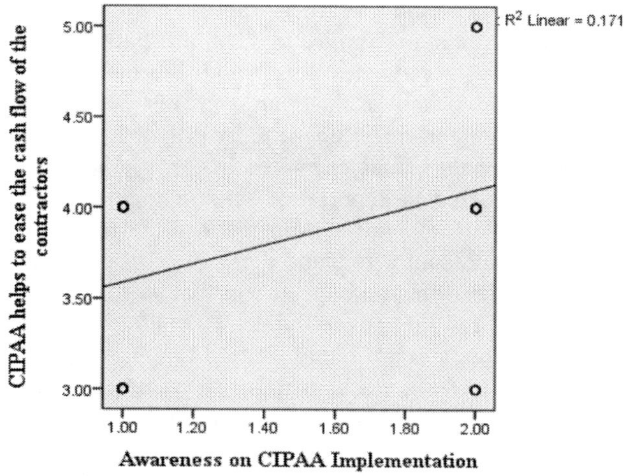

FIGURE 2: Scatter plot of awareness of CIPAA vs CIPAA helps to ease the cash flow of the contractors

Table 2 shows the relationship between the awareness of CIPAA and the importance of CIPAA implementation. Only one correlation was found using the point-biserial correlation test which is CIPAA helps to ease the cash flow of the contractors. The results proved that the lack of awareness of the subcontractors of CIPAA, thus, it shows that subcontractor matter of greatest importance is the security of payment rather than any other importance of CIPAA. Graph in Figure 2 was plotted to analyse further, the relationship between the awareness of CIPAA and CIPAA helps to ease the cash flow of the contractors. The graph indicates that the respondents are aware about CIPAA implementation agree on the importance of CIPAA, that helps to ease the cash flow of the contractors.

6. Conclusion

This paper manages to achieve the main objective which is regarding the perception of subcontractors towards the implementation of CIPAA in Kuching, Sarawak. It is found that lack of awareness of subcontractors towards the implementation of CIPAA in the construction industry. The relevant bodies in the construction industry should identify effective methods in promoting CIPAA in order to ensure smooth application and enforcements in future. It is also found that among seven (7) variables of importance of CIPAA implementation, CIPAA helps to ease the cash flow of the contractors; parties in dispute are able to remain confidential; and CIPAA encourages professionalism and promotes integrity amongst construction parties are the most important features of CIPAA perceived by the subcontractors.

Although this paper managed to provide a quantitative study, further research is recommended to be conducted in terms of increasing number of respondents and expanding the area of the research. It is also recommended research to be done to identify the factors contributing to lack of awareness of the subcontractor on CIPAA implementation. The research will contribute to promote of CIPAA in increasing the awareness of contracting parties in the future.

IOP Conf. Series: Materials Science and Engineering **429** (2018) 012105 doi:10.1088/1757-899X/429/1/012105

References

[1] Karib A S, Shaffii N and Nor N M 2008 *A Report on the Proposal for a Malaysian Construction Industry Payment and Adjudication Act* (Kuala Lumpur)
[2] Fong L C 2005 The Malaysian Construction Industry - The Present Dilemmas of Unpaid Contractor *Master Build.* **80**
[3] Abidin A 2007 *The Profile of Construction Disputes* (Universiti Teknologi Malaysia)
[4] Cheong Y W 2016 *Benefits of The Construction Industry Payment And Adjudication Act 2012 (CIPAA) And Its Impacts To Payment Problem In Construction Industry In Malaysia* (Universiti Tunku Abdul Rahman)
[5] Lip E 2006 *Curing the Ills of Non-Payment in the Construction Industry – the Singapore Experience* (Kuala Lumpur: ISM)
[6] Ali N A N A 2006 A Construction Industry Payment And Adjudication Act: Reducing Payment-Default and Increasing Dispute Resolution Efficiency in Construction *Master Build.* 4–6
[7] Enshassi A and Abuhamra L 2015 International Conference on Construction and Real Estate Management (ICCREM 2015) 567–75
[8] May A L and Siddiqi K 2006 Contingent-Payment Provision Puzzle—Safeguarding against an Unintended Outcome *J. Archit. Eng.* **12** 158–62
[9] Supardi A and Adnan H 2011 Security of Payment in Malaysian Construction Industry : Eradication of Sub-contract's Contingent Payment *International Conference on Energy, Environment, Entrepreneurship, Innovation* 116–21
[10] Mahmood M Z 2014 *Challenges, consenquences and contractual impacts of the construction industry payment and adjudication act 2012* **2012**
[11] Din N M D N and Ismail Z 2014 Construction Industry Payment and Adjudication Act (CIPAA) Remedying Payment Issues: CIDB G7 Contractor's Perspective *J. Technol. Manag. Bus.* **1** 21–38
[12] Azman M N A, Dzulkalnine N, Hamid Z A, Kamar K A M and Nawi M N M 2014 Payment Scenario in the Malaysian Construction Industry Prior to CIPAA Payment Scenario in the Malaysian Construction Industry Prior to CIPAA *Paym. Scenar. Malaysian Constr. Ind. Prior to CIPAA*
[13] Fong L C 2011 *The Malaysian PWD Form of Construction Contract – 2nd Edition* (Kuala Lumpur: Sweet & Maxwell)
[14] Rajoo S 2012 Dispute Resolution for Construction Industry in Malaysia *Newsl. Kuala Lumpur Reg. Cent. Arbitr.* Apr-June 2012 Issue
[15] Mah B Y R 2015 Confidentiality in Construction Adjudication *Malaysia Soc. Adjudicators* 29–31
[16] Sahab S S and Ismail Z 2011 Construction Industry Payment and Adjudication Act; enhancing security of payment in the Malaysian Construction Industry *Int. Conf. Business, Eng. Ind. Appl.* 153–9
[17] Construction Industry Development Board 2016 CIDB Construction Law Report 344
[18] Cushman R F, Carter J D, Gorman P J and Coppi D F 2001 *Construction Disputes: Representing the Contractor* (New York: Aspen Publishers)
[19] Ye K M and Rahman H A 2010 Risk of Late Payment in the Malaysian Construction Industry *World Acad. Sci. Eng. Technol. Int. J. Mech. Ind. Eng.* **4** 503–11
[20] Master Builders 2014 Construction Contract & Management Issues *Master Build. J.* 70–5

AUTHOR INDEX

AUTHOR INDEX

AUTHOR INDEX

AUTHOR INDEX

Institute of Physics
Dirac House, Temple Back
Bristol BS1 6BE UK

ISSN: 1757-8981
ISBN 978-1-5108-7508-1

9 781510 875081

CURRAN ASSOCIATES INC.
proceedings.com